CRC Handbook of
Avian Body Masses
Second Edition

CRC Handbook of

Avian Body Masses

Second Edition

Edited by John B. Dunning, Jr.

CRC Press
Taylor & Francis Group
Boca Raton London New York

CRC Press is an imprint of the
Taylor & Francis Group, an **informa** business

CRC Press
Taylor & Francis Group
6000 Broken Sound Parkway NW, Suite 300
Boca Raton, FL 33487-2742

No claim to original U.S. Government works
Printed in the United States of America on acid-free paper
10 9 8 7 6 5 4 3 2 1

International Standard Book Number-13: 978-1-4200-6444-5 (Hardcover)

Library of Congress Cataloging-in-Publication Data

Dunning, John B. (John Barnard)
CRC handbook of avian body masses / John B. Dunning, Jr. -- 2nd ed.
p. cm.
Includes bibliographical references and index.
ISBN 978-1-4200-6444-5 (alk. paper)
1. Birds--Size--Tables. 2. Body size--Tables. I. Title.

QL697.D86 2008
571.3'18--dc22 2007024376

Visit the Taylor & Francis Web site at
http://www.taylorandfrancis.com

and the CRC Press Web site at
http://www.crcpress.com

To my wife, Beth, for her support and understanding of a little obsession

Contents

The Editor

John B. Dunning, Jr., is an associate professor of wildlife ecology in the Department of Forestry and Natural Resources, Purdue University, West Lafayette, Indiana.

Dr. Dunning graduated in 1978 from Kent State University, Kent, Ohio, with a B.S. in biological sciences (summa cum laude). He then obtained his Ph.D. in ecology from the Department of Ecology and Evolutionary Biology, University of Arizona, Tucson, in 1986. It was during the latter time that he first learned to band birds under the tutelage of Dr. Stephen M. Russell. The compilation of avian body masses started as a necessary preliminary step in a research project for a graduate class; a project that did not yield anything else of interest.

Dr. Dunning received Elective Member status in the American Ornithologists' Union in 1990. In 2006, he was voted the Outstanding Undergraduate Teacher in his department. During his college career, he received the Kent State University Distinguished Scholar Award in 1978, and a National Science Foundation Pre-Doctoral Fellowship from 1978-1981.

Dr. Dunning studies the community and population ecology of sparrows at landscape and local scales. His research is supported by the National Science Foundation, the U.S. Department of Energy, U.S. Forest Service, and The Nature Conservancy, among others. His research focuses on the impacts that human land-use change has had on bird populations across large spatial scales. Dr. Dunning has published over 75 papers and three books. He is a member of Phi Beta Kappa.

Part I

Body Masses of Birds of the World

INTRODUCTION

John B. Dunning, Jr.

Many studies in avian biology require estimates of body size. Ecological and physiological studies, for instance, often report body mass as part of a set of baseline descriptive statistics when a large number of species are being compared (e.g., Diamond 1972, Brown and Bowers 1985). Body masses are also necessary for allometric scaling of metabolic processes (Calder 1984, Colwell 2000), and for some community structure analyses (Schluter and Repasky 1991). For birds the single best estimator of body size is usually adult body mass (Clark 1979, Rising and Somers 1989); therefore, compilations of masses are extremely useful for all of these research programs.

In spite of their utility, body masses were difficult to locate even for relatively common species (Dunning 1984, 1985) until the publication of the first edition of this handbook. Dunning (1992) presented data for about 6,300 species, using both published and unpublished sources. Since that time, additional data have been published for many birds, improving the samples available for many of the species included in the first edition. A variety of regional handbooks and monograph series (e.g., *The Birds of North America* series) that were unfinished (or not begun) at the time of the first edition's publication are now complete and provide fairly complete coverage for avifauna of certain regions. The second edition of this handbook now includes data for over 8,700 species and better coverage of portions of the world that were inadequately represented in the original compilation.

SOURCES OF DATA

The data presented in the first edition were gathered primarily by searching the published literature. Brough (1983) and George Clark's unpublished bibliography of body mass references were invaluable to this effort. In the current compilation, I gathered mass information from a variety of new sources, an effort that was aided substantially by the completion of regional handbooks and monographs such as *The Birds of Africa* (1982–2004), *The Birds of North America* (1992–2002), *The Handbook of Birds of the Western Palaearctic* (1977–1994), and *The Handbook of Australia, New Zealand, and Antarctic Birds* (1990–2006). The *Handbook of Birds of the World* (1992–2005) also presents a comprehensive reference source; however, it presents only mass ranges for the most part, and therefore I preferred other sources when available.

For the initial compilation, I conducted a systematic search in the papers published in major bird journals published from 1960 to the present, including *Ibis*, *Bulletin of the British Ornithologists' Club*, *Auk*, *Wilson Bulletin* (now *Wilson Journal for Ornithology*), *Condor*, *Emu*, *Corella*, *Notornis*, and the *Journal of Field Ornithology*. For the second edition, I extended my search to include earlier volumes of these journals and expanded coverage of additional journals such as *Journal of Avian Biology*, *Journal für Ornithologie*, *Ornitología Neotropical*, and *El Hornero*, among many others.

I included more coverage of museum series and monographs such as *Fieldiana*, *Proceedings of the Philadelphia Academy of Natural Science*, *Bulletin of the American Museum of Natural History*, among others. I also made extensive use of database search engines such as *Wildlife and Ecology Studies Worldwide* to locate articles on individual species. The recent improvement in university interlibrary loan services gave me much greater access to a variety of resources, including articles published in German, Spanish, Chinese, French, and other languages. It is still true, however, that coverage in this second edition is better for English-language literature published in the Western Hemisphere than for European or Asian journals or for articles published in non-English languages.

Many "bird family" handbooks have been published since 1992, and I consulted as many of these sources as possible, such as the handbooks on thrushes (Clement 2000), ratites (Davies 2002), and albatrosses (Tickell 2000), among others. Researchers also recommended specific sources that were missed in the compilation of the first edition (e.g., Humphrey et al. 1970, Hall 1974).

I requested unpublished data when possible from authors of articles on poorly studied species, and also from museum collections and banding (ringing) operations. One particularly useful source was the SAFRING (South African Bird Ringing Unit) summaries of ringing operations in southern Africa, which are available online (http://www.uct.ac.za/depts/stats/adu/safring-index.htm). I was also able to get useful data for many species from the online catalogs of The Field Museum, Chicago, Illinois, and the U.S. National Museum, Washington, DC. For the first edition I visited some museums, including those of the University of Arizona, Louisiana State University, and the Philadelphia Academy of Natural Sciences. Numerous other museums responded to requests for information (see Acknowledgments).

DATA COLLECTION

I included the "best available sample" that I located for each species. This sample was defined as the data from the single source that had either (1) the largest available sample size, or (2) the most complete descriptive statistics, if sample sizes from several sources were equivalent. I avoided combining data from different sources if possible to reduce the potential heterogeneity created when combining samples from different locales or seasons. I did combine data, however, when all available samples were very small (N < 10). In preparing the new edition, when I located new sources with a sample > 10 individuals, I used these in preference to combined samples from the first edition.

I preferred data from breeding birds; however, the lack of published information for many species required the liberal use of data from other seasons. If ages were reported in the original source, I used masses of only adult birds. For two or three very rare species (e.g., *Gallirallus calayanensis*), the only available data were from independent subadults. These data are reported here in the second edition, with a comment in the record that the masses are not from adult individuals.

I present separate means for males and females for sexually dimorphic species. Where the original sources tested for differences between the sexes and found none, a combined mean for both sexes is presented. In cases where the original source did not test for dimorphism, I tested for significant differences with a Student's t-test when this was possible. If enough information was not presented to test for differences, then I report separate means for the sexes if males and females differed by > 10%. A single mean is presented for samples in which sexes were not identified.

Many species of birds vary in size across their geographic range. The first edition contained relatively little data to express geographic variation, both because of space limitations and because such data were relatively rare. In this second edition, I have sought to add data where possible for geographically variable species (e.g., Golden Whistler, *Pachycephala pectoralis*). Such samples are reported by subspecies or by geographic location, whichever was reported in the original source. I did not assign samples to subspecies if the original source did not do so.

In the years since the publication of the original handbook, I have become aware of a few errors in first edition. For instance, the data reported for *Stercorarius* (*Catharacta*) *skua* in the first edition were actually from a sample of *Stercorarius parasiticus,* a substantially smaller bird. (This particular error resulted from my failure to recognize that an Israeli translation into English of a Russian publication would follow European conventions in the use of common names. Thus, "Arctic Skua" in this reference did not refer to *S. skua* [the only skua that breeds in the Arctic] but rather to *S. parasiticus*, which North American ornithologists call the Parasitic Jaeger.) Although few errors were reported to me, I resolved to eliminate as many errors as possible by verifying every entry in this second edition. This includes all sources used in the first edition. I used this verification search to add information where possible, especially location and season of collection, when reported in the original source but not included in the original compilation.

DEFINITIONS

The table presents as much of the following information as was available from the selected source(s): sex, sample size, mean, standard deviation, range, collecting location and season. Species names and order primarily follow the sixth edition of Clements's (1998) *Checklist of World Birds.* Avian taxonomy and systematics are fluid topics, and revisions of the names and species order are published frequently. It is therefore difficult to select a single authority for avian names and sequence. After I had been working on this revision for several years, the updated revision of Howard and Moore's checklist was published (Dickinson 2003). The *Handbook of the Birds of the World* series (del Hoyo et al. 1992–2005) also presents an authoritative taxonomy (although incomplete as of this writing). None of these sources agree with each other in all cases.

I used Clements (1998) because when I started my revision, it was the only updated source with information on all subspecies and their ranges, which was invaluable in identifying data in older references to current species. Also James Clements maintained a web-based list of corrections and additions that allowed me to keep the list up to date. I changed the taxonomic status of forms when new studies were published, for example, Garrido et al. (2005) for West Indian forms of the genus *Icterus.* I checked the spelling of scientific and Latin names using BirdLexi (Santa Barbara Software, Inc., Santa Barbara CA, 2006). This allowed me to conform to changes in the endings of species' names suggested by David and Gosselin (2002). When BirdLexi identified a different spelling than that in Clements (1998), I checked for a consistent spelling in Dickinson (2003) and del Hoyo et al. (1992–2005).

In addition to the above steps, I compared the status of species in Clements (1998) to a variety of published authorities, and attempted to present a modified list that reflects regional authorities. Thus, I followed the American Ornithologists' Union *Check-list* for North and Central America species where that reference disagrees with Clements (1998). I followed the *Handbook of Birds of the Western Palaearctic* for European species, and so on. In cases where authorities disagreed, I presented as much information as possible to allow flexibility to the users. For example, I included data when available for all forms of potential species splits, listing the data either as subspecies of a polymorphic form, or as separate species depending on the available published opinions.

Recently, widespread changes in the order of species have been proposed, for instance placing Galliformes first. I chose not to rearrange the sequence of orders and families to follow these proposals but largely retained the order in Clements (1998). I did this in part because no single sequence has been accepted as of this writing, and the first edition of this handbook followed the Sibley-Ahlquist sequence of orders and families (newly published at the time), which was subsequently not adopted by most authorities. This decision made the first edition hard to use in later years.

All means and ranges are given in grams. Standard deviation, when available, is given for samples > 10 individuals. When necessary (and possible), I calculated standard deviation from standard errors.

The following codes are used in the table. For sex: M = male; F = female; B = both species combined; U = sex unknown. For collecting season: W = winter; B = breeding; PB = postbreeding; Sp = spring migration; F = fall migration; M = spring and fall migration combined, Y = year round. In many sources, the data were not collected during a time period that fit neatly into these categories (i.e., April through November in temperate zones) and I do not indicate a collecting season for these data. I did not assume a season for tropical studies when not specified in the paper, even if the calendar month was given. It is unfortunate that many sources, including many comprehensive handbooks, still do not indicate the season in which the data were collected. Many sources that reported masses from museum specimens also do not indicate the collection localities.

All sources are numbered in the Literature Cited section, including sources of unpublished data. Data in the table are referenced to the original source using these numbers. Of course, the reference numbers have all changed from the first edition, since the number of sources has doubled. After I numbered the sources for the second edition, I dropped some sources as better data became available from another source. I inserted the phrase "reference deleted" for these replaced sources, so as not to imply errors in the numbering scheme. New sources added after the initial numbering are inserted in alphabetical order and given a number such as "219a."

COMPARISON OF THE FIRST AND SECOND EDITIONS

The number of species included in the handbook has increased from 6,283 in the first edition to 8,734 in the present. About 1,221 species listed in Clements (1998) are not yet included in the database. In addition to >2,400 new species, the second edition includes improved samples for many species that were in the first edition. To cite just one example, in the family Procellariidae (shearwaters and petrels), improved samples were found for 28 of the 50 species included in the first edition. Together with data for 24 new species, the current edition has improved information for 52 of the 74 shearwaters and petrels in the database, leaving only 4 species without masses.

I was able to make similar improvements in many other avian families, especially doves, cuckoos, owls, woodpeckers, ovenbirds, larks, bulbuls, and babblers. The largest family for which I found data for all species was the waterfowl (Anatidae), with masses for 158 species including 9 new to this edition. Families with large numbers of species remaining without data include parrots (67 of 329 species, 20%), hummingbirds (53 of 338, 16%), babblers (85 of 266, 32%), and white-eyes (43 of 95, 45%). Smaller families that still have a high percentage of species with no data include the storks (8 of 32 species, 25%), barn owls (7 of 15 species, 47%), and orioles (10 of 29, 34%). Probably the most frustrating were families in which data are needed for only 1 remaining species. These include the cormorants (no mass for *Phalacrocorax nigrogularis*), thick-knees (*Burhinus superciliaris*), skimmers (*Rynchops albicollis*), sandgrouse (*Pterocles personata*), toucans (*Ramphastos citreolaemus*), kinglets (*Regulus teneriffae*), and bowerbirds (*Sericulus ardens*). Given the tendency for splitting among avian taxonomists, complete coverage of a family is often only a temporary milestone anyway.

I made a special effort to locate data for Neotropical species that were not in the first edition, since that region is not yet covered by a handbook series that includes body masses, unlike Africa, Australia, North America, and Europe. I was able to have emails circulated throughout Taiwan and Japan to get better coverage of the birds of those nations, which were poorly covered in the first edition. Areas of the world that still have many species without data include Pacific Is. nations, Southeast Asia, Indonesia, and China. It is possible that better coverage of the scientific literature published in French and German might help fill those holes.

LIMITATIONS AND USES OF MASS DATA

A variety of topics in avian research require a measure of a bird's size. Size can be measured most accurately through multivariate analysis of several mensural characteristics (Freeman and Jackson 1990); however, the data for such an approach are rarely available, especially when large numbers of species are being compared. Also, these multivariate analyses often require sacrifice of the birds under study (when skeletal measurements are used, for instance), making these approaches less useful for ongoing studies of living birds. Some comparative studies have shown that body mass is the single most accurate univariate measure of size in birds. Rising and Somers (1989) determined that body mass correlated more strongly with overall body size (as estimated by multivariate analysis) than did body length, wing chord, or other linear measure. Freeman and Jackson (1990) showed that body mass and tarsal length were the two best univariate measures of overall body size. Tarsal length can be more subject to measurement error than are other linear metrics (Lougheed et al. 1991), and is available for even fewer bird species than body mass.

Body mass is a measure of the overall size of an organism, including both the organism's structural framework and its nutrient reserves (Clark 1979, Piersma and Davidson 1991). The structural size of an organism is the relatively invariant mass associated with skeletal elements and internal organs, while the nutrient reserves are the protein and fat stores that vary with an individual's condition. Studies of structural size can be done most accurately with multivariate estimates of avian skeletal elements (Freeman and Jackson 1990, Piersma and Davidson 1991) or using skeletal volume (Moser and Rusch 1988), but, again, these measures tend to be rarely available.

There are several sources of variation that cannot be adequately addressed in a compilation such as this. First and foremost, many species show geographic variation, including subspecific variation, clines, and differences among Is. and mainland forms. I have attempted to present more samples to represent geographic variation, but, as before, I was often limited in this effort by the lack of data. Another source of variation is seasonality. To be truly comprehensive, I should attempt to present data from a variety of seasons for many species. This would be especially useful for migratory species: one could easily argue that a full presentation of weight variation would include separate samples of breeding, wintering, and migrant individuals. But such a presentation is beyond the scope of this compilation.

A third source of variation is time of day. It is well established that the mass of individual birds can vary by 10% or more in a single day. Small owls, for instance, can vary in their mass by up to 40%, depending on when they last fed (Dunning 1985). The mass of migrant birds will vary during the day as they restock their energy stores during migration, and can also vary according to the length of stay in a particular patch of habitat. For the purposes of this compilation, this type of variation is a source of error that may make a small sample (e.g., masses of single birds) non-representative for a given species. I have tried to reduce the impact of this type of variation by presenting as large a sample as possible. However, for some purposes it may be critical for a researcher to understand this variation, in which case the present compilation may not be a good source of information.

Although body mass can be a useful measure of overall body size, the use of large numbers of masses collected from various sources, such as the compilation presented here, may not be justified in all comparative studies (Clark 1979). The variation introduced by combining samples from different sources may reduce the validity of comparing mass samples for some purposes. Researchers contemplating using the data in this handbook are urged to consider the effect of such variation on their study. When questions arise, the original source of the data should be consulted.

ACKNOWLEDGMENTS

During the preparation of the first edition of the handbook, I was supported by an NSF (National Science Foundation) grant and a postdoctoral fellowship at the Institute of Ecology, University of Georgia. Nora Mays, Rick Bowers, and Eugene Howard helped in the creating of the original database. George Clark, Steve Russell, and Charles Collins helped locate many sources of data.

In compiling data for the second edition, I taxed the patience and the reach of the interlibrary loan staff at Purdue University. I also consulted the libraries of Indiana University, the University of Georgia, University of Arizona, Kent State University, University of Canterbury (Christchurch, New Zealand), MacQuarie University (Sydney, Australia), the Cleveland Museum of Natural History, the Alexander Library of the Edward Grey Institute (Oxford, UK), and the Josselyn Van Tyne Memorial Library of the University of Michigan. The Western Foundation of Vertebrate Zoology (Los Angeles, California) and the Josselyn Van Tyne Library provided hard copies of articles prior to the current era of interlibrary access. Scott Edwards forwarded copies of bulletin issues from Harvard University. Diane Packett and Guofan Shao translated articles in Spanish and Chinese, respectively.

I also thank the curators and directors of the following collections and institutions who provided me with data for the second edition: The Burke Museum, University of Washington; Departamento de Zoologia da Universidade Federal de Minas Gerais, Brazil; Instituto de Ciencias Naturales, Universidad Nacional de Colombia; Museo Nacional, Rio de Janiero, Brazil; Universiteit van Leiden.

The following collections and institutions provided data for the first edition, much of which is included in the current edition: University of Arizona, Royal Ontario Museum, University of Alaska, Texas Cooperative Wildlife Collections (Texas A & M University), Louisiana State University, Philadelphia Academy of Natural Sciences, Field Museum of Natural History (Chicago, Illinois), Cowan Vertebrate Museum (University of British Columbia), California State University – Sacramento, University of Oklahoma, University of Washington, University of Minnesota, Delaware Museum of Natural History, Denver Museum of Natural History, Yale Peabody Museum, Museum of Vertebrate Zoology (University of California – Berkeley), Western Foundation of Vertebrate Zoology (Los Angeles, California), National History Museum (Bulawayo, Zimbabwe), New Zealand National Museum (Wellington, New Zealand), Los Angeles County Museum (California), U.S. National Museum, British Museum of Natural History, and the San Diego Museum of Natural History.

The following observatories and organizations provided unpublished data from their research programs since the publication of the first edition: Africam Safari, Puebla, Mexico; SalvaNatura, El Salvador; the MAPS (Monitoring Avian Productivity and Survivorship) program, United States (Phil Nott, Dave DeSante); Endemic Species Research Institute, Museum of Taiwan. For the first edition, data were provided by: Long Point Bird Observatory (Ontario, Canada), Point Reyes Bird Observatory (California), Manomet Bird Observatory (Massachusetts), Braddock Bay Raptor Research Project (New York), Cape May Bird Observatory and the Cape May Raptor Banding Project (New Jersey), Hastings Reservation (California), National Audubon Society, Powdermill Nature Center (Pennsylvania), Canadian Wildlife Service, and the U.S. Fish and Wildlife Service. The Zoological Society of Philadelphia, Milwaukee County Zoological Park (Wisconsin), Kobi Oji Zoo (Kobe, Japan), New York Zoological Society, and the Audubon Zoological Gardens (New Orleans, Louisiana) provided data from captive birds.

This project benefited from the cooperation of a large number of ornithologists who provided unpublished data from their research files. The predecessors to this volume included Dunning (1984) and the first edition of this handbook; both included much previously unpublished data. The internet has allowed me to contact researchers around the world to a degree that was simply impossible for the previous publication; however, I am sure that additional data remains to be uncovered. I thank all those who responded to my letters and emails with data or suggestions.

John Blake and Bette Loiselle (Panama and Costa Rica), Robert Sutton (Jamaica), Doug Stotz (Brazil), and John Fitzpatrick (Peru) provided extensive unpublished data for the first edition. Charles Collins provided me with his unpublished compilation of mass data for swifts. Although these and other unpublished sources were used in Dunning (1992), I have chosen to continue to cite the original sources as "unpublished compilations" even though, technically, the data were published in my first edition. Retaining the original citation gives credit to the researchers. I have changed the wording of the full citations in the Literature Cited section to reflect when the data were previously reported in Dunning (1992). More recently my requests for mass data were circulated in specific regions by John Innes (New Zealand) and Desmond Allen (Japan), resulting in much better coverage of these avifauna.

Many people have contributed data to the compilation, set me on the path of new sources, or responded to my queries. I thank the following individuals who have contributed to the project since the publication of the first edition: Isa Alpino, Elisabeth Ammon, Leticia Andino, Volker Bahn, Peter Barthel, John Bates, Jim Beckelman, F. Betancourt, Robin Bjork, Rick Bowers, Reed Bowman, Tom Braile, Mike Braun, Mark Brazil, Eliot Brenowitz, Dan Buden, Carlos Cadena, Juan Cesar Canales-Delgadillo, Ya Cheng-te, Dan Christian, Alice Cibois, Charles Collins, Juan Cornejo, Oswaldo Cortes, Andres Cuervo, Robert Curry, James Dean, Jenny Dell, Mike Dickison, Peter Dostine, Andrzej Dyrcz, Scott Edwards, Stephen Ervin, Robert Faucett, Edward Fendick, Chris Filardi, John Fitzpatrick, Jon Fjeldså, Padu Franco, Alejandro Grajal, Russell Greenberg, Joseph Grzybowski, Amanda Hale, Gergo Halmos, Sebastian Herzog, Janet Hinshaw, Richard Holmes, Jon Hornbuckle, Wayne Hsu, Jocelyn Hudon, Juan Carlos Illera, Jim Ingold, K. Ishida, T. Ishizuka, Sue Jackson, Fabian Jaksic, Stephanie Jones, Leo Joseph, K. Kazuto, Howard King, Guy Kirwan, Oliver Komar, Reiko Kurosawa, Adrian Long, N. Krabbe, Andrew Kratter, Thomas Labedz, Steve Latta, Oscar Laverde, Scott Ruey-Shing Lin, Weiting Liu, John Loegering, Michel Louettte, Stephen Loughead, Joao Magahaes, Miguel Marini, Tom Martin, Virginia Mascitti, Mara McDonald, Robert McFarlane, Ross McGregor, John McLaughlin, Martin McNicholl, Chris Mead, Allan Mee, Andy Mitchell, William Moskoff, Jim Mountjoy, H. Nagata, Phil Nott, Hugo Ochoa-Acuna, Jose Ochoa Quintero, Dieter Oschadleus, Luis Padilla, Patricia Parker, Robert Payne, George Powell, Paulo Pulgarin, Van Remsen, Swen Renner, Sam Riffell, Chris Rimmer, James Rising, Frank Rivera, Adriana Rodriguez, Sievert Rohwer, Tom Skovlund Romdal, Diego Rubolini, Doug Runde, Peter Ryan, Paul Salaman, Diego Santiago, Martin Schaefer, Jessie Schillaci, Nathalie Seddon, Chad Seewagen, Shin-Ichi Seki, Marisol Sepulveda, Cagan Sekercioglu, Jui-yun Shen, Jevgeni Shergain, Harry Slack, Beth Slikas, Troy Smith, Justine Stahl, Doug Stotz, Dan Strickland, Hiraoka Takashi, Michael Tarburton, Jean-Claude Thibault, Richard Thomas, Joseph Tobias, Thomas Valqui, Jan van der Brugge, Marcelo Vasconcelos, Heath Wakelee, George Wallace, Eric Walters, Matthias Waltert, Bruno Walther, David Wege, Allison Wells, Ethan White, David Wiedenfeld, James W. Wiley, Stephen Willis, Maiken Winter, Christopher Witt, John Woinarski, Rueven Yosef, and Sun Yue-Hua.

In addition to the individuals cited above, I thank the following people for providing data that were used in the preparation of the first edition (much of which was retained in the current compilation): Keith Arnold, Jon Atwood, Yves Aubry, Tom Bancroft, Luis Baptista, Jon Barlow, Carl Barrentine, John Bates, Range Bayer, Craig Benkman, Byron Berger, Richard Bierregaard, Linda Birch, Sharon Birks, Robin Bjork, Charles Blake, John Blake, David Blockstein, E. Clark Bloom, Peter Bloom, Jeffrey Bouton, Rick Bowers, Roger Boyd, Dawn Breese, H. Brieschke, Kay Burk, Bill Calder, Peter Cannell, Dick Cannings, David Capin, Charles Chase, Jane Church, Roger Clapp, George Clark, William Clark, Mario Cohn-Haft, Charles Collins, Charles Corchran, Malcolm Coulter, Humphrey Crick, Alexander Cruz, Christian Dau, Dave DeSante, James Dick, Edward Diebold, George Divoky, Robert Duncan, Erica Dunn, Bruce Eichhorst, Susan Elbin, John Fitzpatrick, Charles Francis, Daniel Gibson, Robert Gill, David Goldstein, Walter Graul, Ken Graupman, John Groves, Joseph Grzybowski, Jeremy Hatch, Rod Hay, Sue Heath, Paul Hendricks, Joanna Hill, Larry Hood, Robert Hosea, C. Stuart Houston, Peter Hubbell, Steve Hubbell, George Hunt, John Innes, Ross James, Joseph Jehl, Kent Jenson, Kent Johnson, Ned Johnson, Virginia Johnson, Jim Jolly, Philip Kahl, Kenn Kaufman, Paul Kerlinger, Brina Kessel, Lloyd Kiff, Greg Lasley, Peter Lawson, Calvin Lensink, Fred

Lohrer, Bette Loiselle, Annarie Lyles, Carl Marti, Tom Martin, Harold Mayfield, H. Elliott McClure, Bonnie McKinney, Barbara McKnight, Martin McNicholl, Davide Melville, Rita Mesquita, L. Richard Mewaldt, Alex Middleton, George Millikan, Burt Monroe, R. J. Moorhouse, P. J. Mundy, Koichi Murata, Nancy Newfield, David Niles, Robert Norton, Gary Nuechterlein, Yoshika Oniki, Jose Ottenwalder, James Otto, George Page, Dennis Paulson, Donald Payne, Margaret Peterson, Lisa Petit, Kenneth Prescott, Trevor Price, Ronald Pulliam, Elizabeth Pullman, Michael Putnam, William Radke, John Ratti, Amadeo Rea, Roland Redmond, Kerry Reese, J. Van Remsen, Robert Ricklefs, Mark Robbins, Sievert Rohwer, Roland Roth, Ruth Russell, Steve Russell, Andrew Sanders, S. D. Schemnitz, Dolph Schluter, Gary Schnell, Donald Schroeder, Chris Schultz, Peter Shannon, David Shepherd, Jay Sheppard, Doug Stotz, Joseph Strauch, Robert Sutton, John Terborgh, Bernie Tershy, Betsy Trent Thomas, Robert Tintle, Pepper Trail, Jolan Truan, Joan Tweit, Robert Tweit, Jerry Verner, D.L. Walker, George Wallace, Philip Walters, Richard Weatherly, Betsy Webb, John Weske, Pamela Williams, Kerry-Jayne Wilson, David Wingate, J. Wood, D. Scott Woods, and Richard Zusi.

Body Masses of World Birds

Species Name	Common Name	Sex	N	Mean	S.D.	Min	Max	Season	Location	Source #
Family Struthionidae										
Struthio camelus	Ostrich	U	13	111000		86000	145000			256
Family Rheidae										
Rhea americana	Greater Rhea	U		23000		10500	40000			133a
Rhea pennata	Lesser Rhea	M	3	23900		19100	28600			256
Family Casuariidae										
Casuarius casuarius	Southern Cassowary	B		44000		29200	58500			133a
Casuarius bennetti	Dwarf Cassowary	U		35000						291
Casuarius unappendiculatus	Northern Cassowary	M		36600						256
		F	1	58000						
Family Dromaiidae										
Dromaius novaehollandiae	Emu	M	11	31500		17700	46500			256
		F	11	36900		25900	48000			
Family Apterygidae										
Apteryx australis	Brown Kiwi	M	15	2120		1720	2730		New Zealand	197
		F	31	2540		2060	3850			
Apteryx owenii	Little Spotted Kiwi	M	51	1135	119	880	1356		New Zealand	745
		F	41	1351	164	1000	1400			
Apteryx haastii	Great Spotted Kiwi	M	7	1692		1215	2320		New Zealand	745
		F	12	2418	389	1530	2718			
Family Tinamidae										
Tinamus tao	Gray Tinamou	M	113	1565		1325	1863			256
		F	54	1636		1430	2080			
Tinamus solitarius	Solitary Tinamou	M	7	1284		1014	1469			256
		F	11	1497		1300	1710			
Tinamus osgoodi	Black Tinamou	M	1	1285					Peru	370

Body Masses of World Birds (continued)

Species Name	Common Name	Sex	N	Mean	S.D.	Min	Max	Season	Location	Source #
Tinamus major	Great Tinamou	M	16	960		885	1142			256
		F	7	1097		945	1249			
Tinamus guttatus	White-throated Tinamou	M	3	638		623	652			256
		F	4	738		680	800			
Nothocercus bonapartei	Highland Tinamou	B	8	763		455	1050		Panama; Colombia	488, 489, 798
Crypturellus berlepschi	Berlepsch's Tinamou	M				430	527			256
		F				512	615			
Crypturellus cinereus	Cinereous Tinamou	M	7	460		355	527			256, 498
		F	4	558		512	615			
Crypturellus erythropus	Red-legged Tinamou	M	1	485						498
Crypturellus soui	Little Tinamou	M	6	198		165	204		Belize	988, 1027
		F	8	236		213	268			
Crypturellus obsoletus	Brown Tinamou	M	55	421		358	482			256
		F	88	467		395	548			
Crypturellus undulatus	Undulated Tinamou	M	6	513		462	569			256
		F	1	621						
Crypturellus transfasciatus	Pale-browed Tinamou	B	6	283					Peru	1302
Crypturellus strigulosus	Brazilian Tinamou	M	27	412		332	464			256
		F	8	450		388	500			
Crypturellus duidae	Gray-legged Tinamou	B	6	426		370	489		Peru	10
Crypturellus noctivagus	Yellow-legged Tinamou	U		562		533	602			256
Crypturellus atrocapillus	Black-capped Tinamou	M	2	453		450	455		Peru	370
Crypturellus boucardi	Slaty-breasted Tinamou	M	22	418	20.6	375	464		Belize	1027
		F	18	454	45.4	374	526			
Crypturellus variegatus	Variegated Tinamou	B	9	378		354	423			256

Body Masses of World Birds (continued)

Species Name	Common Name	Sex	N	Mean	S.D.	Min	Max	Season	Location	Source #
Crypturellus cinnamomeus	Thicket Tinamou	B	8	422		401	448		Mexico; Belize	910a, 1027
Crypturellus bartletti	Bartlett's Tinamou	U	6	241					Peru	370
Crypturellus parvirostris	Small-billed Tinamou	M	9	180		141	205			256, 657
		F	10	220		176	250			
Crypturellus tataupa	Tataupa Tinamou	M	38	202		169	229			256
		F	53	237		189	281			
Rhynchotus rufescens	Red-winged Tinamou	M	19	803		700	920			256
		F	16	886		580	1040			
Nothoprocta ornata	Ornate Tinamou	M	10	569		444	700			256
		F	10	674		593	761			
Nothoprocta perdicaria	Chilean Tinamou	U	4	458					captive	1328
Nothoprocta cinerascens	Brushland Tinamou	M	6	479		457	493			256
		F	4	573		540	615			
Nothoprocta pentlandii doeringi	Andean Tinamou	U	10	266		211	330			256
Nothoprocta pentlandii mendozae	Andean Tinamou	U	14	340		301	384			256
Nothoprocta curvirostris	Curve-billed Tinamou	U	2	376		372	380		Peru	1042
Nothura boraquira	White-bellied Tinamou	B	7	283		239	337			256
Nothura minor	Lesser Nothura	B				158	174			256
Nothura darwinii darwinii	Darwin's Nothura	M	13	215		200	245		Argentina	151
		F	16	243		209	285			
Nothura darwinii salvadorii	Darwin's Nothura	M	27	250		192	307		Argentina	151
		F	27	274		197	330			
Nothura maculosa annectens	Spotted Nothura	M	90	250		193	297	W	Argentina	151
		F	92	289		192	361			
Nothura maculosa maculosa	Spotted Nothura	M	198	234		152	292		Brazil	256
		F	182	260		164	316			

Body Masses of World Birds (continued)

Species Name	Common Name	Sex	N	Mean	S.D.	Min	Max	Season	Location	Source #
Taoniscus nanus	Dwarf Tinamou	U	1	43						256
Eudromia elegans elegans	Elegant Crested Tinamou	M		680			809		Argentina	88
		F		709			892			
Eudromia elegans albida	Elegant Crested Tinamou	M		676			695		Argentina	88
		F		749			836			
Tinamotis pentlandii	Puna Tinamou	U	1	650						362
Family Spheniscidae										
Aptenodytes patagonicus	King Penguin	M	70	12435		10200	15300	B	Marion Is.	1036
		F	71	11067		8600	13700			
Aptenodytes forsteri	Emperor Penguin	M	9	38200				B	Antarctica	745
		F	7	29500						
Pygoscelis papua	Gentoo Penguin	M	32	6400	700			B	South Georgia	1172a
		F	32	5500	650					
Pygoscelis adeliae	Adelie Penguin	M	15	5000				B	Ross Is.	3
		F	10	4700						
Pygoscelis antarcticus	Chinstrap Penguin	M	19	4435	315			B	South Shetland Is.	230
		F	25	3876	310					
Eudyptes pachyrhynchus	Fiordland Penguin	M	38	4110	390			B	New Zealand	829
		F	24	3710	400					
Eudyptes robustus	Snares Penguin	M	41	3320	347	2450	4300			745
		F	32	2780	300	2300	3400			
Eudyptes sclateri	Erect-crested Penguin	M	22	6382	520			B	Antipodes Is.	745
		F	22	5434	431					
Eudyptes chrysocome	Rockhopper Penguin	M	49	2430	230			B	Antipodes Is.	745
		F	33	2230	140					
Eudyptes schlegeli	Royal Penguin	M		4500				B	Macquaire Is.	536
		F		4000						
Eudyptes chrysolophus	Macaroni Penguin	M	71	4280	120	2300	6000		Crozet Is.	745
		F	52	4700	120	3100	6800			
Megadyptes antipodes	Yellow-eyed Penguin	M	80	5530	340	4880	6350	B	New Zealand	745
		F	80	5130	320	4310	5780			

Body Masses of World Birds (continued)

Species Name	Common Name	Sex	N	Mean	S.D.	Min	Max	Season	Location	Source #
Eudyptula minor novaehollandiae	Little Penguin	M	12278	1172	164	550	2130		Australia	745
		F	10973	1048	159	550	2100			
Eudyptula minor albosignata	Little (Blue) Penguin	M	11	1358	182	1067	1546		New Zealand	745
		F	7	1280		1108	1500			
Eudyptula minor iredalei	Little Penguin	M	11	935	96.6	752	1055		New Zealand	745
		F	6	842		698	956			
Eudyptula minor minor	Little Penguin	M	10	1244	127	1080	1475		New Zealand	745
		F	9	1085		895	1374			
Spheniscus demersus	Jackass Penguin	M	127	3310	260				South Africa	308
		F	127	2960	310					
Spheniscus humboldti	Humboldt Penguin	M	165	4711	406	3450	6000		Peru	1364
		F	123	4047	390	2950	5400			
Spheniscus magellanicus	Magellanic Penguin	M	49	4470	490			B	Argentina	745
		F	49	3770	400					
Spheniscus mendiculus	Galapagos Penguin	M	56	2135		1780	2880	B	Galapagos Is.	1312
		F	45	1730		1500	2000			
Family Gaviidae										
Gavia stellata	Red-throated Loon	U	8	1486		1150	1923		Ontario, Canada	1147a
Gavia arctica	Arctic Loon	M	5	3494		3100	4000		Siberia; Mongolia	926, 1025
		F	3	2665		2255	3075			
Gavia pacifica	Pacific Loon	M	8	1830		1135	2450		Alaska, USA	1149
		F	8	1507		990	2038			
Gavia immer	Common Loon	M	20	5460		4770	6130	B	Ontario, Canada	779
		F	20	4500		3380	4650			
Gavia adamsii	Yellow-billed Loon	M	14			4000	5800			860
		F	13			4025	6400			
Family Podicipedidae										
Tachybaptus ruficollis	Little Grebe	M	5	140		133	146	PB	Netherlands	223
		F	8	130		117	142			
Tachybaptus ruficollis philippensis	Little Grebe	M	14	222		190	250		Philippines	983
		F	11	204		175	235			

Body Masses of World Birds (continued)

Species Name	Common Name	Sex	N	Mean	S.D.	Min	Max	Season	Location	Source #
Tachybaptus novaehollandiae	Australasian Grebe	U	36	219	25	166	281		New South Wales, Australia	745
Tachybaptus pelzelnii	Madagascar Grebe	U	2	184		172	197		Madagascar	68, 957
Tachybaptus dominicus	Least Grebe	M	12	129	11.4				Panama	489
		F	12	116	11.8					
Tachybaptus dominicus	Least Grebe	M	16	139	16.9	112	164		Texas, USA; Costa Rica	1176
		F	16	122	16	94	150			
Tachybaptus dominicus	Least Grebe	M	5	161		134	182		West Indies	1176
		F	4	133		81	167			
Podilymbus podiceps	Pied-billed Grebe	M	36	474	60.6	321	568			827
		F	40	358	51	253	479			
Podilymbus gigas	Atitlan Grebe	M	2	830		804	856		Guatemala	664
		F	2	568		552	584			
Rollandia rolland	White-tufted Grebe	M	14	248	42					699
		F	11	424	18					
Rollandia microptera	Short-winged Grebe	B	19	706						699
Poliocephalus poliocephalus	Hoary-headed Grebe	M	14	258	30.2	202	311			1174
		F	14	223	27.6	190	276			
Poliocephalus rufopectus	New Zealand Grebe	B	15	249		170	291		New Zealand	1174
Podiceps major	Great Grebe	F	1	1646					Tierra del Fuego	546
Podiceps grisegena	Red-necked Grebe	B	6	1023		743	1270		Alaska, USA	894
Podiceps cristatus	Great Crested Grebe	M	7	738		596	813		Netherlands	223
		F	4	609		568	686			
Podiceps cristatus	Great Crested Grebe	F	4	869		630	1000		Tsingtao, China	1088
Podiceps auritus	Horned Grebe	U	47	453		327	528	W	Maryland, USA	525
Podiceps nigricollis	Eared Grebe	M	394	422	39.7	195	521	F	Utah, USA	238
		F	338	374	31.1	281	477			

Body Masses of World Birds (continued)

Species Name	Common Name	Sex	N	Mean	S.D.	Min	Max	Season	Location	Source #
Podiceps nigricollis	Eared Grebe	M	4	475		430	520		Tsingtao, China	1088
		F	7	386		335	440			
Podiceps occipitalis	Silvery Grebe	M	13	335	53					699
		F	7	307						
Podiceps taczanowskii	Junin Grebe	M	6	423						699
		F	7	360						
Podiceps gallardoi	Hooded Grebe	B	4	575		420	740			1173
Aechmophorus occidentalis occidentalis	Western Grebe	M	41	1429	163	1137	1826			1177
		F	21	1199	209	808	1753			
Aechmophorus occidentalis ephemeralis	Western Grebe	M	9	1138		828	1338			1177
		F	4	857		823	912			
Aechmophorus clarkii transitionalis	Clark's Grebe	M	25	1341	162	1001	1685			1177
		F	6	1133		1006	1258			
Aechmophorus clarkii clarkii	Clark's Grebe	M	7	1029		815	1201			1177
		F	9	906		718	1251			
Family Diomedeidae										
Diomedea exulans	Wandering Albatross	M	52	9110		8200	10600		South Georgia	1222
		F	53	7270		6350	8300			
Diomedea exulans gibsoni	Wandering Albatross	M	35	6800		5500	8600		Auckland Is.	1222
		F	36	5800		4600	7300			
Diomedea exulans antipodensis	Wandering Albatross	M	10	7460					Antipodes Is.	1222
		F	15	5840						
Diomedea epomophora	Royal Albatross	M	11	10300	740			B	Campbell Is.	745
		F	7	7700						
Diomedea epomophora sanfordi	Royal Albatross	U				6530	6800	B	Chatham Is.	745
Diomedea amsterdamensis	Amsterdam Is. Albatross	B	33	6270	780	4800	8000	B	Amsterdam Is.	611
Phoebastria albatrus	Short-tailed Albatross	U		4411						339
Phoebastria irrorata	Waved Albatross	M	7	3750		3110	4390			133
		F	13	3040		2700	3380			

Body Masses of World Birds (continued)

Species Name	Common Name	Sex	N	Mean	S.D.	Min	Max	Season	Location	Source #
Phoebastria immutabilis	Laysan Albatross	M	152	3310	284	2700	4100	B	Midway Is.	384
		F	54	2990	264	2500	3600			
Phoebastria nigripes	Black-footed Albatross	M	123	3400		2600	4300		Hawaiian Is.	1222
		F	104	2990		2600	3600			
Thalassarche chrysostoma	Gray-headed Albatross	M	133	3751					South Georgia	943
		F	95	3264						
Thalassarche melanophrys melanophrys	Black-browed Albatross	M	132	3922					South Georgia	943
		F	94	3206						
Thalassarche melanophrys impavida	Black-browed Albatross	M	15	3100		2750	3800		Campbell Is.	1222
		F	18	2700		2200	3150			
Thalassarche bulleri platei	Buller's Albatross	M	18	2840		2500	3300	B	Chatham Is.	1222
		F	18	2430		2150	2800			
Thalassarche bulleri bulleri	Buller's Albatross	M	6	3120		2850	3350	B	Snares Is.	1222
		F	6	2780		2050	3100			
Thalassarche cauta cauta	Shy Albatross	M	18	4350		3900	5100			745
		F	18	3700		3200	4400			
Thalassarche cauta steadi	Shy Albatross	M	6	4430		3300	5300	B	Auckland Is.	1222
		F	7	3450		2600	4200			
Thalassarche cauta salvini	Shy Albatross	M	17	4000		3300	4900	B	Bounty Is.	1222
		F	12	3590		3300	3700			
Thalassarche cauta eremita	Shy Albatross	M	13	4000		3600	4700	B	Chatham Is.	1222
		F	10	3770		3100	3900			
Thalassarche chlororhynchos bassi	Yellow-nosed Albatross	U	33	2060		1750	2600	B	Amsterdam Is.	1222
Thalassarche chlororhynchos	Yellow-nosed Albatross	U	26	2200		1780	2840	B	Gough Is.	1222
Phoebetria fusca	Sooty Albatross	B	176	2500	200	2100	3400	B	Marion Is.	71
Phoebetria fusca	Sooty Albatross	M	18	2730		2200	3250		Iles Crozet	1222
		F	12	2440		2100	2800			

Body Masses of World Birds (continued)

Species Name	Common Name	Sex	N	Mean	S.D.	Min	Max	Season	Location	Source #
Phoebetria palpebrata	Light-mantled Albatross	U	13	3150		2850	3600		Iles Crozet	1222
Family Procellariidae										
Macronectes giganteus	Antarctic Giant Petrel	M	37	4940	410	4100	5800		S. Orkney Is.	745
		F	37	3850	370	3000	4800			
Macronectes giganteus	Antarctic Giant Petrel	M	15	3500	300			B	Patagonia, Argentina	217a
		F	21	2500	200					
Macronectes halli	Hall's Giant Petrel	M	57	4790	330	4100	5450	B	Macquarie Is.	745
		F	49	3580	300	3100	4450			
Macronectes halli	Hall's Giant Petrel	M	56	4902	391	4150	5800	B	South Georgia	745
		F	43	3724	313	3050	4500			
Fulmarus glacialis	Northern Fulmar	M	47	649	52.1	552	773	B	Alaska, USA	495
		F	57	577	52.8	445	710			
Fulmarus glacialoides	Southern Fulmar	M	12	845		740	1030	B	Antarctica	745
		F	9	745		670	855			
Thalassoica antarctica	Antarctic Petrel	M	10	699		610	780		Antarctica	745
		F	10	718		610	805			
Daption capense capense	Cape Petrel	M	23	442		380	550	B	Signy Is.	745
		F	22	407		360	510			
Daption capense	Cape Petrel	M	52	452	40	345	577	B	Snares Is.	745
		F	43	419	31	342	476			
Pagodroma nivea	Snow Petrel	B	52	268	35.9	202	322	B	Antarctica	1032
Pterodroma macroptera gouldii	Great-winged Petrel	M	56	560				B	New Zealand	551
		F	28	505						
Pterodroma aterrima	Mascarene Petrel	M	1	232					Reunion	610
Pterodroma rostrata	Tahiti Petrel	U	141	409	38					1141
Pterodroma lessonii	White-headed Petrel	B	18	698	58.6	580	810	B	Kerguelen Is.	1284
Pterodroma hasitata	Black-capped Petrel	U	1	278					Dominican Republic	888

Body Masses of World Birds (continued)

Species Name	Common Name	Sex	N	Mean	S.D.	Min	Max	Season	Location	Source #
Pterodroma cahow	Bermuda Petrel	U	1	246					Bermuda	1321
Pterodroma incerta	Atlantic Petrel	U		520						2
Pterodroma alba	Phoenix Petrel	U	20	259	28.1			B	Christmas Is.	1301
Pterodroma inexpectata	Mottled Petrel	U	89	316	32.1	247	441	B	New Zealand	969
Pterodroma solandri	Providence Petrel	M	3	518		443	600		Lord Howe Is.	745
		F	2	424		414	433			
Pterodroma ultima	Murphy's Petrel	U	13	360		325	377		Pitcairn Is.	1309
Pterodroma neglecta	Kermadec Petrel	U	9	501		370	590	B	Macauley Is.; Chile	745, 861b
Pterodroma magentae	Magenta Petrel	U	17	465	21.6	423	507			745
Pterodroma arminjoniana	Trinidade Petrel	U	68	394	43	287	460		Round Is.	745
Pterodroma heraldica	Herald Petrel	U	4	266	29					1140
Pterodroma mollis	Soft-plumaged Petrel	B	85	312	34.7	250	380	B	Marion Is.	1059
Pterodroma feae	Cape Verde Petrel	U	17	311	20.6	275	355		Desertas	125a
Pterodroma madeira	Madeira Petrel	B				175	231			1130
Pterodroma baraui	Barau's Petrel	U		400					estimated	123
Pterodroma phaeopygia	Galapagos Petrel	U	37	389	40				Santa Cruz, Galapagos	1226
Pterodroma phaeopygia	Galapagos Petrel	U	17	413	40				Floreana, Galapagos	1226
Pterodroma sandwichensis	Hawaiian Petrel	B	38	434	52.9			B	Hawaiian Is.	1113
Pterodroma externa	Juan Fernandez Petrel	U	207	428	47.5					1141
Pterodroma cervicalis	White-necked Petrel	U	30	445	31.5	380	545		Kermadec Is.	745
Pterodroma occulta	Vanuatu Petrel	U				300	350		estimated	551a

Body Masses of World Birds (continued)

Species Name	Common Name	Sex	N	Mean	S.D.	Min	Max	Season	Location	Source #
Pterodroma cookii	Cook's Petrel	M	3	164		112	250		New Zealand	223
		F	7	193		149	241			
Pterodroma defilippiana	Defilippe's Petrel	U	6	169		143	220			125a, 1148
Pterodroma leucoptera	Gould's Petrel	U	129	159	14.8					1141
Pterodroma leucoptera brevipes	Gould's Petrel	B	20	136		116	158	B	Fiji	1277
Pterodroma hypoleuca	Bonin Petrel	B	168	176	13	150	220	B	Hawaiian Is.	486
Pterodroma nigripennis	Black-winged Petrel	B	30	163	10.9	132	185		Kermadec Is.	745
Pterodroma nigripennis	Black-winged Petrel	B	15	185	19.6	156	228	B	Chatham Is.	745
Pterodroma axillaris	Chatham Petrel	F	1	165					Chatham Is.	745
Pterodroma longirostris	Stejneger's Petrel	U	27	143	9					1140
Pterodroma pycrofti	Pycroft's Petrel	B	80	159	16.8	128	198		New Zealand	745
Pterodroma macgillivrayi	Fiji Petrel	U	1	143					Fiji	1278
Halobaena caerulea	Blue Petrel	U	215	202	17.5	163	251	B	Marion Is.	396a
Pachyptila vittata	Broad-billed Prion	U	29	196				B	New Zealand	969
Pachyptila salvini	Salvin's Prion	B	89	159	13	130	210	B	Crozet Is.	115
Pachyptila desolata	Antarctic Prion	U	118	147	14.1	115	183	B	Kerguelen Is.	1284
Pachyptila belcheri	Slender-billed Prion	U	66	145	12.5	118	180	B	Kerguelen Is.	1284
Pachyptila crassirostris crassirostris	Fulmar Prion	M	9	149		134	172	B	Chatham Is.	745
		F	8	141		110	157			
Pachyptila crassirostris eatoni	Fulmar Prion	U	25	126	10.1	102	144		Heard Is.	745
Pachyptila turtur	Fairy Prion	U	54	137		110	175	B	Crozet Is.	115

Body Masses of World Birds (continued)

Species Name	Common Name	Sex	N	Mean	S.D.	Min	Max	Season	Location	Source #
Bulweria bulwerii	Bulwer's Petrel	U	191	99	13.8	78	130	B	Hawaiian Is.	486
Bulweria fallax	Jouanin's Petrel	U	8	173		150	185		Indian Ocean	31
Procellaria cinerea	Gray Petrel	U	37	1131	133	900	1520	B	Kerguelen Is.	1284
Procellaria aequinoctialis	White-chinned Petrel	U	87	1213	134	980	1885	B	Crozet Is.	611a
Procellaria parkinsoni	Parkinson's Petrel	M	9	723		620	855	B	Little Barrier Is., Australia	745
		F	8	682		587	791			
Procellaria westlandica	Westland Petrel	U	67	1199	109			B	New Zealand	745
Aphrodroma brevirostris	Kerguelen Petrel	U	126	357	43.2	255	451	B	Marion Is.	1059
Calonectris leucomelas	Streaked Shearwater	B	181	580	66.7	435	713		Korea	679
Calonectris diomedea	Cory's Shearwater	U	37	535	60	400	650	B	Crete	1020
Puffinus creatopus	Pink-footed Shearwater	B	19	744	56.3	600	825		Chile	1082
Puffinus carneipes	Flesh-footed Shearwater	B	9	609		533	692	B	New Zealand	745
Puffinus gravis	Greater Shearwater	U	58	849	26.7			PB	Nova Scotia, Canada	141
Puffinus pacificus	Wedge-tailed Shearwater	U	124	388	33.4	320	510	B	Hawaiian Is.	486
Puffinus bulleri	Buller's Shearwater	B	30	407	40.6	339	449		New Zealand	745
Puffinus griseus	Sooty Shearwater	U	100	787		666	978	B	New Zealand	894
Puffinus tenuirostris	Short-tailed Shearwater	U	67	559	53	460	740	B	Australia	745
Puffinus nativitatis	Christmas Shearwater	U	99	356	30	280	415	B	Hawaiian Is.	486
Puffinus puffinus	Manx Shearwater	M	60	468	34.9			B	Wales, UK	124
		F	53	439	30.6					

Body Masses of World Birds (continued)

Species Name	Common Name	Sex	N	Mean	S.D.	Min	Max	Season	Location	Source #
Puffinus mauretanicus	Balearic Shearwater	U	448	497	44.6			B	Spain	16a
Puffinus yelkouan	Levantine Shearwater	B				330	485			1130
Puffinus huttoni	Hutton's Shearwater	U	17	364				B	New Zealand	363
Puffinus opisthomelas	Black-vented Shearwater	B	258	408	44	332	545		Natividad Is., Mexico	624
Puffinus auricularis townsendi	Townsend's Shearwater	B	10	323		290	358			6
Puffinus auricularis newelli	Newell's Shearwater	U	35	381	45	340	411	B	Hawaiian Is.	6
Puffinus gavia	Fluttering Shearwater	M	2	221		200	243		New Zealand	745
		F	8	237		206	302			
Puffinus assimilis	Little Shearwater	U	91	226		170	275	B	Gough Is., S. Atlantic Ocean	1192
Puffinus lherminieri	Audubon's Shearwater	U	78	168		128	211	B	Galapagos Is.	481
Puffinus lherminieri bailloni	Audubon's Shearwater	U	12	226	28.2	165	259		Reunion	610
Puffinus lherminieri nicolae	Audubon's Shearwater	U	62	168	19				Seychelles	114
Puffinus lherminieri colstoni	Audubon's Shearwater	U	8	214					Aldabra	114
Puffinus persicus	Persian Shearwater	U	1	195					Indian Ocean	31
Family Hydrobatidae										
Garrodia nereis	Gray-backed Storm-Petrel	U	38	38.2	3.5	31	44	B	Kerguelen Is.	1284
Oceanites oceanicus	Wilson's Storm-Petrel	U	31	32	3	27	39	B	Crozet Is.	611a
Oceanites oceanicus	Wilson's Storm-Petrel	B	9	28.9		25	35		Chile	1082

Body Masses of World Birds (continued)

Species Name	Common Name	Sex	N	Mean	S.D.	Min	Max	Season	Location	Source #
Oceanites gracilis	White-vented Storm-Petrel	U	3	16.7		16	17		Galapagos Is.	482, 1141a
Pelagodroma marina maoriana	White-faced Storm-Petrel	U	100	47.2	4	40	62		New Zealand	223
Fregetta tropica	Black-bellied Storm-Petrel	U	38	52	3	43	59	B	Crozet Is.	611a
Fregetta grallaria grallaria	White-bellied Storm-Petrel	U	35	52	4.7	45	65	B	Lord Howe Is.	745
Fregetta grallaria segethi	White-bellied Storm-Petrel	U	7	38.9		34	43		Humboldt Current, Pacific	1141a
Fregetta grallaria titan	White-bellied Storm-Petrel	U	1	64					E. Pacific Ocean	1141a
Nesofregetta fuliginosa	Polynesian Storm-Petrel	U	6	70	5					1140
Hydrobates pelagicus	European Storm-Petrel	U	1770	25.2		20.3	31.1	B	Britain	515
Oceanodroma microsoma	Least Storm-Petrel	U	11	20.5	1.1	18.2	21.7		E. Pacific Ocean	227
Oceanodroma tethys	Wedge-rumped Storm-Petrel	U	57	22.6	2.2	19	30	B	Galapagos Is.	482
Oceanodroma castro	Band-rumped Storm-Petrel	U	102	45.3	5.1	33.5	54	B	Galapagos Is.	482
Oceanodroma castro	Band-rumped Storm-Petrel	U	229	44.1	4.4	33	58	B	Azores Is. (hot season)	1116
Oceanodroma castro	Band-rumped Storm-Petrel	U	729	49.2	4.5	36	67	B	Azores Is. (cool season)	1116
Oceanodroma leucorhoa leucorhoa	Leach's Storm-Petrel	B	150	41.4	5.8	24	57.5		Pacific Ocean	227
Oceanodroma leucorhoa beali	Leach's Storm-Petrel	B	124	39.8	5	29	53		E. Pacific Ocean	227
Oceanodroma leucorhoa socorroensis	Leach's Storm-Petrel	B	152	31.7	4.2	22.4	44		Guadalupe Is.	227
Oceanodroma leucorhoa chapmani	Leach's Storm-Petrel	B	78	34.8	2.7	29	42	B	Baja California, Mexico	227

Body Masses of World Birds (continued)

Species Name	Common Name	Sex	N	Mean	S.D.	Min	Max	Season	Location	Source #
Oceanodroma monorhis	Swinhoe's Storm-Petrel	U	4	39.8		39	40			31, 223
Oceanodroma tristrami	Tristram's Storm-Petrel	B	80	92	8.2	71	112	B	Laysan Is., Hawaiian Is.	756
Oceanodroma tristrami	Tristram's Storm-Petrel	U	23	86.2	6	74	94	B	Nihoa Is., Hawaiian Is.	1116
Oceanodroma markhami	Markham's Storm-Petrel	B	40	53.1						125a
Oceanodroma matsudairae	Matsudaira's Storm-Petrel	M	1	62						745
Oceanodroma melania	Black Storm-Petrel	B	112	61.5	5	53	77	B	California, USA	4
Oceanodroma homochroa	Ashy Storm-Petrel	U	20	36.9	2.3	33.3	42.4		E. Pacific Ocean	227
Oceanodroma hornbyi	Ringed Storm-Petrel	B	3	40.6		32.5	46		South Pacific Ocean	1148
Oceanodroma furcata	Fork-tailed Storm-Petrel	U	422	54.4	3.4	45	67.5	B	Washington, USA	87
Oceanodroma furcata	Fork-tailed Storm-Petrel	U	81	65.6	4.7	50	82	B	Alaska, USA	87
Family Pelecanoididae										
Pelecanoides garnotii	Peruvian Diving-Petrel	U	65	202					Peru	1363
Pelecanoides garnotii	Peruvian Diving-Petrel	B	10	239	22	210	270		Chile	861b, 1082
Pelecanoides magellani	Magellanic Diving-Petrel	M	4	134		115	145		Chile	861b, 1082
Pelecanoides georgicus	South Georgia Diving-Petrel	U	71	121	13	90	150	B	Crozet Is.	611a
Pelecanoides urinatrix	Common Diving-Petrel	U	52	141	13	105	165	B	Crozet Is.	611a
Pelecanoides urinatrix exsul	Common Diving-Petrel	U	44	164	11	133	185	B	Kerguelen Is.	745

Body Masses of World Birds (continued)

Species Name	Common Name	Sex	N	Mean	S.D.	Min	Max	Season	Location	Source #
Family Phaethontidae										
Phaethon aethereus	Red-billed Tropicbird	U		750						971
Phaethon rubricauda	Red-tailed Tropicbird	B	411	659	59.4				Johnson Atoll	1063
Phaethon rubricauda	Red-tailed Tropicbird	B	688	685	63.5				Christmas Is.	1063
Phaethon lepturus lepturus	White-tailed Tropicbird	U	59	334					Aldabra Atoll	279
Phaethon lepturus catesbyi	White-tailed Tropicbird	U	110	367	27.8				Bermuda; West Indies	678
Phaethon lepturus dorotheae	White-tailed Tropicbird	U	4	288		251	312		Pacific Ocean	678
Family Pelecanidae										
Pelecanus onocrotalus	Great White Pelican	M	52	11450	1340	9000	15000			591
		F	22	7590	1020	5400	9000			
Pelecanus rufescens	Pink-backed Pelican	M	19	5970	780	4500	7000			591
		F	64	4920	630	3900	6200			
Pelecanus philippensis	Spot-billed Pelican	M	2	5400		5100	5700		captive	591
		F	1	4650						
Pelecanus crispus	Dalmatian Pelican	M	6	10400		9500	12000			591
		F	4	8700		7250	10000			
Pelecanus conspicillatus	Australian Pelican	U	4	5505		4000	6800			745
Pelecanus erythrorhynchos	American White Pelican	M	11	6329		3510	8000			591
		F	6	4970		3630	5900			
Pelecanus thagus	Peruvian Pelican	U		6000					Peru	307
Pelecanus occidentalis	Brown Pelican	M	56	3702	389				Florida, USA	489
		F	47	3174	329					
Family Sulidae										
Morus bassanus	Northern Gannet	M	27	2932		2470	3470		Great Britain	845
		F	27	3067		2570	3610			

Body Masses of World Birds (continued)

Species Name	Common Name	Sex	N	Mean	S.D.	Min	Max	Season	Location	Source #
Morus capensis	Cape Gannet	M	61	2618		2296	2920			745
		F	53	2669		2381	3118			
Morus serrator	Australian Gannet	U	50	2350		2000	2800		New Zealand	745
Sula abbotti	Abbott's Booby	M	3	1503		1370	1620			845
		F	3	1600		1470	1700			
Sula nebouxii	Blue-footed Booby	M	23	1283		1100	1580		Galapagos Is.	845
		F	28	1801		1450	2230			
Sula variegata	Peruvian Booby	M	3	1350		1250	1350		Peru; Chile	845, 861b
		F	2	1520		1500	1540			
Sula dactylatra	Masked Booby	M	195	1623		1300	1980		Christmas Is.	12
		F	196	1803		1375	2250			
Sula dactylatra personata	Masked Booby	M	27	1880		1503	2211	B	Hawaiian Is.	316
		F	27	2095		1616	2353			
Sula dactylatra dactylatra	Masked Booby	M	2	1450		1400	1500		St. Thomas	314
		F	4	1584		1450	1660			
Sula granti	Nazca Booby	M	48	1627		1220	1970		Galapagos Is.	12
		F	37	1881		1470	2350			
Sula sula websteri	Red-footed Booby	M	20	938		850	1160		Galapagos Is.	413
		F	20	1068		850	1210			
Sula sula rubripes	Red-footed Booby	B	10	1223	109				Midway Is.	1064
Sula sula rubripes	Red-footed Booby	B	803	1110	131				Johnson Atoll	1064
Sula sula	Red-footed Booby	B	490	857	105				Christmas Is.	1064
Sula leucogaster leucogaster	Brown Booby	M	4	1078					Virgin Is.	1062
		F	23	1360	143					
Sula leucogaster brewsteri	Brown Booby	M	158	1191	108				San Pedro Martir, Mexico	1062
		F	120	1450	101					
Sula leucogaster plotus	Brown Booby	M	52	1187	126				Johnston Atoll	1062
		F	54	1525	153					

Body Masses of World Birds (continued)

Species Name	Common Name	Sex	N	Mean	S.D.	Min	Max	Season	Location	Source #
Family Phalacrocoracidae										
Phalacrocorax sulcirostris	Little Black Cormorant	M	42	1100	100	800	1300		Australia	745
		F	34	900	100	700	1300			
Phalacrocorax auritus floridanus	Double-crested Cormorant	M	33	1808	224				Florida, USA	489
		F	32	1540	215					
Phalacrocorax auritus auritus	Double-crested Cormorant	M	763	2089	147				Quebec, Canada	494
		F	375	1831	139					
Phalacrocorax fuscicollis	Indian Cormorant	B	5	706		600	790			591
Phalacrocorax brasilianus brasilianus	Neotropic Cormorant	M	10	1260	12.6				Panama	489
		F	3	1070						
Phalacrocorax brasilianus mexicanus	Neotropic Cormorant	M	14	1393		1150	1550			591
		F	8	1256		1100	1450			
Phalacrocorax carbo sinensis	Great Cormorant	M	36	2283		1975	2687	B	Germany	223
		F	17	1936		1673	2174			
Phalacrocorax carbo carbo	Great Cormorant	M	6	3240					Greenland	493
		F	6	2630						
Phalacrocorax carbo lucidus	Great Cormorant	M	3	3090						591
		F	5	2248		1590	2950			
Phalacrocorax capensis	Cape Cormorant	M	24	1330						591
		F	11	1104						
Phalacrocorax neglectus	Bank Cormorant	M	77	2107		1775	2425			591
		F	92	1794		1500	2150			
Phalacrocorax capillatus	Japanese Cormorant	M	6	3140		3000	3350		Japan	1276
		F	7	2500		2300	2700			
Phalacrocorax penicillatus	Brandt's Cormorant	M	3	2570		2382	2682			591
		F	12	1925		1399	2298			
Phalacrocorax aristotelis	European Shag	M	3	1948		1760	2154			591
		F	7	1598		1407	1788			

Body Masses of World Birds (continued)

Species Name	Common Name	Sex	N	Mean	S.D.	Min	Max	Season	Location	Source #
Phalacrocorax pelagicus	Pelagic Cormorant	M	16	2072		1814	2440			591
		F	11	1641		1214	2041			
Phalacrocorax urile	Red-faced Cormorant	M	20	2428	113	2200	2500		Aleutian Is.	176
		F	16	1847	64.9	1650	2100			
Phalacrocorax magellanicus	Rock Shag	M	7	1553		1440	1680			591
		F	10	1417		1330	1550			
Phalacrocorax bougainvillii	Guanay Cormorant	M	7	2485		2100	3222			591, 1082
Phalacrocorax varius	Pied Cormorant	M	60	2196		1814	2472			591
		F	175	1715		1361	2062			
Phalacrocorax fuscescens	Black-faced Cormorant	M	30	1515						591
Phalacrocorax carunculatus	Rough-faced Shag	M	1	2655					New Zealand	745
		F	1	2500						
Phalacrocorax chalconotus	Bronze Shag	M	6	2717		1797	3875		New Zealand	745
		F	8	1814		1447	2356			
Phalacrocorax onslowi	Chatham Is. Shag	M	1	2400					Chatham Is.	745
		F	1	1790						
Phalacrocorax colensoi	Auckland Is. Shag	U		2000					Auckland Is.	503
Phalacrocorax campbelli	Campbell Is. Shag	M	2	2000					Campbell Is.	745
		F	8	1800		1600	1900			
Phalacrocorax ranfurlyi	Bounty Is. Shag	M	7	2500		2300	2900		Bounty Is., New Zealand	745
		F	6	2500		2000	2700			
Phalacrocorax bransfieldensis	Antarctic Shag	M	16	3022	170					745
		F	21	2576	138					
Phalacrocorax georgianus	South Georgia Shag	M	130	2883	255				S. Georgia Is.	745
		F	101	2473	154					
Phalacrocorax atriceps	Imperial Shag	M	10	2716		2104	3341			591, 861b
		F	5	2266		2087	2400			
Phalacrocorax nivalis	Heard Is. Shag	M	4	3157				B	Heard Is.	450
		F	3	2467						

Body Masses of World Birds (continued)

Species Name	Common Name	Sex	N	Mean	S.D.	Min	Max	Season	Location	Source #
Phalacrocorax melanogenis	Crozet Shag	M	5	2450		1700	2700		Possession Is.	745
		F	5	2040		1950	2250			
Phalacrocorax verrucosus	Kerguelen Shag	U	27	2630	373	2100	3300		Nuageuses Is.	1284
Phalacrocorax purpurascens	Macquarie Shag	M	16	2860	145	2650	3200		Macquarie Is.	745
		F	18	2430	152	2180	2700			
Phalacrocorax gaimardi	Red-legged Cormorant	M	6	1379		1200	1550		Chile	1082
		F	4	1338		1250	1400			
Phalacrocorax punctatus	Spotted Shag	M	14	1210	200	850	1670			745
		F	8	1160		770	1610			
Phalacrocorax featherstoni	Pitt Is. Shag	M	2	985		645	1325			745
		F	2	1102		1078	1127			
Phalacrocorax melanoleucos melanoleucos	Little Pied Cormorant	M	28	822		691	963		Australia	591
		F	78	730		567	864			
Phalacrocorax melanoleucos brevirostris	Little Pied Cormorant	M	6	796		715	877		New Zealand	591
		F	5	586		413	704			
Phalacrocorax africanus	Long-tailed Cormorant	B	19	545		435	635			591
Phalacrocorax coronatus	Crowned Cormorant	M	6	798		710	880			591
		F	5	728		670	780			
Phalacrocorax niger	Little Cormorant	U		427		360	525			591
Phalacrocorax pygmeus	Pygmy Cormorant	M	3	743		650	870			223
		F	3	615		565	640			
Phalacrocorax harrisi	Flightless Cormorant	M	32	3500	500	2700	4650		Galapagos Is.	1231
		F	37	2600	300	1950	3650			
Family Anhingidae										
Anhinga anhinga	Anhinga	B	26	1235					Florida, USA	489
Anhinga melanogaster rufa	Darter	M	10	1436		1206	1711			591
		F	9	1357		1055	1660			
Anhinga melanogaster melanogaster	Darter	U	4	1340		1160	1500			591

Body Masses of World Birds (continued)

Species Name	Common Name	Sex	N	Mean	S.D.	Min	Max	Season	Location	Source #
Anhinga melanogaster novaehollandiae	Darter	M	16	1600		1200	2100			745
		F	18	1700		900	2600			
Family Fregatidae										
Fregata aquila	Ascension Is. Frigatebird	U		1620						133
Fregata andrewsi	Christmas Is. Frigatebird	M		1400					Christmas Is.	745
		F		1550						
Fregata magnificens	Magnificent Frigatebird	M	19	1465	105				Barbuda	280
		F	19	1704	89					
Fregata magnificens	Magnificent Frigatebird	M	40	1265	73.7				Isla Isabel, Mexico	1225a
		F	48	1563	64.6					
Fregata minor	Great Frigatebird	M	316	927	111	640	1350	B	Christmas Is.	1062a
		F	312	1183	115	850	1550			
Fregata minor	Great Frigatebird	M	26	1280	102				Hawaiian Is.	792
		F	25	1662	96					
Fregata minor	Great Frigatebird	M	54	1234	103				Johnson Atoll	792
		F	33	1490	135					
Fregata ariel	Lesser Frigatebird	M	29	754		625	875		Aldabra Atoll	279
		F	45	858		760	955			
Family Ardeidae										
Syrigma sibilatrix	Whistling Heron	F	2	463		370	555		Brazil; Argentina	63, 211
Pilherodius pileatus	Capped Heron	M	5	530		444	591		Suriname	489
		F	4	610		562	632			
Ardea cinerea	Gray Heron	B	30	1443		1020	2073		Netherlands	223
Ardea herodias	Great Blue Heron	M	24	2480	290				British Columbia, Canada	158
		F	29	2110	340					
Ardea herodias occidentalis	"Great White" Heron	M	11	3015	255	2680	3390		Florida, USA	76
		F	8	2570		1980	3200			
Ardea cocoi	Cocoi Heron	U		3200					Brazil	1105
Ardea cocoi	Cocoi Heron	M	1	1148					Suriname	498
		F	1	1465						

Body Masses of World Birds (continued)

Species Name	Common Name	Sex	N	Mean	S.D.	Min	Max	Season	Location	Source #
Ardea pacifica	Pacific Heron	U	9	881		600	1220		Australia	745
Ardea melanocephala	Black-headed Heron	U	6	1060		710	1650			133
Ardea sumatrana	Great-billed Heron	F	5	2024		1300	2600			745
Ardea goliath	Goliath Heron	U	3	4468		4310	4750			133
Ardea purpurea	Purple Heron	M				617	1218		Netherlands	223
		F				525	1135			
Ardea purpurea	Purple Heron	M	2	1112		1100	1125		Philippines	983
		F	4	1019		925	1150			
Ardea alba	Great Egret	M	12	935	134					489
		F	9	812						
Egretta rufescens	Reddish Egret	B	7	614		364	869			725
Egretta picata	Pied Heron	U	6	259		210	372		Australia	745
Egretta vinaceigula	Slaty Egret	U	5	288		250	340		Zambia	302
Egretta ardesiaca	Black Heron	U	7	324		270	390		Zambia	302
Egretta tricolor	Tricolored Heron	M	35	415	45.5					489
		F	5	334						
Egretta intermedia	Intermediate Egret	M	7	408		350	500		China	184a
		F	5	516		484	562			
Egretta novaehollandiae	White-faced Heron	M	7	599		505	693		New Zealand	745
		F	4	521		462	559			
Egretta caerulea	Little Blue Heron	M	11	364	47.1					488
		F	8	315						
Egretta thula	Snowy Egret	U	17	371						489
Egretta garzetta	Little Egret	F	9	312		280	344			184a
Egretta gularis	Western Reef-Heron	M	1	400						223
Egretta eulophotes	Chinese Egret	M	2	554		530	578		Tsingtao, China	1088
		F	2	385		380	390			

Body Masses of World Birds (continued)

Species Name	Common Name	Sex	N	Mean	S.D.	Min	Max	Season	Location	Source #
Egretta sacra	Pacific Reef-Heron	B	4	383		341	440		New Britain; Solomon Is.	426, 658
Egretta sacra	Pacific Reef-Heron	B	2	658		620	695		Fiji	1196c
Ardeola ralloides	Squacco Heron	B	23	287		230	370			223
Ardeola grayii	Indian Pond-Heron	M	2	253		230	276		Maldives	223, 740
Ardeola bacchus	Chinese Pond-Heron	M	7	332					Hopei, China	1087
		F	2	280						
Bubulcus ibis	Cattle Egret	M	27	372		296	460		Florida, USA	1207
		F	59	360		270	512			
Butorides striata	Striated Heron	F	1	226					Solomon Is.	658
Butorides striata javanica	Striated Heron	U				153	208		Malaysia	1287
Butorides striata carcinophila	Striated Heron	B	11	184		145	220		Philippines	437, 983
Butorides virescens	Green Heron	U	34	212	5.9				Florida, USA	489
Butorides virescens	Green Heron	B	70	187	17.3	138	220		Caribbean Is.	19
Agamia agami	Agami Heron	B	8	567					Panama; Venezuela; Suriname	489, 498, 1308
Nycticorax nycticorax	Black-crowned Night-Heron	B	10	810		615	1014		Northeast USA; Suriname	498, 894
Nycticorax caledonicus	Rufous Night-Heron	M	4	856		750	1014			658, 786, 1017
Nyctanassa violacea	Yellow-crowned Night-Heron	M	8	716					Florida, USA	488
		F	7	649						
Gorsachius goisagi	Japanese Night-Heron	M	1	527					Philippines	954
Gorsachius melanolophus	Malayan Night-Heron	B	5			377	451	F		1287
Cochlearius cochlearius	Boat-billed Heron	M	11	693		577	770		Mexico; Central America; Suriname	295, 498, 1151
		F	7	597		503	726			

Body Masses of World Birds (continued)

Species Name	Common Name	Sex	N	Mean	S.D.	Min	Max	Season	Location	Source
Tigrisoma mexicanum	Bare-faced Tiger-Heron	M	3	1274					Panama	489
		F	4	1046						
Tigrisoma fasciatum	Fasciated Tiger-Heron	U		850					Costa Rica	1170
Tigrisoma lineatum	Rufescent Tiger-Heron	B	8	813		630	980		Suriname	498
Zebrilus undulatus	Zigzag Heron	F	1	123					Peru	988
Ixobrychus involucris	Stripe-backed Bittern	U	12	80.1		69	104		Brazil; Peru; Argentina Suriname	63, 211, 370, 49
Ixobrychus exilis	Least Bittern	U	20	86.3	4.3					489
Ixobrychus sinensis	Yellow Bittern	B	28	94.3		63	120		Hopei, China	1087
Ixobrychus minutus dubius	Little Bittern	U	15	83.9	16.9	59	120		Australia	745
Ixobrychus minutus	Little Bittern	B	32	118		67	144	B	South Africa	1036
Ixobrychus eurhythmus	Schrenck's Bittern	M	12	139		120	158		Hopei, China	1087
		F	5			141	150			
Ixobrychus cinnamomeus	Cinnamon Bittern	M	4	128		106	148		Philippines; China	437, 95 1087
		F	1	125						
Ixobrychus sturmii	Dwarf Bittern	B	2	140		130	150		Angola	266
Ixobrychus flavicollis	Black Bittern	B	8	321		275	358			658, 81 954
Botaurus pinnatus	Pinnated Bittern	M	4	1047		875	1157		Mexico; Suriname	498, 11
		F	2	630		554	705			
Botaurus lentiginosus	American Bittern	U	16	706	183	520	1072	B	Ontario, Canada	1147a
Botaurus stellaris	Great Bittern	M	11	1209		950	1730		Hopei, China	1087
		F	2	1440		1230	1650			
Botaurus poiciloptilus	Australasian Bittern	M	15	1353	308	875	2085		Australia; New Zealand	745
		F	10	868	221	571	1135			
Family Scopidae										
Scopus umbretta	Hamerkop	U	10	472		420	534		South Africa	1036

Body Masses of World Birds (continued)

Species Name	Common Name	Sex	N	Mean	S.D.	Min	Max	Season	Location	Source #
Family Ciconiidae										
Mycteria americana	Wood Stork	M	9	2702						489, 498
		F	2	2415		2050	2780			
Mycteria ibis	Yellow-billed Stork	M	1	2384						473, 1085a
		F	6	1949		1190	2330			
Mycteria leucocephala	Painted Stork	B	10	3180		3030	3370			473
Anastomus lamelligerus	African Openbill	U	11	1081		628	1400			133, 1156
Ciconia nigra	Black Stork	U	13	2926						473
Ciconia abdimii	Abdim's Stork	B	10	1398	117	1160	1570		captive	800
Ciconia episcopus	Wooly-necked Stork	B	8	2061		1590	2790			473
Ciconia maguari	Maguari Stork	M	9	4200						473
		F	5	3800						
Ciconia ciconia	White Stork	M	41	3571		2610	4400	PB	Germany	223
		F	27	3325		2275	3900			
Ciconia boyciana	Oriental Stork	M	7	5014		4200	5900		captive	644, 828a
		F	8	4687		3400	5200			
Ephippiorhynchus asiaticus	Black-necked Stork	U		4100						133
Ephippiorhynchus senegalensis	Saddle-billed Stork	M	5	6378		5085	7524			473, 1156
		F	3	5947		5000	6840			
Jabiru mycteria	Jabiru	M	8	6892		5902	8100			473, 1085a
		F	4	5217		4300	6356			
Leptoptilos javanicus	Lesser Adjutant	B	7	4651		4000	5230			473, 1085a, 1287
Leptoptilos crumeniferus	Marabou Stork	M	37			5600	8900			473
		F	22			4000	6800			
Family Balaenicipididae										
Balaeniceps rex	Shoebill	M	1	6700						473
		F	4	5268		4360	5900			

Body Masses of World Birds (continued)

Species Name	Common Name	Sex	N	Mean	S.D.	Min	Max	Season	Location	Source #
Family Threskiornithidae										
Threskiornis aethiopicus	Sacred Ibis	M	40	1618	140	1268	1963		Pretoria, South Africa	713
		F	54	1378	126	1131	1718			
Threskiornis melanocephalus	Black-headed Ibis									
Threskiornis molucca	Australian Ibis	M	72	2006	138	1700	2350		Australia	745
		F	84	1606	146	1300	2120			
Threskiornis spinicollis	Straw-necked Ibis	M	3	1465		1400	1570		Australia	745
		F	4	1238		1150	1320			
Pseudibis davisoni	White-shouldered Ibis	M	1	1588						1287
Pseudibis gigantea	Giant Ibis	M	1	3515						1287
Geronticus eremita	Waldrapp	B	9	1202		1000	1350		captive	800, 851
Nipponia nippon	Crested Ibis	U		1900					Japan	339
Bostrychia hagedash	Hadada Ibis	B	5	1238		993	1512			1156, 1212
Theristicus caerulescens	Plumbeous Ibis	M	1	1500					Bolivia	1145
Theristicus caudatus	Buff-necked Ibis	B	5	1726		1550	1950		Brazil; Argentina; Chile	63, 546, 861b, 1082
Theristicus branickii	Andean Ibis	M	1	1492					Peru	1042
Mesembrinibis cayennensis	Green Ibis	B	7	756		670	865		Peru; Suriname	193
Phimosus infuscatus	Bare-faced Ibis	B	4	559		512	600		Brazil; Argentina	63, 804
Eudocimus albus	White Ibis	M	12	1036	105	873	1261	B	Florida, USA	663
		F	16	764	68.4	593	864			
Eudocimus ruber	Scarlet Ibis	M	8	741		710	770		Trinidad	359
		F	13	588		505	640			
Plegadis falcinellus	Glossy Ibis	M				557	768	S	Former USSR	223
		F				530	680			
Plegadis chihi	White-faced Ibis	M	32	697	58.9	563	807	B	Utah, USA	314
		F	35	546	45.3	433	677			

Body Masses of World Birds (continued)

Species Name	Common Name	Sex	N	Mean	S.D.	Min	Max	Season	Location	Source #
Plegadis ridgwayi	Puna Ibis	M	3	657		608	734		Bolivia; Peru	1042, 1145
		F	3	511		478	554			
Platalea leucorodia	Eurasian Spoonbill	B	8	1868		1656	2080			473, 926
Platalea regia	Royal Spoonbill	M	6	1886		1650	2070		Australia	745
		F	4	1575		1400	1800			
Platalea alba	African Spoonbill	B	5	1521		1020	1900			508, 800
Platalea minor	Black-faced Spoonbill	U		1228					Japan	339
Platalea flavipes	Yellow-billed Spoonbill	M	4	1895		1750	2000			743
		F	2	1600		1500	1700			
Platalea ajaja	Roseate Spoonbill	B	30	1490	170	1225	1800		Florida, USA	76
Family Phoenicopteridae										
Phoenicopterus ruber roseus	Greater Flamingo	M	13	3540		3100	4100	B	France	50
		F	12	2530		2100	3300			
Phoenicopterus ruber	Greater Flamingo	M	4	3493		3280	3800		Namibia; South Africa	1036
		F	4	2700		2180	3080			
Phoenicopterus chilensis	Chilean Flamingo	B	5	2277		1790	3200			362, 546, 766, 1042
Phoenicopterus minor	Lesser Flamingo	F	1	1500						1036
Phoenicopterus andinus	Andean Flamingo	M	1	4900					Argentina	766
Phoenicopterus jamesi	Puna Flamingo	U		2000						274
Family Anhimidae										
Anhima cornuta	Horned Screamer	U		3150					Brazil	1105
Chauna chavaria	Northern Screamer	M	1	2117					captive	1143
		F	1	3090						

Body Masses of World Birds (continued)

Species Name	Common Name	Sex	N	Mean	S.D.	Min	Max	Season	Location	Source #
Chauna torquata	Southern Screamer	F	1	4400					Brazil, approximate	63
Family Anatidae										
Anseranas semipalmata	Magpie Goose	M	402	2766	283	1838	3195			745
		F	359	2071	237	1405	2770			
Dendrocygna guttata	Spotted Whistling-Duck	F		800						1012
Dendrocygna eytoni	Plumed Whistling-Duck	M	63	788		600	930			745
		F	65	792		580	1400			
Dendrocygna bicolor	Fulvous Whistling-Duck	M	138	770	70.1	545	958		SW Louisiana, USA	529
		F	148	743	77.2	595	964			
Dendrocygna arcuata	Wandering Whistling-Duck	M	287	866		741	948			745
		F	293	732		453	976			
Dendrocygna javanica	Lesser Whistling-Duck	U				450	600			7
Dendrocygna viduata	White-faced Whistling-Duck	U	10	690	35.9	635	760		South Africa	508
Dendrocygna arborea	West Indian Whistling-Duck	F		1150						1012
Dendrocygna autumnalis	Black-bellied Whistling-Duck	M	61	796	68.6	660	950		Texas, USA	571
		F	59	760	49.9	655	880			
Dendrocygna autumnalis	Black-bellied Whistling-Duck	M	78	741		530	890		Guyana	571
		F	82	726		530	890			
Thalassornis leuconotos	White-backed Duck	B	10			575	840		Angola	266
Cygnus olor	Mute Swan	M	59	11800	890	9200	14300	W	Britain	967
		F	35	9670	640	7600	10600			
Cygnus atratus	Black Swan	M	270	6200		4600	8700			1075
		F	243	5100		3700	7200			
Cygnus melanocoryphus	Black-necked Swan	M	9	5550		4500	6754			546, 107
		F	7	4000		3500	4400			
Cygnus buccinator	Trumpeter Swan	M	152	11900		9100	14500	W		802
		F	120	10300		7000	12500			

Body Masses of World Birds (continued)

Species Name	Common Name	Sex	N	Mean	S.D.	Min	Max	Season	Location	Source #
Cygnus cygnus	Whooper Swan	B	12	9350		7400	14000		Northern Europe	223
Cygnus columbianus	Tundra Swan	M	1447	7200	800			W	Eastern USA	694
		F	1290	6300	700					
Cygnus columbianus bewickii	Tundra (Bewick's) Swan	M	96	6400		4900	7800	W	Britain	515
		F	95	5700		3400	7200			
Coscoroba coscoroba	Coscoroba Swan	M	3	4854		3800	5400			546, 1075
		F	3	3800		3200	4500			
Anser cygnoides	Swan Goose	M	1	3575					Mongolia, China	925, 926
		F	1	3450						
Anser fabalis rossicus	Bean Goose	M	126	2668	233	1970	3390	W	Netherlands	223
		F	117	2374	203	2000	2800			
Anser fabalis fabalis	Bean Goose	M	68	3198	302	2690	4060	W	Netherlands	223
		F	58	2843	274	2220	3490			
Anser brachyrhynchus	Pink-footed Goose	M	750	2770	310	1900	3350	F	Britain	223
		F	796	2520	270	1810	3150			
Anser albifrons frontalis	Greater White-fronted Goose	M	384	2075	215			F	California, USA	334
		F	407	2348	222					
Anser albifrons gambeli	Greater White-fronted Goose	M	54	2700	220			W	California, USA	334
		F	53	3000	220					
Anser erythropus	Lesser White-fronted Goose	M	2	1900		1800	2000		Norway; Denmark; Wales; UK	223
		F	2	1622		1400	1843			
Anser anser	Greylag Goose	M	94	3509	321	2600	4560	W	Scotland, UK	223
		F	75	3108	274	2160	3800			
Anser indicus	Bar-headed Goose	M	1	2460					Mongolia	926
		F	2	1990		1860	2120			
Chen caerulescens caerulescens	Snow Goose	M	467	2744	235			F	James Bay, Canada	214
		F	422	2517	217					
Chen caerulescens atlanticus	Snow Goose	M	92	2847	231			PB	Bylot Is., Canada	822
		F	100	2456	149					
Chen rossii	Ross' Goose	M	25	1632	153	1350	1955	B	Northwest Terr., Canada	1035
		F	20	1640	245	1155	2040			

Body Masses of World Birds (continued)

Species Name	Common Name	Sex	N	Mean	S.D.	Min	Max	Season	Location	Source
Chen canagica	Emperor Goose	M	13	2370	178			PB	Alaska, USA	921a
		F	101	1926	130					
Branta bernicla nigricans	Brant	M	430	1370	350	1080	1790	PB	Northwest Terr., Canada	109
		F	361	1230	400	880	1590			
Branta bernicla hrota	Brant	M	27	1293	99			B	Southhampton Is., Canada	959
		F	41	1224	179					
Branta leucopsis	Barnacle Goose	M	366	1788	163	1350	2230	PB	Spitzbergen	889a
		F	253	1586	122	1210	1950			
Branta hutchinsii hutchinsii	Cackling Goose	M	129	2180	3					823
		F	125	1920	20					
Branta hutchinsii taverneri	Cackling Goose	M	60	2606	267					823
		F	61	2421	238					
Branta hutchinsii leucopareia	Cackling Goose	M	36	1946	137					823
		F	46	1704	157					
Branta hutchinsii minima	Cackling Goose	M	52	1480	29.8	1240	1700		California, USA	956
		F	58	1264	29.4	940	1490			
Branta canadensis canadensis	Canada Goose	M	232	3814			6265			844
		F	159	3314			5902			
Branta canadensis occidentalis	Canada Goose	M	130	3232	261					823
		F	131	2640	202					
Branta canadensis parvipes	Canada Goose	M	70	3266	320					823
		F	59	2854	335					
Branta canadensis moffitti	Canada Goose	M	26	4460	306			W	Minnesota, USA	780
		F	31	3770	334					
Branta canadensis interior	Canada Goose	M	128	4181		3799	4727	W	Illinois, USA	955
		F	121	3514		3062	3912			
Branta canadensis maxima	Canada Goose	M	6	4858	280					823
		F	10	4825	425					
Branta sandvicensis	Hawaiian Goose (Nene)	M		2165		1695	3050		captive	39
		F		1930		1525	2560			
Branta leucopsis	Barnacle Goose	M	366	1788	163	1350	2230	PB	Spitzbergen	889a
		F	253	1586	122	1210	1950			

Body Masses of World Birds (continued)

Species Name	Common Name	Sex	N	Mean	S.D.	Min	Max	Season	Location	Source #
Branta ruficollis	Red-breasted Goose	M	5	1375		1200	1625	W		223
		F	2	1094		1058	1130			
Cereopsis novaehollandiae	Cape Barren Goose	M	14			3700	5100	F		745, 1012
		F		3560						
Stictonetta naevosa	Freckled Duck	M	63	969		747	1130			745
		F	31	842		691	985			
Cyanochen cyanoptera	Blue-winged Goose	F		1520						1012
Chloephaga melanoptera	Andean Goose	U	4	2037		2000	2100			362
Chloephaga picta	Upland Goose	M	8	3105		2752	3533		Argentina	210, 546, 978a
		F	5	2672		2404	2823			
Chloephaga hybrida	Kelp Goose	M	2	2575		2540	2611		Argentina	546, 978a
		F	2	2022		2000	2043			
Chloephaga poliocephala	Ashy-headed Goose	M	3	2289		2270	2327		Tierra del Fuego	546, 978a
		F	1	1760						
Chloephaga rubidiceps	Ruddy-headed Goose	U	2	2086		2043	2128		Tierra del Fuego	546
Neochen jubata	Orinoco Goose	F		1250						1012
Alopochen aegyptiana	Egyptian Goose	M	53	1873		1520	2100			223, 1036
		F				1500	1800			
Tadorna ferruginea	Ruddy Shelduck	M	10	1368		1156	1640		China	184a
		F	8	1115		919	1315			
Tadorna cana	South African Shelduck	B	8	1182		1012	1295		South Africa	508
Tadorna tadornoides	Australian Shelduck	M	67	1559		990	1980			745
		F	185	1291		878	1850			
Tadorna variegata	Paradise Shelduck	M	23	1712	145	1422	2000			745
		F	10	1387	178	1059	1799			
Tadorna tadorna	Common Shelduck	M	11	1261	110	1100	1450	S	Netherlands; Denmark; Kazakhstan	223
		F	5	1043		926	1250			

Body Masses of World Birds (continued)

Species Name	Common Name	Sex	N	Mean	S.D.	Min	Max	Season	Location	Source #
Tadorna radjah	Radjah Shelduck	M	46	934		750	1101			745
		F	49	839		600	1130			
Tachyeres pteneres	Flightless Steamerduck	M	3	5344		4625	6186		Tierra del Fuego	546
		F	1	4200						
Tachyeres leucocephalus	White-headed Steamerduck	F		3013						1012
Tachyeres brachypterus	Falkland Steamerduck	F		3450						1012
Tachyeres patachonicus	Flying Steamerduck	M	7	3062		2892	3178		Tierra del Fuego	546
		F	8	2600		2438	2835			
Plectropterus gambensis	Spur-winged Goose	U	11	3869		2400	5400		South Africa	1036
Cairina moschata	Muscovy Duck	M	4	2858						489
		F	4	2042						
Cairina scutulata	White-winged Duck	M				2945	3855			7
		F				2150	3050			
Sarkidiornis melanotos	Comb Duck	M	1	2610						7
		F				1925	2325			
Pteronetta hartlaubii	Hartlaub's Duck	F		790						1012
Nettapus pulchellus	Green Pygmy-goose	M	47	310		300	430			745
		F	26	304		245	340			
Nettapus coromandelianus	Cotton Pygmy-goose	M	52	403		311	495			745
		F	37	380		255	439			
Nettapus auritus	African Pygmy-goose	U	7	266		220	325			133, 266
Callonetta leucophrys	Ringed Teal	M	2	410		370	450			63, 212a
		F	3	321		285	340			
Aix sponsa	Wood Duck	M	248	681			907	F		894
		F	163	635			907			
Aix galericulata	Mandarin Duck	M		628		571	693			273
		F		512		428	608			
Chenonetta jubata	Maned Duck	M	45	815		700	955			745
		F	26	800		662	984			

Body Masses of World Birds (continued)

Species Name	Common Name	Sex	N	Mean	S.D.	Min	Max	Season	Location	Source #
Amazonetta brasiliensis	Brazilian Teal	U		500					Brazil	1105
Hymenolaimus malacorhynchos	Blue Duck	M	11	897	48.6	820	970		New Zealand	745
		F	14	768	59.4	680	870			
Merganetta armata	Torrent Duck	M	1	482						546, 1012
		F		330						
Salvadorina waigiuensis	Salvadori's Teal	M	1	324					New Guinea	770
Anas sparsa	African Black Duck	M	18	1027		950	1156		South Africa	1036
		F	10	938		820	1055			
Anas penelope	Eurasian Wigeon	M	42	819		610	1073		Denmark	894
		F	24	724		552	962			
Anas americana	American Wigeon	M	65	792	79.1	635	1036	Y	Western Canada	1326
		F	68	719	80.6	512	872			
Anas sibilatrix	Chiloe Wigeon	M	2	1072		1036	1107		Tierra del Fuego	546
		F	3	795		624	965			
Anas falcata	Falcated Duck	M	4	713		590	770		Hopei, China	1087
		F	5	585		422	700			
Anas strepera	Gadwall	M	68	968	9.8	790	1250	W	Alabama, USA	691
		F	27	866	14.4	720	980			
Anas formosa	Baikal Teal	M	12	437		360	520		Hopei, China	1087
		F	8	431		402	505			
Anas crecca carolinensis	Green-winged Teal	M	194	364			454	F		894
		F	81	318			409			
Anas crecca crecca	Green-winged Teal	M	999	320				W	England	622a
		F	498	291						
Anas crecca nimia	Green-winged Teal	M	15	392		310	440	B	Amchitka, Aleutian Is.	622a
		F	4	378		338	418			
Anas flavirostris	Speckled Teal	U	9	434		395	478		Argentina	210
Anas capensis	Cape Teal	U	63	402		342	590		South Africa	223
Anas bernieri	Bernier's Teal	M	7	382		320	405		Madagascar	1358
		F	2	375		365	385			

Body Masses of World Birds (continued)

Species Name	Common Name	Sex	N	Mean	S.D.	Min	Max	Season	Location	Source #
Anas gibberifrons	Sunda Teal	M	218	508	38					701
		F	153	469	36					
Anas albogularis	Andaman Teal	M		400						622a
		F		340						
Anas gracilis	Gray Teal	M	10	525	17.4	500	547		New Zealand	1358
		F	7	420		376	499			
Anas gracilis	Gray Teal	M	210	507		395	670		Australia	745
		F	138	474		350	602			
Anas castanea	Chestnut Teal	M	89	660	65					701
		F	58	590	138					
Anas aucklandica	Auckland Is. Teal	M	10	521	49				Auckland Is.	701
		F	6	410						
Anas nesiotis	Campbell Is. Teal	M	1	426					Campbell Is.	701
Anas chlorotis	Brown Teal	B	42	582					New Zealand	701
Anas platyrhynchos	Mallard	U	5847	1082	129	720	1580	Y	Britain, UK	889
Anas platyrhynchos	Mallard	M	1308	1246	108			W	Mississippi, USA	305
		F	453	1095	106					
Anas platyrhynchos diazi	Mallard (Mexican Duck)	M				960	1060			894
		F				815	990			
Anas laysanensis	Laysan Duck	M	64	456	35.3	348	548		Laysan Is.	819
		F	48	463	37	365	525			
Anas wyvilliana	Hawaiian Duck	M	28	644					Hawaiian Is.	1286
		F	19	585						
Anas fulvigula	Mottled Duck	M	86	1043	86.2	876	1241		Florida, USA	807
		F	71	934	89.3	699	1151			
Anas rubripes	American Black Duck	M	376	1400		900	1800	F	Massachusetts, USA	894
		F	176	1100		900	1500			
Anas undulata	Yellow-billed Duck	U	42	1008	121	660	1220		South Africa	508
Anas melleri	Meller's Duck	M	19	1010	99.8	883	1240		captive	1357
		F	21	911	66.6	832	1140			

Body Masses of World Birds (continued)

Species Name	Common Name	Sex	N	Mean	S.D.	Min	Max	Season	Location	Source #
Anas poecilorhyncha	Spot-billed Duck	M	13	1156		1000	1340		Hopei, China	1087
		F	2	865		750	980			
Anas superciliosa	Pacific Black Duck	M	112	1089	105	765	1276		New Zealand	745
		F	120	981	108	624	1276			
Anas superciliosa	Pacific Black Duck	M	100	1133	110	800	1400		Australia	745
		F	100	1015	180	600	1400			
Anas superciliosa pelewensis	Pacific Black Duck	M	1	642					Solomon Is.	658
		F	2	600		560	640			
Anas luzonica	Philippines Duck	M	7	891		852	977		Philippines	954
		F	9	777		752	818			
Anas specularis	Spectacled Duck	M	1	1461					Tierra del Fuego	546
		F	1	1532						
Anas specularioides	Crested Duck	M	1	1078					Tierra del Fuego	546
		F	1	965						
Anas acuta	Northern Pintail	M	188	1006	96	605	1245	W	California, USA	28
		F	151	887	86	615	1100			
Anas eatoni	Eaton's Pintail	U	38	571	68	455	695		Kerguelen Is.	1284
Anas georgica georgica	Yellow-billed Pintail	M	3	632		610	660	B	South Georgia	1285
		F	2	535		460	610			
Anas georgica spinicauda	Yellow-billed Pintail	M	4	718		624	786		Tierra del Fuego	546, 1042
Anas bahamensis	White-cheeked Pintail	M	8	535		475	590		Suriname	498
		F	6	502		402	633			
Anas bahamensis galapagensis	White-cheeked Pintail	B	2	438		425	450		Galapagos	1286
Anas erythrorhyncha	Red-billed Duck	M	104	587		509	667		Southern Africa	1036
		F	52	532		464	607			
Anas puna	Puna Duck	M	2	554		546	562		Peru	1042
Anas versicolor	Silver Duck	M	1	624					Tierra del Fuego	546
		F	1	610						
Anas hottentota	Hottentot Duck	B	10			235	325		Angola	266

Body Masses of World Birds (continued)

Species Name	Common Name	Sex	N	Mean	S.D.	Min	Max	Season	Location	Source #
Anas querquedula	Garganey	M	100	342					France	223
		F	100	310						
Anas discors	Blue-winged Teal	M	110	380	35.7			W	Mexico	1013
		F	82	340	28.1					
Anas cyanoptera	Cinnamon Teal	M	44	383		315	450			401
		F	69	372		265	470			
Anas platalea	Red Shoveler	M	2	473		472	475		Tierra del Fuego	546
		F	1	458						
Anas smithii	Cape Shoveler	M	5	585						1012, 1036
		F		598						
Anas rhynchotis	Australian Shoveler	B	146	667		545	852		Australia	745
Anas rhynchotis variegata	Australian Shoveler	M	43	637	46.9	540	750		New Zealand	745
		F	32	614	46.2	530	708			
Anas clypeata	Northern Shoveler	M	90	636			908	F		894
		F	71	590			726			
Malacorhynchus membranaceus	Pink-eared Duck	M	77	404		290	480		Australia	745
		F	81	344		272	423			
Marmaronetta angustirostris	Marbled Duck	U		477		420	500	W	Former USSR	273
Rhondonessa caryophyllacea	Pink-headed Duck	U				793	1360			7
Netta rufina	Red-crested Pochard	B	29	1118		990	1300	S	Kazakhstan	223
Netta peposaca	Rosy-billed Pochard	U		1000					Brazil	1105
Netta erythrophthalma	Rosy-billed Pochard	B				700	870		Angola	266
Aythya ferina	Common Pochard	B	321	823		467	1240	W	France	223
Aythya valisineria	Canvasback	M	191	1252	124	850	1600	W	New York, USA	821
		F	54	1154	103	900	1530			

Body Masses of World Birds (continued)

Species Name	Common Name	Sex	N	Mean	S.D.	Min	Max	Season	Location	Source #
Aythya americana	Redhead	M	119	1118	65.4	925	1320	W	Texas, USA	1336
		F	92	1035	76.7	890	1245			
Aythya collaris	Ring-necked Duck	M	285	730			1090			894
		F	151	680			1180			
Aythya nyroca	Ferruginous Pochard	B	13	574		470	740	S	France; former USSR	223
Aythya innotata	Madagascar Pochard	M	1	685					captive	1143
Aythya baeri	Baer's Pochard	M	4	724		625	800		Hopei, China	1087
		F	3	641		594	730			
Aythya australis	White-eyed Duck	M	105	902		525	1100			745
		F	88	838		530	1060			
Aythya fuligula	Tufted Duck	M	343	723		400	950	W	France	223
		F	700	680		450	920			
Aythya novaeseelandiae	New Zealand Scaup	M	13	677	55.6	550	746		New Zealand	745
		F	3	687		600	742			
Aythya marila	Greater Scaup	M	345	1054	105	657	1316	W	Long Is. Sound, USA	631
		F	104	959	108	688	1210			
Aythya affinis	Lesser Scaup	M	112	850		620	1050			894
		F	118	790		540	960			
Somateria mollissima	Common Eider	M	22	2218		1384	2800	B	North Atlantic	62
		F	32	1915		1192	2895			
Somateria spectabilis	King Eider	M	41	1668				F	Alaska, USA	894
		F	141	1567						
Somateria fischeri	Spectacled Eider	M	53	1494	102	1275	1750	Sp		921
		F	11	1623	152	1300	1850			
Polysticta stelleri	Steller's Eider	M	48	773		628	900	B	Russia	894
		F	42	842		651	1000			
Camptorhynchus labradorius	Labrador Duck	M	1	865					extinct	186
		F	1	482						
Histrionicus histrionicus	Harlequin Duck	M	82	610		520	680		Alberta, Canada	1000
		F	23	517		480	571			

Body Masses of World Birds (continued)

Species Name	Common Name	Sex	N	Mean	S.D.	Min	Max	Season	Location	Source #
Clangula hyemalis	Long-tailed Duck	M	661	932				W	Michigan, USA	894
		F	636	814						
Melanitta nigra	Black Scoter	M	34	1117	102				Canada	93
		F	21	987	110					
Melanitta perspicillata	Surf Scoter	M	221	1148	104			W	California, USA	1049
		F	21	1047	101					
Melanitta fusca fusca	White-winged Scoter	B	12	1757		1519	1980	W	Denmark	223
Melanitta fusca deglandi	White-winged Scoter	M	29	1917		1388	2128	W	Alaska, USA	140
		F	10	1732		1566	1946			
Bucephala clangula	Common Goldeneye	M	47	1120	13	910	1329		Utah, USA	320
		F	48	711	14.1	600	760			
Bucephala clangula	Common Goldeneye	M	9	1136		966	1245	W	Denmark	320
		F	9	787		707	860			
Bucephala Is.ica	Barrow's Goldeneye	M	18	1130	79.3	980	1315	W	Washington, USA	321
		F	10	751	76.8	577	857			
Bucephala albeola	Bufflehead	M	29	473	32.8	424	551		Oregon	506
		F	16	334	23.2	297	374			
Mergellus albellus	Smew	M		652		540	825	F	Former USSR	223
		F		568		515	630			
Lophodytes cucullatus	Hooded Merganser	M	12	554		453	652			310
		F	19	680		595	879			
Mergus octosetaceus	Brazilian Merganser	U		983					estimated	700
Mergus serrator	Red-breasted Merganser	M	18	1135			1317			894
		F	17	908			1271			
Mergus merganser	Common Merganser	M	13	1709		1528	2054	F		894
		F	11	1232		1050	1362			
Mergus squamatus	Scaly-sided Merganser	U		1234					estimated	700
Mergus australis	Auckland Is. Merganser	U		898					estimated	700

Body Masses of World Birds (continued)

Species Name	Common Name	Sex	N	Mean	S.D.	Min	Max	Season	Location	Source #
Heteronetta atricapilla	Black-headed Duck	M	11	513		434	580			592
		F	13	565		470	630			
Nomonyx dominicus	Masked Duck	M	19	385						592
		F	17	346						
Oxyura jamaicensis	Ruddy Duck	M	157	629	37.6			PB	Manitoba, Canada	143
		F	9	588	57					
Oxyura ferruginea	Andean Duck	M	5	784		688	850		Argentina	210, 546
Oxyura leucocephala	White-headed Duck	M	3	737		553	865	W	Pakistan	223
		F	3	593		539	631			
Oxyura maccoa	Maccoa Duck	M	1	820						592
		F	3	554		516	580			
Oxyura vittata	Lake Duck	M	8	663		600	850		Argentina	592, 776
		F	1	623						
Oxyura australis	Blue-billed Duck	F	47	798	16.8			Sp		745
Biziura lobata	Musk Duck	M	243	2398		1811	3120		Australia	745
		F	292	1551		993	1844			
Family Cathartidae										
Coragyps atratus atratus	Black Vulture	U	50	2159	130				Texas, USA	148
Coragyps atratus brasiliensis	Black Vulture	U	119	1640	100				Venezuela	148
Cathartes aura septentrionalis	Turkey Vulture	U	124	2006				M	Southern Florida, USA	635
Cathartes aura meridionalis	Turkey Vulture	U	130	1430	228			W	Venezuela	634
Cathartes aura ruficollis	Turkey Vulture	U	65	1220	80.6				Venezuela	634
Cathartes burrovianus	Lesser Yellow-headed Vulture	B	11	935		820	1272		Mexico; Suriname; Brazil	75, 498, 1202
Cathartes melambrotus	Greater Yellow-headed Vulture	B	8	1373		1130	1500		Ecuador, Suriname	498, 747

Body Masses of World Birds (continued)

Species Name	Common Name	Sex	N	Mean	S.D.	Min	Max	Season	Location	Source #
Gymnogyps californianus	California Condor	M	13	8800		7900	9900		captive	1134
		F	15	8100		7000	8900			
Vultur gryphus	Andean Condor	M		12500		10900	13600			1269
		F		10100		9600	11400			
Sarcoramphus papa	King Vulture	B		3400		3100	3700			1269
Family Pandionidae										
Pandion haliaetus	Osprey	M	10	1403		1220	1600			138
		F	14	1568		1250	1900			
Family Accipitridae										
Aviceda cuculoides	African Cuckoo-Hawk	B	8	277		247	302		Ivory Coast	1213
Aviceda jerdoni	Jerdon's Baza	M	1	363					Philippines	954
Aviceda subcristata	Pacific Baza	M	10	307	34.5	259	357		Australia	745
		F	9	347		290	448			
Aviceda subcristata gurneyi	Pacific Baza	M	4	314		285	356		Solomon Is.	658
		F	2	324		319	330			
Aviceda leuphotes	Black Baza	B				168	224			358
Leptodon cayanensis	Gray-headed Kite	B	15	474		414	643			359, 489, 498, 61[illegible] 1027, 1219
Leptodon forbesi	White-collared Kite	B	3	577		550	600		Northeast Brazil	1203
Chondrohierax uncinatus	Hook-billed Kite	M	6	264					Panama; Venezuela; Mexico	296, 489, 640
		F	4	310						
Henicopernis longicauda	Long-tailed Honey-Buzzard	M	1	447						138
		F	1	730						
Pernis apivorus	European Honey-Buzzard	M	10	684		510	800			138
		F	8	832		625	1050			
Pernis celebensis	Barred Honey-Buzzard	B	2	722		692	752		Philippines	954

Body Masses of World Birds (continued)

Species Name	Common Name	Sex	N	Mean	S.D.	Min	Max	Season	Location	Source #
Pernis ptilorhynchus	Oriental Honey-Buzzard	M	4	1069		986	1150		Hopei, China; Thailand	1087, 1287
		F	5	1304		950	1480			
Pernis ptilorhynchus	Oriental Honey-Buzzard	U	9	1066		750	1490		Kazakhstan	273
Lophoictinia isura	Square-tailed Kite	M	1	501					Australia	524, 746
		F	3	670		660	680			
Hamirostra melanosternon	Black-breasted Kite	M	2	1196		1150	1242		Australia	746
Elanoides forficatus	Swallow-tailed Kite	B	14	442		372	510			894
Elanoides forficatus	Swallow-tailed Kite	B	11	392		354	435		Suriname	498
Macheiramphus alcinus	Bat Hawk	U	1	650						139
Gampsonyx swainsonii	Pearl Kite	U	4	92.9		80.5	104			359, 498, 1215
Elanus caeruleus	Black-shouldered Kite	M		241		222	252		Ivory Coast	1213
		F		280		264	290			
Elanus axillaris	Australian Kite	M	11	249	39	181	295		Australia	746
		F	5	293		270	340			
Elanus leucurus	White-tailed Kite	B	33	346	23.3				California, USA	312
Elanus scriptus	Letter-winged Kite	U	45	316	67.4	160	420		Australia	746
Chelictinia riocourii	Scissor-tailed Kite	U	1	110					labeled as "fat"	358
Rostrhamus sociabilis plumbeus	Snail Kite	M	28	394	22.8	360	440		Florida, USA	1194
		F	29	446	47.8	350	570			
Rostrhamus sociabilis sociabilis	Snail Kite	M	7	308		225	383			1194
		F	6	335		268	416			
Rostrhamus hamatus	Slender-billed Kite	M	10	420		377	448		Suriname	498
		F	13	431		367	485			

Body Masses of World Birds (continued)

Species Name	Common Name	Sex	N	Mean	S.D.	Min	Max	Season	Location	Source #
Harpagus bidentatus	Double-toothed Kite	M	6	181		165	210		Costa Rica; Panama; Peru; Belize	370, 379, 489, 1027
		F	2	256		206	305			
Harpagus diodon	Rufous-thighed Kite	F	1	200					Bolivia	1143
Ictinia mississippiensis	Mississippi Kite	M	11	245		216	269	B	Oklahoma, USA	894
		F	5	311		278	339			
Ictinia plumbea	Plumbeous Kite	M	16	243		190	297		Suriname	498
		F	7	257		232	294			
Milvus milvus	Red Kite	M	10	947		802	1052	Sp		223
		F	4	1213		1140	1298			
Milvus migrans	Black Kite	B	96	567	83.8	360	775		Northern Terr., Australia	746
Milvus migrans lineatus	Black Kite	M	4	870		800	920		China	925, 108[illegible]
		F	6	802		680	920			
Haliastur sphenurus	Whistling Kite	U	452	753	94.9	480	1050		Northern Terr., Australia	746
Haliastur indus	Brahminy Kite	B	17	529		434	673			292, 658, 746, 95[illegible]
Haliaeetus leucogaster	White-bellied Sea-Eagle	M	10	2400	220	2120	2900		Australia; Papua New Guinea	746
		F	12	3330	290	3000	3900			
Haliaeetus sanfordi	Solomon Sea-Eagle	M	1	1980					male is subadult	138, 658
		F	3			2300	2500			
Haliaeetus vocifer	African Fish-Eagle	M	3			1928	2497			223
		F	2	3400		3170	3630			
Haliaeetus vociferoides	Madagascar Fish-Eagle	F		3000					Madagascar	1279
Haliaeetus leucoryphus	Pallas' Fish-Eagle	M	10	2597		2040	3278		Mongolia	223, 926
		F	9	3207		2100	3700			
Haliaeetus albicilla	White-tailed Eagle	M	12	4014	755	3075	5430			223
		F	18	5572	980	4080	6920			
Haliaeetus leucocephalus	Bald Eagle	M	35	4130		3637	4819		Alaska, USA	894
		F	37	5350		4631	6400			

Body Masses of World Birds (continued)

Species Name	Common Name	Sex	N	Mean	S.D.	Min	Max	Season	Location	Source #
Haliaeetus pelagicus	Steller's Sea-Eagle	B	3	7757		6800	8970			273
Ichthyophaga humilis	Lesser Fish-Eagle	U	2	782		780	785			358
Ichthyophaga ichthyaetus	Gray-headed Fish-Eagle	M	1	1590						138
		F				2270	2700			
Gypohierax angolensis	Palm-nut Vulture	U	6	1600		1361	1712			133
Necrosyrtes monachus	Hooded Vulture	U	7	2043		1900	2250		Southern Africa	1036
Gypaetus barbatus	Lammergeier	U	10	5680		5000	6750		Europe	273
Gypaetus barbatus	Lammergeier	U	5	5710		5200	6250		South Africa	1036
Neophron percnopterus	Egyptian Vulture	B	7	2082		1829	2400			223
Gyps africanus	White-backed Vulture	U	44	5433		4500	7200		Southern Africa	1036
Gyps bengalensis	White-rumped Vulture	B	29	4385		3500	5560			7
Gyps indicus	Long-billed Griffon	B	5	5515		5540	6250			7
Gyps rueppellii	Rueppell's Griffon	U		7400						538
Gyps himalayensis	Himalayan Griffon	U				8000	12000			138
Gyps fulvus	Eurasian Griffon	B	15	7436		6200	8500		SE Europe	223
Gyps coprotheres	Cape Griffon	U	3	8177		7587	8575		South Africa	647
Aegypius monachus	Cinereous Griffon	M	20			7000	11500		Romania	273
		F	21			7500	12500			
Torgus tracheliotus	Lappet-faced Vulture	U	21	6969		5400	8500		Southern Africa	1036
Trigonoceps occipitalis	White-headed Vulture	U	5	3016		2120	4600		Zimbabwe	1036

Body Masses of World Birds (continued)

Species Name	Common Name	Sex	N	Mean	S.D.	Min	Max	Season	Location	Source #
Sarcogyps calvus	Red-headed Vulture	B				3700	5400			358
Circaetus gallicus	Short-toed Eagle	M	10	1664	293	1180	2000			223
		F	12	1735	350	1304	2324			
Circaetus pectoralis	Black-breasted Snake-Eagle	B		1500		1200	2300			358
Circaetus cinereus	Brown Snake-Eagle	U	26	2048		1540	2465			139
Circaetus fasciolatus	Fasciated Snake-Eagle	M	2	934		908	960			139
		F	1	1100						
Terathopius ecaudatus	Bateleur	U	10	2200		1800	3000			358
Spilornis cheela burmanicus	Crested Serpent-Eagle	U		900						138
Spilornis cheela asturinus	Crested Serpent-Eagle	M	1	420						358
		F	1	565						
Spilornis cheela pallidus	Crested Serpent-Eagle	U				675	925			358
Spilornis holospilus	Philippine Serpent-Eagle	B	7	705		603	858		Philippines	436, 954
Circus aeruginosus	Western Marsh-Harrier	M	213	609	55.1	460	750	Y	Western France	52a
		F	121	814	77.6	640	1030			
Circus ranivorus	African Marsh-Harrier	U	21	507		382	606			133
Circus spilonotus	Eastern Marsh-Harrier	M	2	502		480	525		Northern China	925
		F	2	622		585	660			
Circus approximans	Swamp Harrier	M	54	640	50.7	520	720		New Zealand	746
		F	66	870	64.9	700	1035			
Circus maillardi maillardi	Reunion Harrier	F		573					estimated	1111
Circus maillardi macrosceles	(Madagascar) Reunion Harrier	M		600						1111
		F		850						
Circus buffoni	Long-winged Harrier	M	11	420		390	464		Suriname	358, 498
		F	5	613		579	645			

Body Masses of World Birds (continued)

Species Name	Common Name	Sex	N	Mean	S.D.	Min	Max	Season	Location	Source #
Circus assimilis	Spotted Harrier	M	6	465		412	537		Australia	746
		F	5	671		530	745			
Circus maurus	Black Harrier	U	8	509		350	760		South Africa	1036
Circus cinereus	Cinereous Harrier	M	4	337		315	359		Tierra del Fuego; Chile	546, 570
		F	4	474		427	510			
Circus cyaneus hudsonius	Northern Harrier	M	186	359	39.9	283	472	F	New Jersey, USA	166
		F	174	515	54.6	375	661			
Circus cyaneus cyaneus	Hen (Northern) Harrier	M	3	300		290	320		Northern China	925
		F	1	430						
Circus macrourus	Pallid Harrier	M	4	332		311	374			273
		F	17	445		402	550			
Circus melanoleucos	Pied Harrier	M		290						1111
		F		390						
Circus pygargus	Montagu's Harrier	M	13	261		227	305		Britain, UK	223
		F	6	370		319	445			
Polyboroides typus	African Harrier-Hawk	M		564		500	710		Ivory Coast	1213
		F		711		580	820			
Kaupifalco monogrammicus	Lizard Buzzard	M		251		220	275		Ivory Coast	1213
		F		310		282	340			
Melierax metabates	Dark Chanting-Goshawk	B	35			500	910		Angola	266
Melierax poliopterus	Eastern Chanting-Goshawk	M	7	548		514	581			358
		F	3	724		673	802			
Melierax canorus	Pale Chanting-Goshawk	B	12			550	800		Angola	266
Micronisus gabar	Gabar Goshawk	B	30			125	230		Angola	266
Accipiter trivirgatus	Crested Goshawk	M		199						400
		F		353						
Accipiter toussenelii	Red-chested Goshawk	M	2	164		148	185		DR Congo	707
		F	4	306		257	340			

Body Masses of World Birds (continued)

Species Name	Common Name	Sex	N	Mean	S.D.	Min	Max	Season	Location	Source #
Accipiter griseiceps	Sulawesi Goshawk	M	1	212						358
		F	1	299						
Accipiter tachiro	African Goshawk	M	4	202		192	218		Zimbabwe	562, 563, 566
		F	1	381						
Accipiter castanilius	Chestnut-flanked Sparrowhawk	M	5			135	150			139
		F	7			152	200			
Accipiter badius	Shikra	M		122		100	150		Ivory Coast	1213
		F		141		130	150			
Accipiter brevipes	Levant Sparrowhawk	M	362	168	20.9	108	268	M	Israel	1351
		F	263	207	35.7	110	290			
Accipiter soloensis	Chinese Sparrowhawk	M	2	150		140	159		Philippines	138, 436, 437
		F	2	181		158	204			
Accipiter francesii	France's Goshawk	M	13	117	9.3	104	140		Madagascar	957
		F	4	160		112	185			
Accipiter trinotatus	Spot-tailed Goshawk	U	2	146		135	158		Sulawesi, Indonesia	1272
Accipiter hiogaster dampieri	Variable Goshawk	M	3	183		175	197		New Britain	427
		F	2	334		304	365			
Accipiter hiogaster pulchellus	Variable Goshawk	F	1	280					Guadalcanal	399
Accipiter hiogaster leucosomus	Variable Goshawk	M	1	247						358
		F	1	439						
Accipiter novaehollandiae	Gray Goshawk	M	5	356		238	422		Australia	746
		F	11	720	103	530	894			
Accipiter novaehollandiae	Gray Goshawk	M	2	180		172	187		Solomon Is.	658
		F	1	245						
Accipiter fasciatus fasciatus	Brown Goshawk	M	15	359	27.4	320	425	F	Victoria, Australia	746
		F	30	592	44.4	535	700			
Accipiter fasciatus didimus	Brown Goshawk	M	8	227		200	268		Northwest Australia	746
		F	8	343		310	390			

Body Masses of World Birds (continued)

Species Name	Common Name	Sex	N	Mean	S.D.	Min	Max	Season	Location	Source #
Accipiter melanochlamys	Black-mantled Goshawk	B	3	265		244	294			138, 387, 428
Accipiter albogularis	Pied Goshawk	M	3	221		181	265			138, 399
		F	1	280						
Accipiter haplochrous	New Caledonian Goshawk	M	3	155		152	162		New Caledonia	1017
		F	3	254		227	268			
Accipiter rufitorques	Fiji Goshawk	M	1	209					Fiji	1196c
Accipiter luteoschistaceus	Slaty-mantled Goshawk	M	2	214		205	222		both immatures	427
Accipiter imitator	Imitator Goshawk	M	1	208						138, 658
		F	2	238		225	250			
Accipiter poliocephalus	Gray-headed Goshawk	M	3	205		200	210			138
		F	1	225						
Accipiter superciliosus	Tiny Hawk	M	3	73.9		61.5	86.2		Colombia; Brazil; Peru; Ecuador	370, 747, 1037, 1160
		F	3	134		116	152			
Accipiter erythropus	Red-thighed Sparrowhawk	M	3	84.7		78	94			139
		F	3	146		132	170			
Accipiter minullus	Little Sparrowhawk	M	2	75.7		75.3	76.1		Zimbabwe	564
		F	1	101						
Accipiter gularis	Japanese Sparrowhawk	M	7	104		90	142			138, 241, 436
		F	15	140		111	192			
Accipiter virgatus	Besra	M	10	100		83.3	142			273, 775, 954
		F	3	143		103	192			
Accipiter virgatus confusus	Besra	M				83	99			358
		F				131	140			
Accipiter virgatus affinis	Besra	M	1	112						358
Accipiter erythrauchen	Rufous-necked Sparrowhawk	M	1	156						138
Accipiter cirrocephalus	Collared Sparrowhawk	M	16	126	14.2	101	156		Australia	746
		F	23	218	31.4	162	300			
Accipiter brachyurus	New Britain Sparrowhawk	M	1	142					New Ireland	1148

Body Masses of World Birds (continued)

Species Name	Common Name	Sex	N	Mean	S.D.	Min	Max	Season	Location	Source #
Accipiter rhodogaster	Vinous-breasted Sparrowhawk	M	1	113					juveniles	358
		F	1	264						
Accipiter madagascariensis	Madagascar Sparrowhawk	M	6	158		108	200		Madagascar	957
		F	4	206		141	285			
Accipiter ovampensis	Ovampo Sparrowhawk	U		195		119	305			139
Accipiter nisus	Eurasian Sparrowhawk	M	70	150	8.9			B	Scotland, UK	850
		F	246	325	26.3					
Accipiter rufiventris	Rufous-chested Sparrowhawk	U	2	198		185	210			139
Accipiter striatus	Sharp-shinned Hawk	M	435	103	6.4	82	125	F	Wisconsin, USA	825
		F	487	174	10.4	144	208			
Accipiter striatus venator	Sharp-shinned Hawk	M	13	94.9					Puerto Rico	271
		F	11	171						
Accipiter ventralis	Plain-breasted Hawk	F	2	233		224	242		Venezuela	1042
Accipiter erythronemius	Rufous-thighed Hawk	F	1	125				B	Bolivia	1143
Accipiter cooperii	Cooper's Hawk	M	51	349	19.6	297	380	F	Wisconsin, USA	826
		F	57	529	36.1	460	588			
Accipiter gundlachi	Gundlach's Hawk	F	1	675					Cuba	19
Accipiter bicolor	Bicolored Hawk	M	7	212		150	250			211, 489 498, 65 796, 910a
		F	5	390		300	454			
Accipiter chilensis	Chilean Hawk	F	4	368		275	415		Chile	570
Accipiter melanoleucus	Black Goshawk	U	19	699		476	980			139
Accipiter henstii	Henst's Goshawk	F	2	1050		960	1140		Madagascar	957
Accipiter gentilis	Northern Goshawk	M	77	912	14.9	735	1099	F	Wisconsin, USA	824
		F	103	1137	18.6	845	1364			

Body Masses of World Birds (continued)

Species Name	Common Name	Sex	N	Mean	S.D.	Min	Max	Season	Location	Source #
Leucopternis kuhli	White-browed Hawk	M	2	322		315	330		Peru	988
		F	1	385						
Leucopternis semiplumbeus	Semiplumbeous Hawk	M	1	250					Panama	990
		F	1	325						
Leucopternis albicollis	White Hawk	M	3	619		600	653		Suriname; Panama	498, 990
		F	5	800		650	908			
Leucopternis occidentalis	Gray-backed Hawk	F	1	660					Ecuador	991
Buteogallus aequinoctialis	Rufous Crab-Hawk	M	10	610		503	685		Suriname	498
		F	10	709		652	750			
Buteogallus anthracinus	Common Black-Hawk	M	6	793					Panama	489
		F	4	1199						
Buteogallus gundlachi	Cuban Black-Hawk	M	1	673					Cuba	1306
		F	1	800						
Buteogallus urubitinga	Great Black-Hawk	M	5	1065		853	1210			212, 488, 189, 498, 1027
		F	9	1248		900	1560			
Buteogallus meridionalis	Savanna Hawk	B	8	808					Suriname; Panama	489, 498
Parabuteo unicinctus	Harris' Hawk	M	220	690		550	829		Texas, USA	894
		F	177	998		825	1175			
Parabuteo unicinctus	Harris' Hawk	M	37	725		634	877		Arizona, USA	894
		F	14	1047		918	1203			
Busarellus nigricollis	Black-collared Hawk	M	2	614		502	725			489, 496, 497, 1027
		F	6	956		758	1195			
Geranoaetus melanoleucus	Black-chested Buzzard-Eagle	M	2	2139		1670	2608		Chile	583
		F	4	2751		2381	3170			
Geranoaetus melanoleucus australis	Black-chested Buzzard-Eagle	M	1	1700						358
		F	1	3200						
Harpyhaliaetus solitarius	Solitary Eagle	U		3000					estimated	1170
Harpyhaliaetus coronatus	Crowned Eagle	U		2950					Brazil	1105
Buteo nitidus	Gray Hawk	M	20	429		391	470	B	Arizona, USA	72
		F	10	628		552	688			

Body Masses of World Birds (continued)

Species Name	Common Name	Sex	N	Mean	S.D.	Min	Max	Season	Location	Source #
Buteo magnirostris	Roadside Hawk	B	16	269					Panama	489
Buteo lineatus lineatus	Red-shouldered Hawk	M	17	544		486	582			225
		F	29	670		593	774			
Buteo platypterus	Broad-winged Hawk	M	14	420						1135
		F	13	490						
Buteo leucorrhous	White-rumped Hawk	M	1	290					Peru	1145
Buteo brachyurus	Short-tailed Hawk	M	14	422	30	385	480	Y	Florida, USA	799
		F	18	583	61	495	710			
Buteo swainsoni	Swainson's Hawk	M	69	808		693	936	B	Alberta, Canada	337
		F	50	1109		937	1367			
Buteo albicaudatus	White-tailed Hawk	M	4	844		865	923			498, 1260
		F	6	928						
Buteo galapagoensis	Galapagos Hawk	M	74	903	54				Santiago, Galapagos Is.	91
		F	53	1295	91					
Buteo galapagoensis	Galapagos Hawk	M	9	1137					Espanola, Galapagos Is.	91
		F	10	1578	83					
Buteo galapagoensis	Galapagos Hawk	M	17	844	75				Marchena, Galapagos Is.	91
		F	4	1223						
Buteo polyosoma	Red-backed Hawk	M	16	696		501	822			160
		F	4	880		790	980			
Buteo poecilochrous	Puna Hawk	M	8	840		700	950			160
		F	12	1168		1010	1280			
Buteo albonotatus	Zone-tailed Hawk	M	3	628		607	667		Mexico	1151a
		F	4	886		845	937			
Buteo solitarius	Hawaiian Hawk	M	17	414					Hawaiian Is.	190
		F		606						
Buteo jamaicensis	Red-tailed Hawk	M	108	1028						1135
		F	100	1224						
Buteo jamaicensis calurus	Red-tailed Hawk	M	26	923					Western USA	298b
		F	24	1266						

Body Masses of World Birds (continued)

Species Name	Common Name	Sex	N	Mean	S.D.	Min	Max	Season	Location	Source #
Buteo ventralis	Rufous-tailed Hawk	M	2	1103		1031	1175		Argentina; Chile	210, 570
		F	3	1285		1195	1410			
Buteo buteo	Eurasian Buzzard	M	214	781		427	1183	Y	Britain, UK	223
		F	261	969		486	1364			
Buteo buteo vulpinus	Eurasian Buzzard	B	882	578	87			M	Israel	1355
Buteo oreophilus	Mountain Buzzard	U	1	700						139
Buteo brachypterus	Madagascar Buzzard	U	2	511		342	680		Madagascar	957
Buteo rufinus	Long-legged Buzzard	M	8	1035		590	1281			133
		F	11	1314		945	1760			
Buteo hemilasius	Upland Buzzard	M	3	1168		1075	1280		Mongolia; China	925, 926
		F	7	1581		1275	1850			
Buteo regalis	Ferruginous Hawk	M	30	1163		977	1347	B	Western USA	53
		F	37	1776		1501	2047			
Buteo lagopus	Rough-legged Hawk	M	152	847		600	1128	Y	Britain, UK	515
		F	119	1065		783	1660			
Buteo auguralis	Red-necked Buzzard	M		580		560	620		Ivory Coast	1213
		F		760		660	890			
Buteo augur	Augur Buzzard	M	5			880	1160			358
		F	7			1100	1300			
Buteo rufofuscus	Jackal Buzzard	F	1	1530						138
Morphnus guianensis	Crested Eagle	U		1750					Peru	1210
Harpia harpyja	Harpy Eagle	M		4800					Brazil	1105
		F				7600	9000			
Harpyopsis novaeguineae	New Guinea Eagle	U				1600	2400			358
Pithecophaga jefferyi	Great Philippine Eagle	M		4464					Philippines	400
		F		6000						
Ictinaetus malayensis	Black Eagle	U	2	1028		930	1125		Moluccos; Malaysia	785, 118

Body Masses of World Birds (continued)

Species Name	Common Name	Sex	N	Mean	S.D.	Min	Max	Season	Location	Source #
Aquila pomarina	Lesser Spotted Eagle	M	16	1200		1053	1509			133
		F	21	1540		1195	2160			
Aquila clanga	Greater Spotted Eagle	M	3	1733		1600	2000			273
		F	4	2678		2150	3200			
Aquila rapax rapax	Tawny Eagle	U				2000	2500			137a
Aquila nipalensis	Steppe Eagle	M	3	2333		2250	2450		South Africa	793
		F	6	3158		2950	3450			
Aquila adalberti	Spanish Eagle	B				2500	3500			358
Aquila heliaca	Imperial Eagle	M	5			2450	2720			358
		F	5			3160	4530			
Aquila wahlbergi	Wahlberg's Eagle	M		640		437	845			133
Aquila gurneyi	Gurney's Eagle	F	1	3060					immature	358
Aquila chrysaetos	Golden Eagle	M	34	3900	335	3000	4475		Idaho, USA	645
		F	14	4627	420	4075	5280			
Aquila audax	Wedge-tailed Eagle	M	23	3132	342	2300	3585		New South Wales, Australia	131
		F	23	3800	412	3050	4750			
Aquila verreauxii	Verreaux's Eagle	U	21	4195		3000	7000		Southern Africa	1036
Hieraaetus fasciatus	Bonelli's Eagle	M		1500					India	273
		F		2500						
Hieraaetus spilogaster	African Hawk-Eagle	M	10	1318		1229	1450		Southern Africa	1036
		F	10	1631		1480	1800			
Hieraaetus pennatus	Booted Eagle	M	12	701		510	770			133
		F	11	968		840	1250			
Hieraaetus morphnoides	Little Eagle	M	14	633	99.5	440	810		Australia	746
		F	19	949	95.3	745	1120			
Hieraaetus ayresii	Ayres' Hawk-Eagle	M	1	714						138
		F	2	910		879	940			
Hieraaetus kienerii	Rufous-bellied Eagle	U	3	799		732	932		Philippines	436, 954

Body Masses of World Birds (continued)

Species Name	Common Name	Sex	N	Mean	S.D.	Min	Max	Season	Location	Source #
Accipiter gentilis apache	Northern Goshawk	M	45	704	32.7	631	774	B	Arizona, USA	1158
		F	49	986	51.9	907	1100			
Accipiter gentilis schvedowi	Northern Goshawk	M	15	501		357	600		Hopei, China	1087
		F	1	1170						
Accipiter meyerianus	Meyer's Goshawk	M	1	530					Papua New Guinea	268
		F	1	1100						
Erythrotriorchis buergersi	Chestnut-shouldered Goshawk	M	1	575					Papua New Guinea	268
Erythrotriorchis radiatus	Red Goshawk	M	2	635		630	640			746
		F	2	1240		1110	1370			
Megatriorchis doriae	Doria's Goshawk	F	1	1000						358
Urotriorchis macrourus	Long-tailed Hawk	M	1	492						139
Butastur rufipennis	Grasshopper Buzzard	M		326		310	342		Ivory Coast	1213
		F		344		300	383			
Butastur teesa	White-eyed Buzzard	U	1	325						358
Butastur liventer	Rufous-winged Buzzard	U				336	340			358
Butastur indicus	Gray-faced Buzzard	M	4	397		375	433			138, 400
		F		433						
Geranospiza caerulescens	Crane Hawk	M	5	279		230	325		Panama; Suriname	489, 498
		F	7	328		273	430			
Leucopternis plumbeus	Plumbeous Hawk	U	1	482					Panama	615
Leucopternis schistaceus	Slate-colored Hawk	F	1	1000						138
Leucopternis princeps	Barred Hawk	F	1	1000					NW Ecuador	1147
Leucopternis melanops	Black-faced Hawk	M	6	312		297	329			138, 498
		F	3			303	380			

Body Masses of World Birds (continued)

Species Name	Common Name	Sex	N	Mean	S.D.	Min	Max	Season	Location	Source #
Polemaetus bellicosus	Martial Eagle	M	7	3174		2750	3700			1036
		F	7	4952		4450	5312			
Spizastur melanoleucus	Black-and-white Hawk-Eagle	U		850					Peru	1210
Lophaetus occipitalis	Long-crested Eagle	U	45	1067		805	1308		South Africa	1036
Spizaetus africanus	Cassin's Hawk-Eagle	U	3	1047		938	1153			139
Spizaetus cirrhatus	Changeable Hawk-Eagle	M		1360						400
		F		1600						
Spizaetus nipalensis	Mountain Hawk-Eagle	M	1	2800					Mongolia	926
		F	2	3225		2950	3500			
Spizaetus alboniger	Blyth's Hawk-Eagle	U		830						358
Spizaetus philippensis	Philippine Hawk-Eagle	F	1	1168					Philippines	954
Spizaetus nanus	Wallace's Hawk-Eagle	U				510	610			358
Spizaetus tyrannus	Black Hawk-Eagle	M	1	904					Suriname	498
		F	3	1122		1092	1150			
Spizaetus ornatus	Ornate Hawk-Eagle	M	5	1009		835	1215			489, 498, 553, 910a, 990, 1027
		F	4	1421		950	1760			
Stephanoaetus coronatus	Crowned Hawk-Eagle	U	4	3640		3175	4120			133
Family Sagittariidae										
Sagittarius serpentarius	Secretarybird	B	20	4017		2800	5000		Southern Africa	1036
Family Falconidae										
Daptrius ater	Black Caracara	M	8	339		260	382		Suriname	498
		F	8	365		325	445			

Body Masses of World Birds (continued)

Species Name	Common Name	Sex	N	Mean	S.D.	Min	Max	Season	Location	Source #
Ibycter americanus	Red-throated Caracara	B	3	624		586	665		Suriname; Venezuela; Brazil	496, 861, 1308
Phalcoboenus megalopterus	Mountain Caracara	U	2	788		780	795			138, 362
Phalcoboenus albogularis	White-throated Caracara	M	1	809					Tierra del Fuego	546
Phalcoboenus australis	Striated Caracara	U	1	1187						138
Caracara cheriway	Crested Caracara	M	21	1117	71	995	1305		Florida, USA	813
		F	18	1220	76	1095	1355			
Caracara cheriway	Crested Caracara	M	14	834	133				Panama	489
		F	10	953	63.2					
Caracara plancus	Southern Caracara	B	7	1348		1266	1460		Tierra del Fuego; Chile	546, 570
Milvago chimachima	Yellow-headed Caracara	M	10	302		235	329		Suriname	498
		F	11	329		297	364			
Milvago chimango	Chimango Caracara	B	29	296	30.9				Chile	1346
Herpetotheres cachinnans	Laughing Falcon	M	8	583		544	675		Panama; Belize; Suriname	489, 498, 1027
		F	4	667		590	800			
Micrastur ruficollis	Barred Forest-Falcon	M	6	161					Panama; Belize	489
		F	6	196						
Micrastur plumbeus	Plumbeous Forest-Falcon	M	3	180		172	188			1299
		F	4	191		180	213			
Micrastur mintoni	Cryptic Forest-Falcon	M	8	200		171	238		Brazil	1299
		F	11	219		170	264			
Micrastur gilvicollis	Lined Forest-Falcon	U	96	209		170	262		Amazonas, Brazil	1299
Micrastur mirandollei	Slaty-backed Forest-Falcon	F	2	549		542	556		Suriname; Costa Rica	496, 379

Body Masses of World Birds (continued)

Species Name	Common Name	Sex	N	Mean	S.D.	Min	Max	Season	Location	Source
Micrastur semitorquatus	Collared Forest-Falcon	M	5	523		467	646		Belize; Panama; Brazil; Suriname	489, 498 1027, 1206
		F	4	739		649	900			
Spiziapteryx circumcincta	Spot-winged Forest-Falcon	M	9	141		124	164		Argentina; Bolivia	212a, 65
		F	4	196		176	228			
Polihierax semitorquatus	Pygmy Forest-Falcon	U	137	61.5		42.5	85		Southern Africa	1036
Polihierax insignis	White-rumped Falcon	U				84	112			358
Microhierax caerulescens	Collared Falconet	U				30	50			358
Microhierax fringillarius	Black-thighed Falconet	F	1	42.9						1287
Microhierax latifrons	White-fronted Falconet	M	1	41.2					Borneo	1218
Microhierax erythrogenys	Philippine Falconet	M	7	42.6		38	47		Philippines	138, 43
		F	2	58.6		55.8	61.5			
Microhierax melanoleucos	Pied Falconet	U				55	75			358
Falco naumanni	Lesser Kestrel	M	34	141		90	176			133
		F	25	164		109	208			
Falco tinnunculus	Eurasian Kestrel	M	16	167	34.3	132	257	W		223
		F	20	201	36.1	139	260			
Falco newtoni	Madagascar Kestrel	M	4	105		90	117			183
		F	6	145		131	159			
Falco punctatus	Mauritius Kestrel	M	6	143		123	178		Mauritius; captive	183
		F	5	196		152	240			
Falco araea	Seychelles Kestrel	M	14	72.4	16.3					183
		F	32	87.9	26.8					
Falco cenchroides	Australian Kestrel	M	180	163	8				Australia	864
		F	135	182	12.8					
Falco cenchroides	Australian Kestrel	F	6	167		154	189		New Guinea	281

Body Masses of World Birds (continued)

Species Name	Common Name	Sex	N	Mean	S.D.	Min	Max	Season	Location	Source #
Falco rupicoloides	Greater Kestrel	U	333	261		181	334			138
Falco sparverius	American Kestrel	M	69	111	9.3			Y	California, USA	85
		F	111	120	9.2					
Falco sparverius	American Kestrel	M	18	113	8.9			W	Kentucky, USA	98
		F	17	132	9.1					
Falco sparverius paulus	American Kestrel	M	15	90.8	6.6	84	109	W	Florida, USA	673
		F	14	103	9	88	119			
Falco sparverius	American Kestrel	M	19	90.3	9.3	66	105		Caribbean Is.	19
		F	5	108		88	121			
Falco sparverius	American Kestrel	M	2	143		130	156		Chile	546, 570
		F	5	151		140	163			
Falco alopex	Fox Kestrel	U				250	300			358
Falco ardosiaceus	Gray Kestrel	M		222		205	255		Ivory Coast	1213
		F		247		240	250			
Falco dickinsoni	Dickinson's Kestrel	U	42	209		167	246			138
Falco zoniventris	Banded Kestrel	F	1	174					Madagascar	68
Falco chicquera	Red-necked Falcon	B	9	208		139	305			138
Falco vespertinus	Red-footed Falcon	M	19	135		118	159		China	184a
		F	11	170		130	220			
Falco amurensis	Amur Falcon	M	48	136		97	155		Hopei, China	1087
		F	24	148		111	188			
Falco eleonorae	Eleonora's Falcon	U	20	390		340	450			133
Falco concolor	Sooty Falcon	F	1	250					Madagascar	68
Falco femoralis	Aplomado Falcon	M	15	271		208	334		Mexico	806, 894
		F	15	416		310	500			
Falco columbarius	Merlin	M	145	163	14.1	134	223	F	New Jersey, USA	166
		F	189	218	17.2	134	281			
Falco rufigularis	Bat Falcon	M	14	130		108	148		Suriname	359
		F	9	206		177	242			

Body Masses of World Birds (continued)

Species Name	Common Name	Sex	N	Mean	S.D.	Min	Max	Season	Location	Source #
Falco deiroleucus	Orange-breasted Falcon	M	2	339		338	340		Guatemala; Suriname	359, 498
		F	3	623		595	654			
Falco subbuteo	Eurasian Hobby	M	9	186		105	216		Mongolia	273 , 926
		F	4	233		206	247			
Falco cuvierii	African Hobby	M		166		150	178			1213
		F		200		186	224			
Falco severus	Oriental Hobby	M	1	183					New Britain; Philippines	427, 954
		F	2	220		192	249			
Falco longipennis	Australian Hobby	M	8	213		177	250		Australia	248
		F	14	293		201	340			
Falco novaezeelandiae	New Zealand Falcon	M	6	330		307	342		New Zealand	746
		F	16	531	37	461	594			
Falco berigora	Brown Falcon	M	63	506	46.6	405	590		Australia	746
		F	90	681	74	500	860			
Falco hypoleucos	Gray Falcon	M	1	335					Australia	246
		F	1	624						
Falco subniger	Black Falcon	M	5	664		620	710		Australia	247
		F	3	907		843	1000			
Falco biarmicus	Lanner Falcon	M	20	506	35				South Africa	577
		F	21	726	63					
Falco jugger	Laggar Falcon	F	1	755						138
Falco cherrug	Saker Falcon	M	2	870		750	990		Mongolia	926
		F	2	1062		975	1150			
Falco rusticolus	Gyrfalcon	M	7	1170		960	1304			138
		F	12	1752		1396	2000			
Falco mexicanus	Prairie Falcon	M	123	560	36	484	661	B	Idaho, USA	1159a
		F	148	908	65.9	779	1133			
Falco pelegrinoides	Barbary Falcon	M	2	610		609	610			358
Falco fasciinucha	Taita Falcon	M	1	212						139
		F	1	306						

Body Masses of World Birds (continued)

Species Name	Common Name	Sex	N	Mean	S.D.	Min	Max	Season	Location	Source #
Falco peregrinus pealei	Peregrine Falcon	M		894		740	1058			1293
		F		1201		1005	1595			
Falco peregrinus anatum	Peregrine Falcon	M	23	652	52.4	590	810	B	Yukon River, Alaska, USA	1293
		F	82	977	76.4	760	1194			
Falco peregrinus tundrius	Peregrine Falcon	M	23	598		528	650	B	Greenland	1293
		F	87	959		797	1130			
Falco peregrinus macropus	Peregrine Falcon	M	4	600		560	660			746
		F	7	825		657	965			
Falco peregrinus babylonicus	Peregrine Falcon	M				330	398		Pakistan	999
		F				513	765			
Falco peregrinus minor	Peregrine Falcon	M	13	528	29				South Africa	577
		F	20	771	72					
Falco peregrinus	Peregrine Falcon	M	7	567		525	620		Western Cape, South Africa	1036
		F	13	828		715	920			
Family Megapodiidae										
Alectura lathami	Australian Brush-Turkey	M	37	2520	180	2120	2950		Queensland, Australia	605
		F	23	2160	250	1980	2510			
Aepypodius arfakianus	Wattled Brush-Turkey	M	2	1488		1450	1525			605
		F	2	1440		1350	1530			
Talegalla cuvieri	Red-billed Brush-Turkey	F	1	1785						605
Talegalla fuscirostris	Black-billed Brush-Turkey	M	3	1352		1325	1400		New Guinea	605
		F	3	1278		1000	1560			
Talegalla jobiensis	Brown-collared Brush-Turkey	M	4	1608		1531	1705			605
		F	1	1360						
Leipoa ocellata	Malleefowl	M	2	2035		2020	2050		Australia	746
		F	8	1800		1520	2050			
Macrocephalon maleo	Maleo	M	8	1516		1365	1622			605
		F	4	1612		1430	1758			
Megapodius wallacei	Moluccan Scrubfowl	M	1	510					Molucccas	605
		F	4	500		490	510			

Body Masses of World Birds (continued)

Species Name	Common Name	Sex	N	Mean	S.D.	Min	Max	Season	Location	Source
Megapodius pritchardii	Niaufoou Scrubfowl	U	15	365		295	454		Tonga	605
Megapodius laperouse	Micronesian Scrubfowl	U	10	308		230	372			605
Megapodius nicobariensis	Nicobar Scrubfowl	M	4	756		595	907			605
		F	3	926		850	1021			
Megapodius freycinet	Dusky Scrubfowl	M	4	638		600	700			605
		F	3	647		610	670			
Megapodius geelvinkianus	Biak Scrubfowl	F	1	525						605
Megapodius forstenii	Forsten's Scrubfowl	U	4	689		620	752			605
Megapodius eremita	Melanesian Scrubfowl	M	6	711		615	810			605, 658
		F	11	644		395	925			
Megapodius decollatus	New Guinea Scrubfowl	M	1	715					New Guinea	605
		F	3	621		550	692			
Megapodius reinwardt	Orange-footed Scrubfowl	M	5	1098		990	1200		Northern Terr., Australia	746
		F	6	888		562	1200			
Family Cracidae										
Ortalis vetula	Plain Chachalaca	M	106	584	58	468	794	Y	Texas, USA	754
		F	102	542	52	439	709			
Ortalis cinereiceps	Gray-headed Chachalaca	U	2	493		490	495		Panama	155, 99
Ortalis garrula	Chestnut-winged Chachalaca	B	10	532					Panama	489, 61
Ortalis ruficauda	Rufous-vented Chachalaca	M	19	640		455	800			270
		F	11	537						
Ortalis erythroptera	Rufous-headed Chachalaca	B	2	632		620	645		Peru	672
Ortalis wagleri	Rufous-bellied Chachalaca	U	1	834					Mexico	1153

Body Masses of World Birds (continued)

Species Name	Common Name	Sex	N	Mean	S.D.	Min	Max	Season	Location	Source #
Ortalis poliocephala	West Mexican Chachalaca	M	1	760					Oaxaca, Mexico	1145
Ortalis canicollis	Chaco Chachalaca	U	24	539	60				Argentina	179
Ortalis leucogastra	White-bellied Chachalaca	M	1	439					Guatemala	1145
		F	1	560						
Ortalis motmot	Little Chachalaca	B	11	520		385	620		Venezuela; Brazil; Suriname	293, 296, 498
Ortalis guttata columbiana	Speckled Chachalaca	M	1	600					Colombia	796
		F	1	500						
Penelope argyrotis	Band-tailed Guan	M	8	843		686	981			270, 1042
		F	9	772		625	900			
Penelope montagnii	Andean Guan	B	3	706		460	840			1042, 1288
Penelope marail	Marail Guan	M	4	904		772	1015			270, 293
		F	4	856		770	1000			
Penelope superciliaris	Rusty-margined Guan	M	4	895		750	1150			270
Penelope dabbenei	Red-faced Guan	F	1	1230						270
Penelope purpurascens	Crested Guan	B	5	2060		2000	2150		Panama	540, 990
Penelope albipennis	White-winged Guan	U		1750						275a
Penelope jacquacu	Spix's Guan	M	7	1569		1400	1778		Venezuela	270
		F	7	1411		1270	1716			
Penelope obscura	Dusky-legged Guan	U	10	1770		1400	2100			362
Penelope pileata	White-crested Guan	M	1	1100					Brazil	449
		F	1	1420						
Pipile cumanensis	Blue-throated Piping-Guan	M	2	1325		1300	1350		Venezuela	1308
		F	6	1159		970	1270			
Pipile pipile	Trinidad Piping-Guan	U				1000	3300			275a

Body Masses of World Birds (continued)

Species Name	Common Name	Sex	N	Mean	S.D.	Min	Max	Season	Location	Source
Pipile cujubi nattereri	Red-throated Piping-Guan	U				1100	1300		Brazil	1105
Pipile jacutinga	Black-fronted Piping-Guan	U				1100	1400		Brazil	1105
Aburria aburri	Wattled Guan	M	2	1490		1430	1550			270, 370
		F	5	1325		1195	1470			
Chamaepetes unicolor	Black Guan	B	12	1135					Panama	489
Chamaepetes goudotii	Sickle-winged Guan	M	4	685						601, 963
		F	3	773						
Penelopina nigra	Highland Guan	B	2	890		866	914		Oaxaca, Mexico	1145
Oreophasis derbianus	Horned Guan	M	3	2212		2140	2324		captive	217b
		F	4	1940		1749	2075			
Nothocrax urumutum	Nocturnal Curassow	U		1700						1114
Mitu tomentosum	Crestless Curassow	M	6	2433		2300	2600			270
		F	10	1967		1300	2425			
Mitu salvini	Salvin's Curassow	U		3100						1114
Mitu tuberosum	Razor-billed Curassow	M	4	3306		2500	3860		Peru	270, 37[illegible] 1143
		F	1	2320						
Mitu mitu	Alagoas Curassow	M		2960					captive	275a
		F		2745						
Pauxi pauxi	Helmeted Curassow	M		3600						522
		F		2700						
Pauxi unicornis	Horned Curassow	M	1	3860						270
		F	1	3630						
Crax rubra	Great Curassow	B	3	4133		4050	4225		Belize	1027
Crax daubentoni	Yellow-knobbed Curassow	M	4	2688		1625	3200			270
		F	1	2325						
Crax alector	Black Curassow	M	3			2850	3750		Venezuela	270
		F	7	2872		2400	3425			

Body Masses of World Birds (continued)

Species Name	Common Name	Sex	N	Mean	S.D.	Min	Max	Season	Location	Source #
Crax fasciolata	Bare-faced Curassow	B	3	2600		2280	2800		Brazil	449, 1105
Crax globulosa	Wattled Curassow	U	1	2500						270
Crax blumenbachii	Red-billed Curassow	U		3500					Brazil	1105
Family Meleagrididae										
Meleagris gallopavo silvestris	Wild Turkey	M	140	7800		6100	10400	W	New York, USA	324
		F	182	4300		3200	6500			
Meleagris ocellata	Ocellated Turkey	U	1	5525					Belize	1027
Family Tetraonidae										
Falcipennis canadensis	Spruce Grouse	M	62	492	31.5	400	489	PB	Washington, USA	1367
		F	84	456	27.5	370	513			
Dendragapus fuliginosus	Sooty Grouse	M	482	1273	87.8				British Columbia, Canada	1366
		F	223	839	59.7					
Dendragapus obscurus pallidus	Dusky Grouse	M	246	1212	78.4				Washington, USA	1368
		F	252	905	63.5					
Dendragapus falcipennis	Siberian Grouse	M				625	735			734
		F				650	740			
Lagopus leucura	White-tailed Ptarmigan	M	25	359		298	416	W	Alaska, USA	1283
		F	30	351		279	381			
Lagopus lagopus alascensis	Willow Ptarmigan	M	222	613	38.6			W	Alaska, USA	1290
		F	75	525	35.9					
Lagopus lagopus alba	Willow Ptarmigan	M	94	547	42.4			W	Yukon Territory, Canada	1290
		F	79	473	27.2					
Lagopus lagopus scotica	Willow Ptarmigan	M		680						734
		F		586						
Lagopus muta	Rock Ptarmigan	M	100	521	37.9	450	620		Northwest Terr., Canada	530
		F	38	550	49.9	480	640			
Tetrao parvirostris	Black-billed Capercaillie	M				3350	4580			734
		F				1700	2200			

Body Masses of World Birds (continued)

Species Name	Common Name	Sex	N	Mean	S.D.	Min	Max	Season	Location	Source #
Tetrao urogallus	Eurasian Capercaillie	M	75	4100		3600	5050			273
		F	10	1800		1700	1920			
Tetrao tetrix	Black Grouse	M	26	1255		1000	1400	W		273
		F	35	910		765	1050			
Tetrao mlokosiewiczi	Caucasian Grouse	M	12	865			1005			585
		F	5	766			820			
Bonasa bonasia	Hazel Grouse	B	56	429		340	500	W		273
Bonasa sewerzowi	Severtzov's Grouse	M	3	278			300			585
		F	3	257			290			
Bonasa umbellus	Ruffed Grouse	M	103	566	41.7			B	Alberta, Canada	1024
		F	4	498						
Centrocercus urophasianus urophasianus	Greater Sage-Grouse	M	465	3190	183			B	Colorado, USA	54
		F	221	1745	151					
Centrocercus urophasianus phaios	Greater Sage-Grouse	M	199	2600					California, USA	1067
		F	29	1300						
Centrocercus minimus	Gunnison Sage-Grouse	M	47	2100					Colorado, USA	1067
		F	13	1200						
Tympanuchus phasianellus	Sharp-tailed Grouse	M	236	953			1090			844
		F	247	817			999			
Tympanuchus cupido pinnatus	Greater Prairie-Chicken	M	393	933					Kansas, USA	1066
		F	318	795						
Tympanuchus cupido cupido	Heath Hen	M	14	972					Massachusetts, USA	1066
		F	7	813						
Tympanuchus cupido attwateri	(Atwater's) Prairie-Chicken	M	8	1014		760	1135		Texas, USA	683
		F	6	730		709	786			
Tympanuchus pallidicinctus	Lesser Prairie-Chicken	M	46	762					Colorado, USA	416a
		F	28	730						
Family Odontophoridae										
Dendrortyx barbatus	Bearded Wood-Partridge	B		432		405	459		estimated	589

Body Masses of World Birds (continued)

Species Name	Common Name	Sex	N	Mean	S.D.	Min	Max	Season	Location	Source #
Dendrortyx macroura	Long-tailed Wood-Partridge	B	4	431		374	455		Morelos, Mexico	1274
Dendrortyx leucophrys	Buffy-crowned Wood-Partridge	M		397						734
		F		340						
Oreortyx pictus	Mountain Quail	U	56	233			293			844
Callipepla squamata	Scaled Quail	M	143	191			234			584
		F	132	177			218			
Callipepla douglasii	Elegant Quail	B				160	196			734
Callipepla californica californica	California Quail	M	166	186	15.7	140	230		California, USA	163
		F	102	162	16.5	147	186			
Callipepla californica achrustera	California Quail	B	32	151					Baja California Sur, Mexico	163
Callipepla gambelii	Gambel's Quail	M	145	170			207			844
		F	103	162			193			
Philortyx fasciatus	Banded Quail	B	4	130		125	136		Michoacan, Mexico	911
Colinus virginianus	Northern Bobwhite	U	847	178				W	Illinois, USA	1227
Colinus virginianus ridgwayi	Northern Bobwhite	B	45	166		146	195	PB	Sonora, Mexico	1227
Colinus nigrogularis	Black-throated Bobwhite	M	9	123		109	146		Mexico; Belize; Nicaragua	542, 640, 910a, 1027
		F	6	137		120	152			
Colinus cristatus sonnini	Crested Bobwhite	M	18	138		117	153			498
		F	12	131		113	147			
Colinus leucopogon	Spot-bellied Bobwhite	M		144						734
		F		115						
Odontophorus gujanensis	Marbled Wood-Quail	M	4	331		313	380		Suriname	498
		F	1	298						
Odontophorus capueira	Spot-winged Wood-Quail	M		457					estimated	589
		F		396						
Odontophorus melanotis	Black-eared Wood-Quail	B		335					estimated	734

Body Masses of World Birds (continued)

Species Name	Common Name	Sex	N	Mean	S.D.	Min	Max	Season	Location	Source
Odontophorus erythrops	Rufous-fronted Wood-Quail	B		335					estimated	734
Odontophorus atrifrons	Black-fronted Wood-Quail	B		304		298	311		estimated	589
Odontophorus hyperythrus	Chestnut Wood-Quail	M	4	351		325	380		Colombia	380b
		F	3	328		325	330			
Odontophorus melanonotus	Dark-breasted Wood-Quail	B		322					estimated	589
Odontophorus speciosus	Rufous-breasted Wood-Quail	B		317		302	332		estimated	589
Odontophorus dialeucos	Tacarcuna Wood-Quail	B		260						734
Odontophorus strophium	Gorgeted Wood-Quail	B		302					estimated	589
Odontophorus columbianus	Venezuelan Wood-Quail	B		340		336	343		estimated	589
Odontophorus leucolaemus	Black-breasted Wood-Quail	U	103	295	16				Costa Rica	469
Odontophorus balliviani	Stripe-faced Wood-Quail	M		311						734
		F		324						
Odontophorus stellatus	Starred Wood-Quail	M		358						734
		F		315						
Odontophorus guttatus	Spotted Wood-Quail	M	18	314	21.6				Panama	489
		F	4	294						
Dactylortyx thoracicus	Singing Quail	M				180	266			589
		F				168	206			
Cyrtonyx montezumae	Montezuma Quail	M	46	192	14.2	168	222			1185
		F	23	179	19.5	122	230			
Cyrtonyx ocellatus	Ocellated Quail	M		218					estimated	589
		F		182						
Rhynchortyx cinctus	Tawny-faced Quail	U		150						1170

Body Masses of World Birds (continued)

Species Name	Common Name	Sex	N	Mean	S.D.	Min	Max	Season	Location	Source #
Family Phasianidae										
Lerwa lerwa	Snow Partridge	U				454	709			7
Tetraophasis obscurus	Verreaux's Partridge	M	1	938						589
		F	2	780		720	840			
Tetraophasis szechenyii	Szecheny's Partridge	M	2	1260		1020	1500			589
		F	1	880						
Tetraogallus caucasicus	Caucasian Snowcock	M		1933						223
		F		1734						
Tetraogallus caspius	Caspian Snowcock	M		2684						223
		F		2344						
Tetraogallus altaicus	Altai Snowcock	M		3000					estimated	589
		F		2540						
Tetraogallus tibetanus	Tibetan Snowcock	M				1500	1750			589
		F				1170	1600			
Tetraogallus himalayensis	Himalayan Snowcock	B	14	2428		2000	3100	W		273
Alectoris graeca	Rock Partridge	M				550	850			589
		F				410	650			
Alectoris chukar pubescens	Chukar	M	17	541		300	580		Hopei, China	1087
		F	15	466		380	550			
Alectoris philbyi	Philby's Partridge	B		441					estimated	589
Alectoris magna	Przevalski's Partridge	M				445	710			589
		F				442	615			
Alectoris barbara	Barbary Partridge	M		461					estimated	589
		F		376						
Alectoris rufa	Red-legged Partridge	M		516		500	547			734
		F	1	540						
Alectoris melanocephala	Arabian Partridge	M		724					estimated	589
		F		522						

Body Masses of World Birds (continued)

Species Name	Common Name	Sex	N	Mean	S.D.	Min	Max	Season	Location	Source
Ammoperdix griseogularis	See-see Partridge	M				187	238			734
		F				182	205			
Ammoperdix heyi	Sand Partridge	U				180	200			734
Francolinus francolinus asiae	Black Francolin	M	19	482		283	566		India	557
		F	18	424		227	482			
Francolinus pictus	Painted Francolin	U				242	340			7
Francolinus pintadeanus	Chinese Francolin	M				347	388			589
		F		310						
Francolinus pondicerianus	Gray Francolin	M	114	274						589
		F	91	228						
Francolinus gularis	Swamp Francolin	M	1	510						734
Francolinus coqui	Coqui Francolin	M	4	262		227	284			589
		F				218	259			
Francolinus albogularis	White-throated Francolin	U	8	276						589
Francolinus schlegelii	Schlegel's Francolin	M		251					estimated	589
		F		223						
Francolinus lathami	Forest Francolin	B		269		254	284		estimated	589
Francolinus sephaena	Crested Francolin	M		368		320	417			734
		F		315		280	350			
Francolinus streptophorus	Ring-necked Francolin	M	2	385		364	406			589
Francolinus finschi	Finsch's Francolin	B	3			540	600		Angola	266
Francolinus levaillantii	Red-winged Francolin	M	3	463		359	567			589
		F	4	401		354	454			
Francolinus africanus	Gray-winged Francolin	M	13	423		345	539			589
		F	3	359		354	369			

Body Masses of World Birds (continued)

Species Name	Common Name	Sex	N	Mean	S.D.	Min	Max	Season	Location	Source #
Francolinus psilolaemus	Moorland Francolin	M	2	520		510	530			589
		F	2	440		370	510			
Francolinus shelleyi	Shelly's Francolin	B	17	488		397	600			589
Francolinus levaillantoides	Orange River Francolin	U	10			330	430		Angola	266
Francolinus squamatus	Scaly Francolin	M				372	565			589
		F				377	515			
Francolinus ahantensis	Ahanta Francolin	M		608					estimated	589
		F		487						
Francolinus griseostriatus	Gray-striped Francolin	M	1	410					Angola	266
		F	1	203						
Francolinus nahani	Nahan's Francolin	M	2	310		308	312			589
		F	3			234	260			
Francolinus hartlaubi	Hartlaub's Francolin	M	4	265		245	290		Namibia	526
		F	4	220		210	240			
Francolinus bicalcaratus	Double-spurred Francolin	M	5	507						589
		F	3	381						
Francolinus icterorhynchus	Heuglin's Francolin	M	7	571		504	588			589
		F				420	462			
Francolinus clappertoni	Clapperton's Francolin	M		534	8.4	450	600		Sudan	491
		F		413	13	300	530			
Francolinus harwoodi	Harwood's Francolin	M	1	545						734
		F	1	446						
Francolinus adspersus	Red-billed Francolin	M	12	465		346	635			589
		F	24	394		346	549			
Francolinus capensis	Cape Francolin	M				600	915			589
		F				435	659			
Francolinus natalensis	Natal Francolin	M	10	606		485	723			589
		F	5	426		370	482			
Francolinus hildebrandti	Hildebrandt's Francolin	M	2	622		600	645			734
		F	2	455		430	480			

Body Masses of World Birds (continued)

Species Name	Common Name	Sex	N	Mean	S.D.	Min	Max	Season	Location	Source #
Francolinus leucoscepus	Yellow-necked Francolin	M	173	753		615	896			589
		F	223	545		400	615			
Francolinus rufopictus	Gray-breasted Francolin	M		848		779	964			589
		F		588		439	666			
Francolinus afer benguellensis	Red-necked Francolin	B	18			270	635		Angola	266
Francolinus afer cranchii	Red-necked Francolin	B	73			310	625		Angola	266
Francolinus afer manueli	Red-necked Francolin	B	10			425	627		Angola	266
Francolinus swainsonii	Swainson's Francolin	M	90	706		400	875			589
		F	100	505		340	750			
Francolinus jacksoni	Jackson's Francolin	M	2	1097		1064	1130			734
Francolinus nobilis	Handsome Francolin	M	2	878		862	895			734
		F	3			600	670			
Francolinus camerunensis	Cameroon Francolin	M		593					estimated	589
		F		509						
Francolinus swierstrai	Swierstra's Francolin	B	6			375	565		Angola	266
Francolinus castaneicollis	Chestnut-naped Francolin	M				915	1200			589
		F				550	650			
Francolinus erckelii	Erckel's Francolin	M	3	1263		1050	1590		Ethiopia	557
		F	1	1136						
Francolinus ochropectus	Djibouti Francolin	M		809						734
		F		605						
Perdix perdix	Gray Partridge	M	42	418	28	356	483		North Dakota, USA	170
		F	36	393	34	298	472			
Perdix dauurica	Daurian Partridge	M	7	270		245	290		Hopei, China	1087
		F	8	257		230	285			
Perdix hodgsoniae	Tibetan Partridge	B				294	450			734

Body Masses of World Birds (continued)

Species Name	Common Name	Sex	N	Mean	S.D.	Min	Max	Season	Location	Source #
Rhizothera longirostris	Long-billed Partridge	M		800					estimated	589
		F		697						
Margaroperdix madagarensis	Madagascar Partridge	U	2	245		240	250		Madagascar	957
Melanoperdix niger	Black Partridge	B	2	280		280	281			734
Coturnix japonica	Japanese Quail	M	5	93		85	100		Hopei, China	1087
		F	6	96.6		88	105			
Coturnix coturnix	Common Quail	M	144	90		76	111		Africa	589
		F	90	103		81	122			
Coturnix delegorguei	Harlequin Quail	B	27	76		65	94			589
Coturnix coromandelica	Rain Quail	B				64	85			7
Coturnix pectoralis	Stubble Quail	M	30	97.9	6.6	85	114		New South Wales, Australia	746
		F	28	106	11.5	75	125			
Coturnix novaezelandiae	New Zealand Quail	M		220					estimated (extinct)	746
		F		200						
Coturnix ypsilophora australis	Brown Quail	M	13	98	11.8	80	123		Southeast Australia	746
		F	12	103	15.4	85	140			
Coturnix ypsilophora ypsilophora	Brown Quail	M	7	119		103	140			746
		F	9	112		95	125			
Coturnix chinensis	Blue-breasted Quail	M	13			36	49			1287
		F	9			28	54			
Coturnix adansonii	Blue Quail	B	3	44.5		43	46.6			567, 589
Anurophasis monorthonyx	Snow Mountain Quail	B		401					estimated	589
Perdicula asiatica	Jungle Bush-Quail	U				57	82			7
Perdicula argoondah	Rock Bush Quail	M		62					estimated	589
		F		59						
Perdicula erythrorhyncha erythrorhyncha	Painted Bush-Quail	U				70	85			7

Body Masses of World Birds (continued)

Species Name	Common Name	Sex	N	Mean	S.D.	Min	Max	Season	Location	Source
Perdicula erythrorhyncha blewitti	Painted Bush-Quail	U				50	70			7
Perdicula manipurensis	Manipur Bush-Quail	U				64	78			7
Xenoperdix udzungwensis	Udzungwa Partridge	M	2	230		222	239		Tanzania	297
		F	1	220						
Arborophila torqueola	Hill Partridge	M				325	430			589
		F				261	386			
Arborophila rufipectus	Sichuan Partridge	M				410	470			589
		F				350	380			
Arborophila mandellii	Chestnut-breasted Partridge	B		268					estimated	589
Arborophila gingica	White-necked Partridge	U	1	253						589
Arborophila rufogularis	Rufous-throated Partridge	M				325	430			734
		F				261	386			
Arborophila atrogularis	White-cheeked Partridge	M		256		200	312			734
		F		220		200	312			
Arborophila crudigularis	Taiwan Partridge	M	1	311					female estimated	589
		F		212						
Arborophila ardens	Hainan Partridge	M	1	300						589
		F	1	237						
Arborophila javanica	Chestnut-bellied Partridge	B		272		257	286		estimated	589
Arborophila orientalis	Gray-breasted Partridge	U	2	336		320	353			1287
Arborophila campbelli	Campbell's Hill Partridge	B		271					estimated	734
Arborophila brunneopectus	Bar-backed Partridge	M		317					estimated	589
		F		268						
Arborophila davidi	Orange-necked Partridge	U		241					estimated	589

Body Masses of World Birds (continued)

Species Name	Common Name	Sex	N	Mean	S.D.	Min	Max	Season	Location	Source #
Arborophila cambodiana	Chestnut-headed Partridge	M		318					estimated	589
		F		257						
Arborophila hyperythra	Red-breasted Partridge	B		270					estimated	589
Arborophila rubrirostris	Red-billed Partridge	M		243					estimated	589
		F		209						
Arborophila chloropus	Scaly-breasted Partridge	M				280	300			734
		F		250						
Arborophila merlini	Vietnam Partridge	M		290					estimated	734
		F		250						
Arborophila charltonii	Chestnut-necklaced Partridge	M		290					estimated	734
		F		250						
Caloperdix oculeus	Ferruginous Partridge	M	1	190						589, 1287
		F	1	230						
Haematortyx sanguiniceps	Crimson-headed Partridge	M	1	300						589
Rollulus rouloul	Crested Partridge	M	7	232						589
		F	6	202						
Ptilopachus petrosus	Stone Partridge	U	4	193		187	198		Nigeria	777
Bambusicola fytchii	Mountain Bamboo-Pheasant	M				278	400			734
		F				256	340			
Bambusicola thoracicus	Chinese Bamboo-Pheasant	M				242	297			734
		F				200	342			
Galloperdix spadicea	Red Spurfowl	U				284	454			7
Galloperdix lunulata	Painted Spurfowl	M				255	285			7
		F				226	255			
Galloperdix bicalcarata	Ceylon Spurfowl	M				312	368			7
		F				200	312			
Ithaginis cruentus berezowskii	Blood Pheasant	M	5			520	600			588
		F	3			410	620			

Body Masses of World Birds (continued)

Species Name	Common Name	Sex	N	Mean	S.D.	Min	Max	Season	Location	Source #
Tragopan melanocephalus	Western Tragopan	M				1800	2150			7
		F				1250	1400			
Tragopan satyra	Satyr Tragopan	M				1600	2100			7
		F				1000	1200			
Tragopan blythii	Blyth's Tragopan	M		1930						588
		F				1000	1500			
Tragopan temminckii	Temminck's Tragopan	M	2	1404		1362	1447		captive	588
		F	2	964		907	1021			
Tragopan caboti	Cabot's Tragopan	M	1	1400						588
		F		900						
Pucrasia macrolopha	Koklass Pheasant	M	10	1184						184a
		F	10	932						
Lophophorus impejanus	Himalayan Monal	M				1980	2380			7
		F				1800	2150			
Lophophorus sclateri	Sclater's Monal	M	2	2500						588
		F				2126	2267			
Lophophorus lhuysii	Chinese Monal	B	2	3008		2837	3178			734
Gallus gallus	Red Junglefowl	M	5	988		850	1100		Fiji	1196c
		F	2	820		650	990			
Gallus gallus jabouillei	Red Junglefowl	M	10	844		672	1020		China	184a
		F	2	467		435	500			
Gallus sonneratii	Gray Junglefowl	M				790	1136			7
		F				705	790			
Gallus lafayettii	Ceylon Junglefowl	M				790	1140			7
		F				510	625			
Gallus varius	Green Junglefowl	M				672	1450			734
		F				485	1050			
Lophura leucomelanos hamiltoni	Kalij Pheasant	M				910	1080			7
		F				564	1024			

Body Masses of World Birds (continued)

Species Name	Common Name	Sex	N	Mean	S.D.	Min	Max	Season	Location	Source #
Lophura edwardsi	Edward's Pheasant	B	2	1082		1050	1115			588
Lophura hatinhensis	Vietnamese Pheasant	M	1	1100					Vietnam	1264
Lophura swinhoii	Swinhoe's Pheasant	U	2	1100						588
Lophura hoogerwerfi	Hoogerwerf's Pheasant	B	2	685		670	700		Sumatra	1139
Lophura nycthemera beaulieui	Silver Pheasant	M	9			1500	2000			588
		F	2	1230		1160	1300			
Lophura erythrophthalmus	Crestless Fireback	M	2	1043					captive	588
		F	1	837						
Lophura ignita	Crested Fireback	M	9			1810	2500			1287
		F	4			1350	1600			
Lophura diardi	Siamese Fireback	M	1	1420					captive	588
		F	3	835		680	1025			
Lophura bulweri	Bulwer's Pheasant	M	4	1615		1470	1800			588
		F	2	960		916	1004			
Crossoptilon crossoptilon drouynii	White Eared-Pheasant	M	3			2350	2750			588
		F	7			1400	2050			
Crossoptilon mantchuricum	Brown Eared-Pheasant	M				1650	2475			588
		F				1450	2025			
Crossoptilon auritum	Blue Eared-Pheasant	M				1700	2110			588
		F				1450	1880			
Catreus wallichii	Cheer Pheasant	M				1475	1700			7
		F				1250	1360			
Syrmaticus ellioti	Elliot's Pheasant	M	17	1156		1044	1317			588
		F	35	878		726	1090			
Syrmaticus humiae	Hume's Pheasant	M				975	1080			734
		F				650	850			
Syrmaticus mikado	Mikado Pheasant	M	1	1300					captive	734
		F	1	1015						

Body Masses of World Birds (continued)

Species Name	Common Name	Sex	N	Mean	S.D.	Min	Max	Season	Location	Source
Syrmaticus soemmerringi	Copper Pheasant	M		900					Japan	1242
		F		1300						
Syrmaticus reevesii	Reeve's Pheasant	M	24	1529						588
		F	30	949						
Phasianus colchicus	Ring-necked Pheasant	M	6378	1317			1861			844
		F	759	953			1453			
Phasianus versicolor	Green Pheasant	M				900	1100			734
		F				800	900			
Chrysolophus pictus	Golden Pheasant	M	5			575	710			184a
		F	5			350	665			
Chrysolophus amherstiae	Lady Amherst's Pheasant	M	5			675	850			184a
		F	5			624	804			
Polyplectron chalcurum	Bronze-tailed Peacock-Pheasant	M	2	508		425	590			734
		F	4			238	269			
Polyplectron germaini	Germain's Peacock-Pheasant	M	1	510						588
		F	1	397						
Polyplectron bicalcaratum	Gray Peacock-Pheasant	M	4			660	710			588
		F	2	480		460	500			
Polyplectron bicalcaratum katsumatae	Gray Peacock-Pheasant	M	1	456						734
Polyplectron malacense	Malayan Peacock-Pheasant	U	4	608		581	680			588, 12 1333
Polyplectron napoleonis	Palawan Peacock-Pheasant	M	2	436					captive	588
		F	2	322						
Rheinardia ocellata	Crested Argus									
Argusianus argus	Great Argus	M	6	2361		2040	2725			588
		F	3	1627		1590	1700			
Pavo cristatus	Indian Peafowl	M	17	4766	231	4160	5100			613a
		F	5	3622		3040	4400			
Pavo muticus	Green Peafowl	M				3850	5000			588
		F	3			1060	1160			

Body Masses of World Birds (continued)

Species Name	Common Name	Sex	N	Mean	S.D.	Min	Max	Season	Location	Source #
Afropavo congensis	Congo Peacock	M	2	1418		1361	1475			588
		F	2	1144		1135	1154			
Family Numididae										
Agelastes meleagrides	White-breasted Guineafowl	B	2	815					captive	734
Agelastes niger	Black Guineafowl	B		700						734
Numida meleagris	Helmeted Guineafowl	B	10	1299						489
Guttera plumifera	Plumed Guineafowl	B				750	1000			734
Guttera pucherani	Crested Guineafowl	M		1149		721	1573			1244
Guttera pucherani edouardi	Crested Guineafowl	B				1130	1500			734
Acryllium vulturinum	Vulturine Guineafowl	B	4	1330		1026	1645			734
Family Opisthocomidae										
Opisthocomus hoazin	Hoatzin	U	9	696					Venezuela	298
Family Mesitornithidae										
Mesitornis unicolor	Brown Mesite	F	1	148				B	Madagascar egg in oviduct	1143
Mesitornis benschi	Subdesert Mesite	M	52	134	10.8	112	158	Y	Madagascar	1076
		F	35	142	13.6	115	172			
Family Turnicidae										
Turnix sylvaticus sylvaticus	Small Buttonquail	M		60						734
		F		70						
Turnix sylvaticus lepuranus	Small Buttonquail	M	10	36		32.2	43.5			1244
		F	12	51		39	53.7			
Turnix maculosus	Red-backed Buttonquail	M	13	33.9	3.7	27.5	39		Australia	746
		F	9	45.2		38	49			

Body Masses of World Birds (continued)

Species Name	Common Name	Sex	N	Mean	S.D.	Min	Max	Season	Location	Source
Turnix hottentottus nanus	Hottentot Buttonquail	M	4	42.7		40	46		Angola	981, 1244
		F	2	60		57.5	62.4			
Turnix tanki	Yellow-legged Buttonquail	M	3	45.7		36	59		Hopei & Tsingtao, China	1087, 1088
		F	3	61		43	80			
Turnix ocellatus	Spotted Buttonquail	F	1	110						734
Turnix suscitator leggei	Barred Buttonquail	M				35	52			734
		F				47	68			
Turnix nigricollis	Madagascar Buttonquail	U	6	61.3		27.5	84		Madagascar	957
Turnix melanogaster	Black-breasted Buttonquail	M	2	75.8		64.7	87			746
		F	4	100		80	119			
Turnix castanotus	Chestnut-backed Buttonquail	M	3	68.9		68.2	70		Australia	746
		F	3	99.3		87	124			
Turnix varius	Painted Buttonquail	M	16	74.7	8.8	53	85			746
		F	16	109	16.7	72	130			
Turnix everetti	Sumba Buttonquail	B	2	28		28	28			734
Turnix pyrrhothorax	Red-cheeked Buttonquail	M	13	37.1	4.6	27	46		Australia	746
		F	17	56.2	8.3	43	76			
Turnix velox	Little Buttonquail	M	25	37	4.2	28	44			746
		F	19	53	7.6	39	64			
Ortyxelos meiffrenii	Quail-plover	M	2	17.6		15.7	19.5			1244
Family Gruidae										
Balearica regulorum	Gray Crowned-Crane	F	2	3772		3575	3970			1244
Balearica pavonina	Black Crowned-Crane	U	21	3590		2725	4100			133
Anthropoides virgo	Demoiselle Crane	B	10	2417	287	2100	3060	B		586
Grus paradisea	Blue Crane	M	1	5675					captive	586
		F	1	3633						

Body Masses of World Birds (continued)

Species Name	Common Name	Sex	N	Mean	S.D.	Min	Max	Season	Location	Source #
Bugeranus carunculatus	Wattled Crane	B	3	8159		7225	8966			133, 586
Grus leucogeranus	Siberian Crane	M	8	6387		5100	7400	B		586
		F	4	5475		4900	6000			
Grus canadensis canadensis	Sandhill Crane	M	33	3350	22.3	2700	3700	W	New Mexico, USA	314
		F	31	2982	19	2450	3300			
Grus canadensis tabida	Sandhill Crane	M	61	5797	31.5	5040	6700	W	New Mexico, USA	314
		F	28	5345	29.8	4900	6030			
Grus canadensis pratensis	Sandhill Crane	M	34	4670	340					848
		F	34	4070	340					
Grus canadensis rowani	Sandhill Crane	M	51	4800	390					1195
		F	33	4110	250					
Grus antigone	Sarus Crane	M	3	6667		6500	6900		Queensland, Australia	746
		F	4	5250		5000	5500			
Grus rubicunda	Brolga	M	321	6838	649	4761	8729		Northeast Australia	746
		F	217	5663	560	3628	7255			
Grus vipio	White-naped Crane	U	3	4663					captive	972
Grus grus	Common Crane	U	16	5500	129			W	Spain	950
Grus monacha	Hooded Crane	M	7	3930		3284	4870		captive	586
		F	4	3540		3397	3737			
Grus americana	Whooping Crane	U	3	5826		5448	6356		captive	314
Grus nigricollis	Black-necked Crane	F	1	6000						586
Grus japonensis	Red-crowned Crane	U	14	8786	1200				captive	489
Family Aramidae										
Aramus guarauna	Limpkin	U	31	1080	110	900	1270	PB	Florida, USA	847
Family Psophiidae										
Psophia crepitans	Gray-winged Trumpeter	B	14	1026		790	1100		French Guiana; Suriname	293, 498, 1308

Body Masses of World Birds (continued)

Species Name	Common Name	Sex	N	Mean	S.D.	Min	Max	Season	Location	Source
Psophia viridis	Dark-winged Trumpeter	F	1	1071					Brazil	1105
Psophia leucoptera	Pale-winged Trumpeter	M	7	1378		1280	1440		Peru	1094
		F	9	1256		1180	1320			
Family Rallidae										
Sarothrura pulchra	White-spotted Flufftail	B	18	45.2		39	49			1198
Sarothrura elegans	Buff-spotted Flufftail	M	53	46.4	4.1	39.5	58			1198
		F	34	45.7	4.1	40	60.5			
Sarothrura rufa	Red-chested Flufftail	M	14	38.8	4.1	30	46.5			1198
		F	9	36.6		29.5	42			
Sarothrura boehmi	Streaky-breasted Flufftail	M	4	37.3		33.5	42			1198
		F	2	32.8		30.5	35			
Sarothrura affinis	Striped Flufftail	M	4	27.8		25	30			1198
Sarothrura insularis	Madagascar Flufftail	U	2	26.8		23.5	30		Madagascar	68, 957
Sarothrura ayresi	White-winged Flufftail	M	1	31.8						1198
Sarothrura watersi	Slender-billed Flufftail	M	1	26.5					Madagascar	1198
Himantornis haematopus	Nkulengu Rail	M	3	536		504	595			1198
		F	2	640		390	870			
Canirallus oculeus	Gray-throated Rail	B	3	277		273	279			1198
Canirallus kioloides	Madagascar Wood-Rail	U	4	246		172	280		Madagascar	68, 957
Coturnicops noveboracensis	Yellow Rail	B	71	60.9	3.5	53	70	B	Quebec, Canada	998
Coturnicops notatus	Speckled Rail	U	1	30						1198
Micropygia schomburgkii	Ocellated Crake	U	2	31.2		23.5	38.8			1198

Body Masses of World Birds (continued)

Species Name	Common Name	Sex	N	Mean	S.D.	Min	Max	Season	Location	Source #
Rallina rubra	Chestnut Forest-Rail	M	6	81		71	91		New Guinea	282, 423
Rallina leucospila	White-striped Forest-Rail	M	2	120		114	125		New Guinea	282
Rallina forbesi dryas	Forbes' Rail	M		88.4		78	106		New Guinea	768
		F		82		65	96			
Rallina mayri	Mayr' Rail	M	5	128		123	136		New Guinea	282
		F	3	118		112	123			
Rallina tricolor	Red-necked Crake	B	8	178		109	231			746
Rallina fasciata	Red-legged Crake	U		100		68.8	108			1287
Rallina eurizonoides	Slaty-legged Crake	B	6	109		99	118			1198
Anurolimnas castaneiceps	Chestnut-headed Crake	F	1	126						1198
Anurolimnas viridis	Russet-crowned Crake	M	5	60		55	63		Suriname	498
		F	3	69		63	73			
Laterallus melanophaius	Rufous-sided Crake	U	15	52.1						1198
Laterallus ruber	Ruddy Crake	B	10	44.4	3.2	41.5	48.9		Belize; Nicaragua	542, 640, 1027
Laterallus albigularis	White-throated Crake	M	13	49.7	5.8				Panama	489
		F	12	45	6.3					
Laterallus exilis	Gray-breasted Crake	M	10	32.4		26.5	39			1198
		F	9	34.7		27	43			
Laterallus jamaicensis jamaicensis	Black Rail	M	33	34.6	3.1	28.9	43	B	Eastern USA	327
		F	16	36.2	5	29	44			
Laterallus jamaicensis coturniculus	Black Rail	M	16	27.9	2.7	23	34	B	Arizona, USA	327
		F	18	32.2	6.2	25	46			
Laterallus spilonotus	Galapagos Rail	U				36	45		Galapagos Is.	381
Laterallus leucopyrrhus	Rufous-and-white Crake	U	11	45.5		34	52			1198

Body Masses of World Birds (continued)

Species Name	Common Name	Sex	N	Mean	S.D.	Min	Max	Season	Location	Source #
Laterallus xenopterus	Rufous-faced Crake	U	5	56.9		51	68.5			1198
Nesoclopeus woodfordi immaculatus	Woodford's Rail	M	5	593		457	670		Solomon Is.	658, 1153b
		F	6	516		410	571			
Gallirallus australis scotti	Weka	M	70	1034	134	600	1425		Macquarie Is.	746
		F	26	753	87	500	884			
Gallirallus australis hectori	Weka	M	8	1255		925	1605		New Zealand	746
		F	4	651		525	850			
Gallirallus australis australis	Weka	M	9	978		612	1250			746
		F	6	725		350	1035			
Gallirallus sylvestris	Lord Howe Rail	M	49	536	11	410	780		Lord Howe Is.	746
		F	58	456	9	330	615			
Gallirallus okinawae	Okinawa Rail	U	1	433					Okinawa	746
Gallirallus philippensis	Buff-banded Rail	U	16	180		142	218		Tonga	977
Gallirallus philippensis	Buff-banded Rail	F	1	240					Guadalcanal	399
Gallirallus philippensis lacustris	Buff-banded Rail	M	3	225		210	240		New Guinea	768
		F	1	185						
Gallirallus owstoni	Guam Rail	M	27	241		174	303		Guam	578
		F	20	212		170	274			
Gallirallus torquatus	Barred Rail	B	4	280		241	200		Philippines	954, 98
Gallirallus calayanensis	Calayan Rail	F	1	244					Philippines, subadult	7a
Gallirallus striatus	Slaty-breasted Rail	U	4	116		100	142			7, 1287
Rallus longirostris obsoletus	Clapper Rail	M	109	322	35.5	180	400	F	Louisiana, USA	326
		F	157	257	34.3	180	320			
Rallus longirostris scottii	Clapper Rail	M	11	290	15.5	263	310		Florida, USA	326
		F	14	247	30.1	199	314			
Rallus longirostris yumanensis	Clapper Rail	M	81	266	350				California, USA	325
		F	34	193	132					

Body Masses of World Birds (continued)

Species Name	Common Name	Sex	N	Mean	S.D.	Min	Max	Season	Location	Source #
Rallus elegans elegans	King Rail	M	9	415		340	490		Eastern USA	783
		F	9	306		253	325			
Rallus elegans tenuirostris	King Rail	M	4	306		271	331		Mexico	783
		F	3	248		220	268			
Rallus limicola	Virginia Rail	U	147	84.1	14.5	55	124		Arizona, USA	213
Rallus antarcticus	Austral Rail	F	1	60.4					Tierra del Fuego	546
Rallus aquaticus	Water Rail	M	64	125	15.3	92	166	F	Netherlands	223
		F	51	98	12.9	74	138			
Rallus caerulescens	African Rail	M	66	180		146	205			1198
		F	50	146		120	170			
Rallus madagascariensis	Madagascar Rail	B	3	114		88	148		Madagascar	68, 957
Lewinia pectoralis	Lewin's Rail	M	9	80.8		71	100	Y	Australia	746
		F	8	75.4		63	111			
Lewinia muelleri	Auckland Is. Rail	B	4	93		89	100		Auckland Is.	332
Dryolimnas cuvieri	White-throated Rail	M	32	189		145			Madagascar	68
		F	21	176		138	223			
Crecopsis egregia	African Crake	U	19	119		92	141			1198
Crex crex	Corn Crake	M	28	169	18.7	135	202	B	Netherlands	223
		F	3	142		140	145			
Rougetius rougetii	Rouget's Rail	M	1	220						1198
		F	1	170						
Atlantisia rogersi	Inaccessible Is. Rail	M	6	41.8		38	49		Tristan de Cunda	1033
		F	6	36.9		34	42			
Aramides mangle	Little Wood-Rail	F	2	183		164	202		Brazil	1205
Aramides axillaris	Rufous-necked Wood-Rail	M	4	292		262	327		Honduras; Mexico; Suriname	498, 1143
Aramides cajanea	Gray-necked Wood-Rail	B	18	397					Panama	489

Body Masses of World Birds (continued)

Species Name	Common Name	Sex	N	Mean	S.D.	Min	Max	Season	Location	Source #
Aramides wolfi	Brown Wood-Rail	M	1	454					Ecuador	1143
Aramides ypecaha	Giant Wood-Rail	M	3	737		655	860			1198
		F	2	665		565	765			
Aramides saracura	Slaty-breasted Wood-Rail	M	1	540					Brazil	63
Amaurolimnas concolor	Uniform Crake	M	1	133					Brazil	1202
Gymnocrex plumbeiventris	Bare-eyed Rail	U	2	288		255	320			1198
Amaurornis akool	Brown Crake	M				114	170			7
		F				110	140			
Amaurornis olivacea	Plain Bush-hen	B	4	292		250	312		Philippines	954
Amaurornis phoenicurus	White-breasted Waterhen	B	12	180		158	196		Philippines; Malaysia; China	954, 983, 1088, 1287
Amaurornis moluccana	Rufous-tailed Bush-hen	M	7	196		182	238		New Guinea; Indonesia	283, 292
		F	10	178	19.8	135	205			
Amaurornis flavirostra	Black Crake	B	87	86.2		54	125		Southern Africa	1036
Porzana parva	Little Crake	B	21	49.7		30	72			1198
Porzana pusilla affinis	Baillon's Crake	B	12	35.4		22.9	46		Australia	746
Porzana porzana	Spotted Crake	U	149	87.1		47	147	Y	Europe; Africa	1198
Porzana fluminea	Australian Crake	M	8	64.5		50	75		Australia	746
		F	3	57.3		50	61			
Porzana carolina	Sora	U	86	74.8	13.1	49	126			1198
Porzana albicollis	Ash-throated Crake	B	7	100		90	112			1198
Porzana fusca	Ruddy-breasted Crake	U	18			51.5	63.7		Malaysia; Singapore	1287
Porzana paykullii	Band-bellied Crake	B	6	111		84.8	132			1198, 1287

Body Masses of World Birds (continued)

Species Name	Common Name	Sex	N	Mean	S.D.	Min	Max	Season	Location	Source #
Porzana tabuensis	Spotless Crake	M	7	47.7		39	56		Australia	746
		F	4	37.4		21	58			
Porzana atra	Henderson Is. Crake	M	10	80.1		69	87		Henderson Is.	446, 604
		F	12	74.8		66	88			
Porzana flaviventer woodi	Yellow-breasted Crake	M	6	26.1		24	28			1198
		F	10	23.8		20.2	28.3			
Porzana cinerea	White-browed Crake	U[a]	11	52.1	6.8	43	62.5		Australia	746
Aenigmatolimnas marginalis	Striped Crake	F	2	51.2		41.5	61			1198
Neocrex erythrops	Paint-billed Crake	U	7	58.3		43	70		Colombia; Paraguay	1198
Pardirallus maculatus	Spotted Rail	B	15	171	18.2	140	198			63, 75, 212a, 984, 1147
Pardirallus nigricans	Blackish Rail	M	1	217					Brazil	75
Pardirallus sanguinolentus	Plumbeous Rail	U	9	232		190	318		Chile	1198
Eulabeornis castaneoventris	Chestnut Rail	M	5	746		626	910		Western Australia	1198
		F	7	628		550	710			
Megacrex inepta	New Guinea Flightless Rail	U		1200					New Guinea weight approximate	786
Gallicrex cinerea	Watercock	M	9	503		326	650			954, 1287
		F	5	303		207	434			
Porphyrio porphyrio melanotis	Purple Swamphen	M	148	1091	94.9	785	1310		Australia	746
		F	97	885	98.5	679	1252			
Porphyrio porphyrio porphyrio	Purple Swamphen	M	37	869	71.3	720	1000		Spain	223
		F	35	724	78.7	520	870			
Porphyrio porphyrio madagascariensis	Purple Swamphen	M	11	636		528	687			1244
		F	8	556		480	737			
Porphyrio mantelli	Takahe	M	13	2673	258	2250	3250		New Zealand	746
		F	14	2268	194	1850	2600			

Body Masses of World Birds (continued)

Species Name	Common Name	Sex	N	Mean	S.D.	Min	Max	Season	Location	Source
Porphyrio alleni	Allen's Gallinule	M	6	154		132	172			1198
		F	3	125		112	145			
Porphyrio martinica	Purple Gallinule	M	20	257	27.1					489
		F	8	215						
Porphyrio flavirostris	Azure Gallinule	M	9	93.3		73	111			1198
		F	8	92		79	107			
Gallinula silvestris	San Cristobal Moorhen	M	1	450					San Cristobal	1198
Gallinula nesiotis	Tristan Moorhen	U	3	513		505	530			1198
Gallinula chloropus cachinnans	Common Moorhen	M	27	415		370	456			41
		F	33	349		310	398			
Gallinula chloropus chloropus	Common Moorhen	M	74	339		249	493			41
		F	80	271		192	343			
Gallinula tenebrosa	Dusky Moorhen	M	5	570		490	720			746
		F	7	493		336	684			
Gallinula angulata	Lesser Moorhen	M	4	155		145	164			1198
		F	3	109		92	137			
Gallinula melanops	Spot-flanked Gallinule	U	3	191		154	225			1198
Gallinula ventralis	Black-tailed Native-hen	M	8	410		250	530			746
		F	5	364		322	405			
Gallinula mortierii	Tasmanian Native-hen	M	152	1334	109				Tasmania, Australia	746
		F	120	1251	103					
Fulica cristata	Red-knobbed Coot	U	37	826	96.8	585	1080		South Africa	508
Fulica atra australis	Eurasian Coot	B	16	531	81.8	351	660		SE Australia	746
Fulica atra atra	Eurasian Coot	M	215	902		610	1200			1198
		F	214	770		610	1150			
Fulica americana	American Coot	M	27	724		576	848	F		383
		F	20	560		427	628			
Fulica leucoptera	White-winged Coot	B	7	656		400	1420		Brazil; Argentina	63, 210, 546, 1198

Body Masses of World Birds (continued)

Species Name	Common Name	Sex	N	Mean	S.D.	Min	Max	Season	Location	Source #
Fulica ardesiaca	Slate-colored Coot	U	5	988		900	1100		Chile	1198
Fulica armillata	Red-gartered Coot	U	34	1040		860	1260			362
Fulica rufifrons	Red-fronted Coot	U	20	644		533	872			1198
Fulica gigantea	Giant Coot	M	1	2700						1198
		F	2	2210		2020	2400			
Fulica cornuta	Horned Coot	M	3	2374		2100	2611			916, 1198
		F	4	1876		1660	2010			
Family Heliornithidae										
Podica senegalensis	African Finfoot	U	4	599		338	879			1244
Heliornis fulica	Sungrebe	B	12	132					Panama; Suriname	489, 498, 861
Family Rhynochetidae										
Rhynochetos jubatus	Kagu	M	1	860					captive	443
Family Eurypygidae										
Eurypyga helias	Sunbittern	B	8	210		178	295		Suriname; Peru	370, 498
Family Cariamidae										
Cariama cristata	Red-legged Seriema	U		1400					Brazil	1105
Chunga burmeisteri	Black-legged Seriema	B	4	1298		1224	1395		Bolivia; Argentina	212a, 1054
Family Otididae										
Otis tarda	Great Bustard	M	13	11975		8500	18000	Sp	Germany; Russia	223
		F	4	3816		3480	4025			
Ardeotis arabs	Arabian Bustard	M	2	7850		5700	10000			1244
		F	1	4500						
Ardeotis kori	Kori Bustard	M	20	11300		7200	15200		Namibia	1036
		F	35	5620		4300	6700			

Body Masses of World Birds (continued)

Species Name	Common Name	Sex	N	Mean	S.D.	Min	Max	Season	Location	Source #
Ardeotis kori	Kori Bustard	M	1	10900						1244
struthiunculus		F	2	5900						
Ardeotis nigriceps	Indian Bustard	M				8000	14500			7
		F				3500	6750			
Ardeotis australis	Australian Bustard	M	6	5970		4100	7200			746
		F	8	2930		1980	4000			
Chlamydotis undulata	Houbara Bustard	M	5	1960		1500	2220	B	Mongolia; former USSR	223
		F	5	1150		1100	1250			
Neotis ludwigii	Ludwig's Bustard	M	4	4525		4200	4700		South Africa	323
		F	4	2350		2200	2500			
Neotis denhami	Stanley Bustard	M	4			6676	8010		Angola	266
		F	4			1475	3150			
Neotis heuglinii	Heuglin's Bustard	M	2	6000		4000	8000		Kenya	1244
		F	2	2800		2600	3000			
Neotis nuba	Nubian Bustard	M		5440						223
Eupodotis senegalensis	White-bellied Bustard	M	3			930	1000		Angola	266
		F	1	585						
Eupodotis caerulescens	Blue Bustard	U	2	1366		1120	1612		South Africa	508
Eupodotis vigorsii	Karoo Bustard	M	2	1700		1675	1725		Namibia	526
Eupodotis rueppellii	Rueppell's Bustard	M	8	1191		1000	1368		Angola; Namibia	266, 10[illegible]
		F	4	1029		914	1122			
Eupodotis humilis	Little Brown Bustard	U	1	698						590
Eupodotis gindiana	Buff-crested Bustard	M				680	900			590
Eupodotis ruficrista	Red-crested Bustard	M	6	766		688	858		Namibia; South Africa	1036
		F	14	662		250	834			
Eupodotis afra	Black Bustard	U	102	690		500	878			133
Eupodotis afraoides	White-quilled Bustard	B	33	745		525	875		Namibia; South Africa	1036

Body Masses of World Birds (continued)

Species Name	Common Name	Sex	N	Mean	S.D.	Min	Max	Season	Location	Source #
Lissotis melanogaster	Black-bellied Bustard	B	8			1040	1275		Angola	266
Lissotis hartlaubii	Hartlaub's Bustard	F	1	1190						120
Houbaropsis bengalensis	Bengal Florican	B				1800	2250			7
Sypheotides indica	Lesser Florican	B				510	740			7
Tetrax tetrax	Little Bustard	U	2	825		740	910	Sp		590
			4	842		824	866			
Family Jacanidae										
Microparra capensis	Lesser Jacana	F	1	41.3						584a
Actophilornis africanus	African Jacana	M	8	137		115	156			584a
		F	5	261		251	274			
Actophilornis albinucha	Madagascar Jacana	F	1	239					Madagascar	68
Irediparra gallinacea	Comb-crested Jacana	M	6	75.1		68	84			584a
		F	7	130		120	149			
Hydrophasianus chirurgus	Pheasant-tailed Jacana	M	4	130		120	140		Philippines	983
		F	3	197		190	200			
Metopidius indicus	Bronze-winged Jacana	B	10	155		94	210			584a
Jacana spinosa	Northern Jacana	M	20	78.9	8.5				Panama	489
		F	16	112	10.8					
Jacana jacana	Wattled Jacana	M	292	87.5	7.4				Panama	335
		F	140	129	13.1					
Family Rostratulidae										
Rostratula benghalensis	Greater Painted-Snipe	B	34	121		90	164			584a
Rostratula semicollaris	American Painted-Snipe	B	5	76.6		68	86			584a

Body Masses of World Birds (continued)

Species Name	Common Name	Sex	N	Mean	S.D.	Min	Max	Season	Location	Source
Family Dromadidae										
Dromas ardeola	Crab Plover	F	1	325					Eastern Africa	120
Family Haematopodidae										
Haematopus leucopodus	Magellanic Oystercatcher	B	5	618		585	700			584a
Haematopus ater	Blackish Oystercatcher	B	7	724		585	850			584a, 861b
Haematopus bachmani	Black Oystercatcher	B	14	555	47.9			W		15
Haematopus palliatus	American Oystercatcher	M	29	567		499	657			854
		F	30	638		568	720			
Haematopus moquini	African Oystercatcher	B	170	692		507	850		South Africa	1036
Haematopus ostralegus ostralegus	Eurasian Oystercatcher	U	282	526	45.5	430	675	B	Netherlands	223
Haematopus longirostris	Pied Oystercatcher	M	6	602		485	675		Australia	746
		F	9	626		500	782			
Haematopus finschi	South Is. Oystercatcher	M	52	517	38.9	420	593	B	New Zealand	746
		F	63	554	34.9	480	617			
Haematopus chathamensis	Chatham Oystercatcher	M	8	560				B	Chatham Is.	746
		F	8	640						
Haematopus unicolor	Variable Oystercatcher	B	23	671		430	771		New Zealand	746
Haematopus fuliginosus	Sooty Oystercatcher	U	149	819	56.2	705	980		Victoria, Australia	746
Family Ibidorhynchidae										
Ibidorhyncha struthersii	Ibisbill	B	7	294		270	320			584a
Family Recurvirostridae										
Himantopus himantopus	Black-winged Stilt	B	20	161		138	208		Northern Australia	299

Body Masses of World Birds (continued)

Species Name	Common Name	Sex	N	Mean	S.D.	Min	Max	Season	Location	Source #
Himantopus leucocephalus	White-headed Stilt	B	18	176	18				Australia	928
Himantopus leucocephalus	White-headed Stilt	B	29	193	20.9				New Zealand	928
Himantopus novaezelandiae	Black Stilt	B	7	195					New Zealand	746, 928
Himantopus mexicanus mexicanus	Black-necked Stilt	B	145	170		136	202		Nevada, USA	1006
Himantopus mexicanus knudseni	Black-necked Stilt	B	85	202					Hawaiian Is.	1006
Himantopus melanurus	White-backed Stilt	U	2	218		216	219			1042
Cladorhynchus leucocephalus	Banded Stilt	U	14	216	12.2	190	235		South Australia	746
Recurvirostra avosetta	Pied Avocet	B	19	304	57	228	435	B	Netherlands	223
Recurvirostra americana	American Avocet	M	249	307	20.9	261	362		California, USA	1005
		F	257	302	23.7	215	382			
Recurvirostra novaehollandiae	Red-necked Avocet	B	43	325		270	390		Victoria, Australia	746
Recurvirostra andina	Andean Avocet	B	5	361		315	410			584a
Family Burhinidae										
Burhinus vermiculatus	Water Thick-knee	F	1	320					eastern Africa	120
Burhinus oedicnemus	Eurasian Thick-knee	B	13	459		290	535		France	223
Burhinus senegalensis	Senegal Thick-knee	B	6	326		306	362		Kenya; Senegal	1143
Burhinus capensis	Spotted Thick-knee	U	5	423		400	450			133
Burhinus bistriatus	Double-striped Thick-knee	M	1	787					Oaxaca, Mexico	1145
Burhinus grallarius	Bush Thick-knee	M	10	672	78.9	580	860		Australia	746
		F	6	625		530	710			

Body Masses of World Birds (continued)

Species Name	Common Name	Sex	N	Mean	S.D.	Min	Max	Season	Location	Source
Esacus recurvirostris	Great Thick-knee	U		790						7
Esacus magnirostris	Beach Thick-knee	M	8	1032		870	1130		Australia	746
		F	3	1000		980	1020			
Family Glareolidae										
Pluvianus aegyptius	Egyptian Plover	U	10	82		73.4	90.5	B	Ethiopia	223
Cursorius cursor	Cream-colored Courser	B	16	138	19.4	115	198		India; Afghanistan	223
Cursorius rufus	Burchell's Courser	U	1	75					Namibia	1244
Cursorius temminckii	Temminck's Courser	U	21	69.2		54	82		Southern Africa	1036
Rhinoptilus africanus	Double-banded Courser	U	5	87.5		80	95.5		South Africa; Namibia	508, 526
Rhinoptilus cinctus	Three-banded Courser	U	12	105		89.2	130		South Africa	1036
Rhinoptilus chalcopterus	Bronze-winged Courser	U	85	150		91	220		Southern Africa	1036
Stiltia isabella	Australian Pratincole	B	27	65.5		54	75		Australia	515a
Glareola pratincola	Collared Pratincole	B	18	84.9		76	98		Southern Africa	1036
Glareola maldivarum	Oriental Pratincole	B	211	75.2	6.6	59	95		Java, Indonesia	515a
Glareola nordmanni	Black-winged Pratincole	B	10	97.2		87	105			273
Glareola ocularis	Madagascar Pratincole	M	1	103					Madagascar	68
		F	1	82						
Glareola nuchalis	Rock Pratincole	U	5	62.7		48	82.5		Namibia	1036
Glareola cinerea	Gray Pratincole	F	1	37.4				B	Central African Republic	1143
Glareola lactea	Small Pratincole	U	4	39.8		37	44		India	7

Body Masses of World Birds (continued)

Species Name	Common Name	Sex	N	Mean	S.D.	Min	Max	Season	Location	Source #
Family Charadriidae										
Vanellus vanellus	Northern Lapwing	M	32	211	17.9	140	242	B	Netherlands; Germany	223
		F	40	226	21.8	180	317			
Vanellus crassirostris	Long-toed Lapwing	B	2	170		170	170		Eastern Africa	120
Vanellus armatus	Blacksmith Lapwing	B	264	156		114	211			584a
Vanellus spinosus	Spur-winged Lapwing	M	78	192	9.4			B	Israel	1348
		F	78	184	10.2					
Vanellus duvaucelii	River Lapwing	M	3			164	185			7
		F	1	143						
Vanellus malabaricus	Yellow-wattled Lapwing	U	3	140		108	203			584a
Vanellus tectus	Black-headed Lapwing	F	1	100						584a
Vanellus albiceps	White-headed Lapwing	B	14	192		161	214		Zimbabwe; South Africa	1036
Vanellus lugubris	Senegal Lapwing	U	15	122		109	140		South Africa	1036
Vanellus melanopterus	Black-winged Lapwing	M	12	191		163	214		South Africa	1036
		F	4	179		165	197			
Vanellus coronatus	Crowned Lapwing	U	8	165		148	200		Namibia	526, 584a
Vanellus senegallus	Wattled Lapwing	B	28	248		160	287		Zimbabwe	1036
Vanellus melanocephalus	Spot-breasted Lapwing	U	2	214		199	228			584a
Vanellus superciliosus	Brown-chested Lapwing	U		150					estimated	584a
Vanellus cinereus	Gray-headed Lapwing	B	7	270		236	296		Hopei, China	1087
Vanellus indicus	Red-wattled Lapwing	U	21	181		110	230			584a

Body Masses of World Birds (continued)

Species Name	Common Name	Sex	N	Mean	S.D.	Min	Max	Season	Location	Source #
Vanellus macropterus	Sunda Lapwing	U		325					estimated	584a
Vanellus tricolor	Banded Lapwing	U	9	184		140	216		Victoria, Australia	251
Vanellus miles	Masked Lapwing	U	96	387	26.5	333	440		Victoria, Australia	746
Vanellus gregarius	Sociable Lapwing	M	2	252		245	260			584a
		F	7	200		180	252			
Vanellus leucurus	White-tailed Lapwing	B	7	132		99	198		Iraq; Afghanistan; former USSR	223
Vanellus cayanus	Pied Lapwing	M	6	70		54	80			498, 584
		F	4	82		78	84			
Vanellus chilensis	Southern Lapwing	B	3	327		277	426			63, 155, 546
Vanellus resplendens	Andean Lapwing	B	12	214		193	230			584a
Erythrogonys cinctus	Red-kneed Dotterel	U	114	51.4	3.8	42	60		New South Wales, Australia	746
Pluvialis fulva	Pacific Golden-Plover	M	19	130	5.5	120	139	B	Alaska, USA	597
		F	16	140	14.4	116	168			
Pluvialis dominica	American Golden-Plover	M	14	149	7.6	138	166	B	Alaska, USA	597
		F	11	154	5.6	146	166			
Pluvialis apricaria	Eurasian Golden-Plover	U	402	214	15.5	165	260	Sp	Netherlands	223
Pluvialis squatarola	Black-bellied Plover	U	33	250	23.3	195	283	PB	Netherlands	223
Charadrius obscurus	Red-breasted Dotterel	U	23	146	10.4	129	168		New Zealand	746
Charadrius hiaticula	Common Ringed Plover	M	41	63.3		57.6	72	B	Netherlands	432
		F	34	64.8		55.4	74.5			
Charadrius semipalmatus	Semipalmated Plover	M	26	47.4	5.9	37.6	57.4	Sp	Eastern USA	314
		F	24	46.1	4.7	39.2	56.5			
Charadrius placidus	Long-billed Plover	B	19	62.7		41	70			584a

Body Masses of World Birds (continued)

Species Name	Common Name	Sex	N	Mean	S.D.	Min	Max	Season	Location	Source #
Charadrius dubius	Little Ringed Plover	B	461	38.7		33	48.3	B	Netherlands	223
Charadrius wilsonia	Wilson's Plover	B	39	55.1	7.1	45.2	80	W	Panama	1180
Charadrius vociferus	Killdeer	M	10	92.1	10.4	83.9	109	B	Great Plains, USA	314
		F	6	101		87.7	121			
Charadrius melodus	Piping Plover	U	87	55.2		46.4	63.7	B	New York, USA	1304
Charadrius melodus	Piping Plover	B	69	53.6	2.6	46.5	62	B	Maryland, USA	702
Charadrius thoracicus	Madagascar Plover	U	1	11.5					Madagascar	957
Charadrius pecuarius	Kittlitz's Plover	U	886	34		19	49			133
Charadrius tricollaris	Three-banded Plover	U	11	31.2		25	38			584a
Charadrius marginatus	White-fronted Plover	U	262	48.3						584a
Charadrius pallidus	Chestnut-banded Plover	U	17	34.8						584a
Charadrius alexandrinus	Snowy Plover	B	458	42.3		35	58	B	California, USA	892
Charadrius ruficapillus	Red-capped Plover	B	463	37.4		32	50	Y	Victoria, Australia	746
Charadrius peronii	Malaysian Plover	U		42						133
Charadrius collaris	Collared Plover	B	18	28.3	1.7	25.8	31	W	Panama	1180
Charadrius alticola	Puna Plover	B	7	44.9		41	49			584a, 1040, 1042
Charadrius falklandicus	Two-banded Plover	B	8	65		62	72			210, 546, 584a
Charadrius bicinctus	Double-banded Plover	B	183	59.4	3.9			W	Victoria, Australia	746
Charadrius mongolus	Lesser Sand-Plover	U	46	64	3.3	56	71	B	Victoria, Australia	746
Charadrius leschenaultii	Greater Sand-Plover	U	599	74.8	4.5	63	87	Sp	northwest Australia	746

Body Masses of World Birds (continued)

Species Name	Common Name	Sex	N	Mean	S.D.	Min	Max	Season	Location	Source #
Charadrius asiaticus	Caspian Plover	B	21	77.1	7.9	63	91		Kazakhstan; former USSR	223
Charadrius veredus	Oriental Plover	U	58	95	9			Sp	Northwest Australia	746
Charadrius morinellus	Eurasian Dotterel	M	10	100		88	116	B	Former USSR	432
		F	5	117		98.9	130			
Charadrius modestus	Rufous-chested Dotterel	B	5	78		71	89			584a
Charadrius montanus	Mountain Plover	M	26	95.1	6.5			W	California, USA	642
		F	13	96.4	8.7					
Thinornis rubricollis	Hooded Plover	U	12	100	8.9	79	100		Australia	746
Thinornis novaeseelandiae	Shore Plover	B	24	61	3.3	52	69		Chatham Is.	746
Elseyornis melanops	Black-fronted Dotterel	U	40	33	3.1	27	42		New Zealand	746
Peltohyas australis	Inland Dotterel	B	14	83.5		64	107		Australia	746
Anarhynchus frontalis	Wrybill	M	46	58.6	5.3				New Zealand	746
		F	75	56.7	3.7					
Phegornis mitchellii	Diademed Sandpiper-Plover	B	5	35.5		28	46			584a
Oreopholus ruficollis	Tawny-throated Dotterel	U	10	131		120	145			210, 584
Family Pluvianellidae										
Pluvianellus socialis	Magellanic Plover	M		89		79	102			584a
		F		79.5		69.5	87			
Family Scolopacidae										
Scolopax rusticola	Eurasian Woodcock	M	250	306	26.1	250	410	W	Ireland	223
		F	234	313	25.5	205	420			
Scolopax rusticola	Eurasian Woodcock	U	6	306		242	451		Japan	555
Scolopax mira	Amami Woodcock	U	15	447	35.9	405	541		Japan	555

Body Masses of World Birds (continued)

Species Name	Common Name	Sex	N	Mean	S.D.	Min	Max	Season	Location	Source #
Scolopax saturata	Dusky Woodcock	U	2	205		189	220			387, 584a
Scolopax bukidnonensis	Bukidnon Woodcock	B	4	238		193	310		Mindanao, Philippines	626a
Scolopax minor	American Woodcock	M	390	176			222			844
		F	313	219			278			
Coenocorypha pusilla	Chatham Is. Snipe	M	24	75.9	4.8	69	85	B	New Zealand	515a
		F	24	85.4	5.7	77	94			
Coenocorypha aucklandica	Subantarctic Snipe	M	92	101	4.8	89	118	B	New Zealand	515a
		F	62	116	6.6	98	128			
Lymnocryptes minimus	Jack Snipe	M	17	53.7	6.9	41	63	W	Netherlands	223
		F	8	46.7		33	73			
Gallinago solitaria	Solitary Snipe	M				130	148			584a
		F				126	159			
Gallinago hardwickii	Latham's Snipe	M	36	144	22.4	129	182	W	Australia	515a
		F	41	159	18.4	143	178			
Gallinago nemoricola	Wood Snipe	U				142	198			7
Gallinago stenura	Pintail Snipe	B	472	113					Southern India	7
Gallinago megala	Swinhoe's Snipe	B	9	121		103	135		Australia	515a
Gallinago nigripennis	African Snipe	U	28	112		90	164			133
Gallinago macrodactyla	Madagascar Snipe	F	1	216					Madagascar	68
Gallinago media	Great Snipe	M	143	157	6.4			B	Norway	527
		F	67	184	14.5					
Gallinago gallinago raddei	Common Snipe	M	20	97		74	115		Hopei, China	1087
		F	16	113		90	153			
Gallinago delicata	Wilson's Snipe	M	15	128			156			844
		F	14	116			156			
Gallinago paraguaiae	South American Snipe	M	15	106		85	139		Suriname	498
		F	20	113		85	145			

Body Masses of World Birds (continued)

Species Name	Common Name	Sex	N	Mean	S.D.	Min	Max	Season	Location	Source #
Gallinago andina	Puna Snipe	F	1	108					Peru	900
Gallinago nobilis	Noble Snipe	M	1	188						584a
		F	1	197						
Gallinago undulata	Giant Snipe	M	5	294		270	320			584a
		F	3	332		282	363			
Gallinago stricklandii	Fuegian Snipe	M	1	252					Bolivia	963
Gallinago jamesoni	Andean Snipe	U	8	166		140	224			584a
Limnodromus griseus hendersoni	Short-billed Dowitcher	M	7	100		75	129	F		223
		F	3	117		114	122			
Limnodromus griseus griseus	Short-billed Dowitcher	M	12	110		73	152	F	New Jersey, USA	430
		F	30	116		82.5	154			
Limnodromus scolopaceus	Long-billed Dowitcher	M	28	100		90	114	B	Alaska, USA	554
		F	11	109		93	119			
Limnodromus semipalmatus	Asian Dowitcher	B	14	212	12.7	190	235	Sp	Australia	515a
Limosa limosa limosa	Black-tailed Godwit	M	11	252		235	367			273
		F	11	330		297	362			
Limosa haemastica	Hudsonian Godwit	M	6	222						576
		F	6	289						
Limosa lapponica lapponica	Bar-tailed Godwit	M	69	309	25.2	233	360	W	Britain, UK	223
		F	20	276	25.1	348	455			
Limosa lapponica baueri	Bar-tailed Godwit	M	123	282	24.7	225	380	W	New Zealand	772
		F	173	332	21.7	245	400			
Limosa lapponica menzbieri	Bar-tailed Godwit	M	119	264	22.7				Australia	515a
		F	53	292	17.6					
Limosa fedoa	Marbled Godwit	M	40	326	23	285	396	B	Alberta, Canada	445
		F	43	391	25	352	454			
Numenius borealis	Eskimo Curlew	U				270	454			421
Numenius minutus	Little Curlew	U	338	173	16	118	221	Sp	Australia	515a

Body Masses of World Birds (continued)

Species Name	Common Name	Sex	N	Mean	S.D.	Min	Max	Season	Location	Source #
Numenius phaeopus	Whimbrel	M	29	355	22.1	310	403	B	Manitoba & Ontario, Canada	1115
		F	36	404	29.1	345	459			
Numenius phaeopus variegatus	Whimbrel	M	5	325		305	348		Tsingtao, China	1088
		F	4	379		310	460			
Numenius tahitiensis	Bristle-thighed Curlew	M	38	374	26.4	309	450	B	Alaska, USA	757
		F	35	428	23.4	389	485			
Numenius tenuirostris	Slender-billed Curlew	F	2	308		255	360	F		584a
Numenius arquata	Eurasian Curlew	M	124	742	68.7	540	900	F	Netherlands	223
		F	97	869	68.2	720	1050			
Numenius americanus	Long-billed Curlew	M	12	531	32.7	493	597	B	Idaho, USA	314
		F	24	642	31.5	570	689			
Numenius madagascariensis	Far Eastern Curlew	U	12	792		565	1150	W		584a
Bartramia longicauda	Upland Sandpiper	M	27	154		126	170			537
		F	34	164		127	218			
Xenus cinereus	Terek Sandpiper	U	370	78.8	7.8	59	104	Sp	Northwest Australia	515a
Actitis hypoleucos	Common Sandpiper	B	22	48		40	59	B	Afghanistan; Kazakhstan; Mongolia	223
Actitis macularius	Spotted Sandpiper	U	56	40.4	6.2	29.4	59.8	S	Pennsylvania, USA	314
Tringa ochropus	Green Sandpiper	U	19	71.4	10.4	53	93	M	North Africa	223
Tringa solitaria	Solitary Sandpiper	U	104	48.4	8.6	31.1	65.1	Sp	Venezuela	1216
Tringa brevipes	Gray-tailed Tattler	U	348	127	10.6			F	Northwest Australia	515a
Tringa incana	Wandering Tattler	M	13	101		87	114	B	Alaska, USA	554
		F	16	116		98	130			
Tringa erythropus	Spotted Redshank	U	65	158	13.1	140	210	F	Britain, UK	223
Tringa melanoleuca	Greater Yellowlegs	U	16	153	13			W		333

Body Masses of World Birds (continued)

Species Name	Common Name	Sex	N	Mean	S.D.	Min	Max	Season	Location	Source #
Tringa melanoleuca	Greater Yellowlegs	B	19	171		132	215		Suriname	498
Tringa nebularia	Common Greenshank	U	74	187	17	125	230	Sp	Australia	515a
Tringa semipalmata inornata	Willet	M	26	258	14.3	220	282		Alberta, Canada	724
		F	25	283	16.8	246	327			
Tringa semipalmata semipalmata	Willet	M	19	208	5.4	199	218		New York, USA	724
		F	18	240	13.5	217	263			
Tringa flavipes	Lesser Yellowlegs	U	45	77.5	8.2	63	96	W	Latin America	223
Tringa stagnatilis	Marsh Sandpiper	B	61	77.5		55	120	M	China	223
Tringa guttifer	Nordmann's Greenshank	M	3			136	141			584a
		F	1	158						
Tringa glareola	Wood Sandpiper	M	16	62		52	81	B	White Sea, former USSR	223
		F	11	73		51	89			
Tringa glareola	Wood Sandpiper	U	131	53	6.9			Sp	Israel	1356
Tringa glareola	Wood Sandpiper	U	730	61.8		31	130		southern Africa	1036
Tringa totanus	Common Redshank	B	200	129		107	152	B		584a
Prosobonia cancellata	Tuamotu Sandpiper	B	7	36		32	44			584a
Arenaria interpres	Ruddy Turnstone	M	31	134	10.2			Sp	northwest Australia	515a
		F	32	138	8.7					
Arenaria melanocephala	Black Turnstone	M	50	118	7.7	97	134	B	Alaska, USA	474
		F	42	135	12.4	114	169			
Aphriza virgata	Surfbird	M	48	187	17.9	156	225	Sp	Alaska, USA	1081
		F	19	216	18.7	182	251			
Calidris tenuirostris	Great Knot	U	125	192	33.1	115	261	Sp	Western Australia	515a

Body Masses of World Birds (continued)

Species Name	Common Name	Sex	N	Mean	S.D.	Min	Max	Season	Location	Source #
Calidris canutus	Red Knot	B	608	148	19.2	101	206	F	Massachusetts, USA	480
Calidris canutus	Red Knot	B	518	136	16.9	100	187	F	Western Australia	515a
Calidris alba	Sanderling	B	45	57		47	72.5	B	Greenland	515
Calidris alba	Sanderling	M	17	47.3	3			W	Panama	732
		F	10	51.3	2.4					
Calidris pusilla	Semipalmated Sandpiper	U	300	27.5	2.1			B	Manitoba, Canada	444
Calidris mauri	Western Sandpiper	M	39	26.6	1.9			B	Alaska, USA	884
		F	34	29.1	2.3					
Calidris ruficollis	Red-necked Stint	M	70	26.4	1.6	22.3	31.5	W	Australia	515a
		F	56	27.2	1.8	22.1	31			
Calidris minuta	Little Stint	U	248	21.1		17	27	W	Kenya	515a
Calidris temminckii	Temminck's Stint	B	111	23	3	14.7	36	F	France	223
Calidris subminuta	Long-toed Stint	B	20	30.2		23	37			584a
Calidris minutilla	Least Sandpiper	M	42	22	1.4	19.7	26.4	B	British Columbia, Canada	215
		F	50	23.8	1.4	20.9	26.4			
Calidris fuscicollis	White-rumped Sandpiper	U	367	34.7	4.6	27.5	45.7	M	Venezuela	1216
Calidris fuscicollis	White-rumped Sandpiper	M	56	46.9	6.3			Sp	North Dakota, USA	656a
		F	68	48.9	5.7					
Calidris bairdii	Baird's Sandpiper	M	46	38.6	4			B	Alaska, USA	730
		F	16	43.5	4.9					
Calidris melanotos	Pectoral Sandpiper	M	74	97.8	8			B	Alaska, USA	730
		F	38	65	6.2					
Calidris acuminata	Sharp-tailed Sandpiper	M	161	74.4	5.4	54	92	W	Victoria, Australia	515a
		F	195	58.7	5.2	49	88			
Calidris ferruginea	Curlew Sandpiper	M	205	56.6	4.6			W	Victoria, Australia	515a
		F	74	59.6	4.1					

Body Masses of World Birds (continued)

Species Name	Common Name	Sex	N	Mean	S.D.	Min	Max	Season	Location	Source #
Calidris alpina schinzii	Dunlin	M	92	44.2						576
		F	92	49.6						
Calidris alpina sakhalina	Dunlin	M	267	55.4				B	Alaska, USA	731
		F	177	59.7						
Calidris maritima	Purple Sandpiper	U	35	67.6	6	58	80	W	Maine, USA	903
Calidris maritima	Purple Sandpiper	U	72	62.7	6.9	53	85	W	Scotland, UK	903
Calidris ptilocnemis ptilocnemis	Rock Sandpiper	M	33	86.8	6.7	71.5	105	B	Pribilof Is., Alaska, USA	422
		F	12	106	12.8	75	130			
Calidris ptilocnemis tschuktschorum	Rock Sandpiper	M	17	72.6	4.2	62.4	79	B	Nunivak Is., Alaska, USA	422
		F	11	78.6	7.9	59	87			
Calidris ptilocnemis couesi	Rock Sandpiper	M	65	71.3	8.4	57.4	100	B	Aleutian Is., USA	422
		F	37	77.3	9	60	102			
Calidris himantopus	Stilt Sandpiper	M	24	53.8		48	60.8	B	Manitoba, Canada	641
		F	15	60.9		52	68			
Eurynorhynchus pygmeus	Spoon-billed Sandpiper	M		29.5					male estimated	584a
		F	1	34						
Limicola falcinellus	Broad-billed Sandpiper	U	69	37.1	3.4	29	52	F	Australia	515a
Tryngites subruficollis	Buff-breasted Sandpiper	M	67	68.5	3.6	57.5	77.5	B	Alaska, USA	667
		F	58	56.3	3.7	46	65.3			
Philomachus pugnax	Ruff	M	100	170	21.2	130	210		Africa	515a
		F	1001	102	9.6	78	147			
Phalaropus tricolor	Wilson's Phalarope	M	155	51.8	4.1	40	62	B	Saskatchewan, Canada	206
		F	48	68.1	7.1	55	85			
Phalaropus lobatus	Red-necked Phalarope	M	179	34.5	2.6			B	Hudson Bay, Canada	515a
		F	97	39	4.8					
Phalaropus fulicarius	Red Phalarope	M	93	49.7	3.8			B	Alaska, USA	1228
		F	29	62.2	6.1					

Body Masses of World Birds (continued)

Species Name	Common Name	Sex	N	Mean	S.D.	Min	Max	Season	Location	Source #
Family Pedionomidae										
Pedionomus torquatus	Plains-wanderer	M	108	54	7	39	78		New South Wales, Australia	746
		F	64	72.4	10.3	57	95			
Family Thinocoridae										
Attagis gayi	Rufous-bellied Seedsnipe	B	4	311		283	330		Peru; Bolivia	1145
Attagis malouinus	White-bellied Seedsnipe	M	6	340		312	398		Tierra del Fuego	546
Thinocorus orbignyianus	Gray-breasted Seedsnipe	B	16	115	13	96	140		Peru; Bolivia	1145
Thinocorus rumicivorus	Least Seedsnipe	B	15	53.1	3.5	48	59.9		Argentina	210
Family Chionididae										
Chionis albus	Snowy Sheathbill	M	40	719	39.9	640	810		South Shetland Is.	355
		F	37	635	43.5	560	720			
Chionis minor minor	Black-faced Sheathbill	M	37	625	47.8	520	730		Kerguelen Is.	746
		F	35	551	45.3	480	670			
Chionis minor crozettensis	Black-faced Sheathbill	U	23	350	11	250	460		Crozet Is.	746
Family Stercorariidae										
Stercorarius chilensis	Chilean Skua	M	8	1324		1193	1525		Tierra del Fuego; Chile	546, 861b
		F	3	1413		1275	1532			
Stercorarius maccormicki	South Polar Skua	M	26	1277	95	1120	1440	B	Ross Sea, Antarctica	515a
		F	23	1421	75	1280	1550			
Stercorarius antarcticus hamiltoni	Brown Skua	M	9	1429		1300	1600	B	Gough Is.	397
		F	9	1649		1550	1800			
Stercorarius antarcticus lonnbergi	Brown Skua	M	30	1735	75.4	1610	1960	B	Chatham Is.	515a
		F	32	1935	130	1660	2180			
Stercorarius skua	Great Skua	M	112	1271	69.4			B	Shetland Is.	175
		F	125	1409	75					
Stercorarius pomarinus	Pomarine Jaeger	M	73	648	6.2	542	797		Alaska, USA	737
		F	52	740	11.7	576	917			

Body Masses of World Birds (continued)

Species Name	Common Name	Sex	N	Mean	S.D.	Min	Max	Season	Location	Source #
Stercorarius parasiticus	Parasitic Jaeger	M	219	413		306	523	B	North Atlantic	62
		F	189	478		306	604			
Stercorarius longicaudus	Long-tailed Jaeger	M	38	270	21.6	218	320		Greenland	1307
		F	48	307	39.2	262	444			
Family Laridae										
Larus scoresbii	Dolphin Gull	M	16	544	32.7			B	Argentina	1349
		F	16	489	30					
Larus scoresbii	Dolphin Gull	M	3	558		525	575		Chile	1082
		F	4	506		450	550			
Larus pacificus	Pacific Gull	M	5	1550		1200	1800			515a
		F	4	1077		910	1400			
Larus belcheri	Belcher's Gull	M	2	718		710	725		Chile	1082
		F	2	568		560	575			
Larus atlanticus	Olrog's Gull	M	2	930		900	960		Argentina	278
		F	3	763		750	790			
Larus crassirostris	Black-tailed Gull	B	25	561		420	930		Tsingtao, China	1088
Larus modestus	Gray Gull	B	10	344	39.7	250	380		Chile	1082
Larus heermanni	Heermann's Gull	U	10	500	71.6	371	643		Sonora, Mexico	1150
Larus leucophthalmus	White-eyed Gull	M	10	357		325	415		Egypt	1244
		F	9	303		275	355			
Larus hemprichii	Sooty Gull	M	1	510				B	Eastern Africa	120
		F	1	400						
Larus canus canus	Mew Gull	M	96	432		340	552	B	North Atlantic Ocean	62
		F	72	375		290	530			
Larus canus kamtschatschensis	Mew Gull	F	2	490		394	586			273
Larus canus brachyrhynchus	Mew Gull	M	16	409	25.8				British Columbia, Canada	817
		F	8	368						
Larus audouinii	Audouin's Gull	B	37	535		451	660		Spain	1023

Body Masses of World Birds (continued)

Species Name	Common Name	Sex	N	Mean	S.D.	Min	Max	Season	Location	Source #
Larus delawarensis	Ring-billed Gull	M	48	566	42	495	650	B	Ontario, Canada	1034
		F	51	471	46	375	590			
Larus dominicanus	Kelp Gull	M	13	1050		950	1130		New Zealand	515a
		F	15	832		540	970			
Larus dominicanus	Kelp Gull	U	30	989		760	1230		South Africa; Namibia	1036
Larus californicus albertaensis	California Gull	M	32	841	103	653	1045			288
		F	19	710	91.2	568	903			
Larus californicus californicus	California Gull	M	64	657	85.3	490	885			288
		F	84	556	53.4	432	695			
Larus marinus	Great Black-backed Gull	M	116	1829		1380	2272	B	North Atlantic Ocean	62
		F	93	1488		1033	2085			
Larus glaucescens	Glaucous-winged Gull	M	306	1180	71.5			B		1251
		F	331	946	66.2					
Larus occidentalis occidentalis	Western Gull	M	15	1136	47				Farallon Is., California, USA	930
		F	21	879	78					
Larus occidentalis wymani	Western Gull	M	60	931	136				Santa Clara, California, USA	930
		F	35	780	57.6					
Larus livens	Yellow-footed Gull	U	2	1322		1173	1470		Sonora, Mexico	314
Larus hyperboreus	Glaucous Gull	M	39	1576	121	1280	1820		Iceland	223
		F	26	1249	107	1070	1430			
Larus hyperboreus	Glaucous Gull	M	5	1855		1600	2060		Coats Is., Canada	417
		F	7	1497		1250	1690			
Larus glaucoides	Iceland Gull	M	1	863					Britain	515
Larus glaucoides kumlieni	Iceland Gull	U	1	557				PB	Alaska, USA	230
Larus thayeri	Thayer's Gull	M	5	1021		940	1100		Ellesmere Is., Canada	1126
		F	3	870		820	900			
Larus argentatus	Herring Gull	M	169	1147	77.7	973	1143	B	Massachusetts, USA	931
		F	246	1023	106	718	1385			
Larus argentatus vegae	Herring Gull	M	12	1199		987	1775		Hopei, China	1087
		F	11	1007		688	1250			

Body Masses of World Birds (continued)

Species Name	Common Name	Sex	N	Mean	S.D.	Min	Max	Season	Location	Source #
Larus cachinnans michahellis	Yellow-legged Gull	M	80	1275	98	1040	1500		France	223
		F	80	1033	92	800	1400			
Larus cachinnans mongolicus	Yellow-legged Gull	M	4	1164		1075	1250		Mongolia	926
		F	9	997		850	1150			
Larus armenicus	Armenian Gull	B	3	867		850	900			1130
Larus fuscus graellsii	Lesser Black-backed Gull	M	22	880	61	770	1000		Britain, UK	223
		F	31	755	58	620	908			
Larus fuscus fuscus	Lesser Black-backed Gull	M	52	768	59.9	670	945		Norway	223
		F	64	662	64.4	545	840			
Larus ichthyaetus	Great Black-headed Gull	M	8	1599		1130	2000		Former USSR; Pakistan; Mongolia	223
		F	11	1189		950	1500			
Larus schistisagus	Slaty-backed Gull	B	7	1323		1100	1694		Alaska, USA; Japan	415, 1149, 1275
Larus brunnicephalus	Brown-headed Gull	M	4	633		550	714		Hopei, China	1087
		F	3	512		450	626			
Larus cirrocephalus	Gray-hooded Gull	B	6	312		255	360		Kenya; Tanzania; South Africa	223
Larus cirrocephalus	Gray-hooded Gull	U	2	450		440	460		South America	362
Larus hartlaubii	Hartlaub's Gull	B	18	292		235	340		South Africa	1243a
Larus novaehollandiae	Silver Gull	M	15	313	51.9	220	397		Southeast Australia	515a
		F	14	264	34.1	195	320			
Larus novaehollandiae scopulinus	Red-billed Gull	M	10	292	41.8	237	355		New Zealand	515a
		F	3	217		211	225			
Larus bulleri	Black-billed Gull	M	7	243		195	277		New Zealand	515a
		F	9	218		109	260			
Larus maculipennis	Brown-hooded Gull	B	8	339		275	454		Brazil; Argentina; Chile	63, 546, 1082
Larus ridibundus	Black-headed Gull	U	324	284	24.6	195	327	B	Netherlands	223
Larus genei	Slender-billed Gull	B	18	281		223	350		Iran; Saudi Arabia; former USSR	223

Body Masses of World Birds (continued)

Species Name	Common Name	Sex	N	Mean	S.D.	Min	Max	Season	Location	Source #
Larus philadelphia	Bonaparte's Gull	M	58	222	12.1			F	New Brunswick, Canada	153
		F	84	200	12.6					
Larus saundersi	Saunder's Gull	M	5	198		170	220		Hopei & Tsingtao, China	1087, 1088
Larus serranus	Andean Gull	M	1	478						1042
Larus melanocephalus	Mediterranean Gull	U	2	256		232	280			1130
Larus relictus	Relict Gull	M	5	519						638
		F	6	463						
Larus fuliginosus	Lava Gull	U	1	380					Galapagos Is.	71b
Larus atricilla	Laughing Gull	M	37	327	25	249	366		Florida, USA	152
		F	33	289	41	203	371			
Larus pipixcan	Franklin's Gull	U	40	280	9.7	220	335	B	Minnesota, USA	223
Larus minutus	Little Gull	U	87	118		88	162	Sp	Britain, UK	515
Pagophila eburnea	Ivory Gull	U	14	550	43.1	472	605	PB	Russia	476
Rhodostethia rosea	Ross' Gull	U	19	187	9.5	170	210	PB	Alaska, USA	1149
Xema sabini	Sabine's Gull	B	23	198		176	214	B	Canada	264
Creagrus furcatus	Swallow-tailed Gull	B	17	687		610	780	B	Galapagos Is.	483
Rissa brevirostris	Red-legged Kittiwake	B	188	377	34.6	296	489		Alaska, USA	159a
Rissa tridactyla	Black-legged Kittiwake	M	223	421		305	512	B	North Atlantic Ocean	62
		F	183	394		305	525			
Rissa tridactyla	Black-legged Kittiwake	U	198	434	51	316	580		Alaska, USA	159a
Family Sternidae										
Anous tenuirostris	Lesser Noddy	U	199	112	8.9				Pelsaert Is., Australia	515a
Anous stolidus stolidus	Brown Noddy	M	191	178	12.1	151	221		Culebra, Puerto Rico	180
		F	186	167	11.9	142	201			
Anous stolidus pileatus	Brown Noddy	B	335	189	14.5	154	232		Hawaiian Is.	180

Body Masses of World Birds (continued)

Species Name	Common Name	Sex	N	Mean	S.D.	Min	Max	Season	Location	Source #
Anous minutus	Black Noddy	U	36	116	9.3	98	130	B	Ascension Is.	23
Anous minutus marcusi	Black Noddy	U	158	109	9	88	133		Hawaiian Is.	408
Procelsterna cerulea	Blue Noddy	U	52	53	7.21	46	65		Hawaiian Is.	486
Procelsterna albivitta	Gray Noddy	B	11	72.2		57	77		Kermandec Is.	515a
Gypis alba	White Tern	U	109	111	10.4	92	139		Hawaiian Is.	486
Onychoprion fuscatus	Sooty Tern	B	106	175	13.7	142	224		Florida, USA	29
Onychoprion fuscatus oahuensis	Sooty Tern	U	225	208	22.3				Johnston Atoll, Hawaiian Is.	1061
Onychoprion fuscatus nubilosus	Sooty Tern	U	248	176	19.7				Christmas Is.	1061
Onychoprion lunatus	Gray-backed Tern	U	83	146	9.1	115	177		Hawaiian Is.	486
Onychoprion albostriatus	Black-fronted Tern	B	6	94.7		81	105		New Zealand	515a
Onychoprion anaethetus antarctica	Bridled Tern	U	69	95.6	6.6				Seychelles	223
Onychoprion anaethetus	Bridled Tern	M	31	136	11.9	116	168		oceans off southeast USA	475
		F	38	131	9.7	115	155			
Onychoprion aleuticus[a]	Aleutian Tern	U	16	120	12.4	83.4	140		Alaska, USA	1149
Sternula albifrons	Little Tern	B	30	57		50	63		Scotland, UK	223
Sternula saundersi	Saunder's Tern	F	3	42.3		40	45			1243a
Sternula antillarum antillarum	Least Tern	M	75	46.9	3.2	41	54	B	Massachusetts, USA	1217
		F	73	49.3	3.8	40	62			
Sternula antillarum athalassos	Least Tern	M	40	43.1	2.5	38	49		interior USA	1217
		F	54	44.4	2.6	38	50			
Sternula superciliaris	Yellow-billed Tern	B	29	46.4		40	57		Suriname	359
Phaetusa simplex	Large-billed Tern	B	10	235		212	275		Suriname; Brazil; Argentina	63, 498, 1042

Body Masses of World Birds (continued)

Species Name	Common Name	Sex	N	Mean	S.D.	Min	Max	Season	Location	Source #
Gelochelidon nilotica nilotica	Gull-billed Tern	B	12	233	27.3	189	292	PB	Netherlands	223
Gelochelidon nilotica aranea	Gull-billed Tern	B	6	170		160	184	B	Virginia, USA	314
Gelochelidon nilotica macrotarsa	Gull-billed Tern	U	24	264	32.2	188	360		Australia	515a
Gelochelidon nilotica affinis	Gull-billed Tern	B	21	217		178	320		China	515a
Hydroprogne caspia	Caspian Tern	U	84	655		574	782	B	Texas, USA	951
Larosterna inca	Inca Tern	B	105	219					Peru; Chile	1249
Chlidonias niger	Black Tern	U	36	65.3	4.6	60.3	74.1	B	Ontario, Canada	314
Chlidonias leucopterus	White-winged Tern	U	110	54.2	4.9	42	66		South Africa	223
Chlidonias leucopterus	White-winged Tern	U	110	54.2	4.9	42	66		South Africa	223
Chlidonias hybrida	Whiskered Tern	M	17	90	7.3	80	110		Australia	515a
		F	14	77.8	10.7	60	96			
Sterna dougallii	Roseate Tern	U	160	112	73.4	97	128		New York, USA	433
Sterna sumatrana	Black-naped Tern	B	17	106		97	120		Fiji	515a
Sterna striata	White-fronted Tern	B	17	129		103	160		New Zealand	515a
Sterna hirundinacea	South American Tern	F	4	188		172	199		Brazil; Argentina	63, 546
Sterna hirundo	Common Tern	U	265	120		103	145		New York, USA	677
Sterna hirundo longipennis	Common Tern	B	37	139		115	175		Tsingtao, China	1088
Sterna paradisaea	Arctic Tern	U	261	110	8.3	86	127	B	Northeast USA; Canada	177
Sterna vittata	Antarctic Tern	B	21	139		120	160		Tristan da Cunha	515a
Sterna virgata	Kerguelen Tern	U	9	126		85	170		Kerguelen Is.	515a

Body Masses of World Birds (continued)

Species Name	Common Name	Sex	N	Mean	S.D.	Min	Max	Season	Location	Source #
Sterna forsteri	Forster's Tern	B	12	149	15.8	127	173		Oklahoma, USA	1153a
Sterna trudeaui	Snowy-crested Tern	B	2	153		146	160		Brazil	63
Sterna lorata	Peruvian Tern	U	1	53.8					Ecuador	1143
Sterna nereis	Fairy Tern	B	17	72.3		64	80			515a
Sterna balaenarum	Damara Tern	U		60						309
Sterna repressa	White-cheeked Tern	U	12	90		78	105		Egypt	1243a
Thalasseus bengalensis	Lesser Crested Tern	U	40	209	12.3	184	242	W	Northwest Australia	515a
Thalasseus maximus albididorsalis	Royal Tern	U	22	367	14.6	350	395		Mauritania	223
Thalasseus maximus maximus	Royal Tern	U	28	470	19.7					489
Thalasseus bergii	Great Crested Tern	U	50	328	27.8	269	400		Victoria, Australia	515a
Thalasseus sandvicensis	Sandwich Tern	U	10	208	12.7	193	238	B	Virginia, USA	314
Thalasseus sandvicensis eurygnatha	Sandwich Tern	B	12	192		176	226		Suriname	498
Thalasseus bernsteini	Chinese Crested Tern	B	21	284		240	320		Tsingtao, China	1088
Thalasseus elegans	Elegant Tern	U	113	260	21.3	200	325	B	Isla Rosa, Mexico	154
Family Rynchopidae										
Rynchops niger	Black Skimmer	M	56	349	25.7	260	392	B	Texas, USA	951
		F	73	254	17.9	212	292			
Rynchops flavirostris	African Skimmer	B	10	164		111	204			1243a
Family Alcidae										
Alle alle alle	Dovekie	M	45	175	8.5			B	Baffin Bay, Canada	805
		F	43	167	10.1					

Body Masses of World Birds (continued)

Species Name	Common Name	Sex	N	Mean	S.D.	Min	Max	Season	Location	Source #
Alle alle polaris	Dovekie	U	56	202	56				Franz Josef Land	805
Uria aalge	Common Murre	M	121	1006	80	775	1202	B	Newfoundland, Canada	1222
		F	117	979	76	815	1187			
Uria aalge californica	Common Murre	B	70	991				B	California, USA	5
Uria lomvia	Thick-billed Murre	U	139	964						1157
Alca torda	Razorbill	U	549	726	48.8	570	880		Eastern Canada	523
Cepphus grylle	Black Guillemot	B	58	378		320	440		Kattegat, North Sea	157
Cepphus columba	Pigeon Guillemot	B	19	530	48.1				Alaska, USA	345
Cepphus carbo	Spectacled Guillemot	B	3	490		415	545			273
Brachyramphus marmoratus	Marbled Murrelet	B	74	220		188	269	B	British Columbia, Canada	846
Brachyramphus perdix	Long-billed Murrelet	M	8	296		258	357		Asia	846
Brachyramphus brevirostris	Kittlitz's Murrelet	U	36	234	25.5				Alaska, USA	263
Synthliboramphus hypoleucus	Xantus' Murrelet	B	375	167	13	136	215	B	California, USA	833
Synthliboramphus craveri	Craveri's Murrelet	U	8	151						29
Synthliboramphus antiquus	Ancient Murrelet	B	185	207	12.5	175	235	B	Queen Charlotte Is., Canada	406
Synthliboramphus antiquus	Ancient Murrelet	B	46	229		185	280	B	Tsingtao, China	1088
Synthliboramphus wumizusume	Japanese Murrelet	U	1	198						662
Ptychoramphus aleuticus	Cassin's Auklet	U	520	184				PB	Farallon Is., California, USA	744
Aethia psittacula	Parakeet Auklet	M	21	273	19.9	238	304		Alaska, USA	607
		F	11	267	16.8	242	292			

Body Masses of World Birds (continued)

Species Name	Common Name	Sex	N	Mean	S.D.	Min	Max	Season	Location	Source #
Aethia cristatella	Crested Auklet	M	179	268	17.1	227	322	B	St. Lawrence Is., Alaska, USA	606
		F	229	251	17.6	211	300			
Aethia pygmaea	Whiskered Auklet	U	230	112	7	92	132	B	Alaska, USA	547
Aethia pusilla	Least Auklet	U	457	84	7	72	98			262
Cerorhinca monocerata	Rhinoceros Auklet	M	20	510	49	441	569	B	British Columbia, Canada	407
		F	25	456	45	389	616			
Fratercula arctica	Atlantic Puffin	U	165	381	29.9			B	Britain, UK	484
Fratercula arctica	Atlantic Puffin	U	923	923	480	39		B	Labrador, Canada	723
Fratercula corniculata	Horned Puffin	M	210	556	45.6			B		924
		F	219	517	44.6					
Fratercula cirrhata	Tufted Puffin	M	483	806	61.8			B		924a
		F	493	744	57.5					
Family Pteroclidae										
Syrrhaptes tibetanus	Tibetan Sandgrouse	B				299	395			734
Syrrhaptes paradoxus	Pallas' Sandgrouse	M	8	274		255	300		Mongolia	926
		F	6	240		180	270			
Pterocles alchata	Pin-tailed Sandgrouse	M				250	408			734
		F				207	274			
Pterocles namaqua	Namaqua Sandgrouse	U	32	177		143	193			133
Pterocles exustus	Chestnut-bellied Sandgrouse	M				170	295			734
		F				140	241			
Pterocles senegallus	Spotted Sandgrouse	M	2	272		264	280		Morocco; Iraq	223
		F	1	255						
Pterocles orientalis	Black-bellied Sandgrouse	M	9	428		400	460	F	Kazakhstan	223
		F	11	383		300	420			
Pterocles gutturalis	Yellow-throated Sandgrouse	B	9	338		285	400		Eastern Africa	120

Body Masses of World Birds (continued)

Species Name	Common Name	Sex	N	Mean	S.D.	Min	Max	Season	Location	Source #
Pterocles coronatus	Crowned Sandgrouse	M	1	300					Morocco	223
Pterocles decoratus	Black-faced Sandgrouse	M				160	216			734
		F				149	210			
Pterocles lichtensteinii	Lichtenstein's Sandgrouse	M				175	250			734
		F				190	230			
Pterocles bicinctus	Double-banded Sandgrouse	B	28	237		210	280			1243a
Pterocles quadricinctus	Four-banded Sandgrouse	M	1	160					Senegal	1143
		F	1	200						
Pterocles indicus	Painted Sandgrouse	M				166	208			734
		F				166	176			
Pterocles burchelli	Burchell's Sandgrouse	M				180	200			734
		F				160	185			
Family Columbidae										
Columba livia	Rock Pigeon	M	41	369	38.6			B	Kansas, USA	603
		F	37	340	34.7					
Columba rupestris	Hill Pigeon	M	17	252		180	305		Hopei, China	1087
		F	10	239		201	262			
Columba leuconota leuconota	Snow Pigeon	M	3			277	300			7
		F	3			255	307			
Columba guinea	Speckled Pigeon	B	14	352		307	371	B	Ethiopia	1320
Columba albitorques	White-collared Pigeon	B	2	277		262	292			1243a
Columba oenas	Stock Pigeon	M	12	302	18.8	280	334	B		223
		F	7	280	16.2	254	302			
Columba eversmanni	Pale-backed Pigeon	B	12	202		183	234		Afghanistan; India	223
Columba palumbus	Common Wood-pigeon	B	518	490		284	614	Y		223
Columba trocaz	Trocaz Pigeon	M				420	520		Madeira Is.	414
		F				390	440			

Body Masses of World Birds (continued)

Species Name	Common Name	Sex	N	Mean	S.D.	Min	Max	Season	Location	Source #
Columba unicincta	Afep Pigeon	M	5	423		357	490			1243a
		F	3	358		356	360			
Columba arquatrix	African Olive Pigeon	U	53	400		226	489		South Africa	1036
Columba thomensis	Maroon Pigeon	M	2	525		520	530			1243a
		F	1	350						
Columba albinucha	White-naped Pigeon	U	1	325						120
Columba pulchricollis	Ashy Wood-pigeon	U		330						414
Columba elphinstonii	Nilgiri Wood-pigeon	U	1	379						7
Columba punicea	Pale-capped Pigeon	U				370	510			7
Columba palumboides	Andaman Wood-pigeon	M	1	520					India	7
Columba janthina	Japanese Wood-pigeon	U		397					Japan	339
Columba vitiensis	Metallic Wood-pigeon	B	13	353	39.8	268	415		Fiji	1196c
Columba leucomela	White-headed Wood-pigeon	B	16	420		218	500		Eastern Australia	515a
Columba pallidiceps	Yellow-legged Pigeon	M	1	459					New Britain	427
Patagioenas leucocephala	White-crowned Pigeon	M	24	253	31				Florida, USA	37
		F	30	230	14					
Patagioenas leucocephala	White-crowned Pigeon	B	55	238					Cuba	37
Patagioenas squamosa	Scaly-naped Pigeon	B	11	312	49.5	235	388		Cuba; Puerto Rico; Bonaire	19
Patagioenas speciosa	Scaled Pigeon	M	13	262	11.5				Panama	489
		F	6	255						

Body Masses of World Birds (continued)

Species Name	Common Name	Sex	N	Mean	S.D.	Min	Max	Season	Location	Source #
Patagioenas picazuro	Picazuro Pigeon	B	5	279		210	402		Argentina; Brazil	63, 362, 1040
Patagioenas corensis	Bare-eyed Pigeon	U				262	283			414
Patagioenas maculosa	Spot-winged Pigeon	M	4	347		308	345			63, 841
Patagioenas fasciata fasciata	Band-tailed Pigeon	M	5888	353		270	460		Colorado, USA	111
		F	5291	332		226	424			
Patagioenas fasciata monilis	Band-tailed Pigeon	M	1880	398		300	515		California, USA	111
		F	942	386		300	470			
Patagioenas caribaea	Ring-tailed Pigeon	U		250						348
Patagioenas cayennensis	Pale-vented Pigeon	U	10	229		197	254		Suriname	498
Patagioenas flavirostris	Red-billed Pigeon	B	30	261	31.2	206	343			721
Patagioenas flavirostris	Red-billed Pigeon	B	5	281		234	308		Tres Marias Is., Mexico	440
Patagioenas oenops	Peruvian Pigeon	U				250	280			414
Patagioenas inornata	Plain Pigeon	M	1	365					Cuba	19
Patagioenas plumbea	Plumbeous Pigeon	M	6	170		155	187		Suriname	498
		F	6	188		157	206			
Patagioenas nigrirostris	Short-billed Pigeon	B	5	176		128	236		Panama; Mexico	155, 489, 1197
Patagioenas subvinacea	Ruddy Pigeon	M	7	176						370, 489, 861
		F	4	150						
Columba delegorguei	Delegorgue's Pigeon	F	1	133						120
Columba iriditorques	Bronze-winged Pigeon	M	15	130	8					1243a
		F	18	122	10					
Columba larvata	Lemon Dove	U	3	155		147	163			1014

Body Masses of World Birds (continued)

Species Name	Common Name	Sex	N	Mean	S.D.	Min	Max	Season	Location	Source #
Nesoenas mayeri	Pink Pigeon	M	33	315	43.1	240	410			183
		F	29	291	38.2	213	369			
Streptopelia turtur	Eurasian Turtle-Dove	U	246	132		85	170	F	Portugal	223
Streptopelia lugens	Dusky Turtle-Dove	U	20	155	10.8	136	169		Ethiopia	1238
Streptopelia hypopyrrha	Adamawa Turtle-Dove	U	17	162		135	181		Nigeria	777
Streptopelia orientalis	Oriental Turtle-Dove	M	30	238		175	323		Hopei, China	1087
		F	21	228		192	280			
Streptopelia bitorquata	Is. Turtle-Dove	B	12	153		140	174		Philippines	437, 954, 983
Streptopelia decaocto	Eurasian Collared-Dove	M	87	152	11.6	115	184	Y	India	223
		F	80	146	10.3	113	176			
Streptopelia roseogrisea	African Collared-Dove	B	4	155		135	172		Chad; Niger	223
Streptopelia reichenowi	White-winged Collared-Dove	U	14	119		98	135			1243a
Streptopelia decipiens	African Mourning Dove	B	5	134		125	140			120
Streptopelia semitorquata	Red-eyed Dove	U	20	176		145	275			133
Streptopelia capicola	Ring-necked Dove	U	50	142		92	188			133
Streptopelia vinacea	Vinaceous Dove	B	7	110		85	122			120, 777
Streptopelia tranquebarica	Red Collared-Dove	M	13	103		80	144		India; China	7, 1087, 1088
Streptopelia picturata copperingi	Madagascar Turtle-Dove	M	11	184		160	210		Aldabra Is.	67
		F	7	165		150	185			
Streptopelia picturata picturata	Madagascar Turtle-Dove	U	25	182	20	142	220		Madagascar	957
Streptopelia chinensis	Spotted Dove	U	343	159	11	128	194		California, USA	314

Body Masses of World Birds (continued)

Species Name	Common Name	Sex	N	Mean	S.D.	Min	Max	Season	Location	Source #
Streptopelia senegalensis	Laughing Dove	U	1157	101	9.2	72	139	Y	South Africa	223
Streptopelia senegalensis sokotrae	Laughing Dove	U	2	66.8		66.5	67		Socotra	317
Macropygia unchall	Barred Cuckoo-Dove	U	3	168		153	180		Malaysia	775, 1287
Macropygia phasianella	Brown Cuckoo-Dove	B	17	237		205	272		Eastern Australia	515a
Macropygia amboinensis	Slender-billed Cuckoo-Dove	B	17	237		205	272		Eastern Australia	515a
Macropygia rufipennis	Andaman Cuckoo-Dove	U				230	285			7
Macropygia tenuirostris	Philippine Cuckoo-Dove	U	15	177		157	199		Philippines	436, 437, 954
Macropygia nigrirostris	Black-billed Cuckoo-Dove	B	19	90.1		66	104		New Guinea	283
Macropygia mackinlayi	Mackinlay's Cuckoo-Dove	M	10	91	5	85	100		New Guinea	292
		F	5	81		72	90			
Macropygia ruficeps	Little Cuckoo-Dove	U	8	81.9		70.3	97		Malaysia	775, 785, 1287
Reinwardtoena reinwardtii	Great Cuckoo-Dove	U	9	275		257	305		New Guinea	292
Reinwardtoena browni	Pied Cuckoo-Dove	B	4	304		279	325		New Britain; New Ireland	58, 426
Reinwardtoena crassirostris	Crested Cuckoo-Dove	M	2	259		258	260		Guadalcanal	399
Turacoena manadensis	White-faced Cuckoo-Dove	U				172	253			414
Turtur chalcospilos	Emerald-spotted Wood-Dove	B	11	60.6	5.8	50.2	67.3		Zimbabwe	562, 564
Turtur abyssinicus	Black-billed Wood-Dove	U	35	62.7		55	71		Nigeria	777
Turtur afer	Blue-spotted Wood-Dove	U	10	65.6		57.5	74		Ghana	457

Body Masses of World Birds (continued)

Species Name	Common Name	Sex	N	Mean	S.D.	Min	Max	Season	Location	Source #
Turtur tympanistria	Tambourine Dove	B	15	67.7		54	79		Somalia	1335
Turtur tympanistria	Tambourine Dove	U	18	73.6	7.6	64	89		Tanzania	1014
Turtur brehmeri	Blue-headed Wood-Dove	U	1	116					Liberia	616
Oena capensis	Namaqua Dove	B	138	40.6		32	54	Y	South Africa	223
Chalcophaps indica chrysochlora	Emerald Dove	B	21	136		120	172			515a
Chalcophaps indica longirostris	Emerald Dove	B	18	160		140	192			515a
Chalcophaps indica sandwichensis	Emerald Dove	B	8	117		100	134		Vanuatu	656b
Chalcophaps stephani	Stephan's Dove	U	15	118					New Guinea	61
Chalcophaps stephani mortoni	Stephan's Dove	B	4	141		132	150		Guadalcanal	399
Henicophaps albifrons	New Guinea Bronzewing	F	1	247					New Guinea	61
Phaps chalcoptera	Common Bronzewing	M	51	344	37.3	260	450		Australia	515a
		F	43	318	34.9	250	405			
Phaps elegans	Brush Bronzewing	M	23	219	21.9	170	260		Eastern Australia	515a
		F	25	204	15.8	162	234			
Phaps histrionica	Flock Bronzewing	M	17	293	29.3	235	355		Australia	515a
		F	9	285		260	320			
Ocyphaps lophotes	Crested Pigeon	B	21	204		142	260		Australia	515a
Geophaps plumifera plumifera	Spinifex Pigeon	M	12	105	9.3	94	124		Australia	515a
		F	12	91.2	6.6	84	101			
Geophaps plumifera ferruginea	Spinifex Pigeon	M	21	94.2	6.4	83	105		Australia	515a
		F	25	87.8	5.1	77	98			
Geophaps scripta	Squatter Pigeon	M	25	228	17.7	188	250		Australia	515a
		F	22	215	20.4	180	250			

Body Masses of World Birds (continued)

Species Name	Common Name	Sex	N	Mean	S.D.	Min	Max	Season	Location	Source #
Geophaps smithii	Partridge Pigeon	B	40	194		169	250		Australia	515a
Petrophassa rufipennis	Chestnut-quilled Rock-Pigeon	B	41	159		130	178		Australia	515a
Petrophassa albipennis albipennis	White-quilled Rock-Pigeon	B	47	132		117	156		N. Territories, Australia	515a
Petrophassa albipennis boothi	White-quilled Rock-Pigeon	B	30	117		103	142		Western Australia	515a
Geopelia cuneata	Diamond Dove	B	40	31.6		24	39		Australia	515a
Geopelia striata	Zebra Dove	U	7	56.6		51	62.1		Philippines	437
Geopelia placida	Peaceful Dove	B	38	47.8		35	65		Australia	515a
Geopelia humeralis	Bar-shouldered Dove	M	15	133	8.5	118	149		Australia	515a
		F	12	124	8.5	108	137			
Leucosarcia melanoleuca	Wonga Pigeon	M	6	410		220	500		Eastern Australia	515a
		F	6	448		360	500			
Ectopistes migratorius	Passenger Pigeon	U				256	341		extinct	83
Zenaida macroura	Mourning Dove	M	140	123	1.8			PB	Illinois, USA	478
		F	95	115	1.8					
Zenaida graysoni	Socorro Dove	B	6	192		165	215		captive	43
Zenaida auriculata	Eared Dove	B	43	136		116	155		Argentina	1040
Zenaida auriculata noronha	Eared Dove	B	9	89.3		77	95			835
Zenaida aurita	Zenaida Dove	B	356	156	15.2	125	205		Puerto Rico	986
Zenaida galapagoensis galapagoensis	Galapagos Dove	B	116	97.8	20.5	51	153		Galapagos Is.	1044
Zenaida galapagoensis exsul	Galapagos Dove	B	32	126	27.6	76	177		Galapagos Is.	1044
Zenaida asiatica	White-winged Dove	U	30	153	13.2	125	187	B	Arizona, USA	314
Zenaida meloda	Pacific Dove	U				198	221			414

Body Masses of World Birds (continued)

Species Name	Common Name	Sex	N	Mean	S.D.	Min	Max	Season	Location	Source #
Columbina passerina portoricensis	Common Ground-Dove	U	66	35.4	2.5	30.5	41.2	W	Puerto Rico	349
Columbina passerina passerina	Common Ground-Dove	U	19	37					Florida, USA	108
Columbina passerina exigua	Common Ground-Dove	U	218	28.5	2.4	22.4	38.5		Mona Is., Puerto Rico	349
Columbina passerina	Common Ground-Dove	B	61	32	2.5	27	39		Barbuda	19
Columbina passerina	Common Ground-Dove	B	51	39.5	4.1	31	48		Cuba	19
Columbina passerina	Common Ground-Dove	B	94	36.4	4.5	24	46		Dominican Republic	19
Columbina passerina	Common Ground-Dove	B	49	40.3	2.2	36	43		St. Lucia	19
Columbina minuta	Plain-breasted Ground-Dove	B	21	32.9	4.5	26	42.2			75, 496, 1215
Columbina minuta	Plain-breasted Ground-Dove	U	232	32.3	3.4				Venezuela	94
Columbina buckleyi	Ecuadorian Ground-Dove	U		57.5						1232
Columbina talpacoti	Ruddy Ground-Dove	B	74	46.5		35.5	56.5		Trinidad	1131
Columbina talpacoti	Ruddy Ground-Dove	U	121	45.6	6.1				Venezuela	94
Columbina picui	Picui Ground-Dove	U	30	47		41	51			362
Columbina cruziana	Croaking Ground-Dove	B	5	50.1		43.8	52.1		Peru; Chile	776a, 1042
Columbina inca	Inca Dove	U	125	47.5	4.4	33	57		Arizona, USA	314
Columbina squammata	Scaled Dove	U	232	52.9	4.7				Venezuela	94
Claravis pretiosa	Blue Ground-Dove	B	100	68.2		41	103		Brazil	883

Body Masses of World Birds (continued)

Species Name	Common Name	Sex	N	Mean	S.D.	Min	Max	Season	Location	Source #
Claravis mondetoura	Maroon-chested Ground-Dove	M	5	89.7					Panama; Peru	489, 1288
Metriopelia ceciliae	Bare-faced Ground-Dove	B	7	55.1		49	59.8			776a, 1042
Metriopelia melanoptera	Black-winged Ground-Dove	U	2	100		100	100			362
Metriopelia aymara	Golden-spotted Ground-Dove	B	8	61.8		60	68.2			362, 1042
Uropelia campestris	Long-tailed Ground-Dove	F	1	35.2					Brazil	883
Leptotila verreauxi	White-tipped Dove	U	40	161	31.6	114	276		El Salvador	1041a
Leptotila verreauxi tobagensis	White-tipped Dove	U	34	134		96	157		Tobago	359
Leptotila megalura	White-faced Dove	U		218						1232
Leptotila rufaxilla	Gray-fronted Dove	B	15	157	18.9	131	183			111, 161, 188, 245, 246
Leptotila plumbeiceps	Gray-headed Dove	B	10	170	19.6	139	205		Belize	1027
Leptotila pallida	Pallid Dove	U				96	157			414
Leptotila battyi	Brown-backed Dove	U				150	172			414
Leptotila wellsi	Grenada Dove	U	1	200					Grenada	82
Leptotila jamaicensis	Caribbean Dove	B	9	165		139	190		Mexico; Cayman Is.	640, 868, 910a
Leptotila cassini	Gray-breasted Dove	U	12	159	14.3	132	179		Belize; Panama	155, 1027, 1180
Leptotila ochraceiventris	Ochre-bellied Dove	U		146						1232
Geotrygon lawrencii	Purplish-backed Quail-Dove	U		220					Costa Rica	1170

Body Masses of World Birds (continued)

Species Name	Common Name	Sex	N	Mean	S.D.	Min	Max	Season	Location	Source #
Geotrygon carrikeri	Tuxtla Quail-Dove	U	7	267		249	292		captive	217b
Geotrygon costaricensis	Buff-fronted Quail-Dove	M	2	320		310	330		Panama	489
		F	2	254		225	283			
Geotrygon goldmani	Russet-crowned Quail-Dove	U	1	258					Panama	79
Geotrygon saphirina	Sapphire Quail-Dove	U		203						1232
Geotrygon caniceps	Gray-fronted Quail-Dove	U	1	192					Cuba	19
Geotrygon leucometopia	White-fronted Quail-Dove	U	1	154					Dominican Republic	19
Geotrygon versicolor	Crested Quail-Dove	U		225						185
Geotrygon chiriquensis	Chiriqui Quail-Dove	B	19	308					Panama	489
Geotrygon veraguensis	Olive-backed Quail-Dove	U	1	155					Panama	618
Geotrygon albifacies	White-faced Quail-Dove	U	11	238	62.4	158	339		Honduras; El Salvador	1041a, 1145
Geotrygon linearis	Lined Quail-Dove	B	5	261		230	327		Trinidad, Venezuela	613, 125[illegible]
Geotrygon frenata	White-throated Quail-Dove	B	10	307		265	325		Argentina	1040
Geotrygon chrysia	Key West Quail-Dove	B	108	169	21.4	109	212		Caribbean Is.	19
Geotrygon mystacea	Bridled Quail-Dove	B	27	211	33.8	138	256		Guadelupe; Montserrat; St. Kitts	19
Geotrygon violacea	Violaceous Quail-Dove	U	3	95.2		90	102		Panama; Venezuela	200, 61[illegible] 1254
Geotrygon montana	Ruddy Quail-Dove	U	62	115					Peru	370
Geotrygon montana	Ruddy Quail-Dove	U	14	149					Puerto Rico	895

Body Masses of World Birds (continued)

Species Name	Common Name	Sex	N	Mean	S.D.	Min	Max	Season	Location	Source #
Geotrygon montana	Ruddy Quail-Dove	B	8	140		121	165		Belize	220
Starnoenas cyanocephala	Blue-headed Quail-Dove	U	3	219		200	242		Cuba	19, 801, 1267
Caloenas nicobarica	Nicobar Pigeon	M	2	492		460	525		India; Solomon Is.	7, 658
		F	3	588		490	675			
Gallicolumba luzonica	Luzon Bleeding-heart	B	6	196		181	204		Philippines	954
Gallicolumba crinigera	Mindanao Bleeding-heart	U				181	204			414
Gallicolumba rufigula	Cinnamon Ground-Dove	U	2	130					New Guinea	61
Gallicolumba tristigmata	Sulawesi Ground-Dove	U	1	275					Sulawesi, Indonesia	1272
Gallicolumba jobiensis	White-bibbed Ground-Dove	U	11	135					New Guinea	61
Gallicolumba erythroptera	Polynesian Ground-Dove	U				105	122			414
Gallicolumba xanthonura	White-throated Ground-Dove	M	2	128		117	140		Guam	579
		F	4	95.6		58.5	119			
Gallicolumba stairi	Friendly Ground-Dove	U		171						414
Gallicolumba rubescens	Marquesas Ground-Dove	U		60						414
Gallicolumba beccarii	Bronze Ground-Dove	U	10	65.4	4.1	57.7	70.5		Papua New Guinea	387
Gallicolumba beccarii solomonensis	Bronze Ground-Dove	M	3	94.8		93	98		Solomon Is.	399
Trugon terrestris	Thick-billed Ground-Dove	M	1	400					New Guinea	61
Otidiphaps nobilis	Pheasant Pigeon	U		500					New Guinea	61
Goura cristata	Western Crowned-Pigeon	U		2000						1048

Body Masses of World Birds (continued)

Species Name	Common Name	Sex	N	Mean	S.D.	Min	Max	Season	Location	Source
Goura scheepmakeri	Southern Crowned-Pigeon	M	1	2235					captive	786
Goura victoria	Victoria Crowned-Pigeon	U		2384						414
Didunculus strigirostris	Tooth-billed Pigeon	U		400						414
Phapitreron leucotis	White-eared Dove	B	54	108		89.3	136		Philippines	954
Phapitreron amethystinus	Amethyst Dove	B	40	145		112	178		Philippines	954
Phapitreron cinereiceps	Dark-eared Dove	B	3	146		131	159		Philippines	954
Treron olax	Little Green-Pigeon	U		77					Malaysia	665
Treron vernans	Pink-necked Pigeon	B	5	132		126	141		Philippines	437
Treron fulvicollis	Cinnamon-headed Pigeon	U		167					Malaysia	665
Treron bicinctus	Orange-breasted Pigeon	U	2	174		155	194		India	940
Treron pompadora	Pompadour Green-Pigeon	B	10	234		218	257		Philippines	954
Treron curvirostra	Thick-billed Pigeon	M		138		112	161		Malaysia	1287
		F		131		117	171			
Treron capellei	Large Green-Pigeon	M	1	455						1287
		F	2	354		340	369			
Treron phoenicopterus	Yellow-footed Pigeon	B	6	235		226	248		India	740
Treron waalia	Bruce's Green-Pigeon	U	3	254		241	268		Nigeria; Sudan; Ghana	457, 7[illegible] 1319
Treron australis	Madagascar Green-Pigeon	B	7	229		180	256			120, 9[illegible]
Treron seimundi	Yellow-vented Pigeon	F	4			150	166			1287

Body Masses of World Birds (continued)

Species Name	Common Name	Sex	N	Mean	S.D.	Min	Max	Season	Location	Source #
Treron apicauda	Pin-tailed Pigeon	M				185	255			7
Treron calva uellensis	African Green-Pigeon	M	17	218		160	250			1243a
		F	6	179		130	225			
Treron sphenurus	Wedge-tailed Pigeon	U	3	210		205	214			7, 775
Treron sieboldii	White-bellied Pigeon	U		245					Japan	1242
Treron formosae	Whistling Green-Pigeon	U		140					Taiwan	834
Ptilinopus cinctus	Black-backed Fruit-Dove	U				450	570		estimated	515a
Ptilinopus cinctus alligator	Black-banded Fruit-Dove	B	3	243		234	256		Northern Terr., Australia	470
Ptilinopus porphyreus	Pink-headed Fruit-Dove	F	1	168					captive	1188
Ptilinopus occipitalis	Yellow-breasted Fruit-Dove	B	23	238		185	278		Philippines	954
Ptilinopus marchei	Flame-breasted Fruit-Dove	F	2	326		320	333		Luzon, Philippines	1143
Ptilinopus merrilli	Cream-breasted Fruit-Dove	U		263						414
Ptilinopus fischeri	Red-eared Fruit-Dove	U		177						414
Ptilinopus jambu	Jambu Fruit-Dove	U		135					Malaysia	665
Ptilinopus subgularis	Maroon-chinned Fruit-Dove	U		142						414
Ptilinopus leclancheri	Black-chinned Fruit-Dove	B	8	164		150	200		Philippines	437, 983, 954
Ptilinopus magnificus	Wompoo Fruit-Dove	M	29	191	19.6	152	228		New Guinea	515a
		F	29	180	19	132	220			
Ptilinopus perlatus	Pink-spotted Fruit-Dove	U	7	210					New Guinea	61

Body Masses of World Birds (continued)

Species Name	Common Name	Sex	N	Mean	S.D.	Min	Max	Season	Location	Source
Ptilinopus ornatus	Ornate Fruit-Dove	B	10	163	12.6	142	185		New Guinea	283
Ptilinopus tannensis	Tanna Fruit-Dove	B	4	208		202	221		Vanuatu	656b
Ptilinopus aurantiifrons	Orange-fronted Fruit-Dove	B	5	152		140	173		New Guinea	425, 786
Ptilinopus superbus	Superb Fruit-Dove	B	20	118		99	144		New Guinea	515a
Ptilinopus perousii	Many-colored Fruit-Dove	B	2	93.5		85	102		Tonga	977
Ptilinopus porphyraceus	Crimson-crowned Fruit-Dove	U	11	105		85.4	121		Tonga	977
Ptilinopus pelewensis	Palau Fruit-Dove	U		93						414
Ptilinopus roseicapilla	Mariana Fruit-Dove	B	18	90.4		81	103		Guam	35
Ptilinopus regina	Rose-crowned Fruit-Dove	M	12	110	8.8	94	126		Australia	515a
		F	9	93.8		85	108			
Ptilinopus richardsii	Silver-capped Fruit-Dove	M	15	109		94	117		Solomon Is.	399
		F	6	99.1		94.5	110			
Ptilinopus purpuratus frater	Gray-green Fruit-Dove	U				82	93			414
Ptilinopus purpuratus chrysogaster	Gray-green Fruit-Dove	U				95	125			414
Ptilinopus coralensis	Atoll Fruit-Dove	U				70	99			414
Ptilinopus greyii	Red-bellied Fruit-Dove	M	3			86	95			
		F	3			65	88			
Ptilinopus huttoni	Rapa Fruit-Dove	U		130						414
Ptilinopus dupetithouarsii	White-capped Fruit-Dove	U				90	110			414
Ptilinopus insularis	Henderson Is. Fruit-Dove	M	2	104		102	105		Henderson Is.	446
		F	3			95	105			

Body Masses of World Birds (continued)

Species Name	Common Name	Sex	N	Mean	S.D.	Min	Max	Season	Location	Source #
Ptilinopus coronulatus	Coroneted Fruit-Dove	U	10	76.6					New Guinea	61, 786
Ptilinopus pulchellus	Beautiful Fruit-Dove	U	15	70					New Guinea	61
Ptilinopus rivoli	White-breasted Fruit-Dove	B	16	149	15.2	132	188		New Guinea	283, 424, 426
Ptilinopus solomonensis ocularis	Yellow-bibbed Fruit-Dove	B	10	126		112	142		Guadalcanal	399
Ptilinopus solomonensis solomonensis	Yellow-bibbed Fruit-Dove	B	52	93.4		80	105		Solomon Is.	399
Ptilinopus viridis lewisii	Claret-breasted Fruit-Dove	B	9	121		112	135		Solomon Is.	658
Ptilinopus eugeniae	White-headed Fruit-Dove	B	25	114		93.5	130		Solomon Is.	399
Ptilinopus iozonus	Orange-bellied Fruit-Dove	B	11	114					New Guinea	61, 428
Ptilinopus insolitus	Knob-billed Fruit-Dove	M	1	162					New Guinea	426
		F	1	115						
Ptilinopus hyogaster	Gray-headed Fruit-Dove	B	4	149		105	169		N. Moluccas	979
Ptilinopus melanospilus	Black-naped Fruit-Dove	U		92						414
Ptilinopus nainus	Dwarf Fruit-Dove	U	3	47					New Guinea	61
Drepanoptila holosericea	Cloven-feathered Dove	U	2	215		210	220		New Caledonia	1017
Alectroenas madagascariensis	Madagascar Blue-Pigeon	U	3	173		113	210		Madagascar	957
Alectroenas sganzini	Comoro Blue-Pigeon	M	2	146		134	158		Aldabra Is.	67
		F	1	171						
Alectroenas pulcherrima	Seychelles Blue-Pigeon	B	2	165		161	169		Seychelles	409
Ducula poliocephala	Pink-bellied Imperial-Pigeon	B	5	555		510	622		Philippines	436, 954, 983

Body Masses of World Birds (continued)

Species Name	Common Name	Sex	N	Mean	S.D.	Min	Max	Season	Location	Source #
Ducula forsteni	White-bellied Imperial-Pigeon	U		510						414
Ducula radiata	Gray-headed Imperial-Pigeon	U		295						414
Ducula carola	Spotted Imperial-Pigeon	B	3	289		258	320		Philippines	983, 1143
Ducula aenea	Green Imperial-Pigeon	B	14	545		454	644		Philippines	954, 983
Ducula perspicillata	White-eyed Imperial-Pigeon	U	8	667		620	730		Moluccas	1181, 1182
Ducula concinna	Elegant Imperial-Pigeon	U				700	925			414
Ducula pacifica	Pacific Imperial-Pigeon	U				370	420			414
Ducula rubricera	Red-knobbed Imperial-Pigeon	B	9	652		560	765		New Ireland; Solomon Is.	58, 658
Ducula oceanica	Micronesian Imperial-Pigeon	U				340	406			414
Ducula myristicivora	Spice Imperial-Pigeon	U		535						414
Ducula rufigaster	Purple-tailed Imperial-Pigeon	U	2	436					New Guinea	35
Ducula finschii	Finsch's Imperial-Pigeon	M	1	384					New Britain	219
		F	1	308						
Ducula chalconota	Rufescent Imperial-Pigeon	B	2	502		494	509		New Guinea	770
Ducula pistrinaria	Is.Imperial-Pigeon	B	3	515		501	525		Solomon Is.	658
Ducula rosacea	Pink-headed Imperial-Pigeon	U				360	413			414
Ducula whartoni	Christmas Is.Imperial-Pigeon	M	2	625		550	700		Christmas Is.	515a
		F	3	522		450	575			
Ducula brenchleyi	Chestnut-bellied Imperial-Pigeon	M	1	380					Solomon Is.	399

Body Masses of World Birds (continued)

Species Name	Common Name	Sex	N	Mean	S.D.	Min	Max	Season	Location	Source #
Ducula bakeri	Baker's Imperial-Pigeon	M	1	394					Vanuatu	656b
Ducula goliath	New Caledonian Imperial-Pigeon	F	1	600					New Caledonia	1017
Ducula pinon	Pinon Imperial-Pigeon	U	2	756		710	802		New Guinea	61, 786
Ducula melanochroa	Bismarck Imperial-Pigeon	M	1	665					New Ireland	58
Ducula mullerii	Collared Imperial-Pigeon	M	3	612		570	650		New Guinea	515a
		F	2	540		540	540			
Ducula zoeae	Zoe Imperial-Pigeon	U	3	568					New Guinea	61, 786
Ducula badia	Mountain Imperial-Pigeon	B	5	486		316	665			7, 785, 1287
Ducula bicolor spilorrhoa	Torresian Imperial-Pigeon	M	18	468	36.5	378	540		Australia	515a
		F	9	444		380	530			
Ducula bicolor luctuosa	White Imperial-Pigeon	U		410						414
Lopholaimus antarcticus	Topknot Pigeon	M	1	472					Australia	515a
		F	1	522						
Hemiphaga novaeseelandiae	New Zealand Pigeon	B	9	653		510	795		New Zealand	515a
Gymnophaps albertisii	Papuan Mountain-Pigeon	B	6	259					New Guinea	61
Gymnophaps mada	Long-tail Mountain-Pigeon	U	2	334		330	338		Moluccas	1181
Gymnophaps solomonensis	Pale Mountain-Pigeon	B	3	335		310	385		Guadalcanal	399
Family Cacatuidae										
Probosciger aterrimus	Palm Cockatoo	B	5	841		710	1040		Australia	516
Calyptorhynchus banksii banksii	Red-tailed Black-Cockatoo	B	18	775		480	920		Northeast Australia	516
Calyptorhynchus banksii macrorhynchus	Red-tailed Black-Cockatoo	B	49	740		530	1000		North Australia	516

Body Masses of World Birds (continued)

Species Name	Common Name	Sex	N	Mean	S.D.	Min	Max	Season	Location	Source #
Calyptorhynchus banksii samueli	Red-tailed Black-Cockatoo	B	18	656		555	740		Western Australia	516
Calyptorhynchus lathami	Glossy Black-Cockatoo	B	17	437		335	515		Australia	516
Calyptorhynchus funereus funereus	Yellow-tailed Black-Cockatoo	B	18	750		600	900		Southeast Australia	516
Calyptorhynchus funereus xanthonotus	Yellow-tailed Black-Cockatoo	B	28	607		321	730		Southern Australia & Tasmania	516
Calyptorhynchus latirostris	Slender-billed Black-Cockatoo	B	186	612		480	750			1047
Calyptorhynchus baudinii	White-tailed Black-Cockatoo	B	75	620		540	770			1047
Callocephalon fimbriatum	Gang-gang Cockatoo	B	48	256		165	305		Southeast Australia	516
Eolophus roseicapilla roseicapilla	Galah	B	303	332		255	430		Western Australia	516
Eolophus roseicapilla albiceps	Galah	B	19	354		307	432		Southeast Australia	516
Eolophus roseicapilla kuhli	Galah	M	10	286	17.1	259	312		Australia	516
		F	15	257	19.9	227	305			
Cacatua tenuirostris	Long-billed Corella	M	18	588	49.2	491	676		Victoria, Australia	516
		F	23	546	56.5	357	615			
Cacatua pastinator pastinator	Western Corella	U	11	750					Australia	516
Cacatua pastinator derbyi	Western Corella	M	4	610		560	640		Australia	516
		F	4	551		445	640			
Cacatua sanguinea	Little Corella	M	22	562		466	626		Western Australia	1046
		F	17	488		370	540			
Cacatua goffini	Tanimbar Corella	U	12	275	31				captive	371
Cacatua haematuropygia	Philippine Corella	B	5	288		256	302		Philippines	954
Cacatua sulphurea citrinocristata	Yellow-crested Cockatoo	M	1	324					captive	1143
		F	2	284		272	297			

Body Masses of World Birds (continued)

Species Name	Common Name	Sex	N	Mean	S.D.	Min	Max	Season	Location	Source #
Cacatua ducorpsii	Ducorps' Cockatoo	B	9	394		355	460		Guadalcanal; Solomon Is.	399, 658
Cacatua leadbeateri leadbeateri	Pink Cockatoo	B	12	376		275	460		Eastern Australia	516
Cacatua leadbeateri mollis	Pink Cockatoo	M	19	424	26.3	380	480		Western Australia	516
		F	23	392	28.1	360	455			
Cacatua galerita galerita	Sulphur-crested Cockatoo	B	48	790		502	1020		Southeast Australia	516
Cacatua galerita fitzroyi	Sulphur-crested Cockatoo	B	6	657		600	780		North Australia	516
Cacatua ophthalmica	Blue-eyed Cockatoo	M	1	492					captive	1143
Cacatua moluccensis	Salmon-crested Cockatoo	U	2	835		780	890		Seram, Moluccas	1181
Cacatua alba	White Cockatoo	U		570						1257
Nymphicus hollandicus	Cockatiel	M	57	92.2	6.9	79.5	108		New South Wales, Australia	516
		F	25	97.1	6.4	84	117			
Family Psittacidae										
Chalcopsitta atra	Black Lory	U		195					captive	1117
Chalcopsitta sintillata	Yellow-streaked Lory	U	3	190					New Guinea	61, 786
Chalcopsitta cardinalis	Cardinal Lory	B	14	183		168	205		Guadalcanal	399
Eos bornea	Red Lory	U	3	156		131	179		Moluccas	1181, 1182
Eos cyanogenia	Black-winged Lory	B	2	167		166	168		New Guinea	1143
Eos semilarvata	Blue-eared Lory	U				85	105		Seram, Moluccas	1181
Pseudeos fuscata	Dusky Lory	U	10	149					New Guinea	61
Trichoglossus haematodus rubritorquis	Rainbow Lorikeet	B	27	120		103	140		Australia	516

Body Masses of World Birds (continued)

Species Name	Common Name	Sex	N	Mean	S.D.	Min	Max	Season	Location	Source #
Trichoglossus haematodus massena	Rainbow Lorikeet	M	13	92.3		83	104		Guadalcanal	399
		F	9	84.9		74.5	95.5			
Trichoglossus haematodus micropteryx	Rainbow Lorikeet	M	4	135		127	142		New Guinea	292
		F	3	125		125	126			
Trichoglossus haematodus moluccanus	Rainbow Lorikeet	B	84	133		84	169		Eastern Australia	516
Trichoglossus euteles	Olive-headed Lorikeet	M	1	73.7					captive	1143
Trichoglossus johnstoniae	Mindanao Lorikeet	B	11	55.2		48	62.1		Philippines	954
Trichoglossus chlorolepidotus	Scaly-breasted Lorikeet	B	17	87.1		77	97		Queensland, Australia	516
Psitteuteles versicolor	Varied Lorikeet	B	29	53.5		49	61		Australia	516
Psitteuteles goldiei	Goldie's Lorikeet	U	3	58.7					New Guinea	61, 770
Lorius garrulus	Chattering Lory	B	5	143		106	180		captive	1143
Lorius hypoinochrous	Purple-bellied Lory	F	1	187					New Ireland	58
Lorius domicella	Purple-naped Lory	U	4	182		150	205		Seram, Moluccas	1181
Lorius lory	Black-capped Lory	M	6	183		148	210		New Guinea	281
		F	6	171		158	189			
Lorius albidinucha	White-naped Lory	F	4	132		120	146		New Ireland	58
Lorius chlorocercus	Yellow-bibbed Lory	M	11	187		168	225		Solomon Is.	399
		F	4	161		153	174			
Phigys solitarius	Collared Lory	M	2	91.1		90.2	92		Fiji	1196c
		F	3	76.1		71.2	82			
Vini stepheni	Stephen's Lorikeet	M	1	55					Henderson Is.	446
		F	4			42	51			
Vini peruviana	Blue Lorikeet	F	2	29		28.8	29.2		Cook Is.	1153b
Glossopsitta concinna concinna	Musk Lorikeet	B	21	76.1		57	90		Southeast Australia	516

Body Masses of World Birds (continued)

Species Name	Common Name	Sex	N	Mean	S.D.	Min	Max	Season	Location	Source #
Glossopsitta concinna didimus	Musk Lorikeet	B	12	65.4		58	73		Tasmania	516
Glossopsitta pusilla	Little Lorikeet	B	14	39.4		30	53		Eastern Australia	516
Glossopsitta porphyrocephala	Purple-crowned Lorikeet	B	11	44.7		39	50		Australia	516
Charmosyna rubrigularis	Red-chinned Lorikeet	B	22	37		33	40		New Guinea	292
Charmosyna meeki	Meek's Lorikeet	B	9	23.8		21	25.5		Solomon Is.	399
Charmosyna placentis	Red-flanked Lorikeet	B	17	33.6	4.2	26.5	42		New Guinea	283
Charmosyna margarethae	Duchess Lorikeet	M	6	55.7		52.5	55.7		Solomon Is.	399
		F	2	43.8		43	44.5			
Charmosyna pulchella	Fairy Lorikeet	B	6	34.1					New Guinea	61, 283
Charmosyna josefinae	Josephine's Lorikeet	M	1	74					New Guinea	1143
Charmosyna papou	Papuan Lorikeet	M	6	87.2		74	95		New Guinea	424
		F	4	78.8		77	84			
Oreopsittacus arfaki	Plum-faced Lorikeet	B	13	20.2	1.9	17.8	22.5		New Guinea	283, 424
Neopsittacus musschenbroekii	Yellow-billed Lorikeet	B	13	50.9		49	62		New Guinea	283
Neopsittacus pullicauda	Orange-billed Lorikeet	B	10	33.2	4.9	26.5	44		New Guinea	424, 428, 770
Psittrichas fulgidus	Pesquet's Parrot	U		865					New Guinea	291
Nestor notabilis	Kea	M	91	956	172				New Zealand	92
		F	38	779	136					
Nestor meridionalis	New Zealand Kaka	M	18	458	44	402	555		New Zealand	516
		F	21	400	26.6	358	450			
Strigops habroptila	Kakapo	M	28	2000	410	1600	3000		New Zealand	937
		F	28	1500	140	1300	1900			
Micropsitta keiensis	Yellow-capped Pygmy-Parrot	B	12	12.1	1.2	10	14		New Guinea	455, 535

Body Masses of World Birds (continued)

Species Name	Common Name	Sex	N	Mean	S.D.	Min	Max	Season	Location	Source #
Micropsitta geelvinkiana	Geelvink Pygmy-Parrot	B	2	12.2		12	12.5		New Guinea	1143
Micropsitta pusio	Buff-faced Pygmy-Parrot	B	6	14		13	15		New Guinea	292
Micropsitta bruijnii	Red-breasted Pygmy-Parrot	M	3	14.8		13.5	15.5		New Britain	427
		F	2	12.2		11.5	13			
Micropsitta finschii aolae	Finsch's Pygmy-Parrot	B	6	15.7		14.5	17.5		Guadalcanal	399
Micropsitta finschii finschii	Finsch's Pygmy-Parrot	B	13	16.4		14	18		Solomon Is.	399
Micropsitta finschii nanina	Finsch's Pygmy-Parrot	B	14	13	1.1	11.1	15		Solomon Is.	658
Cyclopsitta gulielmitertii	Orange-breasted Fig-Parrot	B	3	31		27	33		New Guinea	283
Cyclopsitta diophthalma	Double-eyed Fig-Parrot	B	5	48.9		44.5	56		New Guinea	283, 770
Cyclopsitta diophthalma marshalli	Double-eyed Fig-Parrot	B	12	35.8		25	42		Australia	516
Psittaculirostris desmarestii	Large Fig-Parrot	B	3	117		108	126		New Guinea	283
Psittaculirostris salvadorii	Salvadori's Fig-Parrot	U	3	118					captive	1121
Bolbopsittacus lunulatus	Guaiabero	B	10	67.6		60	80		Philippines	436
Prosopeia tabuensis	Red Shining-Parrot	U	3	232		205	280		Tonga; Fiji	978, 1196c
Eunymphicus cornutus uvaeensis	Horned Parakeet	U		130						1004
Cyanoramphus unicolor	Antipodes Parakeet	M	13	167	18.5	134	194		Antipodes Is.	516
		F	40	133	12.4	102	162			
Cyanoramphus novaezelandiae	Red-fronted Parakeet	M	181	82.1	8.8	63	113		New Zealand	516
		F	64	67.9	7.5	50	90			
Cyanoramphus novaezelandiae hochstetteri	Red-fronted Parakeet	M	1	125					Antipodes Is.	452
		F	3	95.1		88.8	102			

Body Masses of World Birds (continued)

Species Name	Common Name	Sex	N	Mean	S.D.	Min	Max	Season	Location	Source #
Cyanoramphus auriceps	Yellow-fronted Parakeet	M	55	51.2	4.1	42	58.5		New Zealand	516
		F	11	40.7	3.8	34	45			
Cyanoramphus malherbi	Malherbe's Parakeet	B	18	46.7		37	55		captive	516
Purpureicephalus spurius	Red-capped Parakeet	B	76	116		90	131		Western Australia	516
Barnardius barnardi zonarius	Port Lincoln Parrot	M	94	140	10.4				Western Australia	516
		F	11	126	14.8	108	150			
Barnardius barnardi	Mallee Ringneck	M	51	145	9.7				Australia	516
		F	2	122		102	141			
Platycercus caledonicus	Green Rosella	M	19	149	15.9	130	198		Tasmania	516
		F	13	118	10.7	106	148			
Platycercus elegans elegans	Crimson Rosella	M	28	140	12.5	115	170		Southeast Australia	516
		F	23	125	15.5	99	170			
Platycercus elegans flaveolus	Yellow Rosella	M	10	127	8.1	112	140		Eastern Australia	516
		F	8	102		75	113			
Platycercus elegans adelaidae	Adelaide Rosella	M	16	135	10.3	120	157		South Australia	516
		F	15	118	10.9	102	135			
Platycercus venustus	Northern Rosella	M	10	92.6	5.1	85	100		Northern Australia	516
		F	9	86.4		77	100			
Platycercus eximius	Eastern Rosella	B	23	104		75	125		Tasmania	516
Platycercus adscitus	Pale-headed Rosella	B	22	111		95	175		Southeast Australia	516
Platycercus icterotis	Western Rosella	M	63	65.4	6.7	42	80		Western Australia	704
		F	52	61.2	6.6	28	71			
Psephotus varius	Mulga Parrot	B	44	61.5		50	71		Australia	516
Psephotus haematonotus	Red-rumped Parrot	B	42	61.4		45	77		Eastern Australia	516
Psephotus dissimilis	Hooded Parrot	B	11	45.8		36.1	56		captive	516
Psephotus chrysopterygius	Golden-shouldered Parrot	B	5	46.8		28	56.1			470, 516

Body Masses of World Birds (continued)

Species Name	Common Name	Sex	N	Mean	S.D.	Min	Max	Season	Location	Source #
Northiella haematogaster haematogaster	Bluebonnet	M	27	93.7	9	65	105		Australia	516
		F	13	79.2	6.6	68	90			
Northiella haematogaster haematorrhous	Bluebonnet	B	17	85.5		72	100		Southeast Australia	516
Neophema bourkii	Bourke's Parrot	B	12	44.4		38	51		Australia	516
Neophema chrysostoma	Blue-winged Parrot	B	17	46.1		33	58		Australia	516
Neophema elegans	Elegant Parrot	B	6	43.4		40	51		Australia	516
Neophema petrophila	Rock Parrot	B	8	53.2		48	56		Australia	516
Neophema chrysogaster	Orange-bellied Parrot	U	18	45.9		41	51		Tasmania	516
Neophema pulchella	Turquoise Parrot	M	12	42.7	2.5	37	46.1		Victoria, Australia	516
		F	18	40.1	2.1	34.9	44.3			
Neophema splendida	Scarlet-chested Parrot	B	4	38.6		37	40.2		Western Australia	516
Lathamus discolor	Swift Parrot	B	51	64.7		45	76		Australia	516
Melopsittacus undulatus	Budgerigar	M	173	29.2	1.9				Eastern Australia	1343
		F	164	28.8	2.1					
Pezoporus wallicus	Ground Parrot	M	6	75.7		68	83		Eastern Australia	516
		F	6	67		59	75			
Psittacella picta	Painted Tiger-Parrot	U	3	62.9		56.6	68		New Guinea	387, 770
Psittacella brehmii	Brehm's Tiger-Parrot	B	8	107	10	87.5	118		Papua New Guinea	387
Psittacella modesta	Modest Tiger-Parrot	U	1	44.7					Papua New Guinea	387
Psittacella madaraszi	Madarasz's Tiger-Parrot	M	2	42.2		41.5	43		New Guinea	283
		F	2	35.7		35.3	36			
Geoffroyus geoffroyi	Red-cheeked Parrot	B	15	164					New Guinea; Moluccas	61, 979
Geoffroyus simplex	Blue-collared Parrot	U	1	161					New Guinea	61

Body Masses of World Birds (continued)

Species Name	Common Name	Sex	N	Mean	S.D.	Min	Max	Season	Location	Source #
Geoffroyus heteroclitus	Singing Parrot	B	5	170		154	180		New Britain; Solomon Is.	426, 658
Prioniturus montanus	Luzon Racquet-tail	M	9	126		116	142		Philippines	954
		F	3	111		102	120			
Prioniturus waterstradti	Mindanao Racquet-tail	U	2	119		97.5	140		Mindanao, Philippines	1143
Prioniturus platenae	Blue-headed Racquet-tail	M	2	112		104	119		Palawan, Philippines	1143
Prioniturus luconensis	Green Racquet-tail	B	13	108	11.2	91.3	127		Luzon, Philippines	1143
Prioniturus discurus	Blue-crowned Racquet-tail	M	5	137		127	149		Philippines	954
		F	3	160		136	176			
Prioniturus platurus	Golden-mantled Racquet-tail	F	1	66.3					captive	1143
Tanygnathus megalorynchos	Great-billed Parrot	F	1	325					New Guinea	1143
Tanygnathus lucionensis	Blue-naped Parrot	M	4	215		199	231		Philippines	954
Tanygnathus sumatranus	Azure-rumped Parrot	M	2	248		230	267		Philippines	954
		F	2	293		252	334			
Eclectus roratus	Eclectus Parrot	B	16	561		480	615		Cape York Peninsula, Australia	516
Eclectus roratus	Eclectus Parrot	M	4	412		395	440		Solomon Is.	658
Alisterus scapularis	Australian King-Parrot	M	11	222	19.8	195	250		Eastern Australia	516
		F	11	243	23.1	213	285			
Alisterus amboinensis	Moluccan King-Parrot	B	3	153		145	163		New Guinea	535
Alisterus chloropterus	Papuan King-Parrot	U	10	156					New Guinea	61
Aprosmictus jonquillaceus	Olive-shouldered Parrot	F	2	119		119	119		captive	1143
Aprosmictus erythropterus	Red-winged Parrot	B	32	136		112	164		Australia	516

Body Masses of World Birds (continued)

Species Name	Common Name	Sex	N	Mean	S.D.	Min	Max	Season	Location	Source #
Polytelis swainsonii	Superb Parrot	B	25	153		134	164		Australia	516
Polytelis anthopeplus	Regent Parrot	B	117	170		127	200		Australia	516
Polytelis alexandrae	Alexandra's Parrot	M F	3 3	96 112		92 107	100 120		captive	516
Psittacula eupatria	Alexandrine Parakeet	F	3	214		198	225		India	7, 740
Psittacula krameri	Rose-ringed Parakeet	M F	2 4	126 107		119 95.8	134 120			183, 740, 1319
Psittacula echo	Mauritius Parakeet	F	1	163						183
Psittacula finschii	Gray-headed Parakeet	M	1	89.6					captive	1143
Psittacula cyanocephala	Plum-headed Parakeet	B	12	66		56	71.5			7
Psittacula columboides	Malabar Parakeet	3	87			73	98		India; captive	1148
Psittacula derbiana	Derbyan Parakeet	F	2	251		251	251		captive	1143
Psittacula alexandri	Red-breasted Parakeet	B	7	147		133	168		China	184a
Psittacula longicauda	Long-tailed Parakeet	B	3	107		96.6	123		captive	1143
Loriculus vernalis	Vernal Hanging-Parrot	F	1	35.7					captive	1143
Loriculus beryllinus	Ceylon Hanging-Parrot	U		30					captive	1117
Loriculus philippensis	Philippine Hanging-Parrot	B	19	34.9		30.7	39.5		Philippines	954
Loriculus galgulus	Blue-crowned Hanging-Parrot	U	1	28						1287
Loriculus aurantiifrons	Papuan Hanging-Parrot	B	4	15.5		15	16		New Guinea	292
Loriculus tener	Green-fronted Hanging-Parrot	U	1	12					New Britain	426

Body Masses of World Birds (continued)

Species Name	Common Name	Sex	N	Mean	S.D.	Min	Max	Season	Location	Source #
Agapornis canus	Gray-headed Lovebird	U		30		28	31			133
Agapornis pullarius	Red-headed Lovebird	B	5	39.3		29.5	50			266, 394
Agapornis taranta	Black-winged Lovebird	M		59		53	66			394
		F		56		49	64			
Agapornis swindernianus	Black-collared Lovebird	M	4			39	41			394
Agapornis roseicollis	Rosy-faced Lovebird	B	29			46	63			394
Agapornis fisheri	Fischer's Lovebird	U	4	48.3		43	53			133
Agapornis personatus	Yellow-collared Lovebird	M	8	49						394
		F	9	56						
Agapornis lilianae	Lilian's Lovebird	B	4	37.3		29	40			394
Coracopsis vasa	Vasa Parrot	U	4	525		500	560		Madagascar	957
Coracopsis nigra	Black Parrot	U	5	141		132	153		Seychelles	409, 790
Coracopsis nigra	Black Parrot	U	5	246		215	293		Madagascar	957
Psittacus erithacus	Gray Parrot	U	15	333	20				captive	371
Poicephalus robustus suahelicus	Brown-necked Parrot	M	3	370		310	401			394
		F	3	318		280	364			
Poicephalus gulielmi	Red-fronted Parrot	M	1	290						394
		F	1	260						
Poicephalus meyeri transvaalensis	Meyer's Parrot	M	18	120		102	134			394
		F	10	115		100	123			
Poicephalus rueppellii	Rueppell's Parrot	B	5	116		105	132			394
Poicephalus cryptoxanthus	Brown-headed Parrot	B	13	140		121	156			394
Poicephalus rufiventris	Red-bellied Parrot	M	1	120					Kenya	120
Poicephalus senegalus	Senegal Parrot	B	5	147		120	161			394

Body Masses of World Birds (continued)

Species Name	Common Name	Sex	N	Mean	S.D.	Min	Max	Season	Location	Source #
Poicephalus flavifrons	Yellow-fronted Parrot	B	21	174		140	205			394
Anodorhynchus hyacinthinus	Hyacinth Macaw	U	13	1331	149				captive	171
Anodorhynchus leari	Lear's Macaw	U		940					Brazil	1105
Cyanopsitta spixii	Spix's Macaw	F	1	200					estimated, captive	1118
Ara ararauna	Blue-and-yellow Macaw	U		1125					Peru	1210
Ara glaucogularis	Blue-throated Macaw	U	15	784	63				captive	171
Ara militaris	Military Macaw	M	1	1134					Mexico	1151
Ara ambiguus	Great Green Macaw	U		1300					Costa Rica	1170
Ara macao	Scarlet Macaw	U		1015					Peru	1210
Ara chloropterus	Red-and-green Macaw	U	11	1214	173				captive	171
Ara rubrogenys	Red-fronted Macaw	U	12	468	45				captive	171
Ara severus	Chestnut-fronted Macaw	B	7	343		307	387		Suriname	498
Orthopsittaca manilata	Red-bellied Macaw	B	8	372		360	390		Suriname	498
Primolius couloni	Blue-headed Macaw	B	2	250		207	294		Peru	370
Primolius maracana	Blue-winged Macaw	B	2	256		246	266		Paraguay	1175
Primolius auricollis	Golden-collared Macaw	U	3	245		240	250			362
Diopsittaca nobilis	Red-shouldered Macaw	M	5	156		143	169		Suriname	498
		F	9	146		129	162			
Rhynchopsitta pachyrhyncha	Thick-billed Parrot	M	56	337					captive	1132
		F	60	313						

Body Masses of World Birds (continued)

Species Name	Common Name	Sex	N	Mean	S.D.	Min	Max	Season	Location	Source #
Rhynchopsitta terrisi	Maroon-fronted Parrot	B	3	442		392	468		Coahuila, Mexico	1243
Ognorhynchus icterotis	Yellow-eared Parakeet	U	1	285					captive	1119
Guarouba guarouba	Golden Parakeet	B	2	194		192	196		captive	1148
Conuropsis carolinensis	Carolina Parakeet	U	1	100						1133
Aratinga acuticaudata	Blue-crowned Parakeet	U	8	171		168	176			362, 1055, 1143
Aratinga holochlora	Green Parakeet	B	6	138		96.2	168		Chiapas, Mexico; captive	1143
Aratinga strenua	Pacific Parakeet	B	4	208		163	237		Nicaragua	1143
Aratinga brevipes	Socorro Parakeet	U	8	177		168	195		captive	217b
Aratinga rubritorquis	Red-throated Parakeet	M	1	212					Nicaragua	1143
Aratinga wagleri	Scarlet-fronted Parakeet	U	7	194		162	217			200, 795, 798, 1042
Aratinga mitrata	Mitred Parakeet	U	6	249		219	275			362, 1040, 1055
Aratinga erythrogenys	Red-masked Parakeet	M	1	151					captive	1143
Aratinga finschi	Crimson-fronted Parakeet	U		150					Costa Rica	1170
Aratinga leucophthalma	White-eyed Parakeet	B	9	158		100	218		Suriname; Peru; Brazil	177, 293, 1175
Aratinga chloroptera	Hispaniolan Parakeet	B	2	146		144	147		Dominican Republic	1143
Aratinga solstitialis	Sun Parakeet	B	15			106	135		captive	269
Aratinga pintoi	Sulphur-breasted Parakeet	F	1	110					Brazil	1110
Aratinga jandaya	Jandaya Parakeet	F	2	89.4		87.7	91		captive	1143
Aratinga auricapilla	Golden-capped Parakeet	M	1	130					Brazil	177

Body Masses of World Birds (continued)

Species Name	Common Name	Sex	N	Mean	S.D.	Min	Max	Season	Location	Source #
Aratinga weddellii	Dusky-headed Parakeet	B	6	108		96	129		Peru; Brazil	370, 1178
Aratinga pertinax	Brown-throated Parakeet	B	16	84		76	102		Suriname	498
Aratinga nana	Olive-throated Parakeet	B	5	76.9		72.4	84		Mexico; Jamaica	231, 640, 1027
Aratinga canicularis	Orange-fronted Parakeet	B	13	85					Guatemala	612
Aratinga aurea	Peach-fronted Parakeet	M	8	86.5					Brazil	1109
		F	5	82.8						
Nandayus nenday	Nanday Parakeet	B	4	128		120	141		Paraguay	1175
Cyanoliseus patagonus	Burrowing Parrot	M	22	286	25.8	253	340	B	Argentina	766a
		F	24	270	22.5	227	304			
Pyrrhura frontalis	Maroon-bellied Parakeet	M	3			82.5	94		Brazil	63
		F	2	72		72	72			
Pyrrhura perlata	Crimson-bellied Parakeet	M	2	75		70	80		Brazil	449
Pyrrhura molinae	Green-cheeked Parakeet	U	11	77.1		62	81		Argentina	362, 1040
Pyrrhura picta	Painted Parakeet	B	22	62.1		52	72		Suriname	498
Pyrrhura egregia	Fiery-shouldered Parakeet	M	3	66.3		66	66.6			917, 1148
Pyrrhura melanura	Maroon-tailed Parakeet	U	5	67.9		66	70.5		Venezuela	1308
Pyrrhura melanura melanura	Maroon-tailed Parakeet	U	9	73.3		64	79			973
Pyrrhura orcesi	El Oro Parakeet	U	14	70.8		65	73.5		Ecuador	973
Pyrrhura rupicola	Black-capped Parakeet	U		75					Peru	1210
Pyrrhura albipectus	White-necked Parakeet	M	1	110					Ecuador	953
		F	1	98						

Body Masses of World Birds (continued)

Species Name	Common Name	Sex	N	Mean	S.D.	Min	Max	Season	Location	Source #
Pyrrhura hoematotis	Red-eared Parakeet	B	9	68.7		62	74.2			200, 1042
Pyrrhura hoffmanni	Sulphur-winged Parakeet	B	34	82.2					Panama	489
Enicognathus ferrugineus	Austral Parakeet	B	5	160		140	188		Argentina	186a, 210
Enicognathus leptorhynchus	Slender-billed Parakeet	M	1	193					captive	1143
Myiopsitta monachus	Monk Parakeet	B	158	120				Y	Argentina	16
Bolborhynchus orbygnesius	Andean Parakeet	B	5	43.1		35.8	50			871, 1042
Bolborhynchus lineola	Barred Parakeet	B	8	56.2		45.2	59.4		Colombia; Peru; Panama	489, 798, 871, 1288
Psilopsiagon aymara	Gray-hooded Parakeet	U	2	30		29	31		Bolivia	1055
Psilopsiagon aurifrons	Mountain Parakeet	B	3	33.6		32.7	34.1			1042
Forpus cyanopygius	Mexican Parrotlet	B	13	33.5		30	36.6		Mexico	440, 1151
Forpus passerinus	Green-rumped Parrotlet	B	28	23		20	28		Suriname	498
Forpus xanthopterygius	Blue-winged Parrotlet	B	2	31		30.5	31.5			871
Forpus conspicillatus	Spectacled Parrotlet	U	7	26.4		24	28.2		Colombia	795, 798
Forpus sclateri	Dusky-billed Parrotlet	B	2	27.2		26.5	28			871
Forpus coelestis	Pacific Parrotlet	M	2	27.9					Peru	1302
		F	2	24.5						
Brotogeris tirica	Plain Parakeet	F	1	63					Brazil	1202
Brotogeris versicolurus	Canary-winged Parakeet	M	15	74.6		64.9	80.8		Florida, USA	116
		F	10	68.6		61.7	74.3			

Body Masses of World Birds (continued)

Species Name	Common Name	Sex	N	Mean	S.D.	Min	Max	Season	Location	Source #
Brotogeris chiriri	Yellow-chevroned Parakeet	B	5	61.6		52.5	68.3		Bolivia; California, USA	116
Brotogeris pyrrhopterus	Gray-cheeked Parakeet	M	1	68					Peru	1302
		F	2	60						
Brotogeris jugularis	Orange-chinned Parakeet	M	14	61	4.4				Panama	489
		F	9	65.5						
Brotogeris cyanoptera	Cobalt-winged Parakeet	B	2	56		55	57			871
Brotogeris sanctithomae	Tui Parakeet	U	16	59	3.5				Peru	370
Brotogeris chrysopterus	Golden-winged Parakeet	B	20	54.5		49	69		Suriname	498
Nannopsittaca panychlora	Tepui Parrotlet	U	7	44.1		42	46.5		Venezuela; Peru	871, 1309
Nannopsittaca dachilleae	Amazonian Parrotlet	B	18	41.5		37.5	46			871
Touit batavicus	Lilac-tailed Parrotlet	U	9	55		52	59.5		Trinidad	359
Touit huetii	Scarlet-shouldered Parrotlet	B	2	60		58	62			871
Touit dilectissimus	Blue-fronted Parrotlet	M	2	65		59	71		Colombia	521
Touit purpuratus	Sapphire-rumped Parrotlet	U	6	59.7		52.6	62.6		Brazil	875
Touit stictopterus	Spot-winged Parrotlet	M	3	78.8		70.5	84		Peru	259
Pionites melanocephala	Black-headed Parrot	M	11	157		142	179		Suriname	498
		F	8	141		130	157			
Pionites leucogaster	White-bellied Parrot	U		155					Peru	1210
Pionopsitta haematotis	Brown-hooded Parrot	B	12	149					Mexico; Panama	489, 910a, 990
Pionopsitta pulchra	Rose-faced Parrot	B	4	150		135	176		Ecuador	1143

Body Masses of World Birds (continued)

Species Name	Common Name	Sex	N	Mean	S.D.	Min	Max	Season	Location	Source #
Pionopsitta barrabandi	Orange-cheeked Parrot	U		140					Peru	1210
Pionopsitta caica	Caica Parrot	B	3	132		121	143		Suriname	498
Pionopsitta pileata	Pileated Parrot	B	6	119		95	135		captive	1120
Hapalopsittaca melanotis	Black-winged Parrot	M	2	158					Bolivia	963
Hapalopsittaca amazonina	Rusty-faced Parrot	M	2	106		97	115			447, 900
Hapalopsittaca pyrrhops	Red-faced Parrot	U	2	103		100	106		Ecuador	1147
Graydidascalus brachyurus	Short-tailed Parrot	B	2	159		158	160		captive	1122
Pionus menstruus	Blue-headed Parrot	B	12	251		209	295		Suriname	498
Pionus sordidus	Red-billed Parrot	B	2	266		259	272			200, 1042
Pionus maximiliani	Scaly-headed Parrot	M	1	293					Brazil	63
Pionus senilis	White-crowned Parrot	B	11	212					Panama	489
Pionus tumultuosus	Speckle-faced Parrot	B	2	239		228	250			1042
Pionus chalcopterus	Bronze-winged Parrot	M	1	210					Peru	1302
Pionus fuscus	Dusky Parrot	B	14	206		179	229		Suriname	498
Amazona leucocephala	Cuban Parrot	B	2	227		222	232		Cayman Is.	1147a
Amazona leucocephala	Cuban Parrot	B	18	244	21.3	180	290		Cuba	19
Amazona collaria	Yellow-billed Parrot	M	1	187					Jamaica	1148
Amazona ventralis	Hispaniolan Parrot	F	1	214					captive	1143
Amazona vittata	Puerto Rican Parrot	U				250	312		Puerto Rico	1136

Body Masses of World Birds (continued)

Species Name	Common Name	Sex	N	Mean	S.D.	Min	Max	Season	Location	Source #
Amazona xantholora	Yellow-lored Parrot	B	10	217	12.4	200	232		Yucatan, Mexico	640, 910a
Amazona albifrons	White-fronted Parrot	B	16	206	17.4	176	242		Mexico; Belize	640, 910a, 1027
Amazona agilis	Black-billed Parrot	M	3	213		203	220		Jamaica	1148
Amazona tucumana	Tucuman Parrot	U	5	260		250	280			362
Amazona pretrei	Red-spectacled Parrot	U				240	310		captive	712
Amazona viridigenalis	Red-crowned Parrot	U	13	316		293	345		captive	338
Amazona finschi	Lilac-crowned Parrot	B	3	302		282	312		Western Mexico	1150, 1151
Amazona autumnalis	Red-lored Parrot	B	16	416					Panama	489
Amazona dufresniana	Blue-cheeked Parrot	B	5	623		481	615		Suriname	498
Amazona rhodocorytha	Red-browed Parrot	U				450	500		captive	743
Amazona festiva	Festive Parrot	M	1	430					captive	1143
Amazona barbadensis	Yellow-shouldered Parrot	F	1	270					Bonaire	19
Amazona aestiva	Blue-fronted Parrot	U	7	451		402	500			362, 1055
Amazona oratrix	Yellow-headed Parrot	M	1	517					Oaxaca, Mexico	75
Amazona ochrocephala	Yellow-crowned Parrot	B	8	440		365	530		Brazil; Suriname	498, 861, 1109
Amazona amazonica	Orange-winged Parrot	B	10	370		298	469		Suriname	498
Amazona mercenaria	Scaly-naped Parrot	F	1	340					Peru	1288
Amazona farinosa	Mealy Parrot	B	13	626		525	777		Suriname	498
Amazona vinacea	Vinaceous Parrot	B	4	254		198	282		Paraguay; captive	1143

Body Masses of World Birds (continued)

Species Name	Common Name	Sex	N	Mean	S.D.	Min	Max	Season	Location	Source #
Amazona versicolor	St. Lucia Parrot	B	6	676		505	1000		captive, wild born in St. Lucia	347
Amazona guildingii	St. Vincent Parrot	B	20	646	110	485	880		captive	574
Deroptyus accipitrinus	Red-fan Parrot	B	14	246		190	277		Suriname	498
Triclaria malachitacea	Blue-bellied Parrot	U		90					Brazil	1240
Family Musophagidae										
Corythaeola cristata	Great Blue Turaco	B	10	965		857	1231			394
Tauraco persa	Guinea Turaco	B	6	306		280	350			266, 394
Tauraco livingstonii	Livingstone's Turaco	F	1	249					Kenya	120
Tauraco schalowi	Schalow's Turaco	B	70			142	300		Angola	266
Tauraco corythaix	Knysna Turaco	U	11	308		260	375		South Africa	1036
Tauraco schuettii	Black-billed Turaco	B	16	235		199	272			394
Tauraco leucolophus	White-crested Turaco	B	5	219		198	226			394
Tauraco fischeri	Fischer's Turaco	B	11	250		227	283			394
Tauraco macrorhynchus	Yellow-billed Turaco	M	4	264		250	272			266, 394
		F	9	225						
Tauraco bannermani	Bannerman's Turaco	U	4	224		200	250			394
Tauraco erythrolophus	Red-crested Turaco	B	16			210	325		Angola	266
Tauraco hartlaubi	Hartlaub's Turaco	B	7	224		200	255			120
Tauraco leucotis	White-cheeked Turaco	U	15	265		200	315			394
Tauraco ruspolii	Prince Ruspoli's Turaco	U	5	263		200	290			394
Tauraco porphyreolophus	Purple-crested Turaco	U	13	259		216	298		South Africa	1036

Body Masses of World Birds (continued)

Species Name	Common Name	Sex	N	Mean	S.D.	Min	Max	Season	Location	Source #
Ruwenzorornis johnstoni	Ruwenzori Turaco	B	3	240		232	247			394
Musophaga violacea	Violet Turaco	U	1	360						394
Musophaga rossae	Ross' Turaco	B	10			275	475		Angola	266
Corythaixoides personatus	Bare-faced Go-away-bird	U	6	250		210	300			394
Corythaixoides concolor	Gray Go-away-bird	U	139	258		147	350		South Africa	1036
Corythaixoides leucogaster	White-bellied Go-away-bird	M	2	198		170	225			394
		F	3	236		225	250			
Crinifer piscator	Western Plantain-eater	U	5	370					Ivory Coast	1212
Crinifer zonurus	Eastern Plantain-eater	U	8	527		392	737			120, 394, 1319
Family Cuculidae										
Clamator jacobinus	Pied Cuckoo	B	30			63	100		Angola	266
Clamator levaillantii	Levaillant's Cuckoo	B	13	122		102	141			394
Clamator coromandus	Chestnut-winged Cuckoo	U	44			66.4	83.8			1287
Clamator glandarius	Great Spotted Cuckoo	M	6	169		153	192	B	Spain	223
		F	1	138						
Clamator glandarius	Great Spotted Cuckoo	M	10	139		128	168		South Africa	908
		F	9	129		118	148			
Pachycoccyx audeberti	Thick-billed Cuckoo	M	1	92						394
		F	4	115		100	120			
Cuculus sparverioides	Large Hawk-Cuckoo	M	3	151		140	160			908
Cuculus bocki	Dark Hawk-Cuckoo	U	2	130		130	131			1287
Cuculus varius	Common Hawk-Cuckoo	B	5	103		93	110			740, 908

Body Masses of World Birds (continued)

Species Name	Common Name	Sex	N	Mean	S.D.	Min	Max	Season	Location	Source #
Cuculus vagans	Moustached Hawk-Cuckoo	B	3	57.7		53	62			908, 1287
Cuculus fugax	Javan Hawk-Cuckoo	B	6	79.7		69.1	93		Malaysia	1218, 1287
Cuculus hyperythrus	Rufous Hawk-Cuckoo	U	3	121		99	148		Japan	908
Cuculus nisicolor	Whistling Hawk-Cuckoo	U				69.1	93	M	Malay Peninsula	908
Cuculus pectoralis	Philippine Hawk-Cuckoo	B	23	77.1		67.3	89.2		Philippines	908
Cuculus solitarius	Red-chested Cuckoo	B	21			64	92		Angola	266
Cuculus clamosus	Black Cuckoo	B	11			69	90		Angola	266
Cuculus micropterus micropterus	Indian Cuckoo	B	7	119		112	129		Russia	849
Cuculus micropterus micropterus	Indian Cuckoo	U	17			77.5	122		Malay Peninsula	1287
Cuculus micropterus concretus	Indian Cuckoo	U	2	66.2		63	69.3			1287
Cuculus canorus	Common Cuckoo	M	84	117	17			B	United Kingdom	223
		F	52	106	14					
Cuculus gularis	African Cuckoo	B	11	103		95	115			266, 394
Cuculus saturatus	Himalayan Cuckoo	M	12	89.5		73.2	103		Taiwan, Philippines	908
		F	5	72.6		65	87			
Cuculus optatus	Oriental Cuckoo	M	13	99		76.4	118		Philippines	908
		F	11	89.1		71.2	136			
Cuculus poliocephalus	Asian Lesser Cuckoo	M	3	54.2		48	59			394
		F	1	40						
Cuculus rochii	Madagascar Lesser Cuckoo	M	3	62.7		59	65			394, 957
Cuculus pallidus	Pallid Cuckoo	B	40	87.7		63	118		Australia	516
Cercococcyx mechowi	Dusky Long-tailed Cuckoo	B	8	55.6		52	62			908

Body Masses of World Birds (continued)

Species Name	Common Name	Sex	N	Mean	S.D.	Min	Max	Season	Location	Source #
Cercococcyx olivinus	Olive Long-tailed Cuckoo	M	4	65.9		64	68.5		Angola	266, 394
Cercococcyx montanus	Barred Long-tailed Cuckoo	B	12	58.2		52	64			908
Cacomantis sonneratii	Banded Bay Cuckoo	U	7	33.8		27.9	38			56
Cacomantis merulinus	Plaintive Cuckoo	B	8	25.3		21.3	27.2		Borneo; Malaysia; Philippines	908
Cacomantis variolosus dumetorum	Brush Cuckoo	B	29	33.3		21.8	38		Australia	516
Cacomantis variolosus sepulcralis	Brush Cuckoo	B	14	33.4		24.4	39.5		Philippines	954
Cacomantis variolosus	Brush Cuckoo	B	14	41.1		35.5	45.5		Solomon Is.	399
Cacomantis castaneiventris	Chestnut-breasted Cuckoo	B	14	33.2		25	38		New Guinea; Australia	283, 516, 424
Cacomantis flabelliformis	Fan-tailed Cuckoo	B	39	49.8		42.6	65		Australia	516
Chrysococcyx osculans	Black-eared Cuckoo	U	15	30.9	3.8	25.5	38.5	W	Australia	516
Chrysococcyx basalis	Horsfield's Bronze-Cuckoo	M	22	22	1.2	17	30		Australia	516
		F	16	24.3	3.2	20	32			
Chrysococcyx lucidus lucidus	Shining Bronze-Cuckoo	B	28	25.1		16.7	32		New Zealand	516
Chrysococcyx lucidus plagosus	Shining Bronze-Cuckoo	B	68	23.8		17	35		Australia	516
Chrysococcyx ruficollis	Rufous-throated Bronze-Cuckoo	M	1	21						908
Chrysococcyx meyeri	White-eared Bronze-Cuckoo	M	6	17.8		15	19.5			908
		F	6	21.2		19	27.4			
Chrysococcyx minutillus	Little Bronze-Cuckoo	B	9	18.6		16.9	21			908
Chrysococcyx russatus	Gould's Bronze-Cuckoo	B	11	18.6		16.9	21		Northern Australia	516

Body Masses of World Birds (continued)

Species Name	Common Name	Sex	N	Mean	S.D.	Min	Max	Season	Location	Source #
Chrysococcyx maculatus	Asian Emerald Cuckoo	M	3	25		23	27			908
		F	1	30						
Chrysococcyx xanthorhynchus	Violet Cuckoo	M	10			19.9	24			1287
		F	7			20.5	25.6			
Chrysococcyx flavigularis	Yellow-throated Cuckoo	B	8			27.5	31			207
Chrysococcyx klaas	Klaas' Cuckoo	M	30	26.2		22.7	32.8		Angola	266
		F	38	28.6		24.7	35.1			
Chrysococcyx cupreus	African Emerald Cuckoo	B	32	37.7		30	46			394
Chrysococcyx caprius	Dideric Cuckoo	M	102	29.3		20.5	37.1		Malawi	908
		F	64	36.7		28.3	44.3			
Rhamphomantis megarhynchus	Long-billed Cuckoo	M	1	31.5						908
Surniculus dicruroides	Fork-tailed Drongo-Cuckoo	U	41			32.5	43.3	M	Malay Peninsula	908
Surniculus lugubris	Square-tailed Drongo-Cuckoo	M	6	29.7		26	31			908
Surniculus velutinus	Philippine Drongo-Cuckoo	M	12	33.2		31.4	42		Philippines	908
		F	7	38.2		36	47			
Caliechthrus leucolophus	White-crowned Koel	M	3	116		110	125		New Guinea	283, 535, 908
		F	1	99						
Microdynamis parva	Dwarf Koel	M	4	43.3		39.3	54			908
		F	2	51.5		49	54			
Eudynamys scolopaceus melanorhynchus	Asian (Black-billed) Koel	M	9	202		133	231			908
		F	4	217		192	244			
Eudynamys scolopaceus	Asian Koel	B	10	233	54.3	133	327		Philippines; New Guinea	426, 954
Eudynamys scolopaceus	Asian Koel	B	10	167		136	190		India	7
Eudynamys scolopaceus alberti	Asian Koel	B	6	165		143	190			399, 769
Eudynamys cyanocephala cyanocephala	Australian Koel	B	39	259		167	340		Eastern Australia	516

Body Masses of World Birds (continued)

Species Name	Common Name	Sex	N	Mean	S.D.	Min	Max	Season	Location	Source #
Eudynamys cyanocephala subcyanocephala	Australian Koel	B	21	227		158	267		Australia	516
Eudynamys taitensis	Long-tailed Koel	B	19	117		74	153		New Zealand	516
Scythrops novaehollandiae	Channel-billed Cuckoo	B	24	684		560	935		Australia	516
Ceuthmochares aereus	Chattering Yellowbill	B	20	65.8		54	80			908
Ceuthmochares australis	Whistling Yellowbill	B	25	62		52	72		Eastern Africa	908
Phaenicophaeus diardi	Black-bellied Malkoha	B	3	62.1		55.8	72.2		Borneo; Malaysia	785, 1218
Phaenicophaeus sumatranus	Chestnut-bellied Malkoha	M	5	86		82	95		Kalimantan	908
		F	3	98.3		90	105			
Phaenicophaeus viridirostris	Blue-faced Malkoha	B	21	67		52	82			7
Phaenicophaeus tristis	Green-billed Malkoha	B	6	117		100	128			740, 908
Phaenicophaeus leschenaultii	Sirkeer Malkoha	M	4	218		174	249			908
		F	2	140		132	148			
Phaenicophaeus chlorophaeus	Raffles' Malkoha	M	14	53.2		45	62			908
		F	7	49.3		41	62			
Phaenicophaeus javanicus	Red-billed Malkoha	B	7	109		97	134		Borneo; Malaysia	785, 1218
Phaenicophaeus curvirostris	Chestnut-breasted Malkoha	B	6	154		111	190		Borneo; Malaysia	785, 1218, 1287
Phaenicophaeus superciliosus	Red-crested Malkoha	B	13	123		105	138		Luzon, Philippines	436
Phaenicophaeus cumingi	Scale-feathered Malkoha	M	18	174	16.9	148	204		Philippines	908
		F	24	167	19.8	121	195			
Carpococcyx radiatus	Bornean Ground-Cuckoo	M	1	455						908
		F	1	540						
Coua gigas	Giant Coua	U	4	424		410	450		Madagascar	957
Coua coquereli	Coquerel's Coua	M	1	160					Madagascar	1143

Body Masses of World Birds (continued)

Species Name	Common Name	Sex	N	Mean	S.D.	Min	Max	Season	Location	Source #
Coua serriana	Red-breasted Coua	M	2	293		288	298		Madagascar	957
Coua reynaudii	Red-fronted Coua	B	5	153		128	175		Madagascar	957
Coua ruficeps	Red-capped Coua	B	6	180		112	204		Madagascar	68, 1143
Coua cursor	Running Coua	F	2	110		103	118		Madagascar	957, 1143
Coua cristata	Crested Coua	U	14	147	10.6	131	168		Madagascar	957
Coua caerulea	Blue Coua	U	8	244		225	270		Madagascar	957
Centropus unirufus	Rufous Coucal	M	6	168		146	188			436, 908
		F	8	201		146	228			
Centropus melanops	Black-faced Coucal	M	10	210		195	237		Philippines	954
		F	16	233		182	265			
Centropus milo	Buff-headed Coucal	M	5	769		742	790			908
Centropus goliath	Goliath Coucal	M	7	401		340	453			908
		F	2	622		605	638			
Centropus violaceus	Violaceous Coucal	F	1	500					New Britain	426
Centropus menbeki	Greater Black Coucal	B	8	519					New Guinea	61, 535, 786
Centropus ateralbus	Pied Coucal	B	3	339		330	345			908
Centropus phasianinus phasianinus	Pheasant Coucal	M	16	302	36.6	200	364		Australia	516
		F	22	445	63.2	242	520			
Centropus phasianinus melanurus	Pheasant Coucal	M	13	328	27.8	289	378		Australia	516
		F	12	483	63.9	390	600			
Centropus bernsteini	Lesser Black Coucal	M	5	146		130	160		New Guinea	908
		F	2	180		160	200			
Centropus rectunguis	Short-toed Coucal	U	1	238						1287
Centropus steerii	Black-hooded Coucal	M	4	184		175	200		Philippines	983
		F	4	214		190	238			
Centropus sinensis	Greater Coucal	M	5	247		208	270			908
		F	5	319		275	380			

Body Masses of World Birds (continued)

Species Name	Common Name	Sex	N	Mean	S.D.	Min	Max	Season	Location	Source #
Centropus viridis	Philippine Coucal	B	16	142		106	172		Philippines	954
Centropus toulou	Madagascar Coucal	M	5	135		125	150		Madagascar	957
		F	3	205		185	220			
Centropus grillii	Black Coucal	M	6	100		94	108			394
		F	1	151						
Centropus bengalensis javanensis	Lesser Coucal	M	4	86.2		82.7	90		Philippines	908
		F	5	148		132	161			
Centropus bengalensis sarasinorum	Lesser Coucal	M	5	146		132	156			908
		F	5	264		227	290			
Centropus leucogaster	Black-throated Coucal	M	2	282		270	293			908
		F	2	336		327	346			
Centropus anselli	Gabon Coucal	M	1	210					Angola	908
Centropus monachus	Blue-headed Coucal	M	5	171		163	177			908
		F	3	237		206	283			
Centropus cupreicaudus	Coppery-tailed Coucal	M	6	272		250	293			394
		F	5	299		245	342			
Centropus senegalensis	Senegal Coucal	B	9	169		160	180			908
Centropus superciliosus superciliosus	White-browed Coucal	M	6	124		110	140			908
		F	7	136		125	170			
Centropus superciliosus loandae	White-browed Coucal	M	14	160		142	213			908
		F	10	180		153	212			
Coccycua melanogaster	Black-bellied Cuckoo	M	4	97		85	110		Suriname	498
		F	5	109		97	118			
Coccycya minuta	Little Cuckoo	B	26	39.3		26	49		Suriname	498
Piaya cayana	Squirrel Cuckoo	M	33	104		73	137			908
		F	28	100		76	129			
Coccyzus melacoryphus	Dark-billed Cuckoo	M	14	45.6		42	54			908
		F	7	54.2		50	66.5			
Coccyzus americanus	Yellow-billed Cuckoo	U	103	64	9.1	50	84.6		Pennsylvania, USA	314

Body Masses of World Birds (continued)

Species Name	Common Name	Sex	N	Mean	S.D.	Min	Max	Season	Location	Source #
Coccyzus euleri	Pearly-breasted Cuckoo	B	8	52.3		45	63			498, 908, 1178
Coccyzus minor	Mangrove Cuckoo	U	18	63.9	10.8	50	85.5	W	Puerto Rico	349
Coccyzus ferrugineus	Cocos Is.Cuckoo	U		70					Cocos Is.	1170
Coccyzus pumilus	Dwarf Cuckoo	U	9	37.1						908, 1216
Coccyzus cinereus	Ash-colored Cuckoo	M	3	37.1		34.8	39			908
		F	5	45.4		43	57			
Coccyzus erythropthalmus	Black-billed Cuckoo	U	34	50.9	7.3	38.3	65.4		Pennsylvania, USA	828
Coccyzus lansbergi	Gray-capped Cuckoo	M	4	50.3		46	56			908
Coccyzus pluvialis	Chestnut-bellied Cuckoo	U		163						348
Coccyzus rufigularis	Bay-breasted Cuckoo	M	1	128						908
Coccyzus vetula	Jamaican Lizard-Cuckoo	U	2	96		86	105			908
Coccyzus vieilloti	Puerto Rican Lizard-Cuckoo	B	30	81.9	10.3	68	104		Puerto Rico	19
Coccyzus merlini	Great Lizard-Cuckoo	M	19	146	11.4	130	170		Cuba	19
		F	14	197	26	165	255			
Coccyzus longirostris	Hispaniolan Lizard-Cuckoo	B	11	100	15.2	83.7	124		Dominican Republic	19
Crotophaga major	Greater Ani	M	16	157	12.6			Y	Panama	489
		F	9	140						
Crotophaga ani	Smooth-billed Ani	M	19	100	40.5	79.6	119	B	Florida, USA	952
		F	10	94.8	49.6	70.9	109			
Crotophaga ani	Smooth-billed Ani	M	43	121	20.8	137	251		Caribbean Is.	19
		F	30	151	4.1	142	159			
Crotophaga ani	Smooth-billed Ani	M	11	111		81.9	133		Panama	908
		F	11	92.6		77.8	116			

Body Masses of World Birds (continued)

Species Name	Common Name	Sex	N	Mean	S.D.	Min	Max	Season	Location	Source #
Crotophaga sulcirostris	Groove-billed Ani	M	16	87.3	4.1	80.6	93		Mexico	1148a
		F	19	77.1	9.6	58.7	96.1			
Guira guira	Guira Cuckoo	B	27	141		113	169			908
Tapera naevia	Striped Cuckoo	B	10	52.1					Panama	244
Tapera naevia	Striped Cuckoo	B	19	45		40	54		Suriname	498
Dromococcyx phasianellus	Pheasant Cuckoo	U	4	84.5					Peru; Panama; Brazil	63, 489, 1288
Dromococcyx pavoninus	Pavonine Cuckoo	U	6	46.4		40.5	50			875, 908
Morococcyx erythropygus	Lesser Ground-Cuckoo	M	15	61.8		53.1	70.5			908
		F	9	67.4		56	86.8			
Geococcyx californianus	Greater Roadrunner	U	23	376	90	221	538			545
Geococcyx velox	Lesser Roadrunner	M	5	186		174	203			908
		F	4	174		163	192			
Neomorphus geoffroyi squamiger	"Scaled" Ground-Cuckoo	M	1	340					Brazil	449
Neomorphus geoffroyi	Rufous-vented Ground-Cuckoo	M	6	351		339	370			908
		F	3	375		349	400			
Neomorphus radiolosus	Banded Ground-Cuckoo	U	2	399		433	365		Ecuador	1369
Neomorphus rufipennis	Rufous-winged Ground-Cuckoo	M	2	435		350	520			908
		F	2	328		315	340			
Neomorphus pucheranii	Red-billed Ground-Cuckoo	M	1	330						908
Family Tytonidae										
Tyto tenebricosa	Greater Sooty-Owl	M	5	492		452	565		Australia	516
		F	4	887		753	1040			
Tyto multipunctata	Lesser Sooty-Owl	M	4	369		330	410		Australia	516
		F	5	558		500	600			
Tyto novaehollandiae novaehollandiae	Australian Masked-Owl	M	12	476	79	352	603		Southeast Australia	516
		F	13	630	62.4	476	706			

Body Masses of World Birds (continued)

Species Name	Common Name	Sex	N	Mean	S.D.	Min	Max	Season	Location	Source #
Tyto novaehollandiae castanops	Australian Masked-Owl	M	4	632		395	805		Tasmania	516
		F	5	841		702	945			
Tyto capensis longimembris	Australasian Grass-Owl	M	6	313		265	390		Australia	516
		F	3	370		312	478			
Tyto capensis	African Grass-Owl	B	8	419		355	520			394
Tyto soumagnei	Madagascar Red Owl	U	2	284		245	323		Madagascar	957, 1220
Tyto alba alba	Barn Owl	M	361	330						1199
		F	445	370						
Tyto alba guttata	Barn Owl	M	60	306						1199
		F	122	357						
Tyto alba pratincola	Barn Owl	M	112	474	32.3				Utah, USA	761
		F	166	566	57					
Tyto alba javanica	Barn Owl	M	13	555						1199
		F	26	612						
Tyto alba delicatula	Barn Owl	B	22	355		250	470		Australia	516
Tyto alba punctatissima	Barn Owl	F	2	264					Galapagos Is.	1199
Phodilus badius	Oriental Bay-Owl	U	5			255	308			1287
Phodilus prigoginei	Congo Bay-Owl	F	1	195				B	Congo	159
Family Strigidae										
Otus sagittatus	White-fronted Scops-Owl	U	9			109	138			1287
Otus rufescens	Reddish Scops-Owl	U	11			70.5	83.1			1287
Otus icterorhynchus	Sandy Scops-Owl	B	7	73.3						625
Otus ireneae	Sokoke Scops-Owl	U	3	50.3						759
Otus spilocephalus	Mountain Scops-Owl	U	16	67.5		60	77		Malaysia	775

Body Masses of World Birds (continued)

Species Name	Common Name	Sex	N	Mean	S.D.	Min	Max	Season	Location	Source #
Otus angelinae	Javan Scops-Owl	U	3	82.2		75	90.6		Java	649
Otus bakkamoena marathae	Collared Scops-Owl	M	1	116					India	759
Otus bakkamoena lempiji	Collared Scops-Owl	U	10	108		90	140			649
Otus bakkamoena lempiji	Collared Scops-Owl	U	27			94.1	119		Malay Peninsula	1287
Otus bakkamoena lettia	Collared Scops-Owl	M	3	108						759
		F	4	142						
Otus semitorques	Japanese Scops-Owl	U		138					Japan	339
Otus megalotis	Philippines Scops-Owl	M	4	172		159	182		Philippines	435, 436, 759
		F	1	310						
Otus megalotis everetti	Philippines Scops-Owl	M	4	125					Philippines	759
		F	3	152						
Otus mirus	Mindanao Scops-Owl	M	2	54.9		54.5	64.5		Philippines	1143
		F	1	68.5						
Otus longicornis	Luzon Scops-Owl	M	8	72.3		69	78		Philippines	436
		F	1	95						
Otus brucei	Pallid Scops-Owl	M	3	107		100	110			7, 223
Otus brucei pamelae	Pallid Scops-Owl	F	2	66.5		62	71		Oman	223
Otus senegalensis	African Scops-Owl	U	66	67		40	100		Southern Africa	1036
Otus scops scops	European Scops-Owl	U	169	92	14.3	64	135	F	France	223
Otus sunia	Oriental Scops-Owl	U	10	72.1	14.3	54.4	95			273, 1287
Otus sunia japonicus	Oriental Scops-Owl	M	5	72.2		67	79		Hopei, China	1087
		F	2	91.5		85	98			
Otus flammeolus	Flammulated Owl	M	27	54.7	3.3	49.5	62	B	Colorado, USA	773
		F	25	65.6	10.8	53.5	96			
Otus magicus	Moluccan Scops-Owl	U	4	165						759

Body Masses of World Birds (continued)

Species Name	Common Name	Sex	N	Mean	S.D.	Min	Max	Season	Location	Source #
Otus mantananensis sibutuensis	Mantanani Scops-Owl	U	11	106	7.8					759
Otus elegans botelensis	Ryukyu Scops-Owl	U	1077	124					Lanyu, Taiwan	681
Otus elegans	Ryukyu Scops-Owl	U				100	107			649
Otus manadensis manadensis	Sulawesi Scops-Owl	U	4	88.3						759
Otus collari	Sangihe Scops-Owl	M	1	76						649
Otus umbra	Simeulue Scops-Owl	U	1	95						759
Otus enganensis	Enggano Scops-Owl	U				100	110			649
Otus pauliani	Comoro Scops-Owl	M	1	69.5					Grand Comoros Is.	513
Otus capnodes	Anjouan Scops-Owl	U	1	119					Anjouan, Comoro Is.	649
Otus moheliensis	Moheli Scops-Owl	M	1	95					Moheli, Comoro Is.	649
		F	1	116						
Otus rutilus	Malagasy Scops-Owl	U	28	101		76	121		Madagascar	957
Otus hartlaubi	Sao Tome Scops-Owl	U		79					estimated	625
Megascops kennicottii macfarlanei	Western Screech-Owl	M	74	192	16.9	143	230		Idaho, USA	165
		F	105	236	27.3	166	305			
Megascops kennicottii aikeni	Western Screech-Owl	M	37	119						165
		F	27	145						
Megascops kennicottii bendirei	Western Screech-Owl	M	43	143						165
		F	19	155						
Megascops kennicottii vinaceus	Western Screech-Owl	M	10	106						165
		F	5	123						
Megascops kennicottii suttoni	Western Screech-Owl	M	25	113						165
		F	14	129						

Body Masses of World Birds (continued)

Species Name	Common Name	Sex	N	Mean	S.D.	Min	Max	Season	Location	Source #
Megascops seductus	Balsas Screech-Owl	M	2	160		158	161		Colima, Mexico	1145
Megascops cooperi	Pacific Screech-Owl	M	3	149		145	153		Mexico; Guatemala	294, 1145
		F	1	181						
Megascops cooperi lambi	Pacific Screech-Owl	U		125					estimated	649
Megascops trichopsis	Whiskered Screech-Owl	M	23	84.5		70	104			322
		F	8	92.2		79	121			
Megascops asio naevius	Eastern Screech-Owl	M	31	167	16.3	140	210	Y	Ohio, USA	505
		F	66	194	16.3	150	235			
Megascops choliba	Tropical Screech-Owl	B	12	132	14.6	114	163			496, 752, 796, 876, 1040, 1131
Megascops koepckeae	Koepcke's Screech-Owl	U				113	121			649
Megascops roboratus pacificus	West Peruvian Screech-Owl	B	17	87.2						595
Megascops roboratus roboratus	West Peruvian Screech-Owl	M	7	144						595
		F	3	162						
Megascops clarkii	Bare-shanked Screech-Owl	M	3	135		123	150			990, 1289
		F	1	186						
Megascops barbarus	Bearded Screech-Owl	F	2	69		69	69		Chiapas, Mexico	1147a, 1289
Megascops ingens	Rufescent Screech-Owl	M	3	151		134	168		Peru	370, 1288
		F	1	181						
Megascops columbianus	Columbian Screech-Owl	M				150	156			521, 649
Megascops petersoni	Cinnamon Screech-Owl	M	1	96					Peru	1143
Megascops marshalli	Cloud-forest Screech-Owl	U		115					estimated	1289
Megascops watsonii	Tawny-bellied Screech-Owl	B	9	129		114	172		Brazil; Venezuela	449, 1178 1308

Body Masses of World Birds (continued)

Species Name	Common Name	Sex	N	Mean	S.D.	Min	Max	Season	Location	Source #
Megascops guatemalae	Guatemalan Screech-Owl	B	14	110		91	128		Peru; Mexico	522, 990, 1289
Megascops vermiculatus	Vermiculated Screech-Owl	B	8	118					Ecuador; Peru	370, 1143
Megascops hoyi	Hoy's Screech-Owl	M	1	110					Argentina	1143
Megascops atricapilla	Variable Screech-Owl	U				114	124		Brazil	1105
Megascops sanctaecatarinae	Long-tufted Screech-Owl	M	7	170		155	194		Brazil	63
		F	8	190		174	211			
Megascops nudipes	Puerto Rican Screech-Owl	M	9	131		103	146		Puerto Rico	19
		F	14	144		103	169			
Megascops albogularis	White-throated Screech-Owl	U	23	159	12.3	137	181		Colombia	324a
Gymnoglaux lawrencii	Cuban Screech-Owl	B	10	95.2	9.2	80	111		Cuba	19
Ptilopsis leucotis	Northern White-faced Owl	B	16	204						625
Ptilopsis granti	Southern White-faced Owl	M		190						526, 649
		F	4	242		240	275			
Bubo virginianus pallescens	Great Horned Owl	M	18	914		724	1257			322
		F	12	1142		801	1550			
Bubo virginianus virginianus	Great Horned Owl	M	18	1154		865	1460			322
		F	18	1555		1112	2046			
Bubo magellanicus	Magellanic Horned Owl	M	2	1007		940	1074		Chile	570
		F	1	1335						
Bubo bubo	Eurasian Eagle-Owl	M	14	2380		1835	2810	W	Norway	223
		F	12	2992		2280	4200			
Bubo bengalensis	Rock Eagle-Owl	M	1	1100						7
Bubo capensis	Cape Eagle-Owl	M	4	929						625
		F	3	1347						
Bubo africanus	Spotted Eagle-Owl	B	12			600	690		Angola	266

Body Masses of World Birds (continued)

Species Name	Common Name	Sex	N	Mean	S.D.	Min	Max	Season	Location	Source #
Bubo cinerascens	Grayish Eagle-Owl	U		500					estimated	649
Bubo poensis	Fraser's Eagle-Owl	M	1	575						625
		F	4	746						
Bubo nipalensis	Spot-bellied Eagle-Owl	U				1300	1500			649
Bubo sumatranus	Barred Eagle-Owl	U		1525		1427	1606			649
Bubo shelleyi	Shelley's Eagle-Owl	F	1	1257						625
Bubo lacteus	Verreaux's Eagle-Owl	M	4	1704						625
		F	6	2625						
Bubo leucostictus	Akun Eagle-Owl	M	2	511						625
		F	3	555						
Bubo scandiacus	Snowy Owl	M	23	1806	144	1606	2041	W	Alberta, Canada	630
		F	21	2279	261	1838	2951			
Ketupa zeylonensis	Brown Fish-Owl	F	1	1105						7
Ketupa flavipes	Tawny Fish-Owl	U	8	2315					Taiwan	1342a
Ketupa ketupu	Buffy Fish-Owl	U		1293		1028	2100			649
Scotopelia peli	Pel's Fishing-Owl	M	1	1725						266, 625
		F	4	2188						
Scotopelia ussheri	Rufous Fishing-Owl	M	1	743						625
		F		834						
Scotopelia bouvieri	Vermiculated Fishing-Owl	F		637					estimated	625
Strix seloputo	Spotted Wood-Owl	M	1	1011						1287
Strix leptogrammica	Brown Wood-Owl	U				800	1100			649
Strix leptogrammica bartelsi	Brown Wood-Owl	U				1000	1300			649

Body Masses of World Birds (continued)

Species Name	Common Name	Sex	N	Mean	S.D.	Min	Max	Season	Location	Source #
Strix leptogrammica newarensis	Brown Wood-Owl	M	1	970						649
Strix aluco	Tawny Owl	M	11	426	34.6	385	500	W	Italy	223
		F	7	524		415	620			
Strix butleri	Hume's Owl	F	5	219					Negev Desert; Sinai	648, 789
Strix occidentalis occidentalis	Spotted Owl	M	218	566	35					464
		F	195	646	45					
Strix occidentalis caurina	Spotted Owl	M	68	579	35					464
		F	65	663	45					
Strix occidentalis lucida	Spotted Owl	M	68	509	33					464
		F	68	569	44					
Strix varia	Barred Owl	M	20	632		468	774			322
		F	24	801		610	1051			
Strix hylophila	Rusty-barred Owl	M	2	302		285	320		Brazil	63
		F	1	395						
Strix rufipes	Rufous-legged Owl	B	13	380		280	430		Argentina; Bolivia; Chile	186a, 210, 570, 657
Strix chacoensis	Chaco Owl	U		330					estimated	649
Strix uralensis macroura	Ural Owl	M	40	706	112	503	950	Y	Romania	223
		F	57	863	137	569	1307			
Strix nebulosa	Great Gray Owl	M	21	890	53.3	825	1050	B		149
		F	63	1267	149	1025	1700			
Strix woodfordii	African Wood-Owl	U	15	350		286	445		South Africa	1036
Ciccaba virgata	Mottled Owl	M	7	240		220	256		Guatemala	412
		F	9	336		308	366			
Ciccaba nigrolineata	Black-and-white Owl	M	1	435					Guatemala; Suriname	293, 412
		F	3	482		443	536			
Ciccaba huhula	Black-banded Owl	U		370					Peru	1210
Lophostrix cristata	Crested Owl	M	2	468		425	510		Mexico; Peru	1145
		F	1	620						

Body Masses of World Birds (continued)

Species Name	Common Name	Sex	N	Mean	S.D.	Min	Max	Season	Location	Source #
Jubula lettii	Maned Owl	F	1	183						648
Pulsatrix perspicillata	Spectacled Owl	M	10	767		550	1075			63, 155, 489, 498, 990, 293 1027
		F	8	908		597	1250			
Pulsatrix perspicillata pulsatrix	Spectacled Owl	M	1	1075						649
		F	1	1250						
Pulsatrix koeniswaldiana	Tawny-browed Owl	F	1	481					Argentina	649
Surnia ulula	Northern Hawk Owl	M	42	301	31.1	242	375		Canada; Minnesota, USA	311
		F	54	340	41.2	250	454			
Glaucidium passerinum	Eurasian Pygmy-Owl	M	5			47	62		Finland	223
		F	12			55	70			
Glaucidium brodiei	Collared Pygmy-Owl	M	3	56		52	58.1			7, 785
		F	1	63						
Glaucidium perlatum	Pearl-spotted Owl	M	12	69						625
		F	13	91						
Glaucidium gnoma californicum	Northern Pygmy-Owl	M	42	61.9		54	74			322
		F	10	73		64	87			
Glaucidium gnoma	Northern Pygmy-Owl	M	10	50.8	2.7				Mexico	996
		F	3	61.1						
Glaucidium gnoma	Northern Pygmy-Owl	B	10	61.5		54	64		Alberta, Canada	531
Glaucidium hoskinsii	Cape Pygmy-Owl	U				50	65			649
Glaucidium costaricanum	Costa Rican Pygmy-Owl	M	3	66					Costa Rica; Panama	996
Glaucidium nubicola	Cloud-forest Pygmy-Owl	M	3	76.1					Colombia	996
		F	1	79						
Glaucidium jardinii	Andean Pygmy-Owl	U	12	66.9	7.3	53.6	79.5		Colombia	324a
Glaucidium palmarum	Colima Pygmy-Owl	M	2	45.7		42.4	49		Jalisco, Colima, Mexico	1145

Body Masses of World Birds (continued)

Species Name	Common Name	Sex	N	Mean	S.D.	Min	Max	Season	Location	Source #
Glaucidium sanchezi	Tamaulipas Pygmy-Owl	M	1	55					San Luis Potosi, Mexico	1145
Glaucidium griseiceps	Central American Pygmy-Owl	M				49.5	51.7			649
Glaucidium parkeri	Subtropical Pygmy-Owl	M	3	61.4		59.1	63.1			649
Glaucidium bolivianum	Yungas Pygmy-Owl	M	20	59.6	4.9				Peru; Bolivia	996
		F	13	66.5	5.6					
Glaucidium hardyi	Amazonian Pygmy-Owl	M	3	58.3		55.5	63			649, 1178
Glaucidium minutissimum	Least Pygmy-Owl	U		50					estimated	649
Glaucidium brasilianum	Ferruginous Pygmy-Owl	M	68	64.3	5.4	53	79		Texas, USA	946
		F	38	77.2	8.1	66	102			
Glaucidium tucumanum	Tucuman Pygmy-Owl	U				52	56			649
Glaucidium mooreorum	Pernambuco Pygmy-Owl	M	1	51					Brazil	1106
Glaucidium peruanum	Peruvian Pygmy-Owl	M	4	60.4		58	62			648
		F	1	64.5						
Glaucidium nanum	Austral Pygmy-Owl	M	14	66.9						582
		F	11	81.2						
Glaucidium siju siju	Cuban Pygmy-Owl	M	16	58.4	6.2	49	67		Cuba	19
		F	9	68.8		54	84			
Glaucidium siju vittatum	Cuban Pygmy-Owl	M	2	66.5		65	68		Isle of Pines, Cuba	984
		F	6	87.2		84	92			
Glaucidium tephronotum	Red-chested Owlet	B	10	87.4						625
Glaucidium sjostedti	Sjostedt's Owlet	F		139					estimated	625
Glaucidium cuculoides whitelyi	Asian Barred Owlet	M	10	187		170	240		China	184a
		F	10	210		176	240			
Glaucidium radiatum	Jungle Owlet	B	7	101		88	114		India	7, 740

Body Masses of World Birds (continued)

Species Name	Common Name	Sex	N	Mean	S.D.	Min	Max	Season	Location	Source #
Glaucidium castanonotum	Chestnut-backed Owlet	U		100					estimated	649
Glaucidium capense	African Barred Owlet	U	18	122		52	392		Southern Africa	1036
Glaucidium castaneum	Chestnut Owlet	U		100					estimated	649
Glaucidium albertinum	Albertine Owlet	F		73					estimated	625
Xenoglaux loweryi	Long-whiskered Owlet	B	3	48		46	51		Peru	869
Micrathene whitneyi	Elf Owl	U	20	41		35.9	44.1	B	Arizona, USA	1271
Athene cunicularia hypugaea	Burrowing Owl	M	38	146	11.7				Colorado, USA	500
		F	31	156	20					
Athene cunicularia floridana	Burrowing Owl	B	273	150					Florida, USA	500
Athene brama	Spotted Owlet	B	2	112		110	114		India	7, 740
Athene blewitti	Forest Owlet	M	1	241					India	7
Athene noctua	Little Owl	B	21	164		146	193	PB	Netherlands	223
Athene noctua	Little Owl	B	49	174		142	210	PB	New Zealand	516
Aegolius funereus	Boreal Owl	M	74	101	4.8	90	113	B	Germany	431
		F	96	167	15.7	126	194			
Aegolius funereus	Boreal Owl	M	10	113	9.8			B	Finland	650
		F	11	179	20.3					
Aegolius funereus	Boreal Owl	M	50	117	9.8	93	139		Idaho, USA	502
		F	53	167	17.9	132	215			
Aegolius acadicus	Northern Saw-whet Owl	M	22	77.4	4.5				British Columbia, Canada	164
		F	36	131	16					
Aegolius ridgwayi	Unspotted Saw-whet Owl	U		80					Costa Rica	1170
Aegolius harrisii	Buff-fronted Owl	B	13	120		104	135		Peru; Bolivia, Colombia	63, 324a, 1145, 900

Body Masses of World Birds (continued)

Species Name	Common Name	Sex	N	Mean	S.D.	Min	Max	Season	Location	Source #
Ninox rufa rufa	Rufous Owl	M	2	1225		1150	1300		Northwest Australia	516
		F	2	978		935	1020			
Ninox rufa queenslandica	Rufous Owl	M	4	950		823	1152		Queensland, Australia	516
		F	2	662		583	740			
Ninox strenua	Powerful Owl	M	13	1447	333	995	2220		Southeast Australia	516
		F	9	1252		1040	1465			
Ninox connivens connivens	Barking Owl	M	6	694		639	740		East Australia	516
		F	12	592	117	380	710			
Ninox connivens peninsularis	Barking Owl	B	12	470		385	580		Northern Australia	516
Ninox sumbaensis	Little Sumba Hawk-Owl	U	1	90					Sumba, Indonesia	862
Ninox novaeseelandiae novaeseelandiae	Morepork	B	40	164		105	215		New Zealand	516
Ninox novaeseelandiae undulata	Morepork	F	1	213					Norfolk Is.	516
Ninox novaeseelandiae leucopus	Morepork	M	22	190	33.6	118	250		Tasmania, Australia	516
		F	29	214	40.5	125	275			
Ninox novaeseelandiae rotiensis	Morepork	U		146						862
Ninox boobook	Southern Morepork	M	20	252	41.8	182	320		Southeast Australia	516
		F	16	316	35	250	365			
Ninox scutulata	Brown Hawk-Owl	B	9	191		146	230			7, 436, 740, 1087, 1287
Ninox superciliaris	White-browed Owl	B	4	231		217	236		Madagascar	68, 957
Ninox philippensis	Philippine Hawk-Owl	M	11	116	8.1	110	135		Philippines	436
		F	1	141						
Ninox burhani	Togian Hawk-Owl	M	2	99		98	100		Sulawesi, Indonesia	552
Ninox ios	Cinnabar Hawk-Owl	U	2	86		78	94		Sulawesi, Indonesia	680

Body Masses of World Birds (continued)

Species Name	Common Name	Sex	N	Mean	S.D.	Min	Max	Season	Location	Source #
Ninox squamipila	Moluccan Hawk-Owl	U	1	210						649
Ninox natalis	Christmas Is.Hawk-Owl	U	12	184	17	165	215		Christmas Is.	516
Ninox theomacha	Jungle Hawk-Owl	U	1	125					New Guinea	770
Ninox odiosa	New Britain Hawk-Owl	F	1	209					New Britain	426
Ninox jacquinoti jacquinoti	Solomon Is.Hawk-Owl	M	2	256		253	260		Isabel, Solomon Is.	658
		F	1	277						
Ninox jacquinoti granti	Solomon Is.Hawk-Owl	B	2	156		154	158		Guadalcanal, Solomon Is.	399
Ninox jacquinoti roseoaxillaris	Solomon Is.Hawk-Owl	M	1	147					San Cristoval, Solomon Is.	399
Sceloglaux albifacies	Laughing Owl	U	1	600					New Zealand (extinct)	516
Pseudoscops grammicus	Jamaican Owl	U	1	335					Jamaica	1148
Pseudoscops clamator	Striped Owl	M	7	406		347	485			488, 498, 1040
		F	5	484		400	546			
Asio stygius	Stygian Owl	B	6	565		408	675		Colombia; Argentina; Brazil; Cuba	19, 324a, 752, 796, 1040
Asio otus	Long-eared Owl	M	55	261		223	304	B	Montana, USA	755
		F	49	337		289	409			
Asio flammeus	Short-eared Owl	M	20	315		206	368			322
		F	27	378		284	475			
Asio flammeus	Short-eared Owl	M	1	282					Puerto Rico	19
Asio capensis	Marsh Owl	U	43	334		220	390		Southern Africa	1036
Nesasio solomonensis	Fearful Owl	F	1	1155					Solomon Is.	658
Family Steatornithidae										
Steatornis caripensis	Oilbird	U	13	408	43.4	345	490			1042, 1131, 1308

Body Masses of World Birds (continued)

Species Name	Common Name	Sex	N	Mean	S.D.	Min	Max	Season	Location	Source #
Family Aegothelidae										
Aegotheles insignis	Feline Owlet-Nightjar	B	44	78.5		58	101		New Guinea	938
Aegotheles tatei	Spangled Owlet-Nightjar	U	1	52					New Guinea	938
Aegotheles crinifrons	Moluccan Owlet-Nightjar	U	2	136		105	167		Moluccas	938
Aegotheles wallacii	Wallace's Owlet-Nightjar	M	2	49.5		48.5	50.5		New Guinea	282, 283
		F	1	52						
Aegotheles archboldi	Archbold's Owlet-Nightjar	B	6	30.6					New Guinea	534
Aegotheles albertisi	Mountain Owlet-Nightjar	B	27	35.2	2.8				New Guinea	534
Aegotheles bennettii affinis	Barred Owlet-Nightjar	M				41.5	53			191
		F		60						
Aegotheles cristatus cristatus	Australian Owlet-Nightjar	B	39	44		21	65		Australia	516
Aegotheles cristatus tasmanicus	Australian Owlet-Nightjar	B	7	41.6		31	60		Tasmania	516
Family Podargidae										
Podargus strigoides brachypterus	Tawny Frogmouth	B	20	278		185	416		Western Australia	516
Podargus strigoides strigoides	Tawny Frogmouth	M	55	354	86.8	178	520		Eastern Australia	516
		F	39	297	91.1	157	555			
Podargus ocellatus plumiferus	Marbled Frogmouth	M	8	262		202	322		Queensland, Australia	516
		F	6	204		147	252			
Podargus ocellatus marmoratus	Marbled Frogmouth	M	1	163						516
		F	2	136		131	142			
Podargus ocellatus ocellatus	Marbled Frogmouth	B	5	140		137	150		New Guinea	534
Podargus papuensis	Papuan Frogmouth	M	6	414		290	507		Australia	516
		F	5	341		300	383			
Batrachostomus auritus	Large Frogmouth	U		206						534

Body Masses of World Birds (continued)

Species Name	Common Name	Sex	N	Mean	S.D.	Min	Max	Season	Location	Source #
Batrachostomus septimus	Philippine Frogmouth	M	6	94.9					Philippines	759, 954
		F	3	79.3		75.2	81.7			
Batrachostomus stellatus	Gould's Frogmouth	U	14			43.5	60.6			1287
Batrachostomus moniliger	Ceylon Frogmouth	M	1	46.5						191
		F	1	59						
Batrachostomus hodgsoni	Hodgson's Frogmouth	U	2	51						759
Batrachostomus javensis continentalis	Javan Frogmouth	U	7	46						759
Batrachostomus javensis affinis	Javan Frogmouth	U	4	46.7						759
Family Nyctibiidae										
Nyctibius grandis	Great Potoo	B	12	547		450	624		Suriname	498
Nyctibius aethereus	Long-tailed Potoo	U				434	447		Brazil	1105
Nyctibius jamaicensis mexicanus	Northern Potoo	M				210	274		Mexico	191
		F	1	251						
Nyctibius jamaicensis abbotti	Northern Potoo	M	1	282					Hispaniola	191
Nyctibius maculosus	Andean Potoo	B				145	200			191, 900
Nyctibius griseus panamensis	Common Potoo	B	21	185					Panama	244
Nyctibius griseus griseus	Common Potoo	B	10	160		145	187		Suriname; Trinidad; Brazil	534
Nyctibius leucopterus	White-winged Potoo	B	3	86.4		78.5	98		Brazil	10, 195, 196
Nyctibius bracteatus	Rufous Potoo	U				46	58			534
Family Caprimulgidae										
Lurocalis semitorquatus	Short-tailed Nighthawk	B	8	75.9		69.5	84		Venezuela	1042
Chordeiles pusillus	Least Nighthawk	F	2	30.6		27.8	33.4		Guyana; Argentina	659, 1143
Chordeiles acutipennis	Lesser Nighthawk	U	13	49.9	7.4	41	64		Texas, USA	1148a

Body Masses of World Birds (continued)

Species Name	Common Name	Sex	N	Mean	S.D.	Min	Max	Season	Location	Source #
Chordeiles acutipennis	Lesser Nighthawk	B	37	47		41	55		Suriname	498
Chordeiles minor	Common Nighthawk	B	78	79.3						934
Chordeiles gundlachii	Antillean Nighthawk	U		50					estimated	465
Podager nacunda	Nacunda Nighthawk	B	6	159		130	188		Brazil; Colombia; Argentina	63, 795, 861, 1040
Nyctiprogne leucopyga	Band-tailed Nighthawk	B	4	27.3		26.6	28.2		Peru	10
Eurostopodus argus	Spotted Nightjar	B	29	93.4		72	132		Australia	516
Eurostopodus mystacalis	White-throated Nightjar	M	5	104		80	138		Eastern Australia	516
		F	3	174		155	188			
Eurostopodus papuensis	Papuan Nightjar	F				80	81			191
Eurostopodus archboldi	Archbold's Nightjar	U	2	77		74	80			191
Eurostopodus temminckii	Malaysian Nightjar	U	5	95.9					Malaysia	1287, 1333
Eurostopodus macrotis	Great Eared-Nightjar	M	3	131		125	140		Philippines	954, 983
		F	1	151						
Nyctidromus albicollis	Pauraque	B	37	53.2					Panama	489
Nyctidromus albicollis yucatanensis	Pauraque	B	16	62.9					Yucatan, Mexico	534
Phalaenoptilus nuttallii	Common Poorwill	M	15	47.2	6				British Columbia, Canada	235
		F	10	49.5	6.3					
Nyctiphrynus mcleodii	Eared Poorwill	M	2	35		33.5	36.4		Mexico	1145
Nyctiphrynus yucatanicus	Yucatan Poorwill	B	2	24.5		21.3	27.7		Mexico; Belize	910a, 1027
Nyctiphrynus rosenbergi	Choco Poorwill	F	1	52					Ecuador	992
Nyctiphrynus ocellatus	Ocellated Poorwill	B	15	39.6	4.5					992

Body Masses of World Birds (continued)

Species Name	Common Name	Sex	N	Mean	S.D.	Min	Max	Season	Location	Source #
Caprimulgus carolinensis	Chuck-will's-widow	B	57	109		66	188			1179
Caprimulgus rufus	Rufous Nightjar	U	3	94.2		89	99		Panama; Brazil; Suriname	392, 498, 618, 883
Caprimulgus cubanensis	Greater Antillean Nightjar	U	8	63.5		60	80		Cuba	403
Caprimulgus salvini	Tawny-collared Nightjar	F	1	56					Mexico	763
Caprimulgus badius	Yucatan Nightjar	B	4	59.8		51.2	65.5		Belize; Yucatan, Mexico	640, 910a, 1027
Caprimulgus sericocaudatus	Silky-tailed Nightjar	M	1	83					Paraguay (very fat)	1175
Caprimulgus ridgwayi	Buff-collared Nightjar	U	6	48		39.8	61		Mexico	105
Caprimulgus vociferus	Whip-poor-will	M	20	50.3	4			B	Kansas, USA	189
		F	20	56.5	4.6					
Caprimulgus noctitherus	Puerto Rican Nightjar	M	7	36.1		33.8	38		Puerto Rico	19
Caprimulgus saturatus	Dusky Nightjar	B	2	52		51	52.9			191
Caprimulgus longirostris	Band-winged Nightjar	B	13	45.6		36.4	54		Brazil; Peru; Venezuela; Colombia	63, 324a, 917, 1288, 1308
Caprimulgus candicans	White-winged Nightjar	U	6	49.3		46	52			705
Caprimulgus parvulus	Little Nightjar	B	12	37	5.9	25	46.5		Brazil; Bolivia; Argentina	63, 177, 210a, 392, 657
Caprimulgus maculicaudus	Spot-tailed Nightjar	B	16	30.3		26	39		Suriname	498
Caprimulgus cayennensis	White-tailed Nightjar	B	33	35.5		32	40		Suriname	498
Caprimulgus anthonyi	Scrub Nightjar	M	10	35.2	2.9				Peru; Ecuador	993
		F	9	36.6						
Caprimulgus nigrescens	Blackish Nightjar	B	12	35.7		32	39		Venezuela	1308

Body Masses of World Birds (continued)

Species Name	Common Name	Sex	N	Mean	S.D.	Min	Max	Season	Location	Source #
Caprimulgus whitelyi	Roraiman Nightjar	M				30	40			191, 917
		F	1	47						
Caprimulgus binotatus	Brown Nightjar	U	1	63						394
Caprimulgus ruficollis	Red-necked Nightjar	U	2	68.5		62	75		Algeria; Morocco	223
Caprimulgus indicus	Gray Nightjar	M	4	97		78	113		Hopei & Tsingtao, China	1087, 1088
		F	6	86.5		73	98			
Caprimulgus europaeus	Eurasian Nightjar	U	36	67	8.1	56	85	F	France	223
Caprimulgus fraenatus	Sombre Nightjar	M	4	65		55.2	71.2			394
		F	4	54.9		46	65			
Caprimulgus rufigena	Rufous-cheeked Nightjar	B	10			45	60		Angola	266
Caprimulgus aegyptius	Egyptian Nightjar	B	7	77.3		68	93			133, 223
Caprimulgus nubicus torridus	Nubian Nightjar	M	1	49.3					Kenya	1143
Caprimulgus mahrattensis	Sykes' Nightjar	B	2	57		56	58			534
Caprimulgus eximius	Golden Nightjar	M	1	66.3						191
Caprimulgus macrurus	Large-tailed Nightjar	B	13	66.2		52	78		Australia	516
Caprimulgus atripennis	Jerdon's Nightjar	U	3	61		55	64		India	534
Caprimulgus manillensis	Philippine Nightjar	B	5	70.3		58	84.5		Philippines	437
Caprimulgus donaldsoni	Donaldson-Smith's Nightjar	U	14	29		21	36			394
Caprimulgus nigriscapularis	Black-shouldered Nightjar	M	2	49		48	50			191
Caprimulgus pectoralis	Fiery-necked Nightjar	B	60			42	63		Angola	266
Caprimulgus poliocephalus	Abyssinian Nightjar	B	4	41.8		40	43		Zambia	301

Body Masses of World Birds (continued)

Species Name	Common Name	Sex	N	Mean	S.D.	Min	Max	Season	Location	Source #
Caprimulgus ruwenzorii	Montane Nightjar	M	10	46.4		44	50			394
		F	9	51.6		47	57			
Caprimulgus asiaticus	Indian Nightjar	B	5	42					India	7, 740
Caprimulgus madagascariensis	Madagascar Nightjar	B	19	41.1	5.3	27	51		Madagascar	957
Caprimulgus natalensis	Swamp Nightjar	M	3	77.7		58	110		Angola	266
Caprimulgus inornatus	Plain Nightjar	B	3	53.3		46	60			120
Caprimulgus inornatus	Plain Nightjar	B	13	47.4		32	61		Kenya	394
Caprimulgus stellatus	Star-spotted Nightjar	B	6	58.2		50	69			394
Caprimulgus affinis	Savanna Nightjar	B	5	58.7		42	78		India; Philippines; Indonesia	437, 516, 740
Caprimulgus tristigma	Freckled Nightjar	B	29	79		69	91			394
Caprimulgus enarratus	Collared Nightjar	B	3	52.8		47.5	57		Madagascar	68, 957
Caprimulgus batesi	Bates' Nightjar	F	2	100		89	112			394
Caprimulgus climacurus	Long-tailed Nightjar	B	33	45.8		40	57	W	Senegambia	394
Caprimulgus clarus	Slender-tailed Nightjar	B	19	42.5		34	53			394
Caprimulgus fossii	Square-tailed Nightjar	M	12	59.2		54	67		Zimbabwe; Zambia; Mozambique	394
		F	34	57.4		47.5	73.2			
Macrodipteryx vexillarius	Pennant-winged Nightjar	M	104	67.7	4.4	59	79	B	Zambia	119
		F	14	77.1	5.8	67	82.5			
Macrodipteryx longipennis	Standard-winged Nightjar	B	32	48		37	65		Chad	394
Uropsalis lyra	Lyre-tailed Nightjar	B	2	71.2		68.5	74		Ecuador	1147
Uropsalis segmentata	Swallow-tailed Nightjar	B	3	42.9					Peru	1288
Hydropsalis climacocerca	Ladder-tailed Nightjar	M	5	46		42	52		Suriname	498

Body Masses of World Birds (continued)

Species Name	Common Name	Sex	N	Mean	S.D.	Min	Max	Season	Location	Source #
Hydropsalis torquata	Scissor-tailed Nightjar	M	6			51	63		Brazil	63
		F	2	58.5		57	60			
Eleothreptus anomalus	Sickle-winged Nightjar	M	1	43.7						191
Family Apodidae										
Cypseloides phelpsi	Tepui Swift	B	11	21.8		17.9	25		Venezuela	296, 1308
Cypseloides niger	Black Swift	U	35	45.6	2.5	41	51.5		California, USA	722
Cypseloides niger	Black Swift	U	13	38.7	2.4	32.8	42		Western Mexico	722
Cypseloides niger	Black Swift	U	19	35.4	1.9	32	40		Costa Rica	722
Cypseloides lemosi	White-chested Swift	F	2	28.1		27.4	28.8		Colombia	329
Cypseloides fumigatus	Sooty Swift	B	4			40	44		Brazil	63
Cypseloides storeri	White-fronted Swift	M	1	39.5					Mexico	839
Cypseloides cherriei	Spot-fronted Swift	U	5	22.1		21.8	22.8		Venezuela	200
Cypseloides cryptus	White-chinned Swift	B	13	35.3	1.9	32.5	39.4		Costa Rica	1154
Cypseloides cryptus	White-chinned Swift	B	4	33.8		31.7	36.5		Colombia	329
Cypseloides senex	Great Dusky Swift	B	33	99.8		86.1	110		Argentina	1155
Cypseloides senex	Great Dusky Swift	B	8	67.8		56	80		Brazil	861
Streptoprocne rutila	Chestnut-collared Swift	B	45	20.2		17.8	24.2		Trinidad	199
Streptoprocne semicollaris	White-naped Swift	U	2	175		170	180			1148b
Streptoprocne zonaris	White-collared Swift	U	19	98.1	5.4	85.8	107		Venezuela	200
Streptoprocne zonaris	White-collared Swift	B	16	76.8		65	84		Brazil	861

Body Masses of World Birds (continued)

Species Name	Common Name	Sex	N	Mean	S.D.	Min	Max	Season	Location	Source #
Streptoprocne biscutata	Biscutate Swift	B	12	108	5	96	116		Brazil	198
Streptoprocne biscutata seridoensis	Biscutate Swift	M	1	85					Brazil	1104
Hydrochous gigas	Waterfall Swift	U	19	35.8					Java, Indonesia	56a
Collocalia esculenta	Glossy Swiftlet	B	8	8		7.2	9		Philippines; New Guinea	292, 954
Collocalia esculenta marginata	Glossy Swiftlet	U	5	7.1		6.3	8.3			1143
Collocalia esculenta natalis	Glossy Swiftlet	B	9	4.9		4	5.5		Christmas Is.	516
Collocalia esculenta uropygialis	Glossy Swiftlet	B	7	5.4		4.8	6		Vanuatu	656b
Collocalia esculenta nitens	Glossy Swiftlet	U	50	6.6	0.4	5.6	7.4	Y	Papua New Guinea	1196a
Collocalia linchi	Cave Swiftlet	U	7	5.9		5.8	6.3		Java, Indonesia	198
Collocalia troglodytes	Pygmy Swiftlet	U	21	5.4		4.5	6.6		Philippines	1143
Aerodramus elaphrus	Seychelles Swiftlet	U	13	10.7		9.5	14.3		Seychelles	181
Aerodramus francicus	Mascarene Swiftlet	B	20	8.9		7.9	11.4		Mauritius; Reunion	183
Aerodramus unicolor	Indian Swiftlet	M	2	11		11	11		India	7
Aerodramus mearnsi	Philippine Swiftlet	B	5	11.3		8.7	14.3		Mindanao, Philippines	1143
Aerodramus hirundinaceus	Mountain Swiftlet	B	24	10.1	0.7			B	Irian Jaya, New Guinea	1196e
Aerodramus spodiopygius leucopygia	White-rumped Swiftlet	B	55	6.8		6	8.5		New Caledonia	1017
Aerodramus spodiopygius assimilis	White-rumped Swiftlet	B	102	8.2	0.6				Fiji	1196c
Aerodramus spodiopygius chillagoensis	White-rumped Swiftlet	U	300	9.3	0.7	7.9	12.1			1196d

Body Masses of World Birds (continued)

Species Name	Common Name	Sex	N	Mean	S.D.	Min	Max	Season	Location	Source #
Aerodramus terraereginae	Australian Swiftlet	U	50	12	0.5	10.5	13		Australia	516
Aerodramus brevirostris	Himalayan Swiftlet	B	5	12.6		10	15.3			7, 1287
Aerodramus vulcanorum	Volcano Swiftlet	U	26	14.3	0.9	12.8	16		Java, Indonesia	198
Aerodramus whiteheadi	Whitehead's Swiftlet	F	1	14					Luzon, Philippines	1148
Aerodramus palawanensis	Palawan Swiftlet	U	15	20.4	1.3	17.9	22.5		Palawan, Philippines	198
Aerodramus vanikorensis yorki	Uniform Swiftlet	B	13	10.7	1.7	7.3	13		New Guinea	516
Aerodramus salangana	Mossy-nest Swiftlet	U		11.3						55
Aerodramus bartschi	Mariana Swiftlet	U	10	7.4		6.5	8		Guam	579
Aerodramus sawtelli	Atiu Swiftlet	U	144	8.6	0.7			B	Atiu, Cook Is.	1196c
Aerodramus maximus	Black-nest Swiftlet	U	3	17		13.6	20			1287
Aerodramus fuciphagus	Edible-nest Swiftlet	U	315	10.7	0.76	8.7	14.8		Sabah, Malaysia	380a
Aerodramus germani	German's Swiftlet	B	2	13.5		13	14		Hainan Is., China	1344
Aerodramus papuensis	Papuan Swiftlet	U	35	19.9	2.3	15.4	25.4	W	Papua New Guinea	1196a
Schoutedenapus myoptilus	Scarce Swift	B	2	23.5		22	25		Kenya; Uganda	1145, 1148b
Mearnsia picina	Philippine Needletail	F	1	42.5					Mindanao, Philippines	1148
Mearnsia novaeguineae	Papuan Needletail	U	7	32.6		29	39		Papua New Guinea	1155
Zoonavena thomensis	Sao Tome Spinetail	U		8					estimated	394
Telacanthura ussheri stictilaema	Mottled Spinetail	B	8	31.2		28.5	34.6			128

Body Masses of World Birds (continued)

Species Name	Common Name	Sex	N	Mean	S.D.	Min	Max	Season	Location	Source #
Telacanthura ussheri benguellensis	Mottled Spinetail	B	14	34.1		31	37.5		Zimbabwe	128
Telacanthura melanopygia	Black Spinetail	F	1	52						394
Rhaphidura leucopygialis	Silver-rumped Needletail	U	10	13.6	1.3	11.9	15.4		Borneo	1218
Rhaphidura sabini	Sabine's Spinetail	B	16	16		15	17		Kenya	1148b
Neafrapus cassini	Cassin's Spinetail	M	1	41.5						394
		F	2	38.5		37	40			
Neafrapus boehmi	Bat-like Spinetail	B	27			13	16		Angola	266
Hirundapus caudacutus	White-throated Needletail	B	31	95.5		47.4	169		Australia	516
Hirundapus cochinchinensis	Silver-backed Needletail	U	4	83.8		76.1	90			204, 1287
Hirundapus giganteus indicus	Brown-backed Needletail	U	3	127		123	134		Thailand	204
Hirundapus celebensis	Purple Needletail	U	22	180		170	203		Philippines	204
Chaetura spinicaudus	Band-rumped Swift	U	68	15.2		13	20		Trinidad	359
Chaetura spinicaudus aetherodroma	Band-rumped Swift	U	22	16.2	0.8					750
Chaetura fumosa	Costa Rican Swift	U	10	19.4	1.3				Costa Rica	750
Chaetura martinica	Lesser Antillean Swift	U	10	12.5					Dominica	949, 1148b
Chaetura cinereiventris	Gray-rumped Swift	U	128	13.9		11.8	18		Trinidad	359
Chaetura egregia	Pale-rumped Swift	U	12	23.4	1.5				Ecuador	750
Chaetura pelagica	Chimney Swift	U	1805	23.6		17	29.8	S	Illinois, USA	1362
Chaetura vauxi	Vaux's Swift	U	72	17.1	1.3	15	20.9		California, USA	314

Body Masses of World Birds (continued)

Species Name	Common Name	Sex	N	Mean	S.D.	Min	Max	Season	Location	Source #
Chaetura chapmani chapmani	Chapman's Swift	U	16	24.7		21.8	28		Trinidad	199a
Chaetura chapmani viridipennis	Chapman's Swift	U	1	23						749
Chaetura brachyura	Short-tailed Swift	U	240	18.3		15.5	22		Trinidad	199a
Chaetura andrei meridionalis	Ashy-tailed Swift	U	9	22.2						392, 749
Aeronautes saxatalis	White-throated Swift	U	20	32.1	2.5	27.8	36		California, USA	314
Aeronautes montivagus	White-tipped Swift	M	50	20.6	0.95	18.2	22.8		Venezuela	200
		F	39	19.7	1.1	17.2	21.8			
Aeronautes andecolus andecolus	Andean Swift	U	1	20						362
Aeronautes andecolus parvulus	Andean Swift	U	10	15.5	1.6	13	18			198
Tachornis phoenicobia	Antillean Palm-Swift	B	12	10.2	0.7	9	11		Dominican Republic	19
Tachornis furcata	Pygmy Swift	U	7	6.5						198
Tachornis squamata	Fork-tailed Palm-Swift	B	27	10.3		8	12		Suriname	498
Tachornis squamata	Fork-tailed Palm-Swift	U	10	11.7	0.9	10.5	13.6		Ecuador	747
Panyptila sanctihieronymi	Great Swallow-tailed Swift	B	3	48.1		46.9	50.3		Oaxaca, Mexico	75
Panyptila cayennensis	Lesser Swallow-tailed Swift	B	10	21.1		15.6	28		Belize; Trinidad; Brazil	359, 1027, 1204
Cypsiurus balasiensis	Asian Palm-Swift	B	17			8.3	10	PB		1287
Cypsiurus parvus	African Palm-Swift	U	61	13.6		10	18		Tropical Africa	223
Tachymarptis melba melba	Alpine Swift	U	43	104		76	120	B	Switzerland	223
Tachymarptis melba africanus	Alpine Swift	U	12	76		67	87		Angola	127

Body Masses of World Birds (continued)

Species Name	Common Name	Sex	N	Mean	S.D.	Min	Max	Season	Location	Source #
Tachymarptis melba tuneti	Alpine Swift	U				95	110			127
Tachymarptis melba maximus	Alpine Swift	U		128						127
Tachymarptis aequatorialis	Mottled Swift	B	26	92.8		83	104			126
Apus apus	Common Swift	U	218	37.6		31	43	B	Britain, UK	223
Apus unicolor	Plain Swift	B	208	22.5	3.1	18.4	30			401a
Apus niansae	Nyanza Swift	B	37	32.5		24	39			126
Apus pallidus	Pallid Swift	U	661	41.9	2.8	34.5	50		Spain	502
Apus barbatus	African Swift	B	18	42.8		39	50			126
Apus berliozi	Forbes-Watson's Swift	B	29	38.2		34	46			126
Apus bradfieldi	Bradfield's Swift	B	35	42.4		33	50		Angola	130
Apus balstoni	Madagascar Swift	B	4	46.1		38	51		Madagascar	68
Apus pacificus	Fork-tailed Swift	B	16	46		37	54	B	Mongolia; China	223, 1088
Apus pacificus	Fork-tailed Swift	U	20	33.8	5.9	22	47		Australia	1196a
Apus affinis bannermani	Little Swift	U	5	23.9		23	25			129
Apus affinis theresae	Little Swift	U	30	24.9		19	30		Angola	129
Apus affinis affinis	Little Swift	B	154	17.9		15	21.5		India	836
Apus affinis kuntzi	Little Swift	M	5	21.1		20	32			516
		F	9	26.1		22	35			
Apus nipalensis	House Swift	U	34			24.5	36.3			1287
Apus horus horus	Horus Swift	B	6			26	29		Angola	266
Apus caffer	White-rumped Swift	U	54	22.1		18	28		Tropical Africa	223
Family Hemiprocnidae										
Hemiprocne coronata	Crested Treeswift	B	6	29		26	34		India; Nepal; Thailand	198. 740, 1148

Body Masses of World Birds (continued)

Species Name	Common Name	Sex	N	Mean	S.D.	Min	Max	Season	Location	Source #
Hemiprocne longipennis	Gray-rumped Treeswift	U	1	42.8					Borneo	1218
Hemiprocne comata comata	Whiskered Treeswift	B	10	18.2	1.7	15.6	21.8			1154
Hemiprocne comata major	Whiskered Treeswift	B	20	23.4	1.5	20.2	25.3			1143
Hemiprocne mystacea	Moustached Treeswift	B	11	70.1	8.1	57	80		New Guinea; Solomon Is.	283, 399, 427, 535, 658
Family Trochilidae										
Ramphodon naevius	Saw-billed Hermit	U	366	7.9	1.2				Brazil	880
Eutoxeres aquila	White-tipped Sicklebill	U	24	10.6	1.4	8.6	14.5		Southwest Colombia	1037
Eutoxeres condamini	Buff-tailed Sicklebill	U	24	9.4	0.3					136
Eutoxeres condamini	Buff-tailed Sicklebill	U	22	12.4	1.6				Colombia	1038a
Glaucis dohrnii	Hook-billed Hermit	U	3	5.8					Brazil	880
Glaucis hirsutus	Rufous-breasted Hermit	B	4	5.9		5	7		Panama	990
Glaucis hirsutus	Rufous-breasted Hermit	U	224	7		5	9.5		Trinidad	1131
Glaucis hirsutus	Rufous-breasted Hermit	U	253	7.5	0.9				Brazil	880
Glaucis aeneus	Bronzy Hermit	B	4	4.8		4	5.4		Ecuador; Costa Rica	1145
Threnetes ruckeri	Band-tailed Barbthroat	U	3	5.2		4.5	6		Belize	580
Threnetes niger leucurus	Pale-tailed Barbthroat	U	42	5.6	0.1					136
Phaethornis yaruqui	White-whiskered Hermit	U	18	5.4	0.7	4	7		Colombia	1037

Body Masses of World Birds (continued)

Species Name	Common Name	Sex	N	Mean	S.D.	Min	Max	Season	Location	Source #
Phaethornis guy	Green Hermit	U	86	6.3		5.5	8		Trinidad	576
Phaethornis guy	Green Hermit	B	23			4	5.5		Panama	990
Phaethornis hispidus	White-bearded Hermit	U	41	5	0.4				Colombia	219
Phaethornis longirostris	Western Long-billed Hermit	U		6					Costa Rica	1178
Phaethornis superciliosus	Eastern Long-billed Hermit	U	54	6.3	0.5				Colombia	219a
Phaethornis malaris	Great-billed Hermit	B	14	5.9	0.5				Colombia	1009
Phaethornis syrmatophorus	Tawny-bellied Hermit	U	34	5.8	1	4.3	10.4		Colombia	1037
Phaethornis koepckeae	Koepcke's Hermit	M	16	5.4		4.7	5.8		Peru	259
		F	10	4.7		4.4	4.9			
Phaethornis philippii	Needle-billed Hermit	B	10	5	0.4					208, 319, 861, 1178
Phaethornis bourcieri	Straight-billed Hermit	B	21	4.3	0.5	3.4	5.5		Venezuela	1308
Phaethornis anthophilus	Pale-bellied Hermit	U	14	5.9	0.9	4	8		Venezuela	1254
Phaethornis eurynome	Scale-throated Hermit	U	155	5.3	0.6				Brazil	880
Phaethornis pretrei	Planalto Hermit	U	124	5.6	0.6				Brazil	880
Phaethornis augusti	Sooty-capped Hermit	U	10	6	0.8	4	6		Venezuela; Argentina	1254
Phaethornis subochraceus	Buff-bellied Hermit	B	10	3.8	0.3	3.5	4.5		Bolivia	1143
Phaethornis squalidus	Dusky-throated Hermit	U	24	3.4	0.5				Brazil	880
Phaethornis rupurumii	Streak-throated Hermit	U		3.7						522

Body Masses of World Birds (continued)

Species Name	Common Name	Sex	N	Mean	S.D.	Min	Max	Season	Location	Source #
Phaethornis longuemareus	Little Hermit	U	58	3	0.4					136
Phaethornis idaliae	Minute Hermit	U	8	2.6						453, 880
Phaethornis nattereri	Cinnamon-bellied Hermit	U	1	3.1						453
Phaethornis ruber	Reddish Hermit	U	13	2.4						136, 219, 875, 876, 880, 1178
Phaethornis stuarti	White-browed Hermit	U	2	2.9					Peru	1288
Phaethornis atrimentalis	Black-throated Hermit	B	5	2.6					Colombia	1009
Phaethornis striigularis	Stripe-throated Hermit	U		3						522
Phaethornis griseogularis	Gray-chinned Hermit	U	9	2.3					Peru; Ecuador; Venezuela	136, 953, 1302, 1308, 1311
Androdon aequatorialis	Tooth-billed Hummingbird	U	10	7.3		5.5	9.1		Panama	136, 990, 1037
Doryfera ludovicae	Green-fronted Lancebill	U	17	6	0.3	5.1	6.1		Colombia	324a
Doryfera johannae guianensis	Blue-fronted Lancebill	M	15	4.4		3	5		Venezuela	296
		F	6	4.7		4	6			
Doryfera johannae	Blue-fronted Lancebill	U	21	5.4	3.3				Colombia	1038a
Phaeochroa cuvierii	Scaly-breasted Hummingbird	U	37	8.9	0.4					136
Campylopterus curvipennis	Wedge-tailed Sabrewing	B	13	6	1.1	4.3	7.4		Belize; Yucatan, Mexico	910a, 1027
Campylopterus excellens	Long-tailed Sabrewing	M	22	10.8	1	9.2	12.7		Veracruz, Mexico	1322
		F	18	6.8	0.5	5.9	8			
Campylopterus largipennis	Gray-breasted Sabrewing	U	29	8.3	0.2					136

Body Masses of World Birds (continued)

Species Name	Common Name	Sex	N	Mean	S.D.	Min	Max	Season	Location	Source #
Campylopterus rufus	Rufous Sabrewing	M	9	7.7		7	9		El Salvador	1041a
		F	4	6.2		6	6.3			
Campylopterus rufus	Rufous Sabrewing	U	25	7.5	1.1	5.8	9		El Salvador	1041a
Campylopterus hemileucurus	Violet Sabrewing	M	18	11	1.8	8.1	16		El Salvador	1041a
		F	60	8.6	1.3	6	14			
Campylopterus hyperythrus	Rufous-breasted Sabrewing	B	5	6.4		5	7		Venezuela	1143
Campylopterus ensipennis	White-tailed Sabrewing	U	7	9.6						136, 359
Campylopterus falcatus	Lazuline Sabrewing	U	10	7.2						136, 200, 633, 1254
Campylopterus villaviscensio	Napo Sabrewing	M	8	8.4		7.4	9.3		Peru	259
		F	16	6.5		5.2	7.4			
Campylopterus duidae	Buff-breasted Sabrewing	M	35	6.8	0.8	5	8		Venezuela	1308
		F	22	5.4	0.7	4.7	7.8			
Campylopterus cirrochloris	Sombre Hummingbird	U	878	7.7	0.7				Brazil	880
Eupetomena macroura	Swallow-tailed Hummingbird	U	474	9	0.9				Brazil	880
Florisuga mellivora	White-necked Jacobin	U	28	7.4	0.6					136
Florisuga fuscus	Black Jacobin	U	2317	8.1	1				Brazil	880
Colibri delphinae	Brown Violet-ear	B	12	6.4	0.8	5.5	8			200, 357, 613, 990, 1027
Colibri thalassinus	Green Violet-ear	U	84	5.9	0.3					136
Colibri coruscans	Sparkling Violet-ear	U	16	6.7	1	3.8	8.3		Colombia	324a
Colibri serrirostris	White-vented Violet-ear	U	321	6.7	0.9				Brazil	880
Anthracothorax viridigula	Green-throated Mango	U	25	7.5	0.5					136

Body Masses of World Birds (continued)

Species Name	Common Name	Sex	N	Mean	S.D.	Min	Max	Season	Location	Source #
Anthracothorax prevostii	Green-breasted Mango	B	14	6.4	0.4	5.7	7		Belize; Colombia	1027, 1029
Anthracothorax nigricollis	Black-throated Mango	U	51	7	0.6					136
Anthracothorax dominicus	Antillean Mango	M	47	5.9	0.5	4.8	7.2		Puerto Rico	19
		F	60	5.1	0.5	4	6.4			
Anthracothorax dominicus	Antillean Mango	B	13	6.4	0.9	4.4	8.2		Dominican Republic	19
Anthracothorax viridis	Green Mango	B	25	6.6	0.5	6	7.8		Puerto Rico	19
Anthracothorax mango	Jamaican Mango	U	7	8.3						78, 136, 1159
Avocettula recurvirostris	Fiery-tailed Awlbill	M	1	4.2					Rondonia, Brazil	1143
Topaza pella	Crimson Topaz	M	19	13.6	1.9	11	18			543
		F	18	10.8	1.3	9	12.5			
Topaza pyra	Fiery Topaz	M	7	14.2		12	17			543
		F	4	11.1		10.5	12			
Eulampis jugularis	Purple-throated Carib	U	19	9.3	1.4	7.5	12		Martinique	1065
Eulampis jugularis	Purple-throated Carib	U	30	8.8					Dominica	949
Eulampis jugularis	Purple-throated Carib	M	30	9.7	0.9	8.3	11.7		St. Vincent	19
		F	50	8.3	0.6	7.1	9.5			
Eulampis jugularis	Purple-throated Carib	M	36	9.3	0.6	8.1	11		St. Lucia	19
		F	46	7.9	0.4	6.9	8.7			
Eulampis jugularis	Purple-throated Carib	M	10	9.6	0.7	8.8	10.7		Montserrat	19
		F	16	8.1	0.4	7.4	9			
Eulampis holosericeus	Green-throated Carib	B	36	5.7	0.5	4.7	6.9		Caribbean Is.	19
Chrysolampis mosquitus	Ruby-topaz Hummingbird	U	20	3.9	0.4					136
Orthorhyncus cristatus	Antillean Crested Hummingbird	B	12	2.7	0.5	2	3.8		Guadalupe	19

Body Masses of World Birds (continued)

Species Name	Common Name	Sex	N	Mean	S.D.	Min	Max	Season	Location	Source #
Orthorhyncus cristatus	Antillean Crested Hummingbird	B	14	3.1	0.4	2.5	4		Montserrat	19
Orthorhyncus cristatus	Antillean Crested Hummingbird	B	11	3.3	0.5	2.6	4.3		St. Vincent	19
Klais guimeti	Violet-headed Hummingbird	B	12	2.6						136, 489, 618
Stephanoxis lalandi	Plovercrest	U	8	4						63, 136, 453, 878
Abeillia abeillei	Emerald-chinned Hummingbird	U	1	2.7						136
Lophornis ornatus	Tufted Coquette	U	11	2.3						136, 359
Lophornis gouldii	Dot-eared Coquette	F	1	2.4					Brazil	449
Lophornis magnificus	Frilled Coquette	U	19	2.7	0.3				Brazil	880
Lophornis delattrei	Rufous-crested Coquette	U		2.8					Costa Rica	1170
Lophornis chalybeus verreauxi	Festive Coquette	F	1	3						453
Lophornis pavoninus	Peacock Coquette	M	1	3					Venezuela	296
Lophornis helenae	Black-crested Coquette	U	6	2.6						136, 208
Lophornis adorabilis	White-crested Coquette	U	4	2.7						136
Popelairia popelairii	Wire-crested Thorntail	U		2.5					Peru	1210
Popelairia langsdorffi	Black-bellied Thorntail	U	5	2.8						136, 208
Discosura conversii	Green Thorntail	U	5	3						136
Discosura longicauda	Racket-tailed Coquette	U	2	3.4						136, 453
Trochilus polytmus	Red-billed Streamertail	M	5	5		4.5	5.3		Jamaica	1159
		F	11	4.7	0.3	4.1	5.1			

Body Masses of World Birds (continued)

Species Name	Common Name	Sex	N	Mean	S.D.	Min	Max	Season	Location	Source #
Trochilus scitulus	Black-billed Streamertail	F	1	4					Jamaica	944
Chlorostilbon notatus	Blue-chinned Sapphire	B	30	4.2		3.5	5		Trinidad	359
Chlorostilbon mellisugus gibsoni	Blue-tailed Emerald	U	8	2.8						136, 208
Chlorostilbon mellisugus pumilus	Blue-tailed Emerald	B	10	2.6	0.2				Colombia	1165
Chlorostilbon mellisugus caribaeus	Blue-tailed Emerald	M	6	2.7						1165
		F	4	2.5						
Chlorostilbon mellisugus napensis	Blue-tailed Emerald	M	4	2.7						1165
		F	3	2.4						
Chlorostilbon auriceps	Golden-crowned Emerald	B	2	2.2		2.2	2.2		Jalisco, Mexico	1151
Chlorostilbon forficatus	Cozumel Emerald	B	3	2.5		2.4	2.6		Cozumel Is., Mexico	910a
Chlorostilbon canivetii	Canivet's Emerald	B	19	2.5					Costa Rica	357
Chlorostilbon assimilis	Garden Emerald	U	2	3		3	3.1			453
Chlorostilbon aureoventris	Glittering-bellied Emerald	U	13	3.5		2.8	4.9			63, 208, 453, 657, 877, 878
Chlorostilbon olivaresi	Chiribiquete Emerald	M	6	3.8					Colombia	1165
		F	11	3.6	0.2					
Chlorostilbon ricordii	Cuban Emerald	B	154	3.2	0.3	2.5	5		Cuba	19
Chlorostilbon bracei	Brace's Emerald	U	9	2.8		2.5	3.1		Bahamas (extinct)	1159
Chlorostilbon swainsonii	Hispaniolian Emerald	B	20	4.3	0.6	2.5	5		Dominican Republic	19
Chlorostilbon maugaeus	Puerto Rican Emerald	B	123	3.1	0.4	2.5	6		Puerto Rico	19
Panterpe insignis	Fiery-throated Hummingbird	U	47	5.7	0.4					136

Body Masses of World Birds (continued)

Species Name	Common Name	Sex	N	Mean	S.D.	Min	Max	Season	Location	Source #
Elvira chionura	White-tailed Emerald	U	16	3	0.2					136
Elvira cupreiceps	Coppery-headed Emerald	U	13	3						136
Eupherusa cyanophrys	Blue-capped Hummingbird	B	7	4.8		4.1	5.4		Mexico	208, 1145
Eupherusa eximia	Stripe-tailed Hummingbird	U	126	4.3	0.3					136
Eupherusa nigriventris	Black-bellied Hummingbird	U		3.5					Costa Rica	1163
Goethalsia bella	Rufous-cheeked Hummingbird	B	4	3.4		3	4			208, 990
Goldmania violiceps	Violet-capped Hummingbird	B	16	4.1		3.2	4.5		Panama	79
Cynanthus sordidus	Dusky Hummingbird	U	4	4		3.4	4.3			74, 208
Cynanthus latirostris	Broad-billed Hummingbird	B	19	3.1	0.4	2.5	4		Western Mexico	1151
Cyanophaia bicolor	Blue-headed Hummingbird	B	9	4.5		4.2	4.8		Dominica	19
Thalurania colombica	Violet-crowned Woodnymph	B	114	4.5	0.5				Costa Rica	80
Thalurania ridgwayi	Mexican Woodnymph	B	7	3.8		3.5	4.2		Jalisco, Mexico	1143
Thalurania furcata	Fork-tailed Woodnymph	U	72	4.4	0.2					136
Thalurania furcata baeri	Fork-tailed Woodnymph	U	11	4	0.2	3.8	4.2		Brazil	392
Thalurania watertonii	Long-tailed Woodnymph	U	3	4.6						136, 453
Thalurania glaucopis	Violet-capped Woodnymph	U	880	4.8	0.6				Brazil	880
Damophila julie	Violet-bellied Hummingbird	U	51	3.3	0.2					136

Body Masses of World Birds (continued)

Species Name	Common Name	Sex	N	Mean	S.D.	Min	Max	Season	Location	Source #
Lepidopyga coeruleogularis	Sapphire-throated Hummingbird	F	3	4.2					Panama	489
Lepidopyga goudoti	Shining-green Hummingbird	U	4	4						136
Hylocharis eliciae	Blue-throated Goldentail	U	13	3.6						136
Hylocharis sapphirina	Rufous-throated Sapphire	U	19	4.4	0.4				Brazil	880
Hylocharis cyanus	White-chinned Sapphire	B	25	3.3		2.7	4		Suriname	498
Hylocharis chrysura	Gilded Sapphire	B	14	4.5						208, 210a, 453, 880
Hylocharis grayi	Blue-throated Sapphire	B	9	5.6		4.6	6.3			208, 798, 1167
Hylocharis humboldtii	Humboldt's Sapphire	B	5	6.2		5.8	6.6		Colombia	1167
Hylocharis xantusii	Xanthus's Hummingbird	B	23	3.4		3.1	4		Baja California, Mexico	541
Hylocharis leucotis	White-eared Hummingbird	M	158	3.6	0.3					314
		F	51	3.2	0.2					
Chrysuronia oenone	Golden-tailed Sapphire	M	28	5.3	0.4	4.7	6.3		Venezuela	200
		F	8	4.8	0.4	4.3	5.3			
Leucochloris albicollis	Wine-throated Hummingbird	U	585	6.3	0.8				Brazil	880
Polytmus guainumbi	White-tailed Goldenthroat	B	9	5.2					Suriname	498
Polytmus milleri	Tepui Goldenthroat	U	1	4.1						453
Polytmus theresiae	Green-tailed Goldenthroat	B	22	3.8		3.3	4.2		Suriname	498
Leucippus fallax	Buffy Hummingbird	M	1	6					Venezuela	1254
Leucippus baeri	Tumbes Hummingbird	B	3	4.5		4.2	5			208, 1143

Body Masses of World Birds (continued)

Species Name	Common Name	Sex	N	Mean	S.D.	Min	Max	Season	Location	Source #
Leucippus taczanowskii	Spot-throated Hummingbird	B	3	6.4		6	7.2			136, 208
Leucippus chlorocercus	Olive-throated Hummingbird	B	4	4.6		3.7	5.3			208, 1143
Leucippus chionogaster	White-bellied Hummingbird	M	9	5.3		4.9	6.7		Bolivia	1055
		F	7	4.8		4.6	5.2			
Leucippus viridicauda	Green-and-white Hummingbird	B	3	6.1		5.6	6.3			208, 390
Leucippus hypostictus	Many-spotted Hummingbird	B	10	7					Peru	390
Amazilia tzacatl	Rufous-tailed Hummingbird	U	105	5	0.3					136
Amazilia tzacatl	Rufous-tailed Hummingbird	U	15	4.6	0.4	4	5.5		Belize	220
Amazilia castaneiventris	Chestnut-bellied Hummingbird	U		5					weight approximate	218
Amazilia amazilia	Amazilia Hummingbird	U	18	4.3	0.6					136
Amazilia alticola	Loja Hummingbird	U	3	4.3		3.9	5		Peru	1143
Amazilia yucatanensis	Buff-bellied Hummingbird	M	8	4.1		3	4.7			208, 587
		F	8	3.7		2.9	4.5			
Amazilia rutila	Cinnamon Hummingbird	B	14	5	0.7	3.3	5.7		Mexico	1151
Amazilia rutila graysoni	Cinnamon Hummingbird	B	8	5.5		4.9	6.1		Tres Marias Is., Mexico	440
Agyrtria leucogaster	Plain-bellied Emerald	B	25	4.7		3.9	5.4		Suriname	498
Agyrtria versicolor	Versicolored Emerald	U	185	4.1	0.4				Brazil	880
Agyrtria brevirostris	White-chested Emerald	U	181	4.3	0.7				Brazil	880
Agyrtria franciae	Andean Emerald	U	7	5						136, 208

Body Masses of World Birds (continued)

Species Name	Common Name	Sex	N	Mean	S.D.	Min	Max	Season	Location	Source #
Agyrtria candida	White-bellied Emerald	B	13	3.8	0.4	2.9	4.3		Mexico; Belize	640, 910a, 1027
Agyrtria cyanocephala	Azure-crowned Hummingbird	U	13	5.1	1	3.7	7		El Salvador	1041a
Agyrtria violiceps	Violet-crowned Hummingbird	U	16	5.1	0.4	4.6	5.8		Arizona, USA	1292
Agyrtria viridifrons	Green-fronted Hummingbird	B	10	6.3						136, 1145
Polyerata lactea	Sapphire-spangled Emerald	U	11	4.6	0.1					136
Polyerata fimbriata	Glittering-throated Emerald	U	76	4.9	0.4				Colombia	219
Polyerata amabilis	Blue-chested Hummingbird	U	18	4.3	0.6					136
Polyerata decora	Charming Hummingbird	U		4.7						886
Polyerata rosenbergi	Purple-chested Hummingbird	U	4	3.7		3.4	4			208, 1037
Polyerata boucardi	Mangrove Hummingbird	U		4.5					Costa Rica	1170
Saucerottia saucerrottei	Steely-vented Hummingbird	U	139	4.5	0.4					136
Saucerottia cyanifrons	Indigo-capped Hummingbird	U		5						902
Saucerottia edward	Snowy-bellied Hummingbird	U	58	4.7	0.2					136
Saucerottia cyanura	Blue-tailed Hummingbird	M	1	3.9					Honduras	1145
Saucerottia beryllina	Berylline Hummingbird	B	49	4.5		3	6		El Salvador	1041a
Saucerottia viridigaster	Green-bellied Hummingbird	M	2	4		3.9	4.1		Venezuela	1308
		F	3	3.5		3.4	3.6			

Body Masses of World Birds (continued)

Species Name	Common Name	Sex	N	Mean	S.D.	Min	Max	Season	Location	Source #
Saucerottia cupreicauda	Copper-tailed Hummingbird	B	13	3.9		3.3	5		Brazil; Venezuela	296, 1143
Saucerottia tobaci	Copper-rumped Hummingbird	U	83	4.7		3.5	6		Trinidad	1131
Microchera albocoronata	Snowcap	U	8	2.7	0.2					136
Chalybura buffonii	White-vented Plumeleteer	M	19	7.1	0.7	6	8		Venezuela	1254
		F	24	6.5	0.9	5	8			
Chalybura urochrysia	Bronze-tailed Plumeleteer	M	7	6.8		6.4	7.2		Panama	79
		F	6	6		5.4	7.2			
Lampornis clemenciae	Blue-throated Hummingbird	M	195	8.4	0.2					314
		F	67	6.8	0.3					
Lampornis amethystinus	Amethyst-throated Hummingbird	U	33	5.2	1	3	7		El Salvador	1041a
Lampornis viridipallens	Green-throated Mountain-gem	M	197	5.7	0.8	3	9		El Salvador	1041a
		F	121	5	0.7	3.3	7			
Lampornis hemileucus	White-bellied Mountain-gem	U		5.5					Costa Rica	1163
Lampornis castaneoventris	White-throated Mountain-gem	U	40	5.6	0.3					136
Lampornis calolaemus	Purple-throated Mountain-gem	M	7	5.6					Costa Rica	357
		F	2	4.2						
Lamprolaima rhami	Garnet-throated Hummingbird	B	3	7.8		7	9.5			208, 1145
Phlogophilus hemileucurus	Ecuadorian Piedtail	U	15	2.8	0.5	2.2	3.7			208, 633, 899
Phlogophilus harterti	Peruvian Piedtail	B	26	2.7					Peru	370
Adelomyia melanogenys	Speckled Hummingbird	U	41	4.6	1.7				Southwest Colombia	1037
Adelomyia melanogenys	Speckled Hummingbird	U	23	5.3					Ecuador	1311
Clytolaema rubricauda	Brazilian Ruby	U	1105	7.9	1.1				Brazil	880

Body Masses of World Birds (continued)

Species Name	Common Name	Sex	N	Mean	S.D.	Min	Max	Season	Location	Source #
Heliodoxa aurescens	Gould's Jewelfront	U	13	6.4						136, 208, 318
Heliodoxa rubinoides	Fawn-breasted Brilliant	M	5	8						208
		F	6	7.6						
Heliodoxa leadbeateri	Violet-fronted Brilliant	U	32	7.4	1.2					136
Heliodoxa xanthogonys	Velvet-browed Brilliant	M	14	7.3		6	9		Venezuela	296
		F	6	6.7		6	7			
Heliodoxa schreibersii	Black-throated Brilliant	M	4	9.9					Peru	370
		F	14	8.5	0.4					
Heliodoxa gularis	Pink-throated Brilliant	B	4	6.3						208, 1038a
Heliodoxa branickii	Rufous-webbed Brilliant	B	5	5.2						208, 1288
Heliodoxa imperatrix	Empress Brilliant	U	20	8.8	0.6	7.2	10.4		Colombia	1037
Heliodoxa jacula	Green-crowned Brilliant	M	6	9.1		8.6	9.5		Panama	79
		F	7	7.2		5.6	8.2			
Eugenes fulgens	Magnificent Hummingbird	M	119	7.7	0.4				Oaxaca, Mexico	587
		F	24	6.4	0.5					
Eugenes fulgens	Magnificent Hummingbird	M	10	9.9	0.3	9.4	10.4		Costa Rica	936
		F	12	8.4	0.3	7.9	9			
Sternoclyta cyanopectus	Violet-chested Hummingbird	U	43	8	1.1	6.2	10		Venezuela	1254
Urochroa bougueri	White-tailed Hillstar	U	10	11.4		8.7	14		Peru; Colombia	881, 899, 1037
Boissonneaua matthewsii	Chestnut-breasted Coronet	U	13	7.7						136, 208, 944, 1042
Boissonneaua flavescens	Buff-tailed Coronet	U	12	8.2	0.4	7.3	8.8		Colombia	324a
Boissonneaua jardini	Velvet-purple Coronet	U	14	9.1	0.8	7.8	11		Colombia	1037
Aglaeactis cupripennis	Shining Sunbeam	U	23	7.6	0.6					136

Body Masses of World Birds (continued)

Species Name	Common Name	Sex	N	Mean	S.D.	Min	Max	Season	Location	Source #
Aglaeactis castelnaudii	White-tufted Sunbeam	B	2	6.7		6.2	7.2			208
Aglaeactis pamela	Black-hooded Sunbeam	B	2	5.5		5.2	5.8			208
Oreotrochilus estella	Andean Hillstar	U	37	8.1	0.4					136
Oreotrochilus chimborazo	Chimborazo Hillstar	B	2	8.2		8	8.4		Colombia	1037
Oreotrochilus melanogaster	Black-breasted Hillstar	M	2	9.1		8.7	9.5		Peru	1103
Lafresnaya lafresnayi	Mountain Velvetbreast	B	40	5.3		4.5	6.7		Ecuador	633
Coeligena coeligena	Bronzy Inca	U	34	6.8	0.1					136
Coeligena wilsoni	Brown Inca	U	32	6.7	0.7	4.7	9.3		Colombia	1037
Coeligena prunellei	Black Inca	U				6.6	7.4			218a
Coeligena torquata	Collared Inca	M	32	7.3	1.3	3.8	12.9		Colombia	324a
		F	6	6.2		5.1	6.9			
Coeligena phalerata	White-tailed Starfrontlet	B	2	6.1		5.4	6.8			208
Coeligena eos	Golden Starfrontlet	U		6.6						522
Coeligena bonapartei	Golden-bellied Starfrontlet	M	124	6.6	0.6	5.3	10.1		Colombia	324a
		F	94	6.4	0.6	3.3	8.5			
Coeligena helianthea	Blue-throated Starfrontlet	M	13	7	0.3				Colombia	653
		F	17	6.3	0.3					
Coeligena lutetiae	Buff-winged Starfrontlet	M	10	7.6	0.6				Colombia	653
		F	12	7	0.5					
Coeligena orina	Dusky Starfrontlet	B	4	7		6.7	7.2		Colombia	653
Coeligena violifer	Violet-throated Starfrontlet	M	44	7.7					Bolivia	963
		F	23	7.1						
Coeligena iris	Rainbow Starfrontlet	B	53	6.9		6	8		Ecuador	633

Body Masses of World Birds (continued)

Species Name	Common Name	Sex	N	Mean	S.D.	Min	Max	Season	Location	Source #
Ensifera ensifera	Sword-billed Hummingbird	B	11	11.2						136, 208, 633, 1037
Pterophanes cyanopterus	Great Sapphirewing	M	3	10.2					Bolivia	963
		F	3	8.4						
Patagona gigas	Giant Hummingbird	U	10	20.2	1.3					136
Sephanoides sephaniodes	Green-backed Firecrown	M	15	5.7					Chile	207
		F	12	4.7						
Sephanoides fernandensis	Juan Fernandez Firecrown	M	15	10.9					Chile	207
		F	2	7						
Heliangelus clarisse	Longuemare's Sunangel	U		5.3						522
Heliangelus mavors	Orange-throated Sunangel	U		4.2						522
Heliangelus amethysticollis	Amethyst-throated Sunangel	M	15	5.3					Bolivia	963
		F	11	4.8						
Heliangelus strophianus	Gorgeted Sunangel	M	4	5.5		4.9	6		Colombia	1037
Heliangelus exortis	Tourmaline Sunangel	M	115	4.8	0.6	3.5	8.5		Colombia	324a
		F	28	4.4	0.3	4	5.2			
Heliangelus viola	Purple-throated Sunangel	B	14	5.4		4.4	6.6			208, 633, 1042
Heliangelus regalis	Royal Sunangel	B	2	3.6		3.5	3.7		Peru	259, 1143
Eriocnemis vestitus	Glowing Puffleg	B	22	4.4		3.9	5		Peru	900
Eriocnemis vestitus	Glowing Puffleg	B	80	4.7		3.3	7.2		Colombia	324a
Eriocnemis luciani	Sapphire-vented Puffleg	B	31	6		5.4	6.4		Ecuador	633
Eriocnemis cupreoventris	Coppery-bellied Puffleg	U	13	5.3	0.3	4.7	5.8		Colombia	324a
Eriocnemis mosquera	Golden-breasted Puffleg	B	11	5.3	1.8		6.2		Colombia	1037

Body Masses of World Birds (continued)

Species Name	Common Name	Sex	N	Mean	S.D.	Min	Max	Season	Location	Source #
Eriocnemis glaucopoides	Blue-capped Puffleg	B	13	3.7	0.5	3	4.5		Bolivia	1143
Eriocnemis alinae	Emerald-bellied Puffleg	U				4	4.5		Colombia	1069
Haplophaedia aureliae	Greenish Puffleg	B	10	5.2		4.7	6		Colombia	1068
Haplophaedia assimilis	Buff-thighed Puffleg	B	8	4.8		4.1	5.5		Peru	1143
Haplophaedia lugens	Hoary Puffleg	U	6	6.9		6	8.8		Colombia	881, 1037
Urosticte benjamini	Purple-bellied Whitetip	B	9	3.9		3.5	4.5		Colombia	259, 944, 1037
Urosticte ruficrissa	Rufous-vented Whitetip	U	7	4.3					Colombia	1038a
Ocreatus underwoodii	Booted Racket-tail	U	19	3	0.1					136
Lesbia victoriae	Black-tailed Trainbearer	B	4	5		4.5	5.3			136, 208, 633
Lesbia nuna	Green-tailed Trainbearer	B	12	3.7						136, 208, 953
Sappho sparganura	Red-tailed Comet	B	10	5.3						208, 362, 1040
Polyonymus caroli	Bronze-tailed Comet	B	6	4.8						136, 208
Ramphomicron microrhynchum	Purple-backed Thornbill	B	6	3.6						208, 633
Oreonympha nobilis	Bearded Mountaineer	B	3	7.7		7	9			136, 208
Oxypogon guerinii	Bearded Helmetcrest	U	4	4.8						136, 1263
Metallura tyrianthina	Tyrian Metaltail	B	93	3.8					Ecuador	632
Metallura aeneocauda	Scaled Metaltail	B	25	5.2					Bolivia	963
Metallura eupogon	Fire-throated Metaltail	U	10	4.6	0.3				Peru	1288

Body Masses of World Birds (continued)

Species Name	Common Name	Sex	N	Mean	S.D.	Min	Max	Season	Location	Source #
Metallura theresiae	Coppery Metaltail	B	5	4.9						136, 208
Metallura odomae	Neblina Metaltail	B	9	5.1		4.7	5.4			28, 900
Metallura baroni	Violet-throated Metaltail	B	9	4.3					Ecuador	632
Metallura williami	Viridian Metaltail	B	12	4.5						79, 319
Metallura phoebe	Black Metaltail	B	5	5.4						136, 208, 1042
Chalcostigma ruficeps	Rufous-capped Thornbill	B	5	3.4						208, 963
Chalcostigma stanleyi	Blue-mantled Thornbill	U	11	5.8						136
Chalcostigma herrani	Rainbow-bearded Thornbill	B	10	5.8		5.1	7		Peru; Colombia	900, 1037
Aglaiocercus kingi	Long-tailed Sylph	M	11	5.1	0.5	4	6		Venezuela	1254
		F	8	4.6		4	4.6			
Aglaiocercus coelestis	Violet-tailed Sylph	M	31	6.3	0.5	5	7.9		Colombia	1037
		F	49	5	0.4	4	6.2			
Aglaiocercus berlepschi	Venezuelan Sylph	M		5.5						522
		F		4.5						
Augastes scutatus	Hyacinth Visorbearer	U	1	3						136
Augastes lumachella	Hooded Visorbearer	U	2	4						136
Augastes geoffroyi	Wedge-billed Hummingbird	U	13	3.6	0.2					136
Heliothryx barroti	Purple-crowned Fairy	B	9	5.5	0.2					136
Heliothryx auritus	Black-eared Fairy	U	11	5.4						136, 208, 1308
Heliactin bilopha	Horned Sungem	B	6	4.4					Brazil	1109
Loddigesia mirabilis	Marvelous Spatuletail	U	1	3						136

Body Masses of World Birds (continued)

Species Name	Common Name	Sex	N	Mean	S.D.	Min	Max	Season	Location	Source #
Heliomaster constantii	Plain-capped Starthroat	B	9	7.3		6.5	9			587
Heliomaster longirostris	Long-billed Starthroat	U	10	6.6						136, 208
Heliomaster squamosus	Stripe-breasted Starthroat	U	3	5.9						136, 880
Heliomaster furcifer	Blue-tufted Starthroat	B	5	5.4		4.9	6.5			63, 136, 362, 657
Rhodopis vesper	Oasis Hummingbird	U	6	3.8						136, 208
Thaumastura cora	Peruvian Sheartail	U	3	1.9						136, 208
Tilmatura dupontii	Sparkling-tailed Hummingbird	U	12	2.2	0.2					136
Doricha enicura	Slender Sheartail	U	1	2.4						136
Doricha eliza	Mexican Sheartail	B	4	2.5		2.3	2.7		Yucatan, Mexico	910a
Calliphlox evelynae	Bahama Woodstar	F	2	2.4		2.2	2.6		Bahamas	1159
Calliphlox bryantae	Magenta-throated Woodstar	U	21	3.3						136
Calliphlox mitchellii	Purple-throated Woodstar	U	3	3.1						136
Calliphlox amethystina	Amethyst Woodstar	U	28	2.8	0.3				Brazil	880
Microstilbon burmeisteri	Slender-billed Woodstar	F	1	2.5					Argentina	1040
Calothorax lucifer	Lucifer Hummingbird	B	9	3.3		2.9	3.5	F	Arizona, USA	1026
Calothorax pulcher	Beautiful Hummingbird	B	5	2.7						136, 208
Mellisuga minima	Vervain Hummingbird	U	6	2.4						136

Body Masses of World Birds (continued)

Species Name	Common Name	Sex	N	Mean	S.D.	Min	Max	Season	Location	Source #
Mellisuga helenae	Bee Hummingbird	B	4	2.2		2	2.6		Cuba	19
Archilochus colubris	Ruby-throated Hummingbird	M	121	2.9	0.2	2.4	3.7		Pennsylvania, USA	828
		F	463	3.3	0.3	2.6	4.1			
Archilochus alexandri	Black-chinned Hummingbird	M	1207	3.2	0.4	2.3	4.8	F	Arizona, USA	1026
		F	733	3.6	0.4	2.5	5.1			
Calypte anna	Anna's Hummingbird	M	431	4.5	0.4	3.6	6.3	F	Arizona, USA	1028
		F	541	4	0.4	3.2	5.6			
Calypte costae	Costa's Hummingbird	M	33	3		2.5	5.2			587
		F	27	3.2		2.5	3.4			
Atthis heloisa	Bumblebee Hummingbird	B	15	2.2		2	2.7			587
Atthis ellioti	Wine-throated Hummingbird	F	6	2.9		2	4		Guatemala; El Salvador	966, 1041a
Stellula calliope	Calliope Hummingbird	M	46	2.5		1.9	3.2			587
		F	26	2.8		2.2	3.2			
Myrtis fanny	Purple-collared Woodstar	U	3	2.3		2.3	2.3			136, 208
Chaetocercus mulsant	White-bellied Woodstar	U	8	3.9						136, 208
Chaetocercus bombus	Little Woodstar	U	1	6.5						785a
Selasphorus scintilla	Scintillant Hummingbird	U	39	2.2	0.2					136
Selasphorus flammula simoni	Volcano Hummingbird	B	9	2.8		2.4	3.2			208, 1161
Selasphorus flammula torridus	Volcano Hummingbird	B	30	2.6		2	3		Costa Rica	1161
Selasphorus platycercus	Broad-tailed Hummingbird	M	97	3.5	0.3			B	Colorado, USA	162
		F	109	3.6	0.4					
Selasphorus rufus	Rufous Hummingbird	B	112	3.5		2.8	4.5	F	Arizona, USA	1026
Selasphorus sasin sasin	Allen's Hummingbird	M	38	3.1		2.5	3.8			587
		F	18	3.2		2.8	3.5			

Body Masses of World Birds (continued)

Species Name	Common Name	Sex	N	Mean	S.D.	Min	Max	Season	Location	Source #
Selasphorus sasin sedentarius	Allen's Hummingbird	M	19	3.5						587
		F	26	3.7						
Family Coliidae										
Colius striatus	Speckled Mousebird	U	102	51.1		36.8	64		Kenya	121
Colius leucocephalus	White-headed Mousebird	M	8	37.8		31	42			394
		F	4	34		28	39			
Colius castanotus	Red-backed Mousebird	B	100			39	82		Angola	266
Colius colius	White-backed Mousebird	U	15	41.4		29	54		South Africa	508
Urocolius macrourus	Blue-naped Mousebird	U	7	45.5		35	52.5			120, 852
Urocolius indicus	Red-faced Mousebird	U	17	52.7		47.5	59.5		South Africa	508
Urocolius indicus lacteifrons	Red-faced Mousebird	B	31			45	65		Angola	266
Family Trogonidae										
Apaloderma narina	Narina Trogon	M	11	65		54	77		South Africa	1036
		F	7	70.5		58	80			
Apaloderma vittatum	Bar-tailed Trogon	B	18	55		49	59			591a
Priotelus temnurus	Cuban Trogon	U	5	53.4		40.3	60		Cuba	591a, 782, 801
Priotelus roseigaster	Hispaniolan Trogon	U	1	74.2					Hispaniola	591a
Trogon melanocephalus	Black-headed Trogon	B	20	85.1		60	95			591a
Trogon citreolus	Citreoline Trogon	B	23	79		66	91			591a
Trogon viridis	White-tailed Trogon	B	31	89.7		77	107			591a
Trogon bairdii	Baird's Trogon	U	2	94.4		93.7	95			591a

Body Masses of World Birds (continued)

Species Name	Common Name	Sex	N	Mean	S.D.	Min	Max	Season	Location	Source #
Trogon violaceus	Violaceous Trogon	B	11	51.5		48	55			359
Trogon violaceus caligatus	Violaceous Trogon	U		42						522
Trogon mexicanus	Mountain Trogon	B	20	71		61.5	85			591a
Trogon comptus	White-eyed Trogon	B	4	114		100	126			591a
Trogon collaris	Collared Trogon	B	47	64.2					Panama	489
Trogon elegans	Elegant Trogon	B	18	70.9	7.7	51.6	80		El Salvador	1041a
Trogon aurantiiventris	Orange-bellied Trogon	B	7	66.7		61.5	72			489, 591a, 1180
Trogon personatus	Masked Trogon	U	10	63.4		57	75			953, 963, 1037, 1288, 1308
Trogon rufus	Black-throated Trogon	B	10	53.8	6.6	46	69.5			79, 155, 293, 615, 990, 1180
Trogon surrucura	Surucua Trogon	B	3	73.3		70	78		Brazil	63
Trogon curucui	Blue-crowned Trogon	B	18	54		38	65			591a
Trogon melanurus	Black-tailed Trogon	B	11	114					Suriname	498
Trogon massena	Slaty-tailed Trogon	B	26	141					Panama	489
Trogon clathratus	Lattice-tailed Trogon	U		130					Costa Rica	1170
Euptilotis neoxenus	Eared Quetzal	M	5	116		103	127			591a
		F	6	129		115	148			
Pharomachrus mocinno costaricensis	Resplendent Quetzal	M	12	206	9	193	221		Costa Rica	76
		F	8	199		184	215			
Pharomachrus antisianus	Crested Quetzal	M	4	170		145	226			591a
		F	2	136		134	138			

Body Masses of World Birds (continued)

Species Name	Common Name	Sex	N	Mean	S.D.	Min	Max	Season	Location	Source #
Pharomachrus fulgidus	White-tipped Quetzal	M	1	160					Venezuela	1254
Pharomachrus auriceps	Golden-headed Quetzal	B	11	180		154	222			591a, 1037
Pharomachrus pavoninus	Pavonine Quetzal	M	11	163		152	180			591a
Harpactes fasciatus	Malabar Trogon	M	1	62						7
Harpactes kasumba	Red-naped Trogon	B	4	96.9		72.1	115		Borneo; Malaysia	1216, 1287
Harpactes diardii	Diard's Trogon	M	6			90.7	111			1287
		F	5			87.5	106			
Harpactes ardens	Philippine Trogon	B	11	96.3		82.6	114		Philippines	954
Harpactes orrhophaeus	Cinnamon-rumped Trogon	M	10			46.8	60.6			1287
		F	17			45.7	60			
Harpactes duvaucelii	Scarlet-rumped Trogon	M	14			36.4	43			1287
		F	8			33.8	43.1			
Harpactes erythrocephalus	Red-headed Trogon	U	6	80.3		75	86.6			775, 1287
Harpactes oreskios	Orange-breasted Trogon	M	1	57.3						1287
Harpactes wardi	Ward's Trogon	B	4	119		115	120		India	7
Family Alcedinidae										
Alcedo atthis atthis	Common Kingfisher	U	15	27	2.4	23	33	PB	Iran	223
Alcedo atthis ispida	Common Kingfisher	U	162	35.8	2.1			B	Britain, UK	223
Alcedo semitorquata	Half-collared Kingfisher	B	9			40	56		Angola	266
Alcedo quadribrachys	Shining-blue Kingfisher	U	11	34.9		28	39.5		Ghana; Angola	266, 457
Alcedo meninting	Blue-eared Kingfisher	U	15			17.9	23.3			1287
Alcedo azurea	Azure Kingfisher	U	40	34.9	3.5	24	40			516

Body Masses of World Birds (continued)

Species Name	Common Name	Sex	N	Mean	S.D.	Min	Max	Season	Location	Source #
Alcedo websteri	Bismarck Kingfisher	M	2	55.5		54	57		New Britain	426
		F	1	67						
Alcedo euryzona	Blue-banded Kingfisher	M	5			40.6	45.2			1287
		F	9			38.7	48.5			
Alcedo cyanopectus	Indigo-banded Kingfisher	F	1	21.5					Luzon, Philippines	436
Alcedo argentata	Silvery Kingfisher	M	12	16.5		13.5	18.2		Philippines	475
		F	14	19.3		17.2	22.4			
Alcedo cristata	Malachite Kingfisher	B	10	15.7	4	11	21.1			457, 508, 562
Alcedo vintsioides	Malagasy Kingfisher	U	19	17.9	3.1	13.5	26		Madagascar	957
Alcedo leucogaster	White-bellied Kingfisher	U	5	14.5		13.5	14.8		Liberia	616
Alcedo pusilla	Little Kingfisher	U	14	13.3	0.9	12	14			516
Ceyx erithacus	Black-backed Kingfisher	U	51	18		15.2	21.5			1287
Ceyx melanurus	Philippine Kingfisher	B	3	22.8		22	23.9		Philippines	436, 954
Ceyx fallax	Sulawesi Kingfisher	B	16	18	1.6	15.5	20.5		Sulawesi, Indonesia	1272
Ceyx rufidorsa	Rufous-backed Kingfisher	U	9	17.6					Malaysia	785, 1333
Ceyx lepidus	Variable Kingfisher	U	10	14					New Guinea	61
Ceyx lepidus nigromaxilla	Variable Kingfisher	B	5	19.4		19	20.5		Guadalcanal	399
Ceyx lepidus gentianus	Variable Kingfisher	M	3	26.8		24.5	28		Solomon Is.	399
		F	4	28.4		27	29.5			
Ispidina madagascariensis	Madagascar Pygmy-Kingfisher	B	20	17.7	2.4	13.5	23.5		Madagascar	957
Ispidina picta	African Pygmy-Kingfisher	U	18	11.4		9.6	13.5		Ghana	457

Body Masses of World Birds (continued)

Species Name	Common Name	Sex	N	Mean	S.D.	Min	Max	Season	Location	Source #
Ispidina picta	African Pygmy-Kingfisher	U	26	14.2	1.6	11	18.5		Tanzania	1014
Ispidina lecontei	Dwarf Kingfisher	U	5	9.5		8	10.5			120, 616
Lacedo pulchella	Banded Kingfisher	M	6			40.8	48.5			1287
		F	7			46.5	54.3			
Dacelo novaeguineae	Laughing Kookaburra	M	51	313	27.1	248	371		Canberra, Australia	516
		F	35	356	26.5	285	417			
Dacelo leachii	Blue-winged Kookaburra	M	21	284	34.4	200	330		Northern Australia	516
		F	13	334	51.3	267	447			
Dacelo tyro archboldi	Spangled Kookaburra	M				128	145			393
		F				148	165			
Dacelo gaudichaud	Rufous-bellied Kookaburra	U	22	143					New Guinea	61
Clytoceyx rex	Shovel-billed Kookaburra	M	2	242		238	245		New Guinea	424
Pelargopsis amauroptera	Brown-winged Kingfisher	M	1	162						1287
Pelargopsis capensis gouldi	Stork-billed Kingfisher	M	4	155		143	180		Philippines	983
		F	3	206		182	225			
Pelargopsis melanorhyncha	Black-billed Kingfisher	M	1	184						393
		F	1	203						
Halcyon coromanda	Ruddy Kingfisher	U	27	77.5		68.8	86.8	F	Malaysia	1287
Halcyon badia	Chocolate-backed Kingfisher	B	13	57.9		47	64.5		Uganda	394
Halcyon smyrnensis	White-throated Kingfisher	B	17	91.4		80.5	108		Philippines	954
Halcyon leucocephala	Gray-headed Kingfisher	B	24	41.8		34	52		Tropical & southern Africa	223
Halcyon pileata	Black-capped Kingfisher	U	25	84	10			W	Malaysia	853
Halcyon cyanoventris	Javan Kingfisher	F	1	93						393

Body Masses of World Birds (continued)

Species Name	Common Name	Sex	N	Mean	S.D.	Min	Max	Season	Location	Source #
Halcyon senegalensis	Woodland Kingfisher	B	30			62	87		Angola	266
Halcyon senegaloides	Mangrove Kingfisher	B	5	61.8		57	66		Uganda	394
Halcyon malimbica	Blue-breasted Kingfisher	U	12	91.8		79	105		Ghana	457
Halcyon albiventris	Brown-hooded Kingfisher	B	9	65.1		59.4	69.4		South Africa	508
Halcyon chelicuti	Striped Kingfisher	B	100			34	55		Angola	266
Todiramphus nigrocyaneus	Blue-black Kingfisher	M				51	54			393
		F				55	57			
Todiramphus winchelli	Rufous-lored Kingfisher	M	2	61.8		59.5	64.1		Philippines	954
		F	2	73.5		67	80			
Todiramphus diops	Blue-and-white Kingfisher	M				41	45			393
		F				42	65			
Todiramphus macleayii	Forest Kingfisher	B	32	37.3		29	43.1		Australia	516
Todiramphus albonotatus	New Britain Kingfisher	M	1	32						393
Todiramphus leucopygius	Ultramarine Kingfisher	M	13	46.3		42.5	52		Guadalcanal; Solomon Is.	399, 658
		F	9	49.6		42	61			
Todiramphus farquhari	Chestnut-bellied Kingfisher	B	6	38.8		37	40.5		Vanuatu	656b
Todiramphus pyrrhopygia	Red-backed Kingfisher	B	23	51.7		41	70		Australia	516
Todiramphus cinnamominus cinnamominus	Micronesian Kingfisher	B	17	58.4	6.2	50.7	71.5		captive	923
Todiramphus cinnamominus reichenbachii	Micronesian Kingfisher	M	27	61.8	4.7				Pohnpei, Micronesia	630a
		F	27	66.7	6.8					
Todiramphus chloris	Collared Kingfisher	U	66	58.9		47.6	70.9		Central Malaysia	1287
Todiramphus chloris sordidus	Collared Kingfisher	U	18	80.5	8	60	98			516

Body Masses of World Birds (continued)

Species Name	Common Name	Sex	N	Mean	S.D.	Min	Max	Season	Location	Source #
Todiramphus chloris solomonis	Collared Kingfisher	M	17	57.4		47.5	61		Solomon Is.	399
		F	7	65		61	71.5			
Todiramphus chloris alberti	Collared Kingfisher	B	11	83.4		67	102		Solomon Is.	399, 658
Todiramphus chloris santoensis	Collared Kingfisher	M	1	56.5						656b
Todiramphus saurophagus	Beach Kingfisher	B	7	126		104	146		Solomon Is.	399, 658
Todiramphus sanctus sanctus	Sacred Kingfisher	U	130	43.7	5.1	34.5	62.5		Australia	516
Todiramphus sanctus vagans	Sacred Kingfisher	U	45	64.2	4.6	55	75		New Zealand	516
Todiramphus ruficollaris	Mangaia Kingfisher	B	4	50.3		40.2	66.5		Cook Is.	1153b
Todiramphus tutus	Chattering Kingfisher	M	15	39.9	2.5	36.7	45.1		Cook Is.	1153b
		F	7	44.9		39.7	50.5			
Melidora macrorrhina	Hook-billed Kingfisher	U	7	97					New Guinea	61
Actenoides bougainvillei excelsus	Moustached Kingfisher	F	2	188		160	215		Guadalcanal	399
Actenoides concretus	Rufous-collared Kingfisher	M	16	71.5		62.3	83.3			1287
		F	16	75.5		59.8	88.1			
Actenoides lindsayi	Spotted Kingfisher	M	10	83.7	6.8	72	96		Luzon, Philippines	436
		F	3	103		95	117			
Actenoides hombroni	Blue-capped Kingfisher	B	9	117		106	147		Philippines	954
Syma torotoro	Yellow-billed Kingfisher	B	13	37.7					New Guinea; Australia	61, 516
Syma megarhyncha	Mountain Kingfisher	M			52	60				393
		F			49	63				
Tanysiptera galatea	Common Paradise-Kingfisher	U	71	50					New Guinea	61

Body Masses of World Birds (continued)

Species Name	Common Name	Sex	N	Mean	S.D.	Min	Max	Season	Location	Source #
Tanysiptera nympha	Red-breasted Paradise-Kingfisher	M	1	57						393
		F				47	57			
Tanysiptera danae	Brown-headed Paradise-Kingfisher	M				37	46			393
		F				42	50			
Tanysiptera sylvia	Buff-breasted Paradise-Kingfisher	B	18	49.6		38	61		Australia	516
Megaceryle maxima	Giant Kingfisher	U	4	325		282	375			457, 508
Megaceryle lugubris	Crested Kingfisher	M	7	255		240	268		Hopei, China	1087
		F	5	288		260	305			
Ceryle alcyon	Belted Kingfisher	U	29	148	20.8	125	215		Pennsylvania, USA	314
Ceryle torquatus	Ringed Kingfisher	B	15	317					Panama; Argentina	155, 210a, 489
Ceryle rudis	Pied Kingfisher	M	189	82.4	6	68	100	B	Kenya	223
		F	96	86.4	7.4	71	110			
Chloroceryle amazona	Amazon Kingfisher	M	11	121	8				Panama	489
		F	6	132						
Chloroceryle americana	Green Kingfisher	U	20	37.5	2.1				Panama	489
Chloroceryle americana	Green Kingfisher	M	8	30.3		28	32.5		Trinidad & Tobago	816
		F	5	33.8		30	36.5			
Chloroceryle inda	Green-and-rufous Kingfisher	B	15	52.1		46.6	60			155, 318, 496, 875
Chloroceryle aenea	American Pygmy Kingfisher	B	12	14.9	1.3	13	16.9		Venezuela	1215
Chloroceryle aenea	American Pygmy Kingfisher	U	9	12.7		11.7	14.4		Brazil	392
Family Todidae										
Todus multicolor	Cuban Tody	B	265	5.9	0.5	4.3	8.5		Cuba	19
Todus subulatus	Broad-billed Tody	U	28	8.7	0.7	7.8	10.2		Dominican Republic	19
Todus angustirostris	Narrow-billed Tody	B	6	7.5		6.5	9		Dominican Republic	19

Body Masses of World Birds (continued)

Species Name	Common Name	Sex	N	Mean	S.D.	Min	Max	Season	Location	Source #
Todus todus	Jamaican Tody	B	18	6.4	0.4	5.6	6.9		Jamaica	19
Todus mexicanus	Puerto Rican Tody	B	246	5.9	0.5	4.8	7.8		Puerto Rico	19
Family Momotidae										
Hylomanes momotula	Tody Motmot	B	9	29.3		20.3	38		Panama; Belize	990, 1027
Aspatha gularis	Blue-throated Motmot	U	20	62.7	11.6	34	97		El Salvador	1041a
Momotus mexicanus	Russet-crowned Motmot	B	6	75.7		68	91		Mexico	911, 1080, 1151
Momotus momota lessonii	Blue-crowned Motmot	B	15	133					Panama	489
Momotus momota conexus	Blue-crowned Motmot	M	11	112	6.6				Panama	489
		F	5	102						
Momotus aequatorialis	Highland Motmot	B	8	158		123	176		Peru; Colombia	370, 798
Baryphthengus martii	Rufous Motmot	U	11	153	10.9				Peru	370
Baryphthengus martii	Rufous Motmot	U	4	178		160	200		Colombia	1037
Baryphthengus ruficapillus	Rufous-capped Motmot	U	8	158					Panama; Brazil; Costa Rica	177, 618, 752, 990
Baryphthengus ruficapillus	Rufous-capped Motmot	U	4	127		120	139		Paraguay	132
Electron carinatum	Keel-billed Motmot	M	1	64.9					Honduras	1145
Electron platyrhynchum	Broad-billed Motmot	U	8	73		53	82		Panama; Colombia	155, 615, 990, 1037, 1180
Eumomota superciliosa	Turquoise-browed Motmot	B	17	62.5	7.4	43.6	74		Yucatan, Mexico	910a, 640
Family Meropidae										
Nyctyornis amictus	Red-bearded Bee-eater	M				68	92			393
		F				61	70			

Body Masses of World Birds (continued)

Species Name	Common Name	Sex	N	Mean	S.D.	Min	Max	Season	Location	Source #
Nyctyornis athertoni	Blue-bearded Bee-eater	M				85	93			393
		F				70	91			
Merops gularis	Black Bee-eater	B	31	27.3		22	33		Liberia; Ivory Coast	394
Merops muelleri	Blue-headed Bee-eater	B	24	22.5		17.4	25			394
Merops bullocki	Red-throated Bee-eater	U	64	23.1	1.3				Nigeria	224
Merops bullockoides	White-fronted Bee-eater	U	39	34.8	1.5	31.5	38		South Africa	508
Merops pusillus	Little Bee-eater	B	7	15.1		11	20.6		Zimbabwe; South Africa	564, 565, 1258
Merops variegatus	Blue-breasted Bee-eater	B	66	22.5		19.5	28.4			394
Merops oreobates	Cinnamon-chested Bee-eater	M	16	26.1		20	38			394
		F	6	22		17	28			
Merops hirundineus	Swallow-tailed Bee-eater	B	10	21.7		19.5	23.6		Ghana; South Africa	457, 1258
Merops breweri	Black-headed Bee-eater	B	10			45	55		Angola	266
Merops revoilii	Somali Bee-eater	M				12	15			393
		F				11	14			
Merops albicollis	White-throated Bee-eater	B	8	25.9		24	28		Eastern Africa	120
Merops orientalis	Green Bee-eater	B	5	14.8		14	16		India	740
Merops boehmi	Boehm's Bee-eater	B	50	16.6		14.4	19.5		Malawi	394
Merops viridis	Blue-throated Bee-eater	U	159	34.8		28.4	45	B		1287
Merops persicus	Blue-cheeked Bee-eater	B	37	49.3		41.3	57	F	Kazakhstan	394
Merops superciliosus	Madagascar Bee-eater	U	32	38.5		32	45			1287

Body Masses of World Birds (continued)

Species Name	Common Name	Sex	N	Mean	S.D.	Min	Max	Season	Location	Source #
Merops philippinus	Blue-tailed Bee-eater	B	6	34		29	38.2			425, 786, 787
Merops ornatus	Rainbow Bee-eater	U	41	29.5	1.7	27	33		Victoria, Australia	516
Merops apiaster	European Bee-eater	B	382	56.6				B	France	692
Merops leschenaulti	Chestnut-headed Bee-eater	U	65	27.2		23	33.1	B		1287
Merops malimbicus	Rosy Bee-eater	B	8			47	54		Angola	266
Merops nubicus	Northern Carmine Bee-eater	U	92	42.4		34	51		Ethiopia	394
Merops nubicoides	Southern Carmine Bee-eater	U	129	60	4.1	52	75		Zimbabwe	118
Family Coraciidae										
Coracias garrulus	European Roller	B	17	146		127	160		Belarus	223
Coracias abyssinicus	Abyssinian Roller	B	3	102		99.5	104		Nigeria; Sudan	223, 1319
Coracias caudatus	Lilac-breasted Roller	U	23	110		51	150		Namibia; South Africa	1036
Coracias spatulatus	Racket-tailed Roller	B	50			71	115		Angola	266
Coracias naevius	Rufous-crowned Roller	B	11	168		140	200		Botswana	394
Coracias benghalensis	Indian Roller	M	2	145		120	166		India	7, 740
		F	2	171		166	176			
Coracias temminckii	Purple-winged Roller	U				144	164			274
Coracias cyanogaster	Blue-bellied Roller	B	22	142		110	178		Ivory Coast	394
Eurystomus glaucurus	Broad-billed Roller	B	13	110		98	112		Tropical Africa	223

Body Masses of World Birds (continued)

Species Name	Common Name	Sex	N	Mean	S.D.	Min	Max	Season	Location	Source #
Eurystomus glaucurus suahelicus	Broad-billed Roller	B	50			85	135		Angola	266
Eurystomus gularis	Blue-throated Roller	B	40	96.3		82	110		Ivory Coast	394
Eurystomus orientalis pacificus	Dollarbird	B	19	129		95.3	180		Australia	516
Eurystomus orientalis pacificus	Dollarbird	B	12	140		110	170		New Guinea	516
Eurystomus orientalis solomonensis	Dollarbird	B	7	162		150	185		Solomon Is.	399
Family Brachypteraciidae										
Brachypteracias leptosomus	Short-legged Ground-Roller	B	5	185		154	217		Madagascar	957
Brachypteracias squamigera	Scaly Ground-Roller	M	1	155					Madagascar	957
Atelornis pittoides	Pitta-like Ground-Roller	U	25	91.5	9	74	114		Madagascar	957
Atelornis crossleyi	Rufous-headed Ground-Roller	U	4	79		72.5	84.9		Madagascar	957
Family Leptosomatidae										
Leptosomus discolor	Cuckoo-Roller	U	6	255		232	301		Madagascar	957
Family Upupidae										
Upupa epops	Eurasian Hoopoe	U	75	61.4	8.1	41	83	Sp	France	223
Upupa epops saturata	Eurasian Hoopoe	M	28	74		62	88		Hopei, China	1087
		F	21	66		56	86			
Upupa marginata	Madagascar Hoopoe	M	3	82.8		70	90.5		Madagascar	957
		F	4	72		57	81			
Family Phoeniculidae										
Phoeniculus purpureus	Green Woodhoopoe	M	5	86.6		72.2	99			457, 1319
		F	3	61.9		56	66.3			

Body Masses of World Birds (continued)

Species Name	Common Name	Sex	N	Mean	S.D.	Min	Max	Season	Location	Source #
Phoeniculus damarensis granti	Violet Woodhoopoe	M	5	85.8		78	90		Kenya	394
		F	3	78		72	82			
Phoeniculus somaliensis	Black-billed Woodhoopoe	U	2	26.5		24	29		Nigeria	777
Phoeniculus bollei	White-headed Woodhoopoe	B	5	54.6		47	64			81, 120
Phoeniculus castaneiceps	Forest Woodhoopoe	B	13	24.4		22.3	27		Liberia; Uganda	394
Rhinopomastus aterrimus	Black Scimitar-bill	B	30			21	35		Angola	266
Rhinopomastus cyanomelas	Common Scimitar-bill	B	60			21	35		Angola	266
Rhinopomastus minor	Abyssinian Scimitar-bill	U	3	21.7		21	23			120
Family Bucerotidae										
Tockus albocristatus	White-crested Hornbill	M	4	297		279	315			394
		F	2	282		276	288			
Tockus hartlaubi	Black Dwarf Hornbill	M	7	110		96	135			394
		F	4	88.2						
Tockus camurus	Red-billed Dwarf Hornbill	M	8	111		101	122			394
		F	7	97		84	115			
Tockus monteiri	Monteiro's Hornbill	U	15			210	400			266
Tockus erythrorhynchus	Red-billed Hornbill	M	75	150		124	172			626
		F	75	128		90	151			
Tockus flavirostris	Eastern Yellow-billed Hornbill	M	3	258		225	275			394
		F	3	182		170	191			
Tockus leucomelas	Southern Yellow-billed Hornbill	M	75	211		153	242			626
		F	75	168		138	211			
Tockus deckeni	Von der Decken's Hornbill	M	7	194		165	212			626
		F	5	145		120	155			
Tockus alboterminatus	Crowned Hornbill	M	31	244		191	332			626
		F	13	205		180	249			

Body Masses of World Birds (continued)

Species Name	Common Name	Sex	N	Mean	S.D.	Min	Max	Season	Location	Source #
Tockus bradfieldi	Bradfield's Hornbill	M	1	200						626
		F	4	193		170	204			
Tockus fasciatus	African Pied Hornbill	M	15	278		250	316			394
		F	7	241		227	260			
Tockus hemprichii	Hemprich's Hornbill	F	1	297				B		394
Tockus nasutus nasutus	African Gray Hornbill	M	5	233		220	258			626
		F	4	183		163	215			
Tockus nasutus epirhinus	African Gray Hornbill	M	34	167		159	234			626
		F	29	146		122	175			
Tockus pallidirostris	Pale-billed Hornbill	M	3	256		248	266			626
		F	2	212		206	217			
Ocyceros griseus	Malabar Gray Hornbill	M	4	292		238	340			626
Ocyceros gingalensis	Ceylon Gray Hornbill	M	1	238						626
Ocyceros birostris	Indian Gray Hornbill	M	1	375						626
Anthracoceros coronatus	Malabar Pied-Hornbill	M	7	862		700	960		China	184a
		F	6	755		600	850			
Anthracoceros albirostris albirostris	Oriental Pied-Hornbill	M				680	795			626
		F				567	680			
Anthracoceros albirostris convexus	Oriental Pied-Hornbill	M	1	907					Sumatra	626
		F	1	879						
Anthracoceros malayanus	Black Hornbill	M	1	1050						626
Buceros rhinoceros	Rhinoceros Hornbill	M	5	2580		2465	2960			626
		F	3	2180		2040	2330			
Buceros bicornis	Great Hornbill	M	7	3006		2600	3400			626
		F	3	2591		2155	3350			
Buceros hydrocorax	Rufous Hornbill	M	18	1474		1185	1652		Philippines	954
		F	15	1348		1016	1662			
Buceros vigil	Helmeted Hornbill	M	1	3060						626
		F	2	2725		2610	2840			

Body Masses of World Birds (continued)

Species Name	Common Name	Sex	N	Mean	S.D.	Min	Max	Season	Location	Source #
Anorrhinus tickelli	Rusty-cheeked Hornbill	M	4	869		854	912			626
		F	7	723		683	797			
Anorrhinus galeritus	Bushy-crested Hornbill	U	3	1172		1134	1247			626
Penelopides manillae	Luzon Hornbill	M	9	450		400	479		Luzon, Philippines	436
		F	2	472		470	475			
Penelopides mindorensis	Mindoro Hornbill	M	2	462		425	500		Philippines	983
Penelopides panini	Tarictic Hornbill	M	16	509		453	584		Philippines	954
		F	13	435		335	514			
Penelopides affinis	Mindanao Hornbill	M	4	523		497	554		Philippines	626
		F	6	450		394	514			
Penelopides exarhatus	Sulawesi Hornbill	F	1	370						626
Aceros comatus	White-crowned Hornbill	F	1	1470						626
		F	1	1470						
Aceros nipalensis	Rufous-necked Hornbill	B	2	2385		2270	2500		India	7
Aceros corrugatus	Wrinkled Hornbill	M	1	1590						626
Aceros leucocephalus	Writhed Hornbill	M	3	1086		1012	1140		Philippines	954
Aceros cassidix	Knobbed Hornbill	M	1	2360						626
Aceros undulatus	Wreathed Hornbill	M	28	2515		1680	3650			626
		F	10	1950		1360	2685			
Aceros narcondami	Narcondam Hornbill	M	4			700	750			7
		F	3			600	750			
Aceros subruficollis	Plain-pouched Hornbill	M	3	2042		1815	2270			626
Aceros plicatus	Blyth's Hornbill	M	7	1803		1500	2000			61, 626, 658, 781
		F	5	1610		1450	2000			
Ceratogymna bucinator	Trumpeter Hornbill	M	9	721		607	941			626
		F	10	567		452	670			

Body Masses of World Birds (continued)

Species Name	Common Name	Sex	N	Mean	S.D.	Min	Max	Season	Location	Source #
Ceratogymna fistulator	Piping Hornbill	M	3	554		463	710			394
		F	3	466		413	500			
Ceratogymna brevis	Silvery-cheeked Hornbill	M	5	1308		1265	1400			394
		F	5	1162		1050	1450			
Ceratogymna subcylindricus	Black-and-white-casqued Hornbill	M	9	1311		1078	1390			394
		F	4	1090		1000	1200			
Ceratogymna cylindricus	Brown-checked Hornbill	F	1	921						626
Ceratogymna albotibialis	White-thighed Hornbill	M	3	1286		1200	1411			626
		F	2	976		908	1043			
Ceratogymna atrata	Black-casqued Hornbill	M	4	1348		1069	1600			626
		F	3	1059		907	1182			
Ceratogymna elata	Yellow-casqued Hornbill	M	1	2100						626
		F	2	1750		1500	2000			
Bucorvus abyssinicus	Abyssinian Ground-Hornbill	M	1	4000					estimated	394
Bucorvus leadbeateri	Southern Ground-Hornbill	M	8	4191		3459	6180			626
		F	8	3344		2230	4580			
Family Galbulidae										
Galbalcyrhynchus leucotis	White-eared Jacamar	U				44	50			274
Galbalcyrhynchus purusianus	Chestnut Jacamar	U	7	50					Peru	370
Brachygalba salmoni	Dusky-backed Jacamar	M	1	18.5					Panama	990
		F	1	16						
Brachygalba lugubris	Brown Jacamar	B	10	15.9		14.7	16.9		Suriname	498
Jacamaralcyon tridactyla	Three-toed Jacamar	M				17.4	18.8			274
		F				17.8	19.3			
Galbula albirostris	Yellow-billed Jacamar	B	11	22.1	1.4	20	24		Brazil	449
Galbula cyanicollis	Blue-cheeked Jacamar	B	25			20	27		Peru	988

Body Masses of World Birds (continued)

Species Name	Common Name	Sex	N	Mean	S.D.	Min	Max	Season	Location	Source #
Galbula ruficauda	Rufous-tailed Jacamar	B	14	26.5		24	33.5		Trinidad	359
Galbula galbula	Green-tailed Jacamar	B	21	23		18	29		Suriname	498
Galbula pastazae	Coppery-chested Jacamar	B	2	31.8		31	32.5		Ecuador	953
Galbula cyanescens	Bluish-fronted Jacamar	U	6	25.3					Peru	1288
Galbula tombacea	White-chinned Jacamar	U				21.5	25			274
Galbula chalcothorax	Purplish Jacamar	U				22	26			274
Galbula leucogastra	Bronzy Jacamar	B	13	15.9		15	18		Suriname	498
Galbula dea	Paradise Jacamar	B	11	27.4	3.6	22	31.5			293, 496
Jacamerops aureus	Great Jacamar	B	11	62.9		57	66		French Guiana; Suriname; Venezuela	293, 498, 1308
Family Bucconidae										
Notharchus macrorhynchos	White-necked Puffbird	B	17	95.9					Panama	489
Notharchus pectoralis	Black-breasted Puffbird	M	1	69.1					Panama	1180
Notharchus ordii	Brown-banded Puffbird	B	2	52.8		51.5	54		Venezuela; Peru	1308
Notharchus tectus	Pied Puffbird	B	31	26.8		21	30		Suriname	498
Bucco macrodactylus	Chestnut-capped Puffbird	U		25					Peru	1210
Bucco tamatia	Spotted Puffbird	B	8	35.5		51.5	54		Suriname	498
Bucco capensis	Collared Puffbird	U	15	54	4				Brazil	73
Nystalus radiatus	Barred Puffbird	U	2	61		59	63		Panama	155, 990
Nystalus chacuru	White-eared Puffbird	B	7	52.3					Brazil	63, 177, 392, 883
Nystalus striolatus	Striolated Puffbird	U		47					Peru	1210

Body Masses of World Birds (continued)

Species Name	Common Name	Sex	N	Mean	S.D.	Min	Max	Season	Location	Source #
Nystalus maculatus	Spot-backed Puffbird	B	11	42		32	49		Brazil; Bolivia	362, 657, 882
Hypnelus ruficollis	Russet-throated Puffbird	U	11	49.8	4	43.3	57		Venezuela	1215
Malacoptila fusca	White-chested Puffbird	B	8	45.4		39	53		French Guiana; Suriname; Peru	293, 498, 785a
Malacoptila semicincta	Semicollared Puffbird	U		44					Peru	1210
Malacoptila striata	Crescent-chested Puffbird	U	6	44.1		41.5	46.5		Brazil	878
Malacoptila fulvogularis	Black-streaked Puffbird	U	2	57.4		49.8	65		Peru; Colombia	1038a, 1288
Malacoptila rufa	Rufous-necked Puffbird	M	10			39	55		Peru	988
		F	8			41	58			
Malacoptila panamensis	White-whiskered Puffbird	U	21	42.6	4.9				Costa Rica; Panama	618
Malacoptila mystacalis	Moustached Puffbird	U	5	49		47	51		Venezuela	1042, 1254
Micromonacha lanceolata	Lanceolated Monklet	U	7	19.8					Colombia; Ecuador	747, 1037, 1038a
Nonnula sclateri	Fulvous-chinned Nunlet	U				15	17			274
Nonnula rubecula	Rusty-breasted Nunlet	U	10	18.4		17	20		Peru; Venezuela; Paraguay; Suriname	132, 498, 988, 1308
Nonnula frontalis	Gray-cheeked Nunlet	U	7	15.7		14.5	19.5		Panama	155, 489, 1180
Nonnula ruficapilla	Rufous-capped Nunlet	U		22					Peru	1210
Nonnula amaurocephala	Chestnut-headed Nunlet	U				15	16			274
Hapaloptila castanea	White-faced Nunbird	B	5	80.6		75	84		Peru	900
Monasa atra	Black Nunbird	B	19	90.4		80	100		Suriname	498

Body Masses of World Birds (continued)

Species Name	Common Name	Sex	N	Mean	S.D.	Min	Max	Season	Location	Source #
Monasa nigrifrons	Black-fronted Nunbird	U	15	80.7	5				Peru	1288
Monasa morphoeus	White-fronted Nunbird	U	11	106	11.2	90.5	122		Panama	155, 990, 1180
Monasa morphoeus	White-fronted Nunbird	U	11	72.9	5.6	63.4	80		Venezuela	1308
Monasa flavirostris	Yellow-billed Nunbird	B	6	62.8		55	68		Peru	1143
Chelidoptera tenebrosa	Swallow-wing	B	27	35.9		29	40		Suriname	498
Family Capitonidae										
Psilopogon pyrolophus	Fire-tufted Barbet	U	18	129		115	149		Malaysia	775
Megalaima virens	Great Barbet	B	25	202		123	265			1102
Megalaima zeylanica	Brown-headed Barbet	B	7	119		108	140			1102
Megalaima lineata	Lineated Barbet	M	11	146		117	205			1102
		F	14	152		134	170			
Megalaima viridis	White-cheeked Barbet	U	16	80.5		71	90		India	7
Megalaima corvina	Brown-throated Barbet	M	1	130						1102
Megalaima chrysopogon	Gold-whiskered Barbet	B	26	151		110	182			1102
Megalaima rafflesii	Red-crowned Barbet	B	22	118		99	150			1102
Megalaima mystacophanos	Red-throated Barbet	B	11	77.4		60.3	95			1102, 1218, 1287
Megalaima javensis	Black-banded Barbet	M	1	115						1102
Megalaima flavifrons	Yellow-fronted Barbet	M	2	58.5		57	60		Sri Lanka	7
Megalaima franklinii	Golden-throated Barbet	U	54	63.5		50	73		Malaysia	775

Body Masses of World Birds (continued)

Species Name	Common Name	Sex	N	Mean	S.D.	Min	Max	Season	Location	Source #
Stactolaema olivacea	Green Barbet	B	34	49.4		40	60			1102
Pogoniulus scolopaceus	Speckled Tinkerbird	U	35			11	18		Angola	1102
Pogoniulus coryphaeus	Western Tinkerbird	B	10	10.7		8	13			1102
Pogoniulus leucomystax	Moustached Tinkerbird	U	24	11.8		10.7	12.9		Malawi; Zambia	1102
Pogoniulus simplex	Green Tinkerbird	B	11	8.8		7	10.5		Kenya; Malawi	1102
Pogoniulus atroflavus	Red-rumped Tinkerbird	M	9	18.5		17	21.5			1102
		F	11	17.9		14	20			
Pogoniulus subsulphureus	Yellow-throated Tinkerbird	B	26	10.5		8.5	12			1102
Pogoniulus bilineatus	Yellow-rumped Tinkerbird	U	72	14.8		11	18.6			1102
Pogoniulus bilineatus leucolaema	Yellow-rumped Tinkerbird	B	30	11.6		7.2	14			1102
Pogoniulus chrysoconus	Yellow-fronted Tinkerbird	U	92	10.8		9.6	12.5		Nigeria	777
Pogoniulus pusillus	Red-fronted Tinkerbird	B	17	9.6		6	12.5			1102
Buccanodon duchaillui	Yellow-spotted Barbet	B	21	41.8		30	50		Liberia	1102
Tricholaema hirsuta ansorgii	Hairy-breasted Barbet	B	40	55.2		43	63			1102
Tricholaema diademata	Red-fronted Barbet	B	16	30.6		22	35			1102
Tricholaema frontata	Miombo Barbet	B	7			24	30		Angola	266
Tricholaema leucomelas	Pied Barbet	B	171	32.2		23	39		Transvaal, South Africa	394
Tricholaema lachrymosa	Spot-flanked Barbet	B	31	22.9		18	27			1102

Body Masses of World Birds (continued)

Species Name	Common Name	Sex	N	Mean	S.D.	Min	Max	Season	Location	Source #
Tricholaema melanocephala	Black-throated Barbet	M	17	21.9		18	25			1102
		F	9	19.1		17	23			
Lybius undatus	Banded Barbet	B	19	38.5		31.8	42			1102
Lybius vieilloti	Vieillot's Barbet	B	16	36		34	38			394
Lybius leucocephalus	White-headed Barbet	M	12	63.8		55	70			394
		F	11	61.8		55	65			
Lybius chaplini	Chaplin's Barbet	M	2	69.5		64	75			1102
Lybius guifsobalito	Black-billed Barbet	B	20	42.2		35	55			1102
Lybius torquatus	Black-collared Barbet	B	30	51.7		47.1	59.3		Zimbabwe; South Africa	508, 564, 565, 566
Lybius melanopterus	Brown-breasted Barbet	B	49	52.4		45	61			1102
Lybius minor	Black-backed Barbet	B	17	46.7		40	50			394
Lybius bidentatus	Double-toothed Barbet	M	9	84.5		72	91			1102
		F	8	74.5		65	91			
Lybius dubius	Bearded Barbet	B	23	91		80	105			1102
Lybius rolleti	Black-breasted Barbet	M	1	96						1102
		F	1	105						
Trachyphonus purpuratus elgonensis	Yellow-billed Barbet	B	38	76.1		65	92			1102
Trachyphonus vaillantii	Crested Barbet	M	17	71.9		64.5	84.9		Zimbabwe	1102
		F	9	67		57.4	74.5			
Trachyphonus margaritatus	Yellow-breasted Barbet	M	1	64						1102
		F	4			44	48			
Trachyphonus erythrocephalus	Red-and-yellow Barbet	B	41	63.5		45	75			1102
Trachyphonus darnaudii boehmi	D'Arnaud's Barbet	B	85	28.6		17	39			1102
Trachyphonus darnaudii usambiro	D'Arnaud's Barbet	B	21	44.7		37	50.5			1102

Body Masses of World Birds (continued)

Species Name	Common Name	Sex	N	Mean	S.D.	Min	Max	Season	Location	Source #
Capito aurovirens	Scarlet-crowned Barbet	B	19	56.2		46	75			1102
Capito wallacei	Scarlet-banded Barbet	B	13	69.5	5.2	63	78		Peru	870
Capito maculicoronatus	Spot-crowned Barbet	M	4	56.8		44	66.4			1102
		F	3	49.1		47	52.2			
Capito squamatus	Orange-fronted Barbet	M	9	61.3		58.5	63.5			1102
		F	7	57.3		53.5	61			
Capito hypoleucus	White-mantled Barbet	M	2	74.6		71	78.3		Colombia	1144
		F	3	67.4		64	70.4			
Capito dayi	Black-girdled Barbet	B	24	67.2		56	74			1102
Capito quinticolor	Five-colored Barbet	M	4	61.4		57	68.5			1102
		F	6	58.7		53.5	62			
Capito niger niger	Black-spotted Barbet	M	29	55.2		41.5	68			1102
		F	10	51.6		38	63			
Capito niger auratus	Black-spotted Barbet	B	16	62.8		55	74			1102
Capito niger punctatus	Black-spotted Barbet	B	26	60.6		47.5	70.5			1102
Capito brunneipectus	Brown-chested Barbet	M	1	57.9						1102
Eubucco richardsoni	Lemon-throated Barbet	B	27	33.9		26	42			1102
Eubucco bourcierii	Red-headed Barbet	B	12	37		33	41			1102
Eubucco tucinkae	Scarlet-hooded Barbet	M	3	41.7		40.5	44			1102
Eubucco versicolor	Versicolored Barbet	B	28	32.8		28	39			1102
Semnornis ramphastinus	Toucan Barbet	B	17	97.7		86	111			1102
Semnornis frantzii	Prong-billed Barbet	U	10	57.3	4.6				Costa Rica	80

Body Masses of World Birds (continued)

Species Name	Common Name	Sex	N	Mean	S.D.	Min	Max	Season	Location	Source #
Family Ramphastidae										
Aulacorhynchus prasinus prasinus	Emerald Toucanet	B	20	187		160	239			1102
Aulacorhynchus prasinus wagleri	Emerald Toucanet	B	22	183		158	202			1102
Aulacorhynchus prasinus caeruleogularis	Emerald Toucanet	B	7	145		118	173			1102
Aulacorhynchus sulcatus	Groove-billed Toucanet	B	15	167		150	200			1102, 1254
Aulacorhynchus derbianus	Chestnut-tipped Toucanet	M	25	221		190	262			1102
		F	18	199		175	250			
Aulacorhynchus haematopygus haematopygus	Crimson-rumped Toucanet	M	5	221		208	232			1102
		F	3			200	210			
Aulacorhynchus haematopygus sexnotatus	Crimson-rumped Toucanet	B	27	171		141	200			1102
Aulacorhynchus huallagae	Yellow-browed Toucanet	M	3	260		250	278		Peru	1074
		F	1	278						
Aulacorhynchus coeruleicinctis	Blue-banded Toucanet	B	11	208		173	257			1102
Pteroglossus inscriptus inscriptus	Lettered Aracari	M	11	121		102	140			1102
		F	8	112		95	130			
Pteroglossus inscriptus humboldtii	Lettered Aracari	M	11	147		111	185			1102
		F	3			122	166			
Pteroglossus viridis	Green Aracari	B	30	135		120	162			1102
Pteroglossus bitorquatus sturmii	Red-necked Aracari	B	21	142		119	171			1102
Pteroglossus azara flavirostris	Ivory-billed Aracari	B	25	148		117	200			1102
Pteroglossus mariae	Brown-mandibled Aracari	B	47	141		100	160			1102
Pteroglossus castanotis castanotis	Chestnut-eared Aracari	B	19	273		230	310			1102

Body Masses of World Birds (continued)

Species Name	Common Name	Sex	N	Mean	S.D.	Min	Max	Season	Location	Source #
Megalaima oorti	Black-browed Barbet	B	80	87.7		67	123			1102
Megalaima asiatica	Blue-throated Barbet	B	21	90.5		78.8	103			1102
Megalaima monticola	Mountain Barbet	M	9	86.3		65	110			1102
		F	10	101		84.2	112			
Megalaima henricii	Yellow-crowned Barbet	B	5	74.1		65	83.2			1102
Megalaima armillaris	Flame-fronted Barbet	M	1	74						1102
		F	1	65						
Megalaima pulcherrima	Golden-naped Barbet	B	16	65.4		55	75			1102
Megalaima australis	Blue-eared Barbet	B	23	33.3		26.3	39.5			1102
Megalaima rubricapillus	Crimson-fronted Barbet	B	5	37.6		32	42.5			1102
Megalaima haemacephala haemacephala	Coppersmith Barbet	B	45	46.7		40.5	52.6			1102
Megalaima haemacephala indica	Coppersmith Barbet	B	10	42.4		30	52.2			1102
Calorhamphus fuliginosus	Brown Barbet	B	14	42.6		38.5	47.7			1102
Gymnobucco calvus	Naked-fronted Barbet	M	9	52.6					Liberia	1102
		F	6	56.5						
Gymnobucco peli	Bristle-nosed Barbet	B	30			43	63			394
Gymnobucco bonapartei	Gray-throated Barbet	M	22	60.7		51	69		Sudan; Uganda; Kenya	1102
		F	18	65.3		55	75			
Stactolaema leucotis kilimensis	White-eared Barbet	B	23	54.9		48	61			1102
Stactolaema anchietae	Anchieta's Barbet	B	14	47.3		42	54			1102
Stactolaema whytii	Whyte's Barbet	B	13	55.6		51.1	62.5			1102

Body Masses of World Birds (continued)

Species Name	Common Name	Sex	N	Mean	S.D.	Min	Max	Season	Location	Source #
Pteroglossus castanotis australis	Chestnut-eared Aracari	B	21	244		220	271			1102
Pteroglossus aracari aracari	Black-necked Aracari	M	11	269		233	309			1102
		F	5	223		177	262			
Pteroglossus aracari atricollis	Black-necked Aracari	B	19	261		200	325			1102
Pteroglossus torquatus torquatus	Collared Aracari	M	25	229		175	254			1102
		F	19	216		183	246			
Pteroglossus torquatus erythrozonus	Collared Aracari	M	12	177		148	198			1102
		F	11	166		147	181			
Pteroglossus frantzii	Fiery-billed Aracari	M	6	253		225	280			1102
Pteroglossus sanguineus	Stripe-billed Aracari	M	4	254		235	275			1102
		F	4	236		190	263			
Pteroglossus erythropygius	Pale-mandibled Aracari	B	6	278		240	310			1102
Pteroglossus pluricinctus	Many-banded Aracari	M	5	283		233	302			1102
		F	7	242		215	265			
Pteroglossus beauharnaesii	Curl-crested Aracari	M	21	210		164	245			1102
		F	9	222		190	280			
Baillonius bailloni	Saffron Toucanet	B	5	146		130	169			781, 878, 1102
Andigena laminirostris	Plate-billed Mountain-Toucan	B	10	311		275	355			1102
Andigena hypoglauca lateralis	Gray-breasted Mountain-Toucan	M	10	341		303	370			1102
		F	7	287		244	345			
Andigena cucullata	Hooded Mountain-Toucan	M	6	324		286	370			1102
		F	7	300		222	380			
Andigena nigrirostris	Black-billed Mountain-Toucan	B	4	350		335	367			1102
Selenidera spectabilis	Yellow-eared Toucanet	B	14	219		175	245			1102
Selenidera reinwardtii langsdorffi	Golden-collared Toucanet	M	32	174		156	200			1102
		F	22	158		134	177			

Body Masses of World Birds (continued)

Species Name	Common Name	Sex	N	Mean	S.D.	Min	Max	Season	Location	Source #
Selenidera reinwardtii reinwardtii	Golden-collared Toucanet	B	16	154		129	178			1102
Selenidera nattereri	Tawny-tufted Toucanet	M	2	164		164	165		Venezuela	1308
		F	2	153		150	156			
Selenidera culik	Guianan Toucanet	B	19	147		129	165			1102
Selenidera maculirostris	Spot-billed Toucanet	B	8	164		136	193			132, 1102
Selenidera gouldii	Gould's Toucanet	M	11	187		168	209			1102
		F	7	175		162	183			
Ramphastos sulfuratus sulfuratus	Keel-billed Toucan	M	10	433		362	537			1102
		F	8	401		304	484			
Ramphastos sulfuratus brevicarinatus	Keel-billed Toucan	M	11	427		410	452			1102
		F	10	380		337	405			
Ramphastos brevis	Choco Toucan	B	4	412		365	482			1102
Ramphastos culminatus	Yellow-ridged Toucan	M	12	395		285	430			1102
		F	14	363		320	400			
Ramphastos vitellinus	Channel-billed Toucan	M	15	363		315	430			1102
		F	6	324						
Ramphastos dicolorus	Red-breasted Toucan	B	8	331		240	400			132, 1102
Ramphastos swainsonii	Chestnut-mandibled Toucan	M	8	709		681	746			1102
		F	3			616	680			
Ramphastos ambiguus	Black-mandibled Toucan	M	3			620	740			1102
		F	1	599						
Ramphastos tucanus	Red-billed Toucan	B	20	595		515	700			1102
Ramphastos cuvieri	Cuvier's Toucan	M	22	702		620	830			1102
		F	14	687		560	820			
Ramphastos toco	Toco Toucan	B	15	618		500	860			1102
Family Indicatoridae										
Indicator maculatus	Spotted Honeyguide	B	36	46.8		42	51			1102

Body Masses of World Birds (continued)

Species Name	Common Name	Sex	N	Mean	S.D.	Min	Max	Season	Location	Source #
Indicator variegatus	Scaly-throated Honeyguide	M	53	47		43.2	50		Kenya	1102
		F	53	50.6		46	55			
Indicator indicator	Greater Honeyguide	B	65	50.4		45.8	57			1102
Indicator archipelagicus	Malaysian Honeyguide	B	6	32.4		23	38.5		Borneo; Malaya	1102
Indicator minor minor	Lesser Honeyguide	B	153	28.2		21	39			1102
Indicator conirostris ussheri	Thick-billed Honeyguide	M	11	31.8		29.8	34.2		Liberia	1102
		F	4			28.7	33.6			
Indicator willcocksi	Willcock's Honeyguide	B	15	15.4		11	20.5			1102
Indicator exilis	Least Honeyguide	M	46	19.8		16	23			1102
		F	33	17.6		15	22.3			
Indicator exilis pachyrhynchus	Least Honeyguide	M	64	18.5		14	22.5		Uganda	1102
		F	52	16		12	20			
Indicator pumilio	Dwarf Honeyguide	M	4	13.9		13	15			1102
		F	10	12.4		12	13.5			
Indicator meliphilus	Pallid Honeyguide	M	5	15.9		15	17			1102
		F	5	12.4		11	14			
Indicator xanthonotus	Yellow-rumped Honeyguide	M	4	32.5		32	34			1102
		F	7	28.4		26	31			
Melichneutes robustus	Lyre-tailed Honeyguide	B	10	54.2		46.9	61.5			266, 1102
Melignomon eisentrauti	Yellow-footed Honeyguide	B	15	24.2		18	29			1102
Melignomon zenkeri	Zenker's Honeyguide	F	3	23.6		21.7	25			1102
Prodotiscus insignis	Cassin's Honeyguide	B	20	10.4		8.9	13			1102
Prodotiscus zambesiae	Green-backed Honeyguide	B	13	10.2		9	12		Angola	266, 394
Prodotiscus regulus	Wahlberg's Honeyguide	U	20	14.1		12	16			1102

Body Masses of World Birds (continued)

Species Name	Common Name	Sex	N	Mean	S.D.	Min	Max	Season	Location	Source #
Family Picidae										
Jynx torquilla tschusii	Eurasian Wryneck	U	115	33.5	5.3	26	50	F	Malta	223
Jynx torquilla torquilla	Eurasian Wryneck	B	13	36.5	2.1	34	40	B	Germany	223
Jynx ruficollis	Rufous-necked Wryneck	B	7	46.4		35	52			120, 266
Picumnus pumilus	Orinoco Piculet	U				9	10			274
Picumnus innominatus	Speckled Piculet	B	5	10.2		8.6	12.5			7, 787, 1287
Picumnus aurifrons	Bar-breasted Piculet	M	6	9.3					Peru; Brazil	370, 861
		F	2	8.6		8.4	8.8			
Picumnus lafresnayi	Lafresnaye's Piculet	U				9	10			1100
Picumnus exilis	Golden-spangled Piculet	B	14	9.3		8	11.7		Suriname	498
Picumnus sclateri	Ecuadorian Piculet	M	2	9.8					Peru	1302
		F	1	12.4						
Picumnus squamulatus	Scaled Piculet	U	11	10.2		7	11.5		Venezuela	1042, 1215, 1254
Picumnus spilogaster	White-bellied Piculet	B	2	13.2		13	13.5		Brazil	1178
Picumnus minutissimus	Guianan Piculet	M	20	13.7		12	15		Suriname	498
		F	13	12.9		11.3	14			
Picumnus pygmaeus	Spotted Piculet	F	1	11					Brazil	1206
Picumnus steindachneri	Speckle-chested Piculet	U	5	9.9		9.2	10.5		Peru	899
Picumnus cirratus	White-barred Piculet	B	12	10.2		8.5	11.5		Bolivia	1055
Picumnus dorbygnianus	Ocellated Piculet	B	9	9.6		7	11		Bolivia	514, 1143
Picumnus temminckii	Ochre-collared Piculet	B	10	11.5		10.5	12.5		Brazil; Paraguay	63, 132, 961

Body Masses of World Birds (continued)

Species Name	Common Name	Sex	N	Mean	S.D.	Min	Max	Season	Location	Source #
Picumnus albosquamatus	White-wedged Piculet	U	6	11.9		11	13		Brazil	177, 752, 883
Picumnus rufiventris	Rufous-breasted Piculet	U		21					Peru	1210
Picumnus nebulosus	Mottled Piculet	B	5	11.6					Brazil	63, 924b
Picumnus castelnau	Plain-breasted Piculet	F	1	11.4					Colombia	962
Picumnus subtilis	Fine-barred Piculet	M	4	10.4					Peru	370
Picumnus olivaceus	Olivaceous Piculet	B	13	10.6					Panama	489
Picumnus granadensis	Grayish Piculet	M	1	12.2					Colombia	798
Sasia africana	African Piculet	U	6	9.4		9	10.5		Cameroon	394
Sasia abnormis	Rufous Piculet	U	12	9.2					Malaysia	1333
Sasia ochracea	White-browed Piculet	B	5	9.7					India; Thailand	7, 787
Nesoctites micromegas	Antillean Piculet	M	5	30.3		26.5	32.6		Dominican Republic	19
		F	11	34.2	2.2	30.8	38.2			
Melanerpes candidus	White Woodpecker	M	10	108		98	130			63, 362
Melanerpes lewis	Lewis' Woodpecker	M	5	113		105	122			1224
		F	4	99		88.3	106			
Melanerpes herminieri	Guadeloupe Woodpecker	U		100						348
Melanerpes portoricensis	Puerto Rican Woodpecker	M	35	64.8	7.7	50	77		Puerto Rico	19
		F	37	55.3	6.3	45.7	73			
Melanerpes erythrocephalus	Red-headed Woodpecker	U	89	71.6	7.6	56.1	90.5	Sp	Ontario, Canada	314
Melanerpes formicivorus	Acorn Woodpecker	M	745	81.8	8.2				California, USA	646
		F	650	77.5	10.2					
Melanerpes chrysauchen	Golden-naped Woodpecker	M	4	58.6					Panama	489
		F	5	47						

Body Masses of World Birds (continued)

Species Name	Common Name	Sex	N	Mean	S.D.	Min	Max	Season	Location	Source #
Melanerpes pucherani	Black-cheeked Woodpecker	M	8	66						539
		F	3	59						
Melanerpes cruentatus	Yellow-tufted Woodpecker	B	27	58.1		51	66		Suriname	498
Melanerpes flavifrons	Yellow-fronted Woodpecker	M	2	63		62	64		Brazil	63
		F	1	53						
Melanerpes cactorum	White-fronted Woodpecker	M	10	37.8		31	42		Bolivia	1055
		F	14	32.3		26.4	35.2			
Melanerpes striatus	Hispaniolan Woodpecker	M	8	69.7		57.5	83		Dominican Republic	19
		F	5	77.3		63.5	90			
Melanerpes radiolatus	Jamaican Woodpecker	U	25	108					Jamaica	231
Melanerpes chrysogenys	Golden-cheeked Woodpecker	M	4	72		60	87.3		Mexico	911, 1151
		F	4	63.7		57.8	70.3			
Melanerpes hypopolius	Gray-breasted Woodpecker	M	1	52.4					Oaxaca, Mexico	1145
Melanerpes pygmaeus	Yucatan Woodpecker	M	10	40.5	2.1	35.8	42.5		Mexico; Belize	46, 640, 910a, 1027
		F	3	37.2		35.1	39.5			
Melanerpes rubricapillus	Red-crowned Woodpecker	M	19	55.9	4.2				Panama	489
		F	12	49	3.8					
Melanerpes hoffmannii	Hoffmann's Woodpecker	U				62	84			1100
Melanerpes uropygialis	Gila Woodpecker	M	20	69.7	6.4	54.6	80.6	Y	Arizona, USA	314
		F	24	60	3.1	53.8	67			
Melanerpes aurifrons	Golden-fronted Woodpecker	M	29	85.4	7	73	99		San Luis Potosi, Mexico	1079
		F	14	76.4	6.4	66	90			
Melanerpes carolinus	Red-bellied Woodpecker	M	391	73.2	9.9				Florida, USA	646
		F	203	66	11.4					
Melanerpes superciliaris caymanensis	West Indian Woodpecker	M	11	84.2	8.3	71.3	99.3		Cayman Is.	234
		F	5	71		67.3	73.5			
Sphyrapicus thyroideus	Williamson's Sapsucker	B	19	47.6	2.8	44.4	55.3	B	Nevada, USA	594

Body Masses of World Birds (continued)

Species Name	Common Name	Sex	N	Mean	S.D.	Min	Max	Season	Location	Source #
Sphyrapicus varius varius	Yellow-bellied Sapsucker	U	52	50.3	4.3	40.7	62.2	M	Pennsylvania, USA	314
Sphyrapicus nuchalis	Red-naped Sapsucker	B	16	45.9	4.8	36	54.9	W	Arizona, USA	1150
Sphyrapicus nuchalis	Red-naped Sapsucker	M	48	50.5	4.7				British Columbia, Canada	1270
		F	49	49.2	4.3					
Sphyrapicus ruber ruber	Red-breasted Sapsucker	M	23	58.3	5.2				British Columbia, Canada	1270
		F	22	57.7	4.5					
Sphyrapicus ruber daggetti	Red-breasted Sapsucker	U	13	48.9	4.4	40.1	54.7	Y	California, USA	456
Xiphidiopicus percussus	Cuban Woodpecker	M	105	65.8	8.2	50.4	97		Cuba	19
		F	100	56.3	8.1	46	88			
Campethera punctuligera	Fine-spotted Woodpecker	B	4	71.8		67	76		Ghana; Nigeria	457, 777
Campethera nubica	Nubian Woodpecker	U	13	59.7		46.5	66.6		Eastern Africa	120
Campethera bennettii	Bennett's Woodpecker	B	22	70.1		61	83.5		Zimbabwe	394
Campethera abingoni	Golden-tailed Woodpecker	B	40			57	80		Angola	266
Campethera mombassica	Mombasa Woodpecker	U				50	71			274
Campethera notata	Knysna Woodpecker	U		62						274
Campethera maculosa	Little Green Woodpecker	U		54						274
Campethera cailliautii	Green-backed Woodpecker	B	6	41.1		35	45.7			120, 1014
Campethera tullbergi	Tullberg's Woodpecker	U	2	50.5		49	52			120
Campethera nivosa	Buff-spotted Woodpecker	B	11	37.8	3.5	33	45.5			266, 616
Campethera caroli	Brown-eared Woodpecker	B	20			30	65		Angola	266

Body Masses of World Birds (continued)

Species Name	Common Name	Sex	N	Mean	S.D.	Min	Max	Season	Location	Source #
Geocolaptes olivaceus	Ground Woodpecker	M	3	120		109	130		South Africa	508
Dendropicos elachus	Little Gray Woodpecker	M	4	18.4		17	21			394, 852
Dendropicos poecilolaemus	Speckle-breasted Woodpecker	U	2	28.1		27	29.2		Kenya	120
Dendropicos abyssinicus	Abyssinian Woodpecker	B	4			23	26			394
Dendropicos fuscescens	Cardinal Woodpecker	B	16	26		21	35.7			133
Dendropicos gabonensis	Gabon Woodpecker	B	3			24	28			394
Dendropicos stierlingi	Stierling's Woodpecker	B	2	28		25	31			394
Dendropicos namaquus	Bearded Woodpecker	B	40			70	95		Angola	266
Dendropicos pyrrhogaster	Fire-bellied Woodpecker	B	6	68.5		65	74			394
Dendropicos xantholophus	Golden-crowned Woodpecker	B	40			70	95		Angola	266
Dendropicos elliotii	Elliot's Woodpecker	B	4	37.2		32	40		Angola; Uganda	266, 394, 1078
Dendropicos goertae	Gray Woodpecker	M	4	47.8		43	52			457, 852, 1319
		F	4	41.4		36	45.5			
Dendropicos spodocephalus	Gray-headed Woodpecker	U				43	52			274
Dendropicos griseocephalus	Olive Woodpecker	M	10	44		36	50			394
		F	6	38		33	42			
Dendropicos obsoletus	Brown-backed Woodpecker	B	6	21.4		18.6	24			394
Dendrocopos maculatus	Philippine Woodpecker	B	11	24.5		20	30.3		Philippines	436, 954, 983
Dendrocopos nanus	Brown-capped Woodpecker	B	2	14		12.9	15		India	7, 740

Body Masses of World Birds (continued)

Species Name	Common Name	Sex	N	Mean	S.D.	Min	Max	Season	Location	Source
Dendrocopos moluccensis	Sunda Woodpecker	U				14.5	18			1100
Dendrocopos canicapillus mitchelli	Gray-capped Woodpecker	M	3			20	24		India	7
		F	3	23.3		21	27			
Dendrocopos canicapillus scintilliceps	Gray-capped Woodpecker	B	46	28.5		21	31		Hopei, China	1087
Dendrocopos kizuki	Pygmy Woodpecker	B	20	19		18	21		Hopei, China	1087
Dendrocopos kizuki	Pygmy Woodpecker	B	39	19.4	3	12	22.9		Japan	555
Dendrocopos minor	Lesser Spotted Woodpecker	U	50	19.8		18	22	Y	Great Britain	515
Dendrocopos auriceps	Brown-fronted Woodpecker	M	3			38	44		India	7
		F	3			38	40			
Dendrocopos macei	Fulvous-breasted Woodpecker	M				43	48	W	India	7
		F				38	52			
Dendrocopos atratus	Stripe-breasted Woodpecker	B				42	52		India	7
Dendrocopos mahrattensis	Yellow-crowned Woodpecker	B	15	34.3		32	38		India	7
Dendrocopos dorae	Arabian Woodpecker	U				19	20			274
Dendrocopos hyperythrus	Rufous-bellied Woodpecker	M	10			42	53		India	7
		F	6			42	50			
Dendrocopos darjellensis	Darjeeling Woodpecker	M	5			70	80		India	7
		F	5			61	73			
Dendrocopos cathpharius	Crimson-breasted Woodpecker	M	1	35					India	7
		F	2	27		26	28			
Dendrocopos medius	Middle Spotted Woodpecker	B	31	59		50	80	Y	Central Europe	223
Dendrocopos leucotos	White-backed Woodpecker	B	16	108	8	99	126			223
Dendrocopos leucotos subcirris	White-backed Woodpecker	M	4	124		119	128		Kuril Is.	223
		F	4	119		107	128			

Body Masses of World Birds (continued)

Species Name	Common Name	Sex	N	Mean	S.D.	Min	Max	Season	Location	Source #
Dendrocopos major	Great Spotted Woodpecker	U	50	81.6		71	83	Y	Great Britain	515
Dendrocopos major pinetorum	Great Spotted Woodpecker	M	9	76		70	87	B	Netherlands	223
		F	7	72.7		68	79			
Dendrocopos major cabanisi	Great Spotted Woodpecker	M	44	76		65	98		Hopei, China	1087
		F	32	69		58	77			
Dendrocopos syriacus	Syrian Woodpecker	M	9	79.5		76	82		Southeast Europe	223
		F	11	74	3.2	70	81			
Dendrocopos leucopterus	White-winged Woodpecker	U		67						1100
Dendrocopos assimilis	Sind Woodpecker	U				42	64			1100
Dendrocopos himalayensis	Himalayan Woodpecker	M	3			72	74		India	7
		F	6			59	67			
Picoides lignarius	Striped Woodpecker	B	9	34		28.1	49.5		Chile; Paraguay	328, 1055
Picoides mixtus	Checkered Woodpecker	M	4	30.6		26	34		Brazil; Bolivia	63, 657
		F	8	27		22	33			
Picoides scalaris	Ladder-backed Woodpecker	U	21	30.3	3.7	25	41.4	W	Western USA; Sonora, Mexico	1150
Picoides scalaris	Ladder-backed Woodpecker	M	8	33.7		29.8	38.1		Tres Marias Is., Mexico	440
		F	3	30.9		28	32.3			
Picoides nuttallii	Nuttall's Woodpecker	M	96	38	2.6	32.4	46.9	B		277
		F	49	36.8	1.8	32.8	40			
Picoides pubescens	Downy Woodpecker	M	17	27.5	0.9	26	29.1		Pennsylvania, USA	828
		F	16	25.9	1.7	22.2	28.5			
Picoides pubescens	Downy Woodpecker	B	224	27.7					Indiana, USA	646
Picoides pubescens	Downy Woodpecker	U	5	21.6					Georgia, USA	568
Picoides villosus	Hairy Woodpecker	M	27	70	3.2	60.8	79.6	Y	Pennsylvania, USA	193
		F	11	62.5	2	59.3	65.9			
Picoides villosus	Hairy Woodpecker	M	13	79	4.8				Montana, USA	569
		F	14	67.5	4.7					

Body Masses of World Birds (continued)

Species Name	Common Name	Sex	N	Mean	S.D.	Min	Max	Season	Location	Source #
Picoides villosus	Hairy Woodpecker	M	17	60	2.9				Baja California, Mexico	569
		F	17	52.4	2.8					
Picoides villosus	Hairy Woodpecker	B	8	51.7		46	57.4		Bahamas	19
Picoides arizonae	Arizona Woodpecker	M	11	49.4	2.3			B	Arizona, USA	693
		F	5	44.1						
Picoides stricklandi	Strickland's Woodpecker	M	3	37.7		35.9	40.3		Mexico	910
		F	1	34						
Picoides borealis	Red-cockaded Woodpecker	M	1160	48.6	6.8				North Carolina, USA	646
		F	783	47.4	8.4					
Picoides albolarvatus	White-headed Woodpecker	M	18	63	3.4	55.6	68	PB	Oregon, USA	314
		F	17	59.2	3.8	52.6	66.4			
Picoides tridactylus	Eurasian Three-toed Woodpecker	M	10	70.1	3.8	65	74		Norway	528
		F	5	61.2		57	66			
Picoides dorsalis	American Three-toed Woodpecker	M	16	58.3	3.5	50	64.5	B	British Columbia, Canada	685
		F	12	52	3.7	47	59			
Picoides arcticus	Black-backed Woodpecker	M	6	72		61.3	88.1		California, USA	456
		F	3	66.6		63.4	71.3			
Veniliornis callonotus	Scarlet-backed Woodpecker	M	3	28.5		23.2	32.9			1042
		F	1	24.6						
Veniliornis dignus	Yellow-vented Woodpecker	B	5	37.6		35.7	40		Colombia	798
Veniliornis nigriceps	Bar-bellied Woodpecker	M	3	45.8					Peru; Ecuador; Bolivia	633, 963, 1288
		F	6	42.2						
Veniliornis fumigatus	Smoky-brown Woodpecker	B	16	36.2	5.2	32.4	49.8		Mexico; Belize; Colombia	798, 910a, 1027
Veniliornis passerinus	Little Woodpecker	B	15	32.1	2.8	27	36.3		Peru; Brazil; Argentina	318, 752, 883, 1099, 1109
Veniliornis frontalis	Dot-fronted Woodpecker	U	9	36.7		33	40		Argentina	362, 1040

Body Masses of World Birds (continued)

Species Name	Common Name	Sex	N	Mean	S.D.	Min	Max	Season	Location	Source #
Veniliornis spilogaster	White-spotted Woodpecker	B	11	40.3	2.9	34.5	45		Brazil; Argentina	63, 211, 878, 961, 1178
Veniliornis sanguineus	Blood-colored Woodpecker	B	33	26.5		23	31		Suriname	498
Veniliornis kirkii	Red-rumped Woodpecker	B	12	28.8	1.5	26	30		Venezuela	1254
Veniliornis chocoensis	Choco Woodpecker	U		30					estimated	274
Veniliornis cassini	Golden-collared Woodpecker	B	20	34.6		27	41		Suriname	498
Veniliornis affinis	Red-stained Woodpecker	M	12	36					Peru; Brazil; Venezuela	370, 1178, 1308
		F	5	33.7		32.5	35			
Piculus simplex	Rufous-winged Woodpecker	U		55					Costa Rica	1170
Piculus leucolaemus	White-throated Woodpecker	U		69					Peru	1210
Piculus flavigula	Yellow-throated Woodpecker	B	27	55		49	65		Suriname	498
Piculus chrysochloros	Golden-green Woodpecker	U		88					Peru	1210
Piculus aurulentus	Yellow-browed Woodpecker	B	5	75		52	104		Brazil	63, 75, 1099, 1178
Piculus rubiginosus	Golden-olive Woodpecker	M	14	56.2		51	68		Trinidad	359
		F	9	55.4		51	61			
Piculus auricularis	Gray-crowned Woodpecker	M	3	65.4		63.9	68.3		Oaxaca, Mexico	1145
Piculus rivolii	Crimson-mantled Woodpecker	B	6	102					Peru; Bolivia; Colombia	963, 324a, 1288
Colaptes atricollis	Black-necked Woodpecker	M	3	78.7		71.4	85.7			1042
		F	1	73.6						
Colaptes punctigula	Spot-breasted Woodpecker	M	11	64		59	70		Suriname	498
		F	7	60		52	70			

Body Masses of World Birds (continued)

Species Name	Common Name	Sex	N	Mean	S.D.	Min	Max	Season	Location	Source #
Colaptes melanochloros	Green-barred Woodpecker	M	7			120	137		Brazil	63
		F	11			114	140			
Colaptes melanolaimus	Golden-breasted Woodpecker	M	1	129					Paraguay	1175
		F	3	113		104	120			
Colaptes auratus cafer	Northern Flicker	U	82	142	2.2	121	167		Oregon, USA	314
Colaptes auratus auratus	Northern Flicker	M	18	128	10.3	106	143	Y	Pennsylvania, USA	828
		F	11	125	9.1	104	137			
Colaptes chrysoides mearnsi	Gilded Flicker	U	100	111	7.4	92.2	129	Y	Arizona, USA	314
Colaptes fernandinae	Fernandina's Flicker	B	8	135		121	157		Cuba	801
Colaptes pitius	Chilean Flicker	U	5	159		156	173		Argentina	210
Colaptes rupicola	Andean Flicker	B	10	174		152	190			362, 900
Colaptes campestris	Campo Flicker	B	11	158		138	176			63, 210a, 752, 1040, 1109
Celeus loricatus	Cinnamon Woodpecker	B	4	76.6		74	79.8		Panama	155, 990, 1180
Celeus brachyurus phaioceps	Rufous Woodpecker	B	2	108		102	113		India	7
Celeus brachyurus badiosus	Rufous Woodpecker	B	3	68.4		65.4	73.2		Borneo	1218
Celeus grammicus	Scaly-breasted Woodpecker	M	4	74.5		67	87		Peru; Venezuela	370, 1308
		F	6	72.4						
Celeus undatus	Waved Woodpecker	B	25	64.5		58	70		Suriname	498
Celeus castaneus	Chestnut-colored Woodpecker	B	10	85.6	7.6	71.9	95		Mexico; Belize	910a, 1027
Celeus elegans	Chestnut Woodpecker	B	9	117		93.5	133		Trinidad	1131
Celeus elegans	Chestnut Woodpecker	M	13	142		127	172		Suriname	498
		F	12	148		131	161			

Body Masses of World Birds (continued)

Species Name	Common Name	Sex	N	Mean	S.D.	Min	Max	Season	Location	Source #
Celeus lugubris	Pale-crested Woodpecker	M	4	137		118	157		Paraguay; Brazil	883, 1175
Celeus flavescens	Blond-crested Woodpecker	B	7	139		122	165		Brazil; Argentina	63, 177, 878, 1099
Celeus flavus	Cream-colored Woodpecker	U		201					Peru	1210
Celeus flavus flavus	Cream-colored Woodpecker	B	21	108		94	131		Suriname	498
Celeus spectabilis	Rufous-headed Woodpecker	U	2	114		111	116		Peru	370, 319
Celeus torquatus	Ringed Woodpecker	U		134					Peru	1210
Dryocopus galeatus	Helmeted Woodpecker	F	1	124					Paraguay	1175
Dryocopus lineatus	Lineated Woodpecker	M	12	194					Panama; Suriname	489, 496
		F	7	173						
Dryocopus pileatus	Pileated Woodpecker	M	2	308		308	309		Pennsylvania, USA	314
		F	4	266		250	284			
Dryocopus pileatus	Pileated Woodpecker	M	16	305		275	328	B	Washington, USA	150
		F	11	270		248	290			
Dryocopus schulzi	Black-bodied Woodpecker	M	1	200					Argentina	212
Dryocopus javensis pectoralis	White-bellied Woodpecker	M	4	290		240	347		Philippines	954
		F	6	254		244	267			
Dryocopus hodgei	Andaman Woodpecker	U		156						1100
Dryocopus martius	Black Woodpecker	B	18	321	30.3	255	361			223
Campephilus haematogaster	Crimson-bellied Woodpecker	B	3	237		225	250		Panama	79, 990
Campephilus rubricollis	Red-necked Woodpecker	M	9	222		197	239		Venezuela; Brazil; Suriname	296, 496, 497, 1178, 1308
		F	3	200		193	206			

Body Masses of World Birds (continued)

Species Name	Common Name	Sex	N	Mean	S.D.	Min	Max	Season	Location	Source #
Campephilus robustus	Robust Woodpecker	U		200					Brazil	1105
Campephilus melanoleucos	Crimson-crested Woodpecker	B	18	256		225	281		Suriname	498
Campephilus gayaquilensis	Guayaquil Woodpecker	U				230	253			1100
Campephilus guatemalensis	Pale-billed Woodpecker	B	17	242					Panama	489
Campephilus leucopogon	Cream-backed Woodpecker	U	14	249		222	281			362
Campephilus magellanicus	Magellanic Woodpecker	B	9	324		275	369		Argentina	210, 546
Campephilus principalis	Ivory-billed Woodpecker	U	2	516		464	568			1196
Picus mineaceus	Banded Woodpecker	U	3	97.3		85	105		Malaysia; Borneo	1287, 1333
Picus chlorolophus	Lesser Yellownape	M	4			57	74		India	7
		F	3			62	70			
Picus puniceus	Crimson-winged Woodpecker	U	3	79.1		77	82.8		Malaysia; Borneo	1287, 1333
Picus flavinucha	Greater Yellownape	M	5			172	198		India	7
		F	3			165	174			
Picus mentalis	Checker-throated Woodpecker	U	10			82.2	121			1287
Picus viridanus	Streak-breasted Woodpecker	U				94	132			1100
Picus vittatus	Laced Woodpecker	M	49			95	126			1287
		F	35			94.2	121			
Picus xanthopygaeus	Streak-throated Woodpecker	B	3	99.7		90	111		India	7, 740
Picus squamatus	Scaly-bellied Woodpecker	B	4	170		159	185			7, 273
Picus awokera	Japanese Woodpecker	U		143					Japan	339

Body Masses of World Birds (continued)

Species Name	Common Name	Sex	N	Mean	S.D.	Min	Max	Season	Location	Source #
Picus viridis	Green Woodpecker	B	69	176		138	190	Y	France	223
Picus vaillantii	Levaillant's Woodpecker	M	1	166				F	Algeria	223
Picus erythropygius	Black-headed Woodpecker	U				100	135			1100
Picus canus	Gray-faced Woodpecker	B	28	137		125	165		Switzerland	223
Dinopium rafflesii	Olive-backed Woodpecker	U	2	102		98	105			1287, 1333
Dinopium shorii	Himalayan Flameback	F	1	101						7
Dinopium javanense	Common Flameback	M	26	76.4		63.9	90.4			1287
		F	29	69.3		63.5	79.1			
Dinopium benghalense benghalense	Black-rumped Flameback	U	11	100		89	121		India	7
Chrysocolaptes festivus	White-naped Woodpecker	U		213						1100
Chrysocolaptes lucidus	Greater Flameback	B	16	142		125	164		Philippines	954
Gecinulus grantia	Pale-headed Woodpecker	U				68	85			1100
Gecinulus viridis	Bamboo Woodpecker	M	5			68.9	74.6			1287
		F	2	69.2		65	73.5			
Blythipicus rubiginosus	Maroon Woodpecker	M	10			73	88			1287
		F	13			76.2	92			
Blythipicus pyrrhotis	Bay Woodpecker	B	6	132		114	170			7, 775
Reinwardtipicus validus	Orange-backed Woodpecker	B	5	149		100	176			785, 1218, 1287
Meiglyptes tristis	Buff-rumped Woodpecker	F	1	31.7					Borneo	1218
Meiglyptes jugularis	Black-and-buff Woodpecker	U				50	57			1100

Body Masses of World Birds (continued)

Species Name	Common Name	Sex	N	Mean	S.D.	Min	Max	Season	Location	Source #
Meiglyptes tukki	Buff-necked Woodpecker	U	24	53.1					Malaysia	1333
Hemicircus concretus	Gray-and-buff Woodpecker	U				27	32			1100
Hemicircus canente	Heart-spotted Woodpecker	U				37	50			1100
Mulleripicus funebris	Sooty Woodpecker	B	10	164	9.3	143	180		Philippines	436, 954
Mulleripicus pulverulentus	Great Slaty Woodpecker	U				360	563			1100
Family Eurylaimidae										
Smithornis capensis	African Broadbill	B	29	23.2	1.9	19	28.5		Tanzania	1014
Smithornis sharpei	Gray-headed Broadbill	U	6	37.5		34	40.3			623
Smithornis rufolateralis	Rufous-sided Broadbill	M	12	20.5	2.7				Liberia	623
		F	6	17.7		10	23.1			
Pseudocalyptomena graueri	Grauer's Broadbill	M	1	32.5						623
		F	4	31.1		29	32.5			
Corydon sumatranus	Dusky Broadbill	U		140					estimated	274
Cymbirhynchus macrorhynchos	Black-and-red Broadbill	M	1	59.4					Malaysia	785
Eurylaimus javanicus	Banded Broadbill	B	3	78.2		74.3	81.8		Borneo; Malaysia	785, 1218
Eurylaimus ochromalus	Black-and-yellow Broadbill	B	5	33.3		32	34.7		Borneo	1218
Eurylaimus steerii	Wattled Broadbill	U				33.7	44.4			274
Eurylaimus samarensis	Visayan Broadbill	B	16	27.1		33.4	41.5		Philippines	954
Psarisomus dalhousiae	Long-tailed Broadbill	F	1	67				W	India	7
Serilophus lunatus	Silver-breasted Broadbill	F	2	34		33	35	W	India	7
Calyptomena viridis	Green Broadbill	F	3	58.5		53.2	67.4		Borneo	398, 1218

Body Masses of World Birds (continued)

Species Name	Common Name	Sex	N	Mean	S.D.	Min	Max	Season	Location	Source #
Calyptomena hosii	Hose's Broadbill	M				102	115			274
		F		92						
Calyptomena whiteheadi	Whitehead's Broadbill	M				142	171			274
		F				150	163			
Family Philepittidae										
Philepitta castanea	Velvet Asity	U	85	35.4	3.7	25.5	46.3		Madagascar	957
Philepitta schlegeli	Schlegel's Asity	B	3	27.9		24.4	30		Madagascar	957
Neodrepanis coruscans	Sunbird Asity	U	37	6.5	0.7	5	8.5		Madagascar	957
Neodrepanis hypoxanthus	Yellow-bellied Asity	U	13	7.1	0.8	5.5	8.5		Madagascar	957
Family Furnariidae										
Geositta poeciloptera	Campo Miner	U				17	19			274
Geositta peruviana	Coastal Miner	B	4	20.8		20	22.2			1043
Geositta maritima	Grayish Miner	U				16	17			274
Geositta cunicularia	Common Miner	U	20	28.5		24.3	31.5		Brazil; Argentina	63, 210, 1040, 1041, 1042
Geositta punensis	Puna Miner	M	4	25.8		24	28		Bolivia	1143
Geositta antarctica	Short-billed Miner	U				34	40			274
Geositta saxicolina	Dark-winged Miner	U				31	35			274
Geositta rufipennis harrisoni	Rufous-banded Miner	B	12	26.6		24	29		Chile	748
Geositta rufipennis fasciata	Rufous-banded Miner	B	18	39.4		36	44		Chile	748
Geositta isabellina	Creamy-rumped Miner	U				38	48			274
Geositta crassirostris	Thick-billed Miner	F	1	55.8						1043

Body Masses of World Birds (continued)

Species Name	Common Name	Sex	N	Mean	S.D.	Min	Max	Season	Location	Source #
Geositta tenuirostris tenuirostris	Slender-billed Miner	B	10	34		29	42.5			651
Geositta tenuirostris kalimayae	Slender-billed Miner	B	2	30.5		30	31		Ecuador	651
Upucerthia dumetaria hypoleuca	Scale-throated Earthcreeper	M	3	47.9		47.4	48.7		Patagonia, Argentina	840
		F	5	48.4		44.4	54.5			
Upucerthia dumetaria dumetaria	Scale-throated Earthcreeper	M	13	50.7		47.2	55.5		Argentina	840
		F	8	47.8		44.3	50.9			
Upucerthia jelskii	Plain-breasted Earthcreeper	U	23	40.1					Peru	1071
Upucerthia validirostris	Buff-breasted Earthcreeper	U	6	33.6		30	39			362, 776a
Upucerthia albigula	White-throated Earthcreeper	U	7	39.9					Peru	1071
Upucerthia serrana	Striated Earthcreeper	M	1	48.7					Peru	1043
Upucerthia ruficauda	Straight-billed Earthcreeper	B	17	36.3		31.9	41			840
Upucerthia andaecola	Rock Earthcreeper	U				28	34			274
Ochetorhynchus certhioides	Chaco Earthcreeper	M	7	21.7		18	24		Bolivia	657
		F	5	22.5		20.9	24			
Ochetorhynchus harterti	Bolivian Earthcreeper	M	3	24.9		23.6	26.2		Bolivia	1055
		F	6	23		21.6	23.8			
Eremobius phoenicurus	Band-tailed Earthcreeper	M	4	31.6		29	34		Argentina	840
		F	2	28.8		27.9	29.6			
Chilia melanura	Crag Chilia	U		40					Chile	582
Cinclodes antarcticus	Blackish Cinclodes	M	4	63.2		59	68.9		Argentina	1148
Cinclodes nigrofumosus	Chilean Seaside Cinclodes	U				63	67			274
Cinclodes taczanowskii	Peruvian Seaside Cinclodes	U	10	64.5	2.9				Peru	71a

Body Masses of World Birds (continued)

Species Name	Common Name	Sex	N	Mean	S.D.	Min	Max	Season	Location	Source #
Cinclodes patagonicus	Dark-bellied Cinclodes	U	15	30.7	2	36.7	44		Argentina	210
Cinclodes oustaleti	Gray-flanked Cinclodes	U	10	30.5	1.9	28	33.4		Argentina	210
Cinclodes olrogi	Olrog's Cinclodes	U				24	32			274
Cinclodes fuscus	Bar-winged Cinclodes	B	13	30		26.7	33			63, 1041, 1043, 1263
Cinclodes comechingonus	Comechingones Cinclodes	U				25	32			274
Cinclodes excelsior	Stout-billed Cinclodes	U				62	66			274
Cinclodes aricomae	Royal Cinclodes	U		50						274
Cinclodes pabsti	Long-tailed Cinclodes	U		53					Brazil	1105
Cinclodes atacamensis	White-winged Cinclodes	U	2	53		51.2	54.7		Peru; Argentina	1041, 1103
Cinclodes palliatus	White-bellied Cinclodes	U				99	109			274
Furnarius leucopus	Pale-legged Hornero	U	11	54.8					Ecuador; Peru	318, 1043, 1302, 1311
Furnarius torridus	Bay Hornero	U				48	57			274
Furnarius figulus	Tail-banded Hornero	M	1	28					Brazil	449
Furnarius minor	Lesser Hornero	U				23	29			274
Furnarius rufus	Rufous Hornero	U	67	57.4	3.8			B	Brazil	1015
Furnarius rufus	Rufous Hornero	U	10	43.7		42	48		Argentina	362
Furnarius rufus	Rufous Hornero	B	8	39.9		31.4	44		Bolivia	1055
Furnarius cristatus	Crested Hornero	U	1	25.5					Argentina	1041
Sylviorthorhynchus desmursii	Des Murs' Wiretail	U		10.9					Chile	84

Body Masses of World Birds (continued)

Species Name	Common Name	Sex	N	Mean	S.D.	Min	Max	Season	Location	Source #
Aphrastura spinicauda	Thorn-tailed Rayadito	U	23	11.5	0.8	9.5	12.8		Argentina	210
Leptasthenura andicola	Andean Tit-Spinetail	U				15	16			274
Leptasthenura striata	Streaked Tit-Spinetail	U				8	10			274
Leptasthenura pileata	Rusty-crowned Tit-Spinetail	B	3	10.3		9.9	10.6			1043
Leptasthenura xenothorax	White-browed Tit-Spinetail	F	1	10.5					Peru	1143
Leptasthenura striolata	Striolated Tit-Spinetail	U		10.5					Brazil	1105
Leptasthenura aegithaloides	Plain-mantled Tit-Spinetail	B	41	9.2	0.8	7.5	11		Chile	328
Leptasthenura aegithaloides pallida	Plain-mantled Tit-Spinetail	B	4	12.5		12.2	12.8		Argentina	210
Leptasthenura platensis	Tufted Tit-Spinetail	B	4	10.6		9.8	11.6		Argentina	841, 1040
Leptasthenura fuliginiceps	Brown-capped Tit-Spinetail	U	4	12.6		12	13		Argentina	1040
Leptasthenura setaria	Araucaria Tit-Spinetail	B	2	11		11	11		Brazil	63
Leptasthenura yanacensis	Tawny Tit-Spinetail	F	1	8.8					Peru	1143
Phleocryptes melanops	Wren-like Rushbird	U	6	14.6		13	16		Brazil; Argentina	63, 210, 1043
Limnornis curvirostris	Curve-billed Reedhaunter	B	3	28.6		27	30		Brazil	63
Limnornis rectirostris	Straight-billed Reedhaunter	B	3	18.5		16	21		Brazil	63
Spartonoica maluroides	Bay-capped Spinetail	B	2	11		10	12		Brazil	63
Schoeniophylax phryganophilus	Chotoy Spinetail	U	6	18.6		15	22.5		Brazil; Argentina	63, 210a

Body Masses of World Birds (continued)

Species Name	Common Name	Sex	N	Mean	S.D.	Min	Max	Season	Location	Source #
Poecilurus candei	White-whiskered Spinetail	M	2	15.3		14.5	16	B	Venezuela	97
Poecilurus scutatus	Ochre-cheeked Spinetail	U	12	15.4	1.7	13.5	19.5		Brazil	177
Synallaxis zimmeri	Russet-bellied Spinetail	U				12	14			274
Synallaxis whitneyi	Bahia Spinetail	B	2	16.2		16	16.5		Brazil	891
Synallaxis erythrothorax	Rufous-breasted Spinetail	B	13	17.1	1.2	15	18.8		Mexico; Belize	910a, 1027, 1145
Synallaxis unirufa	Rufous Spinetail	U	32	17.9	2.1	16.9	26.4		Colombia	324a
Synallaxis castanea	Black-throated Spinetail	U	1	15					Venezuela	1043
Synallaxis fuscorufa	Rusty-headed Spinetail	U				15	17			274
Synallaxis azarae	Azara's Spinetail	U	31	16.9	0.9	15.8	19.3		Ecuador	633
Synallaxis courseni	Apurimac Spinetail	M	2	16.8		16	17.5		Peru	1143
Synallaxis frontalis	Sooty-fronted Spinetail	B	10	14		12.1	16.8		Bolivia	1055
Synallaxis hypospodia	Cinereous-breasted Spinetail	U	17	16.9	1	14.9	18.1		Bolivia	110
Synallaxis albescens	Pale-breasted Spinetail	U	50	11.2	0.9	9.2	14.3		Bolivia	110
Synallaxis spixi	Chicli Spinetail	B	3	12.6		12	13		Brazil	63
Synallaxis albigularis	Dark-breasted Spinetail	U	9	15.4						1288, 1311
Synallaxis brachyura	Slaty Spinetail	B	13	18.3					Panama; Peru	489, 990, 1302, 1311
Synallaxis ruficapilla	Rufous-capped Spinetail	U	18	13.8	0.7				Brazil	961
Synallaxis infuscata	Pinto's Spinetail	B	7	18.2		16	20		Brazil	891

Body Masses of World Birds (continued)

Species Name	Common Name	Sex	N	Mean	S.D.	Min	Max	Season	Location	Source #
Synallaxis cabanisi	Cabanis' Spinetail	U	22	22.7	1.5				Peru	1288
Synallaxis moesta	Dusky Spinetail	U				20	25			274
Synallaxis macconnelli	MacConnell's Spinetail	B	9	19.5		17.6	22		Venezuela; Suriname	296, 498
Synallaxis gujanensis	Plain-crowned Spinetail	U	12	18.1	1.3				Peru	1288
Synallaxis albilora	White-lored Spinetail	U				13	17			274
Synallaxis maranonica	Maranon Spinetail	U	1	13					Southern Ecuador	1311
Synallaxis propinqua	White-bellied Spinetail	U				18	22			274
Synallaxis cinerascens	Gray-bellied Spinetail	B	10	13.1		12	14.5		Brazil; Paraguay	63, 132, 878
Synallaxis tithys	Black-faced Spinetail	B	15	15.8					Peru	1302
Synallaxis rutilans	Ruddy Spinetail	B	13	16.7		15.5	18.5		Suriname	498
Synallaxis cherriei	Chestnut-throated Spinetail	M	1	16					Brazil	449
Synallaxis cinnamomea terrestris	Stripe-breasted Spinetail	B	15	19.4		16.5	23		Tobago	359
Synallaxis cinnamomea carri	Stripe-breasted Spinetail	B	9	17.4		15.5	19		Trinidad	359
Synallaxis hellmayri	Red-shouldered Spinetail	U				25	26			274
Synallaxis stictothorax	Necklaced Spinetail	M	1	10.2					Peru	1143
Siptornopsis hypochondriaca	Great Spinetail	U				23	26			274
Hellmayrea gularis	White-browed Spinetail	B	29	12.4		10.5	14.5		Peru	900

Body Masses of World Birds (continued)

Species Name	Common Name	Sex	N	Mean	S.D.	Min	Max	Season	Location	Source #
Cranioleuca albicapilla	Creamy-crested Spinetail	U	22			18.9	24		Peru	739
Cranioleuca albiceps	Light-crowned Spinetail	U	9	20.4					Peru	1288
Cranioleuca albiceps	Light-crowned Spinetail	B	23	15.1					Bolivia	963
Cranioleuca marcapatae	Marcapata Spinetail	F	2	19.8		19.1	20.5		Peru	370
Cranioleuca baroni baroni	Baron's Spinetail	U	20			24	28		Peru	739
Cranioleuca baroni zaratensis	Baron's Spinetail	U	13			15.5	19.8		Peru	739
Cranioleuca antisiensis	Line-cheeked Spinetail	U	12	15.4	1.8	13	18		Ecuador	953
Cranioleuca antisiensis zaratensis	Line-cheeked Spinetail	M	5	18.5		16.8	19.5		Peru	1043
Cranioleuca curtata	Ash-browed Spinetail	U	4	17.5		16	19.7		Peru; Brazil	785a, 890, 1288
Cranioleuca hellmayri	Streak-capped Spinetail	U				14	16			274
Cranioleuca demissa	Tepui Spinetail	B	10	15.2		14	16.8		Venezuela	1308
Cranioleuca pallida	Pallid Spinetail	F	1	11.5					Brazil	1178
Cranioleuca erythrops	Red-faced Spinetail	B	18	16.9					Panama	489
Cranioleuca semicinerea	Gray-headed Spinetail	B	3	15		14.5	16		Brazil	1204
Cranioleuca subcristata	Crested Spinetail	B	3	14.7		14	15		Venezuela	1043
Cranioleuca obsoleta	Olive Spinetail	B	24	13.4		12	16		Brazil	63
Cranioleuca pyrrhophia	Stripe-crowned Spinetail	B	28	14.9		13	17		Brazil	63
Cranioleuca sulphurifera	Sulphur-bearded Spinetail	B	3	13.5		12.5	14		Brazil	63

Body Masses of World Birds (continued)

Species Name	Common Name	Sex	N	Mean	S.D.	Min	Max	Season	Location	Source #
Cranioleuca vulpina	Rusty-backed Spinetail	B	7	15.7		14	17.4		Brazil; Colombia	392, 876, 1147
Cranioleuca vulpecula	Parker's Spinetail	U				17	21			274
Cranioleuca gutturata	Speckled Spinetail	B	10	14.9					Brazil	1016
Cranioleuca muelleri	Scaled Spinetail	U				14	15			274
Acrobatornis fonsecai	Pink-legged Graveteiro	U	4	13.7		12	15		Bahia, Brazil	890
Certhiaxis mustelina	Red-and-white Spinetail	U				14	16			274
Certhiaxis cinnamomea	Yellow-chinned Spinetail	B	20	15.2		13.6	16.3		Suriname	498
Schizoeaca coryi	Ochre-browed Thistletail	B	4	16.8		15.5	17.5		Venezuela	1263
Schizoeaca perijana	Perija Thistletail	U				16	21			274
Schizoeaca fuliginosa	White-chinned Thistletail	B	9	18.3						1263, 1288
Schizoeaca griseomurina	Mouse-colored Thistletail	B	6	17.6		15.5	19		Peru	900
Schizoeaca palpebralis	Eye-ringed Thistletail	U				16	18			274
Schizoeaca vilcabambae	Vilcabamba Thistletail	B	3	20.2		18.5	22		Peru	1143
Schizoeaca helleri	Puna Thistletail	B	6	15.3		13	17.3		Peru	1143
Schizoeaca harterti	Black-throated Thistletail	M	11	13.2					Bolivia	963
		F	21	12.7						
Oreophylax moreirae	Itatiaia Thistletail	U				10	11			274
Asthenes dorbignyi	Creamy-breasted Canastero	B	11	15.9	1.2	14.5	19		Bolivia	1143
Asthenes steinbachi	Steinbach's Canastero	M	1	19.5					Argentina	1041
Asthenes pyrrholeuca	Lesser Canastero	B	10	13.2		11.6	14.2		Argentina	210, 1040, 1041

Body Masses of World Birds (continued)

Species Name	Common Name	Sex	N	Mean	S.D.	Min	Max	Season	Location	Source #
Asthenes baeri	Short-billed Canastero	B	11	17.8		15	21			63, 841
Asthenes patagonica	Patagonian Canastero	B	4	16.4		15.5	17.1		Argentina	841
Asthenes humicola	Dusky-tailed Canastero	B	57	22.5	1.4	18.5	25		Chile	328
Asthenes humicola goodalli	Dusky-tailed Canastero	B	4	19.3		18	20		Chile	748
Asthenes pudibunda	Canyon Canastero	U				13	17			274
Asthenes ottonis	Rusty-fronted Canastero	U		13						274
Asthenes heterura	Iquico Canastero	U				13	14			274
Asthenes modesta	Cordilleran Canastero	U	7	16.8		13	21.5		Argentina	362, 1040, 1041
Asthenes cactorum	Cactus Canastero	U				16	19			274
Asthenes luizae	Cipo Canastero	U	2	28		25.5	30.5		Brazil	265, 1256
Asthenes humilis	Streak-throated Canastero	B	8	21		19.7	22.5			1043
Asthenes wyatti	Streak-backed Canastero	F	1	15.2					Ecuador	1143
Asthenes sclateri	Puna Canastero	U	2	25.9		25.8	26		Argentina	1041
Asthenes anthoides	Austral Canastero	U	5	22.3		21.3	24.5		Argentina	546, 1262
Asthenes maculicauda	Scribble-tailed Canastero	U		19						274
Asthenes virgata	Junin Canastero	U		22						274
Asthenes urubambensis	Line-fronted Canastero	F	1	16					Bolivia	963
Asthenes flammulata	Many-striped Canastero	B	19	22.5		17	27		Peru	900
Phacellodomus fusciceps	Plain Softtail	U	3	25.2		24	25.9		Brazil	890

Body Masses of World Birds (continued)

Species Name	Common Name	Sex	N	Mean	S.D.	Min	Max	Season	Location	Source #
Phacellodomus berlepschi	Russet-mantled Softtail	U		23						274
Phacellodomus rufifrons	Common Thornbill	B	36	24.6		21.5	27.8		Venezuela	1215
Phacellodomus sibilatrix	Little Thornbill	B	8	15.5		14	16		Bolivia; Brazil	657, 890, 1175
Phacellodomus striaticeps	Streak-fronted Thornbill	B	3	29.1		23.3	40			362, 1055
Phacellodomus ruber	Greater Thornbill	B	10	41		35.8	51		Brazil; Argentina	63, 210a, 752
Phacellodomus striaticollis	Freckle-breasted Thornbill	B	3	25.7		24.5	26.5		Brazil	63
Phacellodomus maculipectus	Spot-breasted Thornbill	U	10	21.9	1.6	19.5	25.5		Argentina	1040
Phacellodomus dorsalis	Chestnut-backed Thornbill	U				33	39			274
Phacellodomus erythrophthalmus	Red-eyed Thornbill	M	2	24.5		24	25		Brazil	63
Clibanornis dendrocolaptoides	Canebrake Groundcreeper	U	3	48.2		42	52.5		Brazil; Paraguay	63, 132, 961
Anumbius annumbi	Firewood-gatherer	U	2	41.5		38	45		Brazil	63
Coryphistera alaudina	Lark-like Brushrunner	B	16	30		27	32		Bolivia	657
Pseudoseisura unirufa	Gray-crested Cacholote	B	3	44.9		43	47.8		Bolivia	1143
Pseudoseisura lophotes	Brown Cacholote	M	5	76		71	83		Bolivia	657
		F	6	68.6		63	77			
Pseudoseisura gutturalis	White-throated Cacholote	B	4	70.4		63.3	78.2		Argentina	841, 1041
Metopothrix aurantiacus	Orange-fronted Plushcrown	U	3	11.2		11	11.5		Peru; Brazil	370, 890
Xenerpestes singularis	Equatorial Graytail	U	4	11.6		11	12		Brazil	890
Xenerpestes minlosi	Double-banded Graytail	F	1	11					Ecuador	1147

Body Masses of World Birds (continued)

Species Name	Common Name	Sex	N	Mean	S.D.	Min	Max	Season	Location	Source #
Siptornis striaticollis	Spectacled Prickletail	F	1	13					Ecuador	1147
Premnornis guttuligera	Rusty-winged Barbtail	U	15	15.9	1.1				Peru	1288
Premnoplex brunnescens	Spotted Barbtail	U	18	16.3	1				Peru	1288
Roraimia adusta mayri	Roraiman Barbtail	M	6	15.8		12	18		Venezuela	296
		F	6	16.7		12	20			
Margarornis rubiginosus	Ruddy Treerunner	U	2	17.8		16.1	19.5		Panama	489
Margarornis stellatus	Fulvous-dotted Treerunner	U				21	22			274
Margarornis bellulus	Beautiful Treerunner	M	1	18.5					Panama	990
Margarornis squamiger	Pearled Treerunner	U	17	17.3	1.4				Peru	1288
Pygarrhichas albogularis	White-throated Treerunner	U	13	24	2.1	19.5	27		Argentina	210
Megaxenops parnaguae	Great Xenops	M	1	25					Northeast Brazil	1205
Xenops milleri	Rufous-tailed Xenops	B	7	12.2					Brazil; Peru; Suriname	73, 498, 988, 1178
Xenops tenuirostris	Slender-billed Xenops	B	8	9.7					Peru; Suriname	370, 498
Xenops minutus	Plain Xenops	U	16	10.6		8.2	11.4		Brazil	875
Xenops rutilans	Streaked Xenops	B	12	12.2					Panama	489
Xenops rutilans	Streaked Xenops	B	5	10.3					Peru	1302
Heliobletus contaminatus	Sharp-billed Treehunter	B	10	14		12.5	15.5		Brazil	63, 878, 961, 1178
Anabacerthia amaurotis	White-browed Foliage-gleaner	U	29	19.2	1.2				Brazil	961
Anabacerthia striaticollis	Montane Foliage-gleaner	U	18	24.8	1.3				Peru	1288

Body Masses of World Birds (continued)

Species Name	Common Name	Sex	N	Mean	S.D.	Min	Max	Season	Location	Source #
Anabacerthia variegaticeps	Scaly-throated Foliage-gleaner	U	3	27.8		25.7	30.6		Colombia	1037
Pseudocolaptes boissonneautii	Streaked Tuftedcheek	B	16	43.9		39.1	51.6		Colombia	324a
Pseudocolaptes lawrencii	Buffy Tuftedcheek	B	7	51					Panama; Ecuador	489, 991
Thripadectes flammulatus	Flammulated Treehunter	B	6	55.9		50	62		Peru; Colombia	900, 1037
Thripadectes scrutator	Buff-throated Treehunter	M	1	66.3					Peru	1288
Thripadectes holostictus	Striped Treehunter	U	28	40.7		32.6	45.6		Colombia	324a
Thripadectes melanorhynchus	Black-billed Treehunter	U	14	44.3					Peru; Ecuador; Colombia	953, 1038a, 1288
Thripadectes virgaticeps	Streak-capped Treehunter	U	2	65.8		65	66.5		Colombia	881
Thripadectes ignobilis	Uniform Treehunter	U	4	47.2		42.2	51.5		Colombia; Ecuador	991, 1037
Thripadectes rufobrunneus	Streak-breasted Treehunter	B	4	56.2		51.5	64.8		Panama	489
Syndactyla subalaris	Lineated Foliage-gleaner	U	14	28.9	1.2				Peru	1288
Syndactyla guttulata	Guttulated Foliage-gleaner	U		36						522
Syndactyla rufosuperciliata	Buff-browed Foliage-gleaner	U	12	25.6	1.4				Brazil	961
Syndactyla ruficollis	Rufous-necked Foliage-gleaner	F	5	31		29	34		Peru	900
Berlepschia rikeri	Point-tailed Palmcreeper	B	5	37		35	38.5		Suriname	498
Simoxenops ucayalae	Peruvian Recurvebill	B	11	49.8					Peru; Brazil	370, 449, 1083

Body Masses of World Birds (continued)

Species Name	Common Name	Sex	N	Mean	S.D.	Min	Max	Season	Location	Source #
Simoxenops striatus	Bolivian Recurvebill	U	6	41.6		37.5	48.5		Bolivia	514
Hyloctistes subulatus	Striped Woodhaunter	U	14	28.8	1.6				Peru	370
Ancistrops strigilatus	Chestnut-winged Hookbill	B	8	35.9					Peru	370
Philydor erythropterum	Chestnut-winged Foliage-gleaner	U		30					Peru	1210
Philydor erythrocercum	Rufous-rumped Foliage-gleaner	U	12	26.4	4.6				French Guiana	1214
Philydor erythrocercum ochrogaster	Rufous-rumped Foliage-gleaner	B	41	24.2					Peru	370
Philydor ruficaudatum	Rufous-tailed Foliage-gleaner	U	10	30.1					Brazil	1016
Philydor rufum	Buff-fronted Foliage-gleaner	B	9	25		22	31		Brazil	63, 252, 961, 1043
Philydor lichtensteini	Ochre-breasted Foliage-gleaner	U	12	21		17	23		Paraguay	132
Philydor dimidiatum	Russet-mantled Foliage-gleaner	U	10	29.4	1.5	27	32		Brazil	177
Philydor pyrrhodes	Cinnamon-rumped Foliage-gleaner	B	6	30.5		25	34.5		Suriname	498
Philydor fuscipenne	Slaty-winged Foliage-gleaner	U				25	28			274
Philydor atricapillus	Black-capped Foliage-gleaner	U	9	22.2		20	24		Brazil; Paraguay	132, 878
Philydor novaesi	Alagoas Foliage-gleaner	F	3	38		30	48		Northeast Brazil	1203
Anabazenops dorsalis	Crested Foliage-gleaner	U	7	35.4					Peru	1288
Automolus rufipileatus	Chestnut-crowned Foliage-gleaner	U	14	34.8		31.4	38		Venezuela; Suriname; French Guiana; Brazil	293, 498, 875, 1308

Body Masses of World Birds (continued)

Species Name	Common Name	Sex	N	Mean	S.D.	Min	Max	Season	Location	Source #
Automolus leucophthalmus	White-eyed Foliage-gleaner	U	12	34.5		31	37.5		Brazil	752, 878
Automolus roraimae	White-throated Foliage-gleaner	M	13	27	2.5	22	31.8		Venezuela	1308
		F	8	28.1		26.5	29			
Automolus ochrolaemus	Buff-throated Foliage-gleaner	U	25	40.2	2.5				Costa Rica; Panama	618
Automolus infuscatus	Olive-backed Foliage-gleaner	U	11	32.9		30.2	36.2		Brazil	875
Automolus melanopezus	Brown-rumped Foliage-gleaner	B	5	31.4		28.7	32.9		Peru	319, 370
Automolus rubiginosus	Ruddy Foliage-gleaner	B	9	39.8		32	44			293, 498, 496, 990
Hylocryptus rectirostris	Chestnut-capped Foliage-gleaner	U	5	48		44	51		Brazil	752
Hylocryptus erythrocephalus	Henna-hooded Foliage-gleaner	M	5	44.8					Peru	1302
		F	5	48.2						
Sclerurus rufigularis	Short-billed Leaftosser	U	10	21.6		20.2	23.1		Brazil; Suriname	293, 496, 875
Sclerurus mexicanus	Tawny-throated Leaftosser	B	9	25.1		21	27.7		Panama; Mexico; Colombia; Suriname	498, 881, 990, 1037
Sclerurus albigularis	Gray-throated Leaftosser	B	13	34.8	3.4	26	39.5		Venezuela	1254
Sclerurus caudacutus olivascens	Black-tailed Leaftosser	B	8	35.3		34.8	37		Peru	1143
Sclerurus caudacutus pallidus	Black-tailed Leaftosser	B	4	37.2		34.5	41.3		Brazil	1143
Sclerurus scansor	Rufous-breasted Leaftosser	B	10	36.9		32	41		Brazil	63, 252, 875, 961, 1205
Sclerurus guatemalensis	Scaly-throated Leaftosser	U	25	34.7	3.3				Panama; Costa Rica	618
Lochmias nematura	Sharp-tailed Streamcreeper	B	13	22		20	24.5		Brazil	63, 752, 878

Body Masses of World Birds (continued)

Species Name	Common Name	Sex	N	Mean	S.D.	Min	Max	Season	Location	Source #
Dendrocincla tyrannina	Tyrannine Woodcreeper	U	7	54.9					Peru; Colombia	1037, 1288
Dendrocincla turdina	Thrush-like Woodcreeper	U	22	39		31	45.3		Brazil	878
Dendrocincla anabatina	Tawny-winged Woodcreeper	M	8	37.4		35.2	42.4		Yucatan, Mexico	910a
		F	6	31.3		28.8	34.1			
Dendrocincla fuliginosa	Plain-brown Woodcreeper	B	98	38.7		31	54.5		Trinidad	359
Dendrocincla merula	White-chinned Woodcreeper	B	20	46.6	3.1				Bolivia	185
Dendrocincla merula	White-chinned Woodcreeper	B	67	47.5	3.4				Peru	1316
Dendrocincla merula	White-chinned Woodcreeper	M		57						274
		F		49.5						
Dendrocincla homochroa	Ruddy Woodcreeper	M	6	45.3					Panama	489
		F	4	36.8						
Dendrocincla homochroa	Ruddy Woodcreeper	U	67	36.8	5.3	25	48		El Salvador	1041a
Deconychura longicauda	Long-tailed Woodcreeper	U	11	23.8	2				Costa Rica; Panama	618
Deconychura stictolaema	Spot-throated Woodcreeper	B	9	18.5		14.5	20.5		Brazil	879, 1178
Sittasomus griseicapillus	Olivaceous Woodcreeper	M	21	14.3	0.6				Panama	489
		F	11	12.9	0.7					
Sittasomus griseicapillus	Olivaceous Woodcreeper	M	3	11.3					Peru	1302
		F	5	10.4						
Sittasomus griseicapillus	Olivaceous Woodcreeper	U	15	18		17	19		Paraguay	132
Glyphorynchus spirurus	Wedge-billed Woodcreeper	U	113	14.6		12.2	16.9		Brazil	875
Drymornis bridgesii	Scimitar-billed Woodcreeper	M	2	78		76	80		Bolivia; Brazil	63, 657, 841
		F	1	110						
Nasica longirostris	Long-billed Woodcreeper	U		92					Peru	1210

Body Masses of World Birds (continued)

Species Name	Common Name	Sex	N	Mean	S.D.	Min	Max	Season	Location	Source #
Dendrexetastes rufigula	Cinnamon-throated Woodcreeper	B	13	69.6		64	77		Suriname	498
Hylexetastes stresemanni	Bar-bellied Woodcreeper	U		111					Peru	617
Hylexetastes perrotii	Red-billed Woodcreeper	U	66	114	7.7				Brazil	73
Hylexetastes perrotii uniformis	Uniform Woodcreeper	B	8	93.2		86	106		Brazil	1143
Xiphocolaptes albicollis	White-throated Woodcreeper	U	4	118		109	131		Brazil; Paraguay	63, 132, 177
Xiphocolaptes falcirostris	Moustached Woodcreeper	U		110						274
Xiphocolaptes major	Great Rufous Woodcreeper	B	11	156		142	163			362, 1175
Xiphocolaptes promeropirhynchus	Strong-billed Woodcreeper	B	10	117		103	126		Colombia	324a
Xiphocolaptes promeropirhynchus emigrans	Strong-billed Woodcreeper	U	3	74		62	80			362
Xiphocolaptes carajaensis	Carajas Woodcreeper	M	1	114					Para, Brazil	1108
Dendrocolaptes sanctithomae	Northern Barred-Woodcreeper	U	5	66.9		58.5	74		Chiapas, Mexico; El Salvador	1041a, 1143
Dendrocolaptes certhia	Amazonian Barred-Woodcreeper	B	15	68.7		62.5	79		Suriname	498
Dendrocolaptes hoffmannsi	Hoffmann's Woodcreeper	B	3	83		74	86		Brazil	861, 1178
Dendrocolaptes picumnus	Black-banded Woodcreeper	M	4	82.5		79	88		Suriname	498
		F	6	92.7		87	98			
Dendrocolaptes picumnus	Black-banded Woodcreeper	B	4	73.8					Northern Andes	1313
Dendrocolaptes picumnus	Black-banded Woodcreeper	B	6	61					Nicaragua	1313

Body Masses of World Birds (continued)

Species Name	Common Name	Sex	N	Mean	S.D.	Min	Max	Season	Location	Source #
Dendrocolaptes platyrostris	Planalto Woodcreeper	B	11	61.7		50.5	69		Brazil	63, 877, 878
Xiphorhynchus picus	Straight-billed Woodcreeper	U	14	41.6	3.2	35.3	46.8		Venezuela	1215
Xiphorhynchus picus	Straight-billed Woodcreeper	M	5	38.8		35.8	40.5		Colombia	1168
Xiphorhynchus kienerii	Zimmer's Woodcreeper	M	2	43.8		43.4	44.3		Colombia	1168
Xiphorhynchus obsoletus	Striped Woodcreeper	U	16	32.6		25.9	37		Suriname	498
Xiphorhynchus ocellatus	Ocellated Woodcreeper	U				32	42			274
Xiphorhynchus chunchotambo	Tschudi's Woodcreeper	U	7	35.1	1.9				Peru	1288
Xiphorhynchus spixii	Spix's Woodcreeper	U	17	31.2		27.2	36.2		Brazil	875
Xiphorhynchus elegans	Spix's Woodcreeper	B	13	34.1	2.6	30.5	38.5		Brazil	1178
Xiphorhynchus guttatus eytoni	Buff-throated Woodcreeper	B	11	58.8	7.2	45	69		Brazil	449, 1178
Xiphorhynchus guttatus polystictus	Buff-throated Woodcreeper	B	13	60.6		55	70		Suriname	498
Xiphorhynchus pardalotus	Chestnut-rumped Woodcreeper	M	12	37	2.5	32	41		Venezuela	1308
		F	6	32.3		26.8	36			
Xiphorhynchus susurrans	Cocoa Woodcreeper	B	51	48.8		36.5	59		Trinidad	359
Xiphorhynchus susurrans nanus	Cocoa Woodcreeper	U	5	48.6		41	52		Panama	156
Xiphorhynchus susurrans nanus	Cocoa Woodcreeper	U	27	43	4.6	35	52		Venezuela	1254
Xiphorhynchus flavigaster	Ivory-billed Woodcreeper	M	9	47.2		43.6	52.6		Yucatan, Mexico	910a
		F	12	40	1.7	35	44.1			
Xiphorhynchus flavigaster	Ivory-billed Woodcreeper	U	84	52.2	7.8	32	75		El Salvador	1041a

Body Masses of World Birds (continued)

Species Name	Common Name	Sex	N	Mean	S.D.	Min	Max	Season	Location	Source #
Xiphorhynchus lachrymosus	Black-striped Woodcreeper	B	4	56.4		50.7	66		Panama	874, 990, 1180
Xiphorhynchus erythropygius	Spotted Woodcreeper	U	13	46.8	2.4	42.6	50		Panama	79
Xiphorhynchus triangularis	Olive-backed Woodcreeper	B	13	46.3		39	51.8		Colombia; Ecuador	798, 953, 1043
Xiphorhynchus fuscus	Lesser Woodcreeper	U	20	21.8	1.4				Rio de Janeiro, Brazil	125
Lepidocolaptes leucogaster	White-striped Woodcreeper	B	12	36	2	32.1	39.2		San Luis Potosi, Mexico	1145
Lepidocolaptes souleyetii	Streak-headed Woodcreeper	B	10	25.7					Peru	1302
Lepidocolaptes angustirostris	Narrow-billed Woodcreeper	B	13	27.8		23.6	45.1		Bolivia	1055
Lepidocolaptes angustirostris	Narrow-billed Woodcreeper	B	8	31.5					Brazil	63, 877, 1109
Lepidocolaptes affinis	Spot-crowned Woodcreeper	M	16	35.4	2.3				Panama	489
		F	20	34.6	2.2					
Lepidocolaptes lacrymiger carabayae	Montane Woodcreeper	B	5	30.6		27	33		Peru	1143
Lepidocolaptes squamatus	Scaled Woodcreeper	U	2	28.1		27	29.2		Brazil; Argentina	211, 877
Lepidocolaptes falcinellus	Scalloped Woodcreeper	U				26	30			274
Lepidocolaptes albolineatus	Lineated Woodcreeper	B	23	20.3		18	24		Suriname	498
Campylorhamphus pucherani	Greater Scythebill	M	1	70.5					Peru	1288
Campylorhamphus trochilirostris	Red-billed Scythebill	B	11	40.8		35	55		Peru; Panama; Argentina; Bolivia; Colombia	156, 319, 657, 795, 1040, 1302
Campylorhamphus pusillus	Brown-billed Scythebill	B	11	40.5		33	48		Peru	900, 1037

Body Masses of World Birds (continued)

Species Name	Common Name	Sex	N	Mean	S.D.	Min	Max	Season	Location	Source #
Sakesphorus luctuosus	Glossy Antshrike	B	8	31.4		28	34		Brazil	392, 449
Thamnophilus doliatus fraterculus	Barred Antshrike	B	32	27.9		25	30.5		Trinidad	359
Thamnophilus doliatus tobagensis	Barred Antshrike	B	12	26.2		22	30		Tobago	359
Thamnophilus zarumae	Chapman's Antshrike	M	2	22.8		22	23.5		Ecuador; Peru	953, 1302
Thamnophilus multistriatus	Bar-crested Antshrike	B	7	27.5		25.6	30.4		Colombia	798
Thamnophilus palliatus	Chestnut-backed Antshrike	U	1	23.3					Brazil	875
Thamnophilus tenuepunctatus	Lined Antshrike	U				22	23			274
Thamnophilus bridgesi	Black-hooded Antshrike	U		27					Costa Rica	1170
Thamnophilus nigriceps	Black Antshrike	U	6	22.9		21.8	26.2		Panama	156
Thamnophilus nigrocinereus	Blackish-gray Antshrike	U	2	30		28	32			274
Thamnophilus cryptoleucus	Castelnau's Antshrike	B	2	27.5		27.5	27.5		Colombia	512
Thamnophilus aethiops	White-shouldered Antshrike	B	14	25.7		22	29.5		Venezuela	1308
Thamnophilus unicolor	Uniform Antshrike	B	7	24.2		22.2	27.2		Colombia	798
Thamnophilus aroyae	Upland Antshrike	U				19	21			274
Thamnophilus schistaceus	Plain-winged Antshrike	B	5	20.3		19	22.5		Brazil; Peru	318, 861
Thamnophilus murinus	Mouse-colored Antshrike	U	13	19.3	1.7				French Guiana	1214
Thamnophilus atrinucha	Western Slaty-Antshrike	U		23.6					Nicaragua	194
Thamnophilus punctatus	Guianan Slaty-Antshrike	U	14	19.4	1.5	16.5	21.5		Brazil	752

Body Masses of World Birds (continued)

Species Name	Common Name	Sex	N	Mean	S.D.	Min	Max	Season	Location	Source #
Thamnophilus stictocephalus	Natterer's Slaty-Antshrike	M	1	21.5					Bolivia	560
Thamnophilus sticturus	Bolivian Slaty-Antshrike	B	18	18.2	2.5	16	27		Bolivia	1143
Thamnophilus pelzelni	Planalto Slaty-Antshrike	B	8	20.9		20	22		Sao Paulo, Brazil	1143
Thamnophilus insignis	Streak-backed Antshrike	M	6	26.3		24.1	30		Venezuela	1308
		F	2	23.5		22	25			
Thamnophilus divisorius	Acre Antshrike	B	6	22.2		21	23		Acre, Brazil	1297
Thamnophilus amazonicus	Amazonian Antshrike	B	10	19.4		15.5	24.4		Brazil; Venezuela	861, 875, 1308
Thamnophilus amazonicus divaricatus	Amazonian Antshrike	B	19	18		15.9	19.7		Suriname	498
Thamnophilus caerulescens	Variable Antshrike	B	12	21.1		17	23		Brazil; Paraguay	132, 878
Thamnophilus torquatus	Rufous-winged Antshrike	U	2	21.1		17.9	24.3		Brazil	392, 1143
Thamnophilus ruficapillus	Rufous-capped Antshrike	B	10	20.4		16.3	21.5		Bolivia	1055
Pygiptila stellaris	Spot-winged Antshrike	B	11	24.1		22	26.1		Suriname	498
Megastictus margaritatus	Pearly Antshrike	B	4	20.2		18.5	21		Rondonia, Brazil	1143
Neoctantes niger	Black Bushbird	U		32					Peru	1210
Clytoctantes atrogularis	Rondonia Bushbird	F	1	31					Brazil	669
Xenornis setifrons	Speckled Antshrike	F	1	24.6					Panama	79
Thamnistes anabatinus	Russet Antshrike	B	10	20.7	1.8	19	23.6		Panama; Belize	489, 615, 990, 1027
Dysithamnus mentalis	Plain Antvireo	B	11	18.9		18	20		Paraguay	132
Dysithamnus mentalis andrei	Plain Antvireo	B	14	12.8		11.5	14.5		Trinidad	359

Body Masses of World Birds (continued)

Species Name	Common Name	Sex	N	Mean	S.D.	Min	Max	Season	Location	Source #
Dysithamnus mentalis cumbreanus	Plain Antvireo	B	16	13.6	0.9	12	15		Venezuela	1254
Dysithamnus striaticeps	Streak-crowned Antvireo	U	1	12					Costa Rica	618
Dysithamnus puncticeps	Spot-crowned Antvireo	U	2	14.9		14.6	15.2		Colombia	1037
Dysithamnus leucostictus	White-streaked Antvireo	U	1	20.2					Colombia	1038a
Dysithamnus plumbeus	Plumbeous Antvireo	B	13	20.5	1.6	18	23		Venezuela	1254
Thamnomanes saturninus	Saturnine Antshrike	B	13	20.4	1	19	22.5		Rondonia, Brazil	1178
Thamnomanes ardesiacus	Dusky-throated Antshrike	U	22	17.7	1.5				French Guiana	1214
Thamnomanes caesius	Cinereous Antshrike	B	26	15.7		12.8	18		Venezuela	1308
Thamnomanes schistogynus	Bluish-slate Antshrike	B	11	17.2	1.3	15.5	20		Peru	319
Myrmotherula brachyura	Pygmy Antwren	B	14	6.4		5	7.8		Suriname	498
Myrmotherula ignota obscura	Moustached (Short-billed) Antwren	M	5	7.1						467
Myrmotherula sclateri	Sclater's Antwren	M	8	8.4						467, 988, 1178
Myrmotherula klagesi	Klages' Antwren	B	3	7.8		7.5	8.3		Brazil	195
Myrmotherula ambigua	Yellow-throated Antwren	B	3	6.7		6	7.6		Brazil; Venezuela	1178, 1308
Myrmotherula cherriei	Cherrie's Antwren	B	11	8.2	0.8	7.3	9.9		Colombia; Peru	10, 1143
Myrmotherula pacifica	Pacific Antwren	M	1	10.1					Ecuador	1143
Myrmotherula multostriata	Amazonian Antwren	B	6	8		7.5	8.6		Venezuela; Brazil	1143

Body Masses of World Birds (continued)

Species Name	Common Name	Sex	N	Mean	S.D.	Min	Max	Season	Location	Source #
Myrmotherula surinamensis	Guianan Antwren	B	16	8.3		7.5	9		Suriname	498
Myrmotherula longicauda	Stripe-chested Antwren	M	5	8.3						467
Myrmotherula hauxwelli	Plain-throated Antwren	U	23	10.7		8	16		Brazil	875
Myrmotherula guttata	Rufous-bellied Antwren	U	10	11	1.3				French Guiana	1214
Myrmotherula gularis	Star-throated Antwren	U	13	11.3		10.6	13		Brazil	879, 961
Myrmotherula gutturalis	Brown-bellied Antwren	U	16	9.2	1				French Guiana	1214
Myrmotherula fulviventris	Checker-throated Antwren	U	19	10.4	0.7				Costa Rica; Panama	618
Myrmotherula leucophthalma	White-eyed Antwren	M	12	9.4	0.9	8	10.7		Brazil	1178
Myrmotherula fjeldsaai	Brown-backed Antwren	B	4	9.8		9	10.7		Ecuador; Peru	654
Myrmotherula haematonota	Stipple-throated Antwren	M	20	8.7						467
Myrmotherula spodionota	Foothill Antwren	U	10	10.7					Ecuador; Colombia	953, 1038a
Myrmotherula ornata	Ornate Antwren	U	12	9.4						449, 912
Myrmotherula erythrura	Rufous-tailed Antwren	U	8	11.4						467, 912
Myrmotherula axillaris	White-flanked Antwren	B	25	8.4					Peru	1288
Myrmotherula axillaris	White-flanked Antwren	B	17	7.8		6.5	9.5		Trinidad	359
Myrmotherula fluminensis	Rio de Janeiro Antwren	U		9						274
Myrmotherula schisticolor	Slaty Antwren	B	16	9.6					Panama	489

Body Masses of World Birds (continued)

Species Name	Common Name	Sex	N	Mean	S.D.	Min	Max	Season	Location	Source #
Myrmotherula longipennis	Long-winged Antwren	U	33	9.4	1.1				French Guiana	1214
Myrmotherula minor	Salvadori's Antwren	F	1	6.4					Brazil	1178
Myrmotherula iheringi	Ihering's Antwren	U		8					Peru	1210
Myrmotherula grisea	Ashy Antwren	M	5	8.4						467
Myrmotherula behni	Plain-winged Antwren	B	2	8		7.9	8		Venezuela	296, 1308
Myrmotherula unicolor	Unicolored Antwren	F	1	11.5					Brazil	961
Myrmotherula snowi	Alagoas Antwren	U	3			8	11.5			274
Myrmotherula menetriesii	Gray Antwren	U	13	8.6	0.5				French Guiana	1214
Myrmotherula assimilis	Leaden Antwren	M	5	9.3						467
Dichrozona cincta	Banded Antwren	U		16					Peru	1210
Myrmorchilus strigilatus	Stripe-backed Antbird	U	10	23	1.7				Argentina	178
Herpsilochmus motacilloides	Creamy-bellied Antwren	U	1	12.5						274
Herpsilochmus atricapillus	Black-capped Antwren	M	4	10.8		10.5	11		Bolivia	514
Herpsilochmus atricapillus	Black-capped Antwren	B	75	9		7	12		Brazil	1298
Herpsilochmus sellowi	Caatinga Antwren	B	3	7.5		7	8		Northeast Brazil	1298
Herpsilochmus pileatus	Pileated Antwren	U	2	8.9		8.5	9.3		Northeast Brazil	1298
Herpsilochmus sticturus	Spot-tailed Antwren	U	4	9.5		8.9	11			498, 1295
Herpsilochmus dugandi	Dugand's Antwren	U	4	10.3		9.3	11.3			1295
Herpsilochmus stictocephalus	Todd's Antwren	U	11	8.8		7.8	9.6			1295
Herpsilochmus gentryi	Ancient Antwren	B	6	10.6		10.2	11		Peru; Ecuador	1295

Body Masses of World Birds (continued)

Species Name	Common Name	Sex	N	Mean	S.D.	Min	Max	Season	Location	Source #
Herpsilochmus dorsimaculatus	Spot-backed Antwren	M	3	9.9		9.4	10.5		Brazil	195
Herpsilochmus roraimae	Roraiman Antwren	M	4		10	13				296
		F	4		11	12				
Herpsilochmus longirostris	Large-billed Antwren	B	4	12.8		12	14		Brazil	752
Herpsilochmus axillaris	Yellow-breasted Antwren	U	3	11.4		11	13			899, 1295
Herpsilochmus rufimarginatus	Rufous-winged Antwren	M	2	11.2		11	11.5		Panama; Brazil	990, 1178
		F	1	10						
Microrhopias quixensis	Dot-winged Antwren	B	11	7.9	0.7	7.2	9.7		Belize	1027
Stymphalornis acutirostris	Parana Antwren	U		10						274
Formicivora iheringi	Narrow-billed Antwren	M	1	8.2						274
Formicivora grisea intermedia	White-fringed Antwren	B	12	9.3		9	10.5		Trinidad	359
Formicivora grisea tobagensis	White-fringed Antwren	B	25	11		9.5	14		Tobago	359
Formicivora grisea griseus	White-fringed Antwren	B	16	10.9		8	13.1		Suriname	498
Formicivora melanogaster	Black-bellied Antwren	B	7	10.5		9.8	12		Bolivia	1143
Formicivora serrana	Serra Antwren	U				10	13			274
Formicivora littoralis	Restinga Antwren	U				14	16			274
Formicivora rufa	Rusty-backed Antwren	B	9	10.8		9.7	12.5		Brazil	392, 883, 1109
Drymophila ferruginea	Ferruginous Antwren	U	5	10.6		9.4	12		Brazil	878, 1178, 1205
Drymophila rubricollis	Bertoni's Antbird	U	1	10						274
Drymophila ochropyga	Ochre-rumped Antbird	U	6	10.5		10.2	10.7		Sao Paulo, Brazil	1143

Body Masses of World Birds (continued)

Species Name	Common Name	Sex	N	Mean	S.D.	Min	Max	Season	Location	Source #
Drymophila devillei	Striated Antbird	M	2	10.2		10	10.3		Peru; Brazil	668
Drymophila malura	Dusky-tailed Antbird	B	10	13	1.4	11.5	16.5		Brazil	961
Drymophila caudata	Long-tailed Antbird	B	5	11.9					Peru	1288
Drymophila squamata	Scaled Antbird	M	2	10.8		10.5	11		Brazil	1202
Terenura maculata	Streak-capped Antwren	M	1	6.5					Northeast Brazil	1205
Terenura sicki	Orange-billed Antwren	F	1	6.4					Northeast Brazil	1204
Terenura callinota	Rufous-rumped Antwren	B	6	7		6.5	8		Panama; Ecuador	498, 990, 991
Terenura humeralis	Chestnut-shouldered Antwren	B	3	7.3		7	7.8		Peru	988
Terenura sharpei	Yellow-rumped Antwren	U				7	8			274
Terenura spodioptila	Ash-winged Antwren	B	6	6.5		5	7.3		Brazil; Suriname	498, 1178
Cercomacra cinerascens	Gray Antbird	B	11	14.3		12.9	15.7		Suriname	498
Cercomacra tyrannina	Dusky Antbird	B	32	16.3		12	18.7		Suriname	498
Cercomacra laeta	Willis' Antbird	U				15	17			274
Cercomacra nigrescens	Blackish Antbird	B	12	16.5	1.9	14	21		Brazil	449
Cercomacra ferdinandi	Bananal Antbird	U	1	15						274
Cercomacra serva	Black Antbird	U		17					Peru	1210
Cercomacra nigricans	Jet Antbird	M	7	17.6					Panama	489
		F	4	15.6						
Cercomacra carbonaria	Rio Branco Antbird	F	1	14.5					Roraima, Brazil	1178
Cercomacra melanaria	Mato Grosso Antbird	U	2	19		18.5	19.5		Paraguay	1175

Body Masses of World Birds (continued)

Species Name	Common Name	Sex	N	Mean	S.D.	Min	Max	Season	Location	Source #
Cercomacra manu	Manu Antbird	M	12	19.7					Peru	369
		F	9	16.6						
Pyriglena leuconota	White-backed Fire-eye	U	78	32.3		26.2	37.5		Brazil	875
Pyriglena leucoptera	White-shouldered Fire-eye	U	21	28.8		25.5	35.6		Brazil	878
Pyriglena atra	Fringe-backed Fire-eye	M	1	32					Northeast Brazil	1205
Rhopornis ardesiacus	Slender Antbird	B	6	26.3		23.5	28		Brazil	1200
Myrmoborus leucophrys	White-browed Antbird	B	28	21.5		17.7	24		Suriname	498
Myrmoborus lugubris	Ash-breasted Antbird	U				20	23			274
Myrmoborus myotherinus	Black-faced Antbird	U	15	18.8		12.3	25		Brazil; Peru	319, 392, 785a
Hypocnemis cantator	Warbling Antbird	B	18	12.6		11.1	14		Venezuela	1308
Hypocnemis cantator	Warbling Antbird	B	22	12		11	13		Suriname	498
Hypocnemis hypoxantha	Yellow-browed Antbird	M	6			10.5	12.5		Peru	988
		F	6			10	13.5			
Hypocnemoides melanopogon	Black-chinned Antbird	B	14	14.1		12	18.2		Suriname	498
Hypocnemoides maculicauda	Band-tailed Antbird	U	2	11.8		11.5	12.1		Brazil	392
Myrmochanes hemileucus	Black-and-white Antbird	U				12	13			274
Gymnocichla nudiceps	Bare-crowned Antbird	B	12	30.5	2.8	27	35		Panama; Belize	156, 990, 1027, 1180
Sclateria naevia	Silvered Antbird	B	26	24.6		21	28		Suriname	498
Percnostola rufifrons rufifrons	Black-headed Antbird	B	9	26.2		23.5	31			167
Percnostola rufifrons jensoni	Black-headed Antbird	B	5	21.8		21	22.5		Peru	167

Body Masses of World Birds (continued)

Species Name	Common Name	Sex	N	Mean	S.D.	Min	Max	Season	Location	Source #
Percnostola arenarum	Allpahuayo Antbird	B	8	23		21.3	24.3		Peru	561
Percnostola leucostigma	Spot-winged Antbird	B	13	25.6		22	28.5			112
Percnostola saturata	Roraiman Antbird	B	10	25.8		24	31.7			112
Percnostola caurensis	Caura Antbird	M	1	39.4					Venezuela	1308
Percnostola lophotes	White-lined Antbird	U		28					Peru	1210
Myrmeciza longipes	White-bellied Antbird	U	10	27.9					Panama	615
Myrmeciza exsul	Chestnut-backed Antbird	B	12	26.5	2.1	22.5	30		Panama; Brazil	79, 156, 874, 1180
Myrmeciza ferruginea	Ferruginous-backed Antbird	B	18	26.1		22	31		Suriname	498
Myrmeciza squamosa	Squamate Antbird	U	4	18.5		16.5	20		Brazil	878
Myrmeciza laemosticta	Dull-mantled Antbird	U		25					Costa Rica	1170
Myrmeciza nigricauda	Esmeraldas Antbird	M	1	24					Colombia	521
Myrmeciza disjuncta	Yapacana Antbird	F	1	15					Venezuela	1308
Myrmeciza pelzelni	Gray-bellied Antbird	U				17	18.5		Venezuela	274
Myrmeciza hemimelaena	Southern Chestnut-tailed Antbird	B	18	16					Peru	1288
Myrmeciza castanea	Northern Chestnut-tailed Antbird	U				16	17			274
Myrmeciza hyperythra	Plumbeous Antbird	U		41					Peru	1210
Myrmeciza melanoceps	White-shouldered Antbird	U	1	38.5					Ecuador	1143
Myrmeciza goeldii	Goeldi's Antbird	U		42					Peru	1210

Body Masses of World Birds (continued)

Species Name	Common Name	Sex	N	Mean	S.D.	Min	Max	Season	Location	Source #
Campylorhamphus falcularius	Black-billed Scythebill	U	8	42.6						64, 125, 878, 961
Campylorhamphus procurvoides	Curve-billed Scythebill	B	7	33.5		29.9	36		French Guiana; Venezuela; Brazil	293, 498, 861, 1308
Family Thamnophilidae										
Cymbilaimus lineatus	Fasciated Antshrike	B	17	35.8		31	44		Suriname	498
Cymbilaimus sanctaemariae	Bamboo Antshrike	B	9	28.7					Peru	370
Hypoedaleus guttatus	Spot-backed Antshrike	F	2	38.8		36	41.5		Paraguay	1175
Batara cinerea	Giant Antshrike	B	8	131		97	156		Brazil; Bolivia	64, 657, 878, 1105
Mackenziaena severa	Tufted Antshrike	B	5	51.8		50	54		Brazil; Paraguay	64, 132, 1178
Mackenziaena leachii	Large-tailed Antshrike	B	6	70.2		59	76.1			64, 211, 252
Frederickena viridis	Black-throated Antshrike	F	2	76		72	80		Suriname	498
Frederickena unduligera	Undulated Antshrike	B	6	83					Peru	370
Taraba major	Great Antshrike	B	29	59.2		50	64.4		Suriname	498
Sakesphorus bernardi	Collared Antshrike	B	4	34.4		28.9	41		Peru	1043, 1302
Sakesphorus canadensis	Black-crested Antshrike	B	26	24.4		22	28		Suriname	498
Sakesphorus cristatus	Silvery-cheeked Antshrike	B	8	17.7		13.3	20.5		Minas Gerais, Brazil	1142b, 1146
Sakesphorus melanonotus	Black-backed Antshrike	B	5	25.5		23	26.5		Venezuela	1254
Sakesphorus melanothorax	Band-tailed Antshrike	B	4	31.8		31	33		Suriname	498

Body Masses of World Birds (continued)

Species Name	Common Name	Sex	N	Mean	S.D.	Min	Max	Season	Location	Source #
Myrmeciza fortis	Sooty Antbird	U	20	45.4	3.3				Bolivia	185
Myrmeciza fortis	Sooty Antbird	U	18	46.5	2.8				Peru	1316
Myrmeciza immaculata	Immaculate Antbird	U	8	47.1		41.1	55.5		Panama; Colombia	618, 990, 1037
Myrmeciza griseiceps	Gray-headed Antbird	M	1	15					Peru	1143
Myrmeciza atrothorax	Black-throated Antbird	B	15	16.1		14.7	18.5		Suriname	498
Pithys albifrons	White-plumed Antbird	U	26	20.4	1.9				Colombia	1038a
Pithys albifrons	White-plumed Antbird	B	20	18.6					Peru; Ecuador	667a
Pithys castaneus	White-masked Antbird	B	11	29.8					Peru	667a
Gymnopithys rufigula	Rufous-throated Antbird	U	834	29.1	2.1				Brazil	73
Gymnopithys leucaspis	Bicolored Antbird	U	32	31.1	1.9	28	37		Panama	79
Gymnopithys leucaspis	Bicolored Antbird	B	10	23.8					Peru; Ecuador	667a
Gymnopithys lunulatus	Lunulated Antbird	U				22	26			274
Gymnopithys salvini	White-throated Antbird	U	70	25.9	2.3				Peru	1316
Myrmornis torquata	Wing-banded Antbird	U	10	46.7					Panama	615, 990
Rhegmatorhina melanosticta	Hairy-crested Antbird	B	20	30.6	3				Bolivia	185
Rhegmatorhina melanosticta	Hairy-crested Antbird	U	41	31.4	8.5				Peru	1316
Rhegmatorhina hoffmannsi	White-breasted Antbird	B	7	31.4		27	34		Brazil	1178
Rhegmatorhina gymnops	Bare-eyed Antbird	U				27	30			274

Body Masses of World Birds (continued)

Species Name	Common Name	Sex	N	Mean	S.D.	Min	Max	Season	Location	Source #
Hylophylax naevioides	Spotted Antbird	U	34	17.8		15.8	20		Panama	874
Hylophylax naevius	Spot-backed Antbird	B	24	14.2					Peru	1288
Hylophylax punctulata	Dot-backed Antbird	B	5	12.4		12	13		Bolivia	49
Hylophylax poecilinotus	Scale-backed Antbird	B	43	18.4		15.8	20.5		Venezuela	1308
Phlegopsis nigromaculata	Black-spotted Bare-eye	U	45	44.5		38	53		Brazil	875
Phlegopsis nigromaculata	Black-spotted Bare-eye	U	64	46	3.9				Peru	1316
Phlegopsis erythroptera	Reddish-winged Bare-eye	B	6	51.1		44	61		Venezuela	1308
Phlegopsis erythroptera	Reddish-winged Bare-eye	B	10	58.3					Peru; Ecuador	667a
Skutchia borbae	Pale-faced Antbird	U		50						274
Phaenostictus mcleannani	Ocellated Antbird	U	12	51.1	2.7				Panama; Costa Rica	618
Family Formicariidae										
Formicarius colma	Rufous-capped Antthrush	B	12	47		42	53		Venezuela	1308
Formicarius nigricapillus	Black-headed Antthrush	B	3	56.2		48	61.5			1037, 1147
Formicarius monileger	Mexican Antthrush	B				56.5	67.8			274
Formicarius analis	Black-faced Antthrush	B	16	62.2	3.5	56.5	67.8		Belize	1027
Formicarius rufifrons	Rufous-fronted Antthrush	B		55.5		54	57			274
Formicarius rufipectus	Rufous-breasted Antthrush	B	5	74.4		68	82		Panama; Colombia	990, 1037
Chamaeza nobilis	Striated Antthrush	U		123					Peru	1210

Body Masses of World Birds (continued)

Species Name	Common Name	Sex	N	Mean	S.D.	Min	Max	Season	Location	Source #
Chamaeza campanisona	Short-tailed Antthrush	U	8	99.7		86	112		Brazil; Argentina; Venezuela	211, 296, 878
Chamaeza campanisona	Short-tailed Antthrush	U	8	82.3					Colombia	1038a
Chamaeza ruficauda	Brazilian Antthrush	B	4	71.2		63.5	76		Brazil	64, 961
Chamaeza meruloides	Such's Antthrush	U		69					Brazil	1314
Chamaeza mollissima	Barred Antthrush	B	3	72		65	80		Peru; Bolivia	1143, 963
Pittasoma michleri	Black-crowned Antpitta	U	1	109					Panama	615
Pittasoma rufopileatum	Rufous-crowned Antpitta	B	2	96.5		96	97			274
Grallaria squamigera	Undulated Antpitta	U	3	136		115	166		Ecuador; Bolivia; Colombia	324a, 963, 1311
Grallaria gigantea	Giant Antpitta	M	4	236		204	266			274
Grallaria excelsa	Great Antpitta	U				218	266			522
Grallaria varia	Variegated Antpitta	B	5	119		90	134			274
Grallaria guatimalensis	Scaled Antpitta	B	7	94.2		77	116			618, 798, 990, 1145
Grallaria alleni	Moustached Antpitta	M	2	70.5		64	77			274
Grallaria haplonota	Plain-backed Antpitta	M	1	85					Colombia	1037
Grallaria dignissima	Ochre-striped Antpitta	M	1	110						274
Grallaria eludens	Elusive Antpitta	M	1	115					Peru	715
Grallaria bangsi	Santa Marta Antpitta	M	1	62					Colombia	1164
Grallaria ruficapilla	Chestnut-crowned Antpitta	B	8	80.6		69.9	89.8		Colombia	324a

Body Masses of World Birds (continued)

Species Name	Common Name	Sex	N	Mean	S.D.	Min	Max	Season	Location	Source #
Grallaria ruficapilla	Chestnut-crowned Antpitta	M	5	84.8		75.6	91.5		Colombia	798
Grallaria kaestneri	Cundinamarca Antpitta	U	3	47.5		45.8	49.4		Colombia	1164
Grallaria watkinsi	Watkins' Antpitta	B				58	84			274
Grallaria andicola	Stripe-headed Antpitta	B				48	66			274
Grallaria rufocinerea	Bicolored Antpitta	U		44.8						274
Grallaria nuchalis	Chestnut-naped Antpitta	B	4	117		110	122		Peru	900
Grallaria ridgelyi	Jocotoco Antpitta	M	3	192		176	204		Ecuador	652
		F	2	167		152	182			
Grallaria carrikeri	Pale-billed Antpitta	B				96	124			274
Grallaria flavotincta	Yellow-breasted Antpitta	U	1	63					Colombia	881
Grallaria hypoleuca	White-bellied Antpitta	F	1	73.7					Peru	900
Grallaria przewalskii	Rusty-tinged Antpitta	B	3	65.2		60.5	74			274
Grallaria capitalis	Bay Antpitta	B	3	73.7		72	77			274
Grallaria erythroleuca	Red-and-white Antpitta	M	1	80					Peru	1143
Grallaria albigula	White-throated Antpitta	F	1	95					Argentina	1040
Grallaria rufula	Rufous Antpitta	B	11	35.7					Peru; Ecuador; Bolivia	953, 963, 1288
Grallaria blakei	Chestnut Antpitta	B				38	47			274
Grallaria erythrotis	Rufous-faced Antpitta	B	2	57		53	61		Bolivia	963
Grallaria quitensis	Tawny Antpitta	B	25	65.7		57	83		Peru	900

Body Masses of World Birds (continued)

Species Name	Common Name	Sex	N	Mean	S.D.	Min	Max	Season	Location	Source #
Grallaria milleri	Brown-banded Antpitta	U		52.5						274
Hylopezus macularius	Spotted Antpitta	B	4	44.2		38	53		French Guiana; Suriname	293, 498
Hylopezus perspicillatus	Streak-chested Antpitta	U	13	43						156, 615, 618
Hylopezus auricularis	Masked Antpitta	M	1	43				B	Bolivia	1143
		F	1	38						
Hylopezus dives	Fulvous-bellied Antpitta	U	1	44						274
Hylopezus fulviventris	White-lored Antpitta	M	2	41		38	44		Panama	990
Hylopezus berlepschi	Amazonian Antpitta	U		48					Peru	1210
Hylopezus ochroleucus	White-browed Antpitta	U		28					Brazil	1105
Hylopezus nattereri	Speckle-breasted Antpitta	M	2	32		31	33			274
Myrmothera campanisona	Thrush-like Antpitta	U		47					Peru	1210
Myrmothera simplex	Tepui Antpitta	B	6	52.4		46.8	61		Venezuela; Guyana	44, 1308
Grallaricula flavirostris	Ochre-breasted Antpitta	U	12	17.7					Peru; Colombia	881, 1288
Grallaricula ferrugineipectus	Rusty-breasted Antpitta	B	43	17.2		13	21		Peru	900
Grallaricula loricata	Scallop-breasted Antpitta	B	20	20.7	11.7	18.4	24.5		Venezuela	1253a
Grallaricula peruviana	Peruvian Antpitta	B	6	18.8		17	21		Peru; Ecuador	900, 953
Grallaricula ochraceifrons	Ochre-fronted Antpitta	M				22.5	23.8			274
Grallaricula nana	Slate-crowned Antpitta	B	9	19.8		17.5	21.5		Peru; Ecuador	900, 953
Grallaricula lineifrons	Crescent-faced Antpitta	B	9	21.1					Ecuador	989

Body Masses of World Birds (continued)

Species Name	Common Name	Sex	N	Mean	S.D.	Min	Max	Season	Location	Source #
Family Conopophagidae										
Conopophaga lineata	Rufous Gnateater	B	17	21.2					Brazil	961
Conopophaga lineata	Rufous Gnateater	U	8	30		28	31		Paraguay	132
Conopophaga aurita	Chestnut-bellied Gnateater	M	14			23	28		Peru	988
		F	7			24	31			
Conopophaga roberti	Hooded Gnateater	U	1	20.8					Brazil	875
Conopophaga peruviana	Ash-throated Gnateater	U		23					Peru	1210
Conopophaga ardesiaca	Slaty Gnateater	B	57	26.3					Peru	370
Conopophaga castaneiceps	Chestnut-crowned Gnateater	B	10	27					Peru; Ecuador	785a, 953, 1288, 1311
Conopophaga melanops	Black-cheeked Gnateater	U	3	20.1		19.2	21.5		Brazil	878
Conopophaga melanogaster	Black-bellied Gnateater	M	2	42.5		42	43		Brazil	449, 1178
Family Rhinocryptidae										
Pteroptochos tarnii	Black-throated Huet-huet	B	4	147		138	153		Argentina	210
Pteroptochos castaneus	Chestnut-throated Huet-huet	M				158	185			274
		F				130	175			
Pteroptochos megapodius	Moustached Turca	M	3	114		95	125			274
Scelorchilus albicollis albicollis	White-throated Tapaculo	M	6	54.8		52	60		Chile	748
		F	2	46.5		45	48			
Scelorchilus albicollis atacamae	White-throated Tapaculo	B	2	41.5		39	44		Chile	748
Scelorchilus rubecula	Chucao Tapaculo	U		76					Chile	84
Scelorchilus rubecula	Chucao Tapaculo	B	7	42.3		34.9	48		Argentina	186a, 210
Rhinocrypta lanceolata	Crested Gallito	B	11	61.9		54	66.7		Bolivia	657

Body Masses of World Birds (continued)

Species Name	Common Name	Sex	N	Mean	S.D.	Min	Max	Season	Location	Source #
Teledromas fuscus	Sandy Gallito	B	4	37.9		35.4	41.8			274, 1041
Liosceles thoracicus	Rusty-belted Tapaculo	M				39	42			274
Melanopareia torquata	Collared Crescent-chest	U	2	17.2		16.5	18		Brazil	752
Melanopareia maximiliani argentina	Olive-crowned Crescent-chest	M	1	18					Bolivia	1143
Melanopareia elegans	Elegant Crescent-chest	F	1	16					Peru	1302
Melanopareia maranonica	Maranon Crescent-chest	F	1	23						274
Psilorhamphus guttatus	Spotted Bamboowren	U		11.3					Brazil	1105
Merulaxis ater	Slaty Bristlefront	M	1	37.2					Brazil	1105
		F	1	33						
Eugralla paradoxa	Ochre-flanked Tapaculo	U		44					Chile	84
Myornis senilis	Ash-colored Tapaculo	B	10	21.2		18.1	24.5		Peru; Ecuador	900, 953
Scytalopus unicolor	Unicolored Tapaculo	M	6	18.5		17.6	19.3		Colombia	324a
		F	8	16.8		15.3	19			
Scytalopus latrans latrans	Blackish Tapaculo	M	15	18.2		16.8	20.9		Ecuador	656
		F	14	16.9		15.5	18			
Scytalopus latrans subcinereus	Blackish Tapaculo	M	4	18.8		15.9	20		Ecuador	656
		F	7	16.7		14.2	19.1			
Scytalopus parvirostris	Trilling Tapaculo	M	7	19.7		18.8	21		Peru	1143
		F	2	16		15	17.1			
Scytalopus macropus	Large-footed Tapaculo	U				32	43			656
Scytalopus femoralis	Rufous-vented Tapaculo	U	8	23.2					Peru; Ecuador; Colombia	953, 1037, 1288
Scytalopus micropterus	Long-tailed Tapaculo	M	9	29.7		27	32.5		Ecuador	656

Body Masses of World Birds (continued)

Species Name	Common Name	Sex	N	Mean	S.D.	Min	Max	Season	Location	Source #
Scytalopus bolivianus	Bolivian Tapaculo	B	13	20.8	1.4	19	22.6		Peru	1143
Scytalopus atratus	White-crowned Tapaculo	M	7	26.2		24.6	32.5		Ecuador	656
		F	1	25.3						
Scytalopus vicinior	Narino Tapaculo	M	14	22.5		20.5	24.6		Ecuador	656
		F	2	23.6		22.9	24.2			
Scytalopus argentifrons	Silvery-fronted Tapaculo	U		17					Costa Rica	1170
Scytalopus latebricola	Brown-rumped Tapaculo	B	4	21.5		18	24		Peru; Ecuador	900, 953
Scytalopus meridanus	Merida Tapaculo	U		20						522
Scytalopus caracae	Caracas Tapaculo	U		24					estimated	274
Scytalopus spillmanni	Spillman's Tapaculo	M	36	25.2		21	30		Ecuador	656
		F	5	24.1		20	29.5			
Scytalopus zimmeri	Zimmer's Tapaculo	M				17.6	19.9			274
		F	1	15.9						
Scytalopus simonsi	Puna Tapaculo	M	3	17.6		15.7	20			274
		F	1	15.7						
Scytalopus urubambae	Vilcabamba Tapaculo	M	1	13						274
Scytalopus affinis	Ancash Tapaculo	M	2	14		13	15			274
		F	1	12.3						
Scytalopus canus opacus	Paramo Tapaculo	M	23	16.2		13.9	17.9		Ecuador	656
		F	7	14.8		13.4	15.5			
Scytalopus canus confusus	Paramo Tapaculo	M	1	25						655
Scytalopus magellanicus	Magellanic Tapaculo	B	3	12.8		10.4	14.2		Argentina	546, 1098
Scytalopus magellanicus	Magellanic Tapaculo	B	11	14.6					Bolivia	963
Scytalopus griseicollis	Matorral Tapaculo	M	4	19.7		18	21.8			274, 324a
		F	1	16.6						

Body Masses of World Birds (continued)

Species Name	Common Name	Sex	N	Mean	S.D.	Min	Max	Season	Location	Source #
Scytalopus griseicollis infasciatus	Colombian Tapaculo	M	2	18		17.5	18.4			274
Scytalopus superciliaris	White-browed Tapaculo	B	5	17.8		16	20		Argentina	1040
Scytalopus acutirostris	Tschudi's Tapaculo	M				17	20			274
		F				16.5	19			
Scytalopus speluncae	Mouse-colored Tapaculo	M	1	13					Brazil	64
Scytalopus novacapitalis	Brasilia Tapaculo	U		18.9					Brazil	1105
Scytalopus psychopompus	Bahia Tapaculo	B	2	17.8		17.5	18		Bahia, Brazil	1201
Scytalopus iraiensis	Wetland Tapaculo	F	1	12.4					Southern Brazil	93a
Scytalopus indigoticus	White-breasted Tapaculo	U	2			14.5	18			1201
Scytalopus schulenbergi	Diademed Tapaculo	M	2	16.1		15.2	17		Bolivia	274, 1294
Scytalopus chocoensis	Choco Tapaculo	M	17	21		19	22.5		Ecuador	656
		F	5	19		17	20.1			
Scytalopus robbinsi	Ecuadorian Tapaculo	M	8	19.6		18.1	21		Ecuador	656
		F	3	19.1		18.7	19.5			
Scytalopus stilesi	Stiles's Tapaculo	M	2	23		21.5	24.5		Colombia	236
		F	5	21.2		20	22			
Scytalopus rodriguezi	Upper Magdalena Tapaculo	M	3	21.6		21.4	21.9		Colombia	655
Scytalopus parkeri	Chusquea Tapaculo	M	13	22.5		21	24.4		Ecuador	656
		F	4	20.1		18.8	22.3			
Acropternis orthonyx	Ocellated Tapaculo	M	3	95		90	100		Peru	900
		F	2	85		81	89			
Family Cotingidae										
Phytotoma raimondii	Peruvian Plantcutter	B	4	39.8		36	44		Peru	1143
Phytotoma rutila	White-tipped Plantcutter	B	4	40.5		36	45.1		Bolivia	657, 1055

Body Masses of World Birds (continued)

Species Name	Common Name	Sex	N	Mean	S.D.	Min	Max	Season	Location	Source #
Phytotoma rara	Rufous-tailed Plantcutter	U	15	47		41.5	52		Chile; Argentina	210, 328
Oxyruncus cristatus	Sharpbill	B	6	42		37	46			259, 990, 1105
Phoenicircus nigricollis	Black-necked Red-Cotinga	M	1	90					Venezuela	1308
		F	3	103		99	107			
Phoenicircus carnifex	Guianan Red-Cotinga	M	2	72.3		70	77			498, 1129, 1230
		F	1	95						
Laniisoma elegans	Shrike-like Cotinga	B	4	47.4		45.8	51			1129, 1288
Laniocera rufescens	Speckled Mourner	B	8	48.1		38.6	56		Panama; Belize	488, 615, 990, 1027, 1180
Laniocera hypopyrra	Cinereous Mourner	U		51					Peru	1210
Phibalura flavirostris	Swallow-tailed Cotinga	B	3	46.5		43.5	52			64, 1129
Tijuca condita	Gray-winged Cotinga	F	1	80						1129
Carpornis cucullatus	Hooded Berryeater	B	15	74.2		66.5	80		Brazil	64, 961, 1129
Carpornis melanocephalus	Black-headed Berryeater	B	3	64.2		62.7	66			1129, 1202
Ampelion rubrocristatus	Red-crested Cotinga	B	21	66.3		45	80			1129
Ampelion rufaxilla	Chestnut-crested Cotinga	B	7	73.9		69	77			1129
Doliornis remseni	Chestnut-bellied Cotinga	B	7	63.1		58.5	72		Ecuador	994
Doliornis sclateri	Bay-vented Cotinga	B	28	60.2		53	69		Peru	994
Zaratornis stresemanni	White-cheeked Cotinga	B	16	52.1		46	57			448. 575
Pipreola riefferii	Green-and-black Fruiteater	B	20	49.3		43	61			881, 953, 1129

Body Masses of World Birds (continued)

Species Name	Common Name	Sex	N	Mean	S.D.	Min	Max	Season	Location	Source #
Pipreola intermedia intermedia	Band-tailed Fruiteater	B	7	47.9		44	58			1129
Pipreola intermedia signata	Band-tailed Fruiteater	B	3	54.7		50.8	58.9			1129
Pipreola arcuata	Barred Fruiteater	B	12	118						953, 963, 1129
Pipreola aureopectus	Golden-breasted Fruiteater	B	2	46		46	46			1129
Pipreola pulchra	Masked Fruiteater	B	7	58.2		51.2	63			1129, 1288
Pipreola chlorolepidota	Fiery-throated Fruiteater	B	2	29.5		28	31			1129
Pipreola frontalis	Scarlet-breasted Fruiteater	B	7	42.4		39.5	45.3			1129
Pipreola formosa	Handsome Fruiteater	B	10	47.2		41	56			1043, 1129
Pipreola whitelyi kathleenae	Red-banded Fruiteater	M	1	50					Venezuela	1143
Ampelioides tschudii	Scaled Fruiteater	B	10	78.4		70	95			1037, 1129
Iodopleura pipra	Buff-throated Purpletuft	M	1	9.6					Northeast Brazil	1206
		F	1	10.5						
Iodopleura isabellae	White-browed Purpletuft	B	3	20		19.8	20.2		Peru	370
Iodopleura fusca	Dusky Purpletuft	M	1	15.3						1129
Lipaugus subalaris	Gray-tailed Piha	B	6	81.2						1038a, 1129
Lipaugus cryptolophus	Olivaceous Piha	M	3	73.6		72.1	75.8			785a
Lipaugus fuscocinereus	Dusky Piha	F	3	138					Peru; Colombia	237, 900
Lipaugus uropygialis	Scimitar-winged Piha	M	1	116						1129
Lipaugus vociferans	Screaming Piha	B	12	82.2		77	87		Peru	1129

Body Masses of World Birds (continued)

Species Name	Common Name	Sex	N	Mean	S.D.	Min	Max	Season	Location	Source #
Lipaugus vociferans	Screaming Piha	B	15	69.2		60	74		Suriname	498
Lipaugus unirufus	Rufous Piha	B	9	82.1		69	87.2			1129
Lipaugus lanioides	Cinnamon-vented Piha	B	12	94.8		85	110			1129
Lipaugus streptophorus	Rose-collared Piha	B	3	57.2		52	60.5		Venezuela	1143
Lipaugus weberi	Chestnut-capped Piha	B	2	72.1		72	72.2		Colombia	237
Porphyrolaema porphyrolaema	Purple-throated Cotinga	U		60					Peru	1210
Cotinga amabilis	Lovely Cotinga	B	5	71.5		66	75			1129
Cotinga ridgwayi	Turquoise Cotinga	M	16	53.6	3.6				Panama	489
		F	6	61						
Cotinga maynana	Plum-throated Cotinga	U		69					Peru	1210
Cotinga cotinga	Purple-breasted Cotinga	B	2	54		53	55			1129
Cotinga maculata	Banded Cotinga	M	1	65						1129
Cotinga cayana	Spangled Cotinga	B	16	61.8		55	73		Suriname	498
Xipholena punicea	Pompadour Cotinga	B	15	68.1		58	76			1129
Xipholena atropurpurea	White-winged Cotinga	M				58	65			275
		F				56	67			
Carpodectes hopkei	Black-tipped Cotinga	F	1	102					Ecuador	1148
Carpodectes antoniae	Yellow-billed Cotinga	U		98					Costa Rica	1170
Carpodectes nitidus	Snowy Cotinga	M	2	116					Panama	539
		F	1	89						
Conioptilon mcilhennyi	Black-faced Cotinga	U		90					Peru	1210

Body Masses of World Birds (continued)

Species Name	Common Name	Sex	N	Mean	S.D.	Min	Max	Season	Location	Source #
Gymnoderus foetidus	Bare-necked Fruitcrow	M	2	345		331	359		Suriname	498
		F	1	220						
Haematoderus militaris	Crimson Fruitcrow	F	1	227				B	Rondonia, Brazil	1143
Querula purpurata	Purple-throated Fruitcrow	M	14	105		91	122		Central America; Guianas	1129
		F	9	96.7		91	105			
Querula purpurata	Purple-throated Fruitcrow	M	5	120		112	133		Peru	1129
		F	6	109		100	114			
Pyroderus scutatus	Red-ruffed Fruitcrow	B	8	357		293	419			1129
Cephalopterus penduliger	Long-wattled Umbrellabird	M	1	338					Ecuador	1147
Cephalopterus ornatus	Amazonian Umbrellabird	F	1	380						1129
Cephalopterus glabricollis	Bare-necked Umbrellabird	M		450					Costa Rica	1170
		F		320						
Perissocephalus tricolor	Capuchinbird	M	5	360		320	395			1129
		F	6	319		267	367			
Procnias tricarunculatus	Three-wattled Bellbird	M	6	210		194	233			1129
Procnias albus	White Bellbird	M	1	210						1129
Procnias averano	Bearded Bellbird	M	4	168		150	178			296, 1129, 1043
		F	5	127		111	135			
Procnias nudicollis	Bare-throated Bellbird	M	4	200		177	225			1129
		F	5	148		140	150			
Rupicola rupicola	Guianan Cock-of-the-Rock	M	6	208		188	225			296, 1308
		F	8	182		140	226			
Rupicola peruviana sanguinolenta	Andean Cock-of-the-Rock	M	3	271		242	306			785a, 1129
		F	4	217		206	226			
Family Pipridae										
Chloropipo unicolor	Jet Manakin	U	27	15.5	0.9				Peru	1288
Chloropipo uniformis	Olive Manakin	M	2	19		19	19		Venezuela	296
		F	8	16.8		15	18			

Body Masses of World Birds (continued)

Species Name	Common Name	Sex	N	Mean	S.D.	Min	Max	Season	Location	Source #
Chloropipo holochlora	Green Manakin	U	15	14.9	0.9	13.5	17.5		Colombia	1037
Chloropipo flavicapilla	Yellow-headed Manakin	F	5	17.8		16.8	19.5		Colombia	798
Manacus candei	White-collared Manakin	M	21	18.8	3.1				Costa Rica	618
		F	8	20.9	1					
Manacus aurantiacus	Orange-collared Manakin	U		15.5					Costa Rica	1170
Manacus vitellinus	Golden-collared Manakin	M	17	19.3	0.2				Panama	489
		F	16	17.1	0.2					
Manacus manucus	White-bearded Manakin	U	44	15		12.2	18.1		Brazil	875
Manacus manucus trinitatis	White-bearded Manakin	M	93	18.5	1.3	16	23		Trinidad	359
		F	106	16.8	1.3	14	21.5			
Corapipo gutturalis	White-throated Manakin	U	65	8.2	1				Brazil	73
Corapipo altera	White-ruffed Manakin	M	87	11.1	0.8				Costa Rica	1018a
		F	30	12.5	1					
Corapipo leucorrhoa	White-bibbed Manakin	M	13	10.7						906
		F	4	12.7						
Chiroxiphia lanceolata	Lance-tailed Manakin	U	27	18.5	1.4				Panama	618
Chiroxiphia linearis	Long-tailed Manakin	M	149	17	2.3	11	26		El Salvador	1041a
		F	116	17.9	2.5	9	23.5			
Chiroxiphia pareola	Blue-backed Manakin	M	13	16.3						906
		F	3	17.4						
Chiroxiphia boliviana	Yungas Manakin	M	9	16.4					Peru	370
		F	20	17.9	1.1					
Chiroxiphia caudata	Blue Manakin	U	43	25.6		20.5	30		Brazil	878
Pipra aureola	Crimson-hooded Manakin	M	20	15.8		14	18		Suriname	498
		F	11	16.5		15	18.5			
Pipra fasciicauda	Band-tailed Manakin	B	130	15.9					Peru	1288

Body Masses of World Birds (continued)

Species Name	Common Name	Sex	N	Mean	S.D.	Min	Max	Season	Location	Source #
Pipra filicauda	Wire-tailed Manakin	B	15	15.4	1.3	14	19		Venezuela	1254
Pipra pipra	White-crowned Manakin	M	11	10.3						906
		F	10	12						
Pipra coronata	Blue-crowned Manakin	M	30	8						906
		F	103	8.7						
Pipra erythrocephala	Golden-headed Manakin	M	223	12.8	1	10.5	17		Trinidad	1131
		F	171	14.1	1	12	16.5			
Pipra mentalis	Red-capped Manakin	U	24	15		12.5	17.2		Panama	874
Pipra rubrocapilla	Red-headed Manakin	U	17	12		10.2	13		Brazil	875
Pipra chloromeros	Round-tailed Manakin	B	16	16.5						455
Pipra cornuta	Scarlet-tailed Manakin	F	3	25.3		21.5	28.5		Venezuela	1308
Pipra iris	Opal-crowned Manakin	U	1	8					Brazil	392
Pipra isidorei	Blue-rumped Manakin	U	25	7.8	0.8				Colombia	1038a
Pipra nattereri	Snow-capped Manakin	M	2	8		7.8	8.1		Brazil	1178
		F	8	8.8		8.1	9.4			
Pipra coeruleocapilla	Cerulean-capped Manakin	M	12	8.9						906
		F	6	10.5						
Lepidothrix suavissima	Tepui Manakin	M	1	8.5					Venezuela	917
Lepidothrix serena	White-fronted Manakin	B	10	11.4		10.8	12		French Guiana	293
Antilophia galeata	Helmeted Manakin	M	39	20.6	1.9	18	26.5		Brazil	751
		F	19	22.4	1.6	18.8	26.2			
Antilophia bokermanni	Araripe Manakin	U				19	20.5			275
Masius chrysopterus	Golden-winged Manakin	M	12	10.8	2.2	6.1	14		Colombia	1037
		F	17	12.3	2	6.4	16			

Body Masses of World Birds (continued)

Species Name	Common Name	Sex	N	Mean	S.D.	Min	Max	Season	Location	Source #
Ilicura militaris	Pin-tailed Manakin	M	23	12.7	2.1	11	19.5		Brazil	753
Machaeropterus pyrocephalus	Fiery-capped Manakin	M	11	9.7	0.4	9	10.5		Peru	319
		F	16	9.3	0.5	8.5	10.5			
Machaeropterus regulus	Striped Manakin	M	15	9.6						906
		F	13	9.1						
Machaeropterus deliciosus	Club-winged Manakin	B	23	12.7	1.7	11	18		Colombia	1037
Xenopipo atronitens	Black Manakin	M	8	15.1		14	17		Suriname	498
		F	8	16		14	18			
Heterocercus flavivertex	Yellow-crested Manakin	B	12	21.3		18.3	24		Venezuela	947, 1308
Heterocercus aurantiivertex	Orange-crested Manakin	U				21	22			275
Heterocercus linteatus	Flame-crested Manakin	U	4	21.2		19.8	23.3		Brazil	392
Neopelma chrysocephalum	Saffron-crested Tyrant-Manakin	B	12	15.5		13.4	17		Suriname; Venezuela; Peru; Brazil	10, 1178, 1308
Neopelma sulphureiventer	Sulphur-crested Tyrant-Manakin	M	2	18					Bolivia	49
		F	2	13.7						
Neopelma pallescens	Pale-bellied Tyrant-Manakin	U	5	18.2		16	20.4		Brazil	392
Neopelma aurifrons	Wied's Tyrant-Manakin	U	5	14		13.5	14.9		Brazil	878, 1178
Tyranneutes stolzmanni	Dwarf Tyrant-Manakin	U	6	7.2		6.9	7.5		Venezuela	1308
Tyranneutes virescens	Tiny Tyrant-Manakin	M	5	6.8		6	7.3		Suriname	498
		F	4	7.6		6	8.5			
Piprites pileatus	Black-capped Piprites	M	1	15					Brazil	64
Piprites griseiceps	Gray-headed Piprites	U	4	16					Costa Rica	80
Piprites chloris	Wing-barred Piprites	M	6	16		15	18.5		Suriname	498

Body Masses of World Birds (continued)

Species Name	Common Name	Sex	N	Mean	S.D.	Min	Max	Season	Location	Source #
Family Tyrannidae										
Ornithion inerme	White-lored Tyrannulet	U		7					Peru	1210
Ornithion semiflavus	Yellow-bellied Tyrannulet	B	10	6.6		5	7.4		Mexico; Costa Rica; Belize	379, 580, 1145
Ornithion brunneicapillus	Brown-capped Tyrannulet	U	3	7.1		6.9	7.2		Panama	1180
Camptostoma imberbe	Northern Beardless-Tyrannulet	B	13	7.4	0.8	6.3	9.5	B	Mexico; Belize	1143
Camptostoma obsoletum	Southern Beardless-Tyrannulet	B	17	8.1	1.2	6	10			497, 613, 798, 1180
Phaeomyias murina	Mouse-colored Tyrannulet	U	37	10		8	12		Trinidad	359
Nesotriccus ridgwayi	Cocos Is.Flycatcher	U		11					Cocos Is.	1170
Capsiempis flaveola	Yellow Tyrannulet	B	12	7.7					Panama; Venezuela	489, 615, 1180, 1215
Tyrannulus elatus	Yellow-crowned Tyrannulet	B	11	7		6.3	7.9		Suriname	498
Myiopagis gaimardii	Forest Elaenia	B	13	12.6		9.5	14		Trinidad	359
Myiopagis gaimardii	Forest Elaenia	M	15	12.2		9.5	17		Suriname	498
		F	11	11.3		9.7	13.3			
Myiopagis olallai	Foothill Elaenia	B	4	12.1		11.1	14		Ecuador; Peru	217
Myiopagis caniceps	Gray Elaenia	B	11	10.5		9.5	12.3		Peru; Brazil	64, 259, 988, 1204
Myiopagis subplacens	Pacific Elaenia	M	1	14.9					Peru	1302
		F	4	16.9						
Myiopagis flavivertex	Yellow-crowned Elaenia	M	9	12.3		10.5	14		Suriname	198
		F	9	11.4		10	13			
Myiopagis cotta	Jamaican Elaenia	M	2	11.6		11.5	11.6		Jamaica	19

Body Masses of World Birds (continued)

Species Name	Common Name	Sex	N	Mean	S.D.	Min	Max	Season	Location	Source #
Myiopagis viridicata viridicata	Greenish Elaenia	B	17	12.3	0.8	11	13.6		Argentina	253
Myiopagis viridicata minima	Greenish Elaenia	B	11	10.1	0.4	9.6	10.8		Isla Tres Marias, Mexico	1151
Myiopagis viridicata jaliscensis	Greenish Elaenia	B	18	12.3	1	11.3	14.6		Nayarit, Mexico	1151
Pseudelaenia leucospodia	Gray-and-white Tyrannulet	M	1	11.9					Peru	1043
Elaenia martinica	Caribbean Elaenia	B	23	22.6		18.5	28.5		Cayman Is.	432
Elaenia martinica	Caribbean Elaenia	B	50	19.4	2.1	17	23.2		Antigua	19
Elaenia martinica	Caribbean Elaenia	B	83	19.6	1.5	16.5	23.6		Barbuda	19
Elaenia martinica	Caribbean Elaenia	B	95	21.5	1.9	18	29.2		Guadalupe	19
Elaenia martinica	Caribbean Elaenia	B	402	19.2	1.6	14	25.7		Puerto Rico	19
Elaenia martinica	Caribbean Elaenia	B	18	22.8	1.4	20.1	25.7		St. Vincent	19
Elaenia spectabilis	Large Elaenia	B	8	27.3					Argentina; Brazil	64, 378, 392, 1040
Elaenia flavogaster	Yellow-bellied Elaenia	B	64	24.8		21	29		Trinidad	359
Elaenia ridleyana	Noronha Elaenia	B	9	33		31	40		Ilha Fernando de Noronha	835
Elaenia albiceps	White-crested Elaenia	U	89	15.5	1	13.4	17.8		Argentina	210
Elaenia parvirostris	Small-billed Elaenia	B	37	13.8		10.9	16.8		Bolivia	1055
Elaenia strepera	Slaty Elaenia	M	7	19.3		13.3	21		Peru; Argentina; Bolivia	370, 1040, 1055
Elaenia mesoleuca	Olivaceous Elaenia	B	17			15.5	20		Brazil	64

Body Masses of World Birds (continued)

Species Name	Common Name	Sex	N	Mean	S.D.	Min	Max	Season	Location	Source #
Elaenia gigas	Mottle-backed Elaenia	M	1	31					Peru	1288
Elaenia cristata	Plain-crested Elaenia	U	17	18.2		16	21.6		Suriname	498
Elaenia ruficeps	Rufous-crowned Elaenia	M	15	19.2		16.6	21.3		Suriname	498
		F	9	18.3		16.4	19.7			
Elaenia chiriquensis	Lesser Elaenia	U	260	15.4	1.9	11.9	22.1		Mato Grosso, Brazil	392
Elaenia frantzii	Mountain Elaenia	U	81	19.6	1.4	14	23		El Salvador	1041a
Elaenia obscura	Highland Elaenia	B	14	23.9	5.8	14.7	34.5		Colombia; Brazil	64, 752, 798, 878
Elaenia pallatangae	Sierran Elaenia	B	14	17.8		14.9	20		Venezuela	1308
Elaenia fallax	Greater Antillean Elaenia	U	9	14.4		13.4	15.2		Dominican Republic	19
Serpophaga cinerea	Torrent Tyrannulet	U	8	8.3						489, 953, 1311
Serpophaga nigricans	Sooty Tyrannulet	B	4	8.5						64, 363
Serpophaga hypoleuca	River Tyrannulet	U				5.5	6.7			275
Serpophaga subcristata	White-crested Tyrannulet	B	7	6.6		5	8.5		Brazil; Argentina	64, 210a, 752, 1040
Serpophaga munda	White-bellied Tyrannulet	U	4	5.4		4.5	6.8			210a, 1055
Mionectes oleagineus	Ochre-bellied Flycatcher	B	260	12.1		9	15		Trinidad	1131
Mionectes oleagineus	Ochre-bellied Flycatcher	B	47	9.6	0.7	7.5	10.8		Peru	167
Mionectes oleagineus	Ochre-bellied Flycatcher	B	31	12	1.2	10	14.8		Bolivia	167
Mionectes striaticollis	Streak-necked Flycatcher	U	89	15	1.7				Peru	1288
Mionectes olivaceus	Olive-striped Flycatcher	B	20	16.4	1.2	14.4	18.5		Panama	79

Body Masses of World Birds (continued)

Species Name	Common Name	Sex	N	Mean	S.D.	Min	Max	Season	Location	Source #
Mionectes olivaceus	Olive-striped Flycatcher	U	40	14.3	1.2	11.7	16.6		Colombia	1037
Mionectes rufiventris	Gray-hooded Flycatcher	U	14	13.3		7.5	16.8		Brazil	64, 878
Mionectes macconnelli	MacConnell's Flycatcher	U	39	13.5	1.3	10	16		Bolivia	167
Leptopogon rufipectus	Rufous-breasted Flycatcher	B	8	13.4		10	18		Peru; Ecuador	900, 953
Leptopogon taczanowskii	Inca Flycatcher	U	2	11.5		11.5	11.5		Peru	1143
Leptopogon amaurocephalus	Sepia-capped Flycatcher	U	27	11.7	0.9				Peru	1288
Leptopogon superciliaris	Slaty-capped Flycatcher	B	30	11.7		9.5	13		Trinidad	359
Pseudotriccus pelzelni	Bronze-olive Pygmy-Tyrant	U	22	10.9	0.8				Peru	1288
Pseudotriccus simplex	Hazel-fronted Pygmy-Tyrant	B	8	9.5					Peru	370
Pseudotriccus ruficeps	Rufous-headed Pygmy-Tyrant	U	14	11.3						953, 963, 1288, 1311
Phylloscartes ophthalmicus	Marble-faced Bristle-Tyrant	B	9	9.2					Peru	370
Phylloscartes venezuelanus	Venezuelan Bristle-Tyrant	U		9						522
Phylloscartes lanyoni	Antioquia Bristle-Tyrant	U		8					estimated	275
Phylloscartes orbitalis	Spectacled Bristle-Tyrant	M	6	7.6					Peru	370
		F	8	6.6						
Phylloscartes poecilotis	Variegated Bristle-Tyrant	U	3	8		7	9.5		Colombia; Peru	798, 1288
Phylloscartes eximius	Southern Bristle-Tyrant	U		7.5					Brazil	1105
Phylloscartes nigrifrons	Black-fronted Bristle-Tyrant	B	5	9.1		8	11.1		Venezuela	1143, 1148, 1153

Body Masses of World Birds (continued)

Species Name	Common Name	Sex	N	Mean	S.D.	Min	Max	Season	Location	Source #
Phylloscartes chapmani	Chapman's Bristle-Tyrant	M	13	8.8	0.6	8	9.5		Venezuela	1308
		F	3	7.3		7	7.5			
Phylloscartes gualaquizae	Ecuadorian Tyrannulet	U		8						275
Phylloscartes flaviventris	Rufous-lored Tyrannulet	U		8						275
Phylloscartes parkeri	Cinnamon-faced Tyrannulet	M	1	8.1					Peru	368
Phylloscartes roquettei	Minas Gerais Tyrannulet	M	1	6					Brazil	637
Phylloscartes paulista	Sao Paulo Tyrannulet	U				7	8			275
Phylloscartes oustaleti	Oustalet's Tyrannulet	B	2	10		9.4	10.5		Brazil	1178
Phylloscartes kronei	Restinga Tyrannulet	B	2	8.3		7.8	8.8		Brazil	1315
Phylloscartes difficilis	Serra do Mar Tyrannulet	U	5	6.9		6	8		Brazil	64, 961
Phylloscartes ceciliae	Alagoas Tyrannulet	B	4	7.6		6.1	9		Northeast Brazil	1204
Phylloscartes ventralis	Mottle-cheeked Tyrannulet	B	11	8.3		7.5	10		Brazil, Peru; Paraguay	64, 132, 961, 1288
Phylloscartes beckeri	Bahia Tyrannulet	U	4	8.7		7.5	9		Bahia, Brazil	434
Phylloscartes flavovirens	Yellow-green Tyrannulet	U	1	8.8					Panama	1180
Phylloscartes virescens	Olive-green Tyrannulet	B	7	8.5		6.9	11		French Guiana; Suriname; Brazil	195, 293, 498
Phylloscartes superciliaris	Rufous-browed Tyrannulet	M	2	8.5		8	9		Ecuador	1147
		F	1	7						
Phylloscartes sylviolus	Bay-ringed Tyrannulet	U		8					estimated	275
Phyllomyias fasciatus	Planalto Tyrannulet	B	5	10.3		8.1	11		Brazil	64, 752

Body Masses of World Birds (continued)

Species Name	Common Name	Sex	N	Mean	S.D.	Min	Max	Season	Location	Source #
Phyllomyias zeledoni	White-fronted Tyrannulet	U	1	10.3					Panama	489
Phyllomyias burmeisteri	Rough-legged Tyrannulet	B	4	11.1		9.8	12		Brazil; Peru; Argentina	64, 259, 1040
Phyllomyias virescens	Greenish Tyrannulet	B	9	8.2		6	12		Argentina; Paraguay	132, 1040
Phyllomyias reiseri urichi	Reiser's Tyrannulet	U				7	8			522
Phyllomyias sclateri	Sclater's Tyrannulet	M	1	9.2					Argentina	1037
Phyllomyias griseiceps	Sooty-headed Tyrannulet	B	2	7.2		7	7.4		Ecuador	1147
Phyllomyias nigrocapillus	Black-capped Tyrannulet	B	17	9.5		8.8	10.4		Colombia	324a
Phyllomyias cinereiceps	Ashy-headed Tyrannulet	U	1	9.5					Colombia	881
Phyllomyias uropygialis	Tawny-rumped Tyrannulet	M	11	9.4					Bolivia	963
		F	8	8.6						
Zimmerius vilissimus	Paltry Tyrannulet	U	15	9.7					Panama	675
Zimmerius improbus	Venezuelan Tyrannulet	U		10.9						522
Zimmerius bolivianus	Bolivian Tyrannulet	B	2	11		10	11.9		Peru	1288
Zimmerius cinereicapillus	Red-billed Tyrannulet	B	6	11.8		10.7	13.8		Ecuador; Peru	9
Zimmerius villarejoi	Mishana Tyrannulet	B	3	6.6		6.2	7		Peru	9
Zimmerius gracilipes	Slender-footed Tyrannulet	M	12	7.6		6.8	7.9		Suriname	498
		F	8	6.7		5.7	7.8			
Zimmerius viridiflavus	Peruvian Tyrannulet	U	2	9.6		9.1	10		Peru	785a, 1302
Zimmerius chrysops	Golden-faced Tyrannulet	B	8	10.4					Colombia	798, 1311
Sublegatus obscurior	Amazonian Scrub-Flycatcher	M	1	16					Roraima, Brazil	1143

Body Masses of World Birds (continued)

Species Name	Common Name	Sex	N	Mean	S.D.	Min	Max	Season	Location	Source #
Sublegatus arenarum	Northern Scrub-Flycatcher	U	29	12.3		10.5	15		Trinidad	359
Sublegatus modestus	Southern Scrub-Flycatcher	B	8	14		11	15.7		Argentina; Suriname; Brazil	498, 883, 1040
Suiriri suiriri	Suiriri Flycatcher	M	20	15.3	1.8	12	17.5			501
		F	14	13.8	1.2	12	15.7			
Suiriri islerorum	Chapada Flycatcher	F	1	22.8					Brazil	1365
Mecocerculus leucophrys	White-throated Tyrannulet	B	14	10.4		9	11.5		Venezuela	1308
Mecocerculus leucophrys	White-throated Tyrannulet	U	12	11.6	1.3	9.6	13.9		Colombia	324a
Mecocerculus poecilocercus	White-tailed Tyrannulet	U	11	7.7		7	8.5		Colombia; Ecuador	953, 1037
Mecocerculus hellmayri	Buff-banded Tyrannulet	B	9	8		6.5	9		Bolivia	1143
Mecocerculus calopterus	Rufous-winged Tyrannulet	U	8	7					Peru; Ecuador	747, 953, 1302
Mecocerculus minor	Sulphur-bellied Tyrannulet	U		11						275
Mecocerculus stictopterus	White-banded Tyrannulet	U	9	9.4		8.2	10.7		Ecuador; Peru	633, 953, 963, 1043
Inezia tenuirostris	Slender-billed Tyrannulet	F	1	5					Venezuela	1143
Inezia inornata	Plain Tyrannulet	U	1	12					Peru	319
Inezia subflava	Pale-tipped Tyrannulet	U	14	8.4	0.6	7.5	10.2		Venezuela	1215
Stigmatura napensis	Lesser Wagtail-Tyrant	U				9.1	10.5			275
Stigmatura budytoides	Greater Wagtail-Tyrant	M	9	11.7		10.9	12.3		Bolivia	1055
		F	14	10.5		9.4	12.6			
Anairetes agilis	Agile Tit-Tyrant	B	7	10.1		6.2	11.7		Colombia	324a

Body Masses of World Birds (continued)

Species Name	Common Name	Sex	N	Mean	S.D.	Min	Max	Season	Location	Source #
Anairetes agraphia	Unstreaked Tit-Tyrant	B	2	9.9		9.9	9.9		Peru	1288
Anairetes alpinus bolivianus	Ash-breasted Tit-Tyrant	F	1	9					Peru	1143
Anairetes reguloides	Pied-crested Tit-Tyrant	B	11	5.9		5	8		Chile	748
Anairetes flavirostris	Yellow-billed Tit-Tyrant	B	8	6		5.3	7.6		Argentina; Peru	841, 1041, 1043
Anairetes parulus	Tufted Tit-Tyrant	U	113	6.2	0.5	5	7.5		Chile	328
Tachuris rubrigastra	Many-colored Rush-Tyrant	B	8	7.8		6.5	9.5			64, 210, 1043
Polystictus pectoralis	Bearded Tachuri	B	10	6.2	0.5	5.3	7.3		Colombia	498, 1143
Polystictus superciliaris	Gray-backed Tachuri	U		6						275
Pseudocolopteryx sclateri	Crested Doradito	B	4	8		6.8	9.9			64, 210a, 1041
Pseudocolopteryx acutipennis	Subtropical Doradito	B	3	6.9		6.2	8.3		Argentina	212
Pseudocolopteryx flaviventris	Warbling Doradito	U	2	7.5		7	8		Brazil	64
Euscarthmus meloryphus	Tawny-crowned Pygmy-Tyrant	U	10	6.8		5.3	8			49, 64, 1215, 1302
Euscarthmus rufomarginatus	Rufous-sided Pygmy-Tyrant	U		6					Brazil	901
Myiornis albiventris	White-bellied Pygmy-Tyrant	U				5	6			275
Myiornis auricularis	Eared Pygmy-Tyrant	B	17	5.3						253, 1288
Myiornis atricapillus	Black-capped Pygmy-Tyrant	B	3	5.7		5.5	6		Panama	990
Myiornis ecaudatus	Short-tailed Pygmy-Tyrant	U		6					Peru	1210
Myiornis ecaudatus	Short-tailed Pygmy-Tyrant	U	2	4.6		4.5	4.8		Trinidad	359

Body Masses of World Birds (continued)

Species Name	Common Name	Sex	N	Mean	S.D.	Min	Max	Season	Location	Source #
Lophotriccus pileatus	Scale-crested Pygmy-Tyrant	M	9	8.1					Panama	489
		F	7	6.9						
Lophotriccus vitiosus	Double-banded Pygmy-Tyrant	B	10	6.7	0.4	5.9	7.3		Brazil	195
Lophotriccus eulophotes	Long-crested Pygmy-Tyrant	F	1	6					Peru	1083
Lophotriccus galeatus	Helmeted Pygmy-Tyrant	B	15	6.6		5.6	8		Suriname	498
Lophotriccus pilaris	Pale-eyed Pygmy-Tyrant	U	6	6.8		5.3	9.1		Panama; Peru; Colombia	615, 795, 1216
Oncostoma cinereigulare	Northern Bentbill	B	38	6.4	1.4	4.3	11		El Salvador	1041a
Oncostoma olivaceum	Southern Bentbill	U	12	6.6					Panama	156, 615, 618, 874
Poecilotriccus ruficeps	Rufous-crowned Tody-Tyrant	U	9	11.1					Peru; Colombia	1037, 1288
Poecilotriccus ruficeps peruvianus	Rufous-crowned Tody-Tyrant	B	4	7					Peru	596
Poecilotriccus sylvia	Slate-headed Tody-Tyrant	U	14	7.1					Panama; Colombia; Venezuela; Belize	489, 615, 618, 795, 1027, 1215
Poecilotriccus luluae	Lulu's Tody-Tyrant	B	17	7.3					Peru	596
Poecilotriccus capitalis	Black-and-white Tody-Tyrant	B	2	8		8	8		Peru	899
Poecilotriccus albifacies	White-cheeked Tody-Tyrant	U		8					Peru	617
Taeniotriccus andrei	Black-chested Tyrant	U	2	9.5		9.4	9.6		Brazil; Suriname	498, 875
Hemitriccus minor	Snethlage's Tody-Tyrant	B	14	7.4		6.5	8		Brazil	49
Hemitriccus josephinae	Boat-billed Tody-Tyrant	U	1	10.8					Suriname	498
Hemitriccus flammulatus	Flammulated Bamboo-Tyrant	B	7	10		8.8	11.7		Peru; Bolivia	49, 370

Body Masses of World Birds (continued)

Species Name	Common Name	Sex	N	Mean	S.D.	Min	Max	Season	Location	Source
Hemitriccus diops	Drab-breasted Bamboo-Tyrant	U	15	10		8	11.5		Brazil; Paraguay	64, 132, 878
Hemitriccus obsoletus	Brown-breasted Bamboo-Tyrant	U	24	11.6	1				Brazil	961
Hemitriccus zosterops	White-eyed Tody-Tyrant	U	14	8.7					Colombia; Suriname	785a, 1038a
Hemitriccus zosterops griseipectus	White-eyed Tody-Tyrant	M	6	9.2		8.7	9.7		Brazil	195
		F	1	8.3						
Hemitriccus minimus	Zimmer's Tody-Tyrant	M	5	7		6.4	7.8		Peru	10
Hemitriccus orbitatus	Eye-ringed Tody-Tyrant	B	3	9.7		9	10.5		Brazil	64
Hemitriccus iohannis	Johannes' Tody-Tyrant	U		10					Peru	1210
Hemitriccus striaticollis	Stripe-necked Tody-Tyrant	B	11	8.6		7.8	10.1		Colombia; Brazil	392, 883, 1143
Hemitriccus nidipendulus	Hangnest Tody-Tyrant	B	2	7.5		7	8		Brazil	1178
Hemitriccus spodiops	Yungas Tody-Tyrant	U				6.9	7.5			275
Hemitriccus margaritaceiventer	Pearly-vented Tody-Tyrant	B	24	8.4	1	6.6	11.5		Paraguay	1175
Hemitriccus inornatus	Pelzeln's Tody-Tyrant	U				7.7	9			275
Hemitriccus granadensis	Black-throated Tody-Tyrant	U	11	7.9					Peru; Ecuador	953, 1288
Hemitriccus rufigularis	Buff-throated Tody-Tyrant	U	3	9.4		9	9.8		Peru; Ecuador	259, 747
Hemitriccus cinnamomeipectus	Cinnamon-breasted Tody-Tyrant	U	7	7.5		6.5	8.5		Peru	397
Todirostrum russatum	Ruddy Tody-Flycatcher	M	10	8	0.6	6.5	9			1143, 1148, 1153

Body Masses of World Birds (continued)

Species Name	Common Name	Sex	N	Mean	S.D.	Min	Max	Season	Location	Source #
Todirostrum plumbeiceps	Ochre-faced Tody-Flycatcher	B	14	5.7	0.3	5.3	6		Brazil; Argentina	64, 253, 878, 961, 1040, 1205
Todirostrum fumifrons	Smoky-fronted Tody-Flycatcher	U	10	6.6		6	8		Brazil; Suriname	392, 498, 1109
Todirostrum maculatum	Spotted Tody-Flycatcher	B	29	7.3		6	8		Suriname	498
Todirostrum nigriceps	Black-headed Tody-Flycatcher	F	1	6.5					Panama	990
Todirostrum pictum	Painted Tody-Flycatcher	B	8	6.8		6	8		Brazil; Suriname	195, 498, 1178
Todirostrum cinereum	Common Tody-Flycatcher	B	23	6.4					Panama	489
Todirostrum cinereum peruanum	Common Tody-Flycatcher	B	3	6.6		6.3	7			650a
Todirostrum cinereum sclateri	Common Tody-Flycatcher	F	1	5.9						650a
Todirostrum chrysocrotaphum	Yellow-browed Tody-Flycatcher	U		7					Peru	1210
Poecilotriccus latirostris	Rusty-fronted Tody-Flycatcher	U	22	8.1	0.7				Peru	1288
Poecilotriccus calopterus	Golden-winged Tody-Flycatcher	M	1	7.1						650a
Poecilotriccus calopterus pulchellus	Golden-winged Tody-Flycatcher	B	2	8		7.2	8.7		Peru	370
Corythopis torquata	Ringed Antpipit	B	14	14.3		12.3	15		Brazil; French Guiana; Venezuela	293, 875, 1178, 1308
Corythopis delalandi	Southern Antpipit	B	12	15		13	16		Brazil; Argentina; Bolivia	64, 132, 253
Cnipodectes subbrunneus	Brownish Flycatcher	U	9	23.2					Panama	618
Ramphotrigon megacephalum	Large-headed Flatbill	U	10	14.2		11	17			319, 747, 878, 897

Body Masses of World Birds (continued)

Species Name	Common Name	Sex	N	Mean	S.D.	Min	Max	Season	Location	Source #
Ramphotrigon fuscicauda	Dusky-tailed Flatbill	U	12	18.6		16.5	20.5		Brazil; Peru	897
Ramphotrigon ruficauda	Rufous-tailed Flatbill	B	13	18.4		15	25.6		Suriname	498
Rhynchocyclus brevirostris	Eye-ringed Flatbill	U	25	24.3	3.1	20	34		El Salvador	1041a
Rhynchocyclus pacificus	Pacific Flatbill	U				23.5	26.5			275
Rhynchocyclus olivaceus	Olivaceous Flatbill	B	11	21.3	1.3	19.8	24		Panama; Suriname	79, 156, 497, 990, 1180
Rhynchocyclus fulvipectus	Fulvous-breasted Flatbill	U	5	23.1					Peru; Colombia	785a, 881, 899, 1288
Tolmomyias sulphurescens	Yellow-olive Flycatcher	U	64	14.3	2.2	9	20		El Salvador	1041a
Tolmomyias assimilis	Yellow-margined Flycatcher	U		17					Peru	1210
Tolmomyias poliocephalus	Gray-crowned Flycatcher	B	21	10.8		9.8	11.8		Suriname	498
Tolmomyias traylori	Orange-eyed Flycatcher	M	1	11.5					Peru	1073
Tolmomyias flaviventris	Yellow-breasted Flycatcher	B	24	12.2		9.5	14.5		Trinidad	359
Platyrinchus saturatus	Cinnamon-crested Spadebill	U	14	10.7		8.8	15		Brazil; Venezuela; Peru	10, 875, 1178, 1308
Platyrinchus cancrominus	Stub-tailed Spadebill	B	21	9.2	1	7.8	10.8		Veracruz, Mexico	1143
Platyrinchus flavigularis	Yellow-throated Spadebill	M	1	10					Ecuador	1148
Platyrinchus coronatus	Golden-crowned Spadebill	U	13	9.4	0.6				Costa Rica	618
Platyrinchus coronatus	Golden-crowned Spadebill	U	14	8.9	0.4				Panama	618

Body Masses of World Birds (continued)

Species Name	Common Name	Sex	N	Mean	S.D.	Min	Max	Season	Location	Source #
Platyrinchus mystaceus	White-throated Spadebill	B	23	9.7		7.5	11		Trinidad	359
Platyrinchus platyrhynchos	White-crested Spadebill	U	4	12		10	13.7			785a, 861
Platyrinchus leucoryphus	Russet-winged Spadebill	U	3	16		14	17		Brazil	64, 132, 961
Onychorhynchus coronatus	Royal Flycatcher	U	13	14		9.7	16.2		Brazil	875
Myiotriccus ornatus	Ornate Flycatcher	U	14	9.9	1.2	8.2	12		Colombia	1037
Myiophobus flavicans	Flavescent Flycatcher	U	15	12.7	0.9				Peru	1288
Myiophobus phoenicomitra	Orange-crested Flycatcher	U	11	11	1	9	12.5		Peru; Ecuador	259, 899, 953
Myiophobus roraimae	Roraiman Flycatcher	M	8	14.4		12.2	18		Venezuela	296, 1308
		F	7	12.9		12	14			
Myiophobus inornatus	Unadorned Flycatcher	M	2	11.2		10.4	12		Peru	1143
Myiophobus pulcher	Handsome Flycatcher	B	3	9.5		9.1	10		Peru; Colombia	324a, 370
Myiophobus lintoni	Orange-banded Flycatcher	M	2	9.8		9.5	10		Peru	900
Myiophobus ochraceiventris	Ochraceous-breasted Flycatcher	M	11	11.2					Bolivia	963
		F	3	9.8						
Myiophobus fasciatus	Bran-colored Flycatcher	B	20	9.9		8.5	12		Trinidad	359
Myiophobus cryptoxanthus	Olive-chested Flycatcher	U	2	9.8		8.5	11		Peru; Ecuador	899, 953
Terenotriccus erythrurus	Ruddy-tailed Flycatcher	U	14	7.4	0.3	6.9	8		Panama	79
Myiobius villosus	Tawny-breasted Flycatcher	B	10	14.1		12	16.1			1037, 1147, 1288
Myiobius sulphureipygius	Sulphur-rumped Flycatcher	U	29	11.9	1.1				Panama; Costa Rica	618

Body Masses of World Birds (continued)

Species Name	Common Name	Sex	N	Mean	S.D.	Min	Max	Season	Location	Source #
Myiobius barbatus	Whiskered Flycatcher	B	24	11.4	1.4	8.8	13.5		Venezuela; Brazil; Peru	1143
Myiobius atricaudus	Black-tailed Flycatcher	U	13	10	1.1				Panama	618
Neopipo cinnamomea	Cinnamon Tyrant	B	9	7.7		6.3	12		Brazil; Ecuador; Colombia; Venezuela	73, 747, 1037, 1038a, 1308
Pyrrhomyias cinnamomea	Cinnamon Flycatcher	U	10	9.9						953, 1040, 1043, 1288
Hirundinea ferruginea	Cliff Flycatcher	U	10	30.6		21.2	42		Brazil; Ecuador; Argentina	210a, 747, 953, 1055
Cnemotriccus fuscatus	Fuscous Flycatcher	U	41	13.6		12	16		Trinidad	359
Lathrotriccus euleri	Euler's Flycatcher	B	28	12		9.5	14		Trinidad	359
Lathrotriccus euleri	Euler's Flycatcher	U	12	10.7	0.8	9.6	12.5		Mato Grosso, Brazil	392
Lathrotriccus griseipectus	Gray-breasted Flycatcher	U		11						275
Aphanotriccus capitalis	Tawny-chested Flycatcher	U	6	11.8					Costa Rica	80
Aphanotriccus audax	Black-billed Flycatcher	B	4			10.5	12		Panama	990
Xenotriccus callizonus	Belted Flycatcher	B	3	11.2		10.5	11.5		Chiapas, Mexico	1143
Xenotriccus mexicanus	Pileated Flycatcher	M	1	13.8					Oaxaca, Mexico	1143
Mitrephanes phaeocercus	Tufted Flycatcher	M	8	8.9					Panama	489
		F	8	8.3						
Mitrephanes olivaceus	Olive Flycatcher	M	1	10					Cuzco, Peru	1143
Contopus cooperi	Olive-sided Flycatcher	B	26	32.1	2.3	26.7	42.2	M	Pennsylvania, USA	193
Contopus pertinax	Greater Pewee	B	27	27.2	3.3	21.6	35.8		Mexico	1148a

Body Masses of World Birds (continued)

Species Name	Common Name	Sex	N	Mean	S.D.	Min	Max	Season	Location	Source #
Contopus lugubris	Dark Pewee	U	23	21.5					Panama	489
Contopus fumigatus	Smoke-colored Pewee	B	6	19.9		14	23.2			200, 296, 798, 1040, 1302
Contopus ochraceus	Ochraceous Pewee	U		23					Costa Rica	1170
Contopus sordidulus	Western Wood-Pewee	B	25	13.1	0.8	11	14.9	B	Western USA	277
Contopus virens	Eastern Wood-Pewee	U	63	13.9	0.7	12.1	15.4		Pennsylvania, USA	828
Contopus cinereus	Tropical Pewee	B	21	11.6		8.5	13.5		Trinidad	359
Contopus nigrescens	Blackish Pewee	B	2	9.5		9	10		Peru	259
Contopus caribaeus	Cuban Pewee	B	100	10.5	0.7	8.5	12		Cuba	19
Contopus pallidus	Jamaican Pewee	M	1	9.3					Jamaica	1159
Contopus hispaniolensis	Hispaniolan Pewee	B	18	11.7	0.9	10.5	13.2		Dominican Republic	19
Contopus portoricensis	Puerto Rican Pewee	B	11	10.8	1	9.2	12.4		Puerto Rico	19
Contopus latirostris	Lesser Antillean Pewee	U	2	11.2		11	11.5		Guadeloupe	19
Contopus oberi	St. Lucia Pewee	U	3	9.1		8	9.9		St. Lucia	19
Contopus albogularis	White-throated Pewee	B	11	9.6		8.3	10.5		Suriname	498
Empidonax flaviventris	Yellow-bellied Flycatcher	B	415	11.8	1	9.2	15.5	M	Pennsylvania, USA	461
Empidonax virescens	Acadian Flycatcher	U	72	12.6	0.7	11.3	13.8		Pennsylvania, USA	828
Empidonax alnorum	Alder Flycatcher	B	53	12.7	1.1	10	14.5	B	Ontario, Canada	1147a
Empidonax traillii	Willow Flycatcher	M	13	13.1	1.4	12	15.7	B	Connecticut, USA	439
		F	11	13.7	1.5	11.3	16.4			
Empidonax albigularis	White-throated Flycatcher	B	3	10.8		10.3	11		Oaxaca, Mexico: Guatemala	1145

Body Masses of World Birds (continued)

Species Name	Common Name	Sex	N	Mean	S.D.	Min	Max	Season	Location	Source #
Empidonax albigularis	White-throated Flycatcher	U	3	8.5		7.9	9.6		Belize	220
Empidonax minimus	Least Flycatcher	U	277	10	0.8	7.8	12.2		Pennsylvania, USA	828
Empidonax hammondii	Hammond's Flycatcher	M	91	10.6	0.6	9.2	12.1	B		1077
		F	44	10.3	0.9	8	12.9			
Empidonax wrightii	Gray Flycatcher	B	15	12.3	0.7	11.3	13.6	B	Nevada, USA	594
Empidonax oberholseri	Dusky Flycatcher	B	16	10.4	0.6	9.3	11.4	M	Arizona, USA	314
Empidonax affinis	Pine Flycatcher	M	1	11.3					Mexico - juvenile	259a
Empidonax difficilis	Pacific-slope Flycatcher	B	606	10.7	0.9	7.7	14.3		California, USA	719
Empidonax occidentalis	Cordilleran Flycatcher	B	69	11.6	1.2	8	15	B		277
Empidonax flavescens	Yellowish Flycatcher	U	69	12.5	2.2	9	20		El Salvador	1041a
Empidonax fulvifrons	Buff-breasted Flycatcher	B	23	7.9	0.7	6.7	9.3	Y	Mexico	1142a
Empidonax atriceps	Black-capped Flycatcher	U	3	9		8.9	9.3		Costa Rica	1143
Sayornis phoebe	Eastern Phoebe	U	108	19.7	1.4	16.5	23.2		Pennsylvania, USA	828
Sayornis nigricans	Black Phoebe	M	78	19.5	0.9	16.9	22	B	California, USA	314
		F	154	17.8	1.1	14.6	20.8			
Sayornis saya	Say's Phoebe	B	13	20.9	2.5	15	24.1	B		314, 277
Pyrocephalus rubinus	Vermilion Flycatcher	B	11	14.4		12	16.1		Nicaragua	542
Ochthoeca jelskii	Jelski's Chat-Tyrant	U	5	11.6		9.5	16.6		Peru; Ecuador	953, 1043
Ochthoeca diadema	Yellow-bellied Chat-Tyrant	U	23	11.5	1	9.5	13.5		Colombia	1037
Ochthoeca pulchella	Golden-browed Chat-Tyrant	U	7	12					Peru; Bolivia	963, 1288
Ochthoeca frontalis	Crowned Chat-Tyrant	M	34	10.4					Bolivia	963
		F	36	9.6						

Body Masses of World Birds (continued)

Species Name	Common Name	Sex	N	Mean	S.D.	Min	Max	Season	Location	Source #
Ochthoeca cinnamomeiventris	Slaty-backed Chat-Tyrant	U	3	12		11	13		Ecuador	953, 1288
Ochthoeca oenanthoides	D'Orbigny's Chat-Tyrant	B	4	17.7						363, 1043
Ochthoeca rufipectoralis	Rufous-breasted Chat-Tyrant	M	4	10.5					Bolivia	963
		F	10	9.5						
Ochthoeca fumicolor	Brown-backed Chat-Tyrant	B	11	16.6		14	19		Venezuela; Ecuador; Peru	953, 1043, 1263
Ochthoeca leucophrys	White-browed Chat-Tyrant	B	5	13.1		12.5	14.5		Argentina; Chile	776a, 1041, 1043
Tumbezia salvini	Tumbes Tyrant	M	2	12.2		11.4	13.1		Peru	1043
Colorhamphus parvirostris	Patagonian Tyrant	U	13	10.6	1	9	12.2		Chile	328
Ochthornis littoralis	Drab Water-Tyrant	U	4	13.4		12.3	14.9		Peru; Suriname	370, 498
Myiotheretes striaticollis	Streak-throated Bush-Tyrant	B	5	61.2		56.7	64		Peru; Argentina	1040, 1043, 1288
Myiotheretes fumigatus	Smoky Bush-Tyrant	U	7	33.4		31.6	35		Ecuador; Colombia	324a, 953, 1311
Myiotheretes fuscorufus	Rufous-bellied Bush-Tyrant	B	6	31.7					Bolivia	963, 1288
Xolmis pyrope	Fire-eyed Diucon	U	14	35.3	3.2	31	42.4		Argentina	210
Xolmis cinerea	Gray Monjita	B	10	57.1	3.7	51.7	61.8			64, 210a, 752, 1109
Xolmis coronata	Black-crowned Monjita	U	3	46.8		43	51		Brazil; Argentina	64, 843
Xolmis irupero	White Monjita	U	11	28.7		29	25			64, 210a, 363
Xolmis rubetra	Rusty-backed Monjita	B	18	36.3		31	39.5		Argentina	186a
Heteroxolmis dominicana	Black-and-white Monjita	U	2	42.8		42	43.5		Brazil	64

Body Masses of World Birds (continued)

Species Name	Common Name	Sex	N	Mean	S.D.	Min	Max	Season	Location	Source #
Neoxolmis rufiventris	Chocolate-vented Tyrant	U		77					Brazil	1105
Agriornis montana	Black-billed Shrike-Tyrant	B	9	63.3		60	68		Argentina	210, 843, 1040, 1043
Agriornis lividus	Great Shrike-Tyrant	M	5	102		96.7	109		Argentina	843
		F	4	96.3		93.8	97.7			
Agriornis microptera	Gray-bellied Shrike-Tyrant	B	5	67.8		60	75		Argentina; Chile	776a, 841, 843, 1040
Agriornis murina	Lesser Shrike-Tyrant	F	3	29.9		27.7	32		Argentina	843, 1040
Polioxolmis rufipennis	Rufous-webbed Tyrant	M	1	34.5					Bolivia	1261
Muscisaxicola maculirostris	Spot-billed Ground-Tyrant	B	13	14.2		12.1	16.6			210, 843, 1043
Muscisaxicola fluviatilis	Little Ground-Tyrant	B	4	14.7		12.6	16		Peru, Brazil	1143
Muscisaxicola maclovianus	Dark-faced Ground-Tyrant	M	11	24.2		22.3	26.9		Argentina	843
		F	6	23.4		20.2	24.8			
Muscisaxicola capistrata	Cinnamon-bellied Ground-Tyrant	B	3	26.6		26.6	26.6		Tierra del Fuego	546
Muscisaxicola rufivertex	Rufous-naped Ground-Tyrant	B	4	20		18.7	21.9		Argentina	843
Muscisaxicola juninensis	Puna Ground-Tyrant	F	1	22.2					Peru	1043
Muscisaxicola albilora	White-browed Ground-Tyrant	M	3	22.6					Argentina	843
		F	3			29	31.7			
Muscisaxicola alpina	Plain-capped Ground-Tyrant	F	1	26.1					Peru	1043
Muscisaxicola cinerea	Cinereous Ground-Tyrant	M	13	18.1		15.6	19.5		Argentina	843
		F	6	19.1		17.2	19.9			
Muscisaxicola flavinucha	Ochre-naped Ground-Tyrant	M	12	35.2		33	37.5		Argentina	843
		F	4	37.3		35.3	38.6			

Body Masses of World Birds (continued)

Species Name	Common Name	Sex	N	Mean	S.D.	Min	Max	Season	Location	Source #
Muscisaxicola frontalis	Black-fronted Ground-Tyrant	M	14	27.7		25	30.4		Argentina	843
		F	6	30.1		28.6	31			
Muscigralla brevicauda	Short-tailed Field-Tyrant	U	2	12.6		11.8	13.5		Peru	1103
Lessonia oreas	Andean Negrito	M	2	13.8		13	14.6			363, 1043
Lessonia rufa	Austral Negrito	U	12	13.4	0.9	11.5	15.5		Argentina	210
Knipolegus striaticeps	Cinereous Tyrant	F	1	11					Bolivia	1143
Knipolegus hudsoni	Hudson's Black-Tyrant	U	6	13.8		12	16.5		Bolivia; Brazil	392, 657
Knipolegus poecilocercus	Amazonian Black-Tyrant	U	3	14.3					Brazil	876
Knipolegus signatus cabanisi	Andean Tyrant	M	3	17.8		17	19.5		Argentina	1040, 1041
Knipolegus cyanirostris	Blue-billed Black-Tyrant	B	4	15.4		14	17		Brazil	64, 877
Knipolegus poecilurus	Rufous-tailed Tyrant	B	8	14.6		13.5	15.3		Colombia; Venezuela	798, 917
Knipolegus orenocensis	Riverside Tyrant	B	2	19		18.3	19.7	B	Colombia	1143
Knipolegus aterrimus	White-winged Black-Tyrant	B	12	20.21	3.1	14.8	23.5		Peru; Bolivia; Argentina	363, 370, 657, 1040, 1041
Knipolegus franciscanus	Caatinga Black-Tyrant	F	1	20.5					Minas Gerais, Brazil	1142b
Knipolegus nigerrimus	Velvety Black-Tyrant	M	3	20.3		20	21		Brazil	64, 1205
Knipolegus lophotes	Crested Black-Tyrant	B	3	31.8		30	34		Brazil	64
Hymenops perspicillatus	Spectacled Tyrant	B	10	22.9	1.1	21.2	24.7		Argentina	210
Fluvicola pica	Pied Water-Tyrant	B	19	12.3						156, 489, 613, 795, 1131

Body Masses of World Birds (continued)

Species Name	Common Name	Sex	N	Mean	S.D.	Min	Max	Season	Location	Source #
Fluvicola albiventer	Black-backed Water-Tyrant	B	6	11.6		10	13		Bolivia	1143
Fluvicola nengeta atripennis	Masked Water-Tyrant	M	1	21					Ecuador	1143
Arundinicola leucocephala	White-headed Marsh-Tyrant	U	12	13.8		12	16.5			64, 210a, 359, 498, 1040
Alectrurus tricolor	Cock-tailed Tyrant	F	1	16					Bolivia	1143
Gubernetes yetapa	Streamer-tailed Tyrant	B	3	68.4		64	71.4		Brazil; Argentina	64, 1099
Satrapa icterophrys	Yellow-browed Tyrant	B	11	21.5		15.5	32			64, 363, 1055
Colonia colonus	Long-tailed Tyrant	B	21	18.3		16	20		Suriname	498
Machetornis rixosa	Cattle Tyrant	B	15	29.6		25	37			64, 210a, 363, 1215
Muscipipra vetula	Shear-tailed Gray Tyrant	B	2	27		26	28		Brazil	64
Attila phoenicurus	Rufous-tailed Attila	U	11	32.3	1.9				Brazil	961
Attila cinnamomeus	Cinnamon Attila	B	21	38.8		33	45		Suriname	498
Attila torridus	Ochraceous Attila	M	1	46					Ecuador	991
Attila citriniventris	Citron-bellied Attila	B	5	33.7		30	38		Venezuela; Peru	988, 1308
Attila spadiceus	Bright-rumped Attila	B	18	39.1	2.9	35	43.9		Mexico; Belize	910a, 988, 1027
Attila bolivianus	Dull-capped Attila	F	1	39.5					Brazil	883
Attila rufus	Gray-hooded Attila	U	8	42.6		36.5	51.5		Brazil	64, 878, 961
Casiornis rufa	Rufous Casiornis	U	15	24.8	2.7				Peru	1288

Body Masses of World Birds (continued)

Species Name	Common Name	Sex	N	Mean	S.D.	Min	Max	Season	Location	Source #
Casiornis fusca	Ash-throated Casiornis	U	1	19.5					Brazil	877
Sirystes sibilator	Sirystes	U	9	32.3		26.2	36		Panama; Suriname; Brazil	38, 87, 497, 611
Rhytipterna holerythra	Rufous Mourner	M	18	36.8					Panama	244
Rhytipterna simplex	Grayish Mourner	B	38	31.8		29	37		Suriname	498
Rhytipterna immunda	Pale-bellied Mourner	B	11	27.4		25	30		Suriname	498
Myiarchus semirufus	Rufous Flycatcher	U	8	22.5		20.7	25			670, 1043
Myiarchus yucatanensis	Yucatan Flycatcher	B	8	19.6		17	23		Yucatan, Mexico; Belize	220, 910a
Myiarchus barbirostris	Sad Flycatcher	B	66	13.4	0.9	11.5	16.2	Y	Jamaica	1191
Myiarchus tuberculifer	Dusky-capped Flycatcher	B	37	20.1		14.2	24			1237
Myiarchus tuberculifer atriceps	Dusky-capped Flycatcher	B	42	21.9		16.4	27			670
Myiarchus tuberculifer olivascens	Dusky-capped Flycatcher	B	36	12.6		10.3	14.3		Tres Marias Is., Mexico	440
Myiarchus swainsoni	Swainson's Flycatcher	B	70	25.1		21.3	28.9			670
Myiarchus venezuelensis	Venezuelan Flycatcher	B	10	29.8		27	33			670
Myiarchus panamensis	Panama Flycatcher	B	16	31.7		27.5	38.5			670
Myiarchus ferox	Short-crested Flycatcher	B	76	27.5		21	34			670
Myiarchus cephalotes cephalotes	Pale-edged Flycatcher	B	13	28.6		26	31.5			670
Myiarchus cephalotes caribbaeus	Pale-edged Flycatcher	B	7	24.1		22.5	27.5			670
Myiarchus phaeocephalus	Sooty-crowned Flycatcher	B	8	26.3		24.9	28			670, 1302
Myiarchus apicalis	Apical Flycatcher	B	23	29.2		26	33.5			670

Body Masses of World Birds (continued)

Species Name	Common Name	Sex	N	Mean	S.D.	Min	Max	Season	Location	Source #
Myiarchus cinerascens	Ash-throated Flycatcher	B	268	28.2		21.2	37.8			168
Myiarchus nuttingi	Nutting's Flycatcher	U	9	23		21	23.9		Mexico	1150, 1151
Myiarchus crinitus	Great Crested Flycatcher	U	62	32.1	3.1	26	40.1		Pennsylvania, USA	828
Myiarchus tyrannulus	Brown-crested Flycatcher	B	59	43.8	3.4	37.9	53.7	B	Arizona, USA	314
Myiarchus tyrannulus bahiae	Brown-crested Flycatcher	B	4	28.7		23.9	32			670
Myiarchus nugator	Grenada Flycatcher	U	1	37					Grenada	19
Myiarchus magnirostris	Galapagos Flycatcher	B	22	14.7	1.7	12	18.5		Galapagos Is.	341
Myiarchus validus	Rufous-tailed Flycatcher	U	4	40.6		38.6	43.2		Jamaica	1159
Myiarchus sagrae sagrae	La Sagra's Flycatcher	B	7	18.6		17	20		Cayman Is.	868
Myiarchus sagrae	La Sagra's Flycatcher	U	170	20	1.1	15.9	23.2		Cuba	19
Myiarchus stolidus	Stolid Flycatcher	B	26	20.8	2	17.8	28		Dominican Republic	19
Myiarchus oberi	Lesser Antillean Flycatcher	U	12	41	2.5	35.3	45		St. Lucia	19
Myiarchus oberi	Lesser Antillean Flycatcher	U	23	29.4	1.9	25	32.8		St. Kitts	19
Myiarchus antillarum	Puerto Rican Flycatcher	B	248	23.4	1.2	20.5	27.5		Puerto Rico	19
Deltarhynchus flammulatus	Flammulated Flycatcher	M	1	17.2					Oaxaca, Mexico	1143
Philohydor lictor	Lesser Kiskadee	B	23	25.5		22	28		Suriname	498
Pitangus sulphuratus	Great Kiskadee	B	15	61	4.1	53.5	67.5		Venezuela	1215
Pitangus sulphuratus	Great Kiskadee	B	15	55				B	Peru	147

Body Masses of World Birds (continued)

Species Name	Common Name	Sex	N	Mean	S.D.	Min	Max	Season	Location	Source #
Pitangus sulphuratus texanus	Great Kiskadee	B	17	74				Y	Southern Texas; NE Mexico	147
Megarynchus pitangua	Boat-billed Flycatcher	M F	4 6	70 77					Panama	539
Megarynchus pitangua	Boat-billed Flycatcher	B	17	63.4		55	72		Suriname	498
Myiozetetes cayanensis	Rusty-margined Flycatcher	U	14	25.9	2.1	22	29.5		Venezuela	1215
Myiozetetes similis	Social Flycatcher	B	24	28					Panama	539
Myiozetetes granadensis	Gray-capped Flycatcher	B	10	29.3					Panama	539
Myiozetetes luteiventris	Dusky-chested Flycatcher	F	1	19					Rondonia, Brazil	1178
Conopias albovittatus	White-ringed Flycatcher	B	4	24.4		23.5	26		Panama	990
Conopias parvus	Yellow-throated Flycatcher	B	16	21		18.5	25		Suriname	498
Conopias cinchoneti	Lemon-browed Flycatcher	F	1	25					Peru	1143
Myiodynastes hemichrysus	Golden-bellied Flycatcher	M	1	43.4					Panama	489
Myiodynastes chrysocephalus	Golden-crowned Flycatcher	B	9	38.3		32.5	43			200, 990, 1037, 1043, 1311
Myiodynastes bairdii	Baird's Flycatcher	F	1	45					Peru	1302
Myiodynastes maculatus	Streaked Flycatcher	U	18	43.2		37	50		Suriname	498
Myiodynastes luteiventris	Sulphur-bellied Flycatcher	B	30	46.9		40.8	57.4			728
Legatus leucophaius	Piratic Flycatcher	B	20	22.2		19	26		Suriname	498
Phelpsia inornata	White-bearded Flycatcher	U	35	29.4	1.7	26.8	33.5		Venezuela	1215

Body Masses of World Birds (continued)

Species Name	Common Name	Sex	N	Mean	S.D.	Min	Max	Season	Location	Source #
Empidonomus varius	Variegated Flycatcher	B	12	27.1					Paraguay	378
Griseotyrannus aurantioatrocristatus	Crowned Slaty Flycatcher	B	11	33		22.5	46			64, 363, 1178
Tyrannopsis sulphurea	Sulphury Flycatcher	B	18	53.6		48	61		Suriname	498
Tyrannus niveigularis	Snowy-throated Kingbird	M	1	34.4					Ecuador	1143
Tyrannus albogularis	White-throated Kingbird	B	13	37.1					Brazil; Suriname	498, 1109
Tyrannus melancholicus	Tropical Kingbird	B	35	37.4	2.9	30.9	43		Western Mexico	1151
Tyrannus couchii	Couch's Kingbird	B	32	39				Y	Texas, USA; NE Mexico	146
Tyrannus vociferans	Cassin's Kingbird	B	14	45.6	3.4	37.4	50			1148a
Tyrannus crassirostris	Thick-billed Kingbird	B	12	55.9		51.7	59			720
Tyrannus verticalis	Western Kingbird	B	17	39.6	2.8	34.9	44.1			314
Tyrannus tyrannus	Eastern Kingbird	M	92	38	2.4	33.3	47		New York, USA	830
		F	105	37.3	2.1	33	44.7			
Tyrannus tyrannus	Eastern Kingbird	M	24	42.8	3.8	39.2	54.7		Kansas, USA	830
		F	32	41.6	2.7	37	45.8			
Tyrannus dominicensis	Gray Kingbird	B	30	46.5	5	38	69		Puerto Rico	19
Tyrannus caudifasciatus	Loggerhead Kingbird	B	13	44.1		34.5	52		Cayman Is.	868
Tyrannus caudifasciatus	Loggerhead Kingbird	B	63	40.7	3.2	35	50		Cuba	19
Tyrannus cubensis	Giant Kingbird	B	5	93.6		88	108		Cuba	19
Tyrannus forficatus	Scissor-tailed Flycatcher	B	106	39.3				B	Oklahoma, USA	960
Tyrannus savana	Fork-tailed Flycatcher	B	23	31.9		28	40		Suriname	498
Sapayoa aenigma	Sapayoa	U	9	20.8					Panama	79, 615, 990

Body Masses of World Birds (continued)

Species Name	Common Name	Sex	N	Mean	S.D.	Min	Max	Season	Location	Source #
Schiffornis major	Greater Schiffornis	U		31					Peru	1210
Schiffornis turdina	Thrush-like Schiffornis	B	13	31.7						906
Schiffornis virescens	Greenish Schiffornis	U	10	25.6	1.2	23	27		Brazil	961
Xenopsaris albinucha	White-naped Xenopsaris	B	2	10		9.8	10.2		Venezuela; Brazil	1205, 1216
Pachyramphus castaneus	Chestnut-crowned Becard	U	7	19.5		18	21		Brazil	64, 878, 1178
Pachyramphus viridis	Green-backed Becard	M	1	21					Brazil	64
Pachyramphus versicolor	Barred Becard	B	7	15.5		13	18.4		Colombia; Ecuador	798, 881, 953, 1037
Pachyramphus cinnamomeus	Cinnamon Becard	B	14	20.3	0.8	18.7	22		Panama; Belize	456, 489, 1027, 1180
Pachyramphus polychopterus	White-winged Becard	B	34	20.8		18	25.5		Suriname	498
Pachyramphus major	Gray-collared Becard	U	3	24.8		22.1	28.4		Belize; Mexico; El Salvador	640, 1027, 1041a
Pachyramphus albogriseus	Black-and-white Becard	B	13	17.3		12.5	20		Peru; Colombia	1037, 1145
Pachyramphus marginaus	Black-capped Becard	M	9	19.8		16	25		Suriname	498
		F	10	17.1		14.5	19			
Pachyramphus surinameus	Glossy-backed Becard	M	4	18.6		17.5	21		Suriname	498
		F	3	20.9		18.9	23			
Pachyramphus rufus	Cinereous Becard	B	31	18.8		17	20		Suriname	498
Pachyramphus spodiurus	Slaty Becard	B	4	18.2					Peru	1302
Pachyramphus minor	Pink-throated Becard	B	24	36.6		31.2	44		Suriname	498
Pachyramphus niger	Jamaican Becard	U	2	40.2		38.9	41.5		Jamaica	1148

Body Masses of World Birds (continued)

Species Name	Common Name	Sex	N	Mean	S.D.	Min	Max	Season	Location	Source #
Pachyramphus aglaiae albiventris	Rose-throated Becard	B	9	29.4		27.2	31.2		Western Mexico	440
Pachyramphus aglaiae yucatanensis	Rose-throated Becard	B	9	30		25.8	33.3		Yucatan, Mexico	910a
Pachyramphus homochrous	One-colored Becard	U	3	37.3						990, 1311
Pachyramphus validus	Crested Becard	B	3	43		39.1	49		Brazil; Bolivia	64, 1055, 1178
Tityra cayana	Black-tailed Tityra	B	20	68.1		60	77		Suriname	498
Tityra semifasciata	Masked Tityra	B	25	79.3					Panama	489
Tityra inquisitor	Black-crowned Tityra	B	16	43.1	3.4	36	47		Panama; Suriname; Belize; Mexico; Brazil	64, 496, 795, 910a, 1027, 1180
Family Pittidae										
Pitta nipalensis	Blue-naped Pitta	B	2	124		124	124			340
Pitta soror	Blue-rumped Pitta	B	9	114		110	120			340
Pitta oatesi castaneiceps	Rusty-naped Pitta	B	8	120		117	124			340
Pitta caerulea	Giant Pitta	M	1	202					Malaysia	785
Pitta cyanea	Blue Pitta	M				99	120			340
Pitta guajana	Banded Pitta	M	1	81.5					Borneo	1218
Pitta gurneyi	Gurney's Pitta	M	1	60.5					captive	1143
Pitta sordida	Hooded Pitta	B	51	64.5	8.6	40	82			340
Pitta maxima	Ivory-breasted Pitta	M	1	190						340
Pitta steerii	Azure-breasted Pitta	B	13	94.8	10.7	68.5	114			340
Pitta kochi	Whiskered Pitta	F	1	116					fat individual	340

Body Masses of World Birds (continued)

Species Name	Common Name	Sex	N	Mean	S.D.	Min	Max	Season	Location	Source #
Pitta erythrogaster erythrogaster	Red-bellied Pitta	B	46	58.5	5.9	46	70.4			340
Pitta erythrogaster macklotii	Red-bellied Pitta	B	16	89.6	6.7	80	104			340
Pitta granatina granatina	Garnet Pitta	F		53						340
Pitta granatina coccinea	Garnet Pitta	F	1	57.8						340
Pitta angolensis	African Pitta	B	4	84.3		45.2	115			340, 1014
Pitta angolensis pulih	African Pitta	M	2	46.7		45.2	48.2		Liberia	623
Pitta brachyura	Indian Pitta	B	25	55.5		47	66		India	7
Pitta nympha	Fairy Pitta	M		109		67.5	155			340
		F	2	71.5		70	73			
Pitta moluccensis	Blue-winged Pitta	U	12	76		63	82		Malaysia	340
Pitta elegans	Elegant Pitta	U	9	56.6		47	77			340
Pitta versicolor	Noisy Pitta	B	28	98.9	13.6	72	128		Australia	517
Pitta versicolor simillima	Noisy Pitta	B	5	80		70	93			517
Pitta anerythra	Black-faced Pitta	M	2	86		83	89		Solomon Is.	658
Pitta iris	Rainbow Pitta	B	9	62.2		55.4	72.2			340
Family Atrichornithidae										
Atrichornis rufescens	Rufous Scrub-bird	U	1	24					Australia; probably immature	90
Atrichornis clamosus	Noisy Scrub-bird	M	36	51.8		46	58		Southwest Australia	517
		F	28	34.3		31.5	37			
Family Menuridae										
Menura alberti	Albert's Lyrebird	M	6	928		668	1215		Australia	517
Menura novaehollandiae	Superb Lyrebird	M	10	1100	54	1032	1200		Australia	517
		F	5	860		721	930			

Body Masses of World Birds (continued)

Species Name	Common Name	Sex	N	Mean	S.D.	Min	Max	Season	Location	Source #
Family Acanthisittidae										
Acanthisitta chloris	Rifleman	M	21	6.3	0.6	5	7		New Zealand	1002
		F	6	7.6		7	8.5			
Xenicus longipes	Bush Wren	M	1	15					New Zealand	517
		F	1	17						
Xenicus gilviventris	South Is.Wren	M	18	15.8	1.1				New Zealand	517
		F	19	19.6	1.6					
Family Alaudidae										
Mirafra passerina	Monotonous Lark	B	10	24.7		21	27.5			623
Mirafra cantillans	Singing Lark	M	24	18.6	1.5				Senegal	623
		F	9	18.8	1.4					
Mirafra javanica	Australasian Lark	B	37	23	1.3	20.5	25.8		Australia	470
Mirafra albicauda	White-tailed Lark	B	4	22.2		20	25			623
Mirafra hova	Madagascar Lark	U	9	18.9		17	23		Madagascar	957
Mirafra pulpa	Friedmann's Lark	U	3	21.9		21.2	23.1		Kenya	623
Mirafra hypermetra	Red-winged Lark	U	2	61.4		55	67.9		Somalia; Kenya	623
Mirafra somalica	Somali Long-billed Lark	U	3	46.6		43.6	50.2			623
Mirafra ashi	Ash's Lark	U	6	36.8		31.3	41.9		Somalia	623
Mirafra angolensis	Angola Lark	M	12	38.8		35	42			623
		F	11	33.8		30	36.5			
Mirafra africana africana	Rufous-naped Lark	M	25	44		41.3	47.5			623
		F	12	40.3		33	49			
Mirafra africana nigrescens	Rufous-naped Lark	M	8	51.9		50	54.5			623
		F	2	44.7		44	45.5			
Mirafra rufocinnamomea	Flappet Lark	M	15	25.7		21	30			623
		F	6	26.9		21	32.4			
Mirafra apiata	Cape Clapper Lark	M	32	29		24	44			623
		F	10	24.9		23	27			

Body Masses of World Birds (continued)

Species Name	Common Name	Sex	N	Mean	S.D.	Min	Max	Season	Location	Source #
Mirafra fasciolata	Eastern Clapper Lark	U				26	44			275
Mirafra erythroptera	Indian Bushlark	U	13	21.3		17	27		India	7
Mirafra gilletti	Gillett's Lark	B	13	22.4		20.1	26.1			623
Mirafra assamica	Rufous-winged Lark	B	11	26.2		20	30		India	7
Calendulalauda africanoides	Fawn-colored Lark	B	45	23.4		20.5	30			623
Calendulalauda alopex	Foxy Lark	U		24					estimated	275
Calendulalauda poecilosterna	Pink-breasted Lark	U	5	24.7		23.5	26.4		Somalia; Kenya	623
Calendulalauda sabota	Sabota Lark	B	87	24.7		21	31			623
Calendulalauda erythrochlamys	Dune Lark	M	8	29.4		26.3	33.1		Namibia	623
		F	10	26.8		24.8	28.9			
Calendulalauda albescens	Karoo Lark	B	7	30.4		27	35		South Africa	623
Calendulalauda barlowi	Barlow's Lark	M				27	36			275
		F				25	31			
Calendulalauda burra	Ferruginous Lark	M	7	40.5		38.5	43		South Africa	623
		F	4	34.5		32	38			
Pinarocorys erythropygia	Rufous-rumped Lark	M	1	30.1						623
Pinarocorys nigricans	Dusky Lark	M	7	39.7		30	46			623
		F	8	36.6		31	41			
Heteromirafra sidamoensis	Sidamo Lark	M	1	30.2					Ethiopia	623
Heteromirafra ruddi	Rudd's Lark	F	2	26.6		26	27.2			623
Certhilauda curvirostris	Cape Lark	B	9	41.9		31	53			266, 623
Certhilauda brevirostris	Algulhas Lark	U				35	48			275
Certhilauda semitorquata	Eastern Long-billed Lark	U				30	48			275

Body Masses of World Birds (continued)

Species Name	Common Name	Sex	N	Mean	S.D.	Min	Max	Season	Location	Source #
Certhilauda subcoronata damarensis	Karoo Long-billed Lark	M	1	45					Angola	1143
		F	1	35						
Certhilauda benguelensis	Benguela Lark	U				35	53			275
Certhilauda chuana	Short-clawed Lark	M	1	44.1						623
Chersomanes beesleyi	Beesley's Lark	M		27						275
		F		22						
Chersomanes albofasciata	Spike-heeled Lark	M	14	26.7		22.5	34			623
		F	7	22.7		19.5	27			
Eremopterix australis	Black-eared Sparrow-Lark	B	8	14.2		12	15			623
Eremopterix leucotis	Chestnut-backed Sparrow-Lark	U	8	16.4		12.3	23.8		South Africa; Nigeria	509, 623
Eremopterix nigriceps	Black-crowned Sparrow-Lark	B	4			12	16		Niger; Chad	223
Eremopterix verticalis	Gray-backed Sparrow-Lark	B	57	17		12.5	21.2			623
Eremopterix signatus	Chestnut-headed Sparrow-Lark	B	3	15.7		15	16			623
Eremopterix leucopareia	Fischer's Sparrow-Lark	U	12	14.5	0.7	12.6	15.4		Kenya	1052
Eremopterix griseus	Ashy-crowned Sparrow-Lark	U	11	16		14	18		India	7
Ammomanes cinctura	Bar-tailed Lark	B	14			14	23.5		Algeria; Niger; Chad	223
Ammomanes phoenicura	Rufous-tailed Lark	B	4	25.6		23	28.5		India	7
Ammomanes deserti	Desert Lark	B	7	23.8		20	26.6			623
Ammomanes grayi	Gray's Lark	M	37	22.7		19.3	26.5			623
		F	35	20.2		17.5	23			
Alaemon alaudipes	Greater Hoopoe-Lark	M	9	47.2		39	51		North Africa; Afghanistan; Saudi Arabia	223
		F	4	40.5		30	47			

Body Masses of World Birds (continued)

Species Name	Common Name	Sex	N	Mean	S.D.	Min	Max	Season	Location	Source #
Ramphocoris clotbey	Thick-billed Lark	M	2	53.5		52	55		Algeria	223
		F	2	45		45	45			
Melanocorypha calandra calandra	Calandra Lark	B	24	61.6		44.2	73			223
Melanocorypha bimaculata	Bimaculated Lark	B	20	54.1		48	62			223
Melanocorypha maxima	Tibetan Lark	U		75						275
Melanocorypha mongolica	Mongolian Lark	M	16	59.4	3.2	54	65		Mongolia	927
		F	4	49		45	54			
Melanocorypha leucoptera	White-winged Lark	B	24	45.3		36	52		former USSR	223
Melanocorypha yeltoniensis	Black Lark	M	44	63.7		56	76		former USSR	223
		F	20	56.4		51	68			
Calandrella brachydactyla	Greater Short-toed Lark	U	22			16	24	Sp	Morocco	223
Calandrella brachydactyla longipennis	Greater Short-toed Lark	M	23	22.3		19	25		Afghanistan	7
		F	9	19.9		19	22			
Calandrella acutirostris	Hume's Lark	M	11			19	23		Afghanistan	7
		F	7			18	21			
Calandrella rufescens	Lesser Short-toed Lark	B	43	24.1	2.2	16	27		Mongolia	927
Calandrella cinerea	Red-capped Lark	B	33	23.7		20	26.3			623
Calandrella cheleensis	Asian Short-toed Lark	B	15	22.2		20	24		Hopei, China	1087
Calandrella raytal	Sand Lark	M	3			18	19		India	7
Spizocorys conirostris	Pink-billed Lark	B	24	14.3		12	16.5			623
Spizocorys fringillaris	Botha's Lark	M	6	17.9		15.7	19.2	B	South Africa	510
		F	3	20.5		20.2	21			
Spizocorys sclateri	Sclater's Lark	U	2	19		17	21			623
Spizocorys obbiensis	Obbia Lark	M	5	14.8		13.7	15.6			623
		F	1	12.3						

Body Masses of World Birds (continued)

Species Name	Common Name	Sex	N	Mean	S.D.	Min	Max	Season	Location	Source #
Spizocorys personata	Masked Lark	U	1	19.7						623
Eremalauda dunni	Dunn's Lark	U	16	20.5					Arabian Desert	1223
Eremalauda starki	Stark's Lark	B	70	18.5		15.5	22.5			623
Chersophilus duponti	Dupont's Lark	U	33	39.4	3.2	32	47		Spain	223
Galerida cristata magna	Crested Lark	M	18	39		31	43		Afghanistan	7
		F	9	41.9		37	51			
Galerida cristata macrorhyncha	Crested Lark	M	16			37	42			623
		F	8	47.6		44	54			
Galerida thekla	Thekla Lark	U	23	36.8		34	41		Spain	223
Galerida thekla superflua	Thekla Lark	M	21	39.6		33	43			623
Galerida modesta	Sun Lark	B	9	20.2		18	22.2			623, 777
Galerida deva	Tawny Lark	B	5			18	22		India	7
Galerida magnirostris	Long-billed Lark	B	7	41.6		35	48			623
Pseudalaemon fremantlii	Short-tailed Lark	M	5	25		23	26			623
		F	4	21.9		20.5	25			
Lullula arborea	Wood Lark	B	961	26.9		23	32		former USSR	223
Alauda arvensis	Sky Lark	M	102	42.7	4.7	32	51	F	Great Britain	223
		F	286	37.2	3.6	29	47			
Alauda arvensis	Sky Lark	M	75	37		27	45		Hopei, China	1087
		F	30	33		29	38			
Alauda gulgula	Oriental Skylark	B	17	26.3		24	30	B	Afghanistan	223
Alauda razae	Razo Skylark	M		24						275
		F		20						
Eremophila alpestris	Horned Lark	B	25	32.2				B	Texas, USA	1303
Eremophila alpestris	Horned Lark	B	21	30.2				B	Washington, USA	1303
Eremophila alpestris flava	Horned Lark	M	16	39.9		36	45	W	Germany	623
		F	21	36.9		30	44			

Body Masses of World Birds (continued)

Species Name	Common Name	Sex	N	Mean	S.D.	Min	Max	Season	Location	Source #
Eremophila alpestris brandti	Horned Lark	M	22	32.4		27	40		Mongolia	927
		F	12	29.6		27	32			
Eremophila bilopha	Temminck's Lark	M	2	38.5		38	39	F	Algeria	223
Family Hirundinidae										
Progne tapera	Brown-chested Martin	B	36	32		27.5	36		Suriname	498
Progne subis	Purple Martin	M	89	53.5	3.8				Texas, USA	133a
		F	98	54.1	4.8					
Progne cryptoleuca	Cuban Martin	M	1	46					Cuba	19
Progne dominicensis	Caribbean Martin	U	5	39.6		38	42		Trinidad	359
Progne sinaloae	Sinaloa Martin	B	3	42.6		42	43.7		Mexico	1153
Progne chalybea	Gray-breasted Martin	U	115	42.9		36	48		Trinidad	359
Progne chalybea	Gray-breasted Martin	B	26	39.9		32	45.5		Suriname	498
Progne elegans	Southern Martin	B	8	50.6		46	54.5		Suriname	498
Tachycineta bicolor	Tree Swallow	B	308	21.2		17	25.5	B		1003
Tachycineta albilinea	Mangrove Swallow	B	22	16.7		15	18		Suriname	498
Tachycineta stolzmanni	Tumbes Swallow	B	6	11.3					Ecuador	995
Tachycineta albiventer	White-winged Swallow	U	6	17.7		14	21			64, 496, 613, 1233
Tachycineta leucorrhoa	White-rumped Swallow	U	24	15.3		12.8	18.4		Argentina	210
Tachycineta meyeni	Chilean Swallow	M	21	14.9	0.6	14	16		Chile	328
Tachycineta thalassina	Violet-green Swallow	M	16	14.4		13	16.3	Sp	California, USA	201
		F	15	13.9		12.5	15.2			
Tachycineta cyaneoviridis	Bahama Swallow	B	8	17.5		16.3	19.5		Bahamas	1159
Pygochelidon cyanoleuca	Blue-and-white Swallow	U	21	9.7	0.6	8.5	10.8		Venezuela	1215

Body Masses of World Birds (continued)

Species Name	Common Name	Sex	N	Mean	S.D.	Min	Max	Season	Location	Source #
Notiochelidon murina	Brown-bellied Swallow	B	14	11.7					Bolivia	963
Notiochelidon flavipes	Pale-footed Swallow	B	11	9.8		8.8	11.1		Peru; Ecuador	900, 1154
Notiochelidon pileata	Black-capped Swallow	B	5	12.2		12	13		Guatemala; El Salvador	1041a, 1145
Atticora fasciata	White-banded Swallow	U	10	13		10.5	14		Ecuador; Venezuela	1154, 1308
Atticora melanoleuca	Black-collared Swallow	B	8	10.8		9.7	12.4		Suriname	498
Neochelidon tibialis	White-thighed Swallow	U	12	10.6	1.7	8.5	14		Panama; Ecuador	615, 1154, 1180
Stelgidopteryx andecola	Andean Swallow	U		17		14	19			1233
Stelgidopteryx fucata	Tawny-headed Swallow	U				13	15			1233
Stelgidopteryx serripennis	Northern Rough-winged Swallow	B	47	15.9	0.6	10.3	18.3	Sp	Pennsylvania, USA	193
Stelgidopteryx serripennis ridgwayi	Northern Rough-winged Swallow	B	6	15.5		14.8	16.2		Yucatan, Mexico	910a
Stelgidopteryx ruficollis	Southern Rough-winged Swallow	U	25	16.1	1.4	14	19		Brazil	752
Cheramoeca leucosternus	White-backed Swallow	B	16	14.1		11	17		Australia	518a
Riparia riparia	Bank Swallow	M	698	13	0.8	10.9	16.4		California, USA	405
		F	1013	13.9	1.3	11.1	18.8			
Riparia riparia riparia	Bank Swallow	U	225	11.3		8.9	17.3	Sp	Morocco	623
Riparia paludicola	Plain Martin	U	111	13.5		11	16			623
Riparia cincta	Banded Martin	M	4	22.3		20.5	24.5		Southern Africa	623
		F	5	24.6		22	28.7			
Phedina borbonica	Mascarene Martin	U	19	23.9	1.8	20.9	25.8		Mauritius	183
Phedina borbonica madagascariensis	Mascarene Martin	B	8	20.6		19.5	22.5		Madagascar	957

Body Masses of World Birds (continued)

Species Name	Common Name	Sex	N	Mean	S.D.	Min	Max	Season	Location	Source #
Phedina brazzae	Brazza's Martin	U		13						1233
Petrochelidon pyrrhonota	Cliff Swallow	B	88	21.6	2	17.5	26.7	B	California, USA	314
Petrochelidon rufocollaris	Chestnut-collared Swallow	U				14	18			275
Petrochelidon fulva pallida	Cave Swallow	B	4388	24.2					New Mexico, USA	1291
Petrochelidon fulva citata	Cave Swallow	B	20	17.8					Yucatan, Mexico	1291
Petrochelidon fulva	Cave Swallow	B	23	15.8	1.1	14.4	19.5		Jamaica; Puerto Rico	19
Hirundo griseopyga	Gray-rumped Swallow	M	10	10.6		9	12.5		Southern Africa	623
		F	11	9.8		7.7	11.5			
Hirundo rupestris	Eurasian Crag-Martin	U	1300	20.4		17.7	33		Gibraltar	623
Hirundo obsoleta	Pale Crag-Martin	U	3	10.2		9	11		Iran; central Africa	852, 1143
Hirundo fuligula	Rock Martin	U	20	21.1		18.5	23.6			623
Hirundo concolor	Dusky Crag-Martin	U	4			12	14		India	7
Hirundo rustica	Barn Swallow	M	1337	16.2		12.1	28.2	Sp	Morocco	20
		F	994	15.8		11	24.8			
Hirundo rustica	Barn Swallow	M	106	18.1	1.1				Nebraska, USA	135
		F	130	19.2	1.5					
Hirundo rustica rustica	Barn Swallow	B	16	18.2		16	22		Afghanistan	7
Hirundo rustica	Barn Swallow	B	825	20.4		11	30.5		Southern Africa	1036
Hirundo lucida	Red-chested Swallow	U				12	14			1233
Hirundo aethiopica	Ethiopian Swallow	U	126	13.3		11	17		Chad; Ethiopia	623
Hirundo angolensis	Angola Swallow	U	9	17.8		15.7	19		Zambia	301
Hirundo albigularis	White-throated Swallow	B	17	21.8		19	29			623
Hirundo tahitica	Pacific Swallow	B	12	17.8	0.6	17	19		Fiji	1196c

Body Masses of World Birds (continued)

Species Name	Common Name	Sex	N	Mean	S.D.	Min	Max	Season	Location	Source #
Hirundo neoxena	Welcome Swallow	U	11	14.7	0.9	13.2	16		Victoria, Australia	518a
Hirundo smithii	Wire-tailed Swallow	U	195	12.4		10	15		Ethiopia	623
Hirundo nigrita	White-throated Blue Swallow	B	9	18.3		15.5	22		Angola; Liberia; Gabon	623
Hirundo nigrorufa	Black-and-rufous Swallow	B	2	14.2		14	14.5		Angola	623
Hirundo atrocaerulea	Blue Swallow	B	9	13.8		13	15		Zambia; South Africa; Zimbabwe	301, 1036
Hirundo megaensis	White-tailed Swallow	M	1	11						623
Hirundo dimidiata	Pearl-breasted Swallow	U	11	12.1		10	13.8			623
Hirundo cucullata	Greater Striped-Swallow	B	47	27.1		21	31		Southern Africa	623
Hirundo abyssinica	Lesser Striped-Swallow	B	6			12.5	14.5		Angola	266
Hirundo semirufa	Rufous-chested Swallow	B	50	30.1		25	40			623
Hirundo senegalensis	Mosque Swallow	M	7	47.4		41	51		South Africa	623
		F	9	43.4		38	47			
Hirundo daurica rufula	Red-rumped Swallow	B	52	22.2		19	29	B	Spain	623
Hirundo striolata	Striolated Swallow	F	1	22					India	7
Hirundo preussi	Preuss' Swallow	B	3	13.3		12	14		Cameroon	623
Hirundo rufigula	Red-throated Swallow	U	1	15.9					Angola	266
Hirundo spilodera	South African Swallow	B	1000	20.6		16	26			623
Hirundo nigricans	Tree Martin	B	9	15.4		14.2	17.4		Australia	470

Body Masses of World Birds (continued)

Species Name	Common Name	Sex	N	Mean	S.D.	Min	Max	Season	Location	Source #
Hirundo fluvicola	Streak-throated Swallow	U	14	9.7		8	12		India	7
Hirundo ariel	Fairy Martin	M	193	10.7	0.6	9.1	13	B	Victoria, Australia	518a
		F	202	11	0.8	9.1	14.1			
Delichon urbicum	House Martin	U	252	14.5		10.3	19.8	Sp	Morocco	20
Delichon dasypus	Asian Martin	U		18						1233
Delichon nipalensis	Nepal Martin	U				14	16			1233
Psalidoprocne nitens	Square-tailed Sawwing	U		9.8		8.3	11			1233
Psalidoprocne fuliginosa	Mountain Sawwing	B	25	12.2		10.5	14			623
Psalidoprocne albiceps	White-headed Sawwing	M	17	11.7		6	14		Uganda	623
		F	6	10.6		7	14			
Psalidoprocne petiti	Petit's Sawwing	B	6	10.8		10	12			623
Psalidoprocne pristoptera	Blue Sawwing	M	11	11.3		10	13		Kenya; Angola	266, 623
		F	5	9.4		7	11			
Psalidoprocne orientalis	Eastern Sawwing	M	1	11.9					Zimbabwe	567
Psalidoprocne holomelas	Black Sawwing	B	14	11.2		10	12		South Africa	623
Psalidoprocne obscura	Fanti Sawwing	U	2	9.4		8.8	10			623
Family Motacillidae										
Dendronanthus indicus	Forest Wagtail	B	15	16.4		13	17		Hopei, China	1037
Motacilla alba alba	White Wagtail	U	93	21	2.1	17.3	25	F	Sweden	223
Motacilla alba personata	White Wagtail	B	21	24.4		21	28	Sp	Afghanistan	223
Motacilla alba personata	White Wagtail	B	20	23.6		19	26		Northern China; Mongolia	925, 927
Motacilla alba baicalensis	White Wagtail	B	18	23.3	1.7	20	26		Mongolia	927
Motacilla alba lugens	White (Black-backed) Wagtail	B	20	27.9		23.8	32.9		Russia	30

Body Masses of World Birds (continued)

Species Name	Common Name	Sex	N	Mean	S.D.	Min	Max	Season	Location	Source #
Motacilla grandis	Japanese Wagtail	M	19	32.5	1.5	29.5	35.2	W	Japan	1239
		F	21	28	1.7	24.6	31.9			
Motacilla madaraspatensis	White-browed Wagtail	B	4	30.5		26	36		India	7
Motacilla aguimp	African Pied Wagtail	U	55	27		23	30.5			133
Motacilla capensis	Cape Wagtail	U	174	20.8		17	25			133
Motacilla flaviventris	Madagascar Wagtail	U	15	22.5	1.8	19.5	25.9		Madagascar	957
Motacilla citreola citreola	Citrine Wagtail	B	25	20.3		17	23	B	Russia	223
Motacilla citreola calcarata	Citrine Wagtail	B	21	18		15	23	Y		7
Motacilla tschutschensis	Eastern Yellow Wagtail	M	129	14.9		12.3	20.4	Sp	Morocco	20
		F	61	13.9		11.2	17.9			
Motacilla flava beema	Western Yellow Wagtail	B	22	16.6		13.4	21		India	7
Motacilla flava flava	Western Yellow Wagtail	B	162	17.6		14	27		Germany	223
Motacilla flava leucocephala	Western Yellow Wagtail	B	17	17.8	1.6	14	20		Mongolia	927
Motacilla flava macronyx	Western Yellow Wagtail	B	16	18.8		17	24		Northern China; Mongolia	925, 927
Motacilla cinerea	Gray Wagtail	M	87	18	1.4	15	22		Germany; Switzerland	223
		F	26	17.2	1.4	14	20			
Motacilla cinerea caspica	Gray Wagtail	B	34	16.3		13	20			7
Motacilla clara	Mountain Wagtail	B	10	17.9		15	24.5		Uganda	623
Tmetothylacus tenellus	Golden Pipit	B	24	20.1		18	22			623
Macronyx croceus	Yellow-throated Longclaw	B	55	47.1		38	55			623
Macronyx fuelleborni	Fuelleborn's Longclaw	M	9	57.9		53	64			623
		F	7	52.6		46	60			

Body Masses of World Birds (continued)

Species Name	Common Name	Sex	N	Mean	S.D.	Min	Max	Season	Location	Source #
Macronyx capensis	Orange-throated Longclaw	M	7	52.5		49.2	55			623
		F	11	47.7		45.2	52.2			
Macronyx ameliae	Rosy-throated Longclaw	M	9	34.4		30.3	39.8			623
		F	4	32.6		30.8	34			
Macronyx aurantiigula	Pangani Longclaw	U	11	39.4		36.6	43			623
Anthus chloris	Yellow-breasted Pipit	B	7	25.1		24	25.8		South Africa	623, 1258
Anthus lineiventris	Striped Pipit	U	12	35		33	38		Zimbabwe; South Africa	623
Anthus crenatus	Yellow-tufted Pipit	B	7	30.7		29	32.5		South Africa	1258
Anthus hoeschi	Mountain Pipit	U	13	27		23.5	31			623
Anthus rufulus	Oriental Pipit	M	9			20.1	22.7			7
		F	6			17.4	22			
Anthus novaeseelandiae lugubris	Australasian Pipit	B	19	24.2		22.2	26.3		Philippines	954
Anthus novaeseelandiae australis	Australasian Pipit	B	39	25.8		20	33		Australia	518a
Anthus novaeseelandiae rufuloides	Australasian Pipit	U	10	23.5		21	27			623
Anthus novaeseelandiae lacuum	Australasian Pipit	U	27	22.9		20	25		Kenya	623
Anthus leucophrys leucophrys	Plain-backed Pipit	U	6	23.2		21	24			623
Anthus leucophrys omoensis	Plain-backed Pipit	U	8	26.9		23	32			623
Anthus longicaudatus	Long-tailed Pipit	B	7	32.2						697, 1258
Anthus richardi	Richard's Pipit	M	19	33.5	2.2	30	37		Mongolia	927
		F	8	30.9		28	37			
Anthus vaalensis	Buffy Pipit	U	18	27		23	31			623
Anthus cinnamomeus	African Pipit	U	38	27		18	31		Ethiopia	623

Body Masses of World Birds (continued)

Species Name	Common Name	Sex	N	Mean	S.D.	Min	Max	Season	Location	Source #
Anthus melindae	Malindi Pipit	U	8	22.5		19	27			623
Anthus pseudosimilis	Kimberly Pipit	M	5	31.6		28	35.2		South Africa	698
Anthus campestris	Tawny Pipit	B	44	23		19	28	B	Kazakhstan	223
Anthus godlewskii	Blyth's Pipit	M	10	24.7	1.6	22	26		Mongolia	927
		F	13	26	2.1	22	28			
Anthus similis	Long-billed Pipit	U	68	25.3		21.5	29			623
Anthus nyassae	Woodland Pipit	U				23.8	25.9			275
Anthus berthelotii	Berthelot's Pipit	U	18	16.9	1.2				Madeira Is.	17
Anthus berthelotii	Berthelot's Pipit	U	92	14.9	1.1				Selvagem Is.	17
Anthus brachyurus	Short-tailed Pipit	B	7	15.3		14	16.7		Angola; South Africa	982, 1258
Anthus caffer	Bush Pipit	U	2	16.8		16	17.5		South Africa	623, 1258
Anthus sokokensis	Sokoke Pipit	U	10	15		12	17			623
Anthus trivialis trivialis	Tree Pipit	M	413	21.7	1.2	18	26	B	Germany	223
		F	23	25.1	1.8	22	29			
Anthus hodgsoni	Olive-backed Pipit	B	51	21.3		18	26		Mongolia	223
Anthus gustavi menzbieri	Pechora Pipit	U	10	19.8	1.2	17	21		Russia	223
Anthus pratensis	Meadow Pipit	U	228	18.4	1	13.9	23.4		Wales, UK	223
Anthus cervinus	Red-throated Pipit	U	101	20.9		16.4	29.3	Sp	Ethiopia	623
Anthus roseatus	Rosy Pipit	B	28			17	25		India	7
Anthus petrosus	Rock Pipit	U	59	22.4		18	27	F	Netherlands	623
Anthus spinoletta	Water Pipit	U	100	23.9	2.1	19.5	24	W	Germany	223
Anthus sylvanus	Upland Pipit	M		31						1259
Anthus rubescens	American Pipit	M	28	21.6		19.2	25.5	B	Alaska, USA	554
		F	8	20.1		18.6	23.2			

Body Masses of World Birds (continued)

Species Name	Common Name	Sex	N	Mean	S.D.	Min	Max	Season	Location	Source #
Anthus rubescens alticola	American Pipit	M	19	21.1	1.4	19	23.5	PB	Wyoming, USA	1253
		F	14	20	0.8	18.7	22.3			
Anthus correndera	Correndera Pipit	B	7	20.9		18	26.6		Brazil; Argentina	64, 546
Anthus antarcticus	South Georgia Pipit	U	12	36.1	2.5	33.7	41.1		South Georgia Is.	518a
Anthus spragueii	Sprague's Pipit	B	19	25.5	2.2	21.4	29.7		Oklahoma & Texas, USA	1153a
Anthus furcatus	Short-billed Pipit	B	29			17.5	23		Brazil	64
Anthus hellmayri	Hellmayr's Pipit	B	33			16.5	21		Brazil	64
Anthus bogotensis	Paramo Pipit	M	3	25.1		23.3	26		Venezuela	1043, 1263
Anthus lutescens	Yellowish Pipit	B	18			14	17		Brazil	64
Anthus lutescens	Yellowish Pipit	B	14	12.6					Panama	489
Anthus nattereri	Ochre-bellied Pipit	M	2	19.5		19	20		Brazil	64
Anthus gutturalis	Alpine Pipit	U	3	36.3		33	40		Papua New Guinea	1148
Family Campephagidae										
Coracina maxima	Ground Cuckoo-shrike	B	3	127		124	130		Australia	518a
Coracina welchmani	North Melanesian Cuckoo-Shrike	U				133	165			275
Coracina caledonica	South Melanesian Cuckoo-shrike	B	38	140		117	180		New Caledonia	1018
Coracina novaehollandiae	Black-faced Cuckoo-shrike	B	16	118		93	148		Australia	518a
Coracina caeruleogrisea	Stout-billed Cuckoo-shrike	M	10	148	11.4	127	170		New Guinea	283, 424, 427
		F	11	129	8.3	116	139			
Coracina striata	Bar-bellied Cuckoo-shrike	B	18	111		98.2	128		Philippines	954
Coracina atriceps	Moluccan Cuckoo-shrike	B	2	139		128	150		Moluccas	979

Body Masses of World Birds (continued)

Species Name	Common Name	Sex	N	Mean	S.D.	Min	Max	Season	Location	Source #
Coracina lineata lineata	Yellow-eyed Cuckoo-shrike	M	1	100					Queensland, Australia	518a
Coracina lineata makirae	Yellow-eyed Cuckoo-shrike	B	14	73.7		65.5	84.5		Solomon Is.	399
Coracina lineata pusilla	Yellow-eyed Cuckoo-shrike	B	15	59.6		50	65.5		Guadalcanal	399
Coracina boyeri	Boyer's Cuckoo-shrike	B	15	66.4		61	74		New Guinea	283
Coracina papuensis	White-bellied Cuckoo-shrike	B	31	65.8	5.7	54.3	75.8		Australia	470
Coracina papuensis elegans	White-bellied Cuckoo-shrike	B	14	61.2		56	65		Guadalcanal	399
Coracina longicauda	Hooded Cuckoo-shrike	B	6	101		92	109		New Guinea	424, 428, 770
Coracina analis	New Caledonian Cuckoo-shrike	M	1	97					New Caledonia	1018
		F	1	70						
Coracina pectoralis	White-breasted Cuckoo-shrike	B	5	54.4		49	58		Cameroon; Angola	266, 623
Coracina azurea	Blue Cuckoo-shrike	B	13	46.1					Liberia	623
Coracina caesia	Gray Cuckoo-shrike	M	12	48.1		42	63		Uganda	623
		F	12	46.5		40	50			
Coracina cinerea	Ashy Cuckoo-shrike	U	12	47.2	4.9	35.5	55.5		Madagascar	957
Coracina typica	Mauritius Cuckoo-shrike	U	1	42.9					Mauritius	183
Coracina tenuirostris	Cicadabird	B	20	67.8		60	76		Australia	518a
Coracina tenuirostris erythropygia	Cicadabird	M	11	58.1		54	63		Solomon Is.	399
		F	5	57.4		52	64.5			
Coracina salomonis	San Cristobal Cicadabird	M	15	64.6		55	75		Solomon Is.	399
		F	12	65.7		60.5	73			
Coracina coerulescens	Blackish Cuckoo-shrike	M	8	65.5		61	70		Philippines	436
		F	7	60.3		53	69			

Body Masses of World Birds (continued)

Species Name	Common Name	Sex	N	Mean	S.D.	Min	Max	Season	Location	Source #
Coracina mindanensis	Black-bibbed Cuckoo-shrike	B	3	58.7		55.6	61.3		Philippines	954
Coracina ceramensis	Pale-gray Cuckoo-shrike	U				45	51		Moluccas	1181, 1182
Coracina incerta	Papuan Cuckoo-shrike	B	7	58.9		50	67		New Guinea	283, 424, 427
Coracina schisticeps poliopsa	Gray-headed Cuckoo-shrike	B	13	52.8		48	57		New Guinea	283
Coracina melas	New Guinean Cuckoo-shrike	U				56	63			275
Coracina montana bicinia	Black-bellied Cuckoo-shrike	M	15	81.1	4.4	72	88		New Guinea	282
		F	13	77.4	4.2	72	85			
Coracina holopolia	Solomon Islands Cuckoo-shrike	B	6	48.9		41.2	63		Solomon Is.	399, 658
Coracina mcgregori	McGregor's Cuckoo-shrike	B	21	42.6		39	47.5		Philippines	954
Coracina polioptera	Indochinese Cuckoo-shrike	U	1	35.8					Thailand	787
Coracina melaschistos	Black-winged Cuckoo-shrike	B	20	43		39	47		China	184a
Coracina fimbriata	Lesser Cuckoo-shrike	U	3	29.3		28.4	30		Borneo	1218
Coracina melanoptera	Black-headed Cuckoo-shrike	U	19			24	36		India	7
Campochaera sloetii	Golden Cuckoo-shrike	B	4	41		36	46		New Guinea	283
Lalage melanoleuca	Black-and-white Triller	B	5	46.8		41.9	49		Philippines	954, 983
Lalage nigra	Pied Triller	B	12	28.4		24	31.3		Philippines	954
Lalage tricolor	White-winged Triller	B	21	25		20.7	31.5		Australia	518a
Lalage aurea	Rufous-bellied Triller	M	1	32					Moluccas	979

Body Masses of World Birds (continued)

Species Name	Common Name	Sex	N	Mean	S.D.	Min	Max	Season	Location	Source #
Lalage leucomela polygrammica	Varied Triller	B	6	26.5		24	29		New Guinea	283
Lalage leucomela	Varied Triller	B	16	32.3		29	36		Australia	518a
Lalage atrovirens	Black-browed Triller	B	2	33		31.5	34.5		New Guinea	427, 428
Lalage maculosa	Polynesian Triller	U	19	30.3					Tonga	977
Lalage leucopyga	Long-tailed Triller	B	8	24.3		21	28		Solomon Is.	399
Campephaga petiti	Petit's Cuckoo-shrike	M	16	32		26	37		Cameroon; Uganda; Kenya	623
		F	3	30.7		28	36			
Campephaga flava	Black Cuckoo-shrike	B	18	32.3		27	40		Southern Africa	623
Campephaga phoenicea	Red-shouldered Cuckoo-shrike	B	88	27.1		23	32.5		Ethiopia	623
Campephaga quiscalina	Purple-throated Cuckoo-shrike	B	20	34.5		26	43			623
Campephaga lobata	Ghana Cuckoo-shrike	U	21	33.1		29	37		Liberia	623
Pericrocotus roseus	Rosy Minivet	B	20	19.2		18	22		China	184a
Pericrocotus divaricatus	Ashy Minivet	B	14	24.6		20	28		China	184a
Pericrocotus cinnamomeus	Small Minivet	U	7	8.6		6	12		India	7, 740
Pericrocotus igneus	Fiery Minivet	B	3	15		14	15.8		Borneo	1218
Pericrocotus erythropygius	White-bellied Minivet	U	2	10.8		10.6	11.1		Myanmar	1148
Pericrocotus ethologus	Long-tailed Minivet	M	36	18		17	21		Hopei, China	1087
		F	45	20		17	23			
Pericrocotus brevirostris	Short-billed Minivet	B				16	17		India	7
Pericrocotus flammeus	Scarlet Minivet	B	20	23.3		21	25		China	184a
Pericrocotus solaris	Gray-chinned Minivet	B	10	14.5		11	17		India; Malaysia	7, 785

Body Masses of World Birds (continued)

Species Name	Common Name	Sex	N	Mean	S.D.	Min	Max	Season	Location	Source #
Hemipus picatus	Bar-winged Flycatcher-shrike	U				8.5	9.5		India	7
Hemipus hirundinaceus	Black-winged Flycatcher-shrike	F	1	10.2					Borneo	1218
Family Pycnonotidae										
Spizixos canifrons	Crested Finchbill	M	1	44					India	7
Spizixos semitorques	Collared Finchbill	U				26	31		Taiwan	834
Spizixos semitorques	Collared Finchbill	B	20	39		30	46		China	184a
Pycnonotus zeylanicus	Straw-headed Bulbul	U				80	93			275
Pycnonotus striatus	Striated Bulbul	B	10			45	60		India	7
Pycnonotus melanoleucus	Black-and-white Bulbul	U	1	31					Malaysia	775
Pycnonotus atriceps	Black-headed Bulbul	B	7	25.5		24	27.9		Palawan, Philippines	1143
Pycnonotus flaviventris	Black-crested Bulbul	M	6			25	34		India	7
		F	3			28	31			
Pycnonotus taivanus	Styan's Bulbul	U				22	31		Taiwan	834
Pycnonotus squamatus	Scaly-breasted Bulbul	B	4	22		18.8	24		Borneo; Malaysia	785, 1218
Pycnonotus cyaniventris	Gray-bellied Bulbul	B	4	22		20.5	24		Borneo; Malaysia	785, 1218
Pycnonotus jocosus	Red-whiskered Bulbul	U	20	27.4		25	31		India	7
Pycnonotus jocosus	Red whiskered Bulbul	U	69	31.7	2.4	27	38		New South Wales, Australia	518a
Pycnonotus xanthorrhous	Brown-breasted Bulbul	B	12	26.9	1.9	24.4	31.1		Thailand	787
Pycnonotus sinensis	Light-vented Bulbul	B	20	34.2		30	46		China	184a

Body Masses of World Birds (continued)

Species Name	Common Name	Sex	N	Mean	S.D.	Min	Max	Season	Location	Source #
Pycnonotus barbatus layardi	Common Bulbul	U	609	38.9		23	52.9			623
Pycnonotus barbatus dodsoni	Common Bulbul	U	29	26.5	2.4	21	33		Kenya	623
Pycnonotus nigricans	Black-fronted Bulbul	U	359	30.8		21.6	39.1			133
Pycnonotus capensis	Cape Bulbul	U	176	38.6		28.2	47.3			623
Pycnonotus xanthopygos	White-spectacled Bulbul	U	56	44		35	46		Israel	223
Pycnonotus xanthopygos	White-spectacled Bulbul	U	18	32.2		28.9	35.8		Yemen	317
Pycnonotus leucotis	White-eared Bulbul	U	10	23		18	28		India	7
Pycnonotus leucotis mesopotamia	White-eared Bulbul	U	5	32		29	35.4		Iraq; Kuwait	223
Pycnonotus leucogenys leucogenys	White-cheeked Bulbul	U	5			34	38		India	7
Pycnonotus cafer bengalensis	Red-vented Bulbul	M	518	45.8	3.5	35.5	58.5			558
		F	554	40.1	3.3	31	51.5			
Pycnonotus aurigaster	Sooty-headed Bulbul	M	10	45.5		42	49		China	184a
		F	10	43.2		40	50			
Pycnonotus eutilotus	Puff-backed Bulbul	U	3	35.3		32	37		Borneo	398
Pycnonotus urostictus	Yellow-wattled Bulbul	B	35	24.8		21.8	28.1		Philippines	954
Pycnonotus finlaysoni	Stripe-throated Bulbul	U		26.3					Malaysia	468
Pycnonotus penicillatus	Yellow-eared Bulbul	F	2	36.5		36	37		Sri Lanka	7
Pycnonotus flavescens	Flavescent Bulbul	B	13	28.9	2.1	27.1	35		Thailand	787
Pycnonotus luteolus	White-browed Bulbul	U	10	34.7		28	43		India	7

Body Masses of World Birds (continued)

Species Name	Common Name	Sex	N	Mean	S.D.	Min	Max	Season	Location	Source #
Pycnonotus goiavier	Yellow-vented Bulbul	B	24	27.8		24.4	34		Philippines	954
Pycnonotus plumosus cinereifrons	Olive-winged Bulbul	M	3	36.7		36	37.8		Palawan, Philippines	1143
		F	3	33.1		30.6	37.7			
Pycnonotus blanfordi	Streak-eared Bulbul	U	8	33.4		31.3	35.7		Thailand	787
Pycnonotus simplex	Cream-vented Bulbul	B	3	25.1		22.4	27.2		Malaysia	785
Pycnonotus brunneus	Red-eyed Bulbul	B	4	28.8		20	37.3		Borneo	398, 1218
Pycnonotus erythrophthalmus	Spectacled Bulbul	B	5	19.2		16.5	20.5		Borneo; Malaysia	785, 1218
Andropadus montanus	Cameroon Mountain Greenbul	B	21	32.6		29	36.7			623
Andropadus masukuensis	Shelly's Greenbul	U	59	26.8	1.9	23	34		Tanzania	1014
Andropadus virens	Little Greenbul	M	14	28.5	2.6	23	32		DR Congo	623
		F	13	26.2	1.9	22	30			
Andropadus virens	Little Greenbul	B	158	25.1	2.1	21	32		Tanzania	1014
Andropadus gracilis	Gray Greenbul	M	36	20.9		18	23			623
		F	15	22.4		19	28			
Andropadus ansorgei	Ansorge's Greenbul	B	15	18.7		17	23		Liberia; Kenya	623
Andropadus curvirostris	Plain Greenbul	U	18	24.9	1.9	20.6	28.3		Uganda	1078
Andropadus gracilirostris	Slender-billed Greenbul	M	43	30.7		24	37			623
		F	32	31.8		26	40			
Andropadus importunus	Sombre Greenbul	B	26	27.4		22	30.9		Kenya	623
Andropadus latirostris	Yellow-whiskered Greenbul	U	224	26.2	2.1	17.6	34		Uganda	1078
Andropadus tephrolaemus bamendae	Western Mountain-Greenbul	U	33	38.3		31.8	46.7		Cameroon	623
Andropadus nigriceps kikuyuensis	Eastern Mountain-Greenbul	U	37	33.1	2	28.5	35.5			623

Body Masses of World Birds (continued)

Species Name	Common Name	Sex	N	Mean	S.D.	Min	Max	Season	Location	Source #
Andropadus nigriceps fusciceps	Eastern Mountain-Greenbul	B	130	37.3		30	42		Malawi	303
Andropadus milanjensis	Stripe-cheeked Greenbul	M	39	38.5	2.5	34.6	45.9		Zimbabwe	563
		F	17	36.3	2.4	32.8	40.8			
Calyptocichla serina	Golden Greenbul	M	6	39.8		36.4	42			623
		F	7	37.7		35	42.7			
Baeopogon indicator	Honeyguide Greenbul	M	13	47.8		44	53			623
		F	6	43.9		40	48			
Ixonotus guttatus	Spotted Greenbul	B	23	35.2		31	40		Liberia; Uganda	623
Chlorocichla simplex	Simple Greenbul	M	8	47.8		40.7	54.5			623
		F	4	45.1		40.4	49.2			
Chlorocichla flavicollis	Yellow-throated Greenbul	U	17	47.3	3	42	53		Uganda; Kenya	623
Chlorocichla flavicollis	Yellow-throated Greenbul	U	19	50.2		37	58		Nigeria	777
Chlorocichla falkensteini	Yellow-necked Greenbul	B	6	31.7		29	37		Cameroon; Angola	266, 623
Chlorocichla flaviventris	Yellow-bellied Greenbul	M	21	42	3.8	34.5	51.2		Zimbabwe	623
		F	12	36.9	2.5	31.9	39.6			
Chlorocichla laetissima	Joyful Greenbul	B	13	50.2		43	55		Kenya	623
Thescelocichla leucopleura	Swamp Greenbul	B	18	62.1		55	70		Liberia; Cameroon; Uganda	623
Phyllastrephus scandens	Leaf-love	M	10	47.2		43	53			623
		F	11	40.7		33	51			
Phyllastrephus cabanisi	Cabanis' Greenbul	M	34	37.1		33	42			623
		F	18	31.9		27	39			
Phyllastrephus fischeri	Fischer's Greenbul	M	25	31.2	1.5	29	34.5		Kenya	623
		F	26	26.4	2.2	22	31.1			
Phyllastrephus fischeri placidus	Fischer's (Placid) Greenbul	M	58	27.5		21	32		Kenya	623
		F	66	23.5		19	31			
Phyllastrephus terrestris	Terrestrial Greenbul	M	51	33.9	2.7	29.5	40.5			623
		F	40	28.5	2.4	24.3	33.3			

Body Masses of World Birds (continued)

Species Name	Common Name	Sex	N	Mean	S.D.	Min	Max	Season	Location	Source #
Phyllastrephus strepitans	Northern Brownbul	U	45	27.1	3.5	21.4	32.9		Somalia	1335
Phyllastrephus fulviventris	Pale-olive Greenbul	M	1	37						623
		F	1	30						
Phyllastrephus cerviniventris	Gray-olive Greenbul	B	7	27		23.5	32.9			623
Phyllastrephus baumanni	Baumann's Greenbul	M	5	30.3		27.3	32.9			623
		F	4	24.9		23.5	26.5			
Phyllastrephus hypochloris	Toro Olive-Greenbul	M	14	25.6		19	30		Uganda	623
		F	11	20.9		17	24			
Phyllastrephus poensis	Cameroon Olive-Greenbul	U	23	28.9		23.8	37.1			623
Phyllastrephus lorenzi	Sassi's Greenbul	F	1	19						623
Phyllastrephus flavostriatus	Yellow-streaked Bulbul	M	11	31.8		28.1	35.8		Malawi	303
		F	14	25.2		22	32.4			
Phyllastrephus poliocephalus	Gray-headed Greenbul	B	4	53		37.8	64			623
Phyllastrephus debilis	Tiny Greenbul	M	32	15.4	0.9	13.5	17			623
		F	25	13.3	1	11.6	15			
Phyllastrephus albigularis	White-throated Greenbul	M	23	27.5		24	31			623
		F	23	21.9		17	28			
Phyllastrephus icterinus	Icterine Greenbul	M	28	20.3		16.5	24.5			623
		F	19	17.6		15	22			
Phyllastrephus xavieri	Xavier's Greenbul	M	28	26.3		23	31			623
		F	31	19.5		16	22			
Phyllastrephus madagascariensis madagascariensis	Long-billed Greenbul	U	49	30	4.4	20.5	39		Madagascar	957
Phyllastrephus madagascariensis inceleber	Long-billed Greenbul	U	8	21.4		19.5	26		Madagascar	957
Phyllastrephus zosterops	Spectacled Greenbul	U	143	6.7	3.2	10	27.5		Madagascar	957

Body Masses of World Birds (continued)

Species Name	Common Name	Sex	N	Mean	S.D.	Min	Max	Season	Location	Source #
Phyllastrephus zosterops andapae	Spectacled Greenbul	U	10	17.6	2.3	15	22		Madagascar	957
Phyllastrephus apperti	Appert's Greenbul	U	23	14.6	1.7	11	17		Madagascar	957
Phyllastrephus tenebrosus	Dusky Greenbul	F	1	21.4					Madagascar	1143
Phyllastrephus cinereiceps	Gray-crowned Greenbul	U	38	18.2	2.3	13.5	24.5		Madagascar	957
Bleda syndactylus	Common Bristlebill	M	47	45.3		36	56			623
		F	44	47		35	48			
Bleda notatus ugandae	Lesser Bristlebill	M	9	42.4		37	48			623
		F	9	36		33	39			
Bleda eximius	Green-tailed Bristlebill	U	4	35.7		34.5	37.5		Gabon	616
Bleda canicapillus	Gray-headed Bristlebill	M	6	42.7		39.4	46.5			623
		F	4	37.9		36	40.5			
Nicator chloris	Yellow-spotted Nicator	M	26	49.1		36	55		Uganda	396
		F	17	43.3		32	51			
Nicator gularis	Eastern Nicator	M	10	55.7		50	63		Kenya; Tanzania	396
		F	11	37.3		33.6	41.3			
Nicator vireo	Yellow-throated Nicator	B	4	23.6		21	26		Uganda; Angola	396, 982
Criniger calurus emini	Red-tailed Greenbul	B	14	30		22	35		Uganda; Nigeria; Central African Rep.	623
Criniger calurus verreauxii	Red-tailed Greenbul	M	9	44.4		41.2	48.1			623
		F	4	40.3		37.9	42.8			
Criniger barbatus	Western Bearded-Greenbul	M	9	44.4		41.2	48.1			623
		F	5	40.8		37.8	43			
Criniger chloronotus	Eastern Bearded-Greenbul	U	4	45		38.8	49.5			616, 623
Criniger olivaceus	Yellow-bearded Greenbul	B	12	28.1		23.8	31.7			623
Criniger ndussumensis	White-bearded Greenbul	B	5	23		20.5	25.6		DR Congo; Central African Rep.	1143

Body Masses of World Birds (continued)

Species Name	Common Name	Sex	N	Mean	S.D.	Min	Max	Season	Location	Source #
Alophoixus finschii	Finsch's Bulbul	B	6	24.2		23.1	25		Borneo; Malaysia	785, 1218
Alophoixus flaveolus	White-throated Bulbul	M	10	48.3		38	54		India	7
		F	4			38	48			
Alophoixus pallidus	Puff-throated Bulbul	U	10	46	3	41	52		Thailand	787
Alophoixus ochraceus	Ochraceous Bulbul	M	7	39.6		36.7	46.5		Malaysia	785
		F	5	43.4		39.6	46.9			
Alophoixus bres	Gray-cheeked Bulbul	U	16	41.9	5.3	32	51		Borneo	398
Alophoixus bres	Gray-cheeked Bulbul	M	1	29.2					Malaysia	785
		F	4	36		34	39.4			
Alophoixus phaeocephalus	Yellow-bellied Bulbul	U	41	32	3.8	24	40		Borneo	398
Alophoixus affinis	Golden Bulbul	B	4	40.5		38	42		Moluccas	979
Setornis criniger	Hook-billed Bulbul	U	4	29.5		26	35		Borneo	398
Tricholestes criniger	Hairy-backed Bulbul	U	7	17.1		16	19		Borneo	398
Iole propinqua	Gray-eyed Bulbul	F	1	25.9					Thailand	787
Acritillas indica	Yellow-browed Bulbul	U	33	30.6		27	34		India	7
Ixos palawanensis	Sulphur-bellied Bulbul	B	9	26.8		24.1	30.9		Palawan, Philippines	1143
Ixos philippinus	Philippine Bulbul	B	44	38.6		32	43.6		Philippines	954
Ixos amaurotis	Brown-eared Bulbul	U	27	70.9	9	60	90.6		Japan	621
Ixos everetti	Yellowish Bulbul	B	33	58.3		51.9	66.9		Philippines	954
Ixos rufigularis	Zamboanga Bulbul	B	5	58 7		47.8	71		Philippines	954
Ixos malaccensis	Streaked Bulbul	U	3	37.3		34	41		Borneo	398, 1218
Hemixos flavala	Ashy Bulbul	B	8	32.5		27.2	35.2		Malaysia	775, 785

Body Masses of World Birds (continued)

Species Name	Common Name	Sex	N	Mean	S.D.	Min	Max	Season	Location	Source #
Ixos mcclellandii	Mountain Bulbul	U	126	32.5		25	40		Malaysia	775
Hypsipetes madagascariensis madagascariensis	Madagascar Bulbul	U	126	45.1	4.1	33	54		Madagascar	957
Hypsipetes madagascariensis	Madagascar Bulbul	B	28	44.3	4.3				Comoro Is.	708
Hypsipetes madagascariensis	Madagascar Bulbul	B	19	48.9	4.5				Anjouan (Indian Ocean)	708
Hypsipetes madagascariensis rostratus	Madagascar Bulbul	B	25	45.8	3.8				Aldabra	708
Hypsipetes ganeesa	Square-tailed Black Bulbul	U	25	42.9		35	49			7
Hypsipetes leucocephalus psaroides	Himalayan Black Bulbul	B	16			44	61			7
Hypsipetes crassirostris	Seychelles Bulbul	B	32	79.5	6.2				Seychelles	183
Hypsipetes parvirostris parvirostris	Comoro Bulbul	B	35	56.9	5.3				Comoro Is.	708
Hypsipetes parvirostris moheliensis	Comoro Bulbul	B	62	62.1	5				Moheli	708
Hypsipetes borbonicus	Reunion Bulbul	B	4	55.3		51	57.2		Reunion	183, 708
Hypsipetes olivaceus	Mauritius Bulbul	U	3	75		70	79		Mauritius	183
Hypsipetes thompsoni	White-headed Bulbul	M	1	41.8					Thailand	787
Neolestes torquatus	Black-collared Bulbul	M	4	21.8		19.9	23.5			396
		F	5	23.2		19.8	27			
Family Regulidae										
Regulus satrapa	Golden-crowned Kinglet	M	261	6.3	0.1	4.9	7.7	M	Pennsylvania, USA	193
		F	147	6.1	0.1	4.5	7.8			
Regulus calendula	Ruby-crowned Kinglet	M	55	6.5	0.3	5.8	7.1	M	Pennsylvania, USA	828
		F	48	5.9	0.3	5.4	6.8			
Regulus regulus	Goldcrest	M	491	5.6	0.4	4.6	7.1		Sweden	223
		F	334	5.5	0.4	4.6	7			

Body Masses of World Birds (continued)

Species Name	Common Name	Sex	N	Mean	S.D.	Min	Max	Season	Location	Source #
Regulus goodfellowi	Flamecrest	F	1	4.1					Taiwan	1143
Regulus ignicapillus	Firecrest	U	44	5.6		4.5	8.2	Y	Great Britain	515
Family Chloropseidae										
Chloropsis flavipennis	Philippine Leafbird	F	1	35.4					Mindanao, Philippines	1143
Chloropsis palawanensis	Yellow-throated Leafbird	M	3	31.7		29.7	34.5		Palawan, Philippines	1143
		F	4	29		28.2	31			
Chloropsis sonnerati	Greater Green Leafbird	M	1	45.8					Borneo	1218
Chloropsis cyanopogon	Lesser Green Leafbird	B	7	22		19.1	25.3		Borneo; Malaysia	398, 785, 1218
Chloropsis cochinchinensis	Blue-winged Leafbird	B	2	24.5		24	25		India; Malaysia	740, 785
Chloropsis aurifrons	Golden-fronted Leafbird	M	3	36		35	37			7
		F	1	29						
Chloropsis media	Sumatran Leafbird	F	1	40					Sumatra	275
Chloropsis hardwickii	Orange-billed Leafbird	M				32	40		India	7
		F				25	34			
Family Aegithinidae										
Aegithina tiphia	Common Iora	B	12	12	1.8	10	15		India	7, 740
Aegithina nigrolutea	White-tailed Iora	B	17			10	14		India	7
Aegithina viridissima	Green Iora	B	5	13.4		12.5	14.8		Borneo	1218
Family Ptilogonatidae										
Phainoptila melanoxantha	Black-and-white Silky-flycatcher	U		56					Costa Rica	1170
Ptilogonys cinereus	Gray Silky-flycatcher	B	6	33.6		32.3	34.9		Mexico	910, 1151
Ptilogonys caudatus	Long-tailed Silky-flycatcher	M	4	39.1					Panama	489
		F	1	33.5						
Phainopepla nitens	Phainopepla	B	303	22.1		17.9	28.1		Arizona, USA	187

Body Masses of World Birds (continued)

Species Name	Common Name	Sex	N	Mean	S.D.	Min	Max	Season	Location	Source #
Family Bombycillidae										
Bombycilla garrulus	Bohemian Waxwing	B	45	56.4	5.2	46.5	69		Colorado, USA	314
Bombycilla garrulus	Bohemian Waxwing	B	53	52.5		31	70		Hopei, China	1087
Bombycilla cedrorum	Cedar Waxwing	M	1829	30.6	2.2	25.6	39.9		Pennsylvania, USA	828
		F	904	32.6	2.3	26.4	39.6			
Bombycilla japonica	Japanese Waxwing	B	3	42.3		40	47		Hopei, China	1087
Family Hypocoliidae										
Hypocolius ampelinus	Hypocolius	B	7	53.3		48	57		Iraq; Afghanistan	7, 223
Family Dulidae										
Dulus dominicus	Palmchat	B	8	48.4		41.4	52.5		Dominican Republic	19
Family Cinclidae										
Cinclus cinclus aquaticus	White-throated Dipper	M	178	64.2	3.8	53	76		Germany	223
		F	180	55.4	5.5	46	72			
Cinclus cinclus gularis	White-throated Dipper	M	124	69.1	3	60	76		Scotland	223
		F	101	58.1	3.3	50	67			
Cinclus pallasii	Brown Dipper	B	11	76	8.6	66	88			7, 273
Cinclus mexicanus	American Dipper	M	21	61	2.2	57	66	B	Utah, USA	314
		F	18	54.6	4.8	43	65			
Cinclus mexicanus	American Dipper	M	9	57.4		55	58.5	B	Oregon, USA	703
		F	18	54	3.9	44	61			
Cinclus leucocephalus	White-capped Dipper	U	8	44		38	59		Ecuador	633, 1043
Cinclus schulzi	Rufous-throated Dipper	B	2	37.6		36.9	38.4		Argentina	1041
Family Troglodytidae										
Donacobius atricapilla	Black-capped Donacobius	B	10	36.8		31	44			156, 318, 496, 1043

Body Masses of World Birds (continued)

Species Name	Common Name	Sex	N	Mean	S.D.	Min	Max	Season	Location	Source #
Campylorhynchus albobrunneus	White-headed Wren	U	6	32.9		27.5	39		Panama	990, 1180
Campylorhynchus zonatus	Band-backed Wren	B	7	34.6		28.3	39.5		Belize; Mexico	1027, 1145
Campylorhynchus megalopterus	Gray-barred Wren	B	3	33.1		32.8	33.5		Mexico	910
Campylorhynchus chiapensis	Giant Wren	B	8	50.9		43.4	57		Chiapas, Mexico	1143
Campylorhynchus rufinucha	Rufous-naped Wren	U	11	31	4.6	24	40		El Salvador	1041a
Campylorhynchus gularis	Spotted Wren	B	3	30.1		28.3	31		Jalisco, Mexico	1080
Campylorhynchus jocosus	Boucard's Wren	B	5	27.6		23.8	29.8		Oaxaca, Mexico	1143
Campylorhynchus yucatanicus	Yucatan Wren	U	2	35.5		31	40		Yucatan, Mexico	640
Campylorhynchus brunneicapillus	Cactus Wren	B	42	38.9		33.4	46.9		Arizona, USA	11
Campylorhynchus griseus	Bicolored Wren	U	6	42.4		37	46.5		Venezuela	1215
Campylorhynchus turdinus	Thrush-like Wren	U	5	32.6		26.5	36.5		Brazil	883
Campylorhynchus nuchalis	Stripe-backed Wren	U	10	23.2	1.6	21.1	25.5		Venezuela	1215
Campylorhynchus fasciatus	Fasciated Wren	B	3	30		24.9	37.1			1043, 1302
Odontorchilus branickii	Gray-mantled Wren	B	2	9.6		9	10.2			521, 1288
Odontorchilus cinereus	Tooth-billed Wren	M	1	11					Rondonia, Brazil	1178
Salpinctes obsoletus	Rock Wren	B	31	16.5	0.6				California, USA	935
Salpinctes obsoletus guadeloupensis	Rock Wren	M	6	15					Guadalupe Is., Mexico	935
		F	6	16.4						
Salpinctes obsoletus tenuirostris	Rock Wren	M	4	16.2					San Benitos Is., Mexico	935
		F	2	17.9						

Body Masses of World Birds (continued)

Species Name	Common Name	Sex	N	Mean	S.D.	Min	Max	Season	Location	Source #
Catherpes mexicanus	Canyon Wren	B	11	12.7	2.8	9	18		Mexico	763, 797, 1150
Catherpes mexicanus conspersus	Canyon Wren	B	55	11.3		9.9	14.8			609
Hylorchilus sumichrasti	Slender-billed Wren	F	1	28.4						1147a
Hylorchilus navai	Nava's Wren	F	1	29.3				B	Chiapas, Mexico	228
Cinnycerthia unirufa	Rufous Wren	U	81	29.4	2.7	23	36.9		Colombia	324a
Cinnycerthia olivascens	Sharpe's Wren	M	15	25.9	4.1				Colombia; Ecuador	144
		F	22	23	2.7					
Cinnycerthia peruana	Peruvian Wren	M	36	20.6	1.7				Peru	144
		F	32	18.4	1.3					
Cinnycerthia fulva	Fulvous Wren	M	6	18.4					Peru; Bolivia	144
		F	10	14.8	1.6					
Thryothorus atrogularis	Black-throated Wren	M	6	26		24.3	27.3		Costa Rica	1143
		F	1	22.5						
Thryothorus fasciatoventris	Black-bellied Wren	U	12	24						156, 489, 618, 990, 1180, 1145
Thryothorus eisenmanni	Inca Wren	M	5	24.6		22	27		Peru	898
		F	4	21.7		19	23			
Thryothorus mystacalis	Whiskered Wren	U		29						522
Thryothorus euophrys atriceps	Plain-tailed Wren	B	5	27.1		26	27.5		Peru	898
Thryothorus euophrys schulenbergi	Plain-tailed Wren	B	4	33.8		29.5	38		Peru	898
Thryothorus genibarbis	Moustached Wren	B	12	19.2		16.2	22.8		Brazil	177
Thryothorus coraya	Coraya Wren	B	16	17.2		14.9	19.2		Suriname	498
Thryothorus felix pallidus	Happy Wren	B	12	14.4	1.3	12.3	16.4		Western Mexico	1151
Thryothorus felix	Happy Wren	M	22	12.2	1.4	9	14.6		Tres Marias Is., Mexico	440
		F	8	11.8	0.8	10.5	13.1			

Body Masses of World Birds (continued)

Species Name	Common Name	Sex	N	Mean	S.D.	Min	Max	Season	Location	Source #
Thryothorus maculipectus	Spot-breasted Wren	M	20	16.2	1.1	13.7	18.5		Veracruz, Mexico	1323
		F	14	13.9	1.5	12.3	15.6			
Thryothorus rutilus	Rufous-breasted Wren	B	30	16.5		13.5	18.5		Trinidad	359
Thryothorus sclateri	Speckle-breasted Wren	U	2	20		20	20		Peru	1143
Thryothorus semibadius	Riverside Wren	U	1	17					Costa Rica	618
Thryothorus nigricapillus	Bay Wren	U	12	23.9		19	27		Panama; Colombia	156, 1037, 1180
Thryothorus thoracicus	Stripe-breasted Wren	U	3	17.6					Costa Rica	618
Thryothorus pleurostictus	Banded Wren	U	14	17.7	2.7	12.7	22		El Salvador	1041a
Thryothorus ludovicianus	Carolina Wren	B	1180	17.2	2.6	10	29	B	Eastern USA	278
Thryothorus ludovicianus	Carolina Wren	M	134	21.2	1.3	18.5	27	Y	Alabama, USA	467a
		F	118	18.7	5.1	16.2	21.5			
Thryothorus rufalbus	Rufous-and-white Wren	M	25	28						352
		F	25	25						
Thryothorus rufalbus	Rufous-and-white Wren	U	89	22	3.2	12	29.9		El Salvador	1041a
Thryothorus sinaloa	Sinaloa Wren	B	24	15.1	1.4	10.9	16.6		Jalisco, Mexico	593
Thryothorus modestus	Plain Wren	M	6	19.7					Panama	489
		F	7	17.9						
Thryothorus modestus	Plain Wren	U	14	15.4	2	12.2	18.3		El Salvador	1041a
Thryothorus zeledoni	Canebrake Wren	U		23					Costa Rica	1170
Thryothorus leucotis	Buff-breasted Wren	M	35	21						352
		F	35	18						
Thryothorus superciliaris	Superciliated Wren	F	1	22.8					Peru	1043
Thryothorus guarayanus	Fawn-breasted Wren	B	2	13.5		13	14		Bolivia	49
Thryothorus longirostris	Long-billed Wren	U	5	21.3		17.7	24.2		Sao Paulo, Brazil	1143

Body Masses of World Birds (continued)

Species Name	Common Name	Sex	N	Mean	S.D.	Min	Max	Season	Location	Source #
Thryomanes bewickii	Bewick's Wren	B	56	9.9	0.8	7.8	11.8	Y	Arizona, USA	314
Troglodytes troglodytes	Winter Wren	B	54	8.9	0.2	7.5	10.5	F	Pennsylvania, USA	193
Troglodytes troglodytes	Winter Wren	M	96	9.4	0.3	8.1	11.6		British Columbia, Canada	504
		F	30	9.1	0.9	7.6	10.9			
Troglodytes troglodytes fridariensis	Winter Wren	U	50	12.2					Fair Is.	223
Troglodytes troglodytes troglodytes	Winter Wren	U	272	9.3	1.1	6	12		Germany	223
Troglodytes troglodytes	Winter Wren	U	50	9.9		8	12.7	Y	Great Britain	515
Troglodytes aedon	House Wren	M	41	10.8	0.6	9.8	12.1		Pennsylvania, USA	828
		F	37	11.1	0.8	9.8	13.1			
Troglodytes aedon	House Wren	U	14	11.6	0.9	10	13		El Salvador	1041a
Troglodytes aedon cahooni	House Wren	B	15	10.4	0.6	9.9	11.8		Coahuila, Mexico	797
Troglodytes aedon	House Wren	B	75	10.4	0.9	8.5	13.4		Chile	328
Troglodytes sissonii	Socorro Wren	B	7	7.8		5.6	9.8		Socorro Is., Mexico	1151
Troglodytes cobbi	Cobb's Wren	B	7	19.1		17	20		Falkland Is.	1338
Troglodytes rufociliatus	Rufous-browed Wren	U	17	10	1.7	6	12		El Salvador	1041a
Troglodytes ochraceus	Ochraceous Wren	U	4	9.4		8	10		Panama	489, 990
Troglodytes solstitialis	Mountain Wren	U	14	11.4					Peru; Ecuador	953, 1288
Troglodytes rufulus	Tepui Wren	M	3	15.5		14	16.5		Venezuela	1308
		F	3	16.7		13.9	19.5			
Cistothorus platensis	Sedge Wren	U	28	9	0.7	7.2	10.3		Mexico; Central America	314
Cistothorus platensis	Sedge Wren	M	10	8.2		7.8	8.9			511
		F	10	7.7		7.1	8.9			
Cistothorus platensis hornensis	Sedge Wren	B	3	11.8		10.6	12.4		Tierra del Fuego	546
Cistothorus apolinari	Apolinar's Wren	M	1	17.7					Colombia	1169

Body Masses of World Birds (continued)

Species Name	Common Name	Sex	N	Mean	S.D.	Min	Max	Season	Location	Source #
Cistothorus palustris	Marsh Wren	M	38	11.9	0.7	10.5	13.5	B	New York, USA	314
		F	38	10.6	1	9	13.5			
Cistothorus palustris	Marsh Wren	M	52	10.6	0.7				Georgia, USA	660
		F	25	9.4	1.1					
Cistothorus palustris	Marsh Wren	M	19	11.7		10.9	12.7		Western USA	660
Uropsila leucogastra	White-bellied Wren	M	10	9.6	0.3	9.8	10.5		Yucatan, Mexico	910a
		F	7	8.5		8	9.1			
Thryorchilus browni	Timberline Wren	U		14					Costa Rica	603a
Henicorhina leucosticta prostheleuca	White-breasted Wood-Wren	M	34	14.8	1	13.5	17		Veracruz, Mexico	1323
		F	20	13.6	0.9	12.2	15.2			
Henicorhina leucophrys	Gray-breasted Wood-Wren	M	17	18.1	1				Panama	489
		F	6	16.4						
Henicorhina leucophrys	Gray-breasted Wood-Wren	U	103	14.2	1.2	12.9	15.2		Colombia	1038
Henicorhina leucophrys brunneiceps	Gray-breasted Wood-Wren	U	142	16.3	0.7	15.7	17.4		Colombia	1038
Henicorhina negreti	Munchique Wood-Wren	M	2	16.4		16.1	16.7		Colombia	1038
		F	1	15.2						
Henicorhina leucoptera	Bar-winged Wood-Wren	M	10			12	16.5		Peru	259
		F	7			12	14.5			
Microcerculus philomela	Nightingale Wren	M	7	18.6		17.4	21.5			1162
		F	4	17		16.4	17.4			
Microcerculus marginatus luscinia	Scaly-breasted Wren	M	5	20.1		18.2	22			1162
		F	4	17.4		17	18			
Microcerculus marginatus	Scaly-breasted Wren	U	17	17.3	0.8				Peru	1288
Microcerculus ustulatus	Flutist Wren	B	15	21		18.5	22		Venezuela	296
Microcerculus bambla	Wing-banded Wren	B	11	19.1		17	21.5		Venezuela; Suriname	498, 1308
Cyphorhinus phaeocephalus	Song Wren	U	21	24.6	2.2				Panama; Costa Rica	618

Body Masses of World Birds (continued)

Species Name	Common Name	Sex	N	Mean	S.D.	Min	Max	Season	Location	Source #
Cyphorhinus thoracicus	Chestnut-breasted Wren	M	5	33.4		30.5	35		Colombia	798
		F	4	29.3		29	29.6			
Cyphorhinus arada	Musician Wren	B	14	20.9		18.3	23.3		Suriname	498
Family Mimidae										
Dumetella carolinensis	Gray Catbird	U	438	35.3	2.5	27.3	43.5		Pennsylvania, USA	828
Melanoptila glabrirostris	Black Catbird	B	7	35.4		31.6	38.1		Yucatan, Mexico	640, 910a
Mimus gundlachii	Bahama Mockingbird	B	10	66.8	8.8	57	85		Bahamas	1159
Mimus polyglottos	Northern Mockingbird	B	221	48.5		36.2	55.7		Florida, USA	364
Mimus gilvus	Tropical Mockingbird	U	16	58.4	4.6	52.5	66		Venezuela	1215
Mimus gilvus	Tropical Mockingbird	B	24	46.3	4.9	35	56		Bonaire	19
Mimus graysoni	Socorro Mockingbird	M	20	67.6	6.5	56	80		Socorro Is., Mexico	765
		F	15	63.5	5.2	55	74			
Mimus saturninus	Chalk-browed Mockingbird	U	10	63.7		51	76		Brazil; Bolivia; Argentina	254, 657, 752
Mimus patagonicus	Patagonian Mockingbird	U	11	57.8	3.4	53.8	64.2		Argentina	210
Mimus dorsalis	Brown-backed Mockingbird	B	7	58.7		52	65		Bolivia	1145
Mimus triurus	White-banded Mockingbird	B	8	47.2		42	54		Argentina; Bolivia; Brazil	64, 210, 363, 657
Mimus longicaudatus	Long-tailed Mockingbird	B	19	66.6	5.7	54	79		Peru	1145
Mimus thenca	Chilean Mockingbird	U		66					Chile	582
Nesomimus parvulus	Galapagos Mockingbird	M	150	56.2	4.9				Galapagos Is.	243
		F	178	51.2	4					
Nesomimus trifasciatus	Charles Mockingbird	M	34	65.7	4.1				Galapagos Is.	243
		F	61	59.8	3.9					

Body Masses of World Birds (continued)

Species Name	Common Name	Sex	N	Mean	S.D.	Min	Max	Season	Location	Source #
Nesomimus macdonaldi	Hood Mockingbird	M	140	76.1	4.7				Galapagos Is.	243
		F	77	64.8	5.3					
Nesomimus melanotis	San Cristobal Mockingbird	M	27	53.2	3.1				Galapagos Is.	243
		F	18	48	3					
Oreoscoptes montanus	Sage Thrasher	B	30	44.2		39.6	50.2		Washington, USA	968
Toxostoma rufum	Brown Thrasher	B	177	68.8		57.6	89		Pennsylvania, USA	193
Toxostoma longirostre sennetti	Long-billed Thrasher	B	95	67.6	5.5			B	Texas, USA	1235
Toxostoma guttatum	Cozumel Thrasher	B	3	52.8		49	59.8		Yucatan, Mexico	640, 910a
Toxostoma cinereum	Gray Thrasher	M	4	63.9		58.6	69.8		Western Mexico	1101, 1151
		F	1	54.4						
Toxostoma bendirei	Bendire's Thrasher	B	23	62.2	5.1	53.5	74.6	Y	Arizona, USA	314
Toxostoma ocellatum	Ocellated Thrasher	U	3	84.3		77.7	88.9		Mexico	1145
Toxostoma curvirostre curvirostre	Curve-billed Thrasher	B	31	85.2	6.1			B	Texas, USA	1234
Toxostoma curvirostre palmeri	Curve-billed Thrasher	M	70	80.9	5	56.8	93.6	B	Arizona, USA	1234
		F	38	75.8	4.4	67.7	84			
Toxostoma redivivum	California Thrasher	B	21	84.4	3.9	78	93		California, USA	314
Toxostoma crissale	Crissal Thrasher	B	34	62.7	4.3	53.2	70		Arizona, USA	314
Toxostoma lecontei	Le Conte's Thrasher	B	172	61.9		54.5	75.5	Y	SW USA; Mexico	1092
Ramphocinclus brachyurus	White-breasted Thrasher	U		50					estimated	348
Melanotis caerulescens caerulescens	Blue Mockingbird	B	16	63.5		57.1	69.7		Western Mexico	440
Melanotis caerulescens longirostris	Blue Mockingbird	B	18	59.7	4.5	50.2	68		Western Mexico	1151
Melanotis hypoleucus	Blue-and-white Mockingbird	U	20	68	5.1				El Salvador	646a

Body Masses of World Birds (continued)

Species Name	Common Name	Sex	N	Mean	S.D.	Min	Max	Season	Location	Source #
Cinclocerthia gutturalis	Gray Trembler	B	5	69.2		65	76.2		St. Lucia	19
Cinclocerthia ruficauda	Brown Trembler	B	56	56.1	3.8	49.4	71		Montserrat	19
Cinclocerthia ruficauda	Brown Trembler	B	21	53.9	2.7	48.8	59.5		St. Vincent	19
Allenia fusca	Scaly-breasted Thrasher	B	95	67.6	8.6	56	120		Montserrat	19
Allenia fusca	Scaly-breasted Thrasher	B	30	74.4	5.7	61.4	88		St. Kitts	19
Allenia fusca	Scaly-breasted Thrasher	B	30	62.7	4.2	54	74		St. Lucia	19
Allenia fusca	Scaly-breasted Thrasher	B	10	64.2	3	58.5	67.5		Barbuda	19
Margarops fuscatus	Pearly-eyed Thrasher	B	737	104	10.3	80	138		Puerto Rico	19
Margarops fuscatus	Pearly-eyed Thrasher	B	219	102	9.3	78	136		Montserrat	19
Margarops fuscatus	Pearly-eyed Thrasher	B	19	89.9	6.6	79	103		Bonaire	19
Family Prunellidae										
Prunella collaris	Alpine Accentor	M	13	46.9	1.4				Japan	837
		F	12	40	2					
Prunella collaris rufilata	Alpine Accentor	B	14	30.7		25	35	Y	Kazakhstan	223
Prunella himalayana	Himalayan Accentor	M	4			25.5	29.8		India	7
		F	2	25.7		24.3	27.1			
Prunella rubeculoides	Robin Accentor	M	5			23.2	27.6		India; China	7, 1183
		F	3			21.2	24.1			
Prunella strophiata	Rufous-breasted Accentor	B	29			16.2	20.2		India	7
Prunella montanella	Siberian Accentor	B	68	17		13	23		Hopei, China	1087
Prunella ocularis	Radde's Accentor	M	2	22.5		20	25		Turkey; Iran	223
Prunella fagani	Yemen Accentor	U	4	21.9		20	23.2		North Yemen	958

Body Masses of World Birds (continued)

Species Name	Common Name	Sex	N	Mean	S.D.	Min	Max	Season	Location	Source #
Prunella fulvescens	Brown Accentor	B	14	18.9	1.5	16	21		Mongolia	927
Prunella atrogularis	Black-throated Accentor	B	15	18.5		14.5	23.6	F	Kazakhstan	223
Prunella koslowi	Mongolian Accentor	B	4	17.5		16	18		Mongolia	927
Prunella modularis modularis	Dunnock	U	527	19.7	1.9	14	26	Y	Germany	223
Prunella modularis occidentalis	Dunnock	U	50	20.8	2.5	13	26		Great Britain	223
Prunella rubida	Japanese Accentor	U		16.7					Japan	339
Prunella immaculata	Maroon-backed Accentor	U	3	20.7		19	22.5	W		7
Family Turdidae										
Neocossyphus fraseri	Rufous Flycatcher-Thrush	B	56	35.7		27	44			623
Neocossyphus finschii	Finsch's Flycatcher-Thrush	M		37.7	2.4					623
		F		36.3	1.8					
Neocossyphus rufus	Red-tailed Ant-Thrush	B	12	65.9		55	72			623
Neocossyphus poensis	White-tailed Ant-Thrush	M	41	53.1		45	60			623
		F	21	50.5		43	60			
Pseudocossyphus sharpei	Forest Rock-Thrush	B	62	24.8	1.9	20.5	29.5		Madagascar	957
Pseudocossyphus bensoni	Benson's Rock-Thrush	U				24	28.5			192
Pseudocossyphus imerinus	Littoral Rock-Thrush	M	5	28.3		26	32.5		Madagascar	1143, 957
Monticola rupestris	Cape Rock-Thrush	F	3	61.4		60	63.7			1245
Monticola brevipes	Short-toed Rock-Thrush	U		33						275
Monticola angolensis	Miombo Rock-Thrush	U	2	44.4		44	44.7			1245

Body Masses of World Birds (continued)

Species Name	Common Name	Sex	N	Mean	S.D.	Min	Max	Season	Location	Source #
Monticola saxatilis	Rufous-tailed Rock-Thrush	B	76	50.3		40	65	B	Afghanistan; Kazakhstan	1245
Monticola rufocinereus	Little Rock-Thrush	B	7	24		20	27		Kenya	1245
Monticola cinclorhynchus	Blue-capped Rock-Thrush	B	20	36		30	41		India	7
Monticola gularis	White-throated Rock-Thrush	B	43	34.6		29	43		Hopei, China	1087
Monticola rufiventris	Chestnut-bellied Rock-Thrush	B	7	53.2		48	61	Y	India	7
Monticola solitarius solitarius	Blue Rock-Thrush	U	26	57	3.5	50.5	63.5		Malta	223
Monticola solitarius philippensis	Blue Rock-Thrush	B	30	50.5		39	60		Hopei, China	1087
Myophonus robinsoni	Malayan Whistling-Thrush	U	5	99.2		87	105		Malaysia	775
Myophonus horsfieldii	Malabar Whistling-Thrush	U	10	117		101	130		India	7
Myophonus caeruleus	Blue Whistling-Thrush	M	19	156		136	188		Hopei, China	1087
		F	11	160		137	179			
Zoothera dumasi	Moluccan Thrush	U				49	50		Buru, Moloccas	1182
Zoothera erythronota	Rusty-backed Thrush	U				52	61			275
Zoothera wardii	Pied Thrush	U	6	58.6		52	72		India	7
Zoothera cinerea	Ashy Thrush	U	1	63					Philippines	344
Zoothera citrina	Orange-headed Thrush	U	20	53.3		46	60		India	7
Zoothera sibirica	Siberian Thrush	U	76	75.5		62	90		Malaysia	775
Zoothera piaggiae	Abyssinian Ground-Thrush	B	39	51.1		43	65		Ethiopia; Kenya; Uganda	1245
Zoothera crossleyi	Crossley's Ground-Thrush	U	14	72		63	82		Cameroon	1245

Body Masses of World Birds (continued)

Species Name	Common Name	Sex	N	Mean	S.D.	Min	Max	Season	Location	Source #
Zoothera gurneyi gurneyi	Orange Ground-Thrush	B	110	54.7		44.5	70.2		Southern Africa	1245
Zoothera gurneyi otomitra	Orange Ground-Thrush	B	40	61.2		54	76		Tanzania	1245
Zoothera cameronensis	Black-eared Ground-Thrush	B	5	44		42	47		Cameroon	133
Zoothera princei	Gray Ground-Thrush	B	14	65.4		59	83		Liberia	1245
Zoothera oberlaenderi	Oberlaender's Ground-Thrush	B	5	43.4		41	48		Uganda	1245
Zoothera guttata	Spotted Ground-Thrush	B	32	57.5		45	65		Kenya	1245
Zoothera spiloptera	Spot-winged Thrush	M	1	70					Sri Lanka	5
Zoothera andromedae	Sunda Thrush	B	4	90.5		81	108		Philippines	344, 436, 983, 954
Zoothera mollissima	Plain-backed Thrush	U	4	98.2		89.6	112		India	7
Zoothera dixoni	Long-tailed Thrush	B	6	90		71.5	103		India	7
Zoothera dauma	Scaly Thrush	B	9	104		92	130		India	7
Zoothera dauma	Scaly Thrush	U	1	173					Japan	490
Zoothera major	Amami Thrush	U	1	172						192
Zoothera lunulata	Olive-tailed Thrush	U	8	130		121	138		Queensland, Australia	390a
Zoothera heinei	Russet-tailed Thrush	B	9	82.8		71	96		Australia	518a
Zoothera margaretae	San Cristobal Thrush	U				60	71			192
Zoothera turipavae	Guadalcanal Thrush	M	1	53.5					Guadalcanal	399
Zoothera monticola	Long-billed Thrush	B	2	126		122	131		India	7

Body Masses of World Birds (continued)

Species Name	Common Name	Sex	N	Mean	S.D.	Min	Max	Season	Location	Source #
Zoothera marginata	Dark-sided Thrush	F	1	81					Thailand	787
Nesocichla eremita	Tristan Thrush	U		91					Tristan de Cunha	1031
Cichlherminia lherminieri	Forest Thrush	B	89	105	8.4	88	128		Montserrat	19
Sialia sialis	Eastern Bluebird	B	90	27.5	3.3	21.3	38		Eastern USA	277
Sialia mexicana	Western Bluebird	M	50	25.8	1.1	23.5	28.5	B	Arizona, USA	463
		F	76	27.1	1.3	24.5	31.5			
Sialia currucoides	Mountain Bluebird	B	42	29.6	3			B	Western USA	507
Myadestes townsendi	Townsend's Solitaire	B	111	33.2	2	28.4	39.1	Y	California, USA	103
Myadestes occidentalis	Brown-backed Solitaire	U	125	36.4	3.2	28	46		El Salvador	1041a
Myadestes elisabeth	Cuban Solitaire	B	56	30	2.7	21.5	43		Cuba	19
Myadestes genibarbis	Rufous-throated Solitaire	B	27	29.2	3.1	24.2	34.8			19
Myadestes melanops	Black-faced Solitaire	B	9	32.1					Panama	489
Myadestes coloratus	Varied Solitaire	M	5			26	34		Panama	990
		F	9			24.5	32			
Myadestes unicolor	Slate-colored Solitaire	U	45	37.9	3.3	29.5	43		El Salvador	1041a
Myadestes ralloides	Andean Solitaire	B	23	29.1	2.6	20	34		Colombia	1037
Myadestes obscurus	Omao	M	72	51	3.7				Hawaiian Is.	1265
		F	24	49.9	4.6					
Myadestes palmeri	Puaiohi	B	11	39.6		37	43		Hawaiian Is.	1127
Cichlopsis leucogenys	Rufous-brown Solitaire	U		61					Brazil	1105
Cichlopsis leucogenys	Rufous-brown Solitaire	U	2	47.4		46.8	47.9		Paraguay	785a
Entomodestes leucotis	White-eared Solitaire	U	10	61.7					Peru; Bolivia	785a, 963, 1288

Body Masses of World Birds (continued)

Species Name	Common Name	Sex	N	Mean	S.D.	Min	Max	Season	Location	Source #
Entomodestes coracinus	Black Solitaire	U	1	56					Colombia	1037
Catharus aurantiirostris	Orange-billed Nightingale-Thrush	U	14	29.8	3.1	25	38		El Salvador	1041a
Catharus fuscater	Slaty-backed Nightingale-Thrush	U	22	33.7	2.2	30.5	38		Colombia	65
Catharus occidentalis	Russet Nightingale-Thrush	M	1	26.2					Veracruz, Mexico	259a
Catharus gracilirostris	Black-billed Nightingale-Thrush	U		21					Costa Rica	1170
Catharus frantzii	Ruddy-capped Nightingale-Thrush	B	270	28.9	2.6	18	39		El Salvador	1041a
Catharus mexicanus	Black-headed Nightingale-Thrush	M	9	33		30.4	34.1		Mexico	1145
Catharus dryas	Spotted Nightingale-Thrush	B	58	37.7	3.6	29	47		El Salvador	1041a
Catharus fuscescens	Veery	U	51	31.9	2.5	26.8	37.1	M	Pennsylvania, USA	828
Catharus minimus	Gray-cheeked Thrush	M	140	32.6	3.9	24.9	44.6	M	Illinois, USA	727
		F	137	30.6	3.3	24.8	43.4			
Catharus bicknelli	Bicknell's Thrush	M	118	27.5	1.5	24	31.9	B	Vermont, USA	975
		F	45	28.1	3.5	23	37			
Catharus ustulatus	Swainson's Thrush	U	394	30.3	1.8	25.1	36.1		Pennsylvania, USA	828
Catharus guttatus	Hermit Thrush	B	2532	30.1	3.1	26.1	36.4	Sp	Ontario, Canada	608
Hylocichla mustelina	Wood Thrush	M	1232	47.8	3.3	37	62	B	Delaware, USA	1019
		F	955	52.5	4.4	40.5	76			
Platycichla leucops	Pale-eyed Thrush	U	9	62.4					Peru; Venezuela	917, 1288, 1308
Platycichla flavipes	Yellow-legged Thrush	B	18	60		52	67		Trinidad	359

Body Masses of World Birds (continued)

Species Name	Common Name	Sex	N	Mean	S.D.	Min	Max	Season	Location	Source #
Platycichla flavipes	Yellow-legged Thrush	M		64					Brazil	1105
		F		72						
Psophocichla litsipsirupa	Groundscraper Thrush	U	19	74		67	82.2		Botswana; Zambia; and South Africa	1245
Turdus menachensis	Yemen Thrush	U				72	79			192
Turdus olivaceus olivaceus	Southern Olive Thrush	B	179	78.1		60	98		South Africa	1245
Turdus olivaceus swynnertoni	Southern Olive Thrush	B	152	66		57	79		Zimbabwe	1245
Turdus abyssinicus abyssinicus	Northern Olive Thrush	B	46	61.4		49	87		Kenya	1245
Turdus abyssinicus bambusicola	Northern Olive Thrush	B	13	64.8		52	70		Uganda	1245
Turdus helleri	Taita Thrush	U				53	89		Kenya	275
Turdus ludoviciae	Somali Thrush	U		62					Somalia	275
Turdus roehli	Usambara Thrush	B	5	72.3		66.5	75.5			1245
Turdus olivaceofuscus	Olivaceous Thrush	B	6	87		77	92		Sao Tome	1245
Turdus bewsheri	Comoro Thrush	U	3	63.3		57	68		Comoro Is.	182
Turdus libonyanus	Kurrichane Thrush	U	65	60.6		51	70		Southern Africa	1245
Turdus pelios	African Thrush	U	84	65.4		56	77		Nigeria	777
Turdus tephronotus	African Bare-eyed Thrush	B	12	49.4		45	55	B	Kenya	1245
Turdus hortulorum	Gray-backed Thrush	B	3	66.7		64	68			273, 1088
Turdus unicolor	Tickell's Thrush	U	15	64.2		57	75.1		India	7
Turdus dissimilis	Black-breasted Thrush	U	1	75						192
Turdus cardis	Japanese Thrush	M	84	56.6	4.6	47	73.2		Japan	556
		F	64	61.3	6.2	46	83			

Body Masses of World Birds (continued)

Species Name	Common Name	Sex	N	Mean	S.D.	Min	Max	Season	Location	Source #
Turdus albocinctus	White-collared Thrush	B	5	99.4		90	107		India	7
Turdus torquatus	Ring Ouzel	U	50	109		88.4	132	Sp	Great Britain	515
Turdus boulboul	Gray-winged Blackbird	B	18	97.1		85	111		India	7
Turdus merula merula	Eurasian Blackbird	B	522	113		82	149	F	Great Britain	223
Turdus merula intermedius	Eurasian Blackbird	B	13	93.4		84	102	B	Kazakhstan; Afghanistan	223
Turdus simillimus nigropileus	Indian Blackbird	U	19	74.9		60	94		India	7
Turdus poliocephalus	Is.Thrush	U	38	61.7	3.9	53.5	69.3		Fiji	668
Turdus poliocephalus rennellianus	Is.Thrush	B	20	58.1		52	67		Solomon Is.	289
Turdus poliocephalus tolokiwae	Is.Thrush	B	16	60.4		52	67		Bismarck Is.	289
Turdus poliocephalus keysseri	Is.Thrush	B	7	72.1		69	78		New Guinea	292
Turdus poliocephalus beehleri	Is.Thrush	U				48	52.5		New Ireland	192
Turdus poliocephalus malindangensis	Is.Thrush	M		75						192
		F				84	86			
Turdus poliocephalus sladeni	Is.Thrush	M	4	61.9		57	67.5		Guadalcanal	399
Turdus poliocephalus nigrorum	Is.Thrush	U				77	79		Negros, Philippines	344
Turdus poliocephalus vanikorensis	Is.Thrush	B	5	46.5		42.5	54		Vanuatu	656b
Turdus rubrocanus	Chestnut Thrush	M	2	92.2		84.6	99.7		India	7
Turdus feae	Gray-sided Thrush	B	13	71		67	77		Hopei, China	1087
Turdus obscurus	Eyebrowed Thrush	U	136	62.6		45	80		Malaysia	775

Body Masses of World Birds (continued)

Species Name	Common Name	Sex	N	Mean	S.D.	Min	Max	Season	Location	Source #
Turdus pallidus	Pale Thrush	B	21	72.1	3.8	66.7	83.2		Japan	490
Turdus chrysolaus	Brown-headed Thrush	B	8	65.4		57	72.2		Philippines; Japan	436, 437, 490
Turdus celaenops	Izu Thrush	B	2	84.5		79	90		Japan	1143
Turdus ruficollis	Dark-throated Thrush	U				63	103			275
Turdus atrogularis	Black-throated Thrush	U				54	110			275
Turdus naumanni naumanni	Naumann's Thrush	B	82	77.9		63	94		Hopei, China	1087
Turdus eunomus	Dusky Thrush	M	15	74		63	88		Hopei, China	1087
		F	11	68		50	80			
Turdus pilaris	Fieldfare	B	69	106		81	128	F	Netherlands	223
Turdus iliacus	Redwing	B	282	61.2		46	80	F	Netherlands	223
Turdus philomelos	Song Thrush	M	129	68.9	6	56	89	F	Netherlands	223
		F	140	66.6	6.6	52	89			
Turdus mupinensis	Chinese Thrush	M	9	59		55	65		Hopei, China	1087
		F	3	66.3		62	70			
Turdus viscivorus	Mistle Thrush	M	14	112	10.4	96	134			223
		F	5	123		106	140			
Turdus plumbeus	Red-legged Thrush	B	276	77.9	6.5	60.1	100		Puerto Rico	19
Turdus plumbeus	Red-legged Thrush	B	266	72.8	6.6	41	98		Cuba	19
Turdus chiguanco	Chiguanco Thrush	U	10	93.3		75	120		Argentina	363
Turdus nigrescens	Sooty Robin	U		96					Costa Rica	1170
Turdus fuscater	Great Thrush	B	10	143		126	162			881, 1037, 1263, 1268
Turdus infuscatus	Black Robin	B	33	72.4	7	55	86		El Salvador	1041a

Body Masses of World Birds (continued)

Species Name	Common Name	Sex	N	Mean	S.D.	Min	Max	Season	Location	Source #
Turdus serranus	Glossy-backed Thrush	U	10	84.9		76	95		Colombia; Peru; Argentina	798, 1037, 1040, 1288
Turdus nigriceps	Andean Slaty-Thrush	B	11	52.7		50	56.5		Argentina	1040
Turdus subalaris	Eastern Slaty-Thrush	U	3	49.5		46	53		Mato Grosso, Brazil	392
Turdus olivater	Black-hooded Thrush	M	51	84.8	5	70	94.8		Venezuela	1308
		F	24	88.5	7.3	76.8	104			
Turdus reevei	Plumbeous-backed Thrush	M	1	61					Peru	1302
Turdus maranonicus	Maranon Thrush	U	1	70					Southern Ecuador	1311
Turdus fulviventris	Chestnut-bellied Thrush	B	3	70		69	71		Ecuador	953
Turdus rufiventris	Rufous-bellied Thrush	M	13	66.8	5.7				Paraguay	378
		F	10	72.2	5					
Turdus falcklandii	Austral Thrush	U	8	93.9		82	114		Chile; Tierra del Fuego	328, 546
Turdus leucomelas	Pale-breasted Thrush	U	44	69.1	4.9	58	79		Brazil	752
Turdus amaurochalinus	Creamy-bellied Thrush	B	21	57.9	6	49	75		Brazil	752
Turdus plebejus	Mountain Robin	M	16	81.1	9.6				Panama	489
		F	5	91.9						
Turdus ignobilis	Black-billed Thrush	M	15	61		52	66.7		Venezuela	296
		F	12	66.7		60	81			
Turdus lawrencii	Lawrence's Thrush	M	1	72.6					Peru	370
Turdus fumigatus	Cocoa Thrush	U	15	80.5	4.6	72	88		St. Vincent	19
Turdus fumigatus	Cocoa Thrush	B	153	71.2		55.5	83		Trinidad	359
Turdus obsoletus	Pale-vented Thrush	U	10	74.8		58	98		Trinidad	79, 861
Turdus hauxwelli	Hauxwell's Thrush	U	44	69	4.7				Peru	370

Body Masses of World Birds (continued)

Species Name	Common Name	Sex	N	Mean	S.D.	Min	Max	Season	Location	Source #
Turdus grayi	Clay-colored Robin	U	108	79.5	7.8	57.3	100		El Salvador	1041a
Turdus nudigenis	Bare-eyed Thrush	B	88	63.9		55	74.5		Trinidad	1131
Turdus maculirostris	Ecuadorian Thrush	U	6	69.6					Southern Ecuador	1311
Turdus haplochrous	Unicolored Thrush	U		84						192
Turdus jamaicensis	White-eyed Thrush	U	7	59.1					Jamaica	231, 1159
Turdus assimilis	White-throated Thrush	U	31	70.2	5.1	61	80		El Salvador	1041a
Turdus albicollis	White-necked Thrush	U	85	54		45	63		Trinidad	359
Turdus rufopalliatus	Rufous-backed Robin	B	8	76.8		72	85		Jalisco, Mexico	1080
Turdus rufopalliatus graysoni	Rufous-backed Robin	B	22	73.3		63.5	83.7		Tres Marias Is., Mexico	440
Turdus rufitorques	Rufous-collared Robin	B	3	72		70	74		Guatemala	1145
Turdus migratorius	American Robin	B	255	78.5	7.4	56	112	B	Western United States	277
Turdus swalesi	La Selle Thrush	U		75						348
Turdus aurantius	White-chinned Thrush	U	8	81.9					Jamaica	231, 1159
Ixoreus naevius naevius	Varied Thrush	M	60	78.4	6.3	70	100			411
		F	33	80.3	6.1	70.5	99.5			
Ixoreus naevius meruloides	Varied Thrush	M	10	73.1	4.9	65.2	81.2			411
		F	16	78.4	5.1	66.9	85.3			
Ridgwayia pinicola	Aztec Thrush	M	1	88					Mexico	1243
Brachypteryx stellata	Gould's Shortwing	M	1	23					Nepal	1143
Brachypteryx leucophrys	Lesser Shortwing	U	15	15.8		13	18		Malaysia	775
Brachypteryx montana	White-browed Shortwing	M	15	16.9	1.9	13	20		Philippines	436
		F	17	18.5	1.3	16	21			

Body Masses of World Birds (continued)

Species Name	Common Name	Sex	N	Mean	S.D.	Min	Max	Season	Location	Source #
Chaetops frenatus	Rufous Rock-jumper	M	4	57.4		53.5	61		South Africa	532
		F	3	52.7		50	57.5			
Chaetops aurantius	Orange-breasted Rock-jumper	M	3	54.1		52.5	55		South Africa	1258
Chaetops pycnopygius	Damara Rock-jumper	B	8	28.3		24	34			396
Alethe poliocephala compsonota	Brown-chested Alethe	U	64	37.3		32.5	46.2		Cameroon	623
Alethe poliocephala nandensis	Brown-chested Alethe	B	40	28.1		25.5	32.5			623
Alethe poliophrys	Red-throated Alethe	B	34	35.2		30	40			623
Alethe choloensis	Cholo Alethe	U	5	41.3					Mozambique	1030a
Alethe fuelleborni	White-chested Alethe	U	70	52	3.3	44	59		Tanzania	1014
Alethe diademata	White-tailed Alethe	M	38	32.6		28	38			623
		F	36	29.8		22	39			
Alethe castanea	Fire-crested Alethe	U	39	32.3		28.4	37.6		Cameroon	623
Family Cisticolidae										
Cisticola erythrops	Red-faced Cisticola	M	29	15.7		13.7	17.8		Zimbabwe	1245
		F	47	13.9		11.2	15.9			
Cisticola cantans	Singing Cisticola	U	34	11.7		9.7	14		Nigeria	1245
Cisticola lateralis	Whistling Cisticola	M	12	18.9		14	21		Cameroon	710
		F	10	14.9		12	19			
Cisticola anonymus	Chattering Cisticola	M	4	15.5		15	16		Cameroon	1245
		F	2	12.5		12	13			
Cisticola woosnami	Trilling Cisticola	F	2	13.5		13	14		Uganda; Kenya	1245
Cisticola bulliens	Bubbling Cisticola	M	1	17.5					Angola	266
		F	1	14.3						
Cisticola chubbi	Chubb's Cisticola	M	24	17.5		12	21		Kenya; Uganda	1245
		F	22	14.7		11	19			

Body Masses of World Birds (continued)

Species Name	Common Name	Sex	N	Mean	S.D.	Min	Max	Season	Location	Source #
Cisticola hunteri	Hunter's Cisticola	M	13	16.2		14	19			1245
		F	11	14.8		12	18			
Cisticola nigriloris	Black-lored Cisticola	M	2	21.8		21.8	21.8			1245
		F	1	17.7						
Cisticola aberrans	Rock-loving Cisticola	U	17	13.8		11.4	15.7		Nigeria	777
Cisticola aberrans	Rock-loving Cisticola	M	12	15.2		13.1	16.9		Southern Africa	1245
		F	9	13.7		12	15.9			
Cisticola bodessa	Boran Cisticola	M	1	21					Kenya	1245
		F	1	14						
Cisticola chiniana	Rattling Cisticola	M	23	18.1		13.2	21.4		Zimbabwe	1245
		F	8	14.4		12.6	18.5			
Cisticola cinereolus	Ashy Cisticola	M	4	15.2		15	15.5		Kenya	1245
		F	1	19						
Cisticola ruficeps	Red-pate Cisticola	U	26	9.7		8.3	12		Nigeria	1245
Cisticola guinea	Dorst's Cisticola	U	16	9.3		6.9	11.3		Nigeria	777
Cisticola rufilatus	Gray Cisticola	B	23	11.3		8	14.7			1245
Cisticola subruficapilla	Red-headed Cisticola	M	11	10.7		9.5	14.7			526, 1245
		F	12	9.2		8	12			
Cisticola lais lais	Wailing Cisticola	M	3	16		15.2	17.3		South Africa	1245
		F	3	11.7		11.2	12.3			
Cisticola lais distinctus	Wailing Cisticola	M	2	17		17	17		Uganda	1245
		F	3	15		13	16			
Cisticola restrictus	Tana River Cisticola	B	4	13.2		12	15		Kenya	1245
Cisticola njombe	Churring Cisticola	M	4	11.6		11	12.1		Zambia; Malawi	301, 1245
		F	2	10.2		9.5	10.9			
Cisticola galactotes	Winding Cisticola	M	50	16.2	1.8	13	19.1		Kenya	122
		F	41	12.9	1.6	11	18			
Cisticola galactotes lulubris	Winding Cisticola	M	2	14		14	14		Ethiopia	1245

Body Masses of World Birds (continued)

Species Name	Common Name	Sex	N	Mean	S.D.	Min	Max	Season	Location	Source #
Cisticola galactotes haematocephalus	Winding Cisticola	F	1	10						1245
Cisticola pipiens	Chirping Cisticola	M	5	14.1		13.3	15.3			1245
		F	6	12.9		12.2	14.3			
Cisticola carruthersi	Carruthers' Cisticola	M	37	11.7		10.1	14.4		Kenya	1245
		F	34	10.8		9.5	12.4			
Cisticola tinniens	Tinkling Cisticola	U	20	11.4		8	18.5		Southern Africa	1245
Cisticola robustus nuchalis	Stout Cisticola	M	5	19.3		16	22		Kenya; Uganda	1245
		F	2	15		14	16.1			
Cisticola robustus angolensis	Stout Cisticola	B	5	21.8		17	25		Congo	1245
Cisticola natalensis	Croaking Cisticola	M	7	20.8		16	23		Kenya; Uganda	1245
		F	3	13		10	15			
Cisticola fulvicapilla	Piping Cisticola	M	26	8.6		6	10.6			1245
		F	16	7.9		7	9.2			
Cisticola aberdare	Aberdare Cisticola	M	1	24						1245
Cisticola melanurus	Slender-tailed Cisticola	M	1	8						1245
		F	1	9.5						
Cisticola brachypterus brachypterus	Siffling Cisticola	U	21	7.2		6	8.8		Nigeria	1245
Cisticola brachypterus isabellinus	Siffling Cisticola	U	6	9.2		7.9	10.7		Malawi	1245
Cisticola brachypterus loanda	Siffling Cisticola	B	14	7.2		6	9		DR Congo	1245
Cisticola rufus	Rufous Cisticola	B	7	7.3		6	9.1		Ghana	457
Cisticola nanus	Tiny Cisticola	F	1	5					Kenya	1245
Cisticola juncidis	Zitting Cisticola	M	11	7.5	0.5	7	8.2		Australia	518a
		F	13	6.2	0.9	5	8.2			
Cisticola haesitatus	Socotra Cisticola	B	6			6	8		Socotra Is.	1245
Cisticola cherina	Madagascar Cisticola	B	9	10		8.2	11.2		Madagascar	69

Body Masses of World Birds (continued)

Species Name	Common Name	Sex	N	Mean	S.D.	Min	Max	Season	Location	Source #
Cisticola aridulus	Desert Cisticola	M	11	8.3		5	11		Botswana	1245
		F	6	10.2		8.8	13			
Cisticola textrix	Cloud Cisticola	M	5	11.9		11	12.5		South Africa	1245
		F	2	9		8	10			
Cisticola eximius	Black-necked Cisticola	U				11	13			1030
Cisticola dambo	Cloud-scraping Cisticola	M	9	9.3		8	11			1245
		F	4	8.7		7	10			
Cisticola brunnescens	Pectoral-patch Cisticola	M	1	8					Kenya	1245
Cisticola cinnamomeus	Pale-crowned Cisticola	M	1	9.5						1030
		F	3	8.5		8.5	8.5			
Cisticola ayresii	Wing-snapping Cisticola	U	6	7		6	8		Congo	1245
Cisticola exilis	Golden-headed Cisticola	M	5	7.4		6.2	8.3		Australia	470
		F	10	6.7	0.4	6.1	7.4			
Rhopophilus pekinensis	White-browed Chinese Warbler	M	12	18.7		16.6	20		China	1087, 1183
		F	6	17.4		15.7	18.4			
Incana incanus	Socotra Warbler	U				9	12		Socotra Is.	1245
Scotocerca inquieta	Streaked Scrub-Warbler	U	10	8.2	1.2	6	10			223
Prinia criniger criniger	Striated Prinia	U	3	15.1		13.2	17			7
Prinia criniger striatula	Striated Prinia	U	3	12.5		12	13			7
Prinia polychroa	Brown Prinia	U				10	12		Taiwan	834
Prinia atrogularis	Hill Prinia	U	11	11.8		8	16			7
Prinia buchanani	Rufous-fronted Prinia	U	3			5	9			7
Prinia rufescens	Rufescent Prinia	B	8			6	7			7
Prinia hodgsonii	Gray-breasted Prinia	B	8	6.4		5	9			7
Prinia gracilis	Graceful Prinia	B	117	7.1				W	Israel	223

Body Masses of World Birds (continued)

Species Name	Common Name	Sex	N	Mean	S.D.	Min	Max	Season	Location	Source #
Prinia sylvatica	Jungle Prinia	U	17	16.1		10	21		India	7
Prinia flaviventris	Yellow-bellied Prinia	M	1	7						7
Prinia socialis	Ashy Prinia	B	12	8		6.7	11		India	7
Prinia subflava	Tawny-flanked Prinia	M	37	9		6.7	11		Zimbabwe	1245
		F	25	8.3		7	9.6			
Prinia inornata	Plain Prinia	B	9	7		5	9		India	7
Prinia somalica	Pale Prinia	M	1	7.5						1245
Prinia flavicans	Black-chested Prinia	M	14	9.3		8.6	10.9		Zimbabwe; Botswana	1245
		F	10	8.2		6	9.8			
Prinia maculosa	Karoo Prinia	B	13	9		7.4	11.1			1245
Prinia hypoxantha	Drakensberg Prinia	B	12	10.5		9.2	11.3			1245
Prinia substriata	Namaqua Prinia	M	5	11.4		9.5	13.2			526, 1245
		F	5	9.9		9	10.8			
Prinia molleri	Sao Tome Prinia	B	2	8.8		8.5	9		Sao Tome	1245
Prinia robertsi	Roberts' Prinia	M	23	10.3		8.7	11.5			1245
		F	19	9.7		9	11.4			
Prinia leontica	Sierra Leone Prinia	B	5	12.4		11.9	13.6			1245
Prinia leucopogon	White-chinned Prinia	B	12	11.7		9.5	15		Zambia	1245
Prinia bairdii	Banded Prinia	U	15	13.4					Kenya	1245
Prinia erythroptera	Red-winged Prinia	B	12	11.5		10.3	12.5		Nigeria	1245
Malcorus pectoralis	Rufous-eared Warbler	M	12	10.7		10.3	11.8			1245, 1258
		F	4	9.6		9.2	10.5			
Drymocichla incana	Red-winged Gray Warbler	B	2	10		10	10		Uganda	1245
Urolais epichlorus	Green Longtail	B	13	11.3		10	14		Cameroon	1245
Spiloptila clamans	Cricket Longtail	B	3	7		6	8			1245

Body Masses of World Birds (continued)

Species Name	Common Name	Sex	N	Mean	S.D.	Min	Max	Season	Location	Source #
Apalis pulchra	Black-collared Apalis	M	16	9.1		7	12		Kenya	1245
		F	8	7.7		6	10			
Apalis ruwenzori	Ruwenzori Apalis	B	27	9.9		7	13		Uganda	1245
Apalis thoracica	Bar-throated Apalis	B	74	12.1		10.2	16		Malawi	303
Apalis thoracica arnoldi	Bar-throated Apalis	M	70	11		9.3	13.1		Zimbabwe	1245
		F	60	10		8.3	12.5			
Apalis thoracica lynesi	Bar-throated Apalis	B	4	11.4					Mozambique	1030a
Apalis thoracica flavigularis	Bar-throated Apalis	M	1	11.3						1030
		F	1	10.8						
Apalis nigriceps	Black-capped Apalis	B	16	8.4		7	10.2		Uganda	1245
Apalis jacksoni	Black-throated Apalis	B	23	8.7		6	10		Uganda	1245
Apalis binotata	Masked Apalis	B	9	8.2		6	10		Uganda	1245
Apalis personata	Black-faced Apalis	B	17	11		9	12		Uganda	1245
Apalis flavida	Yellow-breasted Apalis	M	22	8.5		6.8	9.5		Zimbabwe	1245
		F	14	7.9		7.1	9			
Apalis ruddi fomosa	Rudd's Apalis	M	8	9.9		9.2	10.5			1245
		F	4	9.5		9.3	9.8			
Apalis sharpii	Sharpe's Apalis	M	14	9.1					Liberia	1245
		F	7	8.3						
Apalis rufogularis	Buff-throated Apalis	M	26	8.7		6	10		Uganda	1245
		F	8	8		7	9			
Apalis porphyrolaema	Chestnut-throated Apalis	M	7	8.8		7	11		Uganda	1245
		F	4	8		7	10			
Apalis chapini	Chapin's Apalis	U	4	8.8						1245
Apalis melanocephala	Black-headed Apalis	B	23	8.8		7.5	10.1			1245
Apalis chirindensis	Chirinda Apalis	B	48	8.6		7.3	10.6			1245

Body Masses of World Birds (continued)

Species Name	Common Name	Sex	N	Mean	S.D.	Min	Max	Season	Location	Source #
Apalis cinerea	Gray Apalis	M	10	10.5		8	12.5		Kenya; Uganda	1245
		F	5	9.3		7	11			
Apalis alticola	Brown-headed Apalis	B	13	12.1		10	15		Congo	1245
Apalis karamojae	Karamoja Apalis	U	17	9.2		8	10.5		Tanzania	1086
Urorhipis rufifrons	Red-faced Warbler	U	8	6.7		5	7.5		Kenya	1245
Hypergerus atriceps	Oriole Warbler	B	12	29.8		24.2	36		Nigeria	777
Eminia lepida	Gray-capped Warbler	B	12	20.2		16	24		Kenya; Congo	1245
Camaroptera brachyura brevicaudata	Green-backed Camaroptera	B	11	12		11	13		Cameroon	710
Camaroptera brachyura	Green-backed Camaroptera	U	65	9.3	0.8	7.8	10.5		Kenya	122
Camaroptera brachyura tincta	Green-backed Camaroptera	M	11	10.7		10.1	12		Liberia	1245
		F	3	8.7		8.1	9.1			
Camaroptera brachyura erlangeri	Green-backed Camaroptera	M	54	9.2		4	20		Kenya	1245
		F	24	7.8		5	15			
Camaroptera brachyura brachyura	Green-backed Camaroptera	M	20	10.8		9.2	12.1		Southern Africa	1245
		F	15	10.1		9	11.5			
Camaroptera superciliaris	Yellow-browed Camaroptera	B	22	9.7		7	12			1245
Camaroptera chloronota	Olive-green Camaroptera	U	15	11.4	1	8.7	12.3		Liberia; Ghana; Gabon	616
Calamonastes undosus	Miombo Camaroptera	U	13	13		11	16		Congo	1245
Calamonastes fasciolatus	Barred Camaroptera	U	19	13.1		11.1	14.3		Transvaal, South Africa	1245
Euryptila subcinnamomea	Kopje Warbler	U	2	11.5		10.5	12.5			1245
Family Sylviidae										
Tesia castaneocoronata	Chestnut-headed Tesia	U				8	10		India	7

Body Masses of World Birds (continued)

Species Name	Common Name	Sex	N	Mean	S.D.	Min	Max	Season	Location	Source #
Tesia olivea	Slaty-bellied Tesia	B	3	7		6	9		India	7
Tesia cyaniventer	Gray-bellied Tesia	B	5	9.7		8.4	12			7
Urosphena squameiceps	Asian Stubtail	B	3	9.3		8	10		Tsingtao, China	1088
Cettia diphone	Japanese Bush-Warbler	M	32	15.7	1.1	14	18.5		Japan	490
		F	9	11.3		10	12.6			
Cettia seebohmi	Philippine Bush-Warbler	M	7	12.2		11	13.8		Luzon, Philippines	1143
Cettia parens	Shade Bush-Warbler	F	1	14					Solomon Is.	399
Cettia ruficapilla	Fiji Bush-Warbler	U	18	12.7	1.6	9	15.5		Fiji	668
Cettia fortipes	Brownish-flanked Bush-Warbler	U	13			8	11.5			7
Cettia flavolivacea	Aberrant Bush-Warbler	B	4	8		6	10		India; China	7, 1088
Cettia acanthizoides	Yellowish-bellied Bush-Warbler	U				4.3	6.6		Taiwan	834
Cettia brunnifrons	Gray-sided Bush-Warbler	B	16			6	9			7
Cettia cetti	Cetti's Warbler	M	464	15.2	1	12.8	18.5		France	223
		F	873	11.5	0.8	8.5	14.5			
Bradypterus baboecala	African Bush-Warbler	B	19	11.8		10.4	13.5			1245
Bradypterus baboecala	African Bush-Warbler	U	83	14		11.2	17			1245
Bradypterus grandis	Ja River Scrub-Warbler	U	1	18.5					Burundi	1143
Bradypterus carpalis	White-winged Scrub-Warbler	M	23	23.5	1.6	18.9	26.2		Kenya	122
		F	18	21.3	1	19	23			
Bradypterus graueri	Grauer's Scrub-Warbler	M	9	17.3		16	19		Uganda	1245
		F	6	16.7		15	18			
Bradypterus alfredi	Bamboo Scrub-Warbler	F	1	17.5				B	Uganda	1143

Body Masses of World Birds (continued)

Species Name	Common Name	Sex	N	Mean	S.D.	Min	Max	Season	Location	Source #
Bradypterus lopezi lopezi	Cameroon Scrub-Warbler	U	12	17.7		16	20		Bioko	1245
Bradypterus lopezi mariae	Cameroon Scrub-Warbler	U	21	18.9		15.1	20.8		Tanzania	1245
Bradypterus lopezi bangwaensis	Bangwa Scrub-Warbler	M	5	21		18	27		Cameroon	1245
Bradypterus barratti	African Scrub-Warbler	U	36	18.8		16.1	22.2		Southern Africa	1245
Bradypterus cinnamomeus	Cinnamon Bracken-Warbler	M	23	17.8		15	20		Kenya	1245
		F	7	17		10	25			
Bradypterus victorini	Victorin's Scrub-Warbler	U	1	10					South Africa	1245
Bradypterus thoracicus	Spotted Bush-Warbler	M	3	10.8		9	13.4		India; China	7, 1087, 1183
Bradypterus tacsanowskius	Chinese Bush-Warbler	U	1	14					Northern China	925
Bradypterus seebohmi	Russet Bush-Warbler	U		11.3					Taiwan	834
Bradypterus luteoventris	Brown Bush-Warbler	M	1	13.2					China	1183
Bradypterus palliseri	Ceylon Bush-Warbler	M	1	9					Sri Lanka	7
Bradypterus caudatus	Long-tailed Bush-Warbler	B	7	22.9		21	28		Mindanao & Luzon, Philippines	1143
Bradypterus castaneus	Chestnut-backed Bush-Warbler	U	1	19					Sulawesi, Indonesia	1272
Dromaeocercus brunneus	Brown Emu-tail	U	2	14.2		14	14.5		Madagascar	957
Dromaeocercus seebohmi	Gray Emu-tail	M	3	12.3		11.2	13.2		Madagascar	69
Bathmocercus cerviniventris	Black-capped Rufous-Warbler	U	7	15.8					Liberia	1245
Bathmocercus rufus	Black-faced Rufous-Warbler	M	54	16.6		10	21		South Africa	1245
		F	25	14.3		7	20			

Body Masses of World Birds (continued)

Species Name	Common Name	Sex	N	Mean	S.D.	Min	Max	Season	Location	Source #
Bathmocercus winifredae	Mrs. Moreau's Warbler	M	2	20		20	20		Tanzania	1245
		F	1	11						
Nesillas aldabrana	Aldabra Brush-Warbler	B	4	19.2		18	19.7		Aldabra Is.	948
Nesillas longicaudata	Anjouan Brush-Warbler	U	2	19.2		19	19.5		Anjouan	709
Nesillas typica	Madagascar Brush-Warbler	U	100	18.1	1.9	14	22		Madagascar	957
Nesillas typica moheliensis	Madagascar Brush-Warbler	U	9	21.4		19	24.2		Moheli, Comoro Is.	709
Nesillas lantzii	Subdesert Brush-Warbler	U	11	17.5	1.5	15	20		Madagascar	957
Nesillas brevicaudata	Grand Comoro Brush-Warbler	U	17	16.1		14	19		Grand Comoro	709
Nesillas mariae	Moheli Brush-Warbler	U		16					approximate	706
Thamnornis chloropetoides	Thamnornis	U	4	14.3		14	15		Madagascar	957
Melocichla mentalis	Moustached Grass-Warbler	B	12	33.2		30	37		Ghana	1245
Sphenoeacus afer	Cape Grassbird	B	9	30.4		27	33.7		Zimbabwe; Mozambique; South Africa	1245
Locustella lanceolata	Lanceolated Warbler	M	11	10.6					Malaysia	853
Locustella naevia	Grasshopper Warbler	U	50	13.3		11.4	15.2	B	Great Britain	515
Locustella certhiola	Pallas' Warbler	U	136	14.4	1.5			W	Malaysia	853
Locustella ochotensis	Middendorff's Warbler	U	19	18.5		16	24			276
Locustella fluviatilis	Eurasian River Warbler	B	500	16.1	1.1	13.3	21.2	W	Kenya	223
Locustella luscinioides	Savi's Warbler	U	87	13.8		11.2	20.5		Senegal	1245

Body Masses of World Birds (continued)

Species Name	Common Name	Sex	N	Mean	S.D.	Min	Max	Season	Location	Source #
Locustella luscinioides	Savi's Warbler	U	54	14		11	18.5	F	Sudan	22
Locustella fasciolata	Gray's Warbler	M	24	27.6		24	31		former USSR	223
		F	4	26.7		26	28			
Acrocephalus melanopogon	Moustached Warbler	U	33	10.9		9	13	B	Hungary	1245
Acrocephalus paludicola	Aquatic Warbler	U	47	11.6		9.8	15.5	F	Great Britain	515
Acrocephalus schoenobaenus	Sedge Warbler	U	232	11.2		8.1	17.9		Wales, UK	142
Acrocephalus schoenobaenus	Sedge Warbler	U	451	12.5		8.5	16.5	F	Sudan	22
Acrocephalus bistrigiceps	Black-browed Reed-Warbler	M	16	9.4		8	11			7
		F	8	8		7	10			
Acrocephalus agricola	Paddyfield Warbler	U	11	9.6		8	11			7
Acrocephalus concinens	Blunt-winged Warbler	U	4	8.3		7.7	8.9		Thailand	787
Acrocephalus scirpaceus	Eurasian Reed-Warbler	U	331	12.3	2.7	8	19.7	Sp	Nigeria	304
Acrocephalus baeticatus	African Reed-Warbler	U	387	10.2		7	15.5		Transvaal, South Africa	1245
Acrocephalus dumetorum	Blyth's Reed-Warbler	U	25	11.2		8	16			7
Acrocephalus palustris	Marsh Warbler	U	1200	11.5		9	17.4		Sudan	1245
Acrocephalus arundinaceus	Great Reed-Warbler	B	159	30		23	45.5		Southern Africa	1036
Acrocephalus orientalis	Oriental Reed-Warbler	U	13	25.5		22	29			7
Acrocephalus stentoreus	Clamorous Reed-Warbler	U	13	24.2		21.5	28.5	W	Israel	1245
Acrocephalus griseldis	Basra Reed-Warbler	U	440	16.7		13	24		Kenya	1245
Acrocephalus australis	Australian Reed-Warbler	U	142	17.7	1.7	13.8	22.8		Victoria, Australia	518a

Body Masses of World Birds (continued)

Species Name	Common Name	Sex	N	Mean	S.D.	Min	Max	Season	Location	Source #
Acrocephalus luscinia	Nightingale Reed-Warbler	B	5	29.8		27	31		Guam	35. 1151a
Acrocephalus luscinia	Nightingale Reed-Warbler	B	9	26.3		24	28		Pagan Is.	960a
Acrocephalus luscinia	Nightingale Reed-Warbler	M	5	36.4		35	38		Saipan	960a
Acrocephalus familiaris	Millerbird	B	56	18.3		15	21.5		Nihoa, Hawaiian Is.	811
Acrocephalus aequinoctialis	Christmas Is.Warbler	M	2	25		25	25		Christmas Is.	1143
Acrocephalus rimitarae	Rimitara Reed-Warbler	B	10	26.5	1.8	23	29		Rimitara Is.	1210a
Acrocephalus vaughani	Pitcairn Reed-Warbler	M	1	27					Pitcairn Is.	1309
		F	1	22						
Acrocephalus taiti	Henderson Is.Reed-Warbler	M	21	25.8	2	22	30.5		Henderson Is.	446
		F	19	22.8	1.1	21	25			
Acrocephalus kerearako	Cook Is.Reed-Warbler	B	40	22.5	2.3	17.3	28		Cook Is.	1153b
Acrocephalus rufescens	Greater Swamp-Warbler	M	30	23.9	2	21.7	27.2		Kenya	122
		F	36	21.6	1.9	18	25			
Acrocephalus brevipennis	Cape Verde Swamp-Warbler	F	3	16		15	17		Cape Verde Is.	223
Acrocephalus gracilirostris	Lesser Swamp-Warbler	U	9	14.4		12.5	16.2		Kenya	122
Acrocephalus newtoni	Madagascar Swamp-Warbler	B	4	17		15.5	19.5		Madagascar	69, 957
Acrocephalus aedon	Thick-billed Warbler	B	29	22.4		15	31		Eastern China	1087
Acrocephalus rodericanus	Rodrigues Brush-Warbler	U	2	11.8		11.4	12.2		Rodrigues Is.	183
Acrocephalus sechellensis	Seychelles Brush-Warbler	M	43	16.8	0.8				Seychelles	183
		F	25	15	1					
Hippolais caligata	Booted Warbler	U	34	8.9	0.6	7.7	10.8	M	Siberia	223
Hippolais rama	Sykes' Warbler	B	14	9.3		8	11			7

Body Masses of World Birds (continued)

Species Name	Common Name	Sex	N	Mean	S.D.	Min	Max	Season	Location	Source #
Hippolais pallida	Olivaceous Warbler	U	307	9	0.8	7	11.2	W	Kenya	223
Hippolais languida	Upcher's Warbler	U	241	12	0.8	9.9	14.2	W	Kenya	223
Hippolais olivetorum	Olive-tree Warbler	U	270	17.5	2	13.7	23.1	W	Kenya	223
Hippolais polyglotta	Melodious Warbler	U	54	11	1			Sp	Gibraltar	361
Hippolais icterina	Icterine Warbler	U	211	13.2	1.4	11.1	18.3	B	Kalinigrad (former USSR)	223
Chloropeta natalensis	African Yellow Warbler	U	68	11.7		9	13.6			1245
Chloropeta similis	Mountain Yellow Warbler	B	23	11.1		9	15			1245
Chloropeta gracilirostris	Papyrus Yellow Warbler	U	13	10.8	0.7	10	12.5		Kenya	122
Stenostira scita	Fairy Warbler	B	10	5.9	0.6	5.3	6.7		South Africa; Namibia	509, 526, 1258
Phyllolais pulchella	Buff-bellied Warbler	B	2	6		6	6			1245
Orthotomus metopias	African Tailorbird	B	27	8.4		7.5	9.3		Tanzania	1245
Orthotomus moreaui	Long-billed Tailorbird	B	8	9.1		8.3	10			1245
Orthotomus cuculatus	Mountain Tailorbird	B	6	5.9		5.5	7.1		India; Malaysia	7, 785
Orthotomus sutorius	Common Tailorbird	U	10	7.5		6	10		India	7
Orthotomus heterolaemus	Rufous-headed Tailorbird	B	9	8.3		6.6	9.5		Mindanao, Philippines	1143
Orthotomus atrogularis	Dark necked Tailorbird	M	15	8.1		6.8	8.7		Philippines	954
		F	13	7.3		6.2	8.2			
Orthotomus castaneiceps	Philippine Tailorbird	U	8			11	15		Philippines	344
Orthotomus frontalis	Rufous-fronted Tailorbird	M	39	8.4	0.6	6.8	9.6		Philippines	1143
		F	33	7.6	0.9	6	9.7			

Body Masses of World Birds (continued)

Species Name	Common Name	Sex	N	Mean	S.D.	Min	Max	Season	Location	Source #
Orthotomus derbianus	Gray-backed Tailorbird	B	10	10.7		9	12.4		Philippines	436
Orthotomus sericeus	Rufous-tailed Tailorbird	F	1	10.8					Borneo	1218
Orthotomus sepium	Olive-backed Tailorbird	B	5	8.4		7.8	8.8		Borneo	1218
Orthotomus samarensis	Yellow-breasted Tailorbird	M	5	10.1		10	10.2		Philippines	954
		F	2	7.9		7.4	8.4			
Orthotomus nigriceps	White-browed Tailorbird	B	4	9.8		8	10.9		Mindanao, Philippines	1143
Orthotomus cinereiceps	White-eared Tailorbird	B	2	9		8.8	9.3		Philippines	954
Poliolais lopezi	White-tailed Warbler	B	9	12		10.8	13.4			710, 1245
Graueria vittata	Grauer's Warbler	M	6	15.8		14.5	18			1245
		F	4	15.3		14	17			
Eremomela flavicrissalis	Yellow-vented Eremomela	F	1	4.5					Kenya	1245
Eremomela icteropygialis	Yellow-bellied Eremomela	M	10	7.3		6	8.1		Zimbabwe	1245
		F	7	7.8		7	9			
Eremomela pusilla	Senegal Eremomela	U	38	6.2		5	8.4		Ghana	1245
Eremomela canescens	Green-backed Eremomela	M	1	6					Kenya; Uganda	1245
		F	1	8						
Eremomela scotops	Greencap Eremomela	M	21	8.8		6	11		Congo	1245
		F	12	9.2		8	11			
Eremomela gregalis damarensis	Yellow-rumped Eremomela	B	4	6.8		6	7.8		Namibia	526
Eremomela badiceps	Rufous-crowned Eremomela	M	10	10.3		9	11		Uganda	1245
		F	4	10.9		10	11.5			
Eremomela turneri	Turner's Eremomela	M	1	6					Kenya	1245
		F	1	9						

Body Masses of World Birds (continued)

Species Name	Common Name	Sex	N	Mean	S.D.	Min	Max	Season	Location	Source #
Eremomela atricollis	Black-necked Eremomela	B	14	10.2		9	12		Congo; Angola	266, 1245
Eremomela usticollis	Burnt-neck Eremomela	M	23	8.6		7.7	10		Zimbabwe	1245
		F	16	8.2		6.5	9.6			
Newtonia amphichroa	Dark Newtonia	U	51	12.3	1.8	9	18		Madagascar	957
Newtonia brunneicauda	Common Newtonia	U	99	10.1	1.4	7	14.5		Madagascar	957
Newtonia archboldi	Archbold's Newtonia	M	2	7.8		7.2	8.3		Madagascar	957
Newtonia fanovanae	Red-tailed Newtonia	M	2	12.7		12.5	12.9		Madagascar	957
Cryptosylvicola randriansoloi	Cryptic Warbler	U	9	8.6		7.3	9.8		Madagascar	957
Sylvietta virens	Green Crombec	B	23	8.4		5	10		Uganda	1245
Sylvietta denti	Lemon-bellied Crombec	B	2	8.4		8.1	8.7			1245
Sylvietta leucophrys	White-browed Crombec	M	16	11.7		10	13		Uganda	1245
		F	12	11		8	13.5			
Sylvietta brachyura	Northern Crombec	U	85	8.1		6.5	9.6		Ethiopia	1245
Sylvietta philippae	Short-billed Crombec	B	4	8.6		7.4	10		Somalia	1245
Sylvietta ruficapilla	Red-capped Crombec	B	8	11.1		10	12		Congo	1245
Sylvietta whytii	Red-faced Crombec	B	31	9.9		8	11.5		Zimbabwe	1245
Sylvietta isabellina	Somali Crombec	B	2	10		10	10		Kenya	1245
Sylvietta rufescens	Cape Crombec	U	76	11.6		8	23		Southern Africa	1245
Hemitesia neumanni	Neumann's Warbler	B	6	14.2		11.3	17			81, 1245
Macrosphenus kempi	Kemp's Longbill	M	15	13.4		10	14.4			1245
		F	4	11.9		9.8	14.8			

Body Masses of World Birds (continued)

Species Name	Common Name	Sex	N	Mean	S.D.	Min	Max	Season	Location	Source #
Macrosphenus flavicans	Yellow Longbill	M	18	14.3		12	16			1245
		F	6	12.5		11	14			
Macrosphenus concolor	Gray Longbill	B	15	13.9					Liberia	1245
Macrosphenus kretschmeri	Kretschmer's Longbill	B	3	20		17.5	22			1245
Hylia prasina	Green Hylia	M	48	14.1		10	16.5		Uganda	1245
		F	30	12.1		10	15			
Leptopoecile sophiae	White-browed Tit-Warbler	B	25	6.9	0.7	5.8	7.9	Y	China	1183
Leptopoecile elegans	Crested Tit-Warbler	U	9			5.6	6.9		China	183
Phylloscopus laetus	Red-faced Woodland-Warbler	M	17	9.9		9	11			1245
		F	11	9.2		8	10.5			
Phylloscopus laurae	Laura's Wood-Warbler	U	16	8.5		7.7	9.5		Zambia	1245
Phylloscopus ruficapillus	Yellow-throated Wood-Warbler	U	94	7.7		6.2	9.5		Southern Africa	1245
Phylloscopus budongoensis	Uganda Wood-Warbler	U	9	8.2		7	9		Kenya	1245
Phylloscopus umbrovirens	Brown Woodland-Warbler	U	10	8.6		6.8	10			317, 1245
Phylloscopus herberti	Black-capped Woodland-Warbler	B	9	8.9		8.5	9.5			1245
Phylloscopus trochilus	Willow Warbler	U	4296	8.7	0.9	6.1	13.6	F	Sweden	223
Phylloscopus collybita	Common Chiffchaff	U	383	8.3	1	6.2	11.5	F	Netherlands	223
Phylloscopus brehmii	Iberian Chiffchaff	U		7.7						941
Phylloscopus sindianus	Mountain Chiffchaff	U	45	6.8	0.67	5.7	8.4	F	Kashmir	223
Phylloscopus sindianus lorenzii	Mountain Chiffchaff	U		8.3						941

Body Masses of World Birds (continued)

Species Name	Common Name	Sex	N	Mean	S.D.	Min	Max	Season	Location	Source #
Phylloscopus neglectus	Plain Leaf-Warbler	B	8	5.3		5	6.6		Afghanistan; Iran	223
Phylloscopus bonelli	Western Bonelli's Warbler	M	209	8.5		7	11.5	B	Germany	223
		F	8	8.9		8	11			
Phylloscopus orientalis	Eastern Bonelli's Warbler	U		8.5						941
Phylloscopus sibilatrix	Wood Warbler	U	105	9.2		6.4	14	Sp	Cyprus	223
Phylloscopus fuscatus	Dusky Warbler	M	18	9		6	11		Hopei, China	1087
		F	19	8.5		7	10			
Phylloscopus fuligiventer	Smoky Warbler	U				6.5	12			33
Phylloscopus affinis	Tickell's Leaf-Warbler	B	17			6.3	7.7			7
Phylloscopus subaffinis	Buff-throated Warbler	M	1	6.2					Thailand	787
Phylloscopus griseolus	Sulphur-bellied Warbler	U	6	7.6		7	9			7
Phylloscopus armandii	Yellow-streaked Warbler	B	6	9.4		9	9.9		Thailand; China	787, 1087
Phylloscopus schwarzi	Radde's Warbler	M	15	13		9	17		Hopei, China	1087
		F	11	10		8	12			
Phylloscopus pulcher	Buff-barred Warbler	U	10	6.8		5.4	8.1		India	7
Phylloscopus maculipennis	Ashy-throated Warbler	B	18	5.1	0.5	4.5	6		India	7
Phylloscopus chloronotus	Pale-rumped Warbler	U		5.1						941
Phylloscopus proregulus	Lemon-rumped Warbler	B	50	6		5	7		Hopei, China	1087
Phylloscopus sichuanensis	Chinese Leaf-Warbler	U	1	3.5					China	8
Phylloscopus subviridis	Brook's Leaf-Warbler	B	8			5	6		India	7
Phylloscopus inornatus	Inornate Warbler	U	23	6.4		5.4	7		Great Britain	223

Body Masses of World Birds (continued)

Species Name	Common Name	Sex	N	Mean	S.D.	Min	Max	Season	Location	Source #
Phylloscopus inornatus	Inornate Warbler	U	10	6.8		6.6	7.5		Northern China	925
Phylloscopus humei	Hume's Warbler	B	122	6		5.5	9		Hopei, China	1087
Phylloscopus borealis	Arctic Warbler	B	70	10.6		8	15			7
Phylloscopus borealis	Arctic Warbler	U	12	11.2	0.8	10.1	12.5	F	Japan	490
Phylloscopus trochiloides	Greenish Warbler	U	116	7.1	0.5	5.8	8.7		Siberia	223
Phylloscopus trochiloides nitidus	Greenish Warbler	M	28	8.6		8	11		Hopei, China	1087
		F	17	8		6	10			
Phylloscopus trochiloides plumbeitarsus	Greenish Warbler	U		8.6						941
Phylloscopus tenellipes	Pale-legged Leaf-Warbler	M	2	11.8		10.2	13.5			273, 490
Phylloscopus borealoides	Sakhalin Leaf-Warbler	U		9.8						941
Phylloscopus magnirostris	Large-billed Leaf-Warbler	M	2	11.6		11.3	12			7
Phylloscopus tytleri	Tytler's Leaf-Warbler	U	30	7.2	0.5				India	942
Phylloscopus occipitalis	Western Crowned Leaf-Warbler	B	22	9		8	11		Hopei, China	1087
Phylloscopus coronatus	Eastern Crowned Leaf-Warbler	B	13	10.1		9	12		Hopei, China	1087
Phylloscopus ijimae	Ijima's Leaf-Warbler	U		8.7						941
Phylloscopus reguloides	Blyth's Leaf-Warbler	B	45			6.2	9.4			7
Phylloscopus davisoni	White-tailed Leaf-Warbler	U	9	6.4		5.7	8.1		Thailand	787
Phylloscopus cantator	Yellow-vented Warbler	U	3	6		5	7			7
Phylloscopus cebuensis	Lemon-throated Warbler	B	12	9		7.9	10.5		Philippines	436

Body Masses of World Birds (continued)

Species Name	Common Name	Sex	N	Mean	S.D.	Min	Max	Season	Location	Source #
Phylloscopus trivirgatus	Mountain Warbler	B	20	9		8.1	10		Philippines	954
Phylloscopus poliocephalus becki	Is.Leaf-Warbler	M	2	10.8		10	11.5		Guadalcanal	399
Phylloscopus olivaceus	Philippine Leaf-Warbler	M	12	10.1		9.3	11.6		Philippines	954
		F	7	9.3		8.6	10.2			
Phylloscopus makirensis	San Cristobal Leaf-Warbler	B	4	9.2		9	9.5		Solomon Is.	399
Seicercus burkii	Golden-spectacled Warbler	B	13	7.3		4	9.3			7
Seicercus xanthoschistos	Gray-hooded Warbler	M	27			6	8.5			7
		F	10			6	7.6			
Seicercus affinis	White-spectacled Warbler	U	6			6	8	W		7
Seicercus poliogenys	Gray-cheeked Warbler	B	3	6.3		6	7			7
Seicercus castaniceps	Chestnut-crowned Warbler	U	6	5.3		4	6			7
Seicercus montis	Yellow-breasted Warbler	U	11	6.4		5	7		Malaysia	775, 785
Seicercus albogularis	Rufous-faced Warbler	U				4.7	5		Taiwan	834
Abroscopus superciliaris	Yellow-bellied Warbler	M	3			6	7		India	7
Abroscopus schisticeps	Black-faced Warbler	U	3	4.7		4	6		India	7
Tickellia hodgsoni	Broad-billed Warbler	U	2	4.5		4	5		India	7
Hyliota flavigaster	Yellow-bellied Hyliota	B	11	12.5		11.5	14		Angola	1245
Hyliota australis	Southern Hyliota	B	14	11.2		9.5	14		Angola	266, 1245
Hyliota usambarae	Usambara Hyliota	M	3	10.7		10.2	11.2			1245
		F	1	9.5						

Body Masses of World Birds (continued)

Species Name	Common Name	Sex	N	Mean	S.D.	Min	Max	Season	Location	Source #
Hyliota violacea	Violet-backed Hyliota	B	9	15.8						1245
Megalurus pryeri	Marsh Grassbird	M				13.6	15.6		Japan	33
		F				11.1	17.2			
Megalurus timoriensis	Tawny Grassbird	B	14	30.9	6.2	21	40		Philippines; New Guinea	283, 424, 425, 954
Megalurus timoriensis	Tawny Grassbird	M	17	19.4	2.4	15.6	25.5		Australia	518a
		F	8	16.7		13.5	20			
Megalurus gramineus	Little Grassbird	M	41	13.9	2.3				Southern Australia	778
		F	24	14.3	0.9					
Megalurus palustris	Striated Grassbird	M	13	53		48	56.8		Philippines	954
		F	12	36.7		32.4	41.8			
Megalurus punctatus	Fernbird	M	10	26.4	2.1				New Zealand	518a
		F	7	24						
Cincloramphus cruralis	Brown Songlark	M	21	74.8	4	66.5	81.3		New South Wales, Australia	735
		F	141	32.4	1.9	28.4	36.1			
Cincloramphus mathewsi	Rufous Songlark	M	25	37.6	3	32.7	44		Australia	518a
		F	42	22.9	2.7	18.9	31			
Eremiornis carteri	Spinifexbird	U	14	12.2					Western Australia	1341
Megalurulus mariei	New Caledonian Grassbird	B	14	25	2.7	20	29.2		New Caledonia	1018
Megalurulus llaneae	Bougainville Thicketbird	B	2	25.5		24	27		Bougainville Is.	980
Megalurulus whitneyi	Guadalcanal Thicketbird	M	1	36					Guadalcanal	399
Megalurulus rubiginosus	Rusty Thicketbird	B	2	41.2		39.5	43		New Britain	426
Graminicola bengalensis	Rufous-rumped Grassbird	B	6			13	16.5		India	7
Schoenicola brevirostris	Fan-tailed Grassbird	M	5	15.7		14	17		Uganda	1245
Chamaea fasciata	Wrentit	M	72	15.3	1.1	12.4	18		California, USA	413
		F	43	14.2	1.1	12.4	17			

Body Masses of World Birds (continued)

Species Name	Common Name	Sex	N	Mean	S.D.	Min	Max	Season	Location	Source #
Sylvia atricapilla	Blackcap	U	1518	16.7		8.5	29	Sp	Cyprus	1096
Sylvia borin	Garden Warbler	U	518	18.2		13.5	24.7	F	Kenya	1096
Sylvia communis	Greater Whitethroat	U	1776	15.1	2.1	10.9	24.6	Sp	Kenya	223
Sylvia curruca	Lesser Whitethroat	U	664	11.1		6.8	18	Sp	Cyprus	1096
Sylvia curruca	Lesser Whitethroat	U	845	11.8		8	15.5	F	Sudan	22
Sylvia margelanica	Margelanic Lesser Whitethroat	U				12	16.7	F	China	1096
Sylvia althaea	Hume's Whitethroat	U	74	14.9		11.2	16.3	PB	Kashmir	1096
Sylvia nana	Desert Warbler	U	14	8.7		7	10.5		India	7
Sylvia nisoria	Barred Warbler	U	618	22.5		18	28.8	W	Kenya	1096
Sylvia hortensis	Orphean Warbler	U	322	21.9	1.9	16.2	30.5	F	France	223
Sylvia leucomelaena	Red Sea Warbler	U	29	13.9		12.5	16	F	Sudan	223
Sylvia rueppelli	Rueppell's Warbler	M	56	13.1				Sp	Israel	1096
		F	33	12.6						
Sylvia cantillans	Subalpine Warbler	U	328	9.6		7	13.2	F	Senegal	1096
Sylvia melanocephala	Sardinian Warbler	U	485	11.7		9.6	20		Morocco; Algeria; Niger	1245
Sylvia melanothorax	Cyprus Warbler	B	93	11.4		9.4	15		Cyprus	223
Sylvia mystacea	Menetries' Warbler	U	30	9.9		8	11.5			133
Sylvia conspicillata	Spectacled Warbler	U	17	9		7.8	13.1	Sp	France	1096
Sylvia deserticola	Tristram's Warbler	M	5	9.1		8.5	10	F	Algeria	1096
		F	3	8.1		7.7	8.5			
Sylvia undata	Dartford Warbler	U	52	10.8		9.7	11.8	Y	Great Britain	515

Body Masses of World Birds (continued)

Species Name	Common Name	Sex	N	Mean	S.D.	Min	Max	Season	Location	Source #
Sylvia sarda	Marmora's Warbler	U	12	9.8		7.5	13.5	F	Sardinia; Algeria; Spain	1096
Sylvia layardi	Layard's Warbler	B	12	14.3		13	16	Y	South Africa	1096, 1258
Parisoma subcaeruleum	Rufous-vented Warbler	U	84	14.6		11.6	18.6	Y	South Africa	1096
Parisoma lugens	Brown Warbler	B	18	14.9		12	18		Ethiopia; Kenya	1245
Family Polioptilidae										
Microbates collaris	Collared Gnatwren	U	10	10.2		9	11.5		French Guiana; Venezuela	293, 1308
Microbates cinereiventris	Tawny-faced Gnatwren	U	21	11.9	0.8				Panama; Costa Rica	618
Ramphocaenus melanurus	Long-billed Gnatwren	U	17	9.7	0.9				Panama	618
Polioptila caerulea	Blue-gray Gnatcatcher	U	40	5.8	0.3	5.3	6.6		Pennsylvania, USA	828
Polioptila lembeyei	Cuban Gnatcatcher	B	13	5	0.4	4.5	5.5		Cuba	19
Polioptila californica	California Gnatcatcher	M	44	6.1	0.3	5.2	7	B	California, USA	25
		F	25	5.9	0.3	5.3	6.4			
Polioptila californica	California Gnatcatcher	B	32	5.5		4.7	6.8		Baja California Sur, Mexico	25
Polioptila melanura	Black-tailed Gnatcatcher	B	139	5.2		4.2	6.6			354
Polioptila nigriceps	Black-capped Gnatcatcher	U	9	5.6		4.1	6.6		Western Mexico	1150, 1151
Polioptila albiloris albiventris	White-lored Gnatcatcher	B	9	5.6		5	6.2		Yucatan, Mexico	910a
Polioptila albiloris vanrossemi	White-lored Gnatcatcher	B	6	7		6.5	8.1		Yucatan, Mexico	910a
Polioptila guianensis guianensis	Guianan Gnatcatcher	B	4	6.2		5.4	6.9			1296

Body Masses of World Birds (continued)

Species Name	Common Name	Sex	N	Mean	S.D.	Min	Max	Season	Location	Source #
Polioptila guianensis facilis	Guianan (Rio Negro) Gnatcatcher	M	3	5.9		5.5	6.5			1296
Polioptila guianensis paraensis	Guianan (Para) Gnatcatcher	U	2	5.8		5.8	5.9			1296
Polioptila clementsi	Iquitos Gnatcatcher	M	3	5.6		5.1	6		Peru	1296
Polioptila plumbea	Tropical Gnatcatcher	B	25	6.1	0.7	4.8	7			75, 156, 489, 496, 795, 876, 910a, 1027, 1180, 1288
Polioptila lactea	Creamy-bellied Gnatcatcher	U	7	6.5					Argentina; Brazil	64, 254
Polioptila schistaceigula	Slate-throated Gnatcatcher	F	1	6						1296
Polioptila dumicola	Masked Gnatcatcher	B	14	7		5.8	9.5		Brazil; Argentina; Bolivia	752, 1040, 1055
Family Muscicapidae										
Empidornis semipartitus	Silverbird	M	2	22.5		22	23		Uganda	1245
Bradornis pallidus	Pale Flycatcher	B	20	23.6		20	28		Ethiopia	1245
Bradornis infuscatus	Chat Flycatcher	B	7	32.4		27.5	37			266, 526, 1245
Bradornis mariquensis	Mariqua Flycatcher	M	11	24.3		23	27			1245
		F	9	26.1		22	36			
Bradornis microrhynchus neumanni	African Gray Flycatcher	M	12	19.5		17	22		Ethiopia	1245
		F	6	18.5		17	20			
Bradornis microrhynchus pumilus	African Gray Flycatcher	M	4	20.5		20	22		Ethiopia	1245
Melaenornis fischeri	White-eyed Slaty-Flycatcher	B	28	23.8		16	33			1245
Melaenornis chocolatinus	Abyssinian Slaty-Flycatcher	U	8	23.4		22	27		Ethiopia	1238

Body Masses of World Birds (continued)

Species Name	Common Name	Sex	N	Mean	S.D.	Min	Max	Season	Location	Source #
Melaenornis edolioides	Northern Black-Flycatcher	U	11	30.7		21.6	35.7		Nigeria	777
Melaenornis pammelaina	Southern Black-Flycatcher	U	10	30.2		24.2	32.4		Southern Africa	1245, 1258
Melaenornis ardesiacus	Yellow-eyed Black-Flycatcher	B	16	29		26	35		Uganda	1245
Melaenornis annamarulae	Nimba Flycatcher	B	11	39.9		37	42.3			1245
Fraseria ocreata	African Forest-Flycatcher	B	22	35.5		28	42		Gabon; Uganda	1245
Fraseria cinerascens	White-browed Forest-Flycatcher	B	9	23.1		21	24		Gabon	1245
Sigelus silens	Fiscal Flycatcher	B	40	25.7		22.5	37		South Africa	509
Rhinomyias additus	Buru Jungle-Flycatcher	U				22	23		Buru, Moloccas	1182
Rhinomyias umbratilis	Gray-chested Jungle-Flycatcher	U	47	18.2	1.3	15	21		Borneo	398
Rhinomyias olivaceus	Fulvous-chested Jungle-Flycatcher	U	1	15.7					Malaysia	1148
Rhinomyias ruficauda	Chestnut-tailed Jungle-Flycatcher	B	19	17.3		15.1	20.8		Philippines	954
Rhinomyias gularis	Eyebrowed Jungle-Flycatcher	B	6	25.1		21.2	29.8		Myanmar	1148
Rhinomyias insignis	Rusty-flanked Jungle-Flycatcher	U	14	28.7	3.2	23.5	32		Luzon, Philippines	1143
Rhinomyias goodfellowi	Mindanao Jungle-Flycatcher	B	11	31.7	2.5	27.2	36		Mindanao, Philippines	1143
Muscicapa striata	Spotted Flycatcher	U	595	15.9	1.1	13.1	20.8	F	Sweden	223
Muscicapa gambagae	Gambaga Flycatcher	B	2	13		12	14		Kenya; Ethiopia	1245
Muscicapa griseisticta	Gray-streaked Flycatcher	B	9	15.9		14.4	17.4		Philippines	436, 954
Muscicapa sibirica	Dark-sided Flycatcher	B	61	13.2		10	16		Hopei, China	1087

Body Masses of World Birds (continued)

Species Name	Common Name	Sex	N	Mean	S.D.	Min	Max	Season	Location	Source #
Muscicapa dauurica	Asian Brown Flycatcher	U	13	9.8	0.9				Malaysia	853
Muscicapa ruficauda	Rusty-tailed Flycatcher	U	13	13.3		12	16	B		7
Muscicapa muttui	Brown-breasted Flycatcher	U	13	12		10	14	W	India	7
Muscicapa ferruginea	Ferruginous Flycatcher	F	1	12						7
Muscicapa ussheri	Ussher's Flycatcher	M	3	17.5		16.3	19		Liberia	1245
		F	3	18.2		16.8	19.8			
Muscicapa infuscata	Sooty Flycatcher	B	16	17.7		15	20		Uganda	1245
Muscicapa aquatica	Swamp Flycatcher	M	3	11.5		10	12.5		Eastern Africa	1245
Muscicapa olivascens	Olivaceous Flycatcher	M	8	17.6					Liberia	1245
		F	3	13.5						
Muscicapa adusta	African Dusky Flycatcher	B	6	10.4		9.5	11.1			301, 563, 710, 1014
Muscicapa epulata	Little Gray Flycatcher	B	4	9.6		8.8	10.7		Liberia	1245
Muscicapa sethsmithi	Yellow-footed Flycatcher	B	14	9.1		7.9	10		Cameroon; Gabon; Uganda	1245
Muscicapa comitata	Dusky-blue Flycatcher	B	18	14.1		12	16		Gabon; Uganda	1245
Muscicapa cassini	Cassin's Flycatcher	B	3	17.3		16	19		Congo; Uganda	1245
Muscicapa caerulescens	Ashy Flycatcher	B	20	17.6		16	20		Congo	1245
Myioparus griseigularis	Gray-throated Flycatcher	B	43	12.6		9	15		Uganda	1245
Myioparus plumbeus	Gray Tit-Flycatcher	U	14	12.8		11	15		Congo	1245
Ficedula hypoleuca	European Pied Flycatcher	M	285	12.2	0.5	10.2	13.5	B	Germany	223
		F	349	15.6	0.9	12.7	18.3			

Body Masses of World Birds (continued)

Species Name	Common Name	Sex	N	Mean	S.D.	Min	Max	Season	Location	Source #
Ficedula albicollis	Collared Flycatcher	M	37	12.9		9.2	16	Sp	Cyprus	223
		F	19	12.5		10	14.2			
Ficedula semitorquata	Semicollared Flycatcher	B	13	14.4	1.1	13	17.2		Turkey; former USSR	223
Ficedula zanthopygia	Korean Flycatcher	U	9	11.2				F	Malaysia	853
Ficedula zanthopygia	Korean Flycatcher	B	16	13.1		12	15		Hopei, China	1087
Ficedula narcissina elisae	Narcissus Flycatcher	B	103	13.1		11	15		Hopei, China	1087
Ficedula mugimaki	Mugimaki Flycatcher	U	10	11.7		10	14		Malaysia; Japan	490. 775
Ficedula hodgsonii	Slaty-backed Flycatcher	B	5	10		9	11		India; China	7, 1087
Ficedula strophiata	Rufous-gorgeted Flycatcher	B	11			11	15		India	7
Ficedula albicilla	Taiga Flycatcher	B	53	11.8		10	14		Hopei, China	1087
Ficedula parva	Red-breasted Flycatcher	B	11	9.9	1	8.5	11.5			223
Ficedula subrubra	Kashmir Flycatcher	U	15	10.4		9	12	B	Kashmir	7
Ficedula hyperythra	Snowy-breasted Flycatcher	U	19	8.2		6	12		Malaysia	775
Ficedula monileger	White-gorgeted Flycatcher	F	1	11						7
Ficedula dumetoria	Rufous-chested Flycatcher	B	4	9.4		8.2	11.8		Borneo; Malaysia	785, 1218
Ficedula rufigula	Rufous-throated Flycatcher	B	11	11.6	0.9	10	13		Sulawesi, Indonesia	1272
Ficedula buruensis	Cinnamon-chested Flycatcher	U				15	18		Moluccas	1181, 1182
Ficedula basilanica	Little Slaty Flycatcher	M	3	14.8		14	16.1		Philippines	954
		F	1	12.4						

Body Masses of World Birds (continued)

Species Name	Common Name	Sex	N	Mean	S.D.	Min	Max	Season	Location	Source #
Ficedula platenae	Palawan Flycatcher	U	4			11.2	13		Palawan, Philippines	487
Ficedula crypta	Russet-tailed Flycatcher	B	4	11.5		10.8	12.3		Mindanao, Philippines	1143
Ficedula westermanni	Little Pied Flycatcher	B	13	7.8		6.8	10		Philippines	954
Ficedula superciliaris	Ultramarine Flycatcher	B	2	8		8	8	B		7
Ficedula tricolor	Slaty-blue Flycatcher	M	13	9.1		7.5	10	B	India	7
		F	11	7.5		7	8			
Ficedula nigrorufa	Black-and-rufous Flycatcher	M	5			10	11			7
		F	5			7	10			
Ficedula sapphira	Sapphire Flycatcher	M	2	7.8		7.5	8			7
Cyanoptila cyanomelana	Blue-and-white Flycatcher	B	46	22.5		20	26		Hopei, China	1087
Eumyias thalassina	Verditer Flycatcher	B	27	18.1		15	20			7
Eumyias panayensis	Is. Flycatcher	B	23	20		17.7	21.7		Philippines	954
Eumyias albicaudata	Nilgiri Flycatcher	B	14	16.4		12	19	B	India	7
Niltava grandis	Large Niltava	B	11	30.3		27.5	34.9		Malaysia	785
Niltava macgrigoriae	Small Niltava	B				11	13			7
Niltava sundara	Rufous-bellied Niltava	U	28	21.1		19	24		India	7
Niltava vivida	Vivid Niltava	B	3	33						7
Cyornis herioti	Blue-breasted Flycatcher	M	1	21.4					Philippines	435, 436
		F	1	18						
Cyornis pallipes	White-bellied Blue-Flycatcher	B	21	19		14	23	B	India	7
Cyornis unicolor	Pale Blue-Flycatcher	B	2	21		21	21.1			7

Body Masses of World Birds (continued)

Species Name	Common Name	Sex	N	Mean	S.D.	Min	Max	Season	Location	Source #
Cyornis rubeculoides	Blue-throated Flycatcher	B	28	14.2		10	19	W	India	7
Cyornis banyumas	Hill Blue-Flycatcher	U	5	14.5		12.8	15.3		Thailand	787
Cyornis tickelliae tickelliae	Tickell's Blue-Flycatcher	U	19	14.6		12	17		India	7
Cyornis tickelliae jerdoni	Tickell's Blue-Flycatcher	M	1	18					Sri Lanka	7
Cyornis rufigastra	Mangrove Blue-Flycatcher	U	4	18.3		16.8	22		Philippines	344, 954
Cyornis omissus	Sulawesi Blue-Flycatcher	B	64	18	1.5	14	21		Sulawesi, Indonesia	1272
Culicicapa ceylonensis	Gray-headed Canary-Flycatcher	F	4	7.7		7	8.6		Malaysia	785
Culicicapa helianthea	Citrine Canary-Flycatcher	U	7	8.4		7	9		Philippines, Indonesia	344, 954, 1272
Horizorhinus dohrni	Dohrn's Flycatcher	U				17	22			396
Pogonocichla stellata	White-starred Robin	B	207	18.6		15.5	26.3		Malawi	303
Swynnertonia swynnertoni	Swynnerton's Robin	M	105	15.8		13.8	20.4		Zimbabwe	623
		F	44	16.3		14.4	19.5			
Stiphrornis erythrothorax	Forest Robin	B	39	15.6					Liberia	623
Sheppardia bocagei	Bocage's Akalat	U	17	18.3		16.4	20.9		Cameroon	623
Sheppardia cyornithopsis	Lowland Akalat	M	9	19.6		18	21.2		Liberia	623
		F	2	17.2		17.2	17.3			
Sheppardia aequatorialis	Equatorial Akalat	U	21	16.8					Burundi; Rwanda	81
Sheppardia sharpei	Sharpe's Akalat	B	22	14.7	1.1	12.5	17.5		Tanzania	1014
Sheppardia gunningi	East Coast Akalat	M	8	18		17	19			623
		F	3	16.7		16	17			
Sheppardia gabela	Gabela Akalat	U		11						275
Sheppardia montana	Usambara Akalat	U	95	21.8		15.5	25.5			623

Body Masses of World Birds (continued)

Species Name	Common Name	Sex	N	Mean	S.D.	Min	Max	Season	Location	Source #
Sheppardia lowei	Iringa Akalat	U	22	18.4		16.2	21.2			623
Erithacus rubecula	European Robin	B	50	17.7	1.5	14	22	W	Malta	223
Erithacus akahige	Japanese Robin	U		17					estimated	1242
Erithacus komadori	Ryukyu Robin	B	100	23.9	2.2	19.2	30.6		Tokara Is., Japan	1078a
Luscinia sibilans	Rufous-tailed Robin	F	3	16.7		14	20		Hopei & Tsingtao, China	1087, 1088
Luscinia luscinia	Thrush Nightingale	U	28	23.8	2.9	18.2	34.1	W	Ethiopia	21
Luscinia megarhynchos megarhynchos	Common Nightingale	U	276	18.3		12	24.6	Sp	Morocco	20
Luscinia megarhynchos africana	Common Nightingale	U	207	19.8	1.6	16	25		Kenya	223
Luscinia megarhynchos hafizi	Common Nightingale	U	116	20.8	1.8	16	26	W	Kenya	223
Luscinia calliope	Siberian Rubythroat	B	20	18.5		16	27		China	223
Luscinia pectoralis	White-tailed Rubythroat	B	13			21.5	25.2	PB		7
Luscinia svecica svecica	Bluethroat	B	8	19.3		18	22		Alaska, USA	466
Luscinia svecica svecica	Bluethroat	B	142	18.2		14.7	25.5	B	Russia	223
Luscinia svecica pallidogularis	Bluethroat	M	132	16.1	1.1	14	19	F	Siberia	223
		F	64	15.6	1.2	13	20			
Luscinia ruficeps	Rufous-headed Robin	U	1	18					Malaysia	775
Luscinia brunnea	Indian Blue Robin	B	6	17.5		14	20	B		7
Luscinia cyane	Siberian Blue Robin	U	52	14.8				W	Malaysia	853
Tarsiger cyanurus cyanurus	Red-flanked Bluetail	B	125	14		11	17		Hopei, China	1087
Tarsiger cyanurus rufilatus	Red-flanked Bluetail	U	12	12.9		11	14	B		7

Body Masses of World Birds (continued)

Species Name	Common Name	Sex	N	Mean	S.D.	Min	Max	Season	Location	Source #
Tarsiger chrysaeus	Golden Blue-Robin	B	12	13.8	1.4	11.9	16.4	Y	India	7
Tarsiger indicus	White-browed Blue-Robin	B	6	14.6		13.2	15.5	PB	India	7
Tarsiger hyperythrus	Rufous-vented Blue-Robin	M	1	12				Sp		7
Tarsiger johnstoniae	Collared Blue-Robin	U				9.5	11.6		Taiwan	834
Irania gutturalis	White-throated Robin	U	500	21.6		17.6	26.7	W	Kenya	623
Cossypha isabellae	Mountain Robin-Chat	B	15	24		21	28		Cameroon	710
Cossypha archeri	Archer's Robin-Chat	B	7			20	26			623
Cossypha anomala	Olive-flanked Robin-Chat	B	67	24.5		20.6	27.4		Malawi	303
Cossypha caffra	Cape Robin-Chat	B	58	28.5		25.4	34		Malawi	303
Cossypha humeralis	White-throated Robin-Chat	B	81	22.8		17	28		Zimbabwe; South Africa	623
Cossypha cyanocampter	Blue-shouldered Robin-Chat	U	16	26.7	1.7	24.1	31.1		Uganda	1078
Cossypha polioptera polioptera	Gray-winged Robin-Chat	M	4	18.5		17	20		Kenya	623
		F	3	16		15	17			
Cossypha polioptera nigriceps	Gray-winged Robin-Chat	M	5	21.7		19	23.2	W	Liberia	623
		F	3	19.1		18	20.3			
Cossypha semirufa	Rueppell's Robin-Chat	U	11	27.3		22.5	31		Ethiopia	1238, 1243a
Cossypha heuglini	White-browed Robin-Chat	M	14	37.5	3.7	30.5	44.1		Zimbabwe	623
		F	10	32.3	1.9	29.1	35.9			
Cossypha natalensis	Red-capped Robin-Chat	U	61	34.1	2.7	27	41		Tanzania	1014
Cossypha dichroa	Chorister Robin-Chat	M	28	47.6		42	56		Natal, South Africa	623
		F	34	44.1		37	53			

Body Masses of World Birds (continued)

Species Name	Common Name	Sex	N	Mean	S.D.	Min	Max	Season	Location	Source #
Cossypha heinrichi	White-headed Robin-Chat	M	2	65		61	69			623
		F	2	56		56	56			
Cossypha niveicapilla	Snowy-crowned Robin-Chat	U	51	36.7		30.1	44		Nigeria	777
Cossypha albicapillus	White-crowned Robin-Chat	U	15	58.5		46	74		Nigeria	777
Xenocopsychus ansorgei	Angola Cave-Chat	M	2	41		36	46		Angola	266
		F	1	30						
Cichladusa arquata	Collared Palm-Thrush	B	54	34.2		28.5	38.3			623
Cichladusa ruficauda	Rufous-tailed Palm-Thrush	B	24			22	37		Angola	266
Cichladusa guttata	Spotted Morning-Thrush	M	17	24.4		17	30			623
		F	17	21.9		16	30			
Cercotrichas leucosticta	Forest Scrub-Robin	M	9	26.6		23.2	31.2		Liberia	623
		F	7	24.3		21.5	28			
Cercotrichas quadrivirgata	Bearded Scrub-Robin	B	32	26.1		21.2	31.2		Zimbabwe	623
Cercotrichas signata	Brown Scrub-Robin	U	5	38.2		36	42		Natal, South Africa	623
Cercotrichas hartlaubi	Brown-backed Scrub-Robin	B	8	19.2		17	20			266, 623
Cercotrichas leucophrys	Red-backed Scrub-Robin	M	14	17	2.1	12.9	20.3			623
		F	5	16		15	17.9			
Cercotrichas galactotes	Rufous-tailed Scrub-Robin	U	26	20.3		17.5	24	W	Kenya	223
Cercotrichas paena	Kalahari Scrub-Robin	U	21	19.7		17	23			623
Cercotrichas minor	African Scrub-Robin	B	11	18.9		17	21		Chad; Niger; Nigeria	223
Cercotrichas coryphaeus	Karoo Scrub-Robin	U	9	20.4		19	22.6			623, 1258
Cercotrichas podobe	Black Scrub-Robin	B	7	25.3		24	27		Niger; Chad	223

Body Masses of World Birds (continued)

Species Name	Common Name	Sex	N	Mean	S.D.	Min	Max	Season	Location	Source #
Cercotrichas podobe melanoptera	Black Scrub-Robin	U	2	22.6		22.5	22.7		Yemen	317
Namibornis herero	Herero Chat	F	2	27.2		26.5	28		Namibia	526
Copsychus albospecularis pica	Madagascar Magpie-Robin	U	10	21.2	3.1	18	28		Madagascar	957
Copsychus albospecularis inexpectatus	Madagascar Magpie-Robin	U	55	24.4	2.3	19	30.5		Madagascar	957
Copsychus saularis	Oriental Magpie-Robin	B	24	36		31	42.5		India	7
Copsychus malabaricus	White-rumped Shama	M	32	32		28	35		Hawaiian Is.	997
		F	21	26.7		21	30			
Copsychus sechellarum	Seychelles Magpie-Robin	M		77					Seychelles	275
		F		65						
Copsychus luzoniensis	White-browed Shama	M	2	24		22.8	25.3		Philippines	436
Copsychus niger	White-vented Shama	B	6	28.5		25.5	31.8		Palawan, Philippines	1143
Trichixos pyrropygus	Rufous-tailed Shama	B	7	40.9		34.5	46		Malaysia; Borneo	398, 785
Saxicoloides fulicatus cambaiensis	Indian Robin	B	14	16.6		17	21		India	7
Phoenicurus erythronotus	Rufous-backed Redstart	B	23	18.5		15	22	W	Kazakhstan; Afghanistan	223
Phoenicurus caeruleocephala	Blue-capped Redstart	B	15	14.9		13	17			7, 273
Phoenicurus ochruros	Black Redstart	U	50	16.5	1.5	13	20	W	Malta	223
Phoenicurus phoenicurus	Common Redstart	M	1016	14.7	1.4	10	19	F	Poland	223
		F	1282	14.5	1.4	9	21			
Phoenicurus hodgsoni	Hodgson's Redstart	M	7			16	19.5	W	India	7
		F	6			14.5	18.5			
Phoenicurus schisticeps	White-throated Redstart	B	10	18.1	1.6	16	20.6		India; China	7, 1183
Phoenicurus auroreus	Daurian Redstart	B	71	16.2		11	20		Hopei, China	1087

Body Masses of World Birds (continued)

Species Name	Common Name	Sex	N	Mean	S.D.	Min	Max	Season	Location	Source #
Phoenicurus moussieri	Moussier's Redstart	B	4	14.9		14.5	15	F	Algeria	223
Phoenicurus erythrogastrus	White-winged Redstart	M	367	25.2	2.3	22	29	F	India	223
		F	346	23.9	1.3	21	28			
Phoenicurus frontalis	Blue-fronted Redstart	M	25			14.5	19			7
		F	18			12.5	18			
Chaimarrornis leucocephalus	White-capped Redstart	M	16			30	37			7
		F	12			24	30.6			
Rhyacornis fuliginosa	Plumbeous Redstart	M	11	22		20	24		Hopei, China	1087
		F	5	18.8		18	19			
Rhyacornis bicolor	Luzon Redstart	M	1	23					Luzon, Philippines	1143
Hodgsonius phaenicuroides	White-bellied Redstart	B	137	21.1		18	24		Hopei, China	1087
Cinclidium leucurum	White-tailed Robin	B	6			24	30		India	7
Cinclidium frontale	Blue-fronted Robin	B	4	25.4		25	26.2		India	7
Grandala coelicolor	Grandala	B	5	45.6		38	52		India	7
Enicurus scouleri	Little Forktail	M	2	17.8		17	18.6		India	7
		F	3			12	16			
Enicurus ruficapillus	Chestnut-naped Forktail	M	3	29.1		27.9	30		Malaysia	785
		F	2	27.4		24.8	29.9			
Enicurus immaculatus	Black-backed Forktail	U	2	25.5		25	26		India	7
Enicurus schistaceus	Slaty-backed Forktail	U	13	31		25	38		Malaysia	775
Enicurus leschenaulti	White-crowned Forktail	B	7	33.8		24.8	53.4		India; China; Borneo	7, 398, 785
Enicurus maculatus	Spotted Forktail	B	14			34	48		India	7
Cochoa purpurea	Purple Cochoa	B	2	103		100	106		India	7
Cochoa viridis	Green Cochoa	M	5			88	99		India	7
		F	2	120		117	122			

Body Masses of World Birds (continued)

Species Name	Common Name	Sex	N	Mean	S.D.	Min	Max	Season	Location	Source #
Saxicola rubetra	Whinchat	U	72	16.6	1.5	13	22	Sp	Great Britain	223
Saxicola dacotiae	Canary Is.Chat	U	125	12.2	0.6	10.9	14.1		Canary Is.	549, 550
Saxicola rubicola rubicola	Common Stonechat	U	50	15.3	1.4	13	19	W	Malta	223
Saxicola rubicola indica	Common Stonechat	U	17	11.2		8	16		India	7
Saxicola rubicola maura	Common Stonechat	B	29	12.6		11	15		India	7
Saxicola rubicola stejnegeri	Common Stonechat	U	9	15.4		15	16		Northern China	925
Saxicola torquatus sibillus	African Stonechat	U				15	16.5		Madagascar	183
Saxicola torquatus axillaris	African Stonechat	U	13	15.5		13	20			623
Saxicola torquatus torquatus	African Stonechat	U	7	15.2		13	17.1			623
Saxicola tectes	Reunion Stonechat	M	13	12.3	0.7	11.2	13.8		Reunion	623
		F	8	13		11.9	13.6			
Saxicola caprata	Pied Bushchat	B	19	15.2		13	17		India	7
Saxicola ferreus	Gray Bushchat	M	17			13.8	16		India	7
		F	3	14.7		13.1	16			
Saxicola bifasciatus	Buff-streaked Chat	M	1	37.5						623
		F	3	33.7		32.7	34.8			
Oenanthe leucopyga	White-tailed Wheatear	B	8	27.9		24	39	W	Algeria	223
Oenanthe monacha	Hooded Wheatear	M	1	23					Iran	223
		F	2	20.2		18	22.5			
Oenanthe albonigra	Hume's Wheatear	B	27	25.2		22	28.5		Iran	223
Oenanthe leucura	Black Wheatear	M	2	41		38	44			223
Oenanthe monticola	Mountain Wheatear	B	12	33.1	4	26.5	40		Namibia	526
Oenanthe oenanthe oenanthe	Northern Wheatear	M	22	24	1.5	19	27	Sp	Italy	223
		F	45	22.3	2.5	18	29			

Body Masses of World Birds (continued)

Species Name	Common Name	Sex	N	Mean	S.D.	Min	Max	Season	Location	Source #
Oenanthe oenanthe leucorhoa	Northern Wheatear	B	54	30.6				Sp	Scotland	223
Oenanthe lugens	Mourning Wheatear	M	3	24		22	25	F	Algeria	223
		F	3	21		19	22			
Oenanthe finschii	Finsch's Wheatear	M	29	25.6		21	28.5	Sp	former USSR	223
		F	10	29.3		23.5	32.5			
Oenanthe picata	Variable Wheatear	M	41			21	25		India	7
		F	12			18	23			
Oenanthe moesta	Red-rumped Wheatear	U	4	30.2		28	34		Morocco	1148
Oenanthe pleschanka	Pied Wheatear	M	22	18.2	1.6	16	22		Mongolia	927
		F	14	20.6	2.1	17	25			
Oenanthe cypriaca	Cyprus Wheatear	B				12	23		Cyprus	1130
Oenanthe hispanica	Black-eared Wheatear	M	15	16.4	1.7	13	19	Sp		223
		F	3	13.8		12	15			
Oenanthe xanthoprymna chrysopygia	Red-tailed Wheatear	B	9	22.7		20	27	F	Iran; Afghanistan	223
Oenanthe deserti	Desert Wheatear	M	28	19.6	1.5	17	23		Mongolia	927
		F	14	20.1	1.7	18	23			
Oenanthe pileata	Capped Wheatear	U	4	26.8		23.9	32.5			133, 266, 509
Oenanthe isabellina	Isabelline Wheatear	B	12	29.4	3.4	25	38		Mongolia	927
Oenanthe bottae	Red-breasted Wheatear	U	7	19.9	17	23			Nigeria; Chad	623
Oenanthe heuglini	Heuglin's Wheatear	U				17	23			275
Cercomela sinuata	Sickle-winged Chat	B	7	18.1		15.5	19.7		South Africa	509, 1258
Cercomela schlegelii	Karoo Chat	B	6	23.3		18	29		Angola; Namibia	266, 526
Cercomela tractrac	Tractrac Chat	M	6	20.8		18.5	23		Angola	266, 526
Cercomela familiaris	Familiar Chat	B	19	22		17.3	28		South Africa	509

Body Masses of World Birds (continued)

Species Name	Common Name	Sex	N	Mean	S.D.	Min	Max	Season	Location	Source #
Cercomela familiaris	Familiar Chat	U	31	17.2			20.7		Nigeria	777
Cercomela scotocerca	Brown-tailed Chat	F	1	16						623
Cercomela fusca	Indian Chat	F	1	12.7				F	India	7
Cercomela melanura	Blackstart	B	6	15.2		14	17		Chad; Niger; Yemen	223, 317
Cercomela sordida	Moorland Chat	U	12	21		18	23			623
Myrmecocichla aethiops	Northern Anteater-Chat	M	20	58.5		51	66			623
		F	11	54.7		47	58			
Myrmecocichla formicivora	Southern Anteater-Chat	U	65	41.6		22	49			133
Myrmecocichla nigra	Sooty Chat	B	12	40.3		37	46			623
Myrmecocichla albifrons	White-fronted Black-Chat	B	12	20.6		18	24.9			623, 777
Myrmecocichla arnotti	White-headed Black-Chat	B	2	39.6		35.2	44			266, 623
Thamnolaea cinnamomeiventris	Mocking Cliff-Chat	U	13	39.8	2.4	36	43		Nigeria	777
Pinarornis plumosus	Boulder Chat	M	1	65.8					Zimbabwe	567
Family Platysteiridae										
Megabyas flammulatus aequatorialis	African Shrike-flycatcher	M	12	30.7		29	34		Uganda	1245
		F	12	28.1		22	32			
Megabyas flammulatus flammulatus	African Shrike-flycatcher	M	7	27.6					Liberia	1245
		F	5	25.7						
Bias musicus	Black-and-white Shrike-flycatcher	B	13	21.9		20	25		Uganda	1245
Pseudobias wardi	Ward's Flycatcher	U	3	12.7		12	13.3		Madagascar	957
Platysteira cyanea	Brown-eyed Wattle-eye	B	21	14.6		12	17		Cameroon	1245
Platysteira albifrons	White-fronted Wattle-eye	B	2	12.2		11.7	12.6		Angola	485

Body Masses of World Birds (continued)

Species Name	Common Name	Sex	N	Mean	S.D.	Min	Max	Season	Location	Source #
Platysteira peltata	Black-throated Wattle-eye	B	116	13.7		10	18.8		Southern Africa	485
Platysteira laticincta	Banded Wattle-eye	U	9	14.3		13.2	15.2		Cameroon	485
Platysteira castanea	Chestnut Wattle-eye	M	29	14.3		12	16		Uganda	1245
		F	18	13.9		11	15			
Platysteira tonsa	White-spotted Wattle-eye	M	1	11.5					Gabon	1245
Platysteira blissetti	Red-cheeked Wattle-eye	B	22	11					Liberia	485
Platysteira chalybea	Black-necked Wattle-eye	U	10	11.7		11	13		Gabon	485
Platysteira jamesoni	Jameson's Wattle-eye	B	57	11.5		9	15		Uganda	485
Platysteira concreta	Yellow-bellied Wattle-eye	B	45	11.4					Liberia	485
Batis margaritae	Boulton's Batis	B	24	11.7		9.8	15.5			485
Batis mixta	Short-tailed Batis	B	38	12.2		10.5	13.6		Tanzania	1245
Batis diops	Ruwenzori Batis	B	25	12.7		8	15.5		Uganda	485
Batis capensis	Cape Batis	B	301	11.6		8.9	13.9		Southern Africa	485
Batis capensis dimorpha	Cape (Malawi) Batis	U	3	11.9					Mozambique	1030a
Batis fratrum	Woodward's Batis	M	43	12.2		10.4	13.8		Southern Africa	485
		F	23	11.7		10.3	13.6			
Batis molitor	Chinspot Batis	M	95	11.2		8	14		Southern Africa	485
		F	95	12.1		8	14			
Batis soror	Pale Batis	B	74	9.5		8	13.1		Zimbabwe; Mozambique	485
Batis pririt	Pririt Batis	M	12	9		8	10			485
		F	21	9.7		8	14			
Batis senegalensis	Senegal Batis	M	9	10		9.8	10.5		Nigeria	485
		F	9	9.3		9	10			

Body Masses of World Birds (continued)

Species Name	Common Name	Sex	N	Mean	S.D.	Min	Max	Season	Location	Source #
Batis orientalis	Gray-headed Batis	U	33	11.1		8.8	13.4		Ethiopia	485
Batis minor erlanger	Black-headed Batis	B	4	12.5		11	14		Cameroon	710
Batis perkeo	Pygmy Batis	M	9	6.6		5	8		Kenya; Uganda	1245
		F	4	7.3		6	8			
Batis minima	Verreaux's Batis	B	5	10		8	12		Gabon	1245
Batis poensis	Fernando Po Batis	B	11	9.6		9	10		Bioko	1245
Batis occulta	West African Batis	B	4	9		8.7	9.4		Liberia	485
Batis minulla	Angola Batis	B	3	9.9		9.3	10.4		Angola	485
Lanioturdus torquatus	White-tailed Shrike	U	26	29		22.5	45			485
Family Rhipiduridae										
Rhipidura hypoxantha	Yellow-bellied Fantail	B	10			5	6			7
Rhipidura superciliaris	Blue Fantail	B	47	13.1		9.5	16.8		Philippines	954
Rhipidura cyaniceps	Blue-headed Fantail	B	20	13.6		12	15.6		Philippines	436
Rhipidura nigrocinnamomea	Black-and-cinnamon Fantail	B	22	13		10.5	14.5		Philippines	954
Rhipidura albicollis	White-throated Fantail	U	24	12.9		11	15		Malaysia	775
Rhipidura aureola	White-browed Fantail	B	5	11.2		10	13		India	7
Rhipidura rufiventris	Northern Fantail	B	40	12		10	15		Northern Australia	518a
Rhipidura javanica	Pied Fantail	U	13	12.5	0.9	10.7	13.9		Thailand	787
Rhipidura perlata	Spotted Fantail	B	7	13.4		12	16		Borneo; Malaysia	398, 785, 1218
Rhipidura leucophrys picata	Willie Wagtail	B	13	20.1		14.5	25.5		Australia	518a

Body Masses of World Birds (continued)

Species Name	Common Name	Sex	N	Mean	S.D.	Min	Max	Season	Location	Source #
Rhipidura leucophrys melaleuca	Willie Wagtail	U	24	25.9	2.2	22.2	31.8		Papua New Guinea	318
Rhipidura leucophrys melaleuca	Willie Wagtail	M	7	34.4		31.5	37		Solomon Is.	399
		F	5	31.7		27	36			
Rhipidura cockerelli	White-winged Fantail	B	12	17.5		16.5	18.5		Guadalcanal	399
Rhipidura albolimbata	Friendly Fantail	B	78	10.9	1.2	7.2	14		Papua New Guinea	387
Rhipidura hyperythra	Chestnut-bellied Fantail	B	9	11.2		10	12.5		New Guinea	283, 424
Rhipidura threnothorax	Sooty Thicket-Fantail	B	10	17.6	1.8	14.7	20		New Guinea	283
Rhipidura maculipectus	Black Thicket-Fantail	U	7	18					New Guinea	61
Rhipidura leucothorax	White-bellied Thicket-Fantail	B	8	17.6		11	20		New Guinea	281, 283
Rhipidura atra	Black Fantail	B	14	12.2		10.5	14		New Guinea	283
Rhipidura phasiana	Mangrove Fantail	U	46	6.9		4	7.8		Australia	518a
Rhipidura drownei	Brown Fantail	M	4	12		11.5	12.5		Guadalcanal	399
		F	3	10.3		10	10.5			
Rhipidura tenebrosa	Dusky Fantail	M	5	18.9		18.5	19.5		Solomon Is.	399
		F	4	15.9		15.5	16.5			
Rhipidura fuliginosa	Gray Fantail	U	92	8	1	5.5	10		New Zealand	1002
Rhipidura fuliginosa albicauda	Gray Fantail	B	30	7.1		5.7	8		Australia	518a
Rhipidura spilodera	Streaked Fantail	U	21	12.9	1.1	11.5	15		Fiji	668
Rhipidura brachyrhyncha	Dimorphic Fantail	B	31	10.1	0.8	8.7	11.9		Papua New Guinea	387
Rhipidura teysmanni	Rusty-flanked Fantail	U	21	10.7	1.1	9	14		Sulawesi, Indonesia	1272
Rhipidura dedemi	Streaky-breasted Fantail	U				10	12		Seram, Moluccas	1181
Rhipidura rufidorsa	Rufous-backed Fantail	U	6	9.8					New Guinea	61, 428

Body Masses of World Birds (continued)

Species Name	Common Name	Sex	N	Mean	S.D.	Min	Max	Season	Location	Source #
Rhipidura dahli	Bismarck Fantail	B	4	9.9		9	11		New Britain	426
Rhipidura malaitae	Malaita Fantail	U		12						771
Rhipidura rufifrons intermedia	Rufous Fantail	B	101	11.4	0.9	9.6	14		Queensland, Australia	390a
Rhipidura rufifrons mariae	Rufous Fantail	M	2	8.6		8.3	9		Rota	35
Rhipidura rufifrons uraniae	Rufous Fantail	B	12	9		7.2	10		Guam	35
Rhipidura rufifrons rufofronta	Rufous Fantail	M	8	11.6		11	12		Guadalcanal	399
		F	4	11.9		10.5	14.5			
Rhipidura rufifrons russata	Rufous Fantail	M	13	9.1		8.5	10		Solomon Is.	399
		F	9	8.5		8	9.5			
Rhipidura rufifrons ugiensis	Rufous Fantail	M	3	11		10.5	11.5		Solomon Is.	399
		F	4	10.4		10	10.5			
Family Monarchidae										
Erythrocercus mccallii	Chestnut-capped Flycatcher	B	15	7.2		6	8		Cameroon; Uganda	1245
Erythrocercus holochlorus	Yellow Flycatcher	B	12	6		5	7			1245
Erythrocercus livingstonei	Livingstone's Flycatcher	U	3	5.9		5.6	6			1245
Elminia longicauda	African Blue-Flycatcher	U	18	9.9		8.7	10.9		Nigeria	777
Elminia albicauda	White-tailed Flycatcher	U	16	7.8		6	9		Congo	1245
Elminia albicauda	White-tailed Flycatcher	U	7	8.9		8	10		Tanzania	1014
Elminia nigromitrata	Dusky Crested-Flycatcher	M	12	9.6	0.9				Liberia	1245
		F	12	8.5	0.9					
Elminia albiventris	White-bellied Crested-Flycatcher	U	29	8.9		7	10		Cameroon	1245
Elminia albonotata	White-tailed Crested-Flycatcher	B	44	9.2		7.9	11.6		Malawi	303

Body Masses of World Birds (continued)

Species Name	Common Name	Sex	N	Mean	S.D.	Min	Max	Season	Location	Source #
Trochocercus nitens	Blue-headed Crested-Flycatcher	B	25	11.8		9.8	15		Cameroon; Gabon; Uganda	1245
Trochocercus cyanomelas	African Crested-Flycatcher	B	24	10.2		9.3	12.1		Somalia	1335
Hypothymis helenae	Short-crested Monarch	B	4	9.9		9.2	10.3		Philippines	954
Hypothymis azurea	Black-naped Monarch	B	21	11.1		8.9	13.4		Philippines	954
Hypothymis puella	Pale-blue Monarch	B	14	14.8	0.9	13	16			1272
Hypothymis coelestis	Celestial Monarch	M	1	13.6					Philippines	954
Terpsiphone rufiventer	Black-headed Paradise-Flycatcher	U	10	15.1	1.6	12.5	17.3		Liberia	616
Terpsiphone rufocinerea	Rufous-vented Paradise-Flycatcher	F	1	15.4		15.4	16.5		Congo	1245
Terpsiphone batesi	Bates' Paradise-Flycatcher	U	12	15.2		13	17		Gabon	1245
Terpsiphone viridis	African Paradise-Flycatcher	M	142	13.8		10	20		Kenya	1245
		F	29	12.7		10	16			
Terpsiphone atrochalybeia	Sao Tome Paradise-Flycatcher	M	1	12.2						1245
Terpsiphone mutata	Madagascar Paradise-Flycatcher	M	75	13.4	1.6	8.5	18		Madagascar	957
		F	33	13	1.4	10	17			
Terpsiphone corvina	Seychelles Paradise-Flycatcher	U		18					Seychelles	242
Terpsiphone bourbonnensi desolata	Mascarene Paradise-Flycatcher	M	12	11.7	0.4	11	12.5		Mauritius	183
		F	10	11.3	0.5	10.4	12.1			
Terpsiphone bourbonnensis bourbonnensis	Mascarene Paradise-Flycatcher	M	5	10.6		10.1	11.4		Reunion	183
		F	5	9.7		8.2	10.6			

Body Masses of World Birds (continued)

Species Name	Common Name	Sex	N	Mean	S.D.	Min	Max	Season	Location	Source #
Terpsiphone atrocaudata	Japanese Paradise-Flycatcher	M	2	18.7					Thailand	787
Terpsiphone cyanescens	Blue Paradise-Flycatcher	M	2	26.3		25.7	26.9		Palawan, Philippines	487
		F	1	23.1						
Terpsiphone cinnamomea	Rufous Paradise-Flycatcher	B	10	20.4		18	23.1		Philippines	344, 954, 983
Terpsiphone paradisi paradisi	Asian Paradise-Flycatcher	U	50	18.5		16	21	F	India	7
Terpsiphone paradisi incei	Asian Paradise-Flycatcher	B	36	20		16	23		Hopei, China	1087
Chasiempis sandwichensis ridgwayi	Elepaio	M	41	16.2	1.1	14	18.6		Hawaii, Hawaiian Is.	1246
		F	27	14.4	0.8	12.8	15.9			
Chasiempis sandwichensis ibidis	Elepaio	M	66	13.1	0.8	11.9	15.5		Oahu, Hawaiian Is.	1246
		F	10	12.2	1	11.1	13.8			
Chasiempis sandwichensis sclateri	Elepaio	U	13	12.9	1.1	11	15.5		Kauai, Hawaiian Is.	1246
Pomarea dimidiata	Rarotonga Monarch	B		23					Cook Is.	1045
Pomarea nigra	Tahiti Monarch	U	6	25.4					Tahiti	1210b
Pomarea iphis	Iphis Monarch	U	12	23.9	1.7				Marquesas Is.	1210b
Pomarea mendozae motanensis	Marquesas Monarch	U	8	29.6					Marquesas Is.	1210b
Pomarea whitneyi	Fatuhiva Monarch	U	5	42.5		37	50		Marquesas Is.	1210b
Mayrornis lessoni	Slaty Monarch	U	13	11.2	1.5	9.5	14.5		Fiji	668
Neolalage banksiana	Buff-bellied Monarch	M	6	18.2		16.2	20.2		Vanuatu	656b
		F	1	15.5						
Clytorhynchus pachycephaloides	Southern Shrikebill	M	2	24.8		24.5	25		New Caledonia	1018

Body Masses of World Birds (continued)

Species Name	Common Name	Sex	N	Mean	S.D.	Min	Max	Season	Location	Source #
Clytorhynchus pachycephaloides grisescens	Southern Shrikebill	B	5	27.9		25	30.1		Vanuatu	656b
Clytorhynchus hamlini	Rennell Shrikebill	B	4	34.6		28	38		Rennell, Solomon Is.	1153b
Clytorhynchus vitiensis	Fiji Shrikebill	U	74	29.2	3.3				Fiji	668
Monarcha axillaris	Black Monarch	B	14	15.6		14.3	17		New Guinea	283
Monarcha cinerascens	Is.Monarch	B	18	29.6		24	41		New Guinea	292
Monarcha frater	Black-winged Monarch	B	11	22	1.1	20.8	24		New Guinea	283
Monarcha melanopsis	Black-faced Monarch	B	27	22.7		19	27		Australia	518a
Monarcha castaneiventris castaneiventris	Chestnut-bellied Monarch	M	17	27.1	2	24.5	32		Solomon Is.	658
		F	6	25.6		22	29.3			
Monarcha castaneiventris megarhynchus	Chestnut-bellied Monarch	M	20	26.2		25	28.5		Solomon Is.	399
		F	11	25		22	33			
Monarcha castaneiventris ugiensis	Chestnut-bellied Monarch	M	9	31.6		28.5	33.5		Solomon Is.	399
		F	7	28.9		28	30.5			
Monarcha pileatus	White-naped Monarch	U	1	14					Buru, Moloccas	1182
Monarcha leucotis	White-eared Monarch	B	17	11.4	0.4	10.8	12.2		Australia	470
Monarcha guttulus	Spot-winged Monarch	B	14	16.4		14	18.3		New Guinea	283
Monarcha trivirgatus	Spectacled Monarch	U	21	12.7	0.6	11.3	14.1		Queensland, Australia	470
Monarcha boanensis	Black-chinned Monarch	U	1	17.5					Boano Is., Indonesia	803
Monarcha manadensis	Hooded Monarch	U		22.5					New Guinea	229
Monarcha verticalis	Black-tailed Monarch	B	11	21.7	2.6	19.2	29		New Britain; New Ireland	58, 426
Monarcha viduus	White-collared Monarch	M	11	17.4		16	19.5		Solomon Is.	399
		F	19	16.8		16	18			

Body Masses of World Birds (continued)

Species Name	Common Name	Sex	N	Mean	S.D.	Min	Max	Season	Location	Source #
Monarcha barbatus	Black-and-white Monarch	B	13	21		19	23		Guadalcanal	399
Monarcha chrysomela	Golden Monarch	B	8	15.6					New Guinea; New Ireland	58, 61, 428
Arses telescophthalmus	Frilled Monarch	B	11	17		15.7	19		New Guinea	283
Arses telescophthalmus	Frilled Monarch	U	27	13.7	1.6	12	20.5		Cape York Peninsula, Australia	518a
Arses insularis	Rufous-collared Monarch	M	2	19.3		18.5	20		New Guinea	428
Arses kaupi	Pied Monarch	B	11	13.5	1.1	11.9	15.5		Queensland, Australia	470
Myiagra freycineti	Guam Flycatcher	B	7	11.8		10.5	12.5		Guam	35
Myiagra galeata	Moluccan Flycatcher	M	1	11				B	Moluccas	979
Myiagra rubecula concinna	Leaden Flycatcher	B	23	11		9	13		Australia	518a
Myiagra rubecula yorki	Leaden Flycatcher	M	18	13.5	1.4	11.6	16.5		Queensland, Australia	518a
		F	7	11.8		10	13			
Myiagra ferrocyanea	Steel-blue Flycatcher	B	31	12.8		10	14.5		Guadalcanal	399
Myiagra cervinicauda	Ochre-headed Flycatcher	B	23	11.5		10	13		Solomon Is.	399
Myiagra caledonica	Melanesian Flycatcher	U	11	10.8	0.7	10	12		New Caledonia	1018
Myiagra vanikorensis	Vanikoro Flycatcher	U	7	12.3		10.5	13.5		Fiji	668, 1196c
Myiagra azureocapilla	Blue-crested Flycatcher	B	17	12.9		9.5	16		Fiji	668
Myiagra ruficollis	Broad-billed Flycatcher	B	30	10.1		9.2	11.5		Western Australia	518a
Myiagra cyanoleuca	Satin Flycatcher	B	23	17.5		12	23		Australia	518a
Myiagra inquieta	Restless Flycatcher	B	25	20.9		16.5	25		Eastern Australia	518a

Body Masses of World Birds (continued)

Species Name	Common Name	Sex	N	Mean	S.D.	Min	Max	Season	Location	Source #
Myiagra alecto	Shining Flycatcher	M	13	27.2	4	24	31		New Guinea	292
		F	8	24.6		20	32			
Myiagra alecto melvillensis	Shining Flycatcher	B	22	19.6		17	22.5		Northern Australia	518a
Myiagra hebetior	Dull Flycatcher	B	4	20.2		18.5	21.5		New Britain	426
Lamprolia victoriae victoriae	Silktail	U	8	19		16	21		Fiji	668
Lamprolia victoriae kleinschmidti	Silktail	U	5	11.3		10	12.5		Fiji	668
Machaerirhynchus nigripectus	Black-breasted Boatbill	B	11	12		11	13.3		New Guinea	283, 387, 424
Machaerirhynchus flaviventer	Yellow-breasted Boatbill	U	6	11					New Guinea	61
Machaerirhynchus flaviventer	Yellow-breasted Boatbill	U	22	9.7	0.7	9	12		Cape York Peninsula, Australia	518a
Family Petroicidae										
Amalocichla incerta	Lesser Ground-Robin	M	2	32.7		30.7	34.9		New Guinea	283, 387
Monachella muelleriana	Torrent Flycatcher	B	4	24.5		23	27.5		New Guinea	283
Microeca fascinans fascinans	Jacky-winter	B	55	15.9		13.9	20		Southeast Australia	518
Microeca fascinans pallida	Jacky-winter	B	34	12.3		6.8	17.5		Australia	518
Microeca fascinans assimilis	Jacky-winter	B	30	15.5		14	17.5		Australia	518
Microeca flavigaster tormenti	Lemon-bellied Flycatcher	M	9	12		11	14		Western Australia	518
		F	6	11.6		10	13			
Microeca flavigaster flavigaster	Lemon-bellied Flycatcher	B	17	10.8		8	13		Northern Terr., Australia	518
Microeca flavigaster laetissima	Lemon-bellied Flycatcher	B	8	14.3		13.5	16		Queensland, Australia	518
Microeca griseoceps	Yellow-legged Flycatcher	U	10	10.6		9	11.5		Queensland, Australia	518

Body Masses of World Birds (continued)

Species Name	Common Name	Sex	N	Mean	S.D.	Min	Max	Season	Location	Source #
Microeca flavovirescens	Olive Flyrobin	B	14	15		13	17		New Guinea	283
Microeca papuana	Canary Flycatcher	B	14	14.4	0.5	13.6	15.3		Papua New Guinea	387
Eugerygone rubra	Garnet Robin	B	7	8.4		7.9	9		Papua New Guinea	387
Petroica bivittata	Alpine Robin	U		9					New Guinea	229
Petroica multicolor	Scarlet Robin	U	10	9.6	1	8	11		Fiji	668
Petroica multicolor ambrynensis	Scarlet Robin	B	3	8.9		8.4	9.7		Vanuatu	656b
Petroica multicolor multicolor	Scarlet Robin	B	8	14.5		12	18		Norfolk Is.	518
Petroica multicolor boodang	Scarlet Robin	B	41	12.7		11.8	13.6		Victoria, Australia	518
Petroica multicolor campbelli	Scarlet Robin	B	47	12.3		11	14.5		Western Australia	518
Petroica multicolor dennisi	Scarlet Robin	M	6	10.3		10	11		Guadalcanal	399
		F	3	12.3		11	14.5			
Petroica macrocephala	Tomtit	M	14	12	0.6	11.2	13.2		South Is., New Zealand	518
		F	11	11.6	0.4	11	12			
Petroica goodenovii	Red-capped Robin	B	155	8.9	0.5				Victoria, Australia	300
Petroica phoenicea	Flame Robin	U	22	13	0.9	11.2	14.6		Victoria, Australia	518
Petroica rosea	Rose Robin	B	23	8.2		5	11.5		Southeast Australia	518
Petroica rodinogaster	Pink Robin	M	6	9.9		9.3	10.3		Tasmania, Australia	518
		F	19	10.6	0.8	9.5	12.5			
Petroica australis australis	New Zealand Robin	M	20	37.5	2.9	33.4	43.8		South Is., New Zealand	518
		F	5	34.4		30.8	36			
Petroica australis longipes	New Zealand Robin	M	23	28.1	1.7				North Is., New Zealand	518
		F	7	26.6						
Petroica traversi	Chatham Robin	U	10	23.5		20	25.5		Chatham Is., NZ	18
Melanodryas cucullata cucullata	Hooded Robin	B	23	23.7		21.2	26.1		Victoria, Australia	518

Body Masses of World Birds (continued)

Species Name	Common Name	Sex	N	Mean	S.D.	Min	Max	Season	Location	Source #
Melanodryas cucullata picata	Hooded Robin	B	20	18.8		16	23		Australia	518
Melanodryas vittata	Dusky Robin	U	31	26.6	2.5	22.8	33.3		Tasmania, Australia	518
Tregellasia leucops albigularis	White-faced Robin	M	37	15.9	1.1	13	18.3		Queensland, Australia	518
		F	21	13.8	1.3	12.5	17.3			
Tregellasia leucops wahgiensis	White-faced Robin	M	10	17.4	1	15.3	19.4		New Guinea	518
		F	7	15.4		14	18			
Tregellasia capito	Pale-yellow Robin	U	48	13.9	1.8	10	21		Australia	518
Eopsaltria australis	Yellow Robin	M	40	22.2	2.8	16	30		Southeast Australia	518
		F	19	18.3	2.2	14.8	24			
Eopsaltria griseogularis griseogularis	Gray-breasted Robin	U	14	20.2	2.2	16	22.5		Western Australia	518
Eopsaltria griseogularis rosinae	Gray-breasted Robin	U	187	18.6	2	13.8	26.9		Western Australia	518
Eopsaltria flaviventris	Yellow-bellied Robin	B	3	13		10.5	14.5		New Caledonia	1018
Eopsaltria georgiana	White-breasted Robin	M	156	20.8		19	22.5		Western Australia	518
		F	94	17.5		15	19			
Eopsaltria pulverulenta	Mangrove Robin	M	16	21.3	1.5	19	23.5		Western Australia	518
		F	14	17.3	2.1	15	22			
Poecilodryas brachyura	Black-chinned Robin	U		25.1					New Guinea	229
Poecilodryas hypoleuca	Black-sided Robin	U	6	18					New Guinea	61
Poecilodryas superciliosa	White-browed Robin	M	13	21.8	1.8	20	25.5		Australia	518
		F	10	18	1.5	15.5	20.3			
Poecilodryas placens	Olive-yellow Robin	M	2	27		26	28		New Guinea	283
		F	3	24		23	25			
Poecilodryas albonotata	Black-throated Robin	B	10	38.6	2.9	34.5	43		New Guinea	283, 387, 424
Peneothello sigillata	White-winged Robin	B	33	22.8	1.7	18.5	26.7		Papua New Guinea	387
Peneothello cryptoleucus	Smoky Robin	U		18.9					New Guinea	229

Body Masses of World Birds (continued)

Species Name	Common Name	Sex	N	Mean	S.D.	Min	Max	Season	Location	Source #
Peneothello bimaculatus	White-rumped Robin	M	7	26.3		25	28		New Guinea	283
		F	2	21.5		21	22			
Peneothello cyanus	Blue-gray Robin	M	10	26.9	1.5	24	30		New Guinea	283
		F	10	23.2	1.7	20.7	26			
Heteromyias albispecularis	Ashy Robin	B	437	36.3	4.6				Queensland, Australia	391
Heteromyias cinereifrons	Gray-headed Robin	M	1	40.5					Queensland, Australia	470
Pachycephalopsis hattamensis	Green-backed Robin	U		34.6					New Guinea	229
Pachycephalopsis poliosoma	White-eyed Robin	M	10	40.5	2.6	35	42		New Guinea	283
		F	10	35.6	2.7	32	40			
Drymodes superciliaris	Northern Scrub-Robin	B	7	45.9		36	58			283, 518
Drymodes brunneopygia	Southern Scrub-Robin	M	13	36.7	1.8	34.5	40		Western Australia	518
		F	12	29.6	2.1	26.3	33.8			
Family Pachycephalidae										
Mohoua albicilla	Whitehead	M	20	18.3	1.3	16	21		New Zealand	1002
		F	27	17	2.4	12	20			
Mohoua ochrocephala	Yellowhead	M	19	29.5	1.2	28	32.5		New Zealand	518
		F	13	26	1.9	23	29.5			
Mohoua novaeseelandiae	Pipipi	M	51	13.4	0.7	12	15		New Zealand	240
		F	24	11	0.4	10.5	12			
Falcunculus frontatus	Crested Shrike-tit	M	29	30.5	2.9	26.5	40		Southern Australia	518
		F	19	25.6	2.4	23	33.2			
Oreoica gutturalis gutturalis	Crested Bellbird	B	26	65.3		55	77.5		Australia	518
Oreoica gutturalis pallescens	Crested Bellbird	B	24	60.4		53	67		Australia	518
Rhagologus leucostigma	Mottled Whistler	B	7	26.8		23	29.5		New Guinea	283
Pachycare flavogriseum	Dwarf Whistler	B	17	16.2		12.5	19		New Guinea	283

Body Masses of World Birds (continued)

Species Name	Common Name	Sex	N	Mean	S.D.	Min	Max	Season	Location	Source #
Aleadryas rufinucha	Rufous-naped Whistler	B	15	40.5	3.3	37.1	47.1		Papua New Guinea	387
Pachycephala olivacea	Olive Whistler	B	38	41.2		36.2	44.9		Tasmania, Australia	518
Pachycephala rufogularis	Red-lored Whistler	B	13	36.6		34	40		Australia	518
Pachycephala inornata	Gilbert's Whistler	B	57	30.8		20.5	36		New South Wales, Australia	518
Pachycephala grisola	Mangrove Whistler	B	4	21					India	7
Pachycephala albiventris	Green-backed Whistler	B	8	21.1		17.5	26		Luzon, Philippines	1143
Pachycephala homeyeri	White-vented Whistler	U	9			23	29		Philippines	344
Pachycephala hyperythra	Rusty Whistler	B	8	27.7		25.5	29		New Guinea	283
Pachycephala modesta	Brown-backed Whistler	B	10	18.3	1.3	16	20.1		New Guinea	283, 424, 770
Pachycephala sulfuriventer	Sulphur-bellied Whistler	U	25	21.4	1.4	18	24		Sulawesi, Indonesia	1272
Pachycephala meyeri	Vogelkop Whistler	U	5	19					New Guinea	288
Pachycephala philippinensis	Yellow-bellied Whistler	B	67	22.3		18	28.8		Philippines	954
Pachycephala griseiceps	Gray-headed Whistler	B	20	21.8		20	24.3		New Guinea	283
Pachycephala simplex	Gray Whistler	U	34	17.4	1	15	20		Queensland, Australia	518
Pachycephala pectoralis	Golden Whistler	U	26	32.7	2.4	28.5	38		Fiji	668
Pachycephala pectoralis pectoralis	Golden Whistler	B	44	24.1	1.8	20	27.6		Queensland, Australia	390a
Pachycephala pectoralis youngi	Golden Whistler	B	58	26.5		20	35		Southeast Australia	518
Pachycephala pectoralis glaucura	Golden Whistler	M	18	27.8	2	22.7	31		Tasmania, Australia	518
		F	7	30.4		26.8	37			

Body Masses of World Birds (continued)

Species Name	Common Name	Sex	N	Mean	S.D.	Min	Max	Season	Location	Source #
Pachycephala pectoralis fuliginosa	Golden Whistler	B	52	25.6		23	30		Southern Australia	518
Pachycephala pectoralis	Golden Whistler	M	4	24.4		22	26.5		New Ireland	58
		F	4	22.2		20.4	24			
Pachycephala pectoralis cinnamomea	Golden Whistler	M	18	51.7		48	56		Guadalcanal	399
		F	8	44.9		42	47.5			
Pachycephala pectoralis christophori	Golden Whistler	M	41	33.6		28	36		Solomon Is.	399
		F	21	32.2		28	37.5			
Pachycephala soror	Sclater's Whistler	B	18	24.8		22.5	27.2		New Guinea	283
Pachycephala lorentzi	Lorentz's Whistler	U	5	19.4		18	20.3		New Guinea	424
Pachycephala melanura	Black-tailed Whistler	B	23	23.8		20.5	27.5		Northern Australia	518
Pachycephala caledonica	New Caledonian Whistler	U	17	20.9	1.3	18	23		New Caledonia	1018
Pachycephala caledonica intacta	New Caledonian Whistler	B	7	24.6		23.2	26.2		Vanuatu	656b
Pachycephala schlegelii	Regent Whistler	B	19	22.7	1.8	19.3	25.4		Papua New Guinea	387
Pachycephala implicata	Hooded Whistler	M	7	35.7		33.5	37		Guadalcanal	399
		F	13	33.3		30	39			
Pachycephala griseonota	Drab Whistler	U				23	26		Seram, Moluccas	1181
Pachycephala griseonota	Drab Whistler	U				20	21		Buru, Moluccas	1182
Pachycephala leucogastra	White-bellied Whistler	U	2	18.5		17	20		New Guinea	770
Pachycephala rufiventris falcata	Rufous Whistler	B	29	20.8		17	23.5		Northwest Australia	518
Pachycephala rufiventris rufiventris	Rufous Whistler	B	147	24.3		21	31		New South Wales, Australia	518
Pachycephala lanioides carnaroni	White-breasted Whistler	B	62	40.9		31	52		Western Australia	518
Pachycephala lanioides fretorum	White-breasted Whistler	B	9	36.6		33	42		Northern Terr., Australia	518

Body Masses of World Birds (continued)

Species Name	Common Name	Sex	N	Mean	S.D.	Min	Max	Season	Location	Source #
Colluricincla megarhyncha	Rufous Shrike-Thrush	U	81	32					New Guinea	61
Colluricincla megarhyncha normani	Rufous Shrike-Thrush	U	328	32.7	2.4	22	48		Queensland, Australia	518
Colluricincla boweri	Bower's Shrike-Thrush	B	51	43.8		37	49.6		Queensland, Australia	518
Colluricincla woodwardi	Sandstone Shrike-Thrush	B	28	54.5		46.3	62		Australia	518
Colluricincla harmonica harmonica	Gray Shrike-Thrush	B	65	63.2		47	75		Australia	518
Colluricincla harmonica strigata	Gray Shrike-Thrush	B	18	74.9		70	84		Tasmania, Australia	518
Colluricincla harmonica rufiventris	Gray Shrike-Thrush	B	37	63.9		56	79		Australia	518
Colluricincla harmonica brunnea	Gray Shrike-Thrush	B	31	62		51	79		Australia	518
Colluricincla tenebrosa	Morningbird	U		46					New Guinea	229
Pitohui dichrous	Hooded Pitohui	B	17	92.9	4	85	100		New Guinea	283
Pitohui ferrugineus	Rusty Pitohui	M	10	98	8	86	110		New Guinea	283
		F	7	90		77	99			
Pitohui cristatus	Crested Pitohui	U	3	79					New Guinea	61
Pitohui kirhocephalus	Variable Pitohui	B	20	72		67	79		New Guinea	283
Pitohui nigrescens	Black Pitohui	B	8	76.9		68	86		New Guinea	283, 424
Eulacestoma nigropectus	Wattled Ploughbill	B	24	20	1.5	17	23.2		Papua New Guinea	387
Family Picathartidae										
Picathartes gymnocephalus	White-necked Rockfowl	F	2	196		166	226			396
Picathartes oreas	Gray-necked Rockfowl	U	6	220		202	239			396

Body Masses of World Birds (continued)

Species Name	Common Name	Sex	N	Mean	S.D.	Min	Max	Season	Location	Source #
Family Timaliidae										
Garrulax cinereifrons	Ashy-headed Laughingthrush	M	1	70					Sri Lanka	7
Garrulax perspicillatus	Masked Laughingthrush	M	10	124		102	142		China	184a
		F	10	112		98	130			
Garrulax albogularis	White-throated Laughingthrush	M	17			97	114			7
		F	7			78	105			
Garrulax leucolophus	White-crested Laughingthrush	M	3			123	129		India	7
		F	5			119	123			
Garrulax monileger	Lesser Necklaced Laughingthrush	U	5			77	91		India	7
Garrulax pectoralis	Greater Necklaced Laughingthrush	M	1	156						7
		F	2	135		135	135			
Garrulax striatus vibex	Striated Laughingthrush	B	6			126	148		India	7
Garrulax striatus sikkimensis	Striated Laughingthrush	M	1	92					India	7
		F	3			99	106			
Garrulax ruficollis	Rufous-necked Laughingthrush	B	6			60	73		India	7
Garrulax nuchalis	Chestnut-backed Laughingthrush	F	1	78					Myanmar	1148
Garrulax galbanus	Yellow-throated Laughingthrush	B	3	56		55	57		India	7
Garrulax gularis	Rufous-vented Laughingthrush	M	1	92					India	7
Garrulax davidi	Pere David's Laughingthrush	B	45	56.2		48	63		Hopei, China	1087
Garrulax sukatschewi	Sukatschev's Laughingthrush	F	1	67.8				B	China	1360a
Garrulax cinerascens	Moustached Laughingthrush	M	4			47	51		India	7
Garrulax rufogularis	Rufous-chinned Laughingthrush	B	7			58	68		India	7

Body Masses of World Birds (continued)

Species Name	Common Name	Sex	N	Mean	S.D.	Min	Max	Season	Location	Source #
Garrulax ocellatus	Spotted Laughingthrush	B	5	114		110	121		India	7
Garrulax caerulatus	Gray-sided Laughingthrush	M	4			82	99			7
		F	6			79	84			
Garrulax poecilorhynchus	Rusty Laughingthrush	U				70	74		Taiwan	834
Garrulax mitratus	Chestnut-capped Laughingthrush	U	17	62		51	70		Malaysia	775
Garrulax merulinus	Spot-breasted Laughingthrush	F	2	78.8		75	82.5		India	1148
Garrulax canorus	Hwamei	M	10	58.4		65	75		China	184a
		F	10	67.5		53.8	75			
Garrulax sannio	White-browed Laughingthrush	B	20	67.8		53.8	85		China	184a
Garrulax jerdoni	Gray-breasted Laughingthrush	B	12	45.5	2.6	39	49		India	1148
Garrulax lineatus	Streaked Laughingthrush	U	25	40.7		36	46		India	7
Garrulax virgatus	Striped Laughingthrush	F	1	48					Myanmar	1148
Garrulax subunicolor	Scaly Laughingthrush	M	3			63	69			7
Garrulax austeni	Brown-capped Laughingthrush	M	4			63	74		India	7
		F	1	59						
Garrulax squamatus	Blue-winged Laughingthrush	F	1	84						7
Garrulax elliotii	Elliot's Laughingthrush	U	1	45					China	1148
Garrulax variegatus	Variegated Laughingthrush	U	10	64.5		59	72	B	India	7
Garrulax affinis	Black-faced Laughingthrush	B	14			66	80		India	7
Garrulax morrisonianus	White-whiskered Laughingthrush	B	64	73.1	6.9	58	93		Taiwan	1359

Body Masses of World Birds (continued)

Species Name	Common Name	Sex	N	Mean	S.D.	Min	Max	Season	Location	Source #
Garrulax erythrocephalus	Chestnut-crowned Laughingthrush	U	68	71.7		62	95		Malaysia	775
Liocichla steerii	Steere's Liocichla	M	22	32.4	1.3				Taiwan	771a
		F	21	30.4	2.5					
Liocichla phoenicea	Red-faced Liocichla	U	4	49		45	52		India	7
Modulatrix stictigula	Spot-throat	U	49	30.5		27	34			623
Arcanator orostruthus	Dapple-throat	B	22	32.7		28.5	38			623
Trichastoma celebense	Sulawesi Babbler	B	76	23.6	1.6	20	29		Sulawesi, Indonesia	1272
Trichastoma bicolor	Ferruginous Babbler	U	12	27.2	4.7	19.3	34		Malaysia; Borneo	398, 785
Malacocincla abbotti	Abbott's Babbler	M	5			26	32.5			7
		F	2	26.8		26.5	27			
Malacocincla sepiaria	Horsfield's Babbler	B	11	25.5	2.2	23	29		Borneo	398
Malacocincla malaccensis	Short-tailed Babbler	B	20	21.4	2.3	17	26		Borneo	398
Malacocincla cinereiceps	Ashy-headed Babbler	U	14			22.1	25.9		Palawan, Philippines	487
Pellorneum fuscocapillum	Brown-capped Babbler	M	1	30					Sri Lanka	7
Pellorneum tickelli	Buff-breasted Babbler	B	3	17.1		16.2	17.7		Thailand	787
Pellorneum albiventre	Spot-throated Babbler	M	2	21.5		21	22		India	7
Pellorneum ruficeps	Puff-throated Babbler	U	15	26		21	30		India	7
Pellorneum capistratum	Black-capped Babbler	U	10	25.1	1.5	23	27		Borneo	398
Malacopteron palawanense	Palawan Babbler	U	4			28.1	36.7		Palawan, Philippines	487

Body Masses of World Birds (continued)

Species Name	Common Name	Sex	N	Mean	S.D.	Min	Max	Season	Location	Source #
Malacopteron magnirostre	Moustached Babbler	M	8	20.8		19.5	24.9		Borneo; Malaysia	785, 1218
		F	4	18		16.8	19			
Malacopteron affine	Sooty-capped Babbler	M	1	15.7					Malaysia	785
		F	1	18.6						
Malacopteron cinereum	Scaly-crowned Babbler	U	56	18.1	1.7	15	21		Borneo	398
Malacopteron magnum	Rufous-crowned Babbler	U	26	27.2	3.2	22	34		Borneo	398
Malacopteron albogulare	Gray-breasted Babbler	U	15	16.5	1.2	14	18		Borneo	398
Illadopsis cleaveri	Blackcap Illadopsis	U	15	30.8		28	34		Congo	396
Illadopsis albipectus	Scaly-breasted Illadopsis	U	34	30.6		23	36			396
Illadopsis rufescens	Rufous-winged Illadopsis	U	6	37		34	41		Ivory Coast	396
Illadopsis puveli	Puvel's Illadopsis	B	12	48.4		41	52.5		Uganda	396
Illadopsis rufipennis	Pale-breasted Illadopsis	U	21	22.2		18	27		Kenya	396
Illadopsis fulvescens	Brown Illadopsis	U	60	27.9		20	37		Uganda	396
Illadopsis pyrrhoptera	Mountain Illadopsis	U	66	24.8		20	30		Kenya	396
Illadopsis abyssinica	African Hill Babbler	B	34	18.8		14	25		Kenya	396
Kakamega poliothorax	Gray-chested Illadopsis	U	49	36.4	1.2				Kenya	396
Ptyrticus turdinus	Thrush Babbler	B	3	65.8		59	72.5			396
Pomatorhinus hypoleucos	Large Scimitar-Babbler	U	6	75.6		69	82		Malaysia	775, 785
Pomatorhinus erythrocnemis	Spot-breasted Scimitar-Babbler	M	15			57	70			7
		F	6			47	61			
Pomatorhinus erythrogenys	Rusty-cheeked Scimitar-Babbler	B	4	54.9		52.5	57		Thailand	787

Body Masses of World Birds (continued)

Species Name	Common Name	Sex	N	Mean	S.D.	Min	Max	Season	Location	Source #
Pomatorhinus horsfieldii	Indian Scimitar-Babbler	U	12	43		33	53		India	7
Pomatorhinus schisticeps	White-browed Scimitar-Babbler	U	12	43		33	53		India	7
Pomatorhinus montanus	Chestnut-backed Scimitar-Babbler	M	2	30.6		28.2	33.1		Borneo; Malaysia	785, 1218
		F	1	26.8						
Pomatorhinus ruficollis	Streak-breasted Scimitar-Babbler	M	8	33.1		30	39			7
		F	5	30.2		26	35			
Pomatorhinus ochraceiceps	Red-billed Scimitar-Babbler	F	1	34					India	7
Pomatorhinus ferruginosus	Coral-billed Scimitar-Babbler	F	1	40						7
Xiphirhynchus superciliaris	Slender-billed Scimitar-Babbler	U	3	28		27	30			7
Jabouilleia naungmungensis	Naung Mung Scimitar-Babbler	F	2	35.5		35	36		Myanmar	954a
Rimator malacoptilus	Long-billed Wren-Babbler	M	4			18	21			7
Ptilocichla mindanensis	Striated Wren-Babbler	B	16	29.2		26.3	31		Philippines	954
Kenopia striata	Striped Wren-Babbler	U	8	19.9		18.7	21.1		Borneo	398, 1218
Napothera macrodactyla	Large Wren-Babbler	M	1	48.7					Malaysia	785
Napothera crispifrons	Limestone Wren-Babbler	B	2	27		25.8	28.3		Thailand	1148
Napothera brevicaudata	Streaked Wren-Babbler	U	10	19.5		14	22		Malaysia	775
Napothera crassa	Mountain Wren-Babbler	U	4	26.8		24	29.2		Borneo, Malaysia	1148
Napothera rabori	Luzon Wren-Babbler	U	2	61		57	65		Philippines	436
Napothera epilepidota	Eyebrowed Wren-Babbler	B	5	16		14.5	19		Thailand	787

Body Masses of World Birds (continued)

Species Name	Common Name	Sex	N	Mean	S.D.	Min	Max	Season	Location	Source #
Pnoepyga albiventer	Scaly-breasted Wren-Babbler	M	10	20.9		17	24			7
Pnoepyga immaculata	Immaculate Wren-Babbler	M	1	20.5					Nepal	760
Pnoepyga pusilla	Pygmy Wren-Babbler	U	8	12		11	14			7
Spelaeornis caudatus	Rufous-throated Wren-Babbler	B	2	11		10	12			7
Spelaeornis chocolatinus	Long-tailed Wren-Babbler	B	6			10	14		India	7
Sphenocichla humei	Wedge-billed Wren-Babbler	B	6	34.1		29.2	41		Myanmar	1148
Neomixis tenella orientalis	Common Jery	U	21	7.3	0.5	6.5	8.5		Madagascar	957
Neomixis tenella debilis	Common Jery	U	3	6.5		6.3	6.8		Madagascar	957
Neomixis viridis	Green Jery	U	1	7					Madagascar	957
Neomixis striatigula pallidior	Stripe-throated Jery	U	6	8.3		6.2	9.5		Madagascar	957
Hartertula flavoviridis	Wedge-tailed Jery	B	11	9.7	1	8	11.5		Madagascar	1143
Stachyris ambigua	Buff-chested Babbler	B	3	9.7		9	10			7
Stachyris rufifrons	Rufous-fronted Babbler	M	1	10.6					India	1148
Stachyris ruficeps	Rufous-capped Babbler	U	19	10.3		8	12			7
Stachyris pyrrhops	Black-chinned Babbler	B	9			8	11.5			7
Stachyris chrysaea	Golden Babbler	U	13	9		7	10		Malaysia	775
Stachyris plateni	Pygmy Babbler	B	23	8.7		7.5	10.2		Philippines	954
Stachyris dennistouni	Golden-crowned Babbler	B	9	15.7		13.4	17.4		Luzon, Philippines	1143
Stachyris nigrocapitata	Black-crowned Babbler	B	31	14.1		12.6	16		Philippines	954

Body Masses of World Birds (continued)

Species Name	Common Name	Sex	N	Mean	S.D.	Min	Max	Season	Location	Source #
Stachyris capitalis	Rusty-crowned Babbler	B	7	14.9		13.4	15.8		Philippines	436, 954
Stachyris speciosa	Flame-templed Babbler	U	5			15	18		Philippines	344
Stachyris whiteheadi	Chestnut-faced Babbler	M	32	22.8	2	18	26.7		Philippines	436
		F	35	21	2.1	17	26.2			
Stachyris striata	Luzon Striped-Babbler	B	3	13.9		13.4	14.9		Luzon, Philippines	1143
Stachyris latistriata	Panay Striped-Babbler	M	1	29.5					Panay, Philippines	1148
Stachyris nigrorum	Negros Striped-Babbler	U	31	21.6	2.1	18.5	27		Negros, Philippines	1143
Stachyris nigriceps	Gray-throated Babbler	U	42	15.8		13	20		Malaysia	775
Stachyris poliocephala	Gray-headed Babbler	M	4	24.5		23.3	25.9		Borneo; Malaysia	785, 1218
		F	6	22.2		20.1	25.7			
Stachyris oglei	Snowy-throated Babbler	B	2	33		33	33		Myanmar	1148
Stachyris leucotis	White-necked Babbler	B	6	23.7		21.5	26.5		Malaysia	785
Stachyris nigricollis	Black-throated Babbler	U	13	26.2	2.2	23	31		Borneo	398
Stachyris maculata	Chestnut-rumped Babbler	U	24	29.2	2	26	33		Borneo	398
Stachyris erythroptera	Chestnut-winged Babbler	B	8	12.6		11	15.1		Borneo; Malaysia	785, 1218
Dumetia hyperythra	Tawny-bellied Babbler	U	10	12.9		10	15		India	7
Rhopocichla atriceps	Dark-fronted Babbler	B	3	16.3		16	17		Sri Lanka	7
Macronous gularis	Striped Tit-Babbler	B	9			10	14			7
Macronous striaticeps	Brown Tit-Babbler	B	45	16.5		14.2	20		Philippines	954

Body Masses of World Birds (continued)

Species Name	Common Name	Sex	N	Mean	S.D.	Min	Max	Season	Location	Source #
Macronous ptilosus	Fluffy-backed Tit-Babbler	U	2	18		17	19		Malaysia; Borneo	398, 785
Micromacronus leytensis	Miniature Tit-Babbler	B	2	5.5		5	6		Mindanao, Philippines	1143
Timalia pileata	Chestnut-capped Babbler	B	2	16		15	17			7
Chrysomma sinense hypoleucum	Yellow-eyed Babbler	B	10	16.1		12	20			7
Chrysomma sinense sinense	Yellow-eyed Babbler	U	15	18.3		15	21			7
Turdoides nipalensis	Spiny Babbler	M	2	61		58	64		India	7
Turdoides altirostris	Iraq Babbler	B	9	33						223
Turdoides caudatus caudatus	Common Babbler	M	31	35.2	2.6				India	223
		F	37	32.1	2.3					
Turdoides caudatus huttoni	Common Babbler	M	2	47.5		47	48		Afghanistan; Iran	223
		F	4	43.8		41	46			
Turdoides earlei	Striated Babbler	U	2	47		46	48			7
Turdoides gularis	White-throated Babbler	B	2	45.5		43.1	47.9		Myanmar	1148
Turdoides longirostris	Slender-billed Babbler	F	1	35						7
Turdoides malcolmi	Large Gray Babbler	B	27	75.7		63	92		India	7
Turdoides squamiceps	Arabian Babbler	U				64	83		Israel	223
Turdoides fulvus	Fulvous Chatterer	B	17	61.4		46	70		Northern Africa	223
Turdoides aylmeri	Scaly Chatterer	U	15	35.3		31	42			396
Turdoides rubiginosus	Rufous Chatterer	B	18	40.6		33	49		Kenya; Uganda	396
Turdoides subrufus	Rufous Babbler	B				57	78		India	7
Turdoides striatus orientalis	Jungle Babbler	U	10	66		60	80			7

Body Masses of World Birds (continued)

Species Name	Common Name	Sex	N	Mean	S.D.	Min	Max	Season	Location	Source #
Turdoides reinwardtii	Blackcap Babbler	U	12	78.8		69	91		Ghana	396
Turdoides tenebrosus	Dusky Babbler	M	1	76					Sudan	1143
Turdoides sharpei	Black-lored Babbler	U	12	79.3		66	84		Kenya	396
Turdoides melanops melanops	Black-faced Babbler	M	2	77		76	78			396
Turdoides squamulatus	Scaly Babbler	M	7	72.7		55	85			396
		F	6	66.2		60	78			
Turdoides leucopygia	White-rumped Babbler	U	3	82.4		75	93			396
Turdoides hartlaubii	Hartlaub's Babbler	U	23	86.2		68	105		Namibia; Botswana	1036
Turdoides bicolor	Southern Pied-Babbler	U	36	78.3		55	90		Southern Africa	1036
Turdoides hypoleucus	Northern Pied-Babbler	U	3	74		67	80		Kenya	396
Turdoides hindei	Hinde's Pied-Babbler	U	11	67.6		58	77			396
Turdoides plebejus	Brown Babbler	U	25	62.5		52	70		Kenya	396
Turdoides jardineii	Arrow-marked Babbler	U	201	78.2		58.6	115		Southern Africa	1036
Turdoides gymnogenys	Bare-cheeked Babbler	U	28	77		60	93		Namibia	1036
Leiothrix argentauris	Silver-eared Leiothrix	U	41	28.4		24	36		Malaysia	775
Leiothrix lutea	Red-billed Leiothrix	M	20	21.3	1.1				Hawaiian Is.	741
		F	40	21.5	1.6					
Cutia nipalensis	Cutia	M	9			48	56			7
		F	3			40	46			
Pteruthius rufiventer	Black-headed Shrike-Babbler	M	7			44	48			7
		F	2	43		42	44			

Body Masses of World Birds (continued)

Species Name	Common Name	Sex	N	Mean	S.D.	Min	Max	Season	Location	Source #
Pteruthius flaviscapis	White-browed Shrike-Babbler	B	9			34	44			7
Pteruthius xanthochlorus	Green Shrike-Babbler	B	3	14.3		14	15			7
Pteruthius melanotis	Black-eared Shrike-Babbler	U	7	13.3		12	15		Malaysia	775
Gampsorhynchus rufulus	White-hooded Babbler	M	1	37						7
Actinodura egertoni	Rusty-fronted Barwing	B	8	36		33	38		India	7
Actinodura ramsayi	Spectacled Barwing	B	6	38.4		35.1	41.3		Thailand	787
Actinodura nipalensis	Hoary-throated Barwing	M	5			44	48			7
		F	5			39	45			
Actinodura waldeni	Streak-throated Barwing	B	18			39	56		India	7
Actinodura morrisoniana	Taiwan Barwing	U				23	25		Taiwan	834
Minla cyanouroptera	Blue-winged Minla	U	13	17		15.4	19		Malaysia; Thailand	775, 787
Minla strigula	Chestnut-tailed Minla	U	111	19.2		16	22		Malaysia	775
Minla ignotincta	Red-tailed Minla	B	6	14.3		11	18			7
Alcippe chrysotis	Golden-breasted Fulvetta	M	1	5.5						7
Alcippe cinerea	Yellow-throated Fulvetta	F	2	11		11	11			7
Alcippe castaneceps	Rufous-winged Fulvetta	U	41	12.5		11	20		Malaysia	775
Alcippe vinipectus	White-browed Fulvetta	B	11			11	13			7
Alcippe cinereiceps	Streak-throated Fulvetta	M	2	10		10	10			7
Alcippe brunnea	Dusky Fulvetta	U				16	19			7

Body Masses of World Birds (continued)

Species Name	Common Name	Sex	N	Mean	S.D.	Min	Max	Season	Location	Source #
Alcippe brunneicauda	Brown Fulvetta	B	10	14.3		13.3	15.3		Malaysia	785
Alcippe poioicephala	Brown-cheeked Fulvetta	U	14	20.7		18	23		India	7
Alcippe morrisonia	Gray-cheeked Fulvetta	B	30	14.8	0.7	12.8	16.5		Thailand	787
Alcippe morrisonia	Gray-cheeked Fulvetta	M	165	15.4	0.9				Taiwan	661
		F	157	15.6	0.8					
Alcippe peracensis	Mountain Fulvetta	M	7	14.3		13.1	16.4		Malaysia	785
		F	7			15.5	15.8			
Alcippe nipalensis	Nepal Fulvetta	U	90	15.8		13	20		Malaysia	775
Lioptilus nigricapillus	Bush Blackcap	M	4	28.9		27	31		South Africa	1036
		F	7	30.4		23.9	35.8			
Kupeornis rufocinctus	Red-collared Mountain-Babbler	U	1	47					Central Africa	79
Parophasma galinieri	Abyssinian Catbird	U	1	21.5					Ethiopia	1243a
Phyllanthus atripennis	Capuchin Babbler	M	7	79.6		75	90		Uganda	396
		F	3	92		86	96			
Heterophasia annectens	Rufous-backed Sibia	B	2	24.5		24.2	24.8		Thailand	787
Heterophasia capistrata	Rufous Sibia	U	15	39.4		28	45		India	7
Heterophasia gracilis	Gray Sibia	U				34	42		India	7
Heterophasia melanoleuca	Black-backed Sibia	B	11	32.6	2.2	30.5	36.4		Thailand	787
Heterophasia auricularis	White-eared Sibia	U				36	48.5		Taiwan	834
Heterophasia pulchella	Beautiful Sibia	U				35	47		India	7
Heterophasia picaoides	Long-tailed Sibia	U	29	42.4		35	49		Malaysia	775
Yuhina castaniceps	Striated Yuhina	M	4	11.8		11	12		India	7
Yuhina everetti	Chestnut-crested Yuhina	M	2	12.1		11	13.2		Borneo, Malaysia	1148

Body Masses of World Birds (continued)

Species Name	Common Name	Sex	N	Mean	S.D.	Min	Max	Season	Location	Source #
Yuhina bakeri	White-naped Yuhina	U	4			14	21		India	7
Yuhina flavicollis	Whiskered Yuhina	B	35			13	22			7
Yuhina gularis	Stripe-throated Yuhina	B	18			18	24			7
Yuhina occipitalis	Rufous-vented Yuhina	B	4	13		12	16			7
Yuhina brunneiceps	Taiwan Yuhina	M	13	12.8					Taiwan	1360
		F	14	12.2						
Yuhina nigrimenta	Black-chinned Yuhina	B	6	9.5		8	11			7
Yuhina zantholeuca	White-bellied Yuhina	B	6	11.8		8	17			7
Myzornis pyrrhoura	Fire-tailed Myzornis	B	4	11.9		11	13			7
Oxylabes madagascariensis	White-throated Oxylabes	U	41	23.3	3.5	9.5	30		Madagascar	957
Crossleyia xanthophrys	Yellow-browed Oxylabes	U	7	19.9		15.5	22.5		Madagascar	957
Mystacornis crossleyi	Crossley's Babbler	U	15	24.5	1.8	21.5	28.5		Madagascar	957
Family Pomatostomidae										
Pomatostomus isidorei	New Guinea Babbler	U	11	70					New Guinea	61
Pomatostomus temporalis	Gray-crowned Babbler	B	321	75	4.9				Queensland, Australia	418
Pomatostomus superciliosus superciliosus	White-browed Babbler	U	12	40.4	4	34.4	46		South Australia	518
Pomatostomus superciliosus gilgandra	White-browed Babbler	U	15	39.5	2.4	35.8	43		Victoria, Australia	518
Pomatostomus superciliosus ashbyi	White-browed Babbler	M	13	44.4	1.8	42	49.2		Western Australia	518
		F	35	42	3	37.5	50.2			

Body Masses of World Birds (continued)

Species Name	Common Name	Sex	N	Mean	S.D.	Min	Max	Season	Location	Source #
Pomatostomus halli	Hall's Babbler	B	14	41.8	1.3	40	44		Queensland, Australia	518
Pomatostomus ruficeps	Chestnut-crowned Babbler	U	124	57.8		47	71		South Australia	518
Family Paradoxornithidae										
Panurus biarmicus	Bearded Reedling	M	63	15	1	13	17.5	B	France	223
		F	31	14	1.2	11	16			
Panurus biarmicus	Bearded Reedling	M	61	13.4	0.9	11	16	Y	Spain	929
		F	37	13	1	11.4	16			
Conostoma oemodium	Great Parrotbill	M	1	84					Nepal	1143
Paradoxornis unicolor	Brown Parrotbill	U	2	34		32	36			7
Paradoxornis gularis	Gray-headed Parrotbill	B	2	29		29	29		India	7
Paradoxornis guttaticollis	Spot-breasted Parrotbill	F	1	26						7
Paradoxornis webbianus	Vinous-throated Parrotbill	U	73	10.9	1.5				Taiwan	1085
Paradoxornis fulvifrons	Fulvous Parrotbill	B	5	6.4		6	6.5		Nepal	1143
Paradoxornis nipalensis	Black-throated Parrotbill	B				5	6			7
Paradoxornis verreauxi	Golden Parrotbill	U		6					Taiwan	834
Paradoxornis atrosuperciliaris	Black-browed Parrotbill	F	2	16.8		16.7	17		Myanmar; India	1148
Paradoxornis ruficeps	Rufous-headed Parrotbill	B	2	32		30	34		India	7
Family Orthonychidae										
Orthonyx temminckii	Logrunner	M	11	61.3	4	58	70.5		Queensland, Australia	518
		F	5	55		49	58			
Orthonyx spaldingii spaldingii	Chowchilla	M	9	187		167	208		Australia	518
		F	18	135	6.7	123	148			

Body Masses of World Birds (continued)

Species Name	Common Name	Sex	N	Mean	S.D.	Min	Max	Season	Location	Source #
Orthonyx spaldingii melasmenus	Chowchilla	M	2	164		155	172		Australia	518
		F	3	122		120	125			
Family Cinclosomatidae										
Psophodes olivaceus	Eastern Whipbird	M	25	68	3.1	64	76		Eastern Australia	518
		F	21	57.2	4.5	50.5	89			
Psophodes nigrogularis	Western Whipbird	M	11	46	2.2				Western Australia	1123
		F	12	43.1	2.7					
Psophodes occidentalis	Chiming Wedgebill	M	6	43.3		40	49		Australia	518
		F	5	36.3		27	40			
Psophodes cristatus	Chirruping Wedgebill	M	22	42.4	4.6	27.4	47.5		Australia	518
		F	18	39.1	2.4	36	44.5			
Cinclosoma punctatum	Spotted Quail-thrush	M	15	120	12.1	84	144		Eastern Australia	518
		F	10	107	8.5	88	117			
Cinclosoma castanotus	Chestnut Quail-thrush	M	72	74.6	4.1	67	87		Australia	376
		F	44	72.1	6.4	58	85			
Cinclosoma castaneothorax castaneothorax	Chestnut-breasted Quail-thrush	M	13	69.2	3.2	63	75		Australia	376
		F	14	61.5	3.9	56	67			
Cinclosoma castaneothorax marginatus	Chestnut-breasted Quail-thrush	M	62	60.8	4.4	54	74		Australia	376
		F	41	58.7	4.1	51	70			
Cinclosoma cinnamomeum cinnamomeum	Cinnamon Quail-thrush	M	19	58.8	3.1	53	67		Australia	376
		F	6	58.4		56	62			
Cinclosoma cinnamomeum alisteri	Cinnamon Quail-thrush	M	10	54.9		50	62		Australia	518
		F	6	52.7		50.7	54.5			
Ptilorrhoa leucosticta	Spotted Jewel-babbler	B	4	49.2		47.5	51		New Guinea	282, 283
Ptilorrhoa caerulescens	Blue Jewel-babbler	U	21	49					New Guinea	61
Ptilorrhoa castanonota	Chestnut-backed Jewel-babbler	B	6	69.7		58	74		New Guinea	283
Ifrita kowaldi	Blue-capped Ifrita	B	15	32.1	2.1	28.8	36		New Guinea	283, 424

Body Masses of World Birds (continued)

Species Name	Common Name	Sex	N	Mean	S.D.	Min	Max	Season	Location	Source #
Family Aegithalidae										
Aegithalos caudatus	Long-tailed Tit	U	457	8.6		7.1	10	Sp	former USSR	223
Aegithalos leucogenys	White-cheeked Tit	M	9			6	8		India	7
		F	7			6	7			
Aegithalos concinnus	Black-throated Tit	B	29	6.1		5	7.5		India	7
Aegithalos niveogularis	White-throated Tit	U	3	7.8					India	942
Aegithalos iouschistos	Black-browed Tit	M	5	7		6.5	7.5		India	7
Aegithalos fuliginosus	Sooty Tit	U				4	8			479
Psaltriparus minimus	Bushtit	B	53	5.3	0.4	4.5	6		California, USA	314
Family Maluridae										
Clytomyias insignis	Orange-crowned Fairywren	B	15	12.4		10	14		New Guinea	1022
Sipodotus wallacii	Wallace's Fairywren	B	6	7.7		6.5	8.5			1022
Malurus grayi grayi	Broad-billed Fairywren	M	5	16		14	17		New Guinea	285
		F	7	14.2		12	16			
Malurus grayi campbelli	Broad-billed Fairywren	M	3	11		10	12			1022
		F	3	9.7		8	11			
Malurus alboscapulatus	White-shouldered Fairywren	B	14	10.9		9.3	12.3		New Guinea	283
Malurus alboscapulatus moretoni	White-shouldered Fairywren	M	20	9	1.1	8	11			1022
		F	13	8.3	1.2	7	9			
Malurus melanocephalus	Red-backed Fairywren	B	50	7.9		5	10			1022
Malurus leucopterus	White-winged Fairywren	B	62	7.4		6.3	8.4			1022
Malurus cyaneus cyaneus	Superb Fairywren	M	20	11.3	0.8	10	13.8			1022
		F	31	10.5	0.8	9	12.2			
Malurus cyaneus cyanochlamys	Superb Fairywren	B	61	9.3		8	10.5			1022

Body Masses of World Birds (continued)

Species Name	Common Name	Sex	N	Mean	S.D.	Min	Max	Season	Location	Source #
Malurus splendens splendens	Splendid Fairywren	M	43	10.6	0.4	9.2	11.1			1022
		F	46	9.6	0.4	8.6	10.2			
Malurus splendens callainus	Splendid Fairywren	M	20	9	0.8	8	10			1022
		F	4	8.5		7.6	9			
Malurus lamberti assimilis	Variegated Fairywren	B	55	7.6		5.9	10		Australia	1022
Malurus lamberti lamberti	Variegated Fairywren	B	96	8.2		7	11			1022
Malurus lamberti rogersi	Variegated Fairywren	M	32	8.4	0.6	7.1	9.5		Western Australia	517
		F	26	8	0.4	7.1	9.2			
Malurus amabilis	Lovely Fairywren	M	12	9.1	1	7.5	11.5		Queensland, Australia	517
		F	6	8.2		7.5	9			
Malurus elegans	Red-winged Fairywren	M	468	9.9	0.5	8.5	11.5			1022
		F	323	9.3	0.6	8	11.5			
Malurus pulcherrimus	Blue-breasted Fairywren	M	29	9.4	0.5	8.4	10.3			1022
		F	18	9	0.4	7.8	10.5			
Malurus coronatus	Purple-crowned Fairywren	M	41	11.4	0.6	10.1	13			1022
		F	38	10.8	0.7	9.6	12.6			
Malurus cyanocephalus bonapartii	Emperor Fairywren	M	14	13.3	1.5	10	17			1022
		F	11	11.6	1	9.5	14			
Stipiturus malachurus malachurus	Southern Emuwren	B	67	7.3	0.1				Victoria, Australia	736
Stipiturus ruficeps	Rufous-crowned Emuwren	B	29	5		4	6.3		Australia	517
Stipiturus mallee	Mallee Emuwren	M	1	5.5						1022
Amytornis textilis	Thick-billed Grasswren	M	32	24	1.2	21.6	27.6		Western Australia	517
		F	35	22.1	1.4	19.6	25.6			
Amytornis purnelli purnelli	Dusky Grasswren	M	13	22	1.1	20	23.9		Australia	517
		F	11	20	1.2	17.6	21.5			
Amytornis ballarae	Kalkadoon Grasswren	B	15	20.9		17	25.5		Queensland, Australia	517
Amytornis housei	Black Grasswren	M	8	30.2		29	31			517
		F	9	27.7		23.5	34			

Body Masses of World Birds (continued)

Species Name	Common Name	Sex	N	Mean	S.D.	Min	Max	Season	Location	Source #
Amytornis striatus	Striated Grasswren	M	30	20.2	1	18	22.5		South Australia	619
		F	18	19.1	1.2	17	22			
Amytornis merrotsyi	Short-tailed Grasswren	B	7	22.5		21.5	23.5		South Australia	517
Amytornis woodwardi	White-throated Grasswren	M	8	36.8		33	40			1022
		F	8	33.4		30	37			
Amytornis dorotheae	Carpentarian Grasswren	M	3	24.3		24	24.5		Australia	517
		F	4	22.2		20.5	23.2			
Amytornis barbatus	Gray Grasswren	M	58	18.8	1.4	15	21		Australia	517
		F	69	18	1.2	15	21			
Amytornis goyderi	Eyrean Grasswren	M	10	17		15	18.5		Australia	517
		F	7	16.4		14	17.5			
Family Acanthizidae										
Dasyornis longirostris	Western Bristlebird	U	26	32.7	2.8	26.3	38.5		Australia	518
Dasyornis brachypterus	Eastern Bristlebird	U	23	41.8	3.6	34	49		New South Wales, Australia	518
Dasyornis broadbenti	Rufous Bristlebird	B	15	67.9		59	74.5		South Australia	518
Pycnoptilus floccosus floccosus	Pilotbird	B	10	31.4		25	39		Eastern Australia	518
Pycnoptilus floccosus sandlandi	Pilotbird	U	108	26.6	2.2	19	32		New South Wales, Australia	518
Origma solitaria	Rock Warbler	M	15	14.7	0.6	14	16		New South Wales, Australia	518
		F	10	14	0.9	12.3	15			
Oreoscopus gutturalis	Fernwren	U	105	20	1.3	17.4	23.2		Queensland, Australia	518
Crateroscelis murina	Rusty Mouse-Warbler	B	17	15		13	17		New Guinea	283
Crateroscelis robusta	Mountain Mouse-Warbler	U	35	18.6	1.6	15.5	21		Papua New Guinea	387
Sericornis citreogularis citreogularis	Yellow-throated Scrubwren	M	21	19.1	1.7	16.5	22		Eastern Australia	518
		F	11	17.3	2	14	20.5			

Body Masses of World Birds (continued)

Species Name	Common Name	Sex	N	Mean	S.D.	Min	Max	Season	Location	Source #
Sericornis citreogularis cairnsi	Yellow-throated Scrubwren	M	113	17.4	1	14.6	20.7		Queensland, Australia	390a
		F	77	16.7	1.3	14.1	21			
Sericornis frontalis frontalis	White-browed Scrubwren	M	25	13.3	1.3	11	15.8		Eastern Australia	518
		F	20	12.6	1.1	9.6	14			
Sericornis frontalis harterti	White-browed Scrubwren	M	19	16.4	1.2	14	18		Victoria, Australia	518
		F	13	14.9	1.4	12	17			
Sericornis frontalis mellori	White-browed Scrubwren	M	24	12.6	0.9	10.5	14.5		Southwest Australia	518
		F	11	12.1	1.1	11	14			
Sericornis frontalis balstoni	White-browed Scrubwren	B	29	11.4		8.8	18		Western Australia	518
Sericornis humilis	Brown Scrubwren	M	12	19.6	1.5	16.5	21.2		Tasmania, Australia	518
		F	10	17	1.4	14.5	19			
Sericornis keri	Atherton Scrubwren	M	9	12		10	13		Queensland, Australia	518
		F	11	10.8	1.2	8.9	13			
Sericornis beccarii boreonesioticus	Beccari's Scrubwren	B	42	12.5	1	10.4	14.7		New Guinea	282
Sericornis beccarii imitator	Beccari's Scrubwren	M	3	13.3		12.5	14		New Guinea	282
		F	1	11						
Sericornis beccarii dubius	Beccari's Scrubwren	B	26	10.9		8	14		Queensland, Australia	518
Sericornis virgatus	Perplexing Scrubwren	B	7	12.4		11.5	13.5		New Guinea	282
Sericornis nouhuysi	Large Scrubwren	B	96	17.1	1.4	11.4	21.1		Papua New Guinea	387
Sericornis magnirostris	Large-billed Scrubwren	M	214	9.8	0.8	7.9	11.7		Queensland, Australia	390a
Sericornis perspicillatus	Buff-faced Scrubwren	B	6	9		8	10.2		New Guinea	283
Sericornis papuensis	Papuan Scrubwren	B	100	11.2	0.8	9.5	14.4		Papua New Guinea	387
Sericornis arfakianus	Gray-green Scrubwren	B	20	8.7	0.8	6.5	10.3		New Guinea	283, 424
Sericornis spilodera	Pale-billed Scrubwren	B	13	11.7		10.3	13.2		New Guinea	283

Body Masses of World Birds (continued)

Species Name	Common Name	Sex	N	Mean	S.D.	Min	Max	Season	Location	Source #
Acanthornis magnus	Scrubtit	M	8	10.8		9.8	11.6		Tasmania, Australia	518
		F	7	9.6		9	10			
Pyrrholaemus brunneus	Redthroat	B	50	12.5		11	17		Australia	518
Pyrrholaemus sagittatus	Speckled Warbler	M	53	13.7	0.7	12.1	15		Australia	518
		F	54	12.9	1.3	8	17.5			
Calamanthus campestris campestris	Rufous Fieldwren	B	9	14.9		13.5	16		South Australia	518
Calamanthus campestris isabellinus	Rufous Fieldwren	B	11	13		10	15		South Australia	518
Calamanthus fuliginosus	Striated Fieldwren	B	21	22.1		19	26		Tasmania, Australia	518
Calamanthus fuliginosus albiloris	Striated Fieldwren	M	10	21.7	1.6	18.5	24		New South Wales, Australia	518
		F	9	18.8		16	21			
Hylacola pyrrhopygia	Chestnut-rumped Hylacola	B	9	16.8		13	20		Southwest Australia	518
Hylacola cauta cauta	Shy Hylacola	B	17	14.4		13	16.2		Southwest Australia	518
Hylacola cauta macrorhyncha	Shy Hylacola	M	25	16.1	1	14.5	18.5		New South Wales, Australia	518
		F	13	15.2	0.6	14	16.5			
Acanthiza murina	Papuan Thornbill	U	3	8.4		7.8	9		New Guinea	424
Acanthiza reguloides	Buff-rumped Thornbill	U	64	7.7	0.6	6.2	9.3		Victoria, Australia	518
Acanthiza inornata	Western Thornbill	U	137	7	0.5	5.8	8		Western Australia	518
Acanthiza iredalei iredalei	Slender-billed Thornbill	B	28	5.8		4.6	6.8		Southwest Australia	518
Acanthiza iredalei rosinae	Slender-billed Thornbill	U	23	6.3	0.8	5	8		South Australia	518
Acanthiza katherina	Mountain Thornbill	U	33	7.2	0.6	5.5	8.2		Paluma, Australia	518
Acanthiza pusilla pusilla	Brown Thornbill	U	325	6.8	0.7	5	12		Victoria, Australia	518
Acanthiza pusilla diemenensis	Brown Thornbill	M	21	8.6	0.9	7	11		Tasmania, Australia	518
		F	16	7.7	0.8	6	9			

Body Masses of World Birds (continued)

Species Name	Common Name	Sex	N	Mean	S.D.	Min	Max	Season	Location	Source #
Acanthiza ewingii	Tasmanian Thornbill	B	35	6.9		5.5	10.5		Tasmania, Australia	518
Acanthiza apicalis	Inland Thornbill	B	28	7.5		6.5	8.5		Southwest Australia	518
Acanthiza chrysorrhoa chrysorrhoa	Yellow-rumped Thornbill	U	39	9.2	0.4	8.4	9.9		Western Australia	518
Acanthiza chrysorrhoa leachi	Yellow-rumped Thornbill	B	13	10.4		9	11		Tasmania, Australia	518
Acanthiza chrysorrhoa normantoni	Yellow-rumped Thornbill	B	23	8.4		6.5	11		Australia	518
Acanthiza uropygialis	Chestnut-rumped Thornbill	B	56	6.4		4	8.1		Australia	518
Acanthiza robustirostris	Slaty-backed Thornbill	B	41	6.1		4.8	6.8		Australia	518
Acanthiza nana nana	Yellow Thornbill	B	19	6.1		5	7.2		Southern Australia	518
Acanthiza nana modesta	Yellow Thornbill	U	56	6.3	0.5	5.5	8.1		Victoria, Australia	518
Acanthiza lineata lineata	Striated Thornbill	U	280	7.2	0.5	6	8.8		Victoria, Australia	518
Acanthiza lineata alberti	Striated Thornbill	M	19	7	0.5	6	7.9		Queensland, Australia	518
		F	11	6.4	0.4	5.8	7			
Smicrornis brevirostris	Weebill	B	34	5.3		4.2	6.5		Australia	518
Gerygone cinerea	Mountain Gerygone	B	8	7.6		7	8.5		New Guinea	283, 428
Gerygone chloronotus	Green-backed Gerygone	B	29	6.6		5	8		Northern Terr., Australia	518
Gerygone palpebrosa	Fairy Gerygone	B	44	7.4	0.6	6.3	8.6		Queensland, Australia	470
Gerygone olivacea olivacea	White-throated Gerygone	B	19	7.3		5.5	8.5		Southeast Australia	518
Gerygone olivacea rogersi	White-throated Gerygone	B	29	6.2		5	6.9		Western Australia	518
Gerygone chrysogaster	Yellow-bellied Gerygone	U	6	8					New Guinea	61
Gerygone magnirostris	Large-billed Gerygone	B	28	6.8		5	8.4		Northwest Australia	518

Body Masses of World Birds (continued)

Species Name	Common Name	Sex	N	Mean	S.D.	Min	Max	Season	Location	Source #
Gerygone tenebrosa tenebrosa	Dusky Gerygone	B	23	7.4		5.8	10		Western Australia	518
Gerygone tenebrosa christophori	Dusky Gerygone	M	18	8.3	0.5	7.2	9		Western Australia	518
		F	20	7.4	0.4	6.9	8.2			
Gerygone mouki	Brown Gerygone	B	35	5.2		4.5	7.6		Eastern Australia	518
Gerygone sulphurea	Golden-bellied Gerygone	U	2	5.6		5.5	5.8		Thailand	787
Gerygone ruficollis	Brown-breasted Gerygone	M	1	7					New Guinea	428
Gerygone fusca	Western Gerygone	B	46	5.4		4.5	6.6		Northwest Australia	518
Gerygone levigaster	Mangrove Gerygone	B	55	6.2		5	9.5		Australia	518
Gerygone modesta	Norfolk Gerygone	B	8	7.5		7	8.7		Norfolk Is.	518
Gerygone igata	Gray Gerygone	U	112	6.4	0.7	5	8		New Zealand	1002
Gerygone albofrontata	Chatham Gerygone	M	39	11.1	0.9	9.8	13.8		Chatham Is.	1001
		F	11	9.5	1	8.3	11.5			
Gerygone flavolateralis	Fan-tailed Gerygone	U	12	6.3	0.3	5.5	6.5		New Caledonia	1018
Aphelocephala leucopsis leucopsis	Southern Whiteface	B	50	13.3		8.5	16		Australia	518
Aphelocephala leucopsis castaneiventris	Southern Whiteface	B	33	11.4		10	14		Western Australia	518
Aphelocephala pectoralis	Chestnut-breasted Whiteface	U	31	9	0.4	8.5	9.5		South Australia	518
Aphelocephala nigricincta	Banded Whiteface	B	12	10.4		8.5	12.5		Australia	518
Family Epthianuridae										
Epthianura tricolor	Crimson Chat	B	73	10.7		9	14		Australia	517
Epthianura aurifrons	Orange Chat	B	29	10.5		8	13.5		Australia	517
Epthianura crocea	Yellow Chat	B	43	9.3		7	11		Australia	517
Epthianura albifrons	White-fronted Chat	B	393	13.3		10.5	17.5		Australia	517

Body Masses of World Birds (continued)

Species Name	Common Name	Sex	N	Mean	S.D.	Min	Max	Season	Location	Source #
Poecile atricapillus	Black-capped Chickadee	U	293	10.8	0.8	8.6	12.9	Y	Pennsylvania, USA	828
Poecile gambeli	Mountain Chickadee	M	113	11.7	0.8	10	14.3		California, USA	60
		F	48	11.1	1.1	9.4	14.5			
Poecile gambeli	Mountain Chickadee	U	29	10.8	1				Utah, USA	216
Poecile sclateri eidos	Mexican Chickadee	U	74	10.4	0.8	9	12			360
Poecile sclateri sclateri	Mexican Chickadee	U	13	9.8	1	7.5	11			360
Poecile superciliosa	White-browed Tit	U	7	10.7					Sichuan, China	774
Poecile davidi	Pere David's Tit	U	3	10.3					Sichuan, China	774
Poecile rufescens	Chestnut-backed Chickadee	M	112	9.7	0.5	8.5	11.4		California, USA	250
		F	122	9.1	0.6	7.5	12.1			
Poecile hudsonica	Boreal Chickadee	B	84	9.8	0.8	7	12.4	PB	Ontario, Canada	314
Poecile cincta	Gray-headed Chickadee	B	15	11.4		9	13		Hopei, China	1087
Periparus ater	Coal Tit	U	176	9.2	0.6	7.9	10.7	F	Sweden	223
Periparus rufonuchalis	Black-breasted Tit	M	13			12.3	14.7		India	7
		F	8			11.4	12.4			
Periparus rubidiventris	Rufous-vented Tit	M	16			11.6	13.1		India	7
		F	9			10.5	12.3			
Periparus melanolophus	Black-crested Tit	M	23			8	9.8		India	7
		F	12			7	9.5			
Pardaliparus venustulus	Yellow-bellied Tit	U				9	12.5			479
Pardaliparus elegans	Elegant Tit	M	12	13.5	1.3	10.5	15.9		Luzon, Philippines	436
		F	7	12.2		11.4	13.4			
Pardaliparus amabilis	Palawan Tit	M	1	14.9					Palawan, Philippines	1143
		F	1	10.8						
Lophophanes cristatus	Crested Tit	M	39	11.4		10.6	12.5	W	Norway	223
		F	17	10.7		9.7	11.6			

Body Masses of World Birds (continued)

Species Name	Common Name	Sex	N	Mean	S.D.	Min	Max	Season	Location	Source #
Lophophanes dichrous	Gray-crested Tit	M	20			11.2	14		India	7
Lophophanes dichrous dichroides	Gray-crested Tit	U	4	13.8					Sichuan, China	774
Melaniparus leucomelas	White-winged Tit	U	10	16.1	1.2	13.1	17.5		Ghana	457
Melaniparus niger	Southern Black-Tit	B	22	21.5		17.2	26			396
Melaniparus carpi	Carp's Tit	B	17	18.6		14	20.3		Namibia; Angola	266, 396
Melaniparus albiventris	White-bellied Tit	B	3	21.2		20.3	22		Kenya; Sudan	1143
Melaniparus leuconotus leuconotus	White-backed Black-Tit	U	2	16.7		16.6	16.8			479
Melaniparus leuconotus guineensis	White-backed Black-Tit	U				12.9	21		Ethiopia; Cameroon	479
Melaniparus rufiventris	Rufous-bellied Tit	U	3	18.2		16	19.5		Zambia	915
Melaniparus funereus	Dusky Tit	U				22.2	29			479
Melaniparus fasciiventer	Stripe-breasted Tit	B	4	15.2		14	17.6			79, 396
Melaniparus thruppi	Somali Tit	U	1	12					Uganda	479
Melaniparus griseiventris	Miombo Tit	U	5	15.5		14	16.5		Zambia	915
Melaniparus cinerascens	Ashy Tit	B	8	20		17	21.7		Namibia	396, 526
Melaniparus afer	Gray Tit	U	14	20.2		17.3	22.4			396
Parus major	Great Tit	M	186	18.9	0.9	17	21.5	W	Germany	223
		F	123	17.6	0.8	15.8	19.8			
Parus major cashmirensis	Great Tit	U	33	15		13.2	17.1		India	7
Parus major tibetanus	Great Tit	U	4	14					Sichuan, China	774
Parus bokharensis	Turkestan Tit	U				16.9	18			479
Parus monticolus	Green-backed Tit	M	24			12	16.8		India	7
		F	14			12.3	15.3			
Parus nuchalis	White-winged Tit	U	2	13.8		13	14.5			479

Body Masses of World Birds (continued)

Species Name	Common Name	Sex	N	Mean	S.D.	Min	Max	Season	Location	Source #
Parus xanthogenys	Black-lored Tit	M	11			13.7	16.9		India	7
		F	9			12.9	15.9			
Parus spilonotus	Yellow-cheeked Tit	M	2	18.8		18.3	19.4		India	7
Macholophus holsti	Yellow Tit	U				14	19.9		Taiwan	834
Cyanistes caeruleus	Blue Tit	U	69	10.6	0.7	9	12	F	France	223
Cyanistes caeruleus	Blue Tit	U	8	10.1					Canary Is.	441
Cyanistes cyanus	Azure Tit	M	9	13.8		13	14		Mongolia	223
		F	3	15.5		13	16			
Cyanistes flavipectus	Yellow-breasted Tit	U				10.6	12.3			479
Sittiparus semilarvatus snowi	White-fronted Tit	B	17	18	1.4	15.1	20.2		Luzon, Philippines	1143
Sittiparus varius	Varied Tit	U	11	17		15.9	18.2			133
Baeolophus wollweberi	Bridled Titmouse	B	14	10.4	0.4	9.8	11.5	PB	Arizona, USA	314
Baeolophus inornatus inornatus	Oak Titmouse	M	156	16.4	1	13	19.6		Oregon, California, USA	188
		F	114	15.5	1.1	13.2	21.5			
Baeolophus inornatus affabilis	Oak Titmouse	M	55	17.1	1.1	14.6	20		Southern California, USA	188
		F	42	16.3	1	14	19.2			
Baeolophus inornatus mohavensis	Oak Titmouse	M	26	15.9	1.1	13.5	18.2		California, USA	188
		F	15	15.2	1.7	11	19.1			
Baeolophus inornatus cineraceus	Oak Titmouse	M	18	15.7	0.8	14.1	16.9		Baja California Sur, Mexico	188
		F	11	15	0.8	13.6	16.1			
Baeolophus ridgwayi ridgwayi	Juniper Titmouse	M	23	15.9	0.8	13.5	16.8		Arizona, New Mexico, USA	188
		F	99	16.4	0.8	14.5	18			
Baeolophus ridgwayi zaleptus	Juniper Titmouse	B	86	15.3		13.4	19.5		California, Nevada, USA	188
Baeolophus bicolor bicolor	Tufted Titmouse	B	668	21.6		17.5	26.1	Y	Pennsylvania, USA	674a
Baeolophus atricristatus	Black-crested Titmouse	B	437	17.5	2.7	10.5	26		Texas, USA	277

Body Masses of World Birds (continued)

Species Name	Common Name	Sex	N	Mean	S.D.	Min	Max	Season	Location	Source #
Sylviparus modestus	Yellow-browed Tit	B	21			6	8.5		India	7
Melanochlora sultanea	Sultan Tit	B	4	37.6		35	41		India	7
Family Sittidae										
Sitta castanea	Chestnut-bellied Nuthatch	M	8			17.4	20.9		India	7
		F	5			19.9	20.5			
Sitta europaea	Eurasian Nuthatch	U	177	22.6	1.8	19.3	25.5	F	Sweden	223
Sitta europaea roseillia	Eurasian Nuthatch	M	6	20		19	21		Northern China	925
		F	5	18.6		18	19			
Sitta europaea amurensis	Eurasian Nuthatch	B	22	20.5		19	22		Northern China	925
Sitta nagaensis	Chestnut-vented Nuthatch	U	2	14.7		13.9	15.5		Thailand	787
Sitta cashmirensis	Kashmir Nuthatch	U				17.2	21			479
Sitta himalayensis	White-tailed Nuthatch	B	31			12	17		India	7
Sitta pygmaea	Pygmy Nuthatch	B	267	10.6					Western USA	859
Sitta pusilla	Brown-headed Nuthatch	B	123	10.2					Georgia, USA	859
Sitta whiteheadi	Corsican Nuthatch	F	1	11.8						223
Sitta ledanti	Algerian Nuthatch	B	2	17.2		16.5	18	B	Algeria	223
Sitta krueperi	Krueper's Nuthatch	M	8	12.4		10	14.3	PB		223
Sitta villosa	Snowy-browed Nuthatch	U				8	11			479
Sitta yunnanensis	Yunnan Nuthatch	U				7.5	13			479
Sitta canadensis	Red-breasted Nuthatch	B	310	9.8	0.7	8	12.7	F	New Jersey, USA	832
Sitta leucopsis	White-cheeked Nuthatch	M	10			13.5	15.9		India	7
		F	3			14.7	15.1			

Body Masses of World Birds (continued)

Species Name	Common Name	Sex	N	Mean	S.D.	Min	Max	Season	Location	Source #
Ashbyia lovensis	Desert Chat	B	21	17.5		13.5	20.5		Australia	517
Family Neosittidae										
Daphoenositta miranda	Black Sittella	M	1	17.8					New Guinea	428
		F	2	15.8		15.5	16			
Daphoenositta chrysoptera	Varied Sittella	B	50	11.7	1.1	9.8	14.5		Australia	470
Family Climacteridae										
Cormobates leucophaeus	White-throated Treecreeper	B	61	22		18	24.4		Victoria, Australia	517
Climacteris affinis	White-browed Treecreeper	B	24	21.7		19.5	28.5		Australia	517
Climacteris erythrops	Red-browed Treecreeper	B	31	23.3		20	30		Australia	517
Climacteris picumnus picumnus	Brown Treecreeper	B	111	34.9		26	44.5		New South Wales, Australia	517
Climacteris picumnus melanotus	Brown Treecreeper	M	16	28.7	3.1	24.8	34		Cape York Peninsula, Australia	517
		F	7	24.2		22.5	26.5			
Climacteris melanura	Black-tailed Treecreeper	M	38	33.5	2.3	28.5	38		Australia	517
		F	26	30.3	2.4	26	35.5			
Climacteris rufa	Rufous Treecreeper	M	123	34.6	2	29.5	39.5		Western Australia	517
		F	99	30.8	2	25.7	38.3			
Family Paridae										
Poecile lugubris	Sombre Tit	U	41	15.8	0.7			B	Turkey	223
Poecile palustris	Marsh Tit	U	245	10.9		9	13	Y	Germany	223
Poecile palustris	Marsh Tit	U	27	11.4		10	14		Northern China	925
Poecile hypermelaenus	Black-bibbed Tit	U				9	11.5			479
Poecile montana	Willow Tit	U	166	11.1	0.6	9.6	12.8	F	Sweden	223
Poecile songara	Songar Tit	U				8	12			479
Poecile carolinensis	Carolina Chickadee	M	32	10.1	0.4			B	Pennsylvania, USA	818
		F	30	9.9	0.5					

Body Masses of World Birds (continued)

Species Name	Common Name	Sex	N	Mean	S.D.	Min	Max	Season	Location	Source #
Rhabdornis inornatus	Stripe-breasted Rhabdornis	U	54	39.1		31.6	47.2		Philippines	954
Family Remizidae										
Auriparus flaviceps	Verdin	B	49	6.8	0.7	5.5	8.5		Southwest USA; Mexico	1150
Remiz pendulinus	Eurasian Penduline-Tit	B	40	9.3		8.4	10.4		Cyprus	223
Remiz macronyx	Black-headed Penduline-Tit	B	5	10.1		9.3	10.5		Kazakhstan	223
Remiz coronatus	White-crowned Penduline-Tit	B	10	6.9		5.2	7.8		Kazakhstan	223
Remiz consobrinus	Chinese Penduline-Tit	B	22	9.1		8	10		Hopei, China	1087
Anthoscopus punctifrons	Sennar Penduline-Tit	U	1	7						852
Anthoscopus musculus	Mouse-colored Penduline-Tit	M	2	5.2		5.1	5.2		Kenya	1143
Anthoscopus parvulus	Yellow Penduline-Tit	U	2	5.8		5.6	5.9		Nigeria	777
Anthoscopus caroli	African Penduline-Tit	U	17	6.9		5.5	10		Southern Africa	1036
Anthoscopus minutus	Southern Penduline-Tit	U	49	6.6		5	7.6		South Africa	1036
Cephalopyrus flammiceps	Fire-capped Tit	U	2	7		7	7		India	7
Pholidornis rushiae	Tit-hylia	B	4	6		4	7		Angola	266
Family Nectariniidae										
Chalcoparia singalensis	Ruby-cheeked Sunbird	B	2	8.2		8	8.4		Borneo	1218
Deleornis fraseri	Scarlet-tufted Sunbird	M	28	12.3		11	15			396
		F	16	10.4		8.3	13			
Deleornis axillaris	Gray-headed Sunbird	M	32	12.3		11	15			184
		F	18	10.3		8.3	13			

Body Masses of World Birds (continued)

Species Name	Common Name	Sex	N	Mean	S.D.	Min	Max	Season	Location	Source #
Anthreptes reichenowi reichenowi	Plain-backed Sunbird	M	14	7.7		6.7	8.8			184
		F	13	7.1		6.5	7.8			
Anthreptes reichenowi yokanae	Plain-backed Sunbird	B	13	7.2		5	10			184
Anthreptes anchietae	Anchieta's Sunbird	M	1	8						184
Anthreptes simplex	Plain Sunbird	M	12	9.4		8.3	10.7		Brunei	184
		F	11	8.5		7.1	11.8			
Anthreptes malacensis	Plain-throated Sunbird	M	12	12		11.1	13.5		Brunei	184
		F	8	10.7		7.4	13.1			
Anthreptes rhodolaemus	Red-throated Sunbird	B	7	13.1		12	15		Borneo	1218
Anthreptes longuemarei nyassae	Western Violet-backed Sunbird	M	25	11.8		10.7	13			184
		F	20	11.1		9.8	13.4			
Anthreptes orientalis	Kenya Violet-backed Sunbird	B	24	10.3		8	12			396
Anthreptes neglectus	Uluguru Violet-backed Sunbird	M	7	12.4		10	15.5			184, 396
		F	4	11.3		9	15			
Anthreptes seimundi	Little Green Sunbird	M	16	7.6		6.5	11			396
		F	5	5.7		5	6.5			
Anthreptes rectirostris	Green Sunbird	M	34	11.3		10	12			396
		F	16	9.9		7	12			
Anthreptes rubritorques	Banded Sunbird	M	11	10.7	0.4	10.5	11.9			184
Hedydipna collaris	Collared Sunbird	B	297	7.4		5	17		South Africa	1036
Hedydipna collaris elachior	Collared Sunbird	B	42	6.6		5.3	8.5			184
Hedydipna platura	Pygmy Sunbird	B	43	6.7		5.7	7.5		Nigeria	184
Hedydipna metallica	Nile Valley Sunbird	B	7	7.2		7	7.5		Chad; Yemen	223, 317
Hedydipna pallidigaster	Amani Sunbird	B	9	6.8		6	8			184

Body Masses of World Birds (continued)

Species Name	Common Name	Sex	N	Mean	S.D.	Min	Max	Season	Location	Source #
Hypogramma hypogrammicum nuchale	Purple-naped Sunbird	M	9	10.4		7.8	12.3		Malaysia	785
		F	3	11.1		9.7	12.8			
Hypogramma hypogrammicum hypogrammicum	Purple-naped Sunbird	M	16	12.6		9.8	15.2			184
		F	21	11.7		10	13.5			
Anabathmis reichenbachii	Reichenbach's Sunbird	M	2	11.4		9.8	13			396, 982
Anabathmis hartlaubii	Principe Sunbird	U	1	10.5						396
Anabathmis newtonii	Newton's Sunbird	B	2	6.5		6	7			396
Dreptes thomensis	Sao Tome Sunbird	M	1	26.2					Sao Tome	396
		F	4			18	18.9			
Anthobaphes violacea	Orange-breasted Sunbird	M	409	10		7	13		South Africa	1036
		F	156	8.8		8.8	13			
Cyanomitra verticalis viridisplendens	Green-headed Sunbird	M	45	15.2		13.4	16.6		Kenya; Tanzania	184
		F	20	13.7		11.2	15.5			
Cyanomitra cyanolaema	Blue-throated Brown Sunbird	B	30	16.8		13.5	20		Uganda	396
Cyanomitra alinae tanganjicae	Blue-headed Sunbird	M	23	13.5		11	16			184
		F	23	11.7		10	15			
Cyanomitra oritis	Cameroon Sunbird	M	73	12		9.9	13.3			396
		F	41	10.4		8.7	11.9			
Cyanomitra bannermani	Bannerman's Sunbird	F	1	14						396
Cyanomitra olivacea olivacea	Eastern Olive-Sunbird	U	490	11.8		9	14.7			184
Cyanomitra olivacea changamwensis	Eastern Olive-Sunbird	B	183	7.9		6.2	10.9			184
Cyanomitra obscura vincenti	Western Olive-Sunbird	M	30	10.9		9	12.8		Kenya	396
		F	27	9.6		8.1	11			
Cyanomitra obscura ragazzii	Western Olive-Sunbird	M	65	11.5		9	13.8			184
		F	67	10		7.9	12.2			

Body Masses of World Birds (continued)

Species Name	Common Name	Sex	N	Mean	S.D.	Min	Max	Season	Location	Source #
Cyanomitra veroxii	Mouse-colored Sunbird	U	105	11.5		5	15		South Africa	1036
Cyanomitra adelberti	Buff-throated Sunbird	B	11	9.8		8.3	11		Liberia	396
Chalcomitra fuliginosa	Carmelite Sunbird	F	2	11.9		11.6	12.2			266, 396
Chalcomitra rubescens	Green-throated Sunbird	M	19	11.3		8	14			184
		F	5	9.4		6.5	11			
Chalcomitra amethystina	Amethyst Sunbird	M	178	14.7	1.2	9.5	18.5			184
		F	76	12.7	1.2	8.3	16.9			
Chalcomitra senegalensis senegalensis	Scarlet-chested Sunbird	M	55	10.5		7.5	12.7			184
		F	28	8.9		6.8	11			
Chalcomitra senegalensis gutturalis	Scarlet-chested Sunbird	M	38	14.9		11.5	16.4		Malawi	396
		F	17	13.1		10.2	15.3			
Chalcomitra senegalensis lamberti	Scarlet-chested Sunbird	M	39	13.9		11.4	16			184
		F	12	13.3		12.1	14			
Chalcomitra hunteri	Hunter's Sunbird	B	13	11.8		10	14		Kenya	396
Chalcomitra balfouri	Socotra Sunbird	B	33			8	15			396
Leptocoma zeylonica	Purple-rumped Sunbird	B	17			7	11		India	7
Leptocoma minima	Crimson-backed Sunbird	B	16			4	6		India	7
Leptocoma sperata	Purple-throated Sunbird	B	46	6.4		5.2	7.5		Philippines	954
Nectarinia bocagei	Bocage's Sunbird	M	2	14.8		14.5	15		Angola	982
Nectarinia purpureiventris	Purple-breasted Sunbird	B	44	11.5		10	13.5			184
Nectarinia tacazze	Tacazze Sunbird	M	29	15.7		12.3	18.6		Ethiopia	396
		F	19	14.1		10	16.2			
Nectarinia kilimensis	Bronze Sunbird	M	271	16.7		14	19.9			184
		F	96	14.7		13.3	16.7			
Nectarinia reichenowi	Golden-winged Sunbird	M	65	15.7		12.8	17.5			184
		F	32	13.7		11	15.9			

Body Masses of World Birds (continued)

Species Name	Common Name	Sex	N	Mean	S.D.	Min	Max	Season	Location	Source #
Sitta carolinensis	White-breasted Nuthatch	B	104	21		18.3	23.2	Y	Pennsylvania, USA	828
Sitta neumayer	Rock Nuthatch	B	35	26.5	4.7	20.2	35.5		Greece; Turkey; Iran	223
Sitta tephronota	Persian Nuthatch	B	19	35.2		31	39	B	Afghanistan; former USSR	223
Sitta frontalis	Velvet-fronted Nuthatch	B	14	16.5		15.5	17.9		Luzon, Philippines	436
Sitta solangiae	Yellow-billed Nuthatch	U				14	16			479
Sitta oenochlamys	Sulphur-billed Nuthatch	U	2	16.4		16.3	16.4			479
Sitta magna	Giant Nuthatch	U				36	47			479
Family Tichodromidae										
Tichodroma muraria	Wallcreeper	B	32	17.6		13	19.6		Afghanistan; Nepal; former USSR	223
Family Certhiidae										
Certhia familiaris	Eurasian Treecreeper	U	153	9	0.7	7.6	12.9	F	Sweden	223
Certhia americana	Brown Creeper	U	58	8.1	0.5	7.1	9.5	M	Pennsylvania, USA	828
Certhia brachydactyla	Short-toed Treecreeper	B	24	8.2	0.7	7	9.5	Y	France	223
Certhia himalayana	Bar-tailed Treecreeper	B	10	8.8	0.8	7.8	10.3		India	7
Certhia nipalensis	Rusty-flanked Treecreeper	B	6	11.1		10	12		India	7
Certhia discolor	Brown-throated Treecreeper	B	8	10.4		9.6	11		India	7
Salpornis spilonotus	Spotted Creeper	U	5	14.8		13.5	16		India; Angola; Zambia	7, 266, 915
Family Rhabdornithidae										
Rhabdornis mysticalis	Stripe-sided Rhabdornis	B	21	25.6		22.4	31.5		Philippines	954

Body Masses of World Birds (continued)

Species Name	Common Name	Sex	N	Mean	S.D.	Min	Max	Season	Location	Source #
Nectarinia johnstoni	Red-tufted Sunbird	M	64	16.3	0.8	14.6	17.7		Kenya	342
		F	37	14.1	0.9	12.4	16.9			
Nectarinia famosa famosa	Malachite Sunbird	M	90	17.6		15.5	22.5			184
		F	93	14.5		11.5	17.5			
Nectarinia famosa cupreonitens	Malachite Sunbird	M	91	13.8		12	18.7			184
		F	76	12.1		9.1	17.5			
Cinnyris chloropygius	Olive-bellied Sunbird	B	51	6.2		5	8			184
Cinnyris minullus	Tiny Sunbird	B	15	5.7		4.5	6.5			184
Cinnyris manoensis	Miombo Sunbird	M	50	10	0.9	8.1	12.8			184
		F	49	9	0.9	6	11.2			
Cinnyris chalybeus chalybeus	Southern Double-collared Sunbird	M	51	8.5		6	9.5			184
		F	37	7.2		6	9.5			
Cinnyris chalybeus subalaris	Southern Double-collared Sunbird	M	72	8	0.7	6	10			184
		F	55	7.3	0.7	6	9.5			
Cinnyris neergaardi	Neergaard's Sunbird	M	7	6.5		6.2	7.1			396
		F	2	5.8		5.6	5.9			
Cinnyris stuhlmanni	Stuhlmann's Sunbird	M	6	7		6	8			184
		F	7	6		5	7			
Cinnyris prigoginei	Prigogine's Sunbird	M	4	10.3		9.7	10.9			184
		F	1	8						
Cinnyris ludovicensis	Montane Double-collared Sunbird	M	38	9.3		7.1	10.8			184
		F	14	8.1		7.4	9			
Cinnyris preussi	Northern Double-collared Sunbird	B	63	6.8		5.2	8.5			184
Cinnyris afer	Greater Double-collared Sunbird	M	270	12.2		6.5	18		South Africa	1036
		F	66	10.1		6.5	16			
Cinnyris regius	Regal Sunbird	M	37	7.1		5	8			184
		F	14	6.1		5.5	7			
Cinnyris rockefelleri	Rockefeller's Sunbird	M	3	5.3		5.2	5.5			396
Cinnyris mediocris	Eastern Double-collared Sunbird	M	55	8.6	0.6	7	9.8			184
		F	56	7.8	0.5	6.8	9.2			

Body Masses of World Birds (continued)

Species Name	Common Name	Sex	N	Mean	S.D.	Min	Max	Season	Location	Source #
Cinnyris moreaui	Moreau's Sunbird	M	56	8.5		6.8	9.6			184
		F	43	8		5.1	10.4			
Cinnyris pulchellus	Beautiful Sunbird	M	37	7.5		5.6	10.2			184
		F	27	6.6		5.6	8			
Cinnyris loveridgei	Loveridge's Sunbird	M	25	10.6		9	12			184
		F	24	9		8	10			
Cinnyris mariquensis	Mariqua Sunbird	M	102	12.3		7	17		Southern Africa	1036
		F	75	10.7		7	15			
Cinnyris shelleyi	Shelley's Sunbird	M	2	8					Malawi	184, 1036
		F	6	10.1		9.3	11			
Cinnyris erythrocercus	Red-chested Sunbird	M	56	9.2		8.2	10			396
		F	51	8.4		7.4	9.5			
Cinnyris nectarinioides	Black-bellied Sunbird	B	6	4.8		4	6			396
Cinnyris bifasciatus	Purple-banded Sunbird	M	73	7.6		6	8.9		Southern Africa	396
		F	22	6.8		6	8			
Cinnyris tsavoensis	Tsavo Sunbird	B	15	6.8		6	7.6		Kenya	396
Cinnyris chalcomelas	Violet-breasted Sunbird	B	5	8.2		7	9			396
Cinnyris bouvieri	Orange-tufted Sunbird	M	8	8.7		7	10		Nigeria; Kenya	396
Cinnyris osea	Palestine Sunbird	M	8	7.6					Israel	223
		F	12	6.8	0.7					
Cinnyris habessinicus	Shining Sunbird	M	16	10.5		9.5	11.5			184
		F	29	8.6		7.5	9.5			
Cinnyris coccinigastrus	Splendid Sunbird	M	2	14.3		13.3	15.3		Liberia	396
Cinnyris johannae	Johanna's Sunbird	M	11	13.7		12.6	14.7		Liberia	396
		F	3	12.4		11	15			
Cinnyris superbus	Superb Sunbird	B	14	16.8		13.2	20.6			184
Cinnyris rufipennis	Rufous-winged Sunbird	M	2	9.2		9	9.5		Tanzania	581, 1014
		F	2	8.4		8	8.7			

Body Masses of World Birds (continued)

Species Name	Common Name	Sex	N	Mean	S.D.	Min	Max	Season	Location	Source #
Cinnyris talatala	White-breasted Sunbird	M	357	8.5		6	15		Southern Africa	1036
		F	193	7.7		5.5	13			
Cinnyris venustus	Variable Sunbird	M	87	6.9		5.9	9		Southern Africa	1036
		F	76	6.5		5.6	15.5			
Cinnyris venustus	Variable Sunbird	U	93	6.3		4.5	8.2		Nigeria	777
Cinnyris fuscus	Dusky Sunbird	M	65	8.3		6.5	11		Namibia; South Africa	1036
		F	37	7.4		5.3	10			
Cinnyris ursulae	Ursula's Sunbird	U	48	6.5		5.3	7.3			184
Cinnyris batesi	Bates' Sunbird	U	5	6.3		5.7	7			184
Cinnyris cupreus cupreus	Copper Sunbird	M	63	9		7.5	11		Kenya	184
		F	10	8.1		7.2	8.9			
Cinnyris cupreus chalceus	Copper Sunbird	M	70	9.8		8.4	11.4		Malawi	396
		F	48	8.8		7.3	10.2			
Cinnyris asiaticus	Purple Sunbird	B	18	8.1		5	11		India	7
Cinnyris jugularis	Olive-backed Sunbird	M	22	9.2		7.9	11.9		Philippines	954
		F	18	8.1		6.8	10			
Cinnyris jugularis flavigaster	Olive-backed Sunbird	M	15	10		8.8	11		Guadalcanal; Solomon Is.	399, 658
		F	11	8.8		7.1	9.5			
Cinnyris souimanga	Souimanga Sunbird	M	16	7.6	0.8	5.5	8.5			184
		F	16	6.9	0.6	5.5	7.6			
Cinnyris dussumieri	Seychelles Sunbird	U		10.8						902
Cinnyris notatus notatus	Madagascar Sunbird	M	9	15.4		13.8	17			184
		F	5	13.8		12	16			
Cinnyris humbloti	Humblot's Sunbird	B	38	6.4		5.5	8			184
Cinnyris comorensis	Anjouan Sunbird	M	5	8.4		7.3	9			184
		F	2	7.2		6.8	7.5			
Cinnyris coquerellii	Mayotte Sunbird	M	1	9						184
		F	1	7						

Body Masses of World Birds (continued)

Species Name	Common Name	Sex	N	Mean	S.D.	Min	Max	Season	Location	Source #
Cinnyris lotenius	Long-billed Sunbird	B	15			7	11		India	7
Aethopyga primigenia	Gray-hooded Sunbird	B	13	7	1	5.3	8.3		Mindanao, Philippines	1143
Aethopyga boltoni malindangensis	Mount Apo Sunbird	M	5	8		7.5	8.7		Philippines	628
		F	8	7.1		6.1	8.5			
Aethopyga boltoni tibolii	Mount Apo Sunbird	B	7	6.3		4.9	7.8		Philippines	628
Aethopyga linaraborae	Lina's Sunbird	M	5	7.9		7	8.7		Philippines	628
		F	2	6.5		6	7			
Aethopyga flagrans	Flaming Sunbird	B	6	6.7		6.4	7.3		Luzon, Philippines	436
Aethopyga pulcherrima	Metallic-winged Sunbird	B	46	6.2		4.4	7.5		Philippines	954
Aethopyga shelleyi	Lovely Sunbird	B	10	4		3	5.3		Philippines	954, 983
Aethopyga gouldiae	Gould's Sunbird	M	5			6.5	8			7
		F	2	5		4	6			
Aethopyga nipalensis	Green-tailed Sunbird	M	4	6.9		6.5	8			7
		F	8			5	6.5			
Aethopyga christinae	Fork-tailed Sunbird	U	12	5.2		4.7	5.9			184
Aethopyga saturata	Black-throated Sunbird	U	22	5.3		3	7		Malaysia	775
Aethopyga vigorsii	Western Crimson Sunbird	B	4	7.8		7	9			184
Aethopyga siparaja	Crimson Sunbird	M	8	6.7		4.8	9			184, 1272
Aethopyga mystacalis	Scarlet Sunbird	U	2	6		6	6		Borneo	398, 1218
Aethopyga temminckii	Temminck's Sunbird	B	4	5.5		4.8	6.1			184
Aethopyga ignicauda	Fire-tailed Sunbird	M	2	8.2		7.5	9	W		7

Body Masses of World Birds (continued)

Species Name	Common Name	Sex	N	Mean	S.D.	Min	Max	Season	Location	Source #
Arachnothera crassirostris	Thick-billed Spiderhunter	F	2	16		15.5	16.4			184
Arachnothera flavigaster	Spectacled Spiderhunter	F	1	38.4					Borneo	1218
Arachnothera longirostra	Little Spiderhunter	M	134	14		10.3	16.1		Brunei	184
		F	77	11.1		8.8	13.7			
Arachnothera chrysogenys	Yellow-eared Spiderhunter	U	2	23.2		23	23.5		Malaysia; Borneo	775, 1218
Arachnothera clarae	Naked-faced Spiderhunter	U	17	30.1		27.5	34.8		Philippines	954
Arachnothera modesta	Gray-breasted Spiderhunter	B	9	22.7		20.1	25.5		Malaysia; Brunei	184
Arachnothera affinis	Streaky-breasted Spiderhunter	B	5	26.4		22.2	29.2		Borneo	1218
Arachnothera magna	Streaked Spiderhunter	U	34	30.7		23	35		Malaysia	775
Family Melanocharitidae										
Melanocharis nigra	Black Berrypecker	M	9	13.2		11.5	14.7		New Guinea	283
		F	5	15.2		14	16.5			
Melanocharis longicauda	Lemon-breasted Berrypecker	M	1	11					New Guinea	1143
		F	1	13						
Melanocharis versteri	Fan-tailed Berrypecker	M	10	11.9	1	9.7	13.7		New Guinea	283
		F	10	18.1	0.8	16.7	20			
Melanocharis striativentris	Streaked Berrypecker	M	21	16.1		14.6	19.2		New Guinea	1056
		F	11	19.2		15.5	22.5			
Melanocharis crassirostris	Spotted Berrypecker	U	1	18					New Guinea	424
Toxorhamphus novaeguineae	Yellow-bellied Longbill	M	4	13.9		12	16		New Guinea	427, 535
		F	1	10						
Toxorhamphus poliopterus	Slaty-chinned Longbill	M	10	12.4	1	10.5	14.3		New Guinea	283
		F	10	10.4	0.7	9.3	11.5			
Toxorhamphus iliolophus	Dwarf Honeyeater	U	35	11					New Guinea	61

Body Masses of World Birds (continued)

Species Name	Common Name	Sex	N	Mean	S.D.	Min	Max	Season	Location	Source #
Toxorhamphus pygmaeum	Pygmy Honeyeater	B	2	5.8		5.7	6		New Guinea	283
Family Paramythiidae										
Oreocharis arfaki	Tit Berrypecker	B	12	19.7	1.6	17	21.5		New Guinea	283, 424
Paramythia montium	Crested Berrypecker	M	8	41.1		38.5	43		New Guinea	424
		F	6	42.5		38	49			
Family Dicaeidae										
Prionochilus olivaceus	Olive-backed Flowerpecker	B	20	9.4		7.9	10.8		Philippines	954
Prionochilus maculatus	Yellow-breasted Flowerpecker	U	34	7.1	0.7	6	9		Borneo	398
Prionochilus maculatus oblitus	Yellow-breasted Flowerpecker	M	3	9.1		7.7	11.8			184
		F	2	7.8		7.8	7.9			
Prionochilus plateni	Palawan Flowerpecker	U	8			8	9		Palawan, Philippines	487
Prionochilus xanthopygius	Yellow-rumped Flowerpecker	B	32	7.6		6.5	8.6			184
Prionochilus thoracicus	Scarlet-breasted Flowerpecker	M	1	8.9						184
Dicaeum annae	Golden-rumped Flowerpecker	U	2	10		9.7	10.3			184
Dicaeum agile	Thick-billed Flowerpecker	B	11	9		7.5	11		India	7
Dicaeum proprium	Whiskered Flowerpecker	M	2	10.2		10.2	10.2			184
Dicaeum chrysorrheum	Yellow-vented Flowerpecker	M	1	9						7
Dicaeum aureolimbatum	Yellow-sided Flowerpecker	U	8	8.1		7	10		Sulawesi, Indonesia	1272
Dicaeum nigrilore	Olive-capped Flowerpecker	B	20	11		9.5	12.4		Philippines	954
Dicaeum anthonyi	Flame-crowned Flowerpecker	B	4	11.6		9.6	13			184, 954

Body Masses of World Birds (continued)

Species Name	Common Name	Sex	N	Mean	S.D.	Min	Max	Season	Location	Source #
Dicaeum bicolor	Bicolored Flowerpecker	B	16	9.1		7.8	10		Philippines	436, 954
Dicaeum australe	Red-striped Flowerpecker	B	26	8.7		7.9	10.5		Philippines	954
Dicaeum haematostictum	Red-keeled Flowerpecker	B	6	8.8		8	9.9		Negros, Philippines	1148
Dicaeum retrocinctum	Scarlet-collared Flowerpecker	B	7	6.5		5	7		Philippines	983
Dicaeum trigonostigma	Orange-bellied Flowerpecker	B	30	7.1		5.9	8.4		Philippines	954
Dicaeum erythrorhynchos	Pale-billed Flowerpecker	B	17	6.3		4	8		India	7
Dicaeum concolor	Plain Flowerpecker	U	15	6.2		5	8		India	7
Dicaeum erythrothorax	Flame-breasted Flowerpecker	U				8	10		Buru, Moluccas	1182
Dicaeum hypoleucum	White-bellied Flowerpecker	B	38	8		6.9	9.6		Philippines	954
Dicaeum pygmaeum	Pygmy Flowerpecker	B	5	5		4	5.6		Philippines	436, 954
Dicaeum vulneratum	Ashy Flowerpecker	U				8	10		Seram, Moluccas	1181
Dicaeum pectorale	Olive-crowned Flowerpecker	B	12	7.6	0.5	7	8.5		New Guinea	292, 424, 535
Dicaeum geelvinkianum	Red-capped Flowerpecker	B	12	6.8	0.6	5.3	7.5		New Guinea	283
Dicaeum eximium	Red-banded Flowerpecker	B	4	7.9		7	9.5		New Britain	426
Dicaeum aeneum	Midget Flowerpecker	M	22	9.6		8.5	10.5		Guadalcanal	399
		F	9	8.9		8	9.5			
Dicaeum tristrami	Mottled Flowerpecker	B	22	11.7		10	13		Guadalcanal	399
Dicaeum ignipectum	Fire-breasted Flowerpecker	M	16			4	8			7
		F	5			5.5	6.1			

Body Masses of World Birds (continued)

Species Name	Common Name	Sex	N	Mean	S.D.	Min	Max	Season	Location	Source #
Dicaeum celebicum	Gray-sided Flowerpecker	F	1	6.5					Sulawesi, Indonesia	1272
Dicaeum hirundinaceum	Mistletoebird	M	37	9.3	0.7	7.5	11		New South Wales, Australia	518a
		F	14	8.8	0.9	7.5	10			
Dicaeum cruentatum cruentatum	Scarlet-backed Flowerpecker	M				7	8			184
Dicaeum cruentatum nigrimentum	Scarlet-backed Flowerpecker	M	2	5.6		5.5	5.7			184
Family Pardalotidae										
Pardalotus punctatus	Spotted Pardalote	B	56	8.7		6	15		Southeast Australia	518
Pardalotus quadragintus	Forty-spotted Pardalote	U	52	10.7	0.9				Tasmania	1330
Pardalotus rubricatus	Red-browed Pardalote	B	35	10.9		9	15		Australia	518
Pardalotus striatus striatus	Striated Pardalote	M	140	12					Australia	1331
		F	150	12.4						
Pardalotus striatus melanocephalus	Striated Pardalote	B	61	9	0.9	7.5	12.7		Australia	470
Pardalotus striatus substriatus	Striated Pardalote	U	71	11.8	0.9	9.4	14		Southeast Australia	518
Pardalotus striatus ornatus	Striated Pardalote	U	35	12.4	0.9	9.5	13.5		Australia	518
Family Zosteropidae										
Speirops lugubris	Black-capped Speirops	B	2	14.8		14.7	14.9		Principe	1143
Speirops melanocephalus	Cameroon Speirops	U	8	10.9		9	12.5			396
Speirops brunneus	Fernando Po Speirops	B	14	16.1		14	18.5			396
Speirops leucophoeus	Principe Speirops	B	2	14.8		14.7	14.9		Principe	1143
Zosterops senegalensis	African Yellow White-eye	B	193	10.9		8.9	14.1		Malawi	303

Body Masses of World Birds (continued)

Species Name	Common Name	Sex	N	Mean	S.D.	Min	Max	Season	Location	Source #
Zosterops poliogaster	Broad-ringed White-eye	U	20	11		9	14			396
Zosterops abyssinicus	White-breasted White-eye	U	16	7.8		7.2	9		Kenya	396
Zosterops abyssinicus arabs	White-breasted White-eye	U	12	9.6		8.5	10.4		Yemen	317
Zosterops abyssinicus socotrana	White-breasted White-eye	U	3	11.2		10.7	12		Socotra Is.	317
Zosterops pallidus	Cape White-eye	U	1513	11.1		8	20			396
Zosterops mayottensis	Mayotte White-eye	U	6	9.9					Mayotte Is.	706
Zosterops maderaspatanus maderaspatanus	Madagascar White-eye	B	111	11.2	1.7	8	17.5		Madagascar	957
Zosterops maderaspatanus aldabrensis	Madagascar White-eye	U	29	7.6					Aldabra	706
Zosterops maderaspatanus comorensis	Madagascar White-eye	U	32	8.8					Moheli	706
Zosterops maderaspatanus kirki	Madagascar White-eye	U	28	8.3					Comoro Is.	706
Zosterops mouroniensis	Comoro White-eye	U	10	9.9					Comoro Is.	706
Zosterops griseovirescens	Annobon White-eye	U	2	11		11	11			396
Zosterops borbonicus	Mascarene White-eye	B	13	8.2		7.6	9.2		Mauritius	184
Zosterops olivaceus	Reunion White-eye	M	9	8.6		7.7	9.8		Reunion	184
		F	11	9.7	0.9	8.6	11.4			
Zosterops chloronothos	Mauritius White-eye	B	5	8.1		7.5	9		Mauritius	184
Zosterops modestus	Seychelles White-eye	U	3	12.8					Seychelles	706
Zosterops ceylonensis	Ceylon White-eye	F	1	12.6					Sri Lanka	7

Body Masses of World Birds (continued)

Species Name	Common Name	Sex	N	Mean	S.D.	Min	Max	Season	Location	Source #
Zosterops erythropleurus	Chestnut-flanked White-eye	B	35	11		10	13		Hopei, China	1087
Zosterops palpebrosus	Oriental White-eye	U	20	8.6		6	11		India	7
Zosterops japonicus	Japanese White-eye	B	568	11.1	1.5				Hawaiian Is.	1248
Zosterops japonicus	Japanese White-eye	U	25	11.5	1.1	9.9	13.9	F	Japan	490
Zosterops meyeni	Lowland White-eye	F	1	10					Philippines	174
Zosterops conspicillatus	Bridled White-eye	B	9	7.9					Saipan	114
Zosterops atricapillus	Black-capped White-eye	M	1	9.2					Borneo, Malaysia	1148
Zosterops everetti	Everett's White-eye	B	42	10		8.1	12		Philippines	954
Zosterops nigrorum	Yellowish White-eye	B	10	7.3		6	9.2		Philippines	436, 983
Zosterops montanus	Mountain White-eye	B	20	10.5		9.4	12.6		Philippines	954
Zosterops natalis	Christmas Is. White-eye	M	13	9.3	1.4	7	11.7		Christmas Is.	518a
		F	11	10.7	1.5	8	14			
Zosterops atrifrons	Black-crowned White-eye	U	11	10	1	9	12		Sulawesi, Indonesia	1272
Zosterops stalkeri	Seram White-eye	U				9	12		Seram, Moluccas	1181
Zosterops minor	Black-fronted White-eye	B	8	11.1		10	12		New Guinea	283, 424, 427
Zosterops hypoxanthus	Black-headed White-eye	B	4	12.5		11.5	13		New Britain	426
Zosterops fuscicapillus	Capped White-eye	M	1	11.2					New Guinea	424
Zosterops buruensis	Buru White-eye	U				11	13		Buru, Moluccas	1182
Zosterops novaeguineae	New Guinea White-eye	B	3	12.5		12	13.5		New Guinea	283

Body Masses of World Birds (continued)

Species Name	Common Name	Sex	N	Mean	S.D.	Min	Max	Season	Location	Source #
Chlorocharis emiliae	Mountain Black-eye	U	1	13.9					Borneo, Malaysia	1148
Hypocryptadius cinnamomeus	Cinnamon White-eye	B	22	30		24.7	31.5		Philippines	954
Family Promeropidae										
Promerops gurneyi	Gurney's Sugarbird	M	158	37.4	2.6	30	45			184
		F	195	32.6	3.8	23	43			
Promerops cafer	Cape Sugarbird	M	57	37.5		30.5	43.5			396
		F	46	32.2		26	39			
Family Meliphagidae										
Timeliopsis fulvigula	Olive Straightbill	U	12	18.3		16	20.5		Papua New Guinea	283, 387, 424
Timeliopsis griseigula	Tawny Straightbill	U	6	32					New Guinea	61
Melilestes megarhynchus	Long-billed Honeyeater	M	10	46.6	1.7	43	49.5		New Guinea	283
		F	10	41.1	3.3	36.3	47.5			
Glycichaera fallax	Green-backed Honeyeater	B	12	11.2	2.4	9	18		New Guinea; Australia	517
Lichmera argentauris	Olive Honeyeater	M		14					Seram, Moluccas	1181
		F				12	13			
Lichmera indistincta	Brown Honeyeater	M	37	11.8	1.5	9	15		Australia	517
		F	15	10.6	2.1	8	14			
Lichmera incana	Dark-brown Honeyeater	U	8	13.1		9.5	14.5		New Caledonia	1018
Lichmera alboauricularis	Silver-eared Honeyeater	M	12			13.5	19		New Guinea	425
		F	3	13.3		12.5	14.5			
Lichmera deningeri	Buru Honeyeater	M				18	19		Buru, Moluccas	1182
		F				14	16			
Trichodere cockerelli	White-streaked Honeyeater	M	7	16.9		14	20		Cent. York Pen., Australia	517
		F	3	12.7		12	13.5			
Myzomela blasii	Seram Myzomela	M	8	14.9					Ceram	290
Myzomela albigula	White-chinned Myzomela	M	3	16.3						290
		F	1	13						

Body Masses of World Birds (continued)

Species Name	Common Name	Sex	N	Mean	S.D.	Min	Max	Season	Location	Source #
Myzomela eques eques	Red-throated Myzomela	M	22	13.8	2.4					290
Myzomela eques nymani	Red-throated Myzomela	M	21	15.8	2				New Guinea	290
		F	8	12.4						
Myzomela cineracea	Ashy Myzomela	M	39	15.6	1.6				New Britain	290
		F	19	12.2	1.2					
Myzomela obscura	Dusky Myzomela	M	15	13.2	1	11.8	15.1		Northern Terr., Australia	517
		F	12	10.7	0.7	10	12.3			
Myzomela cruentata	Red Myzomela	M	14	8.6	0.6				New Britain	290
		F	9	7.9						
Myzomela nigrita	Black Myzomela	M	26	10.1	0.5				New Guinea	290
		F	9	8.5						
Myzomela pulchella	New Ireland Myzomela	B	5	10		9	11.5		New Ireland	58
Myzomela erythrocephala	Red-headed Myzomela	M	39	8	0.6	7	9.6		Western Australia	517
		F	58	7.6	0.8	6	9.4			
Myzomela adolphinae	Mountain Myzomela	B	10	7.4					New Guinea	290
Myzomela wakoloensis	Wakolo Myzomela	U				7	9		Moluccas	1181
Myzomela caledonica	New Caledonian Myzomela	M	2	7.8		7.5	8.2		New Caledonia	1148
		F	2	6.2		5.9	6.6			
Myzomela sanguinolenta	Scarlet Myzomela	M	166	8.6	1	6	11		Queensland, Australia	77
		F	66	8.1	0.9	6	10			
Myzomela rubratra	Micronesian Myzomela	U		14					Guam	1305
Myzomela cardinalis	Cardinal Myzomela	M	6	15.4		14.5	16		Solomon Is.	399
		F	6	12.3		11.5	13			
Myzomela sclateri	Scarlet-bibbed Myzomela	B	26	10.5					New Guinea Is.	290
Myzomela pammelaena	Ebony Myzomela	M	28	17	1.2				New Guinea Is.	290
		F	28	14.6	1					
Myzomela pammelaena nigerrima	Ebony Myzomela	U				13.2	20		Bismarck Is.	284

Body Masses of World Birds (continued)

Species Name	Common Name	Sex	N	Mean	S.D.	Min	Max	Season	Location	Source #
Myzomela pammelaena meyeri	Ebony Myzomela	U				8.3	10.8		New Guinea	284
Myzomela lafargei	Scarlet-naped Myzomela	M	4	12.8		12.5	13		Solomon Is.	658
Myzomela melanocephala	Black-headed Myzomela	M	12	12.8		11	14		Guadalcanal	399
		F	5	11.2		10	12.5			
Myzomela tristrami	Sooty Myzomela	M	21	14.4		12.5	16.5		San Cristobal	399
		F	10	12.3		11.5	13			
Myzomela jugularis	Orange-breasted Myzomela	M	10	9	0.6	8	10		Fiji	668
		F	9	8.4		8	9			
Myzomela erythromelas	Black-bellied Myzomela	B	46	8					New Britain	290
Myzomela rosenbergii	Red-collared Myzomela	M	15	11	2.5	9.3	12.7		New Guinea	283
		F	4	9.7		9	10.7			
Certhionyx pectoralis	Banded Honeyeater	B	15	10.1		8.2	13		Australia	517
Certhionyx niger	Black Honeyeater	B	14	9.9		8.5	12		Australia	517
Certhionyx variegatus	Pied Honeyeater	M	11	27.5	3.1	23.5	34		Australia	517
		F	7	25.1		23	28			
Meliphaga montana auga	Forest Honeyeater	M	14	30.7	2.3	27	34		New Guinea	283
		F	13	25.6	1.8	22	29			
Meliphaga mimikae	Spot-breasted Honeyeater	M	25	28.5	2	24	33		New Guinea	283
		F	41	25.4	1.7	22	29			
Meliphaga orientalis	Mountain Honeyeater	M	21	18.7		15.7	20.4		New Guinea	282
		F	31	17.5		14.8	20			
Meliphaga albonotata	Scrub Honeyeater	U	5	22.4					New Guinea	61, 535
Meliphaga aruensis	Puff-backed Honeyeater	M	3	27.8		27	28.7		New Guinea	283
		F	3	24.7		23	26.5			
Meliphaga analoga flavida	Mimic Honeyeater	M	14	21.5		18.8	23.5		New Guinea	282
		F	6	19		17.8	20.2			
Meliphaga analoga analoga	Mimic Honeyeater	M	3	24.9		24	25.5		New Guinea	282
		F	4	21.3		21	22			

Body Masses of World Birds (continued)

Species Name	Common Name	Sex	N	Mean	S.D.	Min	Max	Season	Location	Source #
Meliphaga gracilis	Graceful Honeyeater	M	10	15.6	1.6	14	17		Queensland, Australia	365
		F	20	14.4	1.4	12.5	17			
Meliphaga notata	Yellow-spotted Honeyeater	M	9	27		24	29.5		Queensland, Australia	365
		F	23	25.9	1.8	23.5	30			
Meliphaga flavirictus	Yellow-gaped Honeyeater	U	2	18.5		17	20		New Guinea	283
Meliphaga lewinii	Lewin's Honeyeater	M	69	35.8	2.7	29	43		Australia	517
		F	83	31.5	2.8	26	40			
Meliphaga albilineata	White-lined Honeyeater	M	15	25.4	1.8	22.3	29.9		Northern Terr., Australia	517
		F	6	22.1		19	23.5			
Guadalcanaria inexpectata	Guadalcanal Honeyeater	M	6	46.8		41.5	48.5		Guadalcanal	399
		F	10	40.1		37.5	44.5			
Foulehaio carunculata	Wattled Honeyeater	U	18	37	5.2	25.2	45.4		Fiji	1196c
Lichenostomus subfrenatus	Black-throated Honeyeater	M	9	32.6		29	37		New Guinea	283, 424
		F	9	27		25	29			
Lichenostomus obscurus	Obscure Honeyeater	B	11	26.4	2.3	24	31		New Guinea	283
Lichenostomus frenatus	Bridled Honeyeater	M	7	36.9		34	43		Queensland, Australia	517
		F	9	30.4		28.4	32			
Lichenostomus frenatus	Bridled Honeyeater	U	81	35.4	2.7	28.2	42.1		Queensland, Australia	390a
Lichenostomus hindwoodi	Eungella Honeyeater	M	7	24.5		20	27.2		Queensland, Australia	517
		F	11	21.4	2.1	19.7	24			
Lichenostomus chrysops	Yellow-faced Honeyeater	U	124	17.3	1.5	11.1	20.8		Victoria, Australia	517
Lichenostomus versicolor	Varied Honeyeater	M	1	35.5					Queensland, Australia	517
		F	2	32.8		32.5	33			
Lichenostomus versicolor	Varied Honeyeater	U	13	42.3	4.1	34.8	46.8		Papua New Guinea	318
Lichenostomus fasciogularis	Mangrove Honeyeater	M	10	28.5	3	23.9	32		Queensland, Australia	517
		F	8	25.2		22	30			

Body Masses of World Birds (continued)

Species Name	Common Name	Sex	N	Mean	S.D.	Min	Max	Season	Location	Source #
Zosterops luteus	Australian Yellow White-eye	B	39	9.7		8.4	11.3		Australia	518a
Zosterops metcalfii	Yellow-throated White-eye	B	8	14		12	15		Solomon Is.	658
Zosterops rendovae	Gray-throated White-eye	B	25	15.4		13.5	17		Guadalcanal	399
Zosterops rendovae	Gray-throated White-eye	B	9	19.2		17.5	21.5		Solomon Is.	399
Zosterops xanthochroa	Green-backed White-eye	U	46	10.3	0.9	8.5	12		New Caledonia	1018
Zosterops tenuirostris	Slender-billed White-eye	U	10	17.2	0.6	16	18		Norfolk Is.	518a
Zosterops explorator	Layard's White-eye	U	4	10.5		8.5	11.5		Fiji	668
Zosterops lateralis	Silver-eye	U	1083	12.9	1.1	9.5	16.9		New Zealand	1002
Zosterops lateralis tropicus	Silver-eye	B	5	13.7		12.5	15.4		Vanuatu	656b
Zosterops lateralis tephropleurus	Silver-eye	U	101	12.4	0.9				Lord Howe Is.	518a
Zosterops lateralis lateralis	Silver-eye	U	58	11.8	0.9	9.5	13.8		Tasmania, Australia	518a
Zosterops flavifrons	Yellow-fronted White-eye	B	5	10.7		10	11.9		Vanuatu	656b
Zosterops cinereus ponapensis	Gray White-eye	M	1	11.2					Pohnpei, Caroline Is.	1148
Cleptornis marchei	Golden White-eye	B	4	20.2					Saipan	222
Tephrozosterops stalkeri	Rufescent White-eye	U				15	21		Seram, Moluccas	1181
Lophozosterops pinaiae	Gray-hooded White-eye	U				25	30		Seram, Moluccas	1181
Lophozosterops goodfellowi	Mindanao White-eye	B	18	20		18.2	22.3		Philippines	954
Oculocincta squamifrons	Pygmy White-eye	U	3	6.9		6.5	7.1		Borneo, Malaysia	1148

Body Masses of World Birds (continued)

Species Name	Common Name	Sex	N	Mean	S.D.	Min	Max	Season	Location	Source #
Xanthotis macleayana	Macleay's Honeyeater	M	11	34.2		30	39		Queensland, Australia	517
		F	7	28.1		25.6	31			
Apalopteron familiare	Bonin Honeyeater	U		15					Bonin Is.	620
Melithreptus lunatus	White-naped Honeyeater	M	40	15	2.5	11.5	25		Australia	517
		F	15	13.1	1.2	11.5	15			
Melithreptus affinis	Black-headed Honeyeater	B	66	15	0.9	12.5	16.8		Tasmania	517
Melithreptus albogularis	White-throated Honeyeater	M	14	11.6	0.9	10	13		Australia	517
		F	9	10.5		9	12.5			
Melithreptus gularis	Black-chinned Honeyeater	B	60	21	1.1	18.5	23.2	Y	Victoria, Australia	517
Melithreptus gularis laetior	Black-chinned Honeyeater	B	6	18		15.8	20.2		Western Australia	470
Melithreptus validirostris	Strong-billed Honeyeater	M	27	26.4	2	23.5	30.5		Australia	517
		F	24	23.3	2.9	17.5	28.5			
Melithreptus brevirostris	Brown-headed Honeyeater	B	169	12.6	0.8	10.5	15		Western Australia	517
Notiomystis cincta	Stitchbird	M	54	98.4	2.7	93	103		New Zealand	419
		F	45	88.6	2.5	85	94			
Pycnopygius ixoides	Plain Honeyeater	M	4	29.5		28	33		New Guinea	283
		F	8	25.8		22	31			
Pycnopygius cinereus	Marbled Honeyeater	M	7	48.8		40	58		New Guinea	283
		F	4	41.2		36	46			
Pycnopygius stictocephalus	Streak-headed Honeyeater	M	2	38.5		38	39		New Guinea	535, 786
Melitograis gilolensis	White-streaked Friarbird	M	1	54				B	Moluccas	979
Philemon meyeri	Meyer's Friarbird	U	6	52					New Guinea	61
Philemon citreogularis	Little Friarbird	M	19	62.6	8.1	50	77		Australia	517
		F	18	58.6	7.9	48	76.5			
Philemon subcorniculatus	Seram Friarbird	U	4	194		172	223		Seram, Moluccas	1181

Body Masses of World Birds (continued)

Species Name	Common Name	Sex	N	Mean	S.D.	Min	Max	Season	Location	Source #
Philemon buceroides	Helmeted Friarbird	M	13	101	6.3	88	110		Northern Terr., Australia	517
		F	10	90.7	13.1	74	120			
Philemon buceroides	Helmeted Friarbird	F	4	140		133	161		New Guinea	283, 427, 535
Philemon cockerelli	New Britain Friarbird	M	3	144		139	147		New Britain	426
		F	2	165		153	177			
Philemon eichhorni	New Ireland Friarbird	M	1	116					New Ireland	58
		F	2	88		82	94			
Philemon argenticeps	Silver-crowned Friarbird	M	20	85.4	11.6	70	106		Australia	517
		F	17	75.6	6.7	59	88.5			
Philemon corniculatus	Noisy Friarbird	M	23	106	12.4	88	136		Eastern Australia	517
		F	7	95.7	6.8	85	103			
Philemon diemenensis	New Caledonian Friarbird	M	9	74.5		69.5	84		New Caledonia	1018
		F	6	58.3		53	62			
Ptiloprora plumbea	Leaden Honeyeater	B	4	14.1		12.5	16.5		New Guinea	282
Ptiloprora meekiana	Olive-streaked Honeyeater	M	1	18.5					New Guinea	282
Ptiloprora erythropleura	Rufous-sided Honeyeater	M	8	23		21.5	25		New Guinea	282
		F	5	19.4		18.5	20.5			
Ptiloprora mayri	Mayr's Honeyeater	M	8	34.1		30	37.5		New Guinea	282
		F	7	27.2		24	29			
Ptiloprora guisei acrophila	Rufous-backed Honeyeater	M	15	32.3		29.3	36.3		New Guinea	282
		F	10	27.6		25.5	29.2			
Ptiloprora guisei umbrosa	Rufous-backed Honeyeater	M	19	24		21	27		New Guinea	282
		F	6	20.8		19	22			
Ptiloprora perstriata	Black-backed Honeyeater	B	142	26.6	3.3	20	32.7		Papua New Guinea	387
Melidectes fuscus	Sooty Melidectes	M	1	45					New Guinea	770
Melidectes whitemanensis	Bismarck Melidectes	F	2	48		47	49		New Britain	426
Melidectes princeps	Long-bearded Melidectes	F	1	42					New Guinea	770

Body Masses of World Birds (continued)

Species Name	Common Name	Sex	N	Mean	S.D.	Min	Max	Season	Location	Source #
Melidectes ochromelas	Cinnamon-browed Melidectes	M	1	63					New Guinea	1143
Melidectes leucostephes	Vogelkop Melidectes	M	8	80		63	97		New Guinea	288
		F	5	70.8		65	78			
Melidectes belfordi	Belford's Melidectes	M	15	85.2	5.7	78	92		Papua New Guinea	387
		F	26	62.8	2.8	58.5	69.2			
Melidectes rufocrissalis	Yellow-browed Melidectes	M	20	81		70	96.5		New Guinea	281
		F	21	63.3		56	72.7			
Melidectes torquatus	Ornate Melidectes	M	9	51		40	58		New Guinea	283, 770
		F	4	39.6		34	44			
Melidectes sclateri	San Cristobal Melidectes	M	14	80.2		75.5	88		Solomon Is.	399
		F	9	55.6		51.5	58.5			
Melipotes fumigatus	Smoky Honeyeater	B	134	58.6	7.5	43.6	81.6		Papua New Guinea	387
Myza celebensis	Dark-eared Honeyeater	U	1	20					Sulawesi, Indonesia	1272
Gymnomyza aubryana	Crow Honeyeater	M	4	244		211	284		New Caledonia	1018, 1172
		F	2	156		152	159			
Moho braccatus	Kauai Oo	M		39					estimated	1193
		F		38						
Moho bishopi	Bishop's Oo	M		59					estimated	1193
		F		42						
Moho apicalis	Oahu Oo	M		56					estimated	1193
		F		43						
Moho nobilis	Hawaii Oo	M		67					estimated	1193
		F		43						
Chaetophila angustipluma	Kioea	U		97					estimated	1193
Phylidonyris pyrrhoptera	Crescent Honeyeater	M	119	18.2	1.1	15.4	21.1		Tasmania	517
		F	51	14.6	0.9	12.8	16.5			
Phylidonyris novaehollandiae	New Holland Honeyeater	M	344	21.5	1.9	10	26		Australia	517
		F	320	19	2.1	6	29			
Phylidonyris nigra	White-cheeked Honeyeater	M	42	19.5	1.3	14	22		Australia	517
		F	39	17	1.4	15	20			

Body Masses of World Birds (continued)

Species Name	Common Name	Sex	N	Mean	S.D.	Min	Max	Season	Location	Source #
Phylidonyris albifrons	White-fronted Honeyeater	M	198	19.3	1.3	15	22		Australia	517
		F	151	15.7	1.4	12	19			
Phylidonyris undulata	Barred Honeyeater	F	1	16					New Caledonia	1018
Phylidonyris notabilis	New Hebrides Honeyeater	M	1	29.7					Vanuatu	656b
		F	2	24.8		23.3	26.4			
Phylidonyris melanops	Tawny-crowned Honeyeater	M	15	19.3	2	14.7	22		Australia	517
		F	15	17.2	1.2	14	19			
Ramsayornis modestus	Brown-backed Honeyeater	U	17	12.1	1.4	8.5	14		Australia	738
Ramsayornis fasciatus	Bar-breasted Honeyeater	M	18	13.2	1.4	10.8	15.8		Australia	517
		F	12	12.3	1.2	10	14.4			
Plectorhyncha lanceolata	Striped Honeyeater	M	9	42.4		33	48		Australia	517
		F	11	38.8	3.6	33	45			
Conopophila albogularis	Rufous-banded Honeyeater	U	43	11.8	1.1	8	14		Australia	517
Conopophila rufogularis	Rufous-throated Honeyeater	B	51	10.8		8.2	13.9		Australia	517
Conopophila whitei	Gray Honeyeater	B	13	9.8		7	11		Western Australia	517
Grantiella picta	Painted Honeyeater	B	5	21.5		18	25		Southeast Australia	517
Xanthomyza phrygia	Regent Honeyeater	M	8	43.7		41	45.5		New South Wales, Australia	517
		F	7	36.6		33	41			
Acanthorhynchus tenuirostris tenuirostris	Eastern Spinetail	M	33	11.9	0.7	10.4	13.6		Victoria, Australia	517
		F	45	10.5	0.8	8.8	12.2			
Acanthorhynchus tenuirostris dubius	Eastern Spinetail	M	77	10.6	0.7	9	12.1		Tasmania	517
		F	80	9	0.7	7.9	10.2			
Acanthorhynchus superciliosus	Western Spinetail	M	155	10.5	0.6	8.5	11.8		Perth, Australia	517
		F	86	9	0.5	7.9	11			
Entomyzon cyanotis	Blue-faced Honeyeater	M	16	114	10.5	99	130		Australia	517
		F	15	97.2	8.3	83.5	111			

Body Masses of World Birds (continued)

Species Name	Common Name	Sex	N	Mean	S.D.	Min	Max	Season	Location	Source #
Lichenostomus virescens forresti	Singing Honeyeater	M	13	25.1	2	21.8	29.7		Australia	517
		F	10	21.7	1.9	19.1	26.5			
Lichenostomus flavus	Yellow Honeyeater	U	35	21.4	4.3	14.3	30.3		Australia	517
Lichenostomus unicolor	White-gaped Honeyeater	M	26	32.4	3.3	27.8	40.4		Australia	517
		F	18	29.1	2.4	25	33			
Lichenostomus leucotis	White-eared Honeyeater	M	25	26.5	3.8	18.9	35		Southeast Australia	517
		F	14	21.5	3.1	18	30			
Lichenostomus flavicollis	Yellow-throated Honeyeater	M	42	32.6	2.3	28	40		Australia	517
		F	16	25.3	1.5	21.5	28.5			
Lichenostomus melanops	Yellow-tufted Honeyeater	M	62	34.1	2.7	27.6	40.2		Victoria, Australia	517
		F	50	30.6	2.5	26	36.7			
Lichenostomus cratitius	Purple-gaped Honeyeater	M	20	21.9	1.4	19.5	24.7		Australia	517
		F	17	18.9	2.2	15.5	22.5			
Lichenostomus keartlandi	Gray-headed Honeyeater	M	14	15.7	1.2	13	17		Australia	517
		F	9	14.8		12.5	18			
Lichenostomus flavescens	Yellow-tinted Honeyeater	M	43	12.9	1.1	11	16		Australia	517
		F	20	12	1.7	9.5	15.5			
Lichenostomus fuscus	Fuscous Honeyeater	M	29	19.3	1.2	17.5	23.8		Victoria, Australia	517
		F	59	16.2	1.4	13.8	21.8			
Lichenostomus plumulus	Gray-fronted Honeyeater	B	195	16.5	1.8	13	24		Australia	517
Lichenostomus ornatus	Yellow-plumed Honeyeater	M	29	18.4	1.1	16	22.1		Australia	517
		F	21	16.9	2.4	12.2	21.1			
Lichenostomus penicillatus	White-plumed Honeyeater	M	67	20.6	1.7	12.4	24.6		Victoria, Australia	517
		F	93	18.2	1.6	14.4	23.5			
Xanthotis flaviventer	Tawny-breasted Honeyeater	M	12	47.3	4.2	40	55		New Guinea	283
		F	10	42.7	3.1	38	49			
Xanthotis flaviventer	Tawny-breasted Honeyeater	M	8	39		33	41		Cent. York Pen., Australia	517
		F	5	32.4		26.5	37			
Xanthotis polygrammus	Spotted Honeyeater	M	3	21.1		19.7	23.5		New Guinea	283
		F	3	19		18	19.5			

Body Masses of World Birds (continued)

Species Name	Common Name	Sex	N	Mean	S.D.	Min	Max	Season	Location	Source #
Manorina melanophrys	Bell Miner	M	17	32.3	4.2	24.2	45		Victoria, Australia	517
		F	16	28.6	2.6	25	33			
Manorina melanocephala	Noisy Miner	M	34	63.4	10.7	40	91		Australia	517
		F	19	57.2	8.2	46.7	74			
Manorina flavigula flavigula	Yellow-throated Miner	M	16	63.8	9.6	54	86.5		Eastern Australia	517
		F	6	55.8		50	59			
Manorina flavigula wayensis	Yellow-throated Miner	M	17	55.2	4.9	46.5	65		Australia	517
		F	17	49.7	4.2	43	57			
Manorina melanotis	Black-eared Miner	M	2	54		49	60		South Australia	517
Anthornis melanura	New Zealand Bellbird	M	202	30.7	2.4	21	38		New Zealand	1002
		F	94	23.6	2.2	20	32			
Acanthagenys rufogularis	Spiny-cheeked Honeyeater	B	30	47.8	4.2	39.3	57		Victoria, Australia	517
Anthochaera carunculata	Red Wattlebird	M	12	114	9.7	99	131		Australia	517
		F	6	98.7		84	106			
Anthochaera chrysoptera	Brush Wattlebird	M	26	77	6.6	57	85		Southeast Australia	517
		F	14	59.7	10.5	44	82			
Anthochaera lunulata	Little Wattlebird	M	10	71.2	4.4	65	78		Western Australia	517
		F	4	53.7		47	57			
Anthochaera paradoxa	Yellow Wattlebird	M	18	168	15.5	150	210		Tasmania, Australia	517
		F	12	123	9.7	104	140			
Prosthemadera novaeseelandiae	Tui	M	75	125	10	97	150		New Zealand	1002
		F	46	89.6	6.5	70	105			
Family Oriolidae										
Oriolus forsteni	Seram Oriole	U	3	109		105	113		Seram, Moluccas	1181
Oriolus szalayi	Brown Oriole	B	18	95.2	8.2	79	115		New Guinea	283
Oriolus sagittatus	Olive-backed Oriole	B	27	95.2		78.3	120		Australia	518a
Oriolus flavocinctus	Green Oriole	B	17	109		90.5	125		Australia	518a

Body Masses of World Birds (continued)

Species Name	Common Name	Sex	N	Mean	S.D.	Min	Max	Season	Location	Source #
Oriolus xanthonotus persuasus	Dark-throated Oriole	B	2	45		44.4	45.5		Palawan, Philippines	1143
Oriolus isabellae	Isabela Oriole	B	4	46.1		42.4	50		Luzon, Philippines	1143
Oriolus oriolus	Eurasian Golden Oriole	U	51	79	7.4	59.7	96	F	Egypt	809
Oriolus auratus	African Golden Oriole	B	15	66.2		60	72		DR Congo	396
Oriolus chinensis	Black-naped Oriole	B	76	82.6		65	100		Hopei, China	1087
Oriolus tenuirostris	Slender-billed Oriole	U	5			72	92			7
Oriolus crassirostris	Sao Tome Oriole	B	2	50		50	50			396
Oriolus brachyrhynchus	Western Black-headed Oriole	M	9	50.7						396
		F	9	47.4						
Oriolus larvatus	African Black-headed Oriole	U	9	66.1		59	72.4			396
Oriolus percivali	Black-tailed Oriole	U	1	61						79
Oriolus nigripennis	Black-winged Oriole	B	6	56.7		50.2	60			396, 710
Oriolus xanthornus	Black-hooded Oriole	B	7	56.3		45.7	79		India; Philippines	7, 740, 954
Oriolus traillii	Maroon Oriole	B	10			67	81		India	7
Sphecotheres viridis vieilloti	Green Figbird	B	22	132		128	146		Australia	518a
Sphecotheres viridis flaviventris	Green Figbird	B	5	111		97.1	131		Australia	470
Family Irenidae										
Irena puella	Asian Fairy-bluebird	U	6	64.9		56	70.3		Malaysia; Borneo	398, 121
Irena cyanogaster	Philippine Fairy-bluebird	B	35	80.3		71	96.1		Philippines	954

Body Masses of World Birds (continued)

Species Name	Common Name	Sex	N	Mean	S.D.	Min	Max	Season	Location	Source #
Family Laniidae										
Lanius tigrinus	Tiger Shrike	M	10	28.3		27	29		China	184a
		F	10	32.1		29	37			
Lanius bucephalus	Bull-headed Shrike	F	66	44.8	3.7	37.3	54.4	B	Japan	485
Lanius bucephalus	Bull-headed Shrike	B	41	34.7		29	45		Hopei, China	1087
Lanius collurio	Red-backed Shrike	M	161	27.9	1.8	22.5	33.5	B	Germany	223
		F	174	29	1.9	23	34			
Lanius isabellinus	Rufous-tailed Shrike	B	352	25.2		21	30.3	W	Kenya	223
Lanius cristatus cristatus	Brown Shrike	B	27	33.5	3.2	24	39		Mongolia	927
Lanius cristatus lucionensis	Brown Shrike	B	28	31.4		26	40		Hopei, China	1087
Lanius cristatus	Brown Shrike	U	43	31.7	2	27	36		Taiwan	223
Lanius collurioides	Burmese Shrike	F	1	26					immature	485
Lanius gubernator	Emin's Shrike	U		23.5						485
Lanius souzae	Souza's Shrike	B	23	26.3		21	30		Angola	485
Lanius vittatus	Bay-backed Shrike	U	17	20.8		18	24		India	7
Lanius schach	Long-tailed Shrike	M	10	53.4		51	61		China	184a
		F	10	49.7		43	60			
Lanius tephronotus	Gray-backed Shrike	M	8			39	51			7
		F	4			43	54			
Lanius validirostris	Gray-capped Shrike	B	10	37.5	3.2	34	44.9		Luzon & Mindanao, Philippines	1143
Lanius ludovicianus	Loggerhead Shrike	B	105	47.7	3.5	35	58.5	Y	Florida, USA	1350
Lanius ludovicianus	Loggerhead Shrike	B	129	55.8	0.3				Alberta, Canada	205
Lanius excubitor borealis	Northern Shrike	U	22	63.4	3.3	56	69.5	W	Eastern USA	161

Body Masses of World Birds (continued)

Species Name	Common Name	Sex	N	Mean	S.D.	Min	Max	Season	Location	Source #
Lanius excubitor borealis	Northern Shrike	U	44	66.6	4.6	58.5	79	W	Western USA	161
Lanius excubitor excubitor	Northern Shrike	U	44	63.4	6	48	81		Europe	223
Lanius excubitor	Northern Shrike	B	20	60.4	3.3	55	67		Israel	1349
Lanius meridionalis pallidirostris	Southern Gray Shrike	M	5	61		48	68		Mongolia	927
		F	3	51.7		47	58			
Lanius meridionalis meridionalis	Southern Gray Shrike	U	23	70		53	93		Spain	682
Lanius minor	Lesser Gray Shrike	M	16	45.7	1.5	43	48	B	Germany	223
		F	13	47.3	1.4	45	50			
Lanius sphenocercus	Chinese Gray Shrike	B	5	92.1		87.2	100			485
Lanius excubitoroides	Gray-backed Fiscal	M	42	55.3		47	63.5		Kenya	485
		F	26	52.2	3.3	46.5	59			
Lanius cabanisi	Long-tailed Fiscal	M	5	76.8		70	82			396, 485
		F	3	71		69	74			
Lanius dorsalis	Taita Fiscal	B	5	49.9		45	55			485
Lanius somalicus	Somali Fiscal	B	3	45.3		40	48			485
Lanius mackinnoni	Mackinnon's Shrike	M	9	33.3		30	35		East Africa	396
		F	10	35.3		30	38			
Lanius collaris	Common Fiscal	M	84	41.8		24.7	51		Southern Africa	396
		F	69	39.8		27.7	52			
Lanius collaris smithii	Common Fiscal	M	22	34.1	3.2	27	39		Liberia	485
		F	44	31.1	1.7	27	36			
Lanius newtoni	Newton's Fiscal	F	1	22.4					Sao Tome	485
Lanius nubicus	Masked Shrike	B	264	19.7					Israel	1352
Lanius senator	Woodchat Shrike	M	49	34.6	2.7	28	45	B	Germany	223
		F	44	37.4	3.9	33	52			
Corvinella corvina	Yellow-billed Shrike	U	11	65.4		58	80		Ghana	682
Corvinella melanoleuca	Magpie Shrike	B	33	82.2		54.8	97.1			785

Body Masses of World Birds (continued)

Species Name	Common Name	Sex	N	Mean	S.D.	Min	Max	Season	Location	Source #
Eurocephalus rueppelli	White-rumped Shrike	M	12	48.4		39	70			396
		F	10	52.9		37	64			
Eurocephalus anguitimens	White-crowned Shrike	M	13	68		55.7	76.8			485
		F	9	70.1		59	84			
Family Malaconotidae										
Nilaus afer solivagus	Brubru	B	58	23		17	32			396
Nilaus afer minor	Brubru	B	7	15.8		13	20			396
Dryoscopus gambensis	Northern Puffback	B	22	31.9		28	35			396
Dryoscopus pringlii	Pringle's Puffback	B	13	19.7		18.4	22.6			485
Dryoscopus cubla	Black-backed Puffback	M	125	27.1		19.3	36		Southern Africa	485
		F	98	25.4		21	30.4			
Dryoscopus senegalensis	Red-eyed Puffback	B	6	29.3		23	34		Cameroon; Uganda	485
Dryoscopus angolensis	Pink-footed Puffback	U	10	36		27	40		Cameroon; Uganda	485
Dryoscopus sabini	Large-billed Puffback	M	3	38.5		31	43.9		Liberia	485
		F	3	35.8		32.4	39.2			
Tchagra minuta	Marsh Tchagra	B	21	34.4		30	38		Kenya	396
Tchagra senegala	Black-crowned Tchagra	B	80	53.5		33	59.7		Southern Africa	485
Tchagra australis	Brown-crowned Tchagra	B	159	32.3		25	45.8		Southern Africa	485
Tchagra jamesi	Three-streaked Tchagra	B	16	27.4		22	32			485
Tchagra tchagra	Southern Tchagra	U	28	50.3		40	60		South Africa	1036
Laniarius ruficeps	Red-naped Bushshrike	M	1	35						396
		F	2	30.5		30	31			
Laniarius luehderi	Luehder's Bushshrike	B	39	42.9		30	49		East Africa	396

Body Masses of World Birds (continued)

Species Name	Common Name	Sex	N	Mean	S.D.	Min	Max	Season	Location	Source #
Laniarius brauni	Braun's Bushshrike	B	3	54.3		54	55			396
Laniarius aethiopicus	Tropical Boubou	U	104	50.2		44	66		Southern Africa	485
Laniarius aethiopicus	Tropical Boubou	U	39	48.7		38.8	66		Kenya	485
Laniarius bicolor	Gabon Boubou	M	10			52	65		Angola	266
		F	8			45	57			
Laniarius ferrugineus	Southern Boubou	U	32	46.8	3.9	38.8	54		Kenya	121
Laniarius barbarus	Common Gonolek	U	11	47.1	3.4	43	52		Ghana	457
Laniarius erythrogaster	Black-headed Gonolek	U	44	48.8		35	56		Kenya	485
Laniarius atrococcineus	Crimson-breasted Gonolek	U	32	48.8		40	56.4			485
Laniarius mufumbiri	Papyrus Gonolek	U	12	41.5	3.6	34.4	46		Kenya	122
Laniarius atroflavus	Yellow-breasted Boubou	U	5	43		40.1	46.9			396
Laniarius funebris	Slate-colored Gonolek	U	123	40.4		32.2	50.4		Ethiopia	485
Laniarius leucorhynchus	Sooty Boubou	M	8	53.5		51	57			396
		F	2	47		45	49			
Laniarius fuelleborni fuelleborni	Fuelleborn's Boubou	U	20	46.4		37	53			485
Laniarius poensis	Mountain Sooty Boubou	U	11	44.9		39.3	48.9		Cameroon	396
Rhodophoneus cruentus	Rosy-patched Bushshrike	B	6	53.3		40	62			396
Telophorus zeylonus	Bokmakierie	M	12	65.1		57	71.3			485
		F	11	61.2		48	76.1			
Telophorus zeylonus restrictus	Bokmakierie	B	9	69		63.8	76.1			485
Telophorus bocagei	Gray-green Bushshrike	B	10	27		22	28			396

Body Masses of World Birds (continued)

Species Name	Common Name	Sex	N	Mean	S.D.	Min	Max	Season	Location	Source #
Telophorus sulfureopectus	Sulphur-breasted Bushshrike	B	55	27.1		19.5	30.5			485
Telophorus olivaceus	Olive Bushshrike	M	103	34.5		31.1	38.8			396
		F	56	32.7		24.6	43.6			
Telophorus multicolor	Many-colored Bushshrike	B	20	50.9		43	61		Liberia	396
Telophorus nigrifrons	Black-fronted Bushshrike	U	18	34.8		32.2	39.4		Kenya; South Africa	485
Telophorus viridis	Four-colored Bushshrike	B	26	37.4		29.6	40.5		Southern Africa	485
Telophorus viridis quadricolor	Four-colored Bushshrike	B	8	36.9		29.6	40.5		Zimbabwe	562, 563, 565
Telophorus dohertyi	Doherty's Bushshrike	M	3	39.7		36	43			396, 485
		F	1	30.9						
Malaconotus cruentus	Fiery-breasted Bushshrike	B	11	76.7					Liberia	485
Malaconotus lagdeni	Lagden's Bushshrike	M	3	92.3		84	97		Liberia; Uganda	485
Malaconotus gladiator	Green-breasted Bushshrike	U		99						485
Malaconotus blanchoti	Gray-headed Bushshrike	B	51	78.7		73	112		Southern Africa	1036
Family Prionopidae										
Prionops plumatus	White Helmetshrike	M	44	32.7		27.3	35.8			485
		F	41	34.9		29.7	39.8			
Prionops poliolophus	Gray-crested Helmetshrike	M	1	49						396
Prionops alberti	Yellow-crested Helmetshrike	B	3	61.3		61	62			396
Prionops caniceps	Chestnut-bellied Helmetshrike	B	4	55.8		52	62		Liberia	396
Prionops caniceps rufiventris	Chestnut-bellied Helmetshrike	M	8	42.4		39	44			396
		F	4	47.3		44	50			

Body Masses of World Birds (continued)

Species Name	Common Name	Sex	N	Mean	S.D.	Min	Max	Season	Location	Source #
Prionops retzii retzii	Retz's Helmetshrike	U	41	47.8		38	54		Southern Africa	396
Prionops retzii graculinus	Retz's Helmetshrike	M	15	49.3		33	55		Kenya	396
		F	22	53.4		40	60			
Prionops scopifrons	Chestnut-fronted Helmetshrike	B	39	29.4		26	38		Zimbabwe	396
Tephrodornis gularis	Large Woodshrike	U	4	37.8		38	46			7
Tephrodornis pondicerianus	Common Woodshrike	B	17	20.2		18	27		India	7, 740
Philentoma pyrhoptera	Rufous-winged Philentoma	U	32	16.9						485
Philentoma velata	Maroon-breasted Philentoma	B	2	26.1		25	27.2		Borneo	1218
Family Vangidae										
Calicalicus madagascariensis	Red-tailed Vanga	U	12	16	1.7	14	18.6		Madagascar	957
Calicalicus madagascariensis rufocarpalis	Red-tailed Vanga	U	2	16.8		16.1	17.4		Madagascar	957
Schetba rufa	Rufous Vanga	U	25	37.3	3.4	30.5	44		Madagascar	957
Vanga curvirostris curvirostris	Hook-billed Vanga	U	14	70.4	6.1	53.5	80.5		Madagascar	957
Vanga curvirostris cetera	Hook-billed Vanga	U	5	58.5		53	64		Madagascar	957
Xenopirostris xenopirostris	Lafresnaye's Vanga	B	6	59.5		52	63		Madagascar	957
Xenopirostris polleni	Pollen's Vanga	U	1	61.5					Madagascar	957
Falculea palliata	Sickle-billed Vanga	U	6	111		106	119		Madagascar	957
Artamella viridis	White-headed Vanga	B	6	48.8		44	57		Madagascar	69, 957
Leptopterus chabert	Chabert Vanga	U	13	19.6	2.3	17.6	26.7		Madagascar	957

Body Masses of World Birds (continued)

Species Name	Common Name	Sex	N	Mean	S.D.	Min	Max	Season	Location	Source #
Cyanolanius madagascarinus	Blue Vanga	U	5	22.3		19.3	24.5		Madagascar	957
Oriolia bernieri	Bernier's Vanga	B	2	52		52	52		Madagascar	957
Euryceros prevostii	Helmet Vanga	M	5	109		100	114		Madagascar	1143
		F	3	86.2		84	87.5			
Tylas eduardi	Tylas Vanga	U	22	46.8	4.7	35.5	54		Madagascar	957
Hypositta corallirostris	Coral-billed Nuthatch	B	4	14.1		13.5	15.5		Madagascar	957
Family Dicruridae										
Chaetorhynchus papuensis	Papuan Drongo	M	10	42	2	36	45		New Guinea	283
		F	10	33	4	27	39			
Dicrurus ludwigii	Square-tailed Drongo	U	22	29.4		25	34		Tanzania	396
Dicrurus atripennis	Shining Drongo	B	16	38.9		35	47			396
Dicrurus adsimilis	Fork-tailed Drongo	U	12	40.3		34	48		Chad; Gabon	396
Dicrurus modestus	Velvet-mantled Drongo	F	3	45.6		44	48		Liberia	396
Dicrurus aldabranus	Aldabra Drongo	M	1	51					Aldabra Is.	67
		F	1	46						
Dicrurus forficatus	Crested Drongo	B	35	47.1	4.1	39	55		Madagascar	957
Dicrurus macrocercus	Black Drongo	B	91	48.3		41	59		Hopei, China	1087
Dicrurus leucophaeus	Ashy Drongo	B	40	37.6		32	45	W		7
Dicrurus caerulescens	White-bellied Drongo	U	2	40		39	41		India	7
Dicrurus annectans	Crow-billed Drongo	F	1	44						7
Dicrurus aeneus	Bronzed Drongo	M	5			26	28		India	7
		F	2	26		22	30			
Dicrurus remifer	Lesser Racket-tailed Drongo	U	19	43.1		40	50		Malaysia	775

Body Masses of World Birds (continued)

Species Name	Common Name	Sex	N	Mean	S.D.	Min	Max	Season	Location	Source #
Dicrurus hottentottus	Hair-crested Drongo	B	9	79.2		70	91		New Guinea	786
Dicrurus balicassius	Balicassiao	M	16	71.6	3.7	63.3	78.8		Philippines	436
		F	14	73	4.3	65	80			
Dicrurus montanus	Sulawesi Drongo	U	1	61					Sulawesi, Indonesia	1272
Dicrurus megarhynchus	Ribbon-tailed Drongo	F	2	130		129	130		New Ireland	58
Dicrurus bracteatus bracteatus	Spangled Drongo	B	11	85.7		64.5	105		Southeast Australia	518a
Dicrurus bracteatus meeki	Spangled Drongo	B	8	79.8		73.5	86		Guadalcanal	399
Dicrurus bracteatus longirostris	Spangled Drongo	M	3	91.5		85.5	96.5		Solomon Is.	399
Dicrurus paradiseus	Greater Racket-tailed Drongo	B	4	71		70	73		Sri Lanka	7
Dicrurus paradiseus paradiseus	Greater Racket-tailed Drongo	U	2	79.5		74	85		India	7
Dicrurus paradiseus grandis	Greater Racket-tailed Drongo	M	2	112		100	124		India	7
Family Callaeidae										
Callaeas cinerea	Kokako	M	75	233	14.5				New Zealand	518a
		F	55	218	16.8					
Creadion carunculatus	Saddleback	M	138	80.7	4.6				New Zealand	518a
		F	83	69.2	4					
Heteralocha acutirostris	Huia	U	1	406					New Zealand, extinct	518a
Family Grallinidae										
Grallina cyanoleuca cyanoleuca	Magpie-lark	M	24	91.6	12.1	63.9	118		Australia	518a
		F	17	81.3	7.4	70	94.5			
Grallina cyanoleuca neglecta	Magpie-lark	M	22	82.8	7.9	71	102		Northern Australia	518a
		F	9	75.4		60	90			
Grallina bruijni	Torrent-lark	M	1	38.4					New Guinea	283

Body Masses of World Birds (continued)

Species Name	Common Name	Sex	N	Mean	S.D.	Min	Max	Season	Location	Source #
Family Corcoracidae										
Corcorax melanorhamphos	White-winged Chough	U	191	364	22.1			W	Australia	1021
Struthidea cinerea	Apostlebird	B	94	132	8.7	110	155		New South Wales, Australia	1342
Family Artamidae										
Artamus fuscus	Ashy Woodswallow	B	4	39.8		37	42		India	7
Artamus mentalis	Fiji Woodswallow	B	17	41.9	3.3	33	48		Fiji	1196c
Artamus maximus	Great Woodswallow	B	12	58.2		52	64		New Guinea	283
Artamus leucorynchus	White-breasted Woodswallow	B	11	42.4		36.4	49.6		Philippines	954
Artamus leucorynchus	White-breasted Woodswallow	B	34	41		36	46.5		Australia	518a
Artamus insignis	Bismarck Woodswallow	B	3	51.3		49	54		New Britain	426
Artamus personatus	Masked Woodswallow	M	25	34.6		25.5	41		Australia	518a
Artamus superciliosus	White-browed Woodswallow	B	18	35.5		30	41.1		Australia	518a
Artamus cinereus	Black-faced Woodswallow	B	32	35.6	2.5	30.9	40		Australia	470
Artamus cyanopterus	Dusky Woodswallow	U	116	34.6	4.3	21	47			518a
Artamus minor	Little Woodswallow	B	23	16	1.8	13.1	20.9		Australia	470
Family Pityriaseidae										
Pityriasis gymnocephala	Bornean Bristlebird	M	2	118		115	121		Borneo	1218
		F	1	140						
Family Cracticidae										
Peltops montanus	Mountain Peltops	B	11	31.6	2.6	27	36		New Guinea	283

Body Masses of World Birds (continued)

Species Name	Common Name	Sex	N	Mean	S.D.	Min	Max	Season	Location	Source #
Peltops blainvillii	Lowland Peltops	U	11	30					New Guinea	61
Cracticus mentalis	Black-backed Butcherbird	B	11	77		74	80		Cape York Penin, Australia	518a
Cracticus torquatus leucopterus	Gray Butcherbird	M	20	90.4	5.8	79	99		Australia	518a
		F	18	85.1	4.7	76	95			
Cracticus torquatus colletti	Gray Butcherbird	M	7	81.6		79	83		Australia	518a
		F	6	75.2		68	84			
Cracticus cassicus	Hooded Butcherbird	B	7	142		130	155		New Guinea	283
Cracticus nigrogularis	Pied Butcherbird	B	34	128		105	159		Australia	518a
Cracticus quoyi	Black Butcherbird	B	9	168		148	196		Australia	518a
Gymnorhina tibicen terraereginae	Australian Magpie	B	42	256		212	325		Eastern Australia	518a
Gymnorhina tibicen dorsalis	Australian Magpie	B	10	317		265	360		Western Australia	518a
Strepera graculina	Pied Currawong	M	19	319	31.6	259	385		Australia	518a
		F	14	280	25.6	243	324			
Strepera fuliginosa	Black Currawong	M	15	411	20.9	365	457		Tasmania, Australia	518a
		F	7	343		327	353			
Strepera versicolor	Gray Currawong	M	5	425		390	500		Australia	518a
		F	15	368	48.7	300	500			
Family Paradisaeidae										
Cnemophilus loriae loriae	Loria's Bird-of-paradise	B	15	84.6		76	96			385
Cnemophilus loriae amethystina	Loria's Bird-of-paradise	M	5	93		80	101			385
		F	7	79		69	95			
Cnemophilus loriae inexpectata	Loria's Bird-of-paradise	B	12	80		60	95			385
Cnemophilus macgregorii	Crested Bird-of-paradise	M	6	103		94	120			385
		F	10	93		79	125			
Loboparadisea sericea	Yellow-breasted Bird-of-paradise	M	12	62		50	75			385
		F	2	73		71	75			

Body Masses of World Birds (continued)

Species Name	Common Name	Sex	N	Mean	S.D.	Min	Max	Season	Location	Source #
Macgregoria pulchra	Macgregor's Bird-of-paradise	M	6	256		242	274			385
		F	3	198		190	207			
Lycocorax pyrrhopterus pyrrhopterus	Paradise-crow	M	8	269		242	304			385
		F	7	250		218	276			
Lycocorax pyrrhopterus obiensis	Paradise-crow	M	9	332		300	370			385
		F	5	291		250	316			
Manucodia atra atra	Glossy-mantled Manucode	M	17	224		170	315			385
		F	19	207		155	252			
Manucodia atra subalter	Glossy-mantled Manucode	M	6	275		238	300			385
		F	5	212		190	245			
Manucodia jobiensis	Jobi Manucode	M	8	228		212	257			385
		F	5	180		150	205			
Manucodia chalybatus	Crinkle-collared Manucode	M	36	223		164	265			385
		F	21	209		160	255			
Manucodia comrii	Curl-crested Manucode	M	1	448						385
		F	1	418						
Manucodia keraudrenii purpureoviolaceus	Trumpet Manucode	M	20	181		159	240			385
		F	17	156		134	182			
Manucodia keraudrenii keraudrenii	Trumpet Manucode	M	7	149		130	175			385
		F	7	136		130	140			
Paradigalla carunculata	Long-tailed Paradigalla	F	1	170						385
Paradigalla brevicauda	Short-tailed Paradigalla	M	3	173		160	184			385
		F	2	163		155	170			
Astrapia nigra	Arfak Astrapia	M	1	190					immature	385
Astrapia splendidissima	Splendid Astrapia	M	12	138		120	151			385
		F	16	125		108	151			
Astrapia mayeri	Ribbon-tailed Astrapia	M	12	147		134	164			385
		F	7	132		102	157			
Astrapia stephaniae	Princess Stephanie's Astrapia	M	5	150		145	156			385
		F	15	134		123	148			

Body Masses of World Birds (continued)

Species Name	Common Name	Sex	N	Mean	S.D.	Min	Max	Season	Location	Source #
Astrapia rothschildi	Huon Astrapia	M	5	207		186	225			385
		F	9	159		143	200			
Parotia sefilata	Western Parotia	M	12	192		175	205			385
		F	14	172		140	185			
Parotia carolae clelandiae	Carola's Parotia	M	1	205					New Guinea	423
		F	1	163						
Parotia carolae chrysenia	Carola's Parotia	F	3	130		110	152			385
Parotia lawesii	Lawes' Parotia	M	48	167		153	195			385
		F	60	144		122	180			
Parotia lawesii helenae	Lawes' Parotia	M	5	170		162	177			385
		F	2	166		160	173			
Parotia wahnesi	Wahnes' Parotia	M	3	171		170	172			385
		F	7	146		144	154			
Pteridophora alberti	King-of-Saxony Bird-of-paradise	M	13	87		80	95			385
		F	12	77		68	88			
Ptiloris magnificus magnificus	Magnificent Riflebird	M	20	207		180	230			385
		F	8	142		120	185			
Ptiloris magnificus intercedens	Magnificent Riflebird	M	12	184		155	214			385
		F	15	126		102	149			
Ptiloris magnificus alberti	Magnificent Riflebird	M	4	160		143	171			385
		F	3	104		94	112			
Ptiloris paradiseus	Paradise Riflebird	M	5	136		128	143			518a
		F	7	105		86	112			
Ptiloris victoriae	Victoria's Riflebird	M	22	105		91	119			385
		F	25	86		77	96			
Lophorina superba superba	Superb Bird-of-paradise	M	8	90		81	95			385
		F	5	74		68	85			
Lophorina superba latipennis	Superb Bird-of-paradise	M	39	87		78	99			385
		F	38	67		54	78			
Lophorina superba minor	Superb Bird-of-paradise	M	8	83		77	93			385
		F	5	63		56	68			
Epimachus fastuosus fastuosus	Black Sicklebird	M	7	266		250	280			385
		F	7	191		160	235			

Body Masses of World Birds (continued)

Species Name	Common Name	Sex	N	Mean	S.D.	Min	Max	Season	Location	Source #
Epimachus fastuosus atratus	Black Sicklebird	M	8	281		255	315			385
		F	5	206		160	255			
Epimachus meyeri	Brown Sicklebird	M	7	281		253	310			385
		F	6	173		162	188			
Epimachus albertisi	Black-billed Sicklebird	B	14	110		92	128			385
Epimachus bruijnii	Pale-billed Sicklebird	M	4	161		160	164			385
		F	3	146		144	149			
Cicinnurus magnificus chrysopterus	Magnificent Bird-of-paradise	M	36	102		92	119			385
		F	23	91		81	113			
Cicinnurus magnificus hunsteini	Magnificent Bird-of-paradise	M	33	91		75	105			385
		F	37	77		62	94			
Cicinnurus respublica	Wilson's Bird-of-paradise	M	6	61		53	67			385
		F	6	56		52	60			
Cicinnurus regius gymnorhynchus	King Bird-of-paradise	M	32	53		43	65			385
		F	10	49		38	58			
Cicinnurus regius coccineifrons	King Bird-of-paradise	B	52	53		46	64			385
Semioptera wallacii	Wallace's Standardwing	M	3	163		152	174			385
		F	3	135		126	143			
Seleucidis melanoleuca	Twelve-wired Bird-of-paradise	M	12	189		170	202			385
		F	3	161		160	163			
Paradisaea minor finschi	Lesser Bird-of-paradise	M	21	256		183	300			385
		F	16	164		141	210			
Paradisaea minor jobiensis	Lesser Bird-of-paradise	M	2	296		293	300			676
		F	2	168		152	185			
Paradisaea minor minor	Lesser Bird-of-paradise	M	12	239		185	285			385
		F	7	162		145	186			
Paradisaea apoda	Greater Bird-of-paradise	F	4	172		170	173			385
Paradisaea raggiana salvadorii	Raggiana Bird-of-paradise	M	21	266		240	295			385
		F	24	175		135	210			

Body Masses of World Birds (continued)

Species Name	Common Name	Sex	N	Mean	S.D.	Min	Max	Season	Location	Source #
Paradisaea raggiana augustaevictoriae	Raggiana Bird-of-paradise	M	16	276		234	300			385
		F	10	159		133	195			
Paradisaea decora	Goldie's Bird-of-paradise	M	1	237						676
Paradisaea rubra	Red Bird-of-paradise	M	7	201		158	224			385
		F	10	158		115	208			
Paradisaea guilielmi	Emperor Bird-of-paradise	M	6	256		250	265			385
Paradisaea rudolphi	Blue Bird-of-paradise	M	9	176		158	189			385
		F	4	153		124	166			
Melampitta lugubris	Lesser Melampitta	M	8	44.2		41	47			386
		F	6	41.2		38	45			
Melampitta gigantea	Greater Melampitta	M	1	205					New Guinea	287
Family Ptilonorhynchidae										
Ailuroedus buccoides	White-eared Catbird	U	23	138					New Guinea	61
Ailuroedus melanotis	Spotted Catbird	M	27	183	11.2				Queensland, Australia	390
		F	36	171	13.2					
Ailuroedus crassirostris	Green Catbird	M	19	214	29.1	167	289			390
		F	15	197	11.4	169	211			
Scenopoeetus dentirostris	Tooth-billed Catbird	M	51	158	11.2	132	199			390
		F	3	169		157	182			
Archboldia papuensis papuensis	Archbold's Bowerbird	M	2	173		170	175			390
Archboldia papuensis sanfordi	Sanford's Bowerbird	M	6	184		180	190			390
		F	4	176		163	185			
Amblyornis inornatus	Vogelkop Bowerbird	U	28	124		105	155		New Guinea	288
Amblyornis macgregoriae	Macgregor's Bowerbird	M	37	115	9	100	142			390
		F	18	120	7.2	104	130			
Amblyornis subalaris	Streaked Bowerbird	M	8	104		96	107		New Guinea	1058
		F	3	114		109	122			

Body Masses of World Birds (continued)

Species Name	Common Name	Sex	N	Mean	S.D.	Min	Max	Season	Location	Source #
Amblyornis flavifrons	Golden-fronted Bowerbird	M	1	120					New Guinea, estimated	286
Prionodura newtoniana	Golden Bowerbird	M	80	73	2.7	68	76			390
		F	17	84	8.4	62	96			
Sericulus aureus	Masked Bowerbird	B	5	172		165	180		New Guinea	282
Sericulus bakeri	Fire-maned Bowerbird	M	3	180		178	183		New Guinea	427
		F	5	173		164	184			
Sericulus chrysocephalus	Regent Bowerbird	M	22	90	12.4	76	110			390
		F	36	110	12.2	91	138			
Ptilonorhynchus violaceus	Satin Bowerbird	M	77	225	21.3	189	290			390
		F	89	213	20.8	170	258			
Chlamydera guttata	Western Bowerbird	M	8	134		128	142			390
		F	7	138		122	148			
Chlamydera maculata	Spotted Bowerbird	M	19	139	6.2	125	150			390
		F	13	141	11.7	124	162			
Chlamydera nuchalis	Great Bowerbird	M	9	216		180	265			390
		F	17	186	14.2	164	215			
Chlamydera lauterbachi	Yellow-breasted Bowerbird	M	3	132		128	133			390
		F	3	115		112	120			
Chlamydera cerviniventris	Fawn-breasted Bowerbird	M	44	162	11.5	145	182			390
		F	21	151	13.2	117	170			
Family Turnagridae										
Turnagra capensis	Piopio	U		100					New Zealand, extinct	518a
Family Corvidae										
Platylophus galericulatus	Crested Jay	F	1	77.9					Malaysia	785
Platylophus leucopterus	Black Magpie	B	2	180		178	182		Borneo	1218
Perisoreus infaustus	Siberian Jay	U	56	84.4		72	97			133
Perisoreus internigrans	Sichuan Jay	B	12	102		92	123		China	1361
Perisoreus canadensis canadensis	Gray Jay	M	84	75.8	2.7	68.5	82		Ontario, Canada	1184
		F	91	67.6	2.5	61.7	72.5			

Body Masses of World Birds (continued)

Species Name	Common Name	Sex	N	Mean	S.D.	Min	Max	Season	Location	Source #
Cyanocitta stelleri	Steller's Jay	B	67	128		111	142		Western Canada	1337
Cyanocitta cristata	Blue Jay	U	273	88	5.8	73.1	101		Pennsylvania, USA	828
Calocitta colliei	Black-throated Magpie-Jay	B	4	234		225	251	PB	Jalisco, Mexico	1080
Calocitta formosa	White-throated Magpie-Jay	B	2	210		206	213		Michoacan, Mexico	911
Cyanocorax dickeyi	Tufted Jay	U	4	181					Mexico	226, 1150
Cyanocorax affinis	Black-chested Jay	B	11	212	10.8	194	232			156, 489, 795, 990
Cyanocorax yncas	Green Jay	B	122	78.5	5.6	66	92		Texas, USA	410
Cyanocorax morio	Brown Jay	B	13	204		173	224		Mexico	910a
Cyanocorax melanocyaneus	Bushy-crested Jay	B	16	101		79.8	115		Guatemala; Honduras; El Salvador	1041a, 1145
Cyanocorax sanblasianus	San Blas Jay	U	3	109		101	114		Mexico	1151
Cyanocorax yucatanicus	Yucatan Jay	B	13	118	7	105	128		Mexico; Belize	640, 910a, 1027
Cyanocorax beecheii	Purplish-backed Jay	U	18	193					Western Mexico	1325
Cyanocorax cyanomelas	Purplish Jay	B	11	207					Peru; Brazil; Argentina	177, 254, 370
Cyanocorax caeruleus	Azure Jay	M	1	272					Brazil	64
Cyanocorax violaceus	Violaceous Jay	U		262					Peru	1210
Cyanocorax cristatellus	Curl-crested Jay	U	1	178					Brazil	177
Cyanocorax cayanus	Cayenne Jay	B	13	175		147	196		Suriname	498
Cyanocorax chrysops	Plush-crested Jay	U	21	166		154	170			363
Cyanocorax cyanopogon	White-naped Jay	F	2	146		133	160		Brazil	177
Cyanocorax mystacalis	White-tailed Jay	M	1	160					Peru	1043
		F	1	149						

Body Masses of World Birds (continued)

Species Name	Common Name	Sex	N	Mean	S.D.	Min	Max	Season	Location	Source #
Cyanolyca armillata	Black-collared Jay	U	2	200		190	210			733
Cyanolyca viridicyanus	White-collared Jay	M	1	127					Peru	1288
		F	1	113						
Cyanolyca cucullata	Azure-hooded Jay	M	1	109					Mexico; Costa Rica; Honduras	1145
		F	4	94.2						
Cyanolyca pumilo	Black-throated Jay	M	1	47					Guatemala	1145
Cyanolyca nana	Dwarf Jay	B	2	41		41	41		Oaxaca, Mexico	1145
Cyanolyca argentigula	Silvery-throated Jay	B	2	200		190	210		Panama	489
Cyanolyca mirabilis	White-throated Jay	M	3	52.4		49.9	54		Oaxaca, Mexico	1145
Aphelocoma coerulescens	Florida Scrub-Jay	M	242	79.3	4.6			Y	Florida, USA	1340
		F	248	75	5.1					
Aphelocoma insularis	Is. Scrub-Jay	M	42	123		100	147		California, USA	24
		F	44	109		99	117			
Aphelocoma californica nevadae	Western Scrub-Jay	M	56	80.3		69	98		Western USA	24
		F	37	73.8		65	83			
Aphelocoma californica oocleptica	Western Scrub-Jay	M	16	96.6					California, USA	244
		F	11	82.7						
Aphelocoma californica hypoleuca	Western Scrub-Jay	M	10	84.8					Southern Baja California, Mexico	244
		F	12	76.9						
Aphelocoma californica sumichrasti	Western Scrub-Jay	M	5	100					South-central Mexico	244
		F	6	94.6						
Aphelocoma ultramarina	Mexican Jay	M	40	132	6.1				Arizona, USA	137
		F	43	129	9					
Aphelocoma ultramarina couchii	Mexican Jay	M	3	102		98.5	107		Texas, USA	137
		F	3	95		91	99.5			
Aphelocoma unicolor	Unicolored Jay	M	3	124		121	130		Oaxaca & Chiapas, Mexico	1145

Body Masses of World Birds (continued)

Species Name	Common Name	Sex	N	Mean	S.D.	Min	Max	Season	Location	Source #
Gymnorhinus cyanocephalus	Pinyon Jay	M	66	111	0.7			Y	Arizona, USA	36
		F	66	99	0.7					
Garrulus glandarius	Eurasian Jay	M	107	172		153	197		Germany	396
		F	101	164		129	188			
Garrulus glandarius pekinensis	Eurasian Jay	M	20	157		144	175		Hopei, China	1087
		F	10	146		130	169			
Garrulus lanceolatus	Black-headed Jay	B	25	97.7		84	104			7
Cyanopica cyana	Azure-winged Magpie	M	40	100		80	118		Hopei, China	1087
		F	22	92		76	112			
Urocissa ornata	Ceylon Magpie	F	1	196					Sri Lanka	7
Urocissa caerulea	Formosan Magpie	U				254	260		Taiwan	834
Urocissa flavirostris flavirostris	Gold-billed Magpie	M	5			168	180			7
		F	3			132	143			
Urocissa flavirostris cucullata	Gold-billed Magpie	M	6			130	165			7
		F	5			123	163			
Urocissa erythrorhyncha	Blue Magpie	M	23	164		145	192		Hopei, China	1087
		F	4	140		106	155			
Cissa chinensis	Green Magpie	M	4			130	133			7
		F	3			120	124			
Dendrocitta vagabunda pallida	Rufous Treepie	U	13	100		90	118		India	7
Dendrocitta formosae	Gray Treepie	M	38			90	121			7
		F	21			89	115			
Dendrocitta occipitalis	Sumatran Treepie	U	3	95.3		88	102		Borneo, Malaysia	1148
Dendrocitta leucogastra	White-bellied Treepie	F	1	99.2					India	7
Dendrocitta bayleyi	Andaman Treepie	U				92	113		Andaman Is.	7
Pica hudsonia	Black-billed Magpie	M	81	189	10.3	159	209	Y	Colorado, USA	314
		F	39	166	14.3	135	197			
Pica pica	Eurasian Magpie	M	143	221		185	247		Germany	396
		F	136	191		161	240			

Body Masses of World Birds (continued)

Species Name	Common Name	Sex	N	Mean	S.D.	Min	Max	Season	Location	Source #
Pica pica leucoptera	Eurasian Magpie	M	4	242		214	268		Mongolia	927
		F	5	219		208	232			
Pica nuttalli	Yellow-billed Magpie	M	32	174	8.8	158	189	Y	California, USA	1250
		F	31	144	8.7	126	158			
Podoces hendersoni	Mongolian Ground-Jay	M	10	122	10.6	107	135		Mongolia	927
		F	5	119		104	140			
Podoces panderi	Turkestan Ground-Jay	U	4	91.1		86	96			273
Podoces pleskei	Iranian Ground-Jay	U				85	90			733
Podoces humilis	Tibetan Ground-Jay	M	4			42.5	48.5			7
Nucifraga columbiana	Clark's Nutcracker	M	71	137	0.8				Montana, USA	1225
		F	45	123	1.1					
Nucifraga caryocatactes macrorhynchos	Eurasian Nutcracker	M	40	176		153	190			273
		F	28	169		124	184			
Nucifraga caryocatactes interdicta	Eurasian Nutcracker	M	12	205		185	220		Hopei, China	1087
		F	10	182		170	200			
Nucifraga caryocatactes multipunctata	Eurasian Nutcracker	U				155	177			733
Pyrrhocorax pyrrhocorax	Red-billed Chough	M	9	312		258	375			396
		F	6	263		219	300			
Pyrrhocorax pyrrhocorax brachypus	Red-billed Chough	M	6	288		250	312		Hopei, China	1087
		F	10	248		207	280			
Pyrrhocorax graculus	Yellow-billed Chough	M	176	226	16.2	194	277		France	272
		F	144	197	16.8	160	254			
Ptilostomus afer	Piapiac	U	1	128					Sudan	1318
Corvus monedula	Eurasian Jackdaw	U	196	246				Sp	Great Britain	515
Corvus dauuricus	Daurian Jackdaw	B	45	208		154	275		Hopei, China	1087
Corvus splendens	House Crow	M	16	317		270	371		Mauritius	396
		F	10	270		245	295			

Body Masses of World Birds (continued)

Species Name	Common Name	Sex	N	Mean	S.D.	Min	Max	Season	Location	Source #
Corvus moneduloides	New Caledonian Crow	M	13	309	22.8				New Caledonia	627
		F	9	249	17.8					
Corvus enca	Slender-billed Crow	B	10	251		222	285		Philippines	983, 954
Corvus kubaryi	Mariana Crow	M	6	256		231	270		Guam	35
		F	12	244		205	260			
Corvus woodfordi	Guadalcanal Crow	B	7	467		375	568		Guadalcanal; Solomon Is.	399, 658
Corvus tristis	Gray Crow	M	1	635					New Guinea	61
Corvus capensis	Cape Crow	B	6	553		410	697		Southern Africa	396, 526
Corvus frugilegus	Rook	M	162	489		405	560		Great Britain	396
		F	126	418		325	525			
Corvus brachyrhynchos	American Crow	M	248	538				W	Illinois, USA	1252
		F	245	474						
Corvus brachyrhynchos hesperis	American Crow	M	30	423	33.6				California, USA	1252
		F	20	376	28.8					
Corvus caurinus	NorthWestern Crow	M	19	415		389	486		Washington, USA	602
		F	8	368		315	421			
Corvus minutus	Cuban Palm Crow	M	1	315					Cuba	984
		F	1	263						
Corvus nasicus	Cuban Crow	B	15	401	52.7	330	510		Cuba	19
Corvus jamaicensis	Jamaican Crow	B	2	338		336	339		Jamaica	1148
Corvus imparatus	Tamaulipas Crow	B	5	221		192	237		Mexico	1153
Corvus sinaloae	Sinaloan Crow	U	2	244		229	258		Western Mexico	1150
Corvus ossifragus	Fish Crow	M	20	300	20.3	260	332		Florida, USA	52
		F	19	270	23	195	304			
Corvus hawaiiensis	Hawaiian Crow	M	3	555					Hawaiian Is.	38
		F	5	485						
Corvus cryptoleucus	Chihuahuan Raven	M	76	556	43.6	442	667			51
		F	68	512	42.7	378	607			

Body Masses of World Birds (continued)

Species Name	Common Name	Sex	N	Mean	S.D.	Min	Max	Season	Location	Source #
Corvus corone	Carrion Crow	U	126	570				Sp	Great Britain	515
Corvus macrorhynchos intermedius	Large-billed Crow	M	7			460	582			7
		F	10			392	495			
Corvus macrorhynchos tibetosinensis	Large-billed Crow	M	4			560	650			7
		F	7			450	565			
Corvus orru	Torresian Crow	B	3	433		428	444		New Britain	426
Corvus orru	Torresian Crow	M	22	581	56.9	470	670		Australia	518a
		F	19	521	59	440	650			
Corvus bennetti	Little Crow	U	135	389	40.1	282	525			518a
Corvus coronoides	Australian Raven	M	216	675		540	820		Australia	518a
		F	223	615		500	780			
Corvus mellori	Little Raven	M	138	567		407	660		Australia	518a
		F	153	515		365	660			
Corvus tasmanicus boreus	Forest Raven	M	5	701		645	735		New South Wales, Australia	518a
		F	6	648		610	710			
Corvus tasmanicus tasmanicus	Forest Raven	M	24	712		500	800		Australia	518a
		F	25	647		575	695			
Corvus torquatus	Collared Crow	M	10	492		450	512		China	184a
		F	10	450		347	500			
Corvus albus	Pied Crow	B	26	524		421	700		Southern Africa	396
Corvus ruficollis	Brown-necked Raven	M	2	588		580	595		Algeria; Chad; Niger	396
		F	2	530		500	560			
Corvus edithae	Somali Raven	U	2	442		435	450		Ethiopia	396
Corvus rhipidurus	Fan-tailed Raven	B	7	611		512	734		Niger; Chad	396, 1318
Corvus albicollis	White-necked Raven	B	9	894		762	1157		Tanzania; Zimbabwe	396
Corvus crassirostris	Thick-billed Raven	F	1	1135					Ethiopia	396
Corvus corax	Common Raven	B	145	785	71.9	585	985		California, USA	86
Corvus corax	Common Raven	U	11	1097	127			B	Alaska, USA	86

Body Masses of World Birds (continued)

Species Name	Common Name	Sex	N	Mean	S.D.	Min	Max	Season	Location	Source #
Family Sturnidae										
Aplonis metallica nitida	Metallic Starling	M	39	57.3		49	63.5		Solomon Is.	399
		F	22	55.7		47.5	66.5			
Aplonis cantoroides	Singing Starling	B	18	54.3		50.5	59		Guadalcanal	399
Aplonis cantoroides	Singing Starling	U	10	57.2	3.6	51.8	61.8		Papua New Guinea	318
Aplonis feadensis	Atoll Starling	U		72					Northern Melanesia	771
Aplonis insularis	Rennell Starling	U				66	74		Rennell Is., Solomon Is.	356
Aplonis brunneicapillus	White-eyed Starling	M	24	69.2		59.5	76.5		Guadalcanal	399
		F	18	64.2		59	73.5			
Aplonis grandis	Brown-winged Starling	B	36	134		110	151		Guadalcanal	399
Aplonis dichroa	San Cristobal Starling	M	8	86.9		82.5	90		Solomon Is.	399
		F	9	81.2		73	91.5			
Aplonis striata	Striated Starling	U	3	54.2		49.5	60		New Caledonia	1018
Aplonis panayensis	Asian Glossy Starling	B	10	56.2		52	60.5		Philippines	954
Aplonis minor	Short-tailed Starling	M	3	42		39.5	44.7		Mindanao, Philippines	1148
Aplonis opaca	Micronesian Starling	B	11	83.4		71.5	93		Guam	579
Aplonis tabuensis	Polynesian Starling	B	3	62.5		52	68.6		Tonga; Fiji	977, 1196c
Aplonis atrifusca	Samoan Starling	M	3	146		132	158		American Samoa	1148
Aplonis cinerascens	Rarotonga Starling	B	4	82.9		77.5	88		Rarotonga, Cook Is.	1153a
Mino dumontii	Yellow-faced Myna	U	20	217					New Guinea	61
Mino anais	Golden Myna	U	6	152					New Guinea	61, 427
Mino kreffti	Long-tailed Myna	B	9	190		170	215		Guadalcanal	399

Body Masses of World Birds (continued)

Species Name	Common Name	Sex	N	Mean	S.D.	Min	Max	Season	Location	Source #
Basilornis corythaix	Long-crested Myna	U	3	126		121	132		Ceram, Moluccas	1181
Sarcops calvus	Coleto	B	31	142		120	170		Philippines	954
Saroglossa spiloptera	Spot-winged Starling	M	1	47.5					India	7
Saroglossa aurata	Madagascar Starling	M	1	40.5					Madagascar	69
Ampeliceps coronatus	Golden-crested Myna	M				78	99		India	7
Gracula religiosa	Common Hill Myna	B	9	192		161	229		India	7, 453, 740
Gracula indica	Southern Hill Myna	M	1	126						7
Acridotheres grandis	White-vented Myna	M	42	103	12.3				Singapore	221
		F	27	95	11.4					
Acridotheres cristatellus	Crested Myna	B	17	110					British Columbia, Canada	600
Acridotheres cristatellus	Crested Myna	B	13	123		108	140		China	600
Acridotheres fuscus	Jungle Myna	U	15	82.8		72	98			7
Acridotheres ginginianus	Bank Myna	U	6	72		64	76		India	7
Acridotheres tristis	Common Myna	M	57	120	9.1				Singapore	221
		F	33	112	9.2					
Acridotheres tristis	Common Myna	M	20	127	9	113	140		Hawaiian Is.	614
		F	38	114	7.2	98	129			
Acridotheres tristis	Common Myna	U	22	110		82	138		South Africa	396
Leucopsar rothschildi	Bali Myna	B	7	91.3		70	115		captive	1148
Gracupica nigricollis	Black-collared Starling	M	10	152		136	178		China	184a
		F	10	162		122	210			
Gracupica contra	Asian Pied Starling	M	4	84		76	90		India	740
Sturnia sturnina	Daurian Starling	M	1	49.4					Shetland Is.	518a

Body Masses of World Birds (continued)

Species Name	Common Name	Sex	N	Mean	S.D.	Min	Max	Season	Location	Source #
Sturnia philippensis	Chestnut-cheeked Starling	F	1	28.8					Philippines	436
Sturnia sinensis	White-shouldered Starling	U		61						133
Sturnia malabarica	Chestnut-tailed Starling	U	20	39.6		32	44			7
Temenuchus pagodarum	Brahminy Starling	U	10	48.9		40	54		India	7, 740
Pastor roseus	Rosy Starling	M	13	79.6		71	88			396
		F	6	66.9		60	83			
Sturnus cineraceus	White-cheeked Starling	B	15	83		68	95		Hopei, China	1087
Sturnus vulgaris	European Starling	M	37	87.6	4.8	79	100	W	New York, USA	172
		F	28	84.4	4.2	76.3	92			
Sturnus vulgaris indicus	European Starling	U	15	62.1		55	72			7
Sturnus unicolor	Spotless Starling	U				70	100			356
Creatophora cinerea	Wattled Starling	M	63	74.6		64	85			396
		F	50	71.5		51	83			
Lamprotornis nitens	Cape Glossy-Starling	U	16	85.6		76.3	105		South Africa	396
Lamprotornis chalybaeus	Greater Blue-eared Glossy-Starling	M	191	94		79	106		Kenya	396
		F	109	79		66	96			
Lamprotornis chloropterus	Lesser Blue-eared Glossy-Starling	U	26	73.1		63	86		Gambia	396
Lamprotornis chalcurus	Bronze-tailed Glossy-Starling	M	1	63						396
Lamprotornis splendidus	Splendid Glossy-Starling	M	47			111	155			356
		F	25			96	139			
Lamprotornis ornatus	Principe Glossy-Starling	F	1	130						356
Lamprotornis purpureus	Purple Glossy-Starling	U	3			91	140			356

Body Masses of World Birds (continued)

Species Name	Common Name	Sex	N	Mean	S.D.	Min	Max	Season	Location	Source #
Lamprotornis purpuroptera	Rueppell's Glossy-Starling	M	9	83.7		78	92			396
		F	8	71.6		65	78			
Lamprotornis caudatus	Long-tailed Glossy-Starling	U	12	121		103	133		Nigeria	356
Lamprotornis regius	Golden-breasted Glossy-Starling	U	12			46	63			356
Lamprotornis mevesii	Meves' Glossy-Starling	U	29	64.2		56	77			396
Lamprotornis australis	Burchell's Glossy-Starling	U	14			74	138			396
Lamprotornis acuticaudus	Sharp-tailed Glossy-Starling	B	9	67.6		61	76			356
Lamprotornis corruscus	Black-bellied Glossy-Starling	M	26			51	68			356
		F	13			46	62			
Lamprotornis superbus	Superb Starling	U	36	64.9		52	77			396
Lamprotornis hildebrandti	Hildebrandt's Starling	U	12	55.9		50	69		Kenya	396
Lamprotornis shelleyi	Shelley's Starling	F	1	45						356
Lamprotornis pulcher	Chestnut-bellied Starling	M	39	68		59	76			396
		F	31	66.6		60	74			
Lamprotornis purpureiceps	Purple-headed Glossy-Starling	B	18			50	79			396
Lamprotornis cupreocauda	Copper-tailed Glossy-Starling	B	22			51	66			356
Cinnyricinclus leucogaster	Violet-backed Starling	B	20	44		32.7	55.6		Southern Africa	396
Spreo bicolor	African Pied Starling	B	43	104		93	113		South Africa	396
Spreo fischeri	Fischer's Starling	M	1	51						356
Spreo unicolor	Ashy Starling	B	6			55	66			356
Onychognathus morio	Red-winged Starling	M	33	138	9.1	115	155		South Africa	509
		F	22	128	10.7	110	151			

Body Masses of World Birds (continued)

Species Name	Common Name	Sex	N	Mean	S.D.	Min	Max	Season	Location	Source #
Onychognathus tenuirostris	Slender-billed Starling	M	5			127	142			356
		F	2	118		113	122			
Onychognathus fulgidus	Chestnut-winged Starling	M	13			103	140			356
		F	13			81	121			
Onychognathus walleri	Waller's Starling	M	8			73	92			356
		F	1	93						
Onychognathus blythii	Somali Starling	U	1	100						356
Onychognathus frater	Socotra Starling	U	1	100						356
Onychognathus tristramii	Tristram's Starling	U	15			115	131			356
Onychognathus nabouroup	Pale-winged Starling	M	5			94	122			356
		F	13			80	120			
Onychognathus salvadorii	Bristle-crowned Starling	M	1	160						356
		F	3			121	150			
Onychognathus albirostris	White-billed Starling	B	5			130	145			356
Poeoptera lugubris	Narrow-tailed Starling	B	18			35	43			356
Poeoptera stuhlmanni	Stuhlmann's Starling	M	6			38	46			356
		F	4			35	39			
Poeoptera kenricki	Kenrick's Starling	M	4			46	54			356
		F	5			38	49			
Pholia sharpii	Sharpe's Starling	M	5			40	56			356
		F	2	41		40	42			
Grafisia torquata	White-collared Starling	B	2	64		61	67		Cameroon	710
Speculipastor bicolor	Magpie Starling	M	2	67.5		66	69			356
		F	2	61		61	61			
Neocichla gutturalis	Babbling Starling	B	3	66.7		64	72		Angola	266
Buphagus erythrorhynchus	Red-billed Oxpecker	M	61	51		45	56			396
		F	55	50.5		42	59			

Body Masses of World Birds (continued)

Species Name	Common Name	Sex	N	Mean	S.D.	Min	Max	Season	Location	Source #
Buphagus africanus	Yellow-billed Oxpecker	M	2	58.5		57	60			356
		F	5			53	71			
Family Passeridae										
Passer ammodendri stoliczkae	Saxaul Sparrow	B	21	28.2	1.6	26	31		Mongolia	927
Passer ammodendri nigricans	Saxaul Sparrow	M	7	29.7		27	32		Mongolia	927
Passer domesticus	House Sparrow	M	538	28	1.6	20	34	Y	Pennsylvania, USA	193
		F	469	27.4	2.2	20.1	34.5			
Passer domesticus indicus	House Sparrow	B	143	24.3		21	30.2		Transvaal, South Africa	395
Passer hispaniolensis	Spanish Sparrow	B	200	24.2		18	28	Sp	India	7
Passer castanopterus	Somali Sparrow	M		18						1189
Passer rutilans	Russet Sparrow	M	10	17.4		13	18.5		China	184a
		F	7	19.6		17	20.5			
Passer flaveolus	Plain-backed Sparrow	B		19.4		17	23			1189
Passer moabiticus moabiticus	Dead Sea Sparrow	M		17.5		15	20			1189
Passer moabiticus yatii	Dead Sea Sparrow	B		15.5		14	17			1189
Passer insularis	Socotra Sparrow	B	9	27.3		25	29.2		Socotra Is.	317
Passer motitensis	Rufous Sparrow	B	22			26	35		Angola	266
Passer rufocinctus	Kenya Rufous Sparrow	B	64	27.8		25	32			395
Passer melanurus	Mossie	B	90	29.5		22	38			395
Passer griseus	Gray-headed Sparrow	B	44	23.9		18	26.5		South Africa	509
Passer griseus griseus	Gray-headed Sparrow	U	53	28.1		24.2	33		Nigeria	395
Passer griseus ugandae	Gray-headed Sparrow	U	34	39.2		34	43		Western Kenya	395

Body Masses of World Birds (continued)

Species Name	Common Name	Sex	N	Mean	S.D.	Min	Max	Season	Location	Source #
Passer griseus laeneni	Gray-headed Sparrow	M		27		24	30			1189
		F		28.5		26	31			
Passer griseus luangwae	Gray-headed Sparrow	U		22.9		22.7	23.2			1189
Passer swainsonii	Swainson's Sparrow	U	20	31.6		27.3	35.2		Ethiopia	1243a
Passer gongonensis	Parrot-billed Sparrow	U	23	39.9		33.5	48		Kenya	395
Passer diffusus	Cape Sparrow	U	130	24.2		20	29.6		Transvaal, South Africa	395
Passer simplex	Desert Sparrow	B	4	19.5		18	21	W	Algeria	395
Passer montanus	Eurasian Tree Sparrow	B	136	22	1.2			Y	Missouri, USA	13
Passer montanus	Eurasian Tree Sparrow	U	115	20.8		17	25	W	France	395
Passer luteus	Sudan Golden Sparrow	M	3			14.5	15			852
		F	2	11.2		11	11.5			
Passer euchlorus	Arabian Golden Sparrow	B	27	14		12	16.8		North Yemen	100
Passer eminibey	Chestnut Sparrow	U	13	13.4	0.8	11.9	15		Kenya	1052
Petronia pyrgita	Yellow-spotted Petronia	U	17	24.1		21	28.4		Kenya	395
Petronia xanthocollis	Chestnut-shouldered Petronia	B	21	18.1		14	20			7
Petronia superciliaris	Yellow-throated Petronia	U	10	24		22	26		Zambia	925
Petronia dentata	Bush Petronia	B	57	18.4		16	21.5		Nigeria	777
Petronia petronia	Rock Petronia	B	12	30.2		26	35			395
Carpospiza brachydactyla	Pale Rosefinch	M	4	22.3		21.3	25		Turkey; Armenia	395
Montifringilla nivalis	White-winged Snowfinch	U	20	36.9		32	40			133

Body Masses of World Birds (continued)

Species Name	Common Name	Sex	N	Mean	S.D.	Min	Max	Season	Location	Source #
Montifringilla davidiana	Pere David's Snowfinch	B	8	21.2		20	25		Mongolia	927
Montifringilla ruficollis	Rufous-necked Snowfinch	U		22.8					XiNing, China	276
Montifringilla theresae	Afghan Snowfinch	U	40	24.3	2.5	18	31		Afghanistan	1334
Family Ploceidae										
Bubalornis albirostris	White-billed Buffalo-Weaver	M	6	61.9		56	68.5		Kenya	1052
Bubalornis niger	Red-billed Buffalo-Weaver	M	54	82.1		65	98		Namibia; South Africa; Botswana	1036
		F	46	70.7		59	90			
Dinemellia dinemelli	White-headed Buffalo-Weaver	U	8	63.9		57	71		Kenya	1052
Sporopipes frontalis	Speckle-fronted Weaver	U	50	17.3		14.2	21.6		Nigeria	777
Sporopipes squamifrons	Scaly Weaver	U	66	12.3		10.4	13.7		South Africa	395
Plocepasser mahali mahali	White-browed Sparrow-Weaver	U	811	44		31	59		South Africa	395
Plocepasser mahali melanorhynchus	White-browed Sparrow-Weaver	U	35	39.1		31	43		Kenya	395
Plocepasser superciliosus	Chestnut-crowned Sparrow-Weaver	U	12	37.1		32	45.3		Ghana; Cameroon; Nigeria	457, 710, 777
Plocepasser rufoscapulatus	Chestnut-backed Sparrow-Weaver	B	5	44.6		39	49		Angola	395
Plocepasser donaldsoni	Donaldson-Smith's Sparrow-Weaver	B	7	39.9		35.5	43		Kenya	395, 1143
Histurgops ruficauda	Rufous-tailed Weaver	F	1	69						395
Pseudonigrita arnaudi	Gray-headed Social-Weaver	U	54	20		16.2	25.9		Kenya	395
Pseudonigrita cabanisi	Black-capped Social-Weaver	M	5	22.9		21.5	24		Kenya; Tanzania	395
		F	4	18.5		18	19			

Body Masses of World Birds (continued)

Species Name	Common Name	Sex	N	Mean	S.D.	Min	Max	Season	Location	Source #
Philetairus socius	Social Weaver	B	116	27.4		23.8	32		South Africa	395
Ploceus bannermani	Bannerman's Weaver	U	10	30.8		29	33.2		Cameroon	395, 1143
Ploceus baglafecht	Baglafecht Weaver	U	112	31.6		25.5	36.3		Ethiopia	1243a
Ploceus bertrandi	Bertrand's Weaver	M	5	40.1		37.6	43		Malawi	1036
		F	7	38		32.6	40.6			
Ploceus pelzelni	Slender-billed Weaver	M	16	14.2		11.7	15.1		Kenya	395
		F	15	13.1		11.4	14.9			
Ploceus pelzelni	Slender-billed Weaver	B	13	10.9		10	13		D.R. Congo	395
Ploceus luteolus	Little Weaver	U	26	15.1		10.4	21.8		Nigeria	777
Ploceus intermedius	Lesser Masked-Weaver	M	777	21.4		15	43		Southern Africa	1036
		F	499	19.8		15	32.9			
Ploceus ocularis	Spectacled Weaver	U	38	25.2		24	28.7		Kenya	121
Ploceus nigricollis	Black-necked Weaver	U	71	27.1		23.1	40		Nigeria	777
Ploceus melanogaster	Black-billed Weaver	B	20	22.4		20	25		Cameroon	710
Ploceus alienus	Strange Weaver	U	11	22.2	2.2					81
Ploceus temporalis	Bocage's Weaver	M	3	35.8		34.4	37		Angola	982
Ploceus capensis	Cape Weaver	M	57	47.9		44	52		South Africa	395
		F	46	40.4		36	45			
Ploceus subaureus	African Golden-Weaver	M	22	35		29.5	39		South Africa	395
		F	24	26.4		21.5	31			
Ploceus xanthops	Holub's Golden-Weaver	M	9	44.3		39.5	50		Zambia	395
		F	4	37.3		35	40			
Ploceus princeps	Principe Golden-Weaver	U	1	31.2					Principe Is.	395
Ploceus aurantius	Orange Weaver	B	13	23	3	17.3	25.5		Uganda; Cent. African Rep.	1143

Body Masses of World Birds (continued)

Species Name	Common Name	Sex	N	Mean	S.D.	Min	Max	Season	Location	Source #
Ploceus bojeri	Golden Palm Weaver	M	37	25.5		11	29.2		Kenya	395
		F	14	19.7		17	24.7			
Ploceus castaneiceps	Taveta Golden-Weaver	F	3	18.5		17.8	19			395
Ploceus xanthopterus	Southern Brown-throated Weaver	M	394	25.1		22	31		Malawi	395
		F	342	19.2		16	24			
Ploceus castanops	Northern Brown-throated Weaver	M	28	23.5		19.5	25.4		Kenya	395
		F	13	19.5		18	21.6			
Ploceus burnieri	Kilombero Weaver	M	12	19.6		17	21.3		Tanzania	34
		F	21	16.6		14.2	17.5			
Ploceus galbula	Rueppell's Weaver	B	9	23.8		20.7	27		Yemen	317
Ploceus galbula	Rueppell's Weaver	M	9	23.8		20	29		Sudan	395
		F	5	18.1		16	21			
Ploceus heuglini	Heuglin's Masked-Weaver	U	17	22.8		14.9	28.5		Nigeria	777
Ploceus taeniopterus	Northern Masked-Weaver	M	4	19.9		18.5	21.5		Kenya; Ethiopia	1143
		F	1	15.5						
Ploceus velatus	African Masked-Weaver	M	39	33.6	1.8	28	36.4		South Africa	509
		F	29	28.5	1.9	25.5	34.1			
Ploceus vitellinus	Vitelline Masked-Weaver	M	1	20					Cameroon	395
Ploceus cucullatus nigriceps	Village Weaver	M	229	37.7		33	46		Malawi	395
		F	134	31.4		26	37			
Ploceus cucullatus spilonotus	Village Weaver	M	33	40.5		33	47		Transvaal, South Africa	395
		F	31	34.2		29	40			
Ploceus cucullatus	Village Weaver	U	115	37.3		28.8	48.4		Nigeria	777
Ploceus grandis	Giant Weaver	U	4			62	65			395
Ploceus spekei	Speke's Weaver	M	44	38.3		34	43		Kenya	395
		F	26	32.1		28	36.8			
Ploceus nigerrimus	Vieillot's Weaver	M	6	38.5		34	41		Cameroon	710
		F	11	30.5		27	38			

Body Masses of World Birds (continued)

Species Name	Common Name	Sex	N	Mean	S.D.	Min	Max	Season	Location	Source #
Ploceus weynsi	Weyns' Weaver	M	31	30.9		23.9	35.8		Uganda	395
		F	28	27.3		22.8	33.5			
Ploceus golandi	Clarke's Weaver	M	17	24.6		22.5	27		Kenya	395
		F	16	22.5		20.8	24.6			
Ploceus melanocephalus capitalis	Black-headed Weaver	M	17	22.5		20.1	25		Niger	395
		F	11	19.1		17	21.2			
Ploceus melanocephalus dimidiatus	Black-headed Weaver	M	57	28.7		21	36.3		Kenya	395
		F	24	22.8		19.9	27.5			
Ploceus dichrocephalus	Salvadori's Weaver	F	2	18.5		18	19		Kenya	395
Ploceus jacksoni	Golden-backed Weaver	M	38	26.7		22.3	30		Kenya	395
		F	22	20.5		18.9	22.3			
Ploceus rubiginosus rubiginosus	Chestnut Weaver	M	15	32.4		29.2	37		Kenya	395
		F	8	28		25.3	31.2			
Ploceus rubiginosus trothae	Chestnut Weaver	M	39	32.1		28	37			395
		F	55	27.8		25	30			
Ploceus tricolor	Yellow-mantled Weaver	B	10	36.8		32.3	44		D.R. Congo	395
Ploceus albinucha	Maxwell's Black Weaver	B	4	27.2		25.5	29.5		Cameroon; Uganda; Liberia	395
Ploceus nelicourvi	Nelicourvi Weaver	U	41	23.8	3.2	15	31		Madagascar	957
Ploceus sakalava	Sakalava Weaver	U	26	23.8	2.3	19.5	27.5		Madagascar	957
Ploceus manyar	Streaked Weaver	M	2	17.4		16.4	18.4		Thailand	787
Ploceus philippinus	Baya Weaver	B	23	28.2		24	32		India	7
Ploceus hypoxanthus	Asian Golden Weaver	M	1	18.6					Thailand	787
Ploceus megarhynchus	Yellow Weaver	M	3			34	40		India	7
		F	7			30	34			
Ploceus benghalensis	Bengal Weaver	B	16	19.8		18	22			7
Ploceus bicolor	Forest Weaver	B	26	33.1	2.1	29.4	37.5		Somalia	1335

Body Masses of World Birds (continued)

Species Name	Common Name	Sex	N	Mean	S.D.	Min	Max	Season	Location	Source #
Ploceus bicolor	Forest Weaver	M	11	35.9		31	43.3		Southern Africa	1036
		F	9	34.7		27.8	57			
Ploceus preussi	Preuss' Weaver	B	2	31.6		31.4	31.7		Liberia	395
Ploceus nicolli	Usambara Weaver	M	2	31.8		31	32.5		Tanzania	395
		F	3	28		27	29			
Ploceus olivaceiceps	Olive-headed Weaver	M	11			20.6	24		Mozambique	395
		F	8			17.5	21.6			
Ploceus insignis	Brown-capped Weaver	B	12	27.5		26	29.8		Kenya	395
Ploceus sanctithomae	Sao Tome Weaver	U	3	22		20.5	23.5		Sao Tome Is.	395
Pachyphantes superciliosus	Compact Weaver	B	10	21.4		19	24.8		East Africa	395
Malimbus scutatus	Red-vented Malimbe	B	6	30		28	33		Liberia	395
Malimbus ibadanensis	Ibadan Malimbe	M	5			39.9	43.5		Nigeria	331
		F	6			33.1	39.4			
Malimbus nitens	Gray's Malimbe	M	8	42.6		38	47		Liberia	395
		F	11	33.6		29	36			
Malimbus malimbicus	Crested Malimbe	M	2	38.8		37.5	40.1		Uganda; C.A.R., DR Congo	1143
		F	2	33.6		31.3	35.8			
Malimbus rubricollis	Red-headed Malimbe	U	2	42.2		40.5	44		Kenya	742
Anaplectes rubriceps	Red-headed Weaver	M	32	23.6		20	33		Southern Africa	1036
		F	32	21.6		17	35.5			
Brachycope anomala	Bob-tailed Weaver	M	3	25.3		20	30		D.R. Congo	395
Quelea cardinalis	Cardinal Quelea	M	13	12.5		11.3	13.4		Kenya	395
		F	19	12.9		11.4	15			
Quelea erythrops	Red-headed Quelea	M	36	21.1		17.5	25.5		South Africa	395
		F	63	17.7		15	22			
Quelea quelea	Red-billed Quelea	M	3370	19.1		9	32		Southern Africa	1036
		F	1914	18.1		10	31			

Body Masses of World Birds (continued)

Species Name	Common Name	Sex	N	Mean	S.D.	Min	Max	Season	Location	Source #
Foudia madagascariensis	Red Fody	M	25	18.2	1				Mauritius	183
		F	13	16.3	1.3					
Foudia madagascariensis	Red Fody	M	10	16.8		14.9	18.3		Seychelles	183
		F	9	15.2		13.5	16.5			
Foudia eminentissima	Red-headed Fody	M				24.5	26.3		Aldabra Is.	183
		F				22.7	24.7			
Foudia omissa	Forest Fody	U	59	18.9	1.6	14.5	23.2		Madagascar	957
Foudia rubra	Mauritius Fody	B	17	17.7		16	20.1		Mauritius	183
Foudia seychellarum	Seychelles Fody	B	27	17.1		15.5	19.1		Seychelles	183
Foudia flavicans	Rodrigues Fody	M	26	15.9	0.1	14.7	17		Rodrigues Is.	183
		F	14	14.8	0.2	13.5	16			
Euplectes afer	Yellow-crowned Bishop	M	81	14.7	2.7	11.2	20.5		Ghana	457
		F	44	13.9	1	12.2	16.2			
Euplectes diadematus	Fire-fronted Bishop	B	8	13.9					Kenya	395
Euplectes gierowii	Black Bishop	M	17	29.6		23.2	36.2		Kenya	395
		F	7	27		22.4	30			
Euplectes hordeaceus	Black-winged Bishop	M	12	20.4		18	22.9		Kenya	395
		F	6	18.4		17.2	19.5			
Euplectes franciscanus	Orange Bishop	B	131	16		11.5	22		Nigeria	395
Euplectes orix	Red Bishop	M	704	25.5		21	30		South Africa	395
		F	445	20.7		17	26			
Euplectes nigroventris	Zanzibar Bishop	U	12	12.5		11.1	13.8		Kenya	395
Euplectes capensis	Yellow Bishop	M	202	32.9		14.8	45.4		Southern Africa	1036
		F	275	28.3		15.4	43.5			
Euplectes axillaris	Fan-tailed Widowbird	M	231	27.8		18	36.5		South Africa	1036
		F	193	22.7		16	34.5			
Euplectes macrourus	Yellow-shouldered Widowbird	M	26	25.1		22.3	28.5		Nigeria	395
		F	18	18.7		17.1	20.2			

Body Masses of World Birds (continued)

Species Name	Common Name	Sex	N	Mean	S.D.	Min	Max	Season	Location	Source #
Euplectes albonotatus	White-winged Widowbird	M	31	23.2		20	26.5		South Africa	395
		F	21	19.1		16	24			
Euplectes ardens	Red-collared Widowbird	M	60	18.3	1.4	15.6	22		Zimbabwe	566
		F	92	16	1.2	13.1	19.3			
Euplectes ardens	Red-collared Widowbird	M	187	22.7		20	25.5		South Africa	395
		F	211	19.7		16	21.5			
Euplectes psammocromius	Buff-shouldered Widowbird	M	1	35.5					Zambia	301
Euplectes progne	Long-tailed Widowbird	U	95	35.1		25	46.9			133
Euplectes jacksoni	Jackson's Widowbird	M	227	43	2.1					14
		F	141	31.9	2.1					
Amblyospiza albifrons melanota	Grosbeak Weaver	M	7	38.5		33	45.5		Kenya	395
		F	3	36.3		28	44			
Amblyospiza albifrons unicolor	Grosbeak Weaver	M	177	47.7		39.4	58		Malawi	395
		F	95	37.3		29.9	43.4			
Amblyospiza albifrons albifrons	Grosbeak Weaver	M	48	51.3		34.6	57.1		South Africa	395
		F	98	40.6		34.1	53			
Family Estrildidae										
Parmoptila rubrifrons	Red-fronted Antpecker	M	19	9.2		8	10.5		Liberia	395
		F	13	9.8		8	10.5			
Parmoptila woodhousei	Woodhouse's Antpecker	B	3	9.4		9.1	9.5			395
Parmoptila jamesoni	Jameson's Antpecker	M	1	9.5						395
Nigrita fusconotus	White-breasted Negrofinch	B	12	9.1		7	11		Cameroon; Kenya; Bioko	395
Nigrita bicolor	Chestnut-breasted Negrofinch	B	10	10.5		8.6	11.7		Liberia	395
Nigrita luteifrons	Pale-fronted Negrofinch	B	3	13.2		12	14.5			395
Nigrita canicapillus	Gray-headed Negrofinch	B	11	18.7		17	21		Cameroon; Angola	266, 710

Body Masses of World Birds (continued)

Species Name	Common Name	Sex	N	Mean	S.D.	Min	Max	Season	Location	Source #
Nesocharis ansorgei	White-collared Oliveback	B	10	8.4		8	9.5			395
Nesocharis shelleyi	Fernando Po Oliveback	B	5	7.6		7	9		Cameroon	710
Nesocharis capistrata	Gray-headed Oliveback	U	14	11.6		10.8	13.2		Nigeria	777
Pytilia afra	Orange-winged Pytilia	B	13	16.6		15	18			395
Pytilia phoenicoptera	Red-winged Pytilia	U	5	14.5		13.7	16.1		Ghana	457
Pytilia lineata	Red-billed Pytilia	U	99	14.3		11	16.8			395
Pytilia melba	Green-winged Pytilia	B	564	15.4		12.8	18.4			395
Pytilia hypogrammica	Red-faced Pytilia	F	3	14.8		14.3	15		Ghana; Cameroon	457, 710
Mandingoa nitidula	Green-backed Twinspot	U	17	9.6	0.5	8.5	10		Tanzania	1014
Cryptospiza reichenovii	Red-faced Crimson-wing	U	56	12.4	0.9	10.5	14.5		Tanzania	1014
Cryptospiza salvadorii	Abyssinian Crimson-wing	B	22	12.2	0.9	10.9	14		Sudan; Uganda	1143
Cryptospiza jacksoni	Dusky Crimson-wing	U	8	13					Central Africa	81
Cryptospiza shelleyi	Shelley's Crimson-wing	B	16	17.6	1.1	15	19		Uganda; Burundi	1143
Pyrenestes sanguineus	Crimson Seedcracker	M	8	19.8		16.9	26.6		Liberia	395
		F	6	18.6		16.9	20.3			
Pyrenestes ostrinus	Black-bellied Seedcracker	M	18	20.4	3.4				Cameroon - "small morph"	1125
Pyrenestes ostrinus	Black-bellied Seedcracker	M	14	24.5	2.8				Cameroon - "large morph"	1125
Spermophaga poliogenys	Grant's Bluebill	B	8	19.8		18.7	21.7		Dem. Republic Congo	1143
Spermophaga haematina	Western Bluebill	U	18	23	1.7	20	26		Dem. Republic Congo	395

Body Masses of World Birds (continued)

Species Name	Common Name	Sex	N	Mean	S.D.	Min	Max	Season	Location	Source #
Spermophaga ruficapilla	Red-headed Bluebill	B	45	23.6		21.3	26		Kenya	395
Clytospiza monteiri	Brown Twinspot	B	18	14.9		13.5	16		Cameroon	395
Hypargos niveoguttatus	Peters' Twinspot	B	60	14.6		12.4	16.9		Kenya	395
Hypargos margaritatus	Pink-throated Twinspot	B	3	12.8		12	13.5			395
Euschistospiza dybowskii	Dybowski's Twinspot	B	6	13.4		12	14.6		Cameroon; C. Afr. Republic	395
Euschistospiza cinereovinacea	Dusky Twinspot	B	3	12		9	14			266, 395
Lagonosticta rufopicta	Bar-breasted Firefinch	U	34	9.3		8.2	11.2		Nigeria	777
Lagonosticta nitidula	Brown Firefinch	B	3	10.7		10	11		Cameroon; Angola	266, 710
Lagonosticta senegala	Red-billed Firefinch	B	128	8.3		7	10.7		Kenya	121
Lagonosticta senegala rendalli	Red-billed Firefinch	B	53	8.8		7.2	10		Transvaal, South Africa	395
Lagonosticta senegala brunneiceps	Red-billed Firefinch	U	23	10.7		8	12		Ethiopia	395
Lagonosticta rara	Black-bellied Firefinch	B	266	10.4		7.9	13.1		South Africa	395
Lagonosticta rubricata	African Firefinch	B	16	10.2		8.6	11.7		Zimbabwe	565
Lagonosticta rubricata landanae	Pale-billed Firefinch	M	3	10.5		9.5	11.5		Angola	395
		F	3	9.8		9	10.5			
Lagonosticta rhodopareia	Jameson's Firefinch	B	159	9.3		7.5	13		Zimbabwe	395
Lagonosticta virata	Mali Firefinch	M	3	9.1		8.8	9.3		Mali	395
		F	5	9.5		8.5	10.9			
Lagonosticta sanguinodorsalis	Rock Firefinch	B	81	11.4					Nigeria	777
Lagonosticta larvata	Black-faced Firefinch	M	11	10.4		9.3	12.4		Ethiopia	395
		F	6	9.4		8	10			

Body Masses of World Birds (continued)

Species Name	Common Name	Sex	N	Mean	S.D.	Min	Max	Season	Location	Source #
Uraeginthus angolensis	Blue-breasted Cordonbleu	B	374	9.9		6.3	13		South Africa	395
Uraeginthus bengalus	Red-cheeked Cordonbleu	B	62	9.9		8.9	11		Nigeria	395
Uraeginthus cyanocephalus	Blue-capped Cordonbleu	B	27	9.8		8.5	10.8		Kenya	395
Uraeginthus ianthinogaster	Purple Grenadier	B	31	14.1		12	17.4		Kenya	395
Uraeginthus granatina	Violet-eared Waxbill	B	227	11.8		9.5	13.9		Transvaal, South Africa	395
Estrilda caerulescens	Lavender Waxbill	U	171	9.4		8	15.1		Nigeria	777
Estrilda perreini	Black-tailed Waxbill	B	6	7.4		6.1	9		Zimbabwe	395
Estrilda thomensis	Cinderella Waxbill	B	22	7.6		6.8	8.5			395
Estrilda quartinia	Yellow-bellied Waxbill	B	25	6.7		6	8		Sudan; Kenya	395
Estrilda melanotis	Swee Waxbill	M	3	5.9		5.7	6.1		Zimbabwe; Zambia	301, 563
		F	3	7.2		7	7.5			
Estrilda paludicola	Fawn-breasted Waxbill	U	26	7.6		6.6	8.5		Kenya	395
Estrilda melpoda	Orange-cheeked Waxbill	U	43	7.6	0.7	6.5	9.6		Ghana	457
Estrilda rufibarba	Arabian Waxbill	U	1	8.5					Yemen	317
Estrilda rhodopyga	Crimson-rumped Waxbill	U	51	7.7		6.6	9.5		Kenya	395
Estrilda troglodytes	Black-rumped Waxbill	U	22	7.6		6.9	8.2		Nigeria	777
Estrilda astrild astrild	Common Waxbill	U	73	8.8		6	11		South Africa	395
Estrilda astrild massaica	Common Waxbill	U	19	8		6.8	9.8		Kenya	395
Estrilda astrild nyansae	Common Waxbill	U	42	8.1		6.5	9.2		Kenya	395

Body Masses of World Birds (continued)

Species Name	Common Name	Sex	N	Mean	S.D.	Min	Max	Season	Location	Source #
Estrilda nonnula	Black-crowned Waxbill	U	47	7.7		6.5	8.5		Sudan	395
Estrilda atricapilla	Black-headed Waxbill	U	1	7.8					Gabon	616
Estrilda erythronotos	Black-cheeked Waxbill	B	101	9.2		7.9	10.2			395
Estrilda charmosyna	Red-rumped Waxbill	B	2	7.4		7.3	7.5		Kenya	1143
Amandava amandava	Red Avadavat	B	4	9.6		9.5	9.8		India	42
Amandava amandava flavidiventris	Red Avadavat	B	2	8.4		7.2	9.5		Timor	42
Amandava subflava	Zebra Waxbill	M	287	7.3		5.7	10.3			395
		F	192	7.6		5.2	10.9			
Ortygospiza atricollis	Black-faced Quailfinch	U	14	10.6		9	12.4		Nigeria	777
Ortygospiza fuscocrissa	African Quailfinch	M	4	11.4		10.5	12		Southern Africa	395
		F	8	10.6		10	11.5			
Emblema pictum	Painted Firetail	B	8	10.4		9.6	11		Western Australia	945
Stagonopleura bella	Beautiful Firetail	B	34	14		11	16		Australia	518a
Stagonopleura oculata	Red-eared Firetail	U	18	12.9	0.6	12	14			518a
Stagonopleura guttata	Diamond Firetail	U	84	17.6	1.4	14.1	22.7		Victoria, Australia	518a
Stagonopleura fuliginosus	Mountain Firetail	B	17	18.3	1	17.2	20		Papua New Guinea	387
Neochmia temporalis	Red-browed Firetail	U	136	11.4	1	8.4	14.6		Queensland, Australia	518a
Neochmia phaeton	Crimson Finch	B	152	10	0.8				Western Australia	518a
Neochmia ruficauda	Star Finch	M	54	10.6	0.8	8.5	12.5		Western Australia	518a
		F	33	11.7	0.8	10	13.8			
Neochmia modesta	Plum-headed Finch	M	5	13.2		13	13.8		Queensland, Australia	518a
		F	4	12.3		11.5	13			
Taeniopygia guttata castanotis	Zebra Finch	M	560	11.9	0.8	10	15.2		Northern Terr., Australia	518a
		F	521	12.2	0.9	9.4	15.7			

Body Masses of World Birds (continued)

Species Name	Common Name	Sex	N	Mean	S.D.	Min	Max	Season	Location	Source #
Taeniopygia bichenovii	Double-barred Finch	B	228	9.6	0.8	7	12			518a
Poephila personata	Masked Finch	U	358	13	1	10.5	17		Northern Terr., Australia	1329
Poephila acuticauda	Long-tailed Finch	U	486	14	1	10.9	17.2		Northern Terr., Australia	1329
Poephila cincta	Black-throated Finch	B	13	14.9	1.2	12.9	17.1		Queensland, Australia	470
Erythrura hyperythra	Tawny-breasted Parrotfinch	B	5	11		10	13		Philippines; Indonesia	436, 983, 1272
Erythrura trichroa	Blue-faced Parrotfinch	M	22	14	1.4	12	14		Australia	366
		F	13	14.8	1	13	16			
Erythrura coloria	Red-eared Parrotfinch	B	6	11.3		9.8	12.7		Mindanao, Philippines	1143
Erythrura papuana	Papuan Parrotfinch	B	13	19.4	1.9	17	24		New Guinea	283
Erythrura psittacea	Red-throated Parrotfinch	M	1	11.5					New Caledonia	1018
Erythrura cyaneovirens	Red-headed Parrotfinch	U	49	13	1	10.5	15		Fiji	668
Erythrura kleinschmidti	Pink-billed Parrotfinch	U	1	21					Fiji	1196c
Chloebia gouldiae	Gouldian Finch	U	379	13.8	0.9	10.5	16		Northern Terr., Australia	1329
Lonchura nana	Madagascar Munia	U	6	8.2		7.4	9		Madagascar	69, 957
Lonchura cantans	African Silverbill	U	36	12		10	14		Chad	395
Lonchura malabarica	White-throated Munia	U	13	12		10	14		India	7
Lonchura griseicapilla	Gray-headed Silverbill	U	4	12.8		11.6	14		Kenya	42, 1052
Lonchura cucullata	Bronze Mannikin	U	97	9.2	0.8	7.7	11.8		Kenya	122
Lonchura cucullata	Bronze Mannikin	U	150	8.8			10.6		Nigeria	777

Body Masses of World Birds (continued)

Species Name	Common Name	Sex	N	Mean	S.D.	Min	Max	Season	Location	Source #
Lonchura bicolor	Black-and-white Mannikin	B	15	9.9		8	11		Cameroon	710
Lonchura nigriceps	Brown-backed Mannikin	U	121	9.2		7.3	11.3		Zimbabwe	395
Lonchura fringilloides	Magpie Mannikin	B	12	16.8		13	18.9		Cameroon; Zimbabwe	562, 564, 710
Lonchura striata	White-rumped Munia	U	10	12.3		9.5	13		India	7
Lonchura leucogastroides	Javan Munia	U	3	11.7		10.2	12.5		Timor	42
Lonchura fuscans	Dusky Munia	U	8	9.5					Brunei	1281
Lonchura molucca	Black-faced Munia	F	1	12					Puerto Rico (introduced)	1148
Lonchura kelaarti	Black-throated Munia	U	15	14.1		9.5	17		India	7
Lonchura punctulata	Nutmeg Mannikin	B	13	13.6		12	15		India	7
Lonchura leucogastra	White-bellied Munia	B	7	11.4		10.6	12.3		Philippines	954
Lonchura tristissima	Streak-headed Munia	B	6	8.8		7.5	10		New Guinea	292, 424, 427
Lonchura malacca	Black-headed Munia	B	28	12.6		9.8	14.4		Philippines	954
Lonchura atricapilla formosana	Chestnut Munia	B	22	13.4	1	12	15.8		Luzon, Philippines	1143
Lonchura atricapilla jagori	Chestnut Munia	B	35	12	1	10.1	14.2		Philippines	1143
Lonchura ferruginosa	White-capped Munia	M	1	13					Moluccas	979
Lonchura maja	White-headed Munia	U		12.5		11	14			133
Lonchura grandis	Grand Munia	M	4			11.5	14		New Guinea	283, 426
		F	3	14.3		13	14			
Lonchura nevermanni	Gray-crowned Munia	B	15			9	13.4		New Guinea	786

Body Masses of World Birds (continued)

Species Name	Common Name	Sex	N	Mean	S.D.	Min	Max	Season	Location	Source #
Lonchura spectabilis	Hooded Munia	B	11	11.5		10.3	12.7		New Guinea	283
Lonchura flaviprymna	Yellow-rumped Munia	B	4	11.8		10	13		Australia	518a
Lonchura castaneothorax	Chestnut-breasted Munia	U	26	13.6	1	12	15.5		Queensland, Australia	518a
Lonchura stygia	Black Munia	B	5	11.2		10	12		New Guinea	786
Heteromunia pectoralis	Pictorella Munia	M	14	14.6	1.1	13.2	16.6		Australia	518a
		F	10	15.8	1.4	13	18			
Padda oryzivora	Java Sparrow	B	12	24.8		22.5	27.8			42
Padda fuscata	Timor Sparrow	U	2	20.7		19.2	22.2		Timor	42
Amadina fasciata	Cut-throat	U	95	18		15	32		Chad	395
Amadina erythrocephala	Red-headed Finch	U	192	22.5		17.5	30			133
Family Viduidae										
Anomalospiza imberbis	Parasitic Weaver	F	6	21.4					Southern Africa	904
Anomalospiza imberbis	Parasitic Weaver	B	14	19.7		18	21		Kenya	395
Vidua chalybeata	Village Indigobird	B	14	12.5	0.6	11.3	13.4		Zimbabwe; Malawi	909
Vidua raricola	Jambandu Indigobird	M	21	11.6		10.5	12.8		Cameroon; Nigeria	395
		F	7	12.3		11	15			
Vidua larvaticola	Baka Indigobird	B	18	13.1		12	14		Nigeria; Cameroon	395
Vidua maryae	Jos Plateau Indigobird	U	4	12.8		12	14.5		Nigeria	395
Vidua nigeriae	Quailfinch Indigobird	M	11	11.9		10.4	13.7		Nigeria; Cameroon	395
		F	6	12.2		11	14			
Vidua funerea	Variable Indigobird	M	202	14.9		11	19		Southern Africa	1036
		F	163	13.8		11	18			
Vidua codringtoni	Green Indigobird	B	33	12.8		11.4	14.1		Zimbabwe; Malawi	395
Vidua purpurascens	Purple Indigobird	M	21	12.8	0.7	11.8	13.7		Southern Africa	904, 909
		F	21	13.4						

Body Masses of World Birds (continued)

Species Name	Common Name	Sex	N	Mean	S.D.	Min	Max	Season	Location	Source #
Vidua wilsoni	Pale-winged Indigobird	M	12	12.4		11	14			395
		F	1	13.8						
Vidua camerunensis	Cameroon Indigobird	M	24	12.4		11.3	13.8		Cameroon	395
		F	4	11.8		11	13			
Vidua hypocherina	Steel-blue Whydah	M	11	11.9		11	13		Kenya	395
Vidua fischeri	Straw-tailed Whydah	M	13	13.2		10	17.8		Kenya	395
		F	11	11.3		9	17.8			
Vidua regia	Shaft-tailed Whydah	M	43	15	0.9			B	Namibia	48
		F	64	13.8	1					
Vidua macroura	Pin-tailed Whydah	M	33	16		13.3	18.8		Kenya	395
		F	46	15.5		12.2	16.9			
Vidua interjecta	Long-tailed Paradise-Whydah	M	5	20.6		20	21		Nigeria	395
Vidua paradisaea	Eastern Paradise-Whydah	M	102	21.8	1.7			B	Namibia	48
		F	98	19	1.5					
Vidua orientalis	Northern Paradise-Whydah	M	9	21.3		17	27		Cameroon; Chad; Senegal; Nigeria	395
		F	2	16		15	17			
Vidua obtusa	Broad-tailed Paradise-Whydah	M	8	22.2		19.5	26			395
		F	8	19.3		17.8	21.2			
Family Vireonidae										
Vireo brevipennis	Slaty Vireo	M	1	11.8					Oaxaca, Mexico	1145
Vireo griseus	White-eyed Vireo	B	153	11.4		10	14.3	F	Florida, USA	190
Vireo crassirostris	Thick-billed Vireo	B	75	14.4	0.8	13	15.7		New Providence, Bahamas	19
Vireo crassirostris	Thick-billed Vireo	B	15	11.1	0.2	11	11.5		Cayman Is.	19
Vireo crassirostris	Thick-billed Vireo	U	8	13.8		12.5	16.2		Cuba	636
Vireo pallens	Mangrove Vireo	U	12	11		8.9	13		Yucatan, Mexico; Belize	220, 640, 1027
Vireo bairdi	Cozumel Vireo	B	9	12		11.2	12.7		Mexico	640, 910a, 1145

Body Masses of World Birds (continued)

Species Name	Common Name	Sex	N	Mean	S.D.	Min	Max	Season	Location	Source #
Vireo caribaeus	St. Andrew Vireo	M	4	9.3		8.6	10		San Andres Is.	47, 1029
Vireo modestus	Jamaican Vireo	B	8	9.6		9.2	10.4		Jamaica	1159
Vireo gundlachii	Cuban Vireo	B	418	12.9	0.7	11.1	15.2		Cuba	19
Vireo latimeri	Puerto Rican Vireo	U	156	12	0.9	10.1	14.9		Puerto Rico	19
Vireo nanus	Flat-billed Vireo	U	2	10.5		10.4	10.6		Dominican Republic	19
Vireo bellii	Bell's Vireo	B	33	8.5	0.6	7.4	9.8		Arizona, USA	314
Vireo atricapilla	Black-capped Vireo	M	13	9.2	0.4				Oklahoma & Texas, USA	462
		F	9	8.8	0.4					
Vireo nelsoni	Dwarf Vireo	M	1	9.4					Oaxaca, Mexico	1145
Vireo vicinior	Gray Vireo	B	19	12.8	1.1	11	14.9		Sonora, Mexico	314
Vireo osburni	Blue Mountain Vireo	U	4	19.9					Jamaica	233, 1159
Vireo flavifrons	Yellow-throated Vireo	B	16	18	1.6	15.6	21.4	B	Pennsylvania, USA	314
Vireo plumbeus	Plumbeous Vireo	B	28	16.4	1.5	12	20.3		Colorado, USA	245
Vireo cassinii	Cassin's Vireo	B	75	14.7	1.4	10	20.6	B	Western USA	277
Vireo solitarius	Blue-headed Vireo	B	45	15.3		13.2	17.2	B	Canada	572
Vireo carmioli	Yellow-winged Vireo	U	10	13.1	0.9				Costa Rica	1039
Vireo huttoni	Hutton's Vireo	B	257	11.3	0.8	9	15.1	B		258
Vireo gilvus gilvus	Warbling Vireo	B	75	13.5	1.4	11	18.4			402
Vireo gilvus swainsonii	Warbling Vireo	B	524	11.9	1	9.9	15.8			402
Vireo leucophrys	Brown-capped Vireo	B	11	12.3	0.7	11.3	13.4		Colombia; Brazil; Mexico	75, 798, 883
Vireo philadelphicus	Philadelphia Vireo	U	28	11.5	0.7	9.8	12.7	M	Pennsylvania, USA	828
Vireo olivaceus	Red-eyed Vireo	U	658	16.8	1	14	21		Pennsylvania, USA	828

Body Masses of World Birds (continued)

Species Name	Common Name	Sex	N	Mean	S.D.	Min	Max	Season	Location	Source #
Vireo olivaceus chivi	Red-eyed Vireo	U	7	13.8					Peru	370
Vireo olivaceus chivi	Red-eyed Vireo	U	7	17.9		15.5	21		Trinidad	359
Vireo masteri	Choco Vireo	U	2	11.2		11	11.4		Colombia	1039
Vireo hypochryseus	Golden Vireo	B	21	12.2	0.9	10.6	14.1		Western Mexico	1151
Vireo flavoviridis	Yellow-green Vireo	U	33	17.9	0.7	16	21.1	B	Panama	815
Vireo flavoviridis	Yellow-green Vireo	U	128	17.3	2.4	11	31		El Salvador	1041a
Vireo gracilirostris	Noronha Vireo	B	7	16.5		11.5	20		Ilhu Fernando de Noronha	835
Vireo altiloquus	Black-whiskered Vireo	B	27	17.9				B	Jamaica	233
Vireo altiloquus	Black-whiskered Vireo	B	55	20.3	1.9	17.1	25.7		Puerto Rico	19
Vireo altiloquus	Black-whiskered Vireo	B	14	18.6	1.3	16	20.3		Antigua	19
Vireo altiloquus	Black-whiskered Vireo	B	25	21	1.2	19.4	23.8		Dominican Republic	19
Vireo altiloquus	Black-whiskered Vireo	B	18	17.4	1.4	15.7	21.5		New Providence, Bahamas	19
Vireo altiloquus	Black-whiskered Vireo	B	35	18.2	1.1	15.5	20.3		St. Kitts	19
Vireo altiloquus	Black-whiskered Vireo	B	50	19.7	1.5	17.5	26.2		St. Lucia	19
Vireo magister	Yucatan Vireo	B	12	15.4		13.5	17		Cayman Is.	1159
Hylophilus poicilotis	Rufous-crowned Greenlet	B	9	10.4		9	12		Brazil; Argentina	64, 254, 878, 961
Hylophilus thoracicus	Lemon-chested Greenlet	M	1	13.8					Peru	370
Hylophilus semicinereus	Gray-chested Greenlet	M	1	13					Bolivia	49
Hylophilus pectoralis	Ashy-headed Greenlet	U	5	11.6		11	11.8		Suriname; Brazil	392, 497, 1109

Body Masses of World Birds (continued)

Species Name	Common Name	Sex	N	Mean	S.D.	Min	Max	Season	Location	Source #
Hylophilus sclateri	Tepui Greenlet	M	2	11.2		10	12.5		Venezuela	296
Hylophilus muscicapinus	Buff-cheeked Greenlet	U	3	11.7		10.5	13.5		Brazil; Venezuela	73, 1178, 1308
Hylophilus brunneiceps	Brown-headed Greenlet	B	5	9.9		8	11.5		Brazil; Venezuela	449, 1178, 1308
Hylophilus hypoxanthus	Dusky-capped Greenlet	U		17					Peru	1210
Hylophilus semibrunneus	Rufous-naped Greenlet	B	2	11.4		10.2	12.6		Colombia	798
Hylophilus olivaceus	Olivaceous Greenlet	U	5	12		11	13		Peru; Ecuador	953, 1145
Hylophilus flavipes	Scrub Greenlet	B	14	12.6		11.3	14			795, 1180
Hylophilus flavipes insularis	Scrub Greenlet	B	5	10.8		9.5	12		Trinidad	613
Hylophilus ochraceiceps	Tawny-crowned Greenlet	U	27	11.6	0.7				Peru	370
Hylophilus aurantiifrons	Golden-fronted Greenlet	B	24	9.5		7.5	12		Trinidad	359
Hylophilus decurtatus	Lesser Greenlet	U	9	8.7		7.5	10.3		Belize	580, 1027
Hylophilus decurtatus	Lesser Greenlet	U	22	8.3	0.9	6.6	10		El Salvador	1041a
Vireolanius melitophrys	Chestnut-sided Shrike-Vireo	M	1	34.7					Oaxaca, Mexico	1145
Vireolanius pulchellus	Green Shrike-Vireo	B	4	24		22	26.1		Belize; Mexico	1145, 1027
Vireolanius leucotis	Slaty-capped Shrike-Vireo	U		26					Peru	1210
Cyclarhis gujanensis	Rufous-browed Peppershrike	U	26	28.8		22.5	35		Trinidad	359
Cyclarhis nigrirostris	Black-billed Peppershrike	U	4	31.9		29.6	33.1		Colombia	798, 103
Family Fringillidae										
Fringilla coelebs	Chaffinch	M	91	21.9	1.7	17	28.5		New Zealand	1002
		F	61	20.9	1.5	18	27			

Body Masses of World Birds (continued)

Species Name	Common Name	Sex	N	Mean	S.D.	Min	Max	Season	Location	Source #
Fringilla coelebs	Chaffinch	M	12	28.3	1.3				Canary Is.	442
		F	3	27						
Fringilla coelebs	Chaffinch	M	5	23.6					Morocco	442
		F	4	22.1						
Fringilla teydea	Blue Chaffinch	B	3	30.3					Canary Is.	442
Fringilla montifringilla	Brambling	B	164	22.6		17	28		Hopei, China	1087
Fringilla montifringilla	Brambling	U	19	23.8	1.9	21.8	27.8	F	Japan	490
Linurgus olivaceus	Oriole Finch	B	55	20.3		18.1	24.5		Tanzania	395
Rhynchostruthus socotranus percivali	Golden-winged Grosbeak	U	2	31.5		29.6	33.4		Yemen	317
Rhynchostruthus socotranus socotranus	Golden-winged Grosbeak	U	2	29.2		29	29.5		Socotra Is.	317
Rhynchostruthus louisae	Golden-winged Grosbeak	B	3	24.7		23.3	25.9		Somalia	395
Leucosticte nemoricola	Plain Mountain-Finch	B	19			19.8	25.6	PB		7
Leucosticte brandti	Black-headed Mountain-Finch	B	8	30.4		26.4	34		India; Mongolia	7, 927
Leucosticte arctoa	Asian Rosy-Finch	M	13	32.6		27.4	48			273
		F	5	28.9		22.4	35			
Leucosticte tephrocotis griseonucha	Gray-crowned Rosy-Finch	M	32	45.3	2.7	41.2	51.2	B	Aleutian Is., USA	729
		F	14	48.5	5.3	42.1	59.9			
Leucosticte tephrocotis dawsoni	Gray-crowned Rosy-Finch	B	42	24.6	1			W	California, USA	314
Leucosticte atrata	Black Rosy-Finch	M	117	25.1	1.7	21.9	31.5	B		598
		F	98	25.4	2.3	20.7	32			
Leucosticte australis	Brown-capped Rosy-Finch	B	74	26.6		23	31.7	B		599
Pinicola enucleator	Pine Grosbeak	B	17	56.4	3.2	52	62	F	Saskatchewan, Canada	314
Pinicola subhimachala	Crimson-browed Finch	M	6			42.5	48.2			7
		F	4			44	50			

Body Masses of World Birds (continued)

Species Name	Common Name	Sex	N	Mean	S.D.	Min	Max	Season	Location	Source #
Carpodacus nipalensis	Dark-breasted Rosefinch	M	4			22	23			7
		F	3	22.1		20	23.5			
Carpodacus erythrinus	Common Rosefinch	M	31	25		20	32		Hopei, China	1087
		F	41	23		19	33			
Carpodacus cassinii	Cassin's Finch	B	62	26.5	2.4	20.4	37.8		Arizona, USA	314
Carpodacus purpureus	Purple Finch	B	933	23.3		19.8	28.4		Pennsylvania, USA	828
Carpodacus mexicanus	House Finch	B	220	21.4	1.3	19	25.5	Y	California, USA	314
Carpodacus pulcherrimus	Beautiful Rosefinch	B	8	19.2		17.5	21		China	1183
Carpodacus rhodochrous	Pink-browed Rosefinch	M	9			17	20			7
		F	10			16	18.5			
Carpodacus vinaceus	Vinaceous Rosefinch	U				20.5	23.8		Taiwan	834
Carpodacus edwardsii	Dark-rumped Rosefinch	F	1	26.5				F		7
Carpodacus synoicus	Pale Rosefinch	B	11	20.8		19	24			919
Carpodacus roseus	Pallas' Rosefinch	B	141	28.6		21	35		Hopei, China	1087
Carpodacus rhodopeplus	Spot-winged Rosefinch	M	1	23						7
Carpodacus thura	White-browed Rosefinch	M	14			24	35.6	B		7
		F	8			30	36			
Carpodacus rhodochlamys	Rose-mantled Rosefinch	B	7			31	36		India	7
Carpodacus rubicilloides	Streaked Rosefinch	M	3			40.3	43.8	W	China	1183
Carpodacus rubicilla	Great Rosefinch	M	1	43.2						7
Carpodacus puniceus	Red-fronted Rosefinch	M	11			42.8	51.2			7
		F	5			43	50			
Loxia pytyopsittacus	Parrot Crossbill	U	37	53		44	58.2			133
Loxia scotica	Scottish Crossbill	M	6	44.6		42	49	Y	Scotland	919
		F	5	41.8		36.5	46			

Body Masses of World Birds (continued)

Species Name	Common Name	Sex	N	Mean	S.D.	Min	Max	Season	Location	Source #
Loxia curvirostra	Red Crossbill	B	36	36.4	3.7	29.2	44.9	W	Arizona, USA	1150
Loxia curvirostra	Red Crossbill	U	638	40.6		34	48	PB	Great Britain	515
Loxia curvirostra japonica	Red Crossbill	B	39	38		28	53		Hopei, China	1087
Loxia megaplaga	Hispaniolan Crossbill	B	7	28.8		26.7	30.4		Hispaniola	672
Loxia leucoptera leucoptera	White-winged Crossbill	M	119	26	1.3				Canada	66
		F	100	24.4	1.5					
Loxia leucoptera bifasciata	White-winged Crossbill	M	4	30.7		29.5	31.4			66
		F	8	34.8		29.7	37.9			
Carduelis spinoides	Yellow-breasted Greenfinch	B	9	18.6		16	20.8			7
Carduelis chloris	European Greenfinch	B	46	26		21	31	B	France	395
Carduelis flammea	Common Redpoll	B	30	13		10.1	15	B	Alaska, USA	554
Carduelis cabaret	Lesser Redpoll	M	17	11.9	0.7				Great Britain	343
		F	25	11.6	0.7					
Carduelis hornemanni exilipes	Hoary Redpoll	B	54	12.7		10.4	16.1		Alaska, USA	554
Carduelis hornemanni hornemanni	Hoary Redpoll	B	13	18.2		16.6	20.2		Ellesmere Is., Canada	643
Carduelis spinus	Eurasian Siskin	M	385	13.5		10.1	18.5			395
		F	379	13		10.3	18			
Carduelis pinus	Pine Siskin	B	49	12.7		11.7	14.7		Pennsylvania, USA	828
Carduelis atriceps	Black-capped Siskin	M	1	12.5					Chiapas, Mexico	544
		F	1	14.5						
Carduelis notata	Black-headed Siskin	B	13	10.9	0.6	10	12.2		Belize	1027
Carduelis spinescens	Andean Siskin	U	1	11.3					Colombia	324a
Carduelis magellanica	Hooded Siskin	B	48	13.6	1	11.5	15.5		Chile	776a
Carduelis xanthogastra	Yellow-bellied Siskin	U	10	12.7	0.9	10.8	13.8		Panama	675

Body Masses of World Birds (continued)

Species Name	Common Name	Sex	N	Mean	S.D.	Min	Max	Season	Location	Source #
Carduelis cucullata	Red Siskin	B	2	9.8		9.8	9.8		Puerto Rico (introduced)	1148
Carduelis dominicensis	Antillean Siskin	U	1	12.6					Dominican Republic	19
Carduelis psaltria	Lesser Goldfinch	B	202	9.5		8	11.5	B	California, USA	1090
Carduelis psaltria witti	Lesser Goldfinch	B	29	8.1		6	9		Tres Marias Is., Mexico	440
Carduelis lawrencei	Lawrence's Goldfinch	M	33	10.6	0.8	8.8	12.5	B	California, USA	314
		F	30	11.3	1	9.8	14.3			
Carduelis barbata	Black-chinned Goldfinch	U	161	15.5	1.4	13	19.8		Argentina	210
Carduelis atrata	Black Siskin	M	15	15.8		15	16.8		Argentina	842
		F	5	16.3		16	16.8			
Carduelis uropygialis	Yellow-rumped Siskin	M	1	14					Peru	1043
Carduelis tristis	American Goldfinch	M	1816	13.1	0.9	9.7	16.5		Pennsylvania, USA	828
		F	1381	12.5	0.7	10	15.6			
Carduelis carduelis	European Goldfinch	B	211	16		13	20		Western Europe	395
Carduelis sinica	Oriental Greenfinch	B	27	19		17	22		Hopei, China	1087
Carduelis sinica kittlitzi	Oriental Greenfinch	B	18	18.2	1.5	16.7	22.7		Japan	621
Carduelis flavirostris	Twite	U	50	15.4		13	17.6	W	Great Britain	515
Carduelis cannabina	Eurasian Linnet	M	27	20.2		15.9	25.7		Western Europe	395
		F	52	18.9		15.1	20.5			
Carduelis yemenensis	Yemen Linnet	M	4	14.3		13.4	14.9		North Yemen	101
Carduelis johannis	Warsangli Linnet	M	19	13.4		11.8	15		Somalia	395
Carduelis ankoberensis	Ankober Serin	U	3	14.9		14.6	15		Ethiopia	395
Serinus pusillus	Fire-fronted Serin	B	5			10.5	12.7			7
Serinus serinus	European Serin	U	20	11.2		8.5	14			133
Serinus syriacus	Syrian Serin	B	11	11.7		10	13		Mt. Hermon, Lebanon	919

Body Masses of World Birds (continued)

Species Name	Common Name	Sex	N	Mean	S.D.	Min	Max	Season	Location	Source #
Serinus canaria	Is. Canary	M	12	24.3	3	20.1	29.8		captive	113
Serinus citrinella	Citril Finch	U	1	12						453
Serinus thibetanus	Tibetan Serin	U	6			10	11	W		7
Serinus canicollis	Cape Canary	U	207	15.2	1.1	10	19.7		South Africa	1241
Serinus flavivertex	Yellow-crowned Canary	U	29	14		13	16		Kenya	395
Serinus citrinelloides citrinelloides	African Citril	U	41	12.8		11	14.7		Ethiopia	395
Serinus citrinelloides kikuyuensis	African Citril	U	324	13.2	2.4					395
Serinus frontalis	Western Citril	B	3	12.3		12	13			395
Serinus hypostictus	Southern Citril	B	22	13		10	15.5		Tanzania	395
Serinus koliensis	Papyrus Canary	U	30	13.6		11.3	16.1			133
Serinus scotops	Forest Canary	U	31	14.2	0.9	13	16.4		South Africa	1241
Serinus leucopygius	White-rumped Seedeater	U	14	10.8		9.4	12.7		Ethiopia	395
Serinus rothschildi	Olive-rumped Serin	U	1	14					Yemen	317
Serinus xantholaemus	Salvadori's Serin	B	4	11.8		11	12			395
Serinus atrogularis	Black-throated Canary	U	206	11.4		8	14			133
Serinus reichenowi	Reichenow's Seedeater	U	41	10.9		9.4	13.2		Kenya	395
Serinus citrinipectus	Lemon-breasted Seedeater	B	16	11.2		9.6	12.7		Malawi	1036
Serinus mozambicus	Yellow-fronted Canary	B	71	10.6		9.1	13.5		Ghana	457
Serinus mozambicus	Yellow-fronted Canary	U	110	13.3		8.5	16.2		South Africa	395
Serinus donaldsoni	Northern Grosbeak-Canary	U	3	24		20.7	26.9		Kenya	1052

Body Masses of World Birds (continued)

Species Name	Common Name	Sex	N	Mean	S.D.	Min	Max	Season	Location	Source #
Serinus dorsostriatus maculicollis	White-bellied Canary	U	37	14.5		12	17.3			395
Serinus dorsostriatus dorsostriatus	White-bellied Canary	B	42	15.8		13.7	17.7		Kenya	395
Serinus flaviventris	Yellow Canary	U	229	16.2	1.2	10.3	19.5		South Africa	1241
Serinus sulphuratus	Brimstone Canary	U	335	27.9	1.4	20.3	35.3		South Africa	1241
Serinus reichardi	Reichard's Seedeater	U	11	14.5		11.3	17.2		Malawi	1036
Serinus reichardi striatipectus	Reichard's Seedeater	M	1	16.4					Kenya	1143
Serinus albogularis	White-throated Canary	U	507	27.1		13	34		South Africa; Namibia	1036
Serinus gularis	Streaky-headed Seedeater	B	172	16		11.4	21			395
Serinus mennelli	Black-eared Seedeater	M	6	14.9		13.8	16.3		Zimbabwe; Malawi	1036
		F	4	15.9		14.9	17			
Serinus tristriatus	Brown-rumped Seedeater	U	125	15.5		12.2	19.4		Ethiopia	1243a
Serinus striolatus	Streaky Seedeater	U	35	22.4		19.5	25.5		Ethiopia	1238
Serinus burtoni	Thick-billed Seedeater	U	59	30.5		24	35		Kenya	395
Serinus rufobrunneus	Principe Seedeater	B	14	25.9		21	35		Principe	395
Serinus leucopterus	Protea Canary	U	11	22.1	1.2	18.8	24.8		South Africa	1241
Serinus totta	Cape Siskin	U	103	13.4		10	20.3		South Africa	1036
Serinus symonsi	Drakensberg Siskin	B	3	15.8		15	16.8		South Africa	1258
Serinus alario	Black-headed Canary	B	6	12.2		11	14		Namibia	526
Pyrrhula nipalensis	Brown Bullfinch	B	7	24.6		20	29			7, 775
Pyrrhula leucogenys	White-cheeked Bullfinch	M	1	19					Philippines	954

Body Masses of World Birds (continued)

Species Name	Common Name	Sex	N	Mean	S.D.	Min	Max	Season	Location	Source #
Pyrrhula aurantiaca	Orange Bullfinch	B	11			17	22	B	India	7
Pyrrhula erythrocephala	Red-headed Bullfinch	B	25			18	27.6			7
Pyrrhula erythaca	Gray-headed Bullfinch	B	15	19		18	21		Hopei, China	1087
Pyrrhula pyrrhula	Eurasian Bullfinch	U	34	21.8		21	27	B	Great Britain	515
Pyrrhula pyrrhula cineracea	Eurasian Bullfinch	B	5	27		26	28		Northern China	925
Pyrrhula murina	Azores Bullfinch	B		30					estimated	919
Coccothraustes coccothraustes	Hawfinch	M	99	58		46	72		Central Europe	395
		F	53	55.3		46	65			
Coccothraustes vespertinus	Evening Grosbeak	M	193	60.1	3.6	51.1	71.7		Pennsylvania, USA	828
		F	427	59	3.9	48.2	71.7			
Coccothraustes vespertinus	Evening Grosbeak	M	91	55.7	3.3			W	Utah, USA	36a
		F	131	54.6	3.5					
Coccothraustes abeillei	Hooded Grosbeak	B	4	48.5		47.1	49.7		Mexico	910
Eophona migratoria	Yellow-billed Grosbeak	M	10	48		40	57		China	184a
		F	10	45.8		42	50			
Eophona personata	Japanese Grosbeak	M	6	80		65	99		Hopei, China	1087
Mycerobas icterioides	Black-and-yellow Grosbeak	U	2	67					India	942
Mycerobas affinis	Collared Grosbeak	M	3			69	72			7
		F	1	83						
Mycerobas melanozanthos	Spot-winged Grosbeak	U	2	62		50	74			7
Mycerobas carnipes	White-winged Grosbeak	M	6			56.1	66		India; Nepal; Bhutan	7
		F	4			54.7	59.2			
Pyrrhoplectes epauletta	Gold-naped Finch	B	3	19		19	19			7
Rhodopechys sanguinea	Crimson-winged Finch	M	46	37.5		32.7	47.7		Kazakhstan	395
		F	8	38.8		32.4	45.7			

Body Masses of World Birds (continued)

Species Name	Common Name	Sex	N	Mean	S.D.	Min	Max	Season	Location	Source #
Rhodopechys githagineus	Trumpeter Finch	B	11	19.6	2.7	16	25		Algeria; Niger	395
Rhodopechys mongolica	Mongolian Finch	B	15	21.5	1.7	19	25		Mongolia	927
Rhodopechys obsoleta	Desert Finch	B	2	25.5		25	26		India	7
Uragus sibiricus	Long-tailed Rosefinch	M	4	18.2		16	20		Mongolia; China	273, 927
		F	3	16.7		16	17			
Haematospiza sipahi	Scarlet Finch	B	5	39.5		38	42.5		India	7
Family Drepanididae										
Telespiza cantans	Laysan Finch	M	249	34.4	2.3				Hawaiian Is.	810
		F	170	32.2	2.7					
Telespiza ultima	Nihoa Finch	B	6			21	28		captive, Hawaiian Is.	810
Loxioides bailleui	Palila	M	449	38.1	2.4				Hawaiian Is.	40
		F	246	37.6	2.5					
Pseudonestor xanthophrys	Maui Parrotbill	M	26	24.8	1.3	22.4	28.5		Hawaiian Is.	1112
		F	20	19.7	1.8	16.5	23.5			
Hemignathus virens	Hawaii Amakihi	M	144	14	1.2	11	15.6		Hawaiian Is.	637
		F	88	13.4	0.9	10.5	16.2			
Hemignathus flavus	Oahu Amakihi	M	18	12.8		11.3	17		Hawaiian Is.	695
		F	12	12.5		10.8	15.7			
Hemignathus kauaiensis	Kauai Amakihi	M	16	16.9		15	19.5		Hawaiian Is.	695
		F	7	16.4		15	19.5			
Hemignathus parvus	Anianiau	U	60	9.3	0.8	8	11		Hawaiian Is.	686
Hemignathus ellisianus	Greater Akialoa	M		37					Kauai, estimated, extinct	688
		F		31						
Hemignathus ellisianus	Greater Akialoa	M		34					Oahu, estimated, extinct	688
		F		25						
Hemignathus ellisianus	Greater Akialoa	M		25					Hawaii, USA, estimated, extinct	688
		F		21						
Hemignathus obscurus	Lesser Akialoa	M		29					estimated, extinct	688
		F		26						

Body Masses of World Birds (continued)

Species Name	Common Name	Sex	N	Mean	S.D.	Min	Max	Season	Location	Source #
Hemignathus lucidus lucidus	Nukupuu	U		25					Oahu, Hawaiian Is.	939
Hemignathus lucidus hanapepe	Nukupuu	U		27					Kauai, Hawaiian Is.	939
Hemignathus lucidus affinis	Nukupuu	U		23					Maui, Hawaiian Is.	939
Hemignathus munroi	Akiapolaau	M	38	28.9		19.8	37		Hawaiian Is.	939
		F	16	27.1		17	35			
Oreomystis bairdii	Akikiki	M	6	14		11.5	16		Hawaiian Is.	377
		F	4	14.5		12.5	17			
Oreomystis mana	Hawaii Creeper	B	19	14.6		12.5	16		Hawaiian Is.	690
Paroreomyza montana	Maui Alauahio	M	61	14	1.1	11.4	16.4		Hawaiian Is.	32
		F	25	12.3	1.5	8.9	15.4			
Loxops caeruleirostris	Akekee	U	14	10.8	0.8	9.5	12.5		Hawaiian Is.	689
Loxops coccineus	Akepa	B	75	10.5		9.2	13		Hawaiian Is.	687
Vestiaria coccinea	Iiwi	M	21	19.9	1.7				Hawaiian Is.	351
		F	12	16.7	1.4					
Palmeria dolei	Akohekohe	M	9	28.8		26.5	31		Hawaiian Is.	70
		F	8	23.6		20.6	28			
Himatione sanguinea	Apapane	M	42	16	1.3				Hawaiian Is.	350
		F	29	14.4	1.6					
Melamprosops phaeosoma	Poo-uli	U	1	25.5					Hawaiian Is.	336
Family Peucedramidae										
Peucedramus taeniatus	Olive Warbler	B	16	11	0.6	10.1	12.1	B	Coahuila, Mexico	797
Family Parulidae										
Vermivora pinus	Blue-winged Warbler	B	114	8.9		7.5	10.3	B	West Virginia, USA	420
Vermivora chrysoptera	Golden-winged Warbler	M	63	8.9	0.7	7.8	10.8		Pennsylvania, USA	828
		F	59	8.6	0.6	7.4	9.9			

Body Masses of World Birds (continued)

Species Name	Common Name	Sex	N	Mean	S.D.	Min	Max	Season	Location	Source #
Vermivora peregrina	Tennessee Warbler	B	190	8.9	1.1	6.2	14		El Salvador	1041a
Vermivora celata	Orange-crowned Warbler	B	72	9	0.8	7.3	11.6	M	California, USA	314
Vermivora celata sordida	Orange-crowned Warbler	B	128	9.4	0.8	7	11		Channel Is., California, USA	1137
Vermivora ruficapilla	Nashville Warbler	M	74	8.3	0.5	7.2	9.6	M	Pennsylvania, USA	828
		F	53	7.9	0.4	7.1	9.1			
Vermivora virginiae	Virginia's Warbler	B	78	8.2		6	10.5	B	Arizona, USA	865
Vermivora crissalis	Colima Warbler	B	6	9.3		8	10			1142a, 1147a, 1148a
Vermivora luciae	Lucy's Warbler	B	97	6.6	0.5	5.1	7.9	B	Arizona, USA	314
Parula gutturalis	Flame-throated Warbler	B	7	9.5					Panama	489
Parula superciliosa	Crescent-chested Warbler	B	33	9	1.2	7	13		El Salvador	1041a
Parula americana	Northern Parula	B	44	8.6		7.1	10.2		Florida, USA	364
Parula americana	Northern Parula	M	17	7.7	0.4	7.2	8.4		Pennsylvania, USA	828
		F	25	7.3	0.6	6.5	8.8			
Parula pitiayumi	Tropical Parula	M	10	7.2	0.4				Panama	489
		F	6	6.6						
Parula pitiayumi insularis	Tropical Parula	B	23	6.7		5	8.1		Tres Marias Is., Mexico	440
Dendroica petechia	Yellow Warbler	M	264	9.6	0.5	8.5	11.5		Pennsylvania, USA	828
		F	240	9.2	0.5	7.3	11.1			
Dendroica petechia cruciana	Yellow Warbler	M	10	12.4	0.7	11.2	13.5		Puerto Rico	19
		F	7	10.9		10.4	11.5			
Dendroica petechia	Yellow Warbler	B	10	9.6	0.8	8.4	11		Guadeloupe	19
Dendroica petechia aestiva	Yellow Warbler	U		10					Panama	617

Body Masses of World Birds (continued)

Species Name	Common Name	Sex	N	Mean	S.D.	Min	Max	Season	Location	Source #
Dendroica pensylvanica	Chestnut-sided Warbler	M	52	9.5	0.6	8.4	10.9		Pennsylvania, USA	828
		F	75	9.1	0.5	7.9	10.4			
Dendroica magnolia	Magnolia Warbler	M	467	8.4	0.5	6.9	10.3		Pennsylvania, USA	828
		F	433	7.9	0.5	6.6	9.7			
Dendroica tigrina	Cape May Warbler	M	49	10.3	0.6	9.5	12.7		Pennsylvania, USA	828
		F	51	9.8	0.6	8.8	10.9			
Dendroica caerulescens	Black-throated Blue Warbler	M	89	10.5		8.4	12.4		Florida, USA	364
		F	124	9.8		8.8	12.1			
Dendroica coronata	Yellow-rumped Warbler	M	294	12.2	0.7	10.1	14.7		Pennsylvania, USA	828
		F	288	11.4	0.6	9.8	13.6			
Dendroica coronata auduboni	Yellow-rumped Warbler	M	109	12.3	0.9	10	16	Sp	California, USA	314
		F	79	11.9	0.9	10	14			
Dendroica nigrescens	Black-throated Gray Warbler	B	56	8.7	0.6	6	9.8	B	Oregon & California, USA	277
Dendroica chrysoparia	Golden-cheeked Warbler	B	747	9.9		7	14.7	B	Texas, USA	664
Dendroica virens	Black-throated Green Warbler	M	38	8.9	0.5	8.1	10.1	M	Pennsylvania, USA	828
		F	27	8.5	0.5	7.4	9.5			
Dendroica townsendi	Townsend's Warbler	M	48	9.1	0.7	7.3	10.4	Sp	California, USA	314
		F	48	8.6	0.7	7.3	10.7			
Dendroica occidentalis	Hermit Warbler	B	229	10.4		8.8	12.8	B	Western USA	913
Dendroica fusca	Blackburnian Warbler	M	12	10	2					489
		F	18	9.5	0.6					
Dendroica dominica	Yellow-throated Warbler	M	9	9.9		9.3	10.8	B	Pennsylvania, USA	471
		F	6	9.5		8.8	10.5			
Dendroica pityophila	Olive-capped Warbler	B	16	8		7.2	8.7		Bahamas; Cuba	801, 1159
Dendroica graciae	Grace's Warbler	B	9	8.1		7.5	9.1		Nevada, USA	594
Dendroica adelaidae	Adelaide's Warbler	M	72	7.2	0.6	6	9.2		Puerto Rico	19
		F	61	6.7	0.7	5.3	10			
Dendroica subita	Barbuda Warbler	M	3	7.6		7	8		Barbuda	19
Dendroica delicata	St. Lucia Warbler	B	17	7.4	0.4	6.8	8.2		St. Lucia	19

Body Masses of World Birds (continued)

Species Name	Common Name	Sex	N	Mean	S.D.	Min	Max	Season	Location	Source #
Dendroica pinus	Pine Warbler	B	21	11.9	1.2	9.4	15.1		Minnesota, USA	314
Dendroica pinus	Pine Warbler	U	29	11.2	1.7	7	16		North Carolina, USA	1051
Dendroica pinus	Pine Warbler	U	27	12.3	1.4	10.5	17	Y	Florida, USA	1152
Dendroica kirtlandii	Kirtland's Warbler	M	31	13.6	0.6	12.3	14.6	B	Michigan, USA	767
		F	11	14.3	1.2	12.4	16			
Dendroica discolor	Prairie Warbler	M	149	8	0.9	6.1	10.1	F	Florida, USA	855
		F	110	7.3	0.9	5.7	10.8			
Dendroica vitellina	Vitelline Warbler	B	27	6.7		6.2	7.5		Cayman Is.	868
Dendroica palmarum	Palm Warbler	B	176	10.3	0.5	7	12.9		Pennsylvania, USA	193
Dendroica castanea	Bay-breasted Warbler	B	66	11.8		10.2	13.8		Pennsylvania, USA	828
Dendroica striata	Blackpoll Warbler	M	14	12.3	0.6	11.5	13.3	M	Pennsylvania, USA	828
		F	16	11.4	0.6	10.3	12.5			
Dendroica cerulea	Cerulean Warbler	M	47	9.3	0.6				Pennsylvania, USA	472
		F	34	8.8	0.6					
Dendroica plumbea	Plumbeous Warbler	B	29	10.1	0.7	8.4	11.3		Guadeloupe	19
Dendroica pharetra	Arrow-headed Warbler	B	6	9.9		9.1	10.5		Jamaica	19
Dendroica angelae	Elfin-woods Warbler	B	9	7.7		7.5	8.1		Puerto Rico	19
Catharopeza bishopi	Whistling Warbler	U	18	15.1	1.3	13.5	17.5		St. Vincent	19
Mniotilta varia	Black-and-white Warbler	B	116	10.9		8.8	12.9		Pennsylvania, USA	828
Setophaga ruticilla	American Redstart	M	624	8.4	0.6	6.5	10.2		Pennsylvania, USA	828
		F	483	8.1	0.6	6.7	10			
Protonotaria citrea	Prothonotary Warbler	B	95	14.3	0.1			B	Tennessee, USA	922
Helmitheros vermivorum	Worm-eating Warbler	M	30	13.2	0.7	11.8	14.7	B	Connecticut, USA	477
		F	30	15.2	2.2	13.2	17.4			
Limnothlypis swainsonii	Swainson's Warbler	B	19	18.9	1.7	14.3	20.4	F	Florida, USA	784

Body Masses of World Birds (continued)

Species Name	Common Name	Sex	N	Mean	S.D.	Min	Max	Season	Location	Source #
Seiurus aurocapilla	Ovenbird	B	147	18.8	1.2	15.4	21.4		Pennsylvania, USA	828
Seiurus noveboracensis	Northern Waterthrush	B	65	16.3	1.1	14.5	19.6	M	Pennsylvania, USA	828
Seiurus motacilla	Louisiana Waterthrush	B	164	19.9		16.8	23	B	Pennsylvania, USA	828
Oporornis formosus	Kentucky Warbler	B	165	14		12.1	16.9		Pennsylvania, USA	828
Oporornis agilis	Connecticut Warbler	B	55	13.3		11.2	15.5	F	Pennsylvania, USA	828
Oporornis philadelphia	Mourning Warbler	M	76	12	0.7	9.9	13.6	M	Pennsylvania, USA	828
		F	57	11.5	0.7	9.7	13			
Oporornis tolmiei	MacGillivray's Warbler	B	26	10.4	1.2	8.6	12.6	M	Arizona, USA	314
Geothlypis trichas	Common Yellowthroat	M	528	9.7	0.6	8.1	12.2		Pennsylvania, USA	828
		F	386	9.2	0.6	7.8	11.5			
Geothlypis trichas occidentalis	Common Yellowthroat	M	14	10.1		9.5	10.7		Western USA	59
Geothlypis trichas scirpicola	Common Yellowthroat	M	11	9.2		8.2	10		California, USA	59
Geothlypis beldingi	Belding's Yellowthroat	M	28	15.7		13.8	17.7		Mexico	59
Geothlypis flavovelata	Altamira Yellowthroat	M	5	10.9		10.2	11.5		Baja California, Mexico	1145
Geothlypis rostrata	Bahama Yellowthroat	B	9	16.1		15.1	17.3		Bahamas	1159
Geothlypis semiflava	Olive-crowned Yellowthroat	B	2	17.2		17	17.5		Ecuador; Colombia	1037, 1147
Geothlypis speciosa	Black-polled Yellowthroat	B	2	10.8		10	11.6		Mexico City, Mexico	1145
Geothlypis aequinoctialis	Masked Yellowthroat	B	13	13.1	1.4	11	15		Brazil; Colombia; Trinidad	64, 359, 795, 878
Geothlypis poliocephala	Gray-crowned Yellowthroat	B	13	14.6	1	13.2	16.4		Nicaragua; Mexico; Belize	542, 910a, 1027

Body Masses of World Birds (continued)

Species Name	Common Name	Sex	N	Mean	S.D.	Min	Max	Season	Location	Source #
Geothlypis nelsoni	Hooded Yellowthroat	M	4	11		10.1	11.8		Mexico	1145
Microligea palustris	Green-tailed Warbler	B	33	13.2	1.2	9.5	15.2		Dominican Republic	19
Teretistris fernandinae	Yellow-headed Warbler	B	159	9.9	1.4	6.5	18.5		Cuba	19
Teretistris fornsi	Oriente Warbler	B	680	9.7	0.8	7.4	13		Cuba	19
Wilsonia citrina	Hooded Warbler	M	151	10.8	0.6	9.6	12.5		Pennsylvania, USA	828
		F	261	10.3	0.7	8.6	12.8			
Wilsonia pusilla	Wilson's Warbler	B	21	7.2	0.4	6.5	8		Pennsylvania, USA	828
Wilsonia pusilla	Wilson's Warbler	M	82	7	0.5	6	7	Sp	California, USA	203
		F	40	6.7	0.3	6.1	7.1			
Wilsonia canadensis	Canada Warbler	M	145	10.3	0.5	9	11.7		Pennsylvania, USA	828
		F	125	9.8	0.6	8.5	11.4			
Cardellina rubrifrons	Red-faced Warbler	B	5	9.8		8.2	11.2		Arizona, USA	1150
Ergaticus ruber	Red Warbler	B	5	8.1		7.6	8.7		Mexico	910
Ergaticus versicolor	Pink-headed Warbler	M	1	10					Guatemala	1145
Myioborus pictus	Painted Redstart	M	40	9.7		8.5	11.5		Arizona, USA	45
		F	57	9.9		8.5	11.5			
Myioborus miniatus	Slate-throated Redstart	B	170	9	1.2	6	15		El Salvador	1041a
Myioborus castaneocapillus castaneocapillus	Tepui Redstart	B	5	10.6		9.5	12		Venezuela	1143
Myioborus castaneocapillus maguirei	Tepui Redstart	M	10	11.6	0.4	10.8	12		Venezuela	1308
		F	13	10.6	0.9	8.9	12			
Myioborus brunniceps	Brown-capped Redstart	U	12	9.2		8.5	10.2		Argentina	1040, 1041
Myioborus torquatus	Collared Redstart	B	6	10.5					Panama	489, 874
Myioborus melanocephalus	Spectacled Redstart	U	14	11.3	1	10	13		Colombia	1037

Body Masses of World Birds (continued)

Species Name	Common Name	Sex	N	Mean	S.D.	Min	Max	Season	Location	Source #
Myioborus ornatus	Golden-fronted Redstart	B	12	11.7		10.8	13.1		Colombia	324a
Euthlypis lachrymosa	Fan-tailed Warbler	B	233	25.6	2.2	10	21		El Salvador	1041a
Basileuterus fraseri	Gray-and-gold Warbler	B	18	11.6					Peru	1302
Basileuterus bivittatus	Two-banded Warbler	B	38	14.6					Peru	370
Basileuterus chrysogaster	Golden-bellied Warbler	B	26	11.2					Peru	370
Basileuterus signatus	Pale-legged Warbler	B	37	11.3					Bolivia	963
Basileuterus luteoviridis	Citrine Warbler	M	18	13.7					Bolivia	963
		F	11	12.2						
Basileuterus nigrocristatus	Black-crested Warbler	U	21	13.4	0.8	11.6	14.7		Ecuador	633
Basileuterus nigrocristatus	Black-crested Warbler	B	79	13.9	1.4	11.1	19.8		Colombia	324a
Basileuterus coronatus	Russet-crowned Warbler	B	29	16.3	1.6	13.5	19.5		Colombia	1037
Basileuterus culicivorus	Golden-crowned Warbler	U	22	10.5		9.5	12		Trinidad	359
Basileuterus trifasciatus	Three-banded Warbler	U	9	9.7		8.5	11		Ecuador	953
Basileuterus hypoleucus	White-bellied Warbler	U	33	11.4	0.8	10	13		Brazil	177
Basileuterus rufifrons	Rufous-capped Warbler	B	232	10.4	1.4	7	16		El Salvador	1041a
Basileuterus belli	Golden-browed Warbler	U	38	10.7	1.2	7	13		El Salvador	1041a
Basileuterus melanogenys	Black-cheeked Warbler	B	9	11.8					Panama	489
Basileuterus ignotus	Pirre Warbler	B	8			9.5	12.5		Panama	990

Body Masses of World Birds (continued)

Species Name	Common Name	Sex	N	Mean	S.D.	Min	Max	Season	Location	Source #
Basileuterus tristriatus	Three-striped Warbler	U	22	12.7	1.2				Peru	1288
Basileuterus tristriatus bessereri	Three-striped Warbler	U	15	11.4	0.6	10.4	12.8		Venezuela	1254
Basileuterus leucoblepharus	White-rimmed Warbler	U	26	15.1	0.9				Brazil	961
Basileuterus leucophrys	White-striped Warbler	U	23	18.1	1.7	15	22		Brazil	177, 752
Basileuterus flaveolus	Flavescent Warbler	U	16	13.3	1	12	16		Brazil	752
Basileuterus flaveolus	Flavescent Warbler	U	22	13.1	1.6	9	16		Venezuela	1254
Basileuterus fulvicauda	Buff-rumped Warbler	B	12	14.9					Panama; Costa Rica	489, 618
Basileuterus rivularis	River Warbler	U	11	13.5		11.5	16.5		Brazil	878
Zeledonia coronata	Wrenthrush	U		21						1170
Icteria virens	Yellow-breasted Chat	M	14	24.4	0.9	22.3	25.8		Pennsylvania, USA	808
		F	29	25.4	1.1	23.7	27.7			
Granatellus venustus	Red-breasted Chat	B	12	10.8	0.6	9.6	11.4		Western Mexico	1151
Granatellus sallaei	Gray-throated Chat	B	12	9.9		8.8	11		Mexico; Belize	220, 910a, 1145
Granatellus pelzelni	Rose-breasted Chat	U	4	11.2		10	12.5		Brazil; Venezuela	449, 1178, 1308
Family Coerebidae										
Coereba flaveola	Bananaquit	M	603	10.1	0.7	7.8	12.8		Puerto Rico	19
		F	571	9.3	0.8	7	13			
Coereba flaveola	Bananaquit	B	62	10.8	0.9	8.3	12.8		Dominica	19
Coereba flaveola	Bananaquit	M	27	9.9	0.9	8	12		Dominican Republic	19
		F	44	8.8	1.2	6.4	11			
Coereba flaveola	Bananaquit	M	85	10.7	0.8	8.5	12		Montserrat	19
		F	41	10	0.8	8.8	11.5			
Coereba flaveola	Bananaquit	B	65	11.7	1.2	8.4	14.2		St. Vincent	19

Body Masses of World Birds (continued)

Species Name	Common Name	Sex	N	Mean	S.D.	Min	Max	Season	Location	Source #
Coereba flaveola	Bananaquit	B	91	10.7	1.1	8.1	13.5		Guadeloupe	19
Coereba flaveola	Bananaquit	U	14	8.3	0.6	7.1	9.2		Brazil	392
Coereba flaveola luteola	Bananaquit	M	34	10.5	0.7	9	12		Trinidad	359
		F	24	9.9	0.6	9	11			
Family Thraupidae										
Conirostrum speciosum	Chestnut-vented Conebill	U	10	8.8		8	9		Argentina	254
Conirostrum leucogenys	White-eared Conebill	B	3	7		7	7			795, 990
Conirostrum bicolor	Bicolored Conebill	B	2	10.5		10.5	10.5		Trinidad	359
Conirostrum cinereum	Cinereous Conebill	B	11	8.8	0.9	7.8	10.3		Chile	776a
Conirostrum tamarugense	Tamarugo Conebill	F	2	10.5		10.5	10.5		Chile	776a
Conirostrum ferrugineiventre	White-browed Conebill	U	3	11.5		11	12			1070
Conirostrum rufum	Rufous-browed Conebill	U		11						522
Conirostrum sitticolor	Blue-backed Conebill	U	9	11.9		9.6	13.7		Colombia	324a
Conirostrum albifrons	Capped Conebill	B	4	14.9		11.7	21.5		Colombia	324a, 1037
Oreomanes fraseri	Giant Conebill	U	25	25		22	27			559
Orchesticus abeillei	Brown Tanager	U		31.5					Brazil	559
Schistochlamys ruficapillus	Cinnamon Tanager	U	6	31.2		24.1	39			559, 752
Schistochlamys melanopis	Black-faced Tanager	U	22	33		29	40			559
Neothraupis fasciata	White-banded Tanager	U	11	26.8					Brazil	1109
Cypsnagra hirundinacea	White-rumped Tanager	U	14	31.4					Brazil	1109

Body Masses of World Birds (continued)

Species Name	Common Name	Sex	N	Mean	S.D.	Min	Max	Season	Location	Source #
Conothraupis speculigera	Black-and-white Tanager	U	9	25		23	28			559
Cissopis leverianus	Magpie Tanager	U	13	76		73.8	86			559
Lamprospiza melanoleuca	Red-billed Pied Tanager	U	7	34		31	42			559
Chlorornis riefferii	Grass-green Tanager	U	37	53		42	59			559
Compsothraupis loricata	Scarlet-throated Tanager	M	1	72.5						559
Sericossypha albocristata	White-capped Tanager	U	17	114		95	125			559
Nesospingus speculiferus	Puerto Rican Tanager	B	57	35.2	2.7	28.7	40		Puerto Rico	19
Chlorospingus ophthalmicus bolivianus	Common Bush-Tanager	U	27	16		13.3	18			559
Chlorospingus ophthalmicus hiaticolus	Common Bush-Tanager	U	28	22		16	22.5			559
Chlorospingus ophthalmicus jaqueti	Common Bush-Tanager	U	10	16.8	1.2	15.1	18.5		Venezuela	1254
Chlorospingus ophthalmicus flavopectus	Common Bush-Tanager	M	68	24.4	2.1	19.1	28.4		Colombia	324a
		F	45	22.1	1.7	19.5	26.7			
Chlorospingus tacarcunae	Tacarcuna Bush-Tanager	M	13	18.9		18.2	20		Panama	79
Chlorospingus inornatus	Pirre Bush-Tanager	U	14	28		20.2	36			559
Chlorospingus semifuscus	Dusky Bush-Tanager	U	6	19		17.2	23			559
Chlorospingus pileatus	Sooty-capped Bush-Tanager	U	50	21		16	24.1			559
Chlorospingus parvirostris	Short-billed Bush-Tanager	U	15	24		17.5	28.5			559

Body Masses of World Birds (continued)

Species Name	Common Name	Sex	N	Mean	S.D.	Min	Max	Season	Location	Source #
Chlorospingus flavigularis	Yellow-throated Bush-Tanager	B	11	25.6	2.7	21.5	31		Colombia	1037
Chlorospingus flavovirens	Yellow-green Bush-Tanager	U	1	25						559
Chlorospingus canigularis	Ashy-throated Bush-Tanager	U	12	18		14.5	20.8			559
Cnemoscopus rubrirostris	Gray-hooded Bush-Tanager	U	32	18		13	22.5			559
Hemispingus atropileus	Black-capped Hemispingus	U	60	22		18	26			559
Hemispingus calophrys	Orange-browed Hemispingus	U	34	17		14.5	20.5			559
Hemispingus parodii	Parodi's Hemispingus	M	1	21.8					Peru	1143
Hemispingus superciliaris	Superciliaried Hemispingus	U	41	14		11.4	17.2			559
Hemispingus frontalis	Oleaginous Hemispingus	U	53	17		14	20			559
Hemispingus melanotis	Black-eared Hemispingus	U	52	16		13	21.8			559
Hemispingus rufosuperciliaris	Rufous-browed Hemispingus	U	21	29		26	33			559
Hemispingus verticalis	Black-headed Hemispingus	B	10	14		12	15.7			324a, 559, 953
Hemispingus xanthophthalmus	Drab Hemispingus	U	30	12		10.3	15			559
Hemispingus trifasciatus	Three-striped Hemispingus	U	31	14		12	16			559
Pyrrhocoma ruficeps	Chestnut-headed Tanager	U	10	15.6					Brazil; Paraguay	132, 177, 559, 961
Thlypopsis fulviceps	Fulvous-headed Tanager	U	2	12		11	13		Venezuela	1043
Thlypopsis ornata	Rufous-chested Tanager	U	23	12		9.8	14.9			559

Body Masses of World Birds (continued)

Species Name	Common Name	Sex	N	Mean	S.D.	Min	Max	Season	Location	Source #
Thlypopsis pectoralis	Brown-flanked Tanager	U	8	15		14	17			559
Thlypopsis sordida	Orange-headed Tanager	U	29	17		14	19			559
Thlypopsis inornata	Buff-bellied Tanager	U	8	15		14	17			559
Thlypopsis ruficeps	Rust-and-yellow Tanager	U	16	11		9.6	13.1			559
Hemithraupis guira	Guira Tanager	U	22	12		9.5	14			559
Hemithraupis ruficapillus	Rufous-headed Tanager	M	1	11						877
Hemithraupis flavicollis	Yellow-backed Tanager	B	13	12.8		11	15			559, 1308
Chrysothlypis chrysomelas	Black-and-yellow Tanager	U	6	13		11	15			559
Chrysothlypis salmoni	Scarlet-and-white Tanager	M	6	12.5		9.8	14.5		Colombia	559, 1037
Nemosia pileata	Hooded Tanager	U	25	16		12	20.7			559
Phaenicophilus palmarum	Black-crowned Palm-Tanager	B	94	30.5	2.9	28.7	39.7		Dominican Republic	19
Phaenicophilus poliocephalus	Gray-crowned Palm-Tanager	U		31						348
Calyptophilus tertius	Western Chat-Tanager	U	18	49.2	4.9	40.2	55.3	W	Dominican Republic	974
Calyptophilus frugivorus	Eastern Chat-Tanager	U	9	32.1		26.3	38.6	W	Dominican Republic	974
Rhodinocichla rosea	Rosy Thrush-Tanager	U	14	48		43	51.8			559
Mitrospingus cassinii	Dusky-faced Tanager	U	12	40.4	2.9				Costa Rica	618
Mitrospingus oleagineus	Olive-backed Tanager	B	7	41.3		35.5	46		Venezuela	1143
Chlorothraupis carmioli	Olive Tanager	U	23	39	2.5	34.5	43.4		Panama	79

Body Masses of World Birds (continued)

Species Name	Common Name	Sex	N	Mean	S.D.	Min	Max	Season	Location	Source #
Chlorothraupis carmioli	Olive Tanager	U	13	36.6	2.2				Colombia	1038a
Chlorothraupis olivacea	Lemon-spectacled Tanager	U	4	39		36	41			559
Chlorothraupis stolzmanni	Ochre-breasted Tanager	U	20	38.8	1.9	36.6	43		Colombia	1037
Eucometis penicillata	Gray-headed Tanager	U	52	27		22.5	35			559
Lanio fulvus	Fulvous Shrike-Tanager	U	20	24		19	30			559
Lanio versicolor	White-winged Shrike-Tanager	U	18	17		13	20			559
Lanio aurantius	Black-throated Shrike-Tanager	U	16	35		29.8	45			559
Lanio leucothorax	White-throated Shrike-Tanager	U		40					Costa Rica	1170
Creurgops verticalis	Rufous-crested Tanager	U	6	24		21	27			559
Creurgops dentata	Slaty Tanager	U	6	19		16	19.7			559
Heterospingus rubrifrons	Sulphur-rumped Tanager	U	2	38		36	40			559
Heterospingus xanthopygius berliozi	Scarlet-browed Tanager	M	2	38.8		38	39.5		Ecuador	1143
Tachyphonus cristatus	Flame-crested Tanager	B	16	18.8		16.6	23			559, 1308
Tachyphonus rufiventer	Yellow-crested Tanager	U	9	19		15.5	19.8			559
Tachyphonus surinamus	Fulvous-crested Tanager	U	37	19		15	23			559
Tachyphonus luctuosus	White-shouldered Tanager	U	27	13		11.5	15			559
Tachyphonus delatrii	Tawny-crested Tanager	U	17	18	1.9	14	20.8		Colombia	1037
Tachyphonus coronatus	Ruby-crowned Tanager	U	33	29.3		26	33.1		Brazil	878

Body Masses of World Birds (continued)

Species Name	Common Name	Sex	N	Mean	S.D.	Min	Max	Season	Location	Source #
Tachyphonus rufus	White-lined Tanager	U	172	34.4		25.7	42.5			559
Tachyphonus phoenicius	Red-shouldered Tanager	U	43	21		17	25			559
Trichothraupis melanops	Black-goggled Tanager	U	35	24.3		21.7	29.5		Brazil	878
Habia rubica	Red-crowned Ant-Tanager	M	79	34		27.7	42.9			559
		F	47	31		22.5	37			
Habia fuscicauda	Red-throated Ant-Tanager	M	86	41		32.9	46.5			559
		F	62	36		29	44			
Habia fuscicauda	Red-throated Ant-Tanager	M	9	37.4		32.9	41.6		Yucatan, Mexico	910a
		F	5	33.4		30.8	37.7			
Habia atrimaxillaris	Black-crested Ant-Tanager	M	1	48.9					Costa Rica	559
Piranga roseogularis	Rose-throated Tanager	U	21	24		20.7	29.9			559
Piranga flava	Hepatic Tanager	B	35	37.7	3.9	23.2	45.1	B	Mexico; Belize	1145
Piranga olivacea	Scarlet Tanager	B	216	28.2		23.8	33.5		Pennsylvania, USA	828
Piranga rubra	Summer Tanager	B	30	28.2	3.2			Sp	Louisiana, USA	1011
Piranga rubra	Summer Tanager	U	19	30.1	1.9	25.8	33.6	B	Florida, USA	1152
Piranga ludoviciana	Western Tanager	B	90	28.1		22.5	34.5	Sp	California, USA	1093
Piranga bidentata bidentata	Flame-colored Tanager	B	9	34.7		33.3	39.4		Mexico	1151
Piranga bidentata flammea	Flame-colored Tanager	B	18	37.5	4.1	32	48.4		Mexico	1151
Piranga leucoptera	White-winged Tanager	U	58	16		13.3	20			559
Piranga erythrocephala	Red-headed Tanager	U	3	21.8		19.9	24.5			559, 1151
Piranga rubriceps	Red-hooded Tanager	U	8	35.9		28	42			559, 953

Body Masses of World Birds (continued)

Species Name	Common Name	Sex	N	Mean	S.D.	Min	Max	Season	Location	Source #
Calochaetes coccineus	Vermilion Tanager	U	5	46.4		42	49		Peru; Ecuador	899, 953
Ramphocelus sanguinolentus	Crimson-collared Tanager	U	27	41		34.7	48.1			559
Ramphocelus nigrogularis	Masked Crimson Tanager	U	10	31		27	36			559
Ramphocelus dimidiatus	Crimson-backed Tanager	U	34	28		23.8	34			559
Ramphocelus melanogaster	Huallaga Tanager	M	1	25					Peru	559
Ramphocelus carbo	Silver-beaked Tanager	U	261	28		23.5	37.5			559
Ramphocelus carbo atrosericeus	Silver-beaked Tanager	U	21	24		21.5	27			559
Ramphocelus bresilius	Brazilian Tanager	U	8	32.9		27.9	35.5		Brazil	878
Ramphocelus passerinii	Passerini's Tanager	U	37	32		25.5	37			559
Ramphocelus flammigerus	Flame-rumped Tanager	M	13	33		29.6	35.6			559
Spindalis zena zena	Western Spindalis	B	1186	21.1	1.7	11	28.5		Cuba	19
Spindalis zena benedicti	Western Spindalis	B	15	30.5		26.5	35.6		Cozumel Is., Mexico	404
Spindalis portoricensis	Puerto Rican Spindalis	M	109	30.8	2.8	22.5	37		Puerto Rico	404
		F	42	33.5	3.3	28	41.1			
Spindalis dominicensis	Hispaniolan Spindalis	U	7	29.4		25	33.2		Hispaniola	19, 559
Spindalis nigricephala	Jamaican Spindalis	M	111	43	0.2	38	64.8		Jamaica	404
		F	48	47.2		45.2	53.7			
Thraupis episcopus	Blue-gray Tanager	U	181	35		27	45			559
Thraupis glaucocolpa	Glaucous Tanager	U	13	31.6	2.5	27	36		Venezuela	1254
Thraupis sayaca	Sayaca Tanager	U	27	32		27.9	34.4			559

Body Masses of World Birds (continued)

Species Name	Common Name	Sex	N	Mean	S.D.	Min	Max	Season	Location	Source #
Thraupis sayaca glaucocolpa	Sayaca Tanager	U	12	33		31.3	37.3		Venezuela	1215
Thraupis cyanoptera	Azure-shouldered Tanager	U	6	43.3		40	46			559, 961
Thraupis ornata	Golden-chevroned Tanager	F	1	33					Brazil	1178
Thraupis cyanocephala	Blue-capped Tanager	U	53	36		27	47			559
Thraupis bonariensis	Blue-and-yellow Tanager	U	41	36		28.2	46.5			559
Thraupis abbas	Yellow-winged Tanager	U	42	45		38	55			559
Thraupis palmarum	Palm Tanager	U	82	39		27	48			559
Cyanicterus cyanicterus	Blue-backed Tanager	M	1	34						559
Bangsia arcaei	Blue-and-gold Tanager	U	2	37.2		34.6	39.9		Panama; Costa Rica	79, 80
Bangsia melanochlamys	Black-and-gold Tanager	M	5	38.8		37.7	39.4		Colombia	1166
		F	3	35.6		35	36.3			
Bangsia edwardsi	Moss-backed Tanager	U	7	37		34	39.4		Colombia	1037
Bangsia aureocincta	Gold-ringed Tanager	B	7	41.7		38.3	46.6		Colombia	1166
Buthraupis montana	Hooded Mountain-Tanager	U	35	96		72	116			559
Buthraupis montana montana	Hooded Mountain-Tanager	U	17	79		69	91			559
Buthraupis eximia	Black-chested Mountain-Tanager	B	15	62.7		50	70		Peru	900
Buthraupis aureodorsalis	Golden-backed Mountain-T anager	U	10	85		75	94			559

Body Masses of World Birds (continued)

Species Name	Common Name	Sex	N	Mean	S.D.	Min	Max	Season	Location	Source #
Buthraupis wetmorei	Masked Mountain-Tanager	U	2	62.5		62	63			559
Wetmorethraupis sterrhopteron	Orange-throated Tanager	U	7	50.2		43.6	56			559, 747
Anisognathus melanogenys	Santa Marta Mountain-Tanager	M	1	41					Colombia	559
Anisognathus lacrymosus	Lacrimose Mountain-Tanager	U	80	31		25	38			559
Anisognathus igniventris	Scarlet-bellied Mountain-Tanager	U	98	34		21.5	41.2			559
Anisognathus somptuosus flavinuchus	Blue-winged Mountain-Tanager	U	53	42		33	56			559
Stephanophorus diadematus	Diademed Tanager	U	10	35.4		28.5	42.3		Brazil	878, 961
Iridosornis analis	Yellow-throated Tanager	U	63	26		20	29			559
Iridosornis jelskii	Golden-collared Tanager	U	45	20		16	26			559
Iridosornis rufivertex	Golden-crowned Tanager	U	14	23		18	27.9			559
Iridosornis reinhardti	Yellow-scarfed Tanager	U	73	24		19.5	28			559
Dubusia taeniata	Buff-breasted Mountain-Tanager	U	30	37		31	45			559
Delothraupis castaneoventris	Chestnut-bellied Mountain-Tanager	U	27	28		24.6	33			559
Pipraeidea melanonota	Fawn-breasted Tanager	U	14	21		18	25.2			559
Euphonia jamaica	Jamaican Euphonia	U	4	16.6					Jamaica	19, 231

Body Masses of World Birds (continued)

Species Name	Common Name	Sex	N	Mean	S.D.	Min	Max	Season	Location	Source #
Euphonia plumbea	Plumbeous Euphonia	U	5	8.9		8.7	9.5			559, 1178, 1308
Euphonia affinis	Scrub Euphonia	U	28	10		8.5	12.8			559
Euphonia chlorotica	Purple-throated Euphonia	U	16	11		8	14.3			559
Euphonia luteicapilla	Yellow-crowned Euphonia	U	5	13		11.4	14.5			559
Euphonia trinitatis	Trinidad Euphonia	U	4	11		8.8	14			559
Euphonia concinna	Velvet-fronted Euphonia	U	3	10.3		9	12		Colombia	795
Euphonia finschi	Finsch's Euphonia	U				10	11			559
Euphonia violacea	Violaceous Euphonia	U	87	15		12.5	17			559
Euphonia laniirostris	Thick-billed Euphonia	U	28	15		13	16.5			559
Euphonia hirundinacea	Yellow-throated Euphonia	U	50	14		11.6	17.8			559
Euphonia chalybea	Green-chinned Euphonia	U	4	19		18	20			559
Euphonia elegantissima	Elegant Euphonia	U	28	15		13.1	17			559
Euphonia musica	Antillean Euphonia	B	15	13.7	1.2	12.4	16		Puerto Rico	19
Euphonia musica	Antillean Euphonia	B	8	15.7		13.4	18.5		Guadeloupe	19
Euphonia cyanocephala	Golden-rumped Euphonia	U		14						522
Euphonia imitans	Spot-crowned Euphonia	U		14					Costa Rica	1170
Euphonia fulvicrissa	Fulvous-vented Euphonia	U	10	11		10.1	13			559, 747
Euphonia gouldi	Olive-backed Euphonia	U	20	14		10.9	16			559

Body Masses of World Birds (continued)

Species Name	Common Name	Sex	N	Mean	S.D.	Min	Max	Season	Location	Source #
Euphonia mesochrysa	Bronze-green Euphonia	U	9	13		12	15			559
Euphonia chrysopasta	White-lored Euphonia	U	11	14		11	16.2			559
Euphonia minuta	White-vented Euphonia	U	13	10		7.9	11.5			559
Euphonia anneae	Tawny-capped Euphonia	U	3	14.9		14.4	15.4		Panama	79, 1180
Euphonia xanthogaster	Orange-bellied Euphonia	U	107	13		9	16			559
Euphonia rufiventris	Rufous-bellied Euphonia	U	10	14		13	18			559
Euphonia cayennensis	Golden-sided Euphonia	U	3	14.3		11.8	16			559, 874
Euphonia pectoralis	Chestnut-bellied Euphonia	B	10	14.4						64, 378, 1202
Chlorophonia flavirostris	Yellow-collared Chlorophonia	M	1	11						559
Chlorophonia cyanea	Blue-naped Chlorophonia	U	13	14		11	15			559
Chlorophonia pyrrhophrys	Chestnut-breasted Chlorophonia	U	5	17		16	18			559
Chlorophonia occipitalis	Blue-crowned Chlorophonia	B	9	25.8		23	28.3			1145
Chlorophonia callophrys	Golden-browed Chlorophonia	B	6	25.8					Panama	489
Chlorochrysa phoenicotis	Glistening-green Tanager	U	5	21.8		20	24.5			559, 1037
Chlorochrysa calliparaea	Orange-eared Tanager	U	18	17		14.9	21.5			559
Chlorochrysa nitidissima	Multicolored Tanager	B	8	18.6		17.3	21.6		Colombia	798
Tangara inornata	Plain-colored Tanager	U	13	18		16.4	19.1			559

Body Masses of World Birds (continued)

Species Name	Common Name	Sex	N	Mean	S.D.	Min	Max	Season	Location	Source #
Tangara mexicana	Turquoise Tanager	B	19	20.5		18	23.5		Trinidad	359
Tangara palmeri	Gray-and-gold Tanager	U	10	32.3						559
Tangara chilensis	Paradise Tanager	U	17	23		17	27			559
Tangara chilensis paradisea	Paradise Tanager	U	4	17		16	17			559
Tangara seledon	Green-headed Tanager	B	9	18.7					Paraguay	378, 1205
Tangara cyanocephala	Red-necked Tanager	U	16	18		16	21.6		Sao Paulo, Brazil	878
Tangara desmaresti	Brassy-breasted Tanager	B	5	20.4		19	21.5		Brazil	559, 961, 1178
Tangara cyanoventris	Gilt-edged Tanager	U	2	16.5		16	17		Minas Gerais, Brazil	877
Tangara schrankii	Green-and-gold Tanager	U	35	19		14	23			559
Tangara florida	Emerald Tanager	B	3	19.3		18.5	20.5		Panama	990
Tangara arthus	Golden Tanager	U	21	22		18.7	27.5			559
Tangara icterocephala	Silver-throated Tanager	U	61	22		17.7	24.7			559
Tangara chrysotis	Golden-eared Tanager	U	4	24		23	25.5			559
Tangara xanthocephala	Saffron-crowned Tanager	U	33	19		15	23.6			559
Tangara parzudakii	Flame-faced Tanager	U	21	28		25	31			559
Tangara xanthogastra	Yellow-bellied Tanager	U	17	15		13	18			559
Tangara punctata	Spotted Tanager	U	17	15		13	17			559
Tangara guttata	Speckled Tanager	U	22	18.4		15	20.5		Trinidad	359
Tangara varia	Dotted Tanager	U	1	10						559

Body Masses of World Birds (continued)

Species Name	Common Name	Sex	N	Mean	S.D.	Min	Max	Season	Location	Source #
Tangara rufigula	Rufous-throated Tanager	U	11	19.1	0.8					837a
Tangara gyrola	Bay-headed Tanager	U	168	21		17.5	26.5			559
Tangara lavinia	Rufous-winged Tanager	U		24					Costa Rica	1170
Tangara cayana	Burnished-buff Tanager	U	30	18		15.2	22.5			559
Tangara peruviana	Black-backed Tanager	B	15	22.2		18.5	25.5		Brazil	808
Tangara cucullata	Lesser Antillean Tanager	U	10	27.9	1.3	25.7	30.3		St. Vincent	19
Tangara preciosa	Chestnut-backed Tanager	U	6	22.8		22	24		Argentina	212, 559
Tangara vitriolina	Scrub Tanager	U	15	23		18.4	26.8			559
Tangara meyerdeschauenseei	Green-capped Tanager	B	2	26		25.4	26.5		Peru	1072
Tangara rufigenis	Rufous-cheeked Tanager	U	2	16.5		16	17		Venezuela	1254
Tangara ruficervix	Golden-naped Tanager	U	15	19		16	22.2			559
Tangara labradorides	Metallic-green Tanager	U	14	15		13	16.4			559
Tangara cyanotis	Blue-browed Tanager	U	12	15		12.5	17			559
Tangara cyanicollis	Blue-necked Tanager	U	19	17		14	18.8			559
Tangara larvata	Golden-hooded Tanager	U	35	20		17.1	23.9			559
Tangara nigrocincta	Masked Tanager	U	8	17		15	17.8			559
Tangara fucosa	Green-naped Tanager	U	7	21		18	23			559
Tangara nigroviridis	Beryl-spangled Tanager	U	34	17		14	19.5			559

Body Masses of World Birds (continued)

Species Name	Common Name	Sex	N	Mean	S.D.	Min	Max	Season	Location	Source #
Tangara vassorii	Blue-and-black Tanager	U	63	18		15	21			559
Tangara heinei	Black-capped Tanager	U	6	20.2		18	22.8			559, 1037
Tangara phillipsi	Sira Tanager	B	9	19.9		16.8	21.7		Peru	448, 785a
Tangara viridicollis	Silver-backed Tanager	U	23	21		18.1	24			559
Tangara argyrofenges	Straw-backed Tanager	U	5	19		18	20			559
Tangara cyanoptera	Black-headed Tanager	M	19	20.7	1.7	17.1	24.5		Venezuela	1308
		F	19	22.4	2.2	18	27.5			
Tangara velia	Opal-rumped Tanager	U	11	21		19	23			559
Tangara callophrys	Opal-crowned Tanager	B	5	22.9		21	26.5			370, 559
Iridophanes pulcherrima	Golden-collared Honeycreeper	U	7	15.3		14	17			559, 953
Dacnis albiventris	White-bellied Dacnis	U	2	11.2		11	11.5			559
Dacnis lineata	Black-faced Dacnis	U	20	11		9.5	13			559
Dacnis flaviventer	Yellow-bellied Dacnis	U	3	13		12	14			559
Dacnis nigripes	Black-legged Dacnis	U	10	14		11	15.5			559
Dacnis venusta	Scarlet-thighed Dacnis	U	43	16.1					Panama	489
Dacnis cayana	Blue Dacnis	U	61	13		10	15.5			559
Chlorophanes spiza	Green Honeycreeper	U	108	19		14	23			559
Cyanerpes nitidus	Short-billed Honeycreeper	U	5	9		8	10.2			559
Cyanerpes lucidus	Shining Honeycreeper	B	6	11.4					Panama	489

Body Masses of World Birds (continued)

Species Name	Common Name	Sex	N	Mean	S.D.	Min	Max	Season	Location	Source #
Cyanerpes caeruleus	Purple Honeycreeper	U	53	12		7.9	14			559
Cyanerpes cyaneus	Red-legged Honeycreeper	U	86	14		11	18.3			559
Xenodacnis parina	Tit-like Dacnis	B	11	11.5		10	12			559
Tersina viridis	Swallow-Tanager	U	30	29		25.9	35			559
Catamblyrhynchus diadema	Plush-capped Finch	M	16	14.9					Bolivia	963
		F	8	13.4						
Urothraupis stolzmanni	Black-backed Bush-Tanager	U	5	23.1		19	27		Ecuador	1147
Nephelornis oneilli	Pardusco	M	15	17.5	0.9	16	19		Peru	716
		F	13	14.9	0.7	13.5	15.5			
Family Emberizidae										
Melophus lathami	Crested Bunting	B	23	21.9		18	25			7
Emberiza citrinella	Yellowhammer	B	95	29.7		25	36		former Czechoslovakia	223
Emberiza leucocephalos	Pine Bunting	B	124	28.5		21	37		Hopei, China	1087
Emberiza cirlus	Cirl Bunting	B	63	25.6		23	28	Y	France	223
Emberiza cia	Rock Bunting	M	51	24.2	1.6	21	29	B	Germany	223
		F	38	22.7	1.6	20	27			
Emberiza godlewskii	Godlewski's Bunting	B	71	19		14	23		Hopei, China	1087
Emberiza cioides	Meadow Bunting	B	70	20		17	23		Hopei, China	1087
Emberiza cioides	Meadow Bunting	U	14	22.1	1.5	19.6	24.6	F	Japan	490
Emberiza jankowskii	Rufous-backed Bunting	U	2	22.5		22	23		Mongolia	927
Emberiza buchanani	Gray-hooded Bunting	B	25	21.2		17.2	26		Turkey; Iran	223
Emberiza cineracea	Cinereous Bunting	B	8	24.3		21.1	29.7		Iran	223
Emberiza hortulana	Ortolan Bunting	B	68	19.9				Sp	Israel	1354

Body Masses of World Birds (continued)

Species Name	Common Name	Sex	N	Mean	S.D.	Min	Max	Season	Location	Source #
Emberiza stewarti	Chestnut-breasted Bunting	M	6	17		16	18	Sp		7
		F	11	15		13	17			
Emberiza caesia	Cretzschmar's Bunting	B	14	21.3		18.5	23.5	B	Greece; Turkey; Israel	223
Emberiza caesia	Cretzschmar's Bunting	B	55	18.6				Sp	Israel	1353
Emberiza striolata	House Bunting	B	17	14.8		12	18		Algeria; Chad	223
Emberiza impetuani	Lark-like Bunting	B	26	15.1		13.3	17.2		South Africa	509
Emberiza tahapisi	Cinnamon-breasted Bunting	U	30	13.7		11.6	16		Nigeria	777
Emberiza tahapisi	Cinnamon-breasted Bunting	U	3	12.3		11.6	13.2		Yemen	317
Emberiza socotrana	Socotra Bunting	U				13	16		Socotra	395
Emberiza capensis	Cape Bunting	B	72	20.9	1.3	17.5	26		South Africa; Namibia	1241
Emberiza yessoensis	Ochre-rumped Bunting	U				13	14		Hopei, China	1087
Emberiza tristrami	Tristram's Bunting	B	12	18		14	21		Hopei, China	1087
Emberiza fucata	Chestnut-eared Bunting	B	13	19.8		14	21		India; China	7, 925, 1087
		F								
Emberiza pusilla	Little Bunting	B	141	13.8		12	17		Hopei, China	1087
Emberiza pusilla	Little Bunting	B	164	15.4		12.9	19.3	B	Siberia, Russia	223
Emberiza chrysophrys	Yellow-browed Bunting	M	11	19		16	24		Hopei, China	1087
		F	5	17		15	18			
Emberiza rustica	Rustic Bunting	M	54	21	1.6	17.7	24.3	F	Japan	223
		F	35	20.1	1.3	18	22.9			
Emberiza elegans	Yellow-throated Bunting	U	44	17		14	22		Hopei, China	1087
Emberiza elegans	Yellow-throated Bunting	U	14	16.7	1.1	14.9	19.1	F	Japan	490

Body Masses of World Birds (continued)

Species Name	Common Name	Sex	N	Mean	S.D.	Min	Max	Season	Location	Source #
Emberiza aureola	Yellow-breasted Bunting	B	83	19.6		14	33		Hopei, China	1087
Emberiza flaviventris	Golden-breasted Bunting	M	12	18.8		15.1	28.3		Zimbabwe; Namibia	526, 562, 563, 564
		F	10	17.7		16.1	19			
Emberiza poliopleura	Somali Bunting	U	22	15.8		14.1	17.9		Kenya	395
Emberiza affinis	Brown-rumped Bunting	U	4	15.4		14.5	16		Ghana	457
Emberiza cabanisi	Cabanis' Bunting	U	11	24.9		22.3	27.1		Malawi; Zimbabwe; Ghana	457, 566, 1036
Emberiza rutila	Chestnut Bunting	B	37	17.5		13	24		Hopei, China	1087
Emberiza melanocephala	Black-headed Bunting	M	20	29.7	2.6	25	33	Sp		223
		F	9	27		23	31			
Emberiza bruniceps	Red-headed Bunting	B	46	24.8		18	30.5		Kazakhstan	223
Emberiza sulphurata	Yellow Bunting	M	43	15.2	1.1				Japan	835a
		F	26	14.6	1.2					
Emberiza spodocephala	Black-faced Bunting	U	29	20.4	1.2	17.8	23.5	F	Japan	490
Emberiza spodocephala	Black-faced Bunting	B	42	16.7		13	23		Hopei, China	1087
Emberiza variabilis	Gray Bunting	U	20	23.8	2.5	20.9	30	F	Japan	490
Emberiza pallasi	Pallas' Bunting	M	33	15		12	20		Hopei, China	1087
		F	41	13		10	16			
Emberiza schoeniclus	Reed Bunting	M	153	19.7	1.2	16.1	24.4	F	Siberia, Russia	223
		F	117	17.2	1	14.2	20.6			
Emberiza calandra	Corn Bunting	M	68	53.6	4.1	45	63	W	Great Britain	223
		F	202	43.9	3.1	36	53			
Coryphaspiza melanotis	Black-masked Finch	M	2	15.7		15.6	15.8		Argentina	1099
Saltatricula multicolor	Many-colored Chaco-Finch	B	18	22.2		20	25		Bolivia	657
Coryphospingus pileatus	Pileated Finch	B	33	15.3		12	18		Venezuela	1254

Body Masses of World Birds (continued)

Species Name	Common Name	Sex	N	Mean	S.D.	Min	Max	Season	Location	Source #
Coryphospingus cucullatus	Red-crested Finch	B	32	14.9					Paraguay	378
Coryphospingus cucullatus	Red-crested Finch	B	13	13.7		12	15.8		Bolivia	657
Rhodospingus cruentus	Crimson-breasted Finch	U	10	11.6					Ecuador	1302, 1311
Phrygilus atriceps	Black-hooded Sierra-Finch	U	5	24.3						363, 776a
Phrygilus punensis	Peruvian Sierra-Finch	M	2	37.2		35.9	38.5		Peru	1043
Phrygilus gayi	Gray-hooded Sierra-Finch	M	37	24.9		21.9	30.7		Argentina	842
		F	8	26.3		23.4	28			
Phrygilus patagonicus	Patagonian Sierra-Finch	U	35	22.6	1.9	19.1	27.4		Argentina	210
Phrygilus fruticeti	Mourning Sierra-Finch	B	13	38.8		35.1	41.8		Argentina	841
Phrygilus unicolor	Plumbeous Sierra-Finch	B	6	21.9						1041, 1263, 1288
Phrygilus dorsalis	Red-backed Sierra-Finch	M	4	35.8		35.4	36		Argentina	842
		F	1	40.5						
Phrygilus carbonarius	Carbonated Sierra-Finch	B	3	17.9		17.8	18		Argentina	1040, 1041
Phrygilus alaudinus	Band-tailed Sierra-Finch	B	64	22.7	1.6	18.5	26.8		Chile	328
Phrygilus alaudinus bracki	Band-tailed Sierra-Finch	U	2	19		17	21		Peru	872
Phrygilus alaudinus bipartitus	Band-tailed Sierra-Finch	U	2	21.5						872
Phrygilus alaudinus excelsus	Band-tailed Sierra-Finch	U	6	27.4						872
Phrygilus alaudinus venturii	Band-tailed Sierra-Finch	M	13	28		26.2	30		Argentina	842
		F	4	26.4		25.7	28			
Phrygilus plebejus	Ash-breasted Sierra-Finch	B	25	15.3		13.3	17.4		Argentina	842

Body Masses of World Birds (continued)

Species Name	Common Name	Sex	N	Mean	S.D.	Min	Max	Season	Location	Source #
Phrygilus plebejus	Ash-breasted Sierra-Finch	B	21	14.1	0.6	13	15		Chile	776a
Melanodera melanodera	Canary-winged Finch	B	7	28.7		26.6	32		Tierra del Fuego	546
Melanodera xanthogramma	Yellow-bridled Finch	M	4	36.4		35.5	37.3		Tierra del Fuego	546
		F	2	33.8		30.1	37.3			
Lophospingus pusillus	Black-crested Finch	M	9	14.9		19	20.5		Bolivia	657
		F	1	14						
Lophospingus griseocristatus	Gray-crested Finch	B	25	17.9		13.1	20.2		Bolivia	1055
Donacospiza albifrons	Long-tailed Reed-Finch	M	1	16.5					Brazil	64
		F	4			14	16.5			
Nesospiza acunhae	Nightingale Finch	M	11	29.2	2.6	24.5	34		Tristan da Cunha	382
		F	5	27.9		26.5	29			
Nesospiza wilkinsi	Wilkin's Finch	M	6	44.2		41	44.5		Tristan da Cunha	382
		F	4	42.3		41.8	43.5			
Diuca diuca	Common Diuca-Finch	U	21	36.8	2.7	31	41		Chile	328
Idiopsar brachyurus	Short-tailed Finch	M	1	43					Bolivia	1261
Xenospingus concolor	Slender-billed Finch	F	2	21		20	22		Chile	776a
Incaspiza pulchra	Great Inca-Finch	B	13	27.6	2	25.5	32			1145
Incaspiza personata	Rufous-backed Inca-Finch	M	3	31.5		29.5	33			1145
		F	1	35						
Incaspiza ortizi	Gray-winged Inca-Finch	M	10	34.5	2.8	29	38			1145
		F	3	30.5		29	32			
Incaspiza laeta	Buff-bridled Inca-Finch	B	24	22	1.1	19.5	23.5			1145
Incaspiza watkinsi	Little Inca-Finch	B	9	20.4		17	22.5			1145
Poospiza thoracica	Bay-chested Warbling-Finch	B	3	11.9		11	12.8		Brazil	64, 878

Body Masses of World Birds (continued)

Species Name	Common Name	Sex	N	Mean	S.D.	Min	Max	Season	Location	Source #
Poospiza hypochondria	Rufous-sided Warbling-Finch	B	5	21.2		18	24.5		Bolivia	1143
Poospiza ornata	Cinnamon Warbling-Finch	U		11.8					Argentina	1053
Poospiza erythrophrys	Rusty-browed Warbling-Finch	U	2	14.1		13.2	15			363
Poospiza nigrorufa	Black-and-rufous Warbling-Finch	B	11	17.4		13	20.7		Argentina; Bolivia	210a, 1055
Poospiza whitii	Black-and-chestnut Warbling-Finch	U	1	17						363
Poospiza lateralis	Red-rumped Warbling-Finch	B	12	19.2		16.1	21		Brazil; Argentina	255, 878, 961
Poospiza baeri	Tucuman Mountain-Finch	B	4	32.5		31.5	34.5		Argentina	1041
Poospiza hispaniolensis	Collared Warbling-Finch	M	1	16.5					Ecuador	1143
Poospiza torquata	Ringed Warbling-Finch	B	16	11		10	13.2		Argentina	842
Poospiza melanoleuca	Black-capped Warbling-Finch	B	13	13.1		9	15		Argentina, Paraguay; Bolivia	210a, 363, 918, 1055
Poospiza cinerea	Cinereous Warbling-Finch	B	10	11.6	2	8.4	15.2		Paraguay; Brazil	177
Volatinia jacarina	Blue-black Grassquit	B	41	9.7		8	12		Trinidad	1131
Volatinia jacarina	Blue-black Grassquit	U	17	10.2					Ecuador	1311
Sporophila schistacea	Slate-colored Seedeater	B	10	12.7	0.6	12	13.7		Brazil; Peru	10, 1178
Sporophila plumbea	Plumbeous Seedeater	B	11	9.7		8.8	10.8		Suriname; Brazil	392, 498
Sporophila intermedia	Gray Seedeater	B	15	12.1		11	13.5		Venezuela	1215
Sporophila americana	Wing-barred Seedeater	B	16	12		10	13.5		Suriname	498

Body Masses of World Birds (continued)

Species Name	Common Name	Sex	N	Mean	S.D.	Min	Max	Season	Location	Source #
Sporophila corvina	Variable Seedeater	U	14	11.4	1.1				Costa Rica	618
Sporophila corvina	Variable Seedeater	U	30	9.9	1				Panama	618
Sporophila torquata	White-collared Seedeater	B	41	8.7	1	6.3	12	B	Mexico; Belize, Honduras; Guatemala	1145
Sporophila collaris	Rusty-collared Seedeater	B	5			13	14		Brazil	64
Sporophila bouvronides	Lesson's Seedeater	M	6	9.1		8	10.3		Venezuela; Suriname	498
Sporophila lineola	Lined Seedeater	B	26	9.8		7.5	12		Trinidad	1131
Sporophila luctuosa	Black-and-white Seedeater	U	4	12.5					Peru	1288
Sporophila nigricollis	Yellow-bellied Seedeater	B	12	9.6	0.8	8.5	11.2		Venezuela	1215
Sporophila caerulescens	Double-collared Seedeater	M	11	10.2		9.1	11.6		Bolivia	1055
		F	16	9.3		7.4	11.5			
Sporophila albogularis	White-throated Seedeater	U		9.7					Brazil	1053
Sporophila leucoptera	White-bellied Seedeater	M	3	15.5		15	16		Paraguay	1175
Sporophila peruviana	Parrot-billed Seedeater	U	6	12.6					Peru; Ecuador	1043, 1311
Sporophila bouvreuil	Capped Seedeater	B	22	8.6		7.6	10		Argentina	255
Sporophila minuta	Ruddy-breasted Seedeater	B	19	7.8		7	9		Trinidad	1131
Sporophila ruficollis	Dark-throated Seedeater	B	7	8.6		7	9.7		Bolivia	1099, 1143
Sporophila palustris	Marsh Seedeater	M	1	9					Brazil	64
		F	1	7.5						
Sporophila castaneiventris	Chestnut-bellied Seedeater	U	12	7.8					Peru; Suriname; French Guiana	293, 498, 1288

Body Masses of World Birds (continued)

Species Name	Common Name	Sex	N	Mean	S.D.	Min	Max	Season	Location	Source #
Sporophila hypochroma	Gray-and-chestnut Seedeater	M	1	9.7					Argentina; subadult	255
Sporophila melanogaster	Black-bellied Seedeater	B	3	9.3		8	10.5		Brazil	64
Sporophila telasco	Chestnut-throated Seedeater	M	2	9.6		8.4	10.9		Peru; Chile	776a, 1302
Oryzoborus nuttingi	Nicaraguan Seed-Finch	U		24					Costa Rica	628
Oryzoborus crassirostris	Large-billed Seed-Finch	B	2	20.9		19.8	22		Venezuela; Suriname	498, 1215
Oryzoborus atrirostris gigantirostris	Black-billed Seed-Finch	M	2	26		23	29		Bolivia; Peru	1143
Oryzoborus maximiliani	Great-billed Seed-Finch	M	4	24.6		22	28.4		Peru; Colombia	370, 1037
Oryzoborus angolensis	Chestnut-bellied Seed-Finch	B	12	13		11.4	14.5		Brazil; Suriname; Trinidad; Ecuador	177, 497, 953, 1131
Oryzoborus funereus	Thick-billed Seed-Finch	U	24	12.3	0.9				Panama	618
Amaurospiza moesta	Blackish-blue Seedeater	B	8	13.6		13	14.5		Brazil; Paraguay	64, 132, 961
Amaurospiza concolor	Blue Seedeater	B	6	13.1		12	15		El Salvador	1041a
Amaurospiza carrizalensis	Carrizal Blue-black Seedeater	B	3	13.3		12	14		Venezuela	684
Melopyrrha nigra taylori	Cuban Bullfinch	B	19	16.5		14.5	18.5		Cayman Is.	868
Melopyrrha nigra	Cuban Bullfinch	M	418	14.6	1	9	18.4		Cuba	19
		F	434	13.7	1	10.2	18.3			
Dolospingus fringilloides	White-naped Seedeater	M	10	13.1	0.7				Guyana	987
		F	3	12.3						
Catamenia analis	Band-tailed Seedeater	B	9	12		10.3	16.6			633, 776a, 1041, 1043, 1055
Catamenia inornata	Plain-colored Seedeater	U	16	13.4		12.3	14.5		Ecuador	633

Body Masses of World Birds (continued)

Species Name	Common Name	Sex	N	Mean	S.D.	Min	Max	Season	Location	Source #
Catamenia homochroa	Paramo Seedeater	M	3	12.2					Ecuador; Bolivia	953, 963
Tiaris obscurus	Dull-colored Grassquit	U	29	11.2	0.8				Peru	1288
Tiaris canorus	Cuban Grassquit	B	21	8.1	1	6	9.5		Cuba	19
Tiaris olivacea	Yellow-faced Grassquit	B	318	8.5	0.6	6	10		Cuba	19
Tiaris bicolor	Black-faced Grassquit	B	172	9.8	0.7	7	11.6		Puerto Rico	19
Tiaris fuliginosus	Sooty Grassquit	B	22	13.3		11	16		Trinidad	1131
Loxipasser anoxanthus	Yellow-shouldered Grassquit	B	12	11.4	0.7	10.5	12.5		Jamaica	1159
Euneornis campestris	Orangequit	U	260	16		13.2	19.2			559
Melanospiza richardsoni	St. Lucia Black Finch	B	6	20.9		18	23		St. Lucia	106, 1147a, 1229
Loxigilla portoricensis	Puerto Rican Bullfinch	M	389	34.8	2.6	26.7	44.8		Puerto Rico	19
		F	332	29.5	2.7	23.4	45.5			
Loxigilla violacea violacea	Greater Antillean Bullfinch	M	2	21.6		20.8	22.5		Bahamas	1159
		F	5	19.2		18	20.5			
Loxigilla violacea	Greater Antillean Bullfinch	B	97	26.3	2.2	19.5	32.4		Dominican Republic	19
Loxigilla noctis	Lesser Antillean Bullfinch	B	244	18.5	1.5	15.4	23.4		Dominica	19
Loxigilla noctis	Lesser Antillean Bullfinch	B	62	16.1	1	13.4	18.1		Antigua	19
Loxigilla noctis	Lesser Antillean Bullfinch	B	71	15.6	0.9	13.2	18.1		Barbuda	19
Loxigilla noctis	Lesser Antillean Bullfinch	B	118	16.8	1.4	13.8	20.8		St. Kitts	19
Loxigilla noctis	Lesser Antillean Bullfinch	B	303	16.3	1.3	13.1	23.4		St. Lucia	19
Pinaroloxias inornata	Cocos Is. Finch	U	20	13.1	0.9				Cocos Is.	1124

Body Masses of World Birds (continued)

Species Name	Common Name	Sex	N	Mean	S.D.	Min	Max	Season	Location	Source #
Haplospiza rustica	Slaty Finch	U	13	15.6	1				Peru	1288
Haplospiza unicolor	Uniform Finch	M	8	15.9		14	17.5		Brazil	961
		F	10	14.7	1.4					
Acanthidops bairdii	Peg-billed Finch	U		16					Costa Rica	1170
Diglossa baritula	Cinnamon-bellied Flowerpiercer	B	18	9.2	1.7	6	15		El Salvador	1041a
Diglossa plumbea	Slaty Flowerpiercer	U	4	10		9.3	10.1			559, 618
Diglossa sittoides	Rusty Flowerpiercer	U	13	9		7	10.7			559
Diglossa gloriosissima	Chestnut-bellied Flowerpiercer	U	2	14		14	14		Colombia	950a
Diglossa albilatera	White-sided Flowerpiercer	M	118	10.2	0.8	8.5	12.7		Colombia	324a
		F	87	9.5	0.8	8	11.4			
Diglossa lafresnayii	Glossy Flowerpiercer	U	29	16		13.1	19			559
Diglossa mystacalis albilinea	Moustached Flowerpiercer	B	4	16.2		16	16.5		Peru	1143
Diglossa gloriosa	Merida Flowerpiercer	U		11						522
Diglossa humeralis	Black Flowerpiercer	M	11	13.3	1	12	15.4		Colombia	324a
		F	4	13		12.4	13.6			
Diglossa brunneiventris	Black-throated Flowerpiercer	U	72	12		9.8	15			559
Diglossa carbonaria	Gray-bellied Flowerpiercer	U	22	11		9	14.8			559
Diglossa duidae	Scaled Flowerpiercer	M	20	15.4	1.7	11.9	18.5		Venezuela	1308
		F	12	14.3	0.8	12.9	16			
Diglossa major disjuncta	Greater Flowerpiercer	B	4	22.2		17.5	25		Venezuela	1143
Diglossopis indigotica	Indigo Flowerpiercer	U	12	13.7		10.5	18		Colombia	1037

Body Masses of World Birds (continued)

Species Name	Common Name	Sex	N	Mean	S.D.	Min	Max	Season	Location	Source #
Diglossopis glauca	Deep-blue Flowerpiercer	U	37	12		9.5	13			559
Diglossopis caerulescens	Bluish Flowerpiercer	M	23	14.6	0.6	13.5	15.8		Colombia	324a
		F	12	14.3	1.1	12.9	15.6			
Diglossopis cyanea	Masked Flowerpiercer	M	51	17.4	1.1	14.9	19.6		Colombia	324a
		F	23	16.7	1	15.4	19			
Sicalis flaveola	Saffron Finch	U	21	19.7	1.5	17	23.4		Venezuela	1215
Sicalis flaveola	Saffron Finch	B	28	15.4		13.1	17.8		Bolivia	1055
Sicalis flaveola pelzelni	Saffron Finch	U	7	15.9		15	16.8			255, 363
Sicalis luteola	Grassland Yellow-Finch	U	11	15.9	0.9	14.6	17.9		Argentina	210
Sicalis citrina	Stripe-tailed Yellow-Finch	M	4	11.6		10.7	12.9		Suriname	498
Sicalis uropygialis	Bright-rumped Yellow-Finch	B	5	22.5		21	24.4		Peru	1043
Sicalis auriventris	Greater Yellow-Finch	U	10	55		50	59			363
Sicalis olivascens	Greenish Yellow-Finch	U	8	21.8		17.5	24.2			363, 776a
Sicalis lebruni	Patagonian Yellow-Finch	U	4	25.3		23.1	26.6		Tierra del Fuego	546
Sicalis columbiana	Orange-fronted Yellow-Finch	B	21	13.4	1.1	11.6	16.1		Colombia	1143
Sicalis raimondii	Raimondi's Yellow-Finch	F	1	15.8					Peru	1043
Emberizoides herbicola herbicola	Wedge-tailed Grass-Finch	M	9	30.1		26.2	36.6		Argentina	255
		F	5	27.7		25.8	29.4			
Emberizoides herbicola sphenurus	Wedge-tailed Grass-Finch	B	18	25.1		21	28		Suriname	498
Emberizoides ypiranganus	Lesser Grass-Finch	B	5	20.1		18.5	22			330

Body Masses of World Birds (continued)

Species Name	Common Name	Sex	N	Mean	S.D.	Min	Max	Season	Location	Source #
Embernagra platensis	Great Pampa-Finch	B	5	45.5		45	46.2			255, 363, 1041
Gubernatrix cristata	Yellow Cardinal	B	3	47.6		46	50.8		Brazil; Argentina	64, 255
Paroaria coronata	Red-crested Cardinal	M	4	39.9		34.6	44		Bolivia; Brazil	64, 657
		F	2	35.8		34.5	37			
Paroaria dominicana	Red-cowled Cardinal	U		33.2					Brazil	1053
Paroaria gularis	Red-capped Cardinal	U	9	23.6		20.8	31		Venezuela; Peru	318, 1215
Paroaria capitata	Yellow-billed Cardinal	U	4	22.3		22	23.2		Argentina; Brazil	210a, 883
Lysurus crassirostris	Sooty-faced Finch	M	1	37					Eastern Ecuador	1147
		F	1	40.5						
Lysurus castaneiceps	Olive Finch	M	3	36.9		34.5	39		Peru; Colombia	370, 1037
Pselliophorus tibialis	Yellow-thighed Finch	U	10	30	2.3	26.8	34.8		Panama	675
Pezopetes capitalis	Large-footed Finch	B	2	55.8		55.8	55.9		Panama	489
Atlapetes albinucha	White-naped Brush-Finch	B	5	33.1		31	36		Mexico; El Salvador	1041a, 1145
Atlapetes albinucha gutturalis	White-naped Brush-Finch	U	10	34.6	1.9	31.7	37.2		Panama	675
Atlapetes pallidinucha	Pale-naped Brush-Finch	M	19	36.1	4.2	21.4	40.2		Colombia	324a
		F	6	33.1	4.9	24.9	38.3			
Atlapetes rufinucha	Rufous-naped Brush-Finch	M	9	22.2					Bolivia	963
		F	16	20.9						
Atlapetes latinuchus elaeoprorus	Yellow-breasted Brush-Finch	M	1	27.5						298a
Atlapetes latinuchus nigrifrons	Yellow-breasted Brush-Finch	U	1	31						298a
Atlapetes latinuchus yariguierum	Yellow-breasted Brush-Finch	M	1	22.6					Colombia	298a
		F	1	16						

Body Masses of World Birds (continued)

Species Name	Common Name	Sex	N	Mean	S.D.	Min	Max	Season	Location	Source #
Atlapetes pileatus	Rufous-capped Brush-Finch	U	12	24	2.3	21.5	27.5	B	Oaxaca, Mexico	831
Atlapetes melanocephalus	Santa Marta Brush-Finch	B	6	25						298a
Atlapetes fuscoolivaceus	Dusky-headed Brush-Finch	M	2	32.5						298a
Atlapetes tricolor	Tricolored Brush-Finch	U	8	36.5		29.5	40		Peru; Colombia	1037, 1288
Atlapetes albofrenatus albofrenatus	Moustached Brush-Finch	B	6	30					Colombia	298a
Atlapetes schistaceus	Slaty Brush-Finch	B	140	29.8	2.6	22.6	43.9		Colombia	324a
Atlapetes schistaceus tamae	Slaty Brush-Finch	B	7	26.4						298a
Atlapetes nationi	Rusty-bellied Brush-Finch	B	4	41.6		40.5	43.4		Peru	1043
Atlapetes leucopterus	White-winged Brush-Finch	B	10	20.8					Peru; Ecuador	953, 1302
Atlapetes semirufus	Ochre-breasted Brush-Finch	M	3	30.7		29	33			1043
Atlapetes fulviceps	Fulvous-headed Brush-Finch	U	1	28.2						363
Atlapetes personatus	Tepui Brush-Finch	M	18	33.8	2.2	30	37		Venezuela	1308
		F	8	31.6	1.8	29.2	33.7			
Atlapetes citrinellus	Yellow-striped Brush-Finch	M	1	28					Argentina	1040
Buarremon brunneinucha	Chestnut-capped Brush-Finch	U	14	46.6	3.3				Peru	1288
Buarremon brunneinucha	Chestnut-capped Brush-Finch	B	19	41.5	4	32.5	47		El Salvador	1041a
Buarremon virenticeps	Green-striped Brush-Finch	M	1	41						970
Buarremon torquatus	Stripe-headed Brush-Finch	M	47	40.5	2.9	32.2	44.9		Colombia	324a
		F	16	40.9	3.1	35.9	45.3			
Buarremon torquatus atricapillus	Stripe-headed Brush-Finch	U	9	44.6		40	49		Panama	79

Body Masses of World Birds (continued)

Species Name	Common Name	Sex	N	Mean	S.D.	Min	Max	Season	Location	Source #
Arremon aurantiirostris	Orange-billed Sparrow	U	40	34.5	2.3				Costa Rica	618
Arremon taciturnus	Pectoral Sparrow	U	42	24.8		22	27.8		Brazil	875
Arremon schlegeli	Golden-winged Sparrow	B	10	27.7	2.8	23	32		Venezuela	1254
Arremon abeillei	Black-capped Sparrow	B	10	25.9					Peru	1302
Arremon flavirostris	Saffron-billed Sparrow	U	11	30.9	1.2	29	33		Brazil	177
Arremon flavirostris dorbignii	Saffron-billed Sparrow	U	6	22.1		20	25			363
Arremon franciscanus	San Francisco Sparrow	M	5	24.9		21	25		Minas Gerais, Brazil	1142b, 1146
Arremonops rufivirgatus	Olive Sparrow	M	47	23.2	2.8	17	30		South Texas, USA	145
		F	20	21.7	2.6	15	26			
Arremonops chloronotus	Green-backed Sparrow	B	11	27.4	2.3	24.2	31.2		Belize	1027
Arremonops conirostris	Black-striped Sparrow	M	7	42.3					Panama	489
		F	10	37.2	2.2					
Arremonops conirostris	Black-striped Sparrow	U	18	29.4	1.9	26	32		Venezuela	1254
Melozone kieneri	Rusty-crowned Ground-Sparrow	B	11	38.7		35	41		Mexico	1080
Melozone biarcuata	Prevost's Ground-Sparrow	U	6	30.6		24.8	32		El Salvador	1041a
Melozone leucotis	White-eared Ground-Sparrow	U	5	39.9					Costa Rica; El Salvador	618, 1041a
Pipilo chlorurus	Green-tailed Towhee	B	68	29.4	3.3	21.5	37.3		Arizona, USA	314
Pipilo ocai	Collared Towhee	M	4	64.5		61	68	B	Oaxaca, Mexico	831
		F	5	57.6		54.5	62.5			
Pipilo socorroensis	Socorro Towhee	F	1	28.1					Mexico	1151
Pipilo erythrophthalmus	Eastern Towhee	M	94	41.3	2.4	34	46.6		Pennsylvania, USA	828
		F	60	38.8	2.3	34.8	44.8			

Body Masses of World Birds (continued)

Species Name	Common Name	Sex	N	Mean	S.D.	Min	Max	Season	Location	Source #
Pipilo maculatus	Spotted Towhee	U	21	39	3.6	30.6	45.5	Y	Arizona, USA	313
Pipilo maculatus falcifer	Spotted Towhee	M	22	40.9	2.1	35.9	43.8	F	California, USA	209
		F	16	38	2.2	34.3	41.4			
Pipilo crissalis crissalis	California Towhee	M	43	53.9		48.6	61.2	Y	California, USA	257
		F	29	51.8		46.3	61.2			
Pipilo fuscus	Canyon Towhee	B	98	44.4	3.2	36.6	52.5	Y	Arizona, USA	314
Pipilo aberti	Abert's Towhee	M	37	47.1	3.3	40	54.1		Arizona, USA	314
		F	25	44.8	2.9	39.5	51			
Pipilo albicollis	White-throated Towhee	B	8	46.4		42	50		Mexico	257
Aimophila mystacalis	Bridled Sparrow	M	2	22.2		21.9	22.5		Oaxaca, Mexico	75
Aimophila humeralis	Black-chested Sparrow	B	8	23.9		18	26		Mexico	75, 1145, 1043
Aimophila ruficauda	Stripe-headed Sparrow	M	5	28.2		26.6	30		Mexico	1080, 1151
Aimophila sumichrasti	Cinnamon-tailed Sparrow	M	1	30.3					Oaxaca, Mexico	584
		F	1	28.6						
Aimophila strigiceps	Stripe-capped Sparrow	B	10	22.2		20.5	23.7		Bolivia	657
Aimophila stolzmanni	Tumbes Sparrow	M	1	27.6					Peru	1043
		F	1	23.7						
Aimophila aestivalis	Bachman's Sparrow	B	105	18.4	2.2				South Carolina, USA	1171
Aimophila aestivalis	Bachman's Sparrow	B	49	20.8	5.2				Arkansas, USA	316
Aimophila botterii	Botteri's Sparrow	B	52	19.9	1.9	15.7	25.5	B	Arizona, USA	314
Aimophila cassinii	Cassin's Sparrow	B	92	18.9	1.5	14	23.5	Y	Arizona, USA	314
Aimophila ruficeps	Rufous-crowned Sparrow	M	59	19.3	1.5	16	23.3			1332
		F	39	18.1	1.3	15.2	20.3			
Aimophila carpalis	Rufous-winged Sparrow	B	34	15.3	1	13.3	17.4		Arizona, USA	27

Body Masses of World Birds (continued)

Species Name	Common Name	Sex	N	Mean	S.D.	Min	Max	Season	Location	Source #
Aimophila quinquestriata	Five-striped Sparrow	M	45	20.3	1.2				Arizona, USA	458, 1332
		F	4	18.8		17.9	19.5			
Aimophila notosticta	Oaxaca Sparrow	M	1	27.1					Oaxaca, Mexico	1145
Aimophila rufescens	Rusty Sparrow	B	16	34.3		28.2	38.9		Nicaragua; Belize; Mexico	542
Oriturus superciliosus	Striped Sparrow	B	14	41.4		36.9	53.7		Mexico	75, 259a, 910
Torreornis inexpectata	Zapata Sparrow	B	128	25.5	1.8	19.7	29.3		Cuba	19
Spizella arborea	American Tree Sparrow	M	64	18.6	2	14.9	27.7		Canada	838
		F	40	17.1	1.8	12.6	20.8			
Spizella passerina	Chipping Sparrow	U	195	12.2	0.8	10.3	14.5		Pennsylvania, USA	828
Spizella pallida	Clay-colored Sparrow	B	375	11.2	1.1	9	15	B	Minnesota, USA	1324
Spizella breweri	Brewer's Sparrow	B	83	10.9	0.7			B	Idaho, USA	920
Spizella pusilla	Field Sparrow	U	183	12.5	0.8	11	15.3	Y	Pennsylvania, USA	828
Spizella wortheni	Worthen's Sparrow	M	15	13	1.1	11	15	W	Coahuila, Mexico	163b
		F	24	11.8	1.2	9.5	14.5			
Spizella atrogularis	Black-chinned Sparrow	B	79	11.3	1.1	9	14.8			1209
Pooecetes gramineus	Vesper Sparrow	M	28	26.5				Sp	Michigan, USA	260
		F	15	24.9						
Chondestes grammacus	Lark Sparrow	B	49	29	1.9	24.7	33.3		California, USA	314
Amphispiza bilineata	Black-throated Sparrow	B	89	13.5	1.2	10.2	16.4	W	Arizona, USA	314
Amphispiza belli nevadensis	Sage Sparrow	M	50	18.5	0.2	15.9	21.9		Oregon, USA	762
		F	27	17.7	0.3	15.3	20.9			
Amphispiza belli belli	Sage Sparrow	B	48	15		12.7	16.8	B	California, USA	762
Calamospiza melanocorys	Lark Bunting	B	40	37.6	3.7	29.5	51.5		Arizona, USA	314
Passerculus sandwichensis	Savannah Sparrow	M	71	20.6	1.4			B	Manitoba, Canada	1280
		F	35	19.5	2.3					

Body Masses of World Birds (continued)

Species Name	Common Name	Sex	N	Mean	S.D.	Min	Max	Season	Location	Source #
Passerculus sandwichensis oblitus	Savannah Sparrow	M	417	18	1.4	13	23	B	Minnesota, USA	1324
		F	231	17.1	1.6	13.5	23			
Passerculus sandwichensis	Savannah Sparrow	F	10	17	0.8	15.1	17.8	B	California, USA	1310
Passerculus sandwichensis princeps	Savannah Sparrow	B	103	26				B	Sable Is., Canada	1280
Passerculus sandwichensis beldingi	Savannah Sparrow	M	20	20.4	0.8	19.2	22.5	B	Baja California Norte, Mexico	1292a
Passerculus sandwichensis atratus	Savannah Sparrow	M	18	22.7	1.1	21	25.2	B	Sonora, Mexico	1292a
Ammodramus maritimus maritimus	Seaside Sparrow	M	165	24.3	1.4	19.1	28.7	B	New York, USA	932
		F	119	22.2	1.7	19.4	28.5			
Ammodramus maritimus peninsulae	Seaside Sparrow	M	17	22.6	1.4	20	25.8	B	Florida, USA	932
		F	10	19.8	1.1	18.9	23.2			
Ammodramus nelsoni	Nelson's Sharp-tailed Sparrow	M	13	14.8	0.8	13.7	16.2		North Dakota, USA	454
		F	6	13.9		12.8	15.2			
Ammodramus nelsoni	Nelson's Sharp-tailed Sparrow	B	27	16.8		15.8	20		Ontario, Canada	454
Ammodramus caudacutus	Saltmarsh Sharp-tailed Sparrow	M	33	20.7		18	23.1	B	New Jersey, USA	1339
		F	14	17.8		15.3	19			
Ammodramus lecontei	Le Conte's Sparrow	B	91	13	1.2	10	16	B	Minnesota, USA	1324
Ammodramus henslowii	Henslow's Sparrow	B	107	12.8		10.6	15			512
Ammodramus bairdii	Baird's Sparrow	B	21	17.5	1.4	15	20.3	Sp	Arizona, USA	1150
Ammodramus bairdii	Baird's Sparrow	M	57	18.9					Saskatchewan, Canada	451
		F	13	17.8						
Ammodramus savannarum	Grasshopper Sparrow	B	60	17	2.8	13.4	28.4	W	Arizona, USA	594
Ammodramus savannarum floridanus	Grasshopper Sparrow	M	25	17.2	1			B	Florida, USA	1255
		F	5	18.4						

Body Masses of World Birds (continued)

Species Name	Common Name	Sex	N	Mean	S.D.	Min	Max	Season	Location	Source #
Ammodramus savannarum pratensis	Grasshopper Sparrow	B	9	17.9				B	Connecticut, USA	1255
Ammodramus humeralis	Grassland Sparrow	M	13	16.5		16	18		Suriname	498
		F	7	15.5		13	19			
Ammodramus aurifrons	Yellow-browed Sparrow	U	9	17.8		14.5	19			73, 879, 881, 953, 1311
Passerella iliaca	Fox Sparrow	U	91	32.3		21.7	42.1	W	California, USA	696
Passerella iliaca iliaca	Fox Sparrow	U	85	35.3	2.1	30.2	39.2	M	Pennsylvania, USA	828
Passerella iliaca unalaschensis	Fox Sparrow	B	34	36.1		31	44	B		1282
Passerella iliaca schistacea	Fox Sparrow	B	28	30.4		26	35	B		1282
Passerella iliaca megarhyncha	Fox Sparrow	B	11	32.5		29	38	B		1282
Xenospiza baileyi	Sierra Madre Sparrow	M	1	17.4					Morelos, Mexico	1145
Melospiza melodia	Song Sparrow	M	50	20.5	1.1	18	22.6	Y	Pennsylvania, USA	828
		F	120	19.5	1.3	17.1	23.4			
Melospiza melodia maxima	Song Sparrow	B	4	42.4		40.2	45.7		Alaska, USA	415
Melospiza melodia gouldii	Song Sparrow	U	39	20	0.9				California, USA	758
Melospiza melodia maxillaris	Song Sparrow	U	27	20.8	0.9				California, USA	758
Melospiza melodia pusillula	Song Sparrow	U	37	18.6	0.8				California, USA	758
Melospiza melodia samuelis	Song Sparrow	U	17	18.8	1.2				California, USA	758
Melospiza melodia micronyx	Song Sparrow	U		21.6					California, USA; calculated	1138
Melospiza lincolnii	Lincoln's Sparrow	B	65	16.6	1.2	14.3	19.4	M	Pennsylvania, USA	828
Melospiza georgiana	Swamp Sparrow	U	184	16.1	1.2	13.1	19.2	M	Pennsylvania, USA	828

Body Masses of World Birds (continued)

Species Name	Common Name	Sex	N	Mean	S.D.	Min	Max	Season	Location	Source #
Zonotrichia leucophrys leucophrys	White-crowned Sparrow	B	162	29.4	1	21.6	38.5	M	Pennsylvania, USA	193
Zonotrichia leucophrys gambelii	White-crowned Sparrow	B	50	25.5	1.7	21	28.5	W	California, USA	314
Zonotrichia leucophrys nuttalli	White-crowned Sparrow	B	50	32	2.2	27	35.5	W	California, USA	314
Zonotrichia leucophrys pugetensis	White-crowned Sparrow	B	50	25.3	1.8	21.4	29.1	W	California, USA	314
Zonotrichia leucophrys oriantha	White-crowned Sparrow	B	50	28.4	1.9	23.3	33.7	B	Oregon, USA	314
Zonotrichia albicollis	White-throated Sparrow	B	704	24.4	1.7	19.2	30.4		Pennsylvania, USA	828
Zonotrichia atricapilla	Golden-crowned Sparrow	M	27	31.7	2.2			B	British Columbia, Canada	857
		F	14	32.3	0.9					
Zonotrichia capensis australis	Rufous-collared Sparrow	M	17	21.9		19	25		Argentina	842
		F	15	20.3		17.2	23.2			
Zonotrichia capensis choraules	Rufous-collared Sparrow	M	14	20.3		18	24.7		Argentina	842
		F	4	19		17.2	21.2			
Zonotrichia capensis	Rufous-collared Sparrow	M	22	20.8		16.8	24		Bolivia	1055
		F	14	20		17.1	31			
Zonotrichia capensis	Rufous-collared Sparrow	U	85	20	1.4				El Salvador	646a
Zonotrichia querula	Harris' Sparrow	M	41	37.4	2			B	Northwest Terr., Canada	858
		F	40	33.7	2.2					
Junco hyemalis	Dark-eyed Junco	M	1288	19.3	1	15.6	24.1		Pennsylvania, USA	828
		F	711	17.9	1	15.2	23			
Junco hyemalis cismontanus	Dark-eyed Junco	M	7	19	1	17.9	20.2		Pennsylvania, USA	828
		F	15	17.5	0.9	15.8	19			
Junco hyemalis carolinensis	Dark-eyed Junco	B	48	20.8		18	23.5		Pennsylvania, USA	828
Junco hyemalis mearnsi	Dark-eyed Junco	B	221	18.2	1.3	15.5	23.5	W	Arizona, USA	314
Junco hyemalis dorsalis	Dark-eyed Junco	B	170	21.8	1.4	18	26	W	Arizona, USA	314

Body Masses of World Birds (continued)

Species Name	Common Name	Sex	N	Mean	S.D.	Min	Max	Season	Location	Source #
Junco hyemalis caniceps	Dark-eyed Junco	B	40	19.6	1.1	18	23	W	Arizona, USA	314
Junco hyemalis aikeni	Dark-eyed Junco	M	19	22						856
Junco phaeonotus	Yellow-eyed Junco	B	57	20.4	1.1	18.3	22		Arizona, USA	314
Junco vulcani	Volcano Junco	U		28					Costa Rica	1170
Calcarius mccownii	McCown's Longspur	M	87	26.7				B	Saskatchewan, Canada	1327
		F	48	24.7						
Calcarius lapponicus	Lapland Longspur	M	47	27.4	1.5	24.7	30.7	B	Nunavut, Canada	548
		F	52	28.3	2.9	22.5	35			
Calcarius pictus	Smith's Longspur	M	39	28.1		21	32		Manitoba, Canada	117
		F	29	25.3		19.7	29			
Calcarius ornatus	Chestnut-collared Longspur	B	232	20.3		19.3	22.1	B	Saskatchewan, Canada	519
Plectrophenax nivalis	Snow Bunting	B	35	42.2	5.5	34	56	W	Alaska, USA	314
Plectrophenax hyperboreus	McKay's Bunting	B	11	54.5	7.1	38	62	W	Alaska, USA	314
Geospiza magnirostris	Large Ground-Finch	U	24	34.5		27	39		Galapagos Is.	107
Geospiza fortis	Medium Ground-Finch	U	85	24		18	32		Galapagos Is.	107
Geospiza fuliginosa	Small Ground-Finch	U	33	14.5		12	17		Galapagos Is.	107
Geospiza difficilis	Sharp-beaked Ground-Finch	U		12.3					Galapagos Is.	1
Geospiza scandens	Common Cactus-Finch	U	5	22.7		20	24		Galapagos Is.	107
Geospiza conirostris	Large Cactus-Finch	U		27.6					Galapagos Is.	1
Camarhynchus crassirostris	Vegetarian Finch	U	30	35.5		29	40		Galapagos Is.	107
Camarhynchus heliobates	Mangrove Finch	U		18					Galapagos Is.	173
Camarhynchus psittacula	Large Tree-Finch	U	42	18.7		15	21		Galapagos Is.	107

Body Masses of World Birds (continued)

Species Name	Common Name	Sex	N	Mean	S.D.	Min	Max	Season	Location	Source #
Camarhynchus parvulus	Small Tree-Finch	U	64	13.2		11	16		Galapagos Is.	107
Camarhynchus pauper	Medium Tree-Finch	U		16					Galapagos Is.	173
Camarhynchus pallidus	Woodpecker Finch	U	16	23.1		20	31		Galapagos Is.	107
Certhidea olivacea	Warbler Finch	U	14	9.5		8	12		Galapagos Is.	107
Family Cardinalidae										
Periporphyrus erythromelas	Red-and-black Grosbeak	M	1	48					Suriname	498
Saltator albicollis	Lesser Antillean Saltator	U	25	36.9		30	44		Trinidad	359
Saltator albicollis	Lesser Antillean Saltator	B	20	46.3	3.7	39.8	51.6		St. Lucia	19
Saltator striatipectus	Streaked Saltator	B	2	39		36.5	41.5		Peru; Colombia	1143
Saltator coerulescens	Grayish Saltator	U	21	54.9	5	49	67		Venezuela	1215
Saltator maximus	Buff-throated Saltator	M	17	46.2	1.7				Panama	489
		F	13	49.1	3					
Saltator atriceps	Black-headed Saltator	B	17	79.7		68.2	91.2		Yucatan, Mexico	910a
Saltator atriceps	Black-headed Saltator	U	16	88.1	13.6	63	116		El Salvador	1041a
Saltator grossus	Slate-colored Saltator	B	24	44.2		41	53		Venezuela; Suriname	498, 1308
Saltator atripennis	Black-winged Saltator	U	10	55		44	60		Colombia	798, 1037
Saltator similis	Green-winged Saltator	U	21	43.3	4.3	36	54		Brazil	177
Saltator orenocensis	Orinocan Saltator	U	6	33.7						95, 1215
Saltator aurantiirostris	Golden-billed Saltator	M	9	39		33.1	44.8		Bolivia	1055
		F	11	45		38	54.3			

Body Masses of World Birds (continued)

Species Name	Common Name	Sex	N	Mean	S.D.	Min	Max	Season	Location	Source #
Saltator maxillosus	Thick-billed Saltator	B	4	50.8		48	54		Brazil	64
Saltator cinctus	Masked Saltator	B	7	48.6		43	53		Peru	873
Saltator atricollis	Black-throated Saltator	M	1	53.8					Brazil	752
Caryothraustes poliogaster	Black-faced Grosbeak	B	11	41.8	3.2	34.9	46.7		Belize; Mexico	910a, 1027
Caryothraustes canadensis	Yellow-green Grosbeak	B	6	34.5		31	36			497, 990, 1178
Parkerthraustes humeralis	Yellow-shouldered Grosbeak	F	1	37					Brazil	1178
Rhodothraupis celaeno	Crimson-collared Grosbeak	F	1	60					Mexico	1153
Cardinalis cardinalis	Northern Cardinal	M	38	43.5	2.9	37.1	52.2		Pennsylvania, USA	828
		F	52	41.8	2.6	36.2	47.5			
Cardinalis sinuatus	Pyrrhuloxia	M	191	35.8	2.4	30.1	44.2		Arizona, USA	1236
		F	131	34.6	2.6	24.4	43.2			
Pheucticus chrysopeplus	Yellow Grosbeak	F	1	77.6					Mexico	1151
Pheucticus chrysogaster	Golden-bellied Grosbeak	B	6	55.8		54	59.7		Peru	1143, 1043, 1288, 1302
Pheucticus tibialis	Black-thighed Grosbeak	B	12	62.3					Panama	489
Pheucticus aureoventris	Black-backed Grosbeak	B	19	49		43.8	57.2		Bolivia	1055
Pheucticus ludovicianus	Rose-breasted Grosbeak	B	359	42		34.4	50.6		Pennsylvania, USA	828
Pheucticus melanocephalus	Black-headed Grosbeak	M	151	46.1	3.3				New Mexico, USA	520
		F	156	48.2	4.8					
Cyanocompsa brissonii	Ultramarine Grosbeak	U	18	27.5		27	28			363
Cyanocompsa parellina	Blue Bunting	M	85	16.1	2.1	11	24		El Salvador	1041a
		F	73	15.2	2.7	9.8	21			

Body Masses of World Birds (continued)

Species Name	Common Name	Sex	N	Mean	S.D.	Min	Max	Season	Location	Source #
Cyanocompsa cyanoides	Blue-black Grosbeak	U	44	32.5						618
Cyanoloxia glaucocaerulea	Glaucous-blue Grosbeak	U	3	18.2		16	19.5		Brazil	64, 878
Passerina caerulea	Blue Grosbeak	M	219	27.8	2.3	22	40.5	B		277
		F	84	27	2	23	32.5			
Passerina amoena	Lazuli Bunting	M	58	16		13	19.5	Sp	California, USA	1091
		F	25	15		12.7	16.9			
Passerina cyanea	Indigo Bunting	M	1111	15	0.7	12.5	17.5	B	Michigan, USA	907
		F	398	14.4	9.6	11.9	18.5			
Passerina rositae	Rose-bellied Bunting	B	2	20		19.5	20.5		Chiapas, Mexico	1145
Passerina leclancherii	Orange-breasted Bunting	B	7	14		12	15.1		Mexico	1145
Passerina versicolor	Varied Bunting	M	104	13	1.8					459
		F	53	12.8	1.7					
Passerina ciris	Painted Bunting	M	116	16.1		13.3	19		Florida, USA	364
		F	131	15		12.9	19			
Spiza americana	Dickcissel	M	22	28.5	0.7			B		1208
		F	9	25.2						
Spiza americana	Dickcissel	M	1616	28.1	1.8			W	Venezuela	1208
		F	1169	23.3	1.5					
Family Icteridae										
Dolichonyx oryzivorus	Bobolink	M	142	33.9	2.1			B	New York, USA	764
		F	130	29.2	2.1					
Agelaius flavus	Saffron-cowled Blackbird	M	1	43					Brazil	64
Agelaius thilius	Yellow-winged Blackbird	M	20	35.2	2.2			B	Buenos Aires, Argentina	766b
		F	16	28.2	2.8					
Agelaius xanthophthalmus	Pale-eyed Blackbird	U	24	45.1					Peru	370
Agelaius cyanopus	Unicolored Blackbird	M	1	43						64, 210a
		F	1	32.2						

Body Masses of World Birds (continued)

Species Name	Common Name	Sex	N	Mean	S.D.	Min	Max	Season	Location	Source #
Agelaius phoeniceus	Red-winged Blackbird	M	122	65.3	3.6	55.4	72.1		Pennsylvania, USA	828
		F	15	39.5	2.3	34.5	42.7			
Agelaius assimilis	Red-shouldered Blackbird	M	23	54.4	2.4				Cuba	1300
		F	9	41.8						
Agelaius tricolor	Tricolored Blackbird	M	52	68.2	3.6	60	79	B	California, USA	314
		F	5	49.2		46	54.5			
Agelaius icterocephalus	Yellow-hooded Blackbird	M	27	35.4	2	31.5	40		Trinidad	359
		F	17	26.6		24	31			
Agelaius humeralis	Tawny-shouldered Blackbird	M	10	38.3	1.9				Cuba	1300
		F	8	34.4						
Agelaius xanthomus	Yellow-shouldered Blackbird	B	8	38.4					Puerto Rico	1306
Agelaius ruficapillus	Chestnut-capped Blackbird	M	6			39	44		Brazil	64
		F	2	32		32	32			
Nesopsar nigerrimus	Jamaican Blackbird	U		39						348
Sturnella militaris	Red-breasted Blackbird	M	9	46.2					Panama	489
		F	4	35.5						
Sturnella superciliaris	White-browed Blackbird	M	3	52		50	53		Brazil; Argentina	64, 212
		F	1	39.5						
Sturnella bellicosa	Peruvian Meadowlark	U	1	58					Southern Ecuador	1311
Sturnella defilippii	Pampas Meadowlark	M	1	67.5					Argentina	1097
Sturnella loyca	Long-tailed Meadowlark	U		113					Chile	582
Sturnella magna	Eastern Meadowlark	M	20	102	11.2				Florida, USA	488
		F	9	76						
Sturnella magna lilianae	Eastern Meadowlark	M	7	112		104	131			671
		F	5	90.3		86	94.2			
Sturnella magna alticola	Eastern Meadowlark	M	24	98.3		86	110		Mexico	671
		F	24	77.5		65.6	88.7			

Body Masses of World Birds (continued)

Species Name	Common Name	Sex	N	Mean	S.D.	Min	Max	Season	Location	Source #
Sturnella neglecta	Western	M	51	112				B	Great Plains, USA	1303
	Meadowlark	F	32	89.4						
Xanthocephalus xanthocephalus	Yellow-headed	M	3	79.7		72.5	85.5		Oregon, USA	314
	Blackbird	F	10	49.3	4	42.4	56			
Dives atroviolaceus	Cuban Blackbird	M	6	90.3		85	94		Cuba	19
		F	10	77.4		72	80			
Dives dives	Melodious	M	3	100		98.3	102		Belize; Mexico	259a, 1027
	Blackbird	F	1	83.4						
Dives warszewiczi	Scrub Blackbird	U	3	73.3					Southern Ecuador	1311
Euphagus carolinus	Rusty Blackbird	M	91	64.3	2.4	45.9	80.4	M	Pennsylvania, USA	193
		F	105	55.2	3.6	47	76.5			
Euphagus cyanocephalus	Brewer's	M	19	67.2	3.2	60	73		Oregon, USA	314
	Blackbird	F	15	58.1	4.9	50.6	67			
Quiscalus major	Boat-tailed	M	143	206	14.3	154	239	B	South Carolina, USA	933
	Grackle	F	86	111	8.9	93	147			
Quiscalus quiscula	Common Grackle	M	298	120	7.1	91	142		Illinois, USA	914
		F	289	92.2	7.3	74	124			
Quiscalus mexicanus	Great-tailed	M	29	222	18.5	190	253			593
	Grackle	F	16	116	9.1	102	137			
Quiscalus nicaraguensis	Nicaraguan	M	2	84		82	86		Nicaragua	1148
	Grackle	F	5	47.8		44	50			
Quiscalus niger	Greater Antillean	M	89	86.4	5.8				Puerto Rico	1306
	Grackle	F	95	61.7	5					
Quiscalus niger	Greater Antillean	M	37	93.5	3.2	86.6	100		Cuba	19
	Grackle	F	13	65.9	3.4	60	72.5			
Quiscalus niger caymanensis	Greater Antillean	M	10	75.1	7.1	67.5	88		Cayman Is.	868
	Grackle	F	11	57.4	4.3	51	64			
Quiscalus lugubris	Carib Grackle	M	8	72.5		66	80		Martinique; Trinidad	359, 1065
		F	8	54.8		49	58			
Molothrus badius	Bay-winged	M	11	47.4		39.3	61.5		Bolivia	1055
	Cowbird	F	10	43.2		39.9	45.6			

Body Masses of World Birds (continued)

Species Name	Common Name	Sex	N	Mean	S.D.	Min	Max	Season	Location	Source #
Molothrus rufoaxillaris	Screaming Cowbird	M	4			56	65		Brazil	64
		F	2	47.5		38	57			
Molothrus bonariensis	Shiny Cowbird	B	149	40	6.5	22	53		Dominican Republic	19
Molothrus bonariensis minimus	Shiny Cowbird	M	479	38.7	1.1				Puerto Rico	1306
		F	670	31.9	1					
Molothrus bonariensis bonariensis	Shiny Cowbird	M	69	55.5	7.6				Argentina	726
		F	78	44.9	4.3					
Molothrus aeneus aeneus	Bronzed Cowbird	M	144	68.9	4.2			B	Texas, USA	718
		F	220	56.9	5.1					
Molothrus ater	Brown-headed Cowbird	M	222	48.7	2.6	43.1	57.3		Pennsylvania, USA	828
		F	20	38.1	2.5	32.3	42.3			
Molothrus ater artemisiae	Brown-headed Cowbird	M	283	47.5	3.7				California, USA	717
		F	352	37.6	3.6					
Molothrus ater obscurus	Brown-headed Cowbird	M	63	40.2	3.4				California, USA	717
		F	35	32	3.8					
Molothrus oryzivorus	Giant Cowbird	M	9	212		174	242			156, 212, 293, 359, 1027
		F	3	156		144	167			
Icterus chrysocephalus	Moriche Oriole	B	4	41.2		38	44		Suriname	496, 497
Icterus cayanensis	Epaulet Oriole	U	7	29.7					Paraguay; Bolivia	657, 918, 1055
Icterus cayanensis	Epaulet Oriole	B	3	42.3		40	44.5		Brazil	1109
Icterus chrysater	Yellow-backed Oriole	B	15	53.4	4.9	46.2	61.5		Colombia; Nicaragua; Belize; Panama	542, 795, 1027, 1180
Icterus nigrogularis	Yellow Oriole	B	16	40.2		31	50		Suriname, Venezuela, Trinidad	359, 496, 497, 1215
Icterus leucopteryx	Jamaican Oriole	B	4	36.9		34	41.7			1027, 1159
Icterus auratus	Orange Oriole	B	9	32.1		26.7	36.1		Yucatan, Mexico	640, 910a
Icterus mesomelas	Yellow-tailed Oriole	B	13	39.9	4.8	32.5	47.7		Belize; Mexico	640, 910a, 1027, 1145

Body Masses of World Birds (continued)

Species Name	Common Name	Sex	N	Mean	S.D.	Min	Max	Season	Location	Source #
Icterus auricapillus	Orange-crowned Oriole	M	2	32.2		31	33.5		Panama; Colombia	795, 1180
Icterus pectoralis	Spot-breasted Oriole	M	5	46.5		42.5	52		Western Mexico; El Salvador	1041a, 1151
Icterus gularis yucatanensis	Altamira Oriole	M	7	59.1		56.4	64.3	B	Yucatan, Mexico	910a
		F	3	51.5		47.4	53.6			
Icterus pustulatus microstictus	Streak-backed Oriole	B	38	37.8		31.8	41.6		Western Mexico	440
Icterus pustulatus graysoni	Streak-backed Oriole	B	26	35.8		29	39.4		Tres Marias Is., Mexico	440
Icterus cucullatus	Hooded Oriole	B	56	24.3	2	20.6	33.2	B	Arizona, USA	314
Icterus icterus	Troupial	M	23	77.4	6	65.5	88		Puerto Rico	19
		F	20	67.2	3.4	60.9	73.1			
Icterus icterus jamaicaii	Troupial	B	3	43.4		40.3	48		Bolivia	657
Icterus galbula	Baltimore Oriole	M	35	33.9	2	30.5	38.9		Pennsylvania, USA	828
		F	14	31.8	1.6	27.9	35.2			
Icterus bullockii	Bullock's Oriole	B	73	37.9		32.4	42.5	B	Oklahoma, USA	985
Icterus abeillei	Black-backed Oriole	B	2	33		31	35		Mexico	1153
Icterus spurius	Orchard Oriole	B	130	19.9		16.1	28		Nebraska, USA	1050
Icterus spurius fuertesi	Orchard Oriole	M	2	19		18.8	19.2		Mexico	1147a
Icterus melanopsis	Cuban Oriole	M	34	38.7	1.9	35	42		Cuba	404a
		F	25	35.4	2	30	38.5			
Icterus dominicensis	Hispaniolan Oriole	M	10	36.5	1.1	35	38.3		Hispaniola	404a
		F	9	35		33	40			
Icterus portoricensis	Puerto Rican Oriole	M	27	41	1.6	37.2	44.9		Puerto Rico	404a
		F	22	36.6	1.2	34	38.9			
Icterus prosthemelas	Black-cowled Oriole	M	9	29.5		26.2	32.1		Belize	1027
		F	6	26.5		22.9	29.1			
Icterus wagleri	Black-vented Oriole	B	9	41.8		36.7	46.3		Mexico	104, 1080, 1151

Body Masses of World Birds (continued)

Species Name	Common Name	Sex	N	Mean	S.D.	Min	Max	Season	Location	Source #
Icterus laudabilis	St. Lucia Oriole	B	5	36.9		34	40.5		St. Lucia	19
Icterus bonana	Martinique Oriole	U	1	29					Martinique	19
Icterus oberi	Montserrat Oriole	M	7	38		36	39		Montserrat	404a
		F	6	33.7		31.5	35.8			
Icterus graduacauda	Audubon's Oriole	B	25	42.2	4.9	31	52	B	Mexico	1145
Icterus maculialatus	Bar-winged Oriole	B	12	37.4	7.3	31	59		El Salvador	1041a
Icterus parisorum	Scott's Oriole	B	39	37.4	2.3	32.1	41		Arizona, USA	314
Icterus parisorum	Scott's Oriole	B	80	35.6				B	Texas, USA	372
Amblycercus holosericeus	Yellow-billed Cacique	M	1	86					Brazil	64
		F	1	75.5						
Amblycercus holosericeus	Yellow-billed Cacique	U	13	53.8		30	68		Belize	220, 580
Cacicus cela	Yellow-rumped Cacique	M	126	106	7.5				Peru	1007
		F	225	68.9	3.6					
Cacicus haemorrhous	Red-rumped Cacique	M	10	102						64, 293, 378, 496
		F	3	68.7		62	74			
Cacicus uropygialis	Scarlet-rumped Cacique	M	4	69.3		62.5	71.7		Panama	1180
		F	5	54		50	57.6			
Cacicus chrysopterus	Golden-winged Cacique	M	8	41		34.6	47.2		Brazil; Argentina	64, 211
		F	9	31.9		30.2	34			
Cacicus sclateri	Ecuadorian Cacique	F	1	49				B	Colombia	99
Cacicus solitarius	Solitary Cacique	M	8	86.8					Peru; Argentina	212, 370
		F	3	73.3		59.8	85			
Cacicus melanicterus	Yellow-winged Cacique	M	2	94.8		82.6	107		Mexico	911, 1151
		F	2	69.8		68.6	70.9			
Psarocolius oseryi	Casqued Oropendola	M	5	190					Peru	674
		F	4	99.8						

Body Masses of World Birds (continued)

Species Name	Common Name	Sex	N	Mean	S.D.	Min	Max	Season	Location	Source #
Psarocolius decumanus	Crested Oropendola	M	9	280						156, 211, 359, 489, 496, 497
		F	11	152		124	174			
Psarocolius viridis	Green Oropendola	M	3	383		330	460		French Guiana; Suriname	293, 496
		F	1	215						
Psarocolius atrovirens	Dusky-green Oropendola	F	1	152					Peru	370
Psarocolius angustifrons	Russet-backed Oropendola	M	2	266		254	278		Venezuela	200, 1043
		F	6	171		157	184			
Psarocolius angustifrons	Russet-backed Oropendola	M	1	440					Peru	1007
		F	6			195	215			
Psarocolius wagleri	Chestnut-headed Oropendola	M	9	214					Panama	489
		F	7	113						
Ocyalus latirostris	Band-tailed Oropendola	M	1	114					Colombia	963
		F	1	81						
Gymnostinops montezuma	Montezuma Oropendola	M	4	423		353	528		Belize; Panama	156, 1027, 1180
		F	6	227		198	254			
Gymnostinops bifasciatus yuracares	Amazonian Oropendola	M	1	490					Peru	370
		F	1	230						
Gymnostinops mexicanus	Oriole Blackbird	U	2	98		93	103		Venezuela	1043, 1215
Pseudoleistes guirahuro	Yellow-rumped Marshbird	M	6	91.2						380
		F	3	81.9						
Pseudoleistes virescens	Brown-and-yellow Marshbird	M	22	81.3	4.2				Argentina	791
		F	21	78.6	8.6					
Amblyramphus holosericeus	Scarlet-headed Blackbird	M	13	71	4.4				Panama	489
		F	4	56.4						
Amblyramphus holosericeus	Scarlet-headed Blackbird	M	1	54					Peru	1302
		F	2	48						
Curaeus curaeus	Austral Blackbird	U	14	83.7	3.7	76.9	89.9		Argentina	210
Gnorimopsar chopi	Chopi Blackbird	B	10	65.9	4.8	59	75		Mato Grosso, Brazil	883

Body Masses of World Birds (continued)

Species Name	Common Name	Sex	N	Mean	S.D.	Min	Max	Season	Location	Source #
Oreopsar bolivianus	Bolivian Blackbird	M	3	74.9		72.1	78		Bolivia	1143
		F	2	71		69	73.1			
Macroagelaius imthurni	Golden-tufted Grackle	M	4	83.2		74	89		Venezuela	1143
		F	3	72.7		64.5	80.5			

Part II

Body Masses and Composition of Migrant Birds in the Eastern United States

INTRODUCTION

Eugene P. Odum

In the 1960s, my students, colleagues, and I undertook an extensive study of lipid storage and use by migrating birds. As part of these studies, we compiled an extensive data set on avian body mass and composition, collected mostly from birds killed during migration. Although we have published numerous papers from this study (e.g., Connell et al. 1960, Odum 1960, Odum et al. 1961, Marshall and Baker 1964, Hicks 1967), the full data set has never been published. Since the data are relatively unique, and still requested by researchers, I present them in the following table.

A majority of the bird specimens that were analyzed in this study were spring and fall nocturnal migrants that were killed in collision with a television tower located in Leon County, Florida, between Thomasville, Georgia, and Tallahassee, Florida, near the Gulf coast. During the fall of 1956, Herbert Stoddard searched the grounds of the TV tower at dawn each morning from August to December, collecting the birds that had been killed the previous night. All specimens were immediately frozen, and then transported to the University of Georgia where the study was conducted.

Data for some of the species were collected from other locations. A large series was collected by L. D. Caldwell from a TV tower in Michigan (Caldwell et al. 1964). W. H. Drury donated a fall series of Blackpoll Warblers *(Dendroica striata)* from Massachusetts, all in premigratory and very fat condition. Chimney Swifts *(Chaetura pelagica)* were collected from a chimney roost in Athens, Georgia. Wintering and premigratory Savannah Sparrows *(Passerculus sandwichensis)* and fall premigratory hummingbirds were collected at the Savannah River Ecology Laboratory, Aiken County, South Carolina. Some individuals of other species were collected from TV towers in the southeastern United States; however, most specimens for these species came from the Leon County tower.

Upon receipt at the University of Georgia, the specimens were classified according to species, age and sex, and then weighed and measured. This first mass is referred to as the "wet mass." Aging was done by removing the skin on the top of the head and examining the degree of skull ossification. Specimens were sexed by plumage or gonadal inspection. After weighing, the birds were placed in a vacuum oven, which was set at a constant 40°C, and allowed to dry for 36–108 h, depending on the size of the specimen. When the bird was completely dry, it was weighed again, producing the "dry mass" (Marshall and Baker 1964).

To estimate the amount of fat in each specimen, each bird was then ground in a blender or meat chopper containing about 5 cm of 95% ethyl alcohol (Connell et al. 1960). The pulverized mixture was boiled in a beaker placed in a water bath, and then strained through a very fine mesh sieve. This boiling procedure was repeated 3 times with petroleum ether substituted for the alcohol, after which the residue was dried for 12 h. The residue was weighed again, and the difference between this last measurement and the dry mass was the amount of fat removed by the petroleum ether. The "fat-free mass" was calculated by subtracting the amount of fat from the wet mass. An alternate fat extraction method using chloroform was used on some specimens.

The nonfat dry residue was then burned in a muffle furnace at 600°C until burning eliminated all organic matter, reducing the residue to the ash content of the specimen (Odum et al. 1965). The "ash-free mass" was calculated by subtracting the mass of ash from the fat-free mass. This step was not done on all specimens.

The following table reports wet mass, dry mass, fat-free mass, and ash-free mass for 43 species. For each age and sex class, I report the sample size, mean, standard deviation, and range. Birds in their first calendar year (immatures) are classified as HY (hatching year) in the table. Adult birds are classified as AHY (after hatching year). In addition, I have included statistics for the entire sample for each species, including individual birds that were of unknown sex or age class. The addition of these individuals to the total sample results in slightly larger sample sizes in the total sample than can be summed from the individual age and sex classes.

I would like to express my special appreciation to Shirley Marshall who managed my lipid extraction laboratory with great efficiency during the period when most of these data were obtained.

Body Mass and Composition of North American Migrant Birds

	N	Mean	S.D.	Min	Max
Porzana carolina					
Male HY					
wet	6	71.14	9.85	59.40	84.13
dry	6	32.34	5.35	27.57	40.60
fat-free	6	57.44	8.78	45.17	64.78
Male AHY					
wet	4	68.56		58.21	73.20
dry	4	31.34		20.89	38.96
fat-free	4	54.84		50.84	61.91
Female HY					
wet	8	66.90	13.13	48.67	89.87
dry	8	35.68	9.58	24.41	49.21
fat-free	8	47.42	6.95	36.70	61.24
Female AHY					
wet	10	65.49	8.81	52.70	78.10
dry	10	31.09	9.13	17.50	44.21
fat-free	10	49.99	2.30	46.20	54.89
All individuals					
wet	28	67.54	9.94	48.67	89.87
dry	28	32.70	8.32	17.50	49.21
fat-free	28	51.55	6.83	36.70	64.78
Cuculus americanus					
Male HY					
wet	12	72.09	10.88	51.60	86.34
dry	12	41.25	10.34	22.03	55.68
fat-free	12	46.67	2.93	42.16	52.08
ash-free	11	44.71	2.85	40.49	49.40
Male AHY					
wet	19	77.28	11.31	54.97	88.91
dry	19	46.50	13.03	22.20	59.07
fat-free	19	46.04	4.29	37.00	55.88
ash-free	17	44.14	4.36	35.51	53.84
Female HY					
wet	13	68.56	14.62	50.50	89.23
dry	13	36.64	13.97	19.70	59.04
fat-free	13	45.19	6.27	31.56	52.29
ash-free	9	46.23	2.81	43.06	50.24
Female AHY					
wet	14	75.07	15.64	50.34	93.50
dry	14	41.13	16.25	17.27	62.02
fat-free	14	49.84	3.38	42.62	55.13
ash-free	14	47.88	3.27	40.95	53.22
All individuals					
wet	77	71.51	13.49	49.25	93.50
dry	77	39.71	13.88	17.27	62.02
fat-free	77	46.80	4.49	31.56	55.88
ash-free	63	45.56	3.86	35.51	53.84

Body Mass and Composition of North American Migrant Birds (continued)

	N	Mean	S.D.	Min	Max
Chaetura pelagica					
Male HY					
wet	13	30.37	1.23	28.00	32.41
dry	13	16.42	1.37	13.10	18.62
fat-free	13	19.71	0.84	18.03	21.00
Female HY					
wet	8	30.34	1.29	27.87	31.65
dry	8	15.96	0.88	14.89	17.59
fat-free	8	20.08	1.39	17.64	22.22
All individuals					
wet	21	30.36	1.22	27.87	32.41
dry	21	16.24	1.20	13.10	18.62
fat-free	21	19.85	1.06	17.64	22.22
Archilochus colubris					
Male AHY					
wet	9	3.50	0.77	2.38	4.99
dry	9	1.64	0.62	1.12	2.93
fat-free	9	2.50	0.32	1.84	2.89
Female HY					
wet	5	3.42	0.52	2.76	4.18
dry	5	1.44	0.40	1.07	1.96
fat-free	5	2.71	0.30	2.33	3.00
ash-free	1	2.27			
Female AHY					
wet	7	3.99	0.98	2.91	5.65
dry	7	1.96	0.79	1.21	3.33
fat-free	7	2.72	0.42	2.25	3.37
All individuals					
wet	21	3.64	0.80	2.38	5.65
dry	21	1.70	0.64	1.07	3.33
fat-free	21	2.62	0.35	1.84	3.37
ash-free	1	2.27			
Vireo griseus					
Male HY					
wet	34	10.12	0.77	8.28	11.99
dry	34	4.31	0.57	3.29	5.52
fat-free	34	8.46	0.67	7.06	10.02
Female HY					
wet	52	9.98	0.80	7.56	11.59
dry	52	4.36	0.65	3.16	5.84
fat-free	52	8.18	0.62	5.99	9.66
All individuals					
wet	86	10.04	0.79	7.56	11.99
dry	86	4.34	0.62	3.16	5.84
fat-free	86	8.29	0.65	5.99	10.02

Body Mass and Composition of North American Migrant Birds (continued)

	N	Mean	S.D.	Min	Max
Vireo philadelphicus					
Male HY					
wet	2	14.58		13.95	15.20
dry	2	6.56		6.36	6.76
fat-free	2	10.64		9.76	11.51
Male AHY					
wet	2	11.58		10.95	12.21
dry	2	5.09		4.95	5.23
fat-free	2	9.04		8.45	9.62
Female HY					
wet	1	14.51			
dry	1	7.30			
fat-free	1	9.85			
Female AHY					
wet	1	13.75			
dry	1	6.59			
fat-free	1	10.07			
All individuals					
wet	6	13.43	1.57	10.95	15.20
dry	6	6.20	0.92	4.95	7.30
fat-free	6	9.88	0.98	8.45	11.51
Vireo olivaceus					
Male HY					
wet	39	20.25	3.60	15.47	27.46
dry	39	10.03	3.37	5.50	16.36
fat-free	39	14.84	1.19	12.78	17.16
ash-free	11	13.87	0.68	12.66	14.93
Male AHY					
wet	113	19.32	2.99	13.93	26.37
dry	113	8.90	2.62	4.93	15.12
fat-free	113	14.99	1.33	11.13	19.79
ash-free	12	14.67	1.94	12.34	19.28
Female HY					
wet	55	18.65	2.62	14.12	24.78
dry	55	8.67	2.40	5.15	14.57
fat-free	55	14.54	1.19	11.64	17.11
ash-free	23	13.56	0.87	12.53	16.12
Female AHY					
wet	111	18.07	2.76	13.95	26.20
dry	111	8.42	2.34	5.10	15.29
fat-free	111	14.09	1.26	11.38	18.03
ash-free	9	13.36	1.15	11.93	15.23
All individuals					
wet	323	18.88	3.01	13.93	27.46
dry	323	8.82	2.62	4.93	16.36
fat-free	323	14.59	1.31	11.13	19.79
ash-free	57	13.88	1.26	11.93	19.28

Body Mass and Composition of North American Migrant Birds (continued)

	N	Mean	S.D.	Min	Max
Catharus fuscescens					
Male HY					
wet	41	42.83	4.02	32.09	53.56
dry	41	24.93	4.06	12.38	33.48
fat-free	41	27.75	3.85	22.54	38.46
Male AHY					
wet	3	43.28		42.40	44.32
dry	3	24.84		24.51	25.50
fat-free	3	33.26		32.88	33.55
Female HY					
wet	40	39.59	4.41	28.52	48.58
dry	40	22.42	4.66	11.59	30.01
fat-free	40	25.27	2.65	21.24	34.33
Female AHY					
wet	6	39.72	2.54	36.74	43.40
dry	6	23.92	3.27	20.46	29.20
fat-free	6	26.75	4.27	22.91	34.33
All individuals					
wet	100	41.15	4.26	28.52	53.56
dry	100	23.68	4.29	11.59	33.48
fat-free	100	26.66	3.61	21.24	38.46
Catharus minimus					
Male HY					
wet	43	31.29	4.88	23.90	47.14
dry	43	13.88	5.68		29.51
fat-free	43	25.25	2.94	20.09	34.93
ash-free	42	24.21	2.95	19.21	33.90
Male AHY					
wet	11	38.38	5.82	28.45	47.91
dry	11	19.53	5.92	10.56	29.15
fat-free	11	27.30	2.36	24.38	31.01
ash-free	11	26.26	2.31	23.31	29.88
Female HY					
wet	59	29.99	3.60	24.66	42.17
dry	59	13.00	4.10		24.29
fat-free	59	24.42	4.08	14.46	41.84
ash-free	58	23.63	3.86	19.00	40.74
Female AHY					
wet	15	35.58	5.92	27.17	45.20
dry	15	17.35	5.72	9.63	25.87
fat-free	15	26.12	1.94	22.78	30.49
ash-free	12	25.26	2.09	21.86	29.49
All individuals					
wet	133	32.08	5.43	23.90	47.91
dry	133	14.62	5.54		29.51
fat-free	133	25.20	3.45	14.46	41.84
ash-free	127	24.28	3.36	19.00	40.74

Body Mass and Composition of North American Migrant Birds (continued)

	N	Mean	S.D.	Min	Max
fat-free	24	43.32	2.46	38.57	49.09
ash-free	22	41.58	2.52	36.85	47.21
Male AHY					
wet	19	59.08	6.61	47.65	72.14
dry	19	30.07	8.39	19.23	48.65
fat-free	19	42.06	4.02	34.31	50.13
ash-free	15	41.47	3.34	36.04	48.28
Female HY					
wet	18	54.03	6.63	44.06	65.76
dry	18	25.86	6.33	16.76	36.38
fat-free	18	40.85	3.26	34.51	48.06
ash-free	16	39.53	3.33	32.89	46.36
Female AHY					
wet	28	55.68	8.36	41.94	70.86
dry	28	26.60	7.96	16.00	39.91
fat-free	28	41.89	3.19	35.85	49.77
ash-free	26	40.48	3.10	34.52	48.00
All individuals					
wet	105	56.92	7.03	41.94	73.98
dry	105	27.70	7.33	16.00	48.65
fat-free	105	42.21	3.32	34.13	50.13
ash-free	95	40.84	3.15	32.89	48.28
Dumetella carolinensis					
Male HY					
wet	20	36.89	2.66	32.52	41.90
dry	20	15.23	1.98	11.57	19.43
fat-free	20	31.62	2.66	27.15	35.20
ash-free	20	30.38	2.62	25.96	33.82
Male AHY					
wet	20	37.50	2.83	32.53	42.32
dry	20	14.88	1.62	11.63	17.35
fat-free	20	32.28	2.18	28.76	36.78
ash-free	20	31.08	2.13	27.70	35.54
Female HY					
wet	37	36.37	3.81	30.29	50.14
dry	37	14.74	3.37	11.44	29.71
fat-free	37	31.36	1.89	27.15	34.37
ash-free	37	30.13	1.84	26.02	33.04
Female AHY					
wet	20	38.62	3.99	33.91	50.51
dry	20	16.25	3.77	12.24	27.33
fat-free	20	32.01	1.83	26.90	35.64
ash-free	20	30.80	1.77	25.89	34.28
All individuals					
wet	104	37.17	3.42	30.29	50.51
dry	104	15.13	2.87	11.44	29.71
fat-free	104	31.80	2.13	26.90	36.78
ash-free	104	30.58	2.08	25.89	35.54

Body Mass and Composition of North American Migrant Birds (continued)

	N	Mean	S.D.	Min	Max
Cistothorus palustris					
Male HY					
wet	42	10.43	1.05	7.46	12.76
dry	42	3.78	0.64	2.90	5.99
fat-free	42	9.28	0.77	6.48	11.13
ash-free	27	8.76	0.71	6.19	9.89
Male AHY					
wet	23	9.98	1.28	8.07	13.21
dry	23	4.01	1.06	2.57	6.60
fat-free	23	8.43	0.72	7.17	9.61
ash-free	12	8.09	0.66	6.89	8.79
Female HY					
wet	30	9.47	1.14	7.04	11.21
dry	30	3.64	0.82	2.34	5.55
fat-free	30	8.16	0.89	6.59	9.89
ash-free	14	8.04	0.79	6.86	9.48
Female AHY					
wet	22	8.85	0.93	7.41	10.45
dry	22	3.37	0.49	2.51	4.64
fat-free	22	7.73	0.83	6.36	8.78
ash-free	17	7.31	0.84	6.09	8.40
All individuals					
wet	127	9.77	1.25	7.04	13.21
dry	127	3.72	0.81	2.34	6.60
fat-free	127	8.48	1.01	6.36	11.13
ash-free	77	8.09	0.96	6.09	9.89
Passer domesticus					
Male HY					
wet	8	27.22	3.31	21.00	32.55
dry	8	10.21	2.38	7.29	15.39
fat-free	8	25.04	2.50	20.34	27.75
Male AHY					
wet	5	25.58	3.47	21.35	29.36
dry	5	9.32	2.49	7.07	12.41
fat-free	5	24.15	2.74	20.73	27.61
Female HY					
wet	14	24.82	2.80	19.99	31.01
dry	14	8.73	1.78	6.99	13.22
fat-free	14	23.00	2.42	19.45	28.32
Female AHY					
wet	5	23.65	3.80	19.78	28.82
dry	5	9.73	3.61	6.86	15.67
fat-free	5	21.42	1.81	19.13	23.89
All individuals					
wet	32	25.36	3.27	19.78	32.55
dry	32	9.35	2.34	6.86	15.67
fat-free	32	23.44	2.60	19.13	28.32

Body Mass and Composition of North American Migrant Birds (continued)

	N	Mean	S.D.	Min	Max
Vermivora peregrine					
Male HY					
wet	13	13.26	1.02	11.72	14.71
dry	13	7.78	0.71	6.59	8.82
fat-free	13	7.99	0.55	6.96	8.99
ash-free	3	7.27		6.65	7.63
Male AHY					
wet	17	13.17	1.26	10.52	15.37
dry	17	7.79	1.26	4.43	9.93
fat-free	17	7.97	0.49	6.84	8.81
Female HY					
wet	27	12.86	1.25	10.36	15.83
dry	27	7.71	1.15	5.71	10.72
fat-free	27	7.70	0.45	6.75	8.94
ash-free	7	7.11	0.43	6.52	7.56
Female AHY					
wet	36	12.63	1.15	9.25	14.15
dry	36	7.25	1.24	3.22	8.80
fat-free	36	7.87	0.50	6.73	9.35
All individuals					
wet	95	12.90	1.18	9.25	15.83
dry	95	7.57	1.16	3.22	10.72
fat-free	95	7.85	0.49	6.73	9.35
ash-free	12	7.19	0.41	6.52	7.63
Parula americana					
Male HY					
wet	75	8.09	1.23	5.51	11.17
dry	75	3.87	1.20	2.26	7.07
fat-free	75	6.22	0.46	4.88	7.24
ash-free	56	6.11	0.37	5.13	6.96
Male AHY					
wet	54	8.07	1.20	5.73	10.64
dry	54	3.87	1.17	2.20	6.30
fat-free	54	6.20	0.38	5.28	7.11
ash-free	54	5.97	0.37	5.09	6.86
Female HY					
wet	96	6.92	1.03	5.56	10.08
dry	96	3.10	0.98	2.11	5.56
fat-free	96	5.58	0.41	4.40	6.78
ash-free	49	5.52	0.39	4.20	6.52
Female AHY					
wet	39	7.45	1.08	5.81	10.14
dry	39	3.50	1.15	2.33	6.11
fat-free	39	5.82	0.34	5.03	6.56
ash-free	39	5.60	0.32	4.85	6.31
All individuals					
wet	304	7.60	1.25	5.51	11.17

Body Mass and Composition of North American Migrant Birds (continued)

	N	Mean	S.D.	Min	Max
dry	304	3.56	1.17	2.11	7.07
fat-free	304	5.93	0.49	4.40	7.24
ash-free	236	5.82	0.43	4.20	6.96
Dendroica pensylvanica					
Male HY					
wet	11	13.22	0.71	12.08	13.99
dry	11	7.85	0.65	6.65	8.69
fat-free	11	8.03	0.39	7.30	8.53
Male AHY					
wet	11	12.92	1.40	9.21	14.46
dry	11	7.17	1.30	3.69	8.89
fat-free	11	8.33	0.40	7.41	8.78
Female HY					
wet	20	12.17	1.20	9.23	13.72
dry	20	6.89	1.27	3.44	8.36
fat-free	20	7.84	0.43	6.97	8.44
Female AHY					
wet	26	13.04	0.92	11.16	15.31
dry	26	7.61	0.74	6.01	9.41
fat-free	26	7.97	0.42	7.17	9.01
All individuals					
wet	73	12.79	1.22	9.05	15.57
dry	73	7.35	1.16	3.16	9.41
fat-free	73	8.01	0.45	6.97	9.07
Dendroica magnolia					
Male HY					
wet	21	9.68	1.55	7.43	12.51
dry	21	4.88	1.51	2.89	7.22
fat-free	21	6.95	0.47	6.00	7.70
ash-free	19	6.74	0.42	6.07	7.42
Male AHY					
wet	2	9.31		8.30	10.32
dry	2	4.38		3.04	5.72
fat-free	2	7.06		6.66	7.45
ash-free	2	6.80		6.40	7.20
Female HY					
wet	9	9.56	1.37	7.49	11.25
dry	9	5.02	1.38	2.78	6.33
fat-free	9	6.68	0.35	6.31	7.44
ash-free	6	6.38	0.22	6.07	6.54
Female AHY					
wet	3	9.69		9.30	10.04
dry	3	4.92		4.58	5.19
fat-free	3	7.35		6.84	8.32
All individuals					
wet	35	9.63	1.39	7.43	12.51

Body Mass and Composition of North American Migrant Birds (continued)

	N	Mean	S.D.	Min	Max
Catharus ustulatus					
Male HY					
wet	94	32.40	7.59	22.70	53.41
dry	94	15.61	6.62	7.12	34.96
fat-free	94	24.33	1.90	20.58	29.57
ash-free	88	23.20	1.85	19.72	28.39
Male AHY					
wet	52	36.52	7.87	22.00	49.51
dry	52	18.58	7.05	7.14	30.28
fat-free	52	25.53	2.10	20.79	28.95
ash-free	39	24.26	2.18	19.89	27.93
Female HY					
wet	68	28.74	6.45	20.56	45.45
dry	68	13.17	5.44	6.99	26.83
fat-free	68	22.54	1.89	19.56	27.70
ash-free	61	21.39	1.72	18.77	26.39
Female AHY					
wet	50	34.76	6.79	20.90	44.60
dry	50	17.78	6.00	7.00	27.74
fat-free	50	24.32	2.31	18.99	28.52
ash-free	38	23.11	2.35	18.18	27.18
All individuals					
wet	299	32.66	7.51	20.56	53.41
dry	299	15.85	6.51	6.99	34.96
fat-free	299	24.18	2.20	18.99	29.57
ash-free	239	22.96	2.18	18.18	28.39
Catharus guttatus					
Male HY					
wet	2	30.03		27.58	32.48
dry	2	11.74		10.58	12.90
fat-free	2	26.77		24.84	28.70
ash-free	2	25.74		23.87	27.62
Female HY					
wet	6	28.58	0.84	27.09	29.44
dry	6	11.38	1.00	9.95	13.03
fat-free	6	25.53	1.68	23.12	27.18
ash-free	6	24.57	1.63	22.24	26.15
All individuals					
wet	12	29.13	1.55	27.09	32.48
dry	12	11.82	1.07	9.95	13.17
fat-free	12	25.62	1.60	23.12	28.70
ash-free	12	24.66	1.55	22.24	27.62
Catharus (Hylochichla) mustelina					
Male HY					
wet	24	59.87	5.93	48.66	73.98
dry	24	29.84	6.70	17.24	43.34

Body Mass and Composition of North American Migrant Birds (continued)

	N	Mean	S.D.	Min	Max
dry	35	4.89	1.39	2.78	7.22
fat-free	35	6.92	0.49	6.00	8.32
ash-free	27	6.67	0.41	6.07	7.42
Dendroica caronata					
Male HY					
wet	25	11.75	0.89	10.16	13.77
dry	25	4.38	0.43	3.67	5.34
fat-free	25	10.56	0.82	9.18	12.13
Male AHY					
wet	1	14.59			
dry	1	7.37			
fat-free	1	10.43			
ash-free	1	10.10			
Female HY					
wet	91	11.97	1.35	8.63	14.99
dry	91	5.16	1.04	3.54	8.08
fat-free	91	10.00	0.84	6.99	12.32
All individuals					
wet	120	11.94	1.28	8.63	14.99
dry	120	5.00	1.01	3.54	8.08
fat-free	120	10.13	0.86	6.99	12.32
ash-free	1	10.10			
Dendroica fusca					
Male HY					
wet	6	14.01	0.86	12.83	15.16
dry	6	8.33	0.49	7.84	9.08
fat-free	6	8.34	0.70	7.45	9.46
ash-free	6	8.01	0.68	7.12	9.08
Male AHY					
wet	12	13.31	1.61	10.35	16.41
dry	12	7.91	1.60	4.42	10.34
fat-free	12	7.86	0.41	7.37	8.72
ash-free	12	7.56	0.39	7.08	8.37
Female HY					
wet	17	12.46	1.25	10.19	14.66
dry	17	7.50	1.22	5.66	10.69
fat-free	17	7.30	0.47	6.47	8.08
ash-free	17	7.02	0.46	6.16	7.76
Female AHY					
wet	9	13.03	0.62	11.88	13.90
dry	9	7.82	0.38	7.16	8.27
fat-free	9	7.60	0.63	6.83	8.37
ash-free	9	7.31	0.61	6.55	8.07
All individuals					
wet	52	12.98	1.28	10.19	16.41
dry	52	7.81	1.13	4.42	10.69

Body Mass and Composition of North American Migrant Birds (continued)

	N	Mean	S.D.	Min	Max
fat-free	52	7.58	0.61	6.47	9.46
ash-free	51	7.30	0.59	6.16	9.08
Dendroica dominica					
Male HY					
wet	4	9.71		9.31	10.27
dry	4	3.98		3.58	4.36
fat-free	4	8.20		7.88	8.47
ash-free	4	7.86		7.57	8.10
Female HY					
wet	9	9.16	0.69	7.96	10.12
dry	9	3.68	0.45	3.05	4.37
fat-free	9	7.93	0.44	7.18	8.37
ash-free	9	7.61	0.41	6.87	7.98
Female AHY					
wet	1	15.26			
dry	1	8.79			
fat-free	1	9.37			
All individuals					
wet	14	9.76	1.70	7.96	15.26
dry	14	4.13	1.41	3.05	8.79
fat-free	14	8.11	0.53	7.18	9.37
ash-free	13	7.69	0.38	6.87	8.10
Dendroica palmarum					
Male HY					
wet	30	10.43	1.23	8.27	12.51
dry	30	4.60	1.19	3.25	7.02
fat-free	30	8.69	0.47	7.46	9.65
ash-free	29	8.36	0.46	7.19	9.29
Male AHY					
wet	14	11.34	1.09	9.68	13.60
dry	14	5.48	1.02	4.16	7.62
fat-free	14	8.73	0.43	7.88	9.67
ash-free	14	8.42	0.42	7.55	9.32
Female HY					
wet	34	9.76	1.37	7.75	12.90
dry	34	4.24	1.27	2.70	6.91
fat-free	34	8.13	0.54	7.29	9.36
ash-free	34	7.84	0.52	7.00	9.01
Female AHY					
wet	4	9.53	0.64	8.78	10.22
dry	4	4.23	0.65	3.55	5.11
fat-free	4	7.92	0.21	7.71	8.19
ash-free	4	7.64	0.23	7.43	7.94
All individuals					
wet	100	10.25	1.29	7.75	13.30
dry	100	4.62	1.18	2.70	7.62

Body Mass and Composition of North American Migrant Birds (continued)

	N	Mean	S.D.	Min	Max
fat-free	100	8.38	0.58	7.10	9.67
ash-free	99	8.07	0.56	6.82	9.32
Dendroica castanea					
Male HY					
wet	2	14.52		12.99	16.04
dry	2	8.44		6.78	10.11
fat-free	2	9.56		9.42	9.69
Male AHY					
wet	3	16.84		16.50	17.46
dry	3	8.87		7.55	10.19
fat-free	3	11.24		10.90	11.79
Female HY					
wet	2	14.89		14.56	15.22
dry	2	8.09		7.77	8.41
fat-free	2	9.92		9.81	10.03
Female AHY					
wet	6	14.60	1.16	12.80	16.10
dry	6	7.71	0.96	6.28	8.95
fat-free	6	10.10	0.63	9.24	10.96
All individuals					
wet	14	14.99	1.47	12.80	17.46
dry	14	8.15	1.13	6.28	10.19
fat-free	14	10.17	0.79	9.03	11.79
Dendroica striata					
Male HY					
wet	6	16.20	3.85	11.13	19.97
dry	6	8.73	4.02	3.87	12.21
fat-free	6	10.77	0.89	9.29	11.84
ash-free	5	10.66	0.54	9.96	11.38
Male AHY					
wet	7	13.90	5.33	9.34	24.08
dry	7	6.88	4.79	3.22	15.87
fat-free	7	10.11	0.90	8.82	11.80
ash-free	5	9.66	0.29	9.19	9.95
Female HY					
wet	3	15.98		11.92	18.14
dry	3	8.81		4.66	11.07
fat-free	3	10.36		10.20	10.52
ash-free	3	9.93		9.66	10.14
Female AHY					
wet	3	11.30		10.96	11.59
dry	3	4.30		4.04	4.67
fat-free	3	9.99		9.71	10.25
ash-free	3	9.64		9.36	9.91
All individuals					
wet	19	14.54	4.26	9.34	24.08

Body Mass and Composition of North American Migrant Birds (continued)

	N	Mean	S.D.	Min	Max
dry	19	7.36	4.03	3.22	15.87
fat-free	19	10.34	0.78	8.82	11.84
ash-free	16	10.02	0.57	9.19	11.38
Setophaga ruticilla					
Male HY					
wet	23	9.89	1.19	7.11	11.83
dry	23	5.41	0.91	3.03	6.64
fat-free	23	6.66	0.53	5.50	7.94
ash-free	23	6.42	0.52	5.19	7.67
Male AHY					
wet	27	8.81	1.20	6.99	11.42
dry	27	4.30	1.26	2.51	6.95
fat-free	27	6.62	0.37	5.93	7.35
ash-free	26	6.39	0.37	5.71	7.10
Female HY					
wet	18	8.93	1.50	6.69	10.85
dry	18	4.57	1.37	2.52	6.34
fat-free	18	6.39	0.33	5.59	7.00
ash-free	18	6.15	0.31	5.37	6.75
Female AHY					
wet	15	8.91	1.08	7.00	10.79
dry	15	4.54	1.10	2.64	6.47
fat-free	15	6.38	0.23	6.07	6.90
ash-free	15	6.16	0.22	5.85	6.68
All individuals					
wet	102	9.11	1.30	6.60	11.83
dry	102	4.71	1.22	2.48	6.95
fat-free	102	6.49	0.42	5.50	7.94
ash-free	101	6.26	0.41	5.19	7.67
Protonotaria citrea					
Male HY					
wet	47	13.62	1.31	11.90	18.64
dry	47	6.21	1.16	4.86	11.56
fat-free	47	11.12	0.67	10.04	12.75
ash-free	46	10.65	0.65	9.61	12.24
Male AHY					
wet	2	17.66		17.09	18.23
dry	2	10.6		8.96	11.17
fat-free	2	11.36		10.64	12.07
ash-free	2	10.92		10.23	11.61
Female HY					
wet	20	13.77	1.35	11.39	16.78
dry	20	6.57	1.32	4.63	9.90
fat-free	20	10.68	0.52	9.68	11.86
ash-free	19	10.25	0.50	9.27	11.40

Body Mass and Composition of North American Migrant Birds (continued)

	N	Mean	S.D.	Min	Max
Female AHY					
wet	3	17.61		17.12	18.13
dry	3	10.19		9.46	11.45
fat-free	3	11.10		10.23	11.63
ash-free	3	10.66		9.78	11.22
All individuals					
wet	72	13.94	1.62	11.39	18.64
dry	72	6.58	1.55	4.63	11.56
fat-free	72	11.01	0.66	9.68	12.75
ash-free	70	10.55	0.64	9.27	12.24
Helmitheros vermivorus					
Male HY					
wet	6	16.71	2.65	11.38	18.30
dry	6	9.30	2.49	4.40	10.85
fat-free	6	11.16	0.67	10.25	12.03
ash-free	3	10.27		9.80	10.97
Female HY					
wet	6	16.43	2.51	11.62	18.47
dry	6	9.00	2.11	4.99	10.78
fat-free	6	11.10	0.62	10.17	11.89
ash-tree	5	10.47	0.54	9.71	11.11
Female AHY					
wet	6	15.57	2.54	11.13	18.22
dry	6	8.60	2.24	4.98	10.55
fal-free	6	10.58	0.73	9.58	11.54
ash-free	5	9.98	0.69	9.15	11.05
All individuals					
wet	28	15.63	2.40	11.01	18.47
dry	28	8.47	2.18	4.40	10.85
fat-free	28	10.79	0.78	9.58	12.46
ash-free	20	10.11	0.62	9.15	11.11
Seiurus aurocapillus					
Male HY					
wet	10	23.92	2.08	20.94	26.68
dry	10	13.12	1.41	10.68	14.81
fat-free	10	15.78	0.96	14.65	17.21
ash-free	8	15.33	0.93	14.04	16.54
Male AHY					
wet	2	22.86		22.50	23.23
dry	2	11.96		11.66	12.27
fat-free	2	15.72		15.62	15.82
ash-free	2	15.12		15.04	15.19
Female HY					
wet	9	21.90	1.38	20.74	24.85
dry	9	11.31	0.95	10.61	13.50
fat-free	9	15.57	0.69	14.98	16.45
ash-free	4	15.41		14.48	15.82

Body Mass and Composition of North American Migrant Birds (continued)

	N	Mean	S.D.	Min	Max
Female AHY					
wet	11	22.05	0.59	20.82	22.99
dry	11	11.72	0.47	11.09	12.51
fat-free	11	15.05	0.77	13.89	16.43
ash-free	2	13.91		13.30	14.53
All individuals					
wet	33	22.68	1.63	20.74	26.68
dry	33	12.04	1.19	10.61	14.81
fat-free	33	15.52	0.87	13.89	17.21
ash-free	17	15.23	0.92	13.30	16.57
Seiurus noveboracensis					
Male HY					
wet	7	17.17	4.13	13.22	25.69
dry	7	8.55	3.23	6.04	15.39
fat-free	7	12.94	1.71	9.94	14.95
ash-free	5	12.09	1.95	9.41	14.38
Male AHY					
wet	7	21.86	3.91	14.06	25.49
dry	7	11.53	2.93	5.80	14.13
fat-free	7	14.91	1.41	12.24	16.55
ash-free	6	14.37	1.53	11.69	15.94
Female HY					
wet	15	20.30	3.72	13.17	24.25
dry	15	11.17	2.90	6.12	14.32
fat-free	15	13.67	1.72	8.76	15.80
ash-free	15	13.12	1.69	8.27	14.18
Female AHY					
wet	33	20.77	2.34	14.92	24.56
dry	33	11.55	2.37	5.44	14.72
fat-free	33	13.61	1.11	11.71	17.00
ash-free	31	13.07	1.12	11.18	16.43
All individuals					
wet	89	20.37	3.41	13.17	25.69
dry	89	11.13	2.96	4.94	15.49
fat-free	89	13.68	1.30	8.76	17.00
ash-free	82	13.14	1.32	8.27	16.43
Oporomis formosus					
Male HY					
wet	69	16.69	2.64	10.89	20.31
dry	69	8.95	2.70	4.21	16.28
fat-free	69	11.52	1.16	5.29	13.43
ash-free	65	11.11	1.13	4.76	12.89
Male AHY					
wet	10	15.68	2.71	12.12	19.84
dry	10	7.77	2.70	4.52	11.62
fat-free	10	11.48	0.51	10.66	12.27

Body Mass and Composition of North American Migrant Birds (continued)

	N	Mean	S.D.	Min	Max
ash-free	10	11.04	0.52	10.21	11.88
Female HY					
wet	48	16.24	2.42	11.82	19.77
dry	48	8.60	2.37	4.25	11.91
fat-free	48	11.16	0.82	9.19	12.84
ash-free	48	10.72	0.80	8.77	12.31
Female AHY					
wet	10	16.58	2.60	12.00	19.30
dry	10	8.76	2.41	4.34	10.80
fat-free	10	11.38	0.52	10.46	12.26
ash-free	10	10.91	0.50	9.99	11.68
All individuals					
wet	156	16.46	2.54	10.83	20.31
dry	156	8.76	2.53	4.21	16.28
fat-free	156	11.36	0.96	5.29	13.43
ash-free	151	10.93	0.93	4.76	12.89
Geothlypis trichas					
Male HY					
wet	9	11.41	1.37	9.98	13.67
dry	9	5.35	1.35	3.75	6.90
fat-free	9	8.67	0.54	7.90	9.46
Male AHY					
wet	8	9.94	0.30	9.61	10.35
dry	8	4.34	0.31	3.98	4.68
fat-free	8	8.07	0.42	7.55	8.65
All individuals					
wet	19	10.78	1.19	9.61	13.67
dry	19	4.95	1.06	3.75	6.90
fat-free	19	8.36	0.54	7.55	9.46
Wilsonia citrina					
Male HY					
wet	74	10.59	1.99	8.17	14.99
dry	74	4.84	1.79	3.24	9.01
fat-free	74	8.41	0.82	4.06	9.94
ash-free	37	8.46	0.94	3.75	9.57
Male AHY					
wet	8	12.76	2.80	9.93	16.42
dry	8	6.30	2.44	3.84	9.46
fat-free	8	9.22	0.79	8.54	10.76
ash-free	8	8.98	0.77	8.24	10.37
Female HY					
wet	52	9.26	1.53	7.46	13.87
dry	52	4.04	1.34	2.79	8.74
fat-free	52	7.66	0.61	6.14	9.26
ash-free	12	7.92	0.60	6.82	8.92

Body Mass and Composition of North American Migrant Birds (continued)

	N	Mean	S.D.	Min	Max
All individuals					
wet	153	10.29	2.02	7.46	16.42
dry	153	4.69	1.73	2.79	9.46
fat-free	153	8.20	0.84	4.06	10.76
ash-free	74	8.35	0.84	3.75	10.37
Melospiza georgiana					
Male HY					
wet	4	17.22		16.20	18.06
dry	4	7.35		6.70	7.81
fat-free	4	14.76		13.99	15.79
ash-free	4	14.17		13.41	15.21
Female HY					
wet	3	14.87		13.72	15.93
dry	3	6.04		5.35	6.71
fat-free	3	12.97		12.23	13.63
ash-free	3	12.48		11.76	13.11
All individuals					
wet	7	16.22	1.54	13.72	18.06
dry	7	6.79	0.88	5.35	7.81
fat-free	7	13.99	1.21	12.23	15.79
ash-free	7	13.45	1.16	11.76	15.21
Zonotrichia albicollis					
Male HY					
wet	7	25.41	2.58	23.28	29.27
dry	7	10.42	1.86	8.46	13.92
fat-free	7	21.85	1.50	20.11	24.69
ash-free	2	22.76		21.37	24.14
Male AHY					
wet	5	24.54	2.04	21.80	27.14
dry	5	9.44	1.79	7.36	12.29
fat-free	5	21.99	1.16	20.49	23.20
Female HY					
wet	26	23.42	1.65	20.90	28.13
dry	26	9.18	1.06	7.53	12.02
fat-free	26	20.78	1.01	19.04	23.23
ash-free	5	21.15	1.02	20.32	22.31
Female AHY					
wet	2	21.94		21.57	22.32
dry	2	8.30		7.72	8.88
fat-free	2	19.50		18.86	20.14
All individuals					
wet	40	23.84	2.00	20.90	29.27
dry	40	9.39	1.37	7.36	13.92
fat-free	40	21.06	1.25	18.86	24.69
ash-free	7	21.61	1.39	20.32	24.14

Body Mass and Composition of North American Migrant Birds (continued)

	N	Mean	S.D.	Min	Max
Passerculus sandwichensis					
Male HY					
wet	55	18.39	1.64	15.00	23.30
dry	55	7.09	0.92	5.47	9.30
fat-free	55	16.54	1.31	13.69	20.14
Male AHY					
wet	75	17.48	1.43	14.44	22.60
dry	75	6.70	0.91	5.28	10.72
fat-free	75	15.81	1.22	13.12	20.34
Female HY					
wet	51	17.11	1.34	14.00	20.65
dry	51	6.36	0.76	5.12	8.14
fat-free	51	15.51	1.20	12.61	18.40
Female AHY					
wet	130	16.39	1.23	13.54	20.30
dry	130	6.39	0.93	4.99	9.94
fat-free	130	14.64	1.07	12.00	16.96
All individuals					
wet	311	17.12	1.55	13.54	23.30
dry	311	6.58	0.93	4.99	10.72
fat-free	311	15.40	1.37	12.00	20.34
Ammodramus maritimus					
Male HY					
wet	22	22.85	1.51	19.65	25.56
dry	22	8.23	0.68	7.08	9.68
fat-free	22	20.46	1.46	17.15	22.84
ash-free	16	19.43	1.57	16.49	21.89
Male AHY					
wet	4	20.62		18.75	23.00
dry	4	7.35		6.95	8.19
fat-free	4	18.61		16.96	20.68
ash-free	4	17.84		16.24	19.86
Female HY					
wet	16	21.12	1.38	19.20	24.25
dry	16	7.72	0.51	6.96	8.74
fat-free	16	18.83	1.33	16.55	21.77
ash-free	16	18.03	1.30	15.85	20.99
Female AHY					
wet	6	20.87	1.63	18.47	23.00
dry	6	7.31	0.73	6.41	8.32
fat-free	6	18.81	1.53	16.69	20.91
ash-free	4	17.71		15.99	19.98
All individuals					
wet	50	21.81	1.73	18.47	25.56
dry	50	7.84	0.72	6.41	9.68
fat-free	50	19.55	1.60	16.55	22.84
ash-free	42	18.54	1.58	15.85	21.89

Body Mass and Composition of North American Migrant Birds (continued)

	N	Mean	S.D.	Min	Max
Piranga rubra					
Male HY					
wet	47	34.96	6.27	25.50	46.58
dry	47	17.89	6.81	10.04	28.63
fat-free	47	25.29	2.10	19.40	30.82
ash-free	32	24.63	2.13	18.43	29.75
Male AHY					
wet	8	39.22	3.42	35.82	44.84
dry	8	23.51	2.80	20.01	26.84
fat-free	8	23.45	1.68	20.78	25.79
ash-free	1	21.71			
Female HY					
wet	23	36.12	5.68	27.36	47.63
dry	23	19.90	6.23	10.51	28.34
fat-free	23	24.37	2.02	20.01	27.70
ash-free	9	24.41	1.14	22.81	26.09
Female AHY					
wet	20	39.44	4.81	31.04	45.66
dry	20	23.06	5.06	12.18	28.85
fat-free	20	24.33	2.58	16.05	27.88
ash-free	4	25.56		23.58	26.94
All individuals					
wet	98	36.50	5.91	25.50	47.63
dry	98	19.88	0.44	10.04	28.85
fat-free	98	24.73	2.21	16.05	30.82
ash-free	46	24.60	1.95	18.43	29.75
Piranga olivacea					
Male HY					
wet	23	39.45	3.01	31.28	44.12
dry	23	23.48	2.39	18.55	28.41
fat-free	23	24.10	2.12	19.28	27.98
ash-free	14	23.38	2.38	18.43	26.95
Male AHY					
wet	11	40.19	5.29	30.43	44.85
dry	11	23.85	6.66	10.05	30.57
fat-free	11	24.56	2.36	19.28	28.59
ash-free	5	15.07	1.69	23.55	27.66
Female HY					
wet	19	39.10	4.73	24.50	44.05
dry	19	23.12	4.72	8.34	29.28
fat-free	19	24.13	1.62	21.44	27.59
ash-free	11	23.42	1.79	20.97	26.59
Female AHY					
wet	19	37.00	5.60	25.46	46.80
dry	19	19.71	6.76	8.41	29.72
fat-free	19	25.19	2.68	20.99	31.41
ash-free	10	25.00	2.71	20.81	30.34

Body Mass and Composition of North American Migrant Birds (continued)

	N	Mean	S.D.	Min	Max
All individuals					
wet	74	38.68	4.73	24.50	46.80
dry	74	22.27	5.42	8.34	30.57
fat-free	74	24.50	2.21	19.28	31.41
ash-free	40	24.01	2.31	18.43	30.34
Passerina cyanea					
Male HY					
wet	23	15.25	1.78	12.89	19.61
dry	23	6.65	1.80	4.99	10.99
fat-free	23	12.68	0.91	10.47	14.27
ash-free	11	11.72	0.79	10.04	12.78
Male AHY					
wet	56	15.58	2.07	12.92	22.18
dry	56	6.89	2.12	4.62	13.05
fat-free	56	12.70	0.78	11.01	14.52
ash-free	31	11.88	0.67	10.54	13.16
Female HY					
wet	15	16.04	2.88	13.31	21.36
dry	15	7.84	2.91	5.08	12.36
fat-free	15	11.94	0.87	9.62	13.12
ash-free	14	11.41	0.84	9.19	12.63
Female AHY					
wet	59	15.24	1.98	11.91	20.63
dry	59	6.97	1.99	4.38	12.29
fat-free	59	12.00	0.97	8.13	14.81
ash-free	42	11.40	1.06	7.72	14.44
All individuals					
wet	155	15.45	2.07	11.91	22.18
dry	155	6.99	2.11	4.38	13.05
fat-free	155	12.34	0.95	8.13	14.81
ash-free	100	11.59	0.90	7.72	14.44
Icterus spurius					
Male HY					
wet	2	22.75		21.20	24.30
dry	2	9.70		8.80	10.60
fat-free	2	19.00		18.10	19.90
ash-free	2	18.23		17.34	19.12
Male AHY					
wet	1	25.50			
dry	1	12.00			
fat-free	1	19.60			
ash-free	1	18.78			
Female HY					
wet	5	21.58	2.18	20.20	25.40
dry	5	9.04	1.40	8.20	11.50
fat-free	5	18.22	1.09	17.30	20.00

Body Mass and Composition of North American Migrant Birds (continued)

	N	Mean	S.D.	Min	Max
ash-free	5	17.44	1.10	16.53	19.24
Female AHY					
wet	3	22.90		21.30	24.60
dry	3	10.47		9.30	11.60
fat-free	3	18.20		16.30	20.50
ash-free	3	17.46		15.66	19.65
All individuals					
wet	11	22.51	2.07	20.20	25.50
dry	11	9.82	1.45	8.20	12.00
fat-free	11	18.48	1.33	16.30	20.50
ash-free	11	17.71	1.30	15.66	19.65
Agelaius phoeniceus					
Male unknown age					
wet	2	58.76		53.80	63.72
dry	2	22.37		22.25	22.49
fat-free	2	54.24		48.15	60.34
Female unknown age					
wet	10	42.72	2.51	39.33	47.17
dry	10	17.29	1.59	14.97	19.97
fat-free	10	37.73	2.62	33.91	43.38
All individuals					
wet	12	45.39	6.98	39.33	63.72
dry	12	18.14	2.44	14.97	22.49
fat-free	12	40.48	7.33	33.91	60.34
Dolichonyx orizivorus					
Male HY					
wet	88	35.00	7.32	19.04	51.23
dry	88	19.12	5.94	10.46	33.71
fat-free	88	24.23	3.30	13.13	30.16
ash-free	77	22.92	3.41	11.92	28.80
Male AHY					
wet	19	44.29	5.29	35.18	54.10
dry	19	26.19	4.48	18.59	34.45
fat-free	19	26.82	2.62	22.22	31.80
ash-free	11	26.28	2.49	21.32	30.71
Female HY					
wet	52	27.91	5.75	19.62	40.29
dry	52	14.17	4.62	7.53	23.97
fat-free	52	20.56	2.51	16.54	27.63
ash-free	41	19.80	2.50	15.68	26.49
Female AHY					
wet	18	37.93	5.01	30.20	49.20
dry	18	22.08	4.16	15.00	30.71
fat-free	18	23.46	2.55	19.86	29.05
ash-free	6	23.36	2.93	18.97	28.05

Body Mass and Composition of North American Migrant Birds (continued)

	N	Mean	S.D.	Min	Max
All individuals					
wet	178	34.20	8.09	19.04	54.10
dry	178	18.73	6.36	7.53	34.45
fat-free	178	23.33	3.54	13.13	31.80
ash-free	136	22.24	3.57	11.92	30.71

Part III

Literature Cited

Literature Cited

1. Abbott, I., L. K. Abbott, and P. R. Grant. 1977. Comparative ecology of Galapagos ground finches (*Geospiza* Gould): evaluation of the importance of floristic diversity and interspecific competition. *Ecological Monograph* 47:151–184.
2. Abrams, R. W., and L. G. Underhill. 1986. Relationships of pelagic seabirds with the Southern Ocean environmental assessed by correspondence analysis. *Auk* 103:221–225.
3. Ainley, D. G., and W. Emison. 1972. Sexual size dimorphism in Adelie Penguins. *Ibis* 114:267–271.
4. Ainley, D. G., and W. T. Everett. 2001. Black Storm-Petrel (*Oceanodroma melania*). In *The Birds of North America,* A. Poole and F. Gill (editors). The Birds of North America, Inc., Philadelphia, PA. Number 577.
5. Ainley, D. G., D. N. Nettleship, H. R. Carter, and A. E. Storey. 2002. Common Murre (*Uria aalge*). In *The Birds of North America*, A. Poole and F. Gill (editors). The Birds of North America, Inc., Philadelphia, PA. Number 666.
6. Ainley, D. G., T. C. Telfer, and M. H. Reynolds. 1997. Townsend's and Newell's Shearwater (*Puffinus auricularis*). In *The Birds of North America*, A. Poole and F. Gill (editors). The Academy of Natural Sciences, Philadelphia, PA, and The American Ornithologists' Union, Washington, DC. Number 297.
7. Ali, S., and S. D. Ripley. 1983. *Handbook of the birds of India and Pakistan*. Oxford University Press, Oxford, UK.

7a. Allen, D., C. Oliveros, C. Espanola, G. Broad, and J. C. T. Gonzales. 2004. A new species of *Gallirallus* from Calayon Island, Philippines. *Forktail* 20:1–7.

8. Alström, P., U. Olsson, and P. R. Colston. 1992. A new species of *Phylloscopus* warbler from central China. *Ibis* 134:329–334.
9. Alverez A. J., and B. M. Whitney. 2001. A new *Zimmerius* tyrannulet (Aves: Tyrannidae) from white sand forests of northern Amazonian Peru. *Wilson Bulletin* 113:1–9.
10. Alverez A. J., and B. M. Whitney. 2003. New distributional records of birds from white-sand forests of the northern Peruvian Amazon, with implications for biogeography of northern South America. *Condor* 105:552–566.
11. Anderson, A. H., and A. Anderson. 1973. *The cactus wren*. University of Arizona Press, Tucson, AZ.
12. Anderson, D. J. 1993. Masked Booby (*Sula dactylatra*). In *The Birds of North America*, A. Poole and F. Gill (editors). The Academy of Natural Sciences, Philadelphia, PA, and The American Ornithologists' Union, Washington, DC. Number 73.
13. Anderson, T. D. 1978. Population studies of European sparrows in North America. *Occasional Papers of the Museum of Natural History, University of Kansas* 70:1–58.
14. Andersson, S. 1993. Sexual dimorphism and modes of sexual selection in lekking Jackson's widowbirds *Euplectes jacksoni* (Ploceinae). *Biological Journal of the Linnean Society* 49:1–17.
15. Andres, B. A., and G. A. Falxa. 1995. Black Oystercatcher (*Haematopus bachmani*). In *The Birds of North America*, A. Poole and F. Gill (editors). The Academy of Natural Sciences, Philadelphia, PA, and The American Ornithologists' Union, Washington, DC. Number 155.
16. Aramburu, R. M. 1995. Ciclo annual de muda, peso corporal y gonadas en la Cotorro Comun (*Myiopsitta monachus monachus*). *Ornitologia Neotropical* 6:81–85.

16a. Arcos, J. M., and D. Oro. 2002. Significance of fisheries discards for a threatened Mediterranean seabird, the Balearic shearwater *Puffinus mauretanicus*. *Marine Ecology Progress Series* 239:209–220.

17. Arctander, P., O. Folmer, and J. Fjeldsa. 1996. The phylogenetic relationships of Berthelot's Pipit *Anthus berthelotii* illustrated by DNA sequence data, with remarks on the genetic distance between Rock and Water Pipits *Anthus spinoletta*. *Ibis* 138:263–272.
18. Ardern, S. L., I. G. McLean, S. Anderson, R. Maloney, and D. M. Lambert. 1994. The effects of blood sampling on the behavior and survival of the endangered Chatham Island Black Robin (*Petroica traversi*). *Conservation Biology* 8:857–862.
19. Arendt, W. J., J. Faaborg, G. E. Wallace, and O. H. Garrido. 2004. Biometrics of birds throughout the Greater Caribbean Basin. *Proceedings of the Western Foundation of Vertebrate Zoology* 8:1–33.
20. Ash, J. S. 1969. Spring weights of trans-Saharan migrants in Morocco. *Ibis* 111:1–10.
21. Ash, J. S. 1973. *Luscinia megarhynchos* and *L. luscinia* in Ethiopia. *Ibis* 115:267–269.
22. Ash, J. S., and G. Nikolaus. 1992. Birds in the hot season along the Nile in the Sudanese eastern Sahara. *Scopus* 16:22–28.
23. Ashmole, N. P. 1962. The Black Noddy *Anous tenuirostris* on Ascension Island. Part 1. General biology. *Ibis* 103b:235–273.
24. Atwood, J. L. 1979. Body weights of the Santa Cruz Island Scrub Jay. *North American Bird Bander* 4:148–153.
25. Atwood, J. L., and D. R. Bontrager. 2001. California Gnatcatcher (*Polioptila californica*). In *The Birds of North America*, A. Poole and F. Gill (editors). The Birds of North America, Inc., Philadelphia, PA. Number 574.
26. Audubon Zoological Garden, New Orleans, LA, unpublished data (P. Shannon).
27. Austin, G. T., and R. E. Ricklefs. 1977. Growth and development of the Rufous-winged Sparrow (*Aimophila carpalis*). *Condor* 79:37–50.
28. Austin, J. E., and M. R. Miller. 1995. Northern Pintail (*Anas acuta*). In *The Birds of North America*, A. Poole and F. Gill (editors). The Academy of Natural Sciences, Philadelphia, PA, and The American Ornithologists' Union, Washington, DC. Number 163.
29. Austin, O., Jr., unpublished data from the files of the Florida Museum of Natural History, *fide* T. Webber. Previously included in Dunning (1992).
30. Badyaev, A. V., D. D. Gibson, and B. Kissel. 1996. White Wagtail (*Motacilla alba*) and Black-backed Wagtail (*Motacilla lugens*). In *The Birds of North America*, A. Poole and F. Gill (editors). The Academy of Natural Sciences, Philadelphia, PA, and The American Ornithologists' Union, Washington, DC. Number 236–237.
31. Bailey, R. 1966. The sea-birds of the southeast coast of Arabia. *Ibis* 108:224–264.

32. Baker, H., and P. E. Baker. 2000. Maui 'Alauahio (*Paroreomyza montana*). In *The Birds of North America*, A. Poole and F. Gill (editors). The Birds of North America, Inc., Philadelphia, PA. Number 504.
33. Baker, K. 1997. *Warblers of Europe, Asia, and North Africa*. Princeton University Press, Princeton, NJ.
34. Baker, N. E., and E. M. Baker. 1990. A new species of weaver from Tanzania. *Bulletin of the British Ornithologists' Club* 110:51–58.
35. Baker, R. H. 1951. The avifauna of Micronesia: its origin, evolution, and distribution. *University of Kansas Publications of the Museum of Natural History* 3:1–359.
36. Balda, R. P. 2002. Pinyon Jay (*Gymnorhinus cyanocephalus*). In *The Birds of North America*, A. Poole and F. Gill (editors). The Birds of North America, Inc., Philadelphia, PA. Number 605.
36a. Balph, M. H. 1976. Some physical characteristics of Evening Grosbeaks wintering in northern Utah. *North American Bird Bander* 1:114–115.
37. Bancroft, G. T., and R. Bowman. 2001. White-crowned Pigeon (*Columba leucocephala*). In *The Birds of North America*, A. Poole and F. Gill (editors). The Birds of North America, Inc., Philadelphia, PA. Number 596.
38. Banko, P. C., D. L. Ball, and W. E. Banko. 2002. Hawaiian Crow (*Corvus hawaiiensis*). In *The Birds of North America*, A. Poole and F. Gill (editors). The Birds of North America, Inc., Philadelphia, PA. Number 648.
39. Banko, P. C., J. M. Black, and W. E. Banko. 1999. Hawaiian Goose (Nene) (*Branta sandvicensis*). In *The Birds of North America*, A. Poole and F. Gill (editors). The Birds of North America, Inc., Philadelphia, PA. Number 434.
40. Banko, P. C., L. Johnson, G. D. Lindsey, S. G. Fancy, T. K. Pratt, J. D. Jacobi, and W. E. Banko. 2002. Palila (*Loxioides bailleui*). In *The Birds of North America*, A. Poole and F. Gill (editors). The Birds of North America, Inc., Philadelphia, PA. Number 679.
41. Bannor, B. K., and E. Kiviat. 2002. Common Moorhen (*Gallinula chloropus*). In *The Birds of North America*, A. Poole and F. Gill (editors). The Birds of North America, Inc., Philadelphia, PA. Number 685.
42. Baptista, L. F. Unpublished data included in Dunning (1992).
43. Baptista, L. F., W. J. Boarman, and P. Kandianidis. 1983. Behavior and taxonomic status of Grayson's Dove. *Auk* 100:907–919.
44. Barber, B. R., and M. B. Robbins. 2002. Nest and eggs of the Tepui Antpitta (*Myrmothera simplex*). *Wilson Bulletin* 114:287–288.
45. Barber, D. R., P. M. Barber, and P. G. Jablonski. 2000. Painted Redstart (*Myioborus pictus*). In *The Birds of North America*, A. Poole and F. Gill (editors). The Birds of North America, Inc., Philadelphia, PA. Number 528.
46. Barlow, J. C., J. A. Dick, D. H. Baldwin, and R. A. Davis. 1969. New records of birds from British Honduras. *Ibis* 111:399–402.
47. Barlow, J. C., and S. V. Nash. 1985. Behavior and nesting biology of the St. Andrew Vireo. *Wilson Bulletin* 97:265–272.
48. Barnard, P. 1995. Timing of ornament growth, phenotypic variation, and size dimorphism in two promiscuous African whydahs (Ploceidae: *Vidua*). *Biological Journal of the Linnean Society* 55:129–141.
49. Bates, J. M., M. C. Garvin, D. C. Schmitt, and C. G. Schmitt. 1989. Notes on bird distribution in northeastern Dpto. Santa Cruz, Bolivia, with 15 species new to Bolivia. *Bulletin of the British Ornithologists' Club* 109:236–244.
50. Bauer, K. M., and U. N. Glutz von Blotzheim. 1966. *Handbuch der Vögel Mitteleuropas*. Akademische Verlagsgesellschaft, Frankfurt, Germany.
51. Baumel, J. J. 1953. Individual variation in the White-necked Raven. *Condor* 55:26–32.
52. Baumel, J. J. 1957. Individual variation in the Fish Crow, *Corvus ossifragus*. *Auk* 74:73–78.
52a. Bavoux, C., G. Burneleau, and V. Bretagnolle. 2006. Gender determination in the Western Marsh Harrier (*Circus aeruginosus*) using morphometrics and discriminant analysis. *Journal of Raptor Research* 40:57–64.
53. Bechard, M. J., and J. K. Schmutz. 1995. Ferruginous Hawk (*Buteo regalis*). In *The Birds of North America*, A. Poole and F. Gill (editors). The Academy of Natural Sciences, Philadelphia, PA, and The American Ornithologists' Union, Washington, DC. Number 172.
54. Beck, T. D. I., and C. E. Braun. 1978. Weights of Colorado sage grouse. *Condor* 80:241–243.
55. Becking, J.-H. 1971. The breeding of *Collocalia gigas*. *Ibis* 113:330–334.
56. Becking, J.-H. 1975. The ultrastructure of the avian eggshell. *Ibis* 117:143–151.
56a. Becking, J.-H. 2006. Notes on Waterfall Swift *Hydrochous gigas*: II. Nestling plumage and phylogenetic relationships. *Bulletin of the British Ornithologists' Club* 126:123–136.
57. Bedard, J. 1969. Adaptive radiation in Alcidae. *Ibis* 111:189–198.
58. Beehler, B. 1978. Notes on the mountain birds of New Ireland. *Emu* 78:65–70.
59. Behle, W. H. 1950. Clines in the Yellow-throats of western North America. *Condor* 52:193–219.
60. Behle, W. H. 1956. A systematic review of the Mountain Chickadee. *Condor* 58:51–70.
61. Bell, H. L. 1982. A bird community of lowland rainforest in New Guinea. I. Composition and density of the avifauna. *Emu* 82:24–41.
62. Belopol'skii, L. O. 1957. *Ecology of sea colony birds of the Barents Sea*. Izdatel'stvo Akademii Nauk SSSR, Moscow, Russia.
63. Belton, W. 1984. Birds of the Rio Grande do Sul, Brazil. Part 1. Rheidae through Furnariidae. *Bulletin of the American Museum of Natural History* 178:369–636.
64. Belton, W. 1985. Birds of the Rio Grande do Sul, Brazil. Part 2. Formicariidae through Corvidae. *Bulletin of the American Museum of Natural History* 180:1–242.
65. Beltrán, W., and G. Kattan. 2001. First record of the Slaty-backed Nightingale-Thrush in the Central Andes of Colombia, with notes on its ecology and geographical variation. *Wilson Bulletin* 113:134–139.
66. Benkman, C. W. 1992. White-winged Crossbill (*Loxia leucoptera*). In *The Birds of North America*, A. Poole, P. Stettenheim, and F. Gill (editors). The Academy of Natural Sciences, Philadelphia, PA, and The American Ornithologists' Union, Washington, DC. Number 27.
67. Benson, C. W., and M. J. Penny. 1971. The land birds of Aldabra. *Philosophical Transactions of the Royal Society of London B* 260:417–527.

68. Benson, C. W., J. F. R. Colebrook-Robjent, and A. Williams. 1976. Contribution a l'ornithologie de Madagascar. *L'Oiseau et R.F.O* 46:103–134, 209–242; 46:367–386.
69. Benson, C. W., J. F. R. Colebrook-Robjent, and A. Williams. 1977. Contribution a l'ornithologie de Madagascar. *L'Oiseau et R.F.O* 47:41–64, 167–191.
70. Berlin, K. E., and E. M. VanGelder. 1999. 'Akohekohe (*Palmeria dolei*). In *The Birds of North America*, A. Poole and F. Gill (editors). The Birds of North America, Inc., Philadelphia, PA. Number 400.
71. Berruti, A. 1979. The breeding biologies of the Sooty Albatrosses *Phoebetria fusca* and *P. palpebrata*. *Emu* 79:161–175.
71a. Bertolero, A., and C. Zavalaga. 2003. Observaciones sobre la biometría y la muda del Churrete Marisquero (*Cinclodes taczanowskii*) en Punta San Juan, Costa Sur del Perú. *Ornitologia Neotropical* 14:469–475.
71b. Betancourt, F. unpublished data.
72. Bibles, B. D., R. L. Glinski., and R. R. Johnson. 2002. Gray Hawk (*Astruina nitida*). In *The Birds of North America*, A. Poole and F. Gill (editors). The Birds of North America, Inc., Philadelphia, PA. Number 652.
73. Bierregaard, R. O. 1988. Morphological data from understory birds in terra firme forest in the Central Amazonian basin. *Revista Brasileira de Biologia* 48:169–178.
74. Binford, L. C. 1985. Re-evaluation of the "hybrid" hummingbird *Cynanthus sordidus* X *C. latirostris* from Mexico. *Condor* 87:148–150.
75. Binford, L. C. 1989. A distributional study of the birds of the Mexican state of Oaxaca. *Ornithological Monographs* 43.
76. Bjork, R., and G. V. N. Powell. National Audubon Society and RARE Center for Tropical Bird Conservation, Philadelphia, PA. Unpublished data.
77. Blaber, S. J. M. 1995. Abundance, site fidelity, morphometrics and sex ratios of Scarlet Honeyeater *Myzomela sanguinolenta* at a site in south-east Queensland. *Corella* 19:55–60.
78. Blake, C. H. 1956. Some bird weights from Jamaica. *Bird-Banding* 27:174–178.
79. Blake, J. G., unpublished database included in Dunning (1992).
80. Blake, J. G., and B. A. Loiselle, unpublished database included in Dunning (1992).
81. Blake, J. G., B. A. Loiselle, and J. P. Vande weghe. 1990. Weights and measurements of some central African birds. *Le Gerfaut* 80:3–11.
82. Blockstein, D. E., unpublished data included in Dunning (1992).
83. Blockstein, D. E. 2002. Passenger Pigeon (*Ectopistes migratorius*). In *The Birds of North America*, A. Poole and F. Gill (editors). The Birds of North America, Inc., Philadelphia, PA. Number 611.
84. Blondel, J., F. Vuilleumier, L. F. Marcus, and E. Terouanne. 1984. Is there ecomorphological convergence among Mediterranean bird communities of Chile, California, and France? *Evolutionary Biology* 18:141–157.
85. Bloom, P. H. 1973. Seasonal variation in body weight of Sparrow Hawks in California. *Western Bird Bander* 48:17–19.
86. Boarman, W. I., and B. Heinrich. 1999. Common Raven (*Corvus corax*). In *The Birds of North America*, A. Poole and F. Gill (editors). The Birds of North America, Inc., Philadelphia, PA. Number 476.
87. Boersma, P. D., and M. C. Silva. 2001. Fork-tailed Storm-Petrel (*Oceanodroma furcata*). In *The Birds of North America*, A. Poole and F. Gill (editors). The Birds of North America, Inc., Philadelphia, PA. Number 569.
88. Bohl, W. H. 1970. *A study of the crested tinamou of Argentina*. U.S. Fish and Wildlife Service, Bureau of Sport Fisheries and Wildlife. Special Scientific Report—Wildlife, number 131.
89. Boles, W. E. 1988. *The robins and flycatchers of Australia*. Angus & Robertson Publishers, North Ryde, NSW, Australia.
90. Boles, W. E., and B. Tynan. 1995. Low altitude record of Rufous Scrub-bird *Atrichornis rufescens*. *Corella* 19:91.
91. Bollmer, J. L., T. Sanchez, M. Donaghy Cannon, D. Sanchez, B. Cannon, J. C. Bednarz, T. de Vries, M. S. Struve, and P. G. Parker. 2003. Variation in morphology and mating system among island populations of Galapagos Hawks. *Condor* 105:428–438.
92. Bond, A. B., K.-J. Wilson, and J. Diamond. 1991. Sexual dimorphism in the Kea *Nestor notabilis*. *Emu* 91:12–19.
93. Bordage, D., and J.-P. Savard. 1995. Black Scoter (*Melanitta nigra*). In *The Birds of North America*, A. Poole and F. Gill (editors). The Academy of Natural Sciences, Philadelphia, PA, and The American Ornithologists' Union, Washington, DC. Number 177.
93a. Bornschein, M. R., B. L. Reinert, and M. Picorim. 1998. Descrição, ecologia e conservação de um novo *Scytalopus* (Rhinocryptidae) do sul do Brasil, com comentários sobre a morfologia da familia. *Ararajuba* 6:3–36.
94. Bosque, C., M. Andreína Pacheco, and M. A. García-Amada. 2004. The annual cycle of *Columbina* ground-doves in seasonal savannas of Venezuela. *Journal of Field Ornithology* 75:1–17.
95. Bosque, C., M. Andreína Pacheco, and R. B. Siegel. 1999. Maintenance energy costs of two partially folivorous tropical passerines. *Auk* 116:246–252.
96. Bosque, C., and M. T. Bosque. 1995. Nest predation as a selective factor in the evolution of developmental rates in altricial birds. *American Naturalist* 145:234–260.
97. Bosque, C., and M. Lentino. 1987. The nest, eggs, and young of the White-whiskered Spinetail (*Synallaxis* [*Poecilurus*] *candei*). *Wilson Bulletin* 99:104–106.
98. Bostrom, M. R., and G. Ritchison. 2006. Possible relationships between morphology, territory quality, and skin color of American Kestrels. *Journal of Field Ornithology* 77:392–398.
99. Botero, C. A. 2001. First specimen of the Ecuadorian Cacique from Colombia, with notes on its nesting behavior. *Wilson Bulletin* 113:327–328.
100. Bowden, C. G. R. 1987. The Arabian Golden Sparrow in North Yemen. *Sandgrouse* 9:94–97.
101. Bowden, C. G. R., and D. J. Brooks. 1987. The Yemen Linnet in North Yemen. *Sandgrouse* 9:111–114.
102. Reference deleted.
103. Bowen, R. V. 1997. Townsend's Solitaire (*Myadestes townsendi*). In *The Birds of North America*, A. Poole and F. Gill (editors). The Academy of Natural Sciences, Philadelphia, PA, and The American Ornithologists' Union, Washington, DC. Number 269.

104. Bowers, R. K. Unpublished banding data included in Dunning (1992).

105. Bowers, R. K., and J. B. Dunning, Jr. 1997. Buff-collared Nightjar (*Caprimulgus ridgwayi*). In *The Birds of North America*, A. Poole and F. Gill (editors). The Birds of North America, Inc., Philadelphia, PA, and The American Ornithologists' Union, Washington, DC. Number 267.

106. Bowman, R. Unpublished data communicated in letter from P. W. Trail, included in Dunning (1992).

107. Bowman, R. I. 1961. Morphological differentiation and adaptation in the Galapagos finches. *University of California Publications in Zoology* 58.

108. Bowman, R. 2002. Common Ground-Dove (*Columbina passerina*). In *The Birds of North America*, A. Poole and F. Gill (editors). The Birds of North America, Inc., Philadelphia, PA. Number 645.

109. Boyd, H., and L. S. Maltby. 1980. Weights and primary growth of Brent Geese *Branta bernicla* moulting in the Queen Elizabeth Islands, N.W.T., Canada, 1973–1975. *Ornis Scandinavica* 11:135–141.

110. Brace, R. C., and J. Hornbuckle. 1998. Distributional records of and identification notes on birds of the Beni Biological Station, Beni, Bolivia. *Bulletin of the British Ornithologists' Club* 118:36–47.

111. Braun, C. E. 1976. Banding worksheet for western birds—Band-tailed Pigeon. Western Bird Banding Association.

112. Braun, M. J., M. L. Isler, P. R. Isler, J. M. Bates, and M. B. Robbins. 2005. Avian speciation in the Pantepui: the case of the Roraiman Antbird (*Percnostola [Schistocichla] "leucostigma" saturata*). *Condor* 107:327–341.

113. Brenowitz, E., and T. Smith. Unpublished data.

114. Bretagnolle, V., C. Attie, and F. Mougeot. 2000. Audubon's Shearwaters *Puffinus lherminieri* on Reunion Island, Indian Ocean: behaviour, census, distribution, biometrics and breeding biology. *Ibis* 142:399–412.

115. Bretagnolle, V., R. Zotier, and P. Jouventin. 1990. Comparative population biology of four prions (genus *Pachyptila*) from the Indian Ocean and consequences for their taxonomic status. *Auk* 197:305–316.

116. Brightsmith, D. 1999. White-winged Parakeet (*Brotogeris versicolurus*) and Yellow-chevroned Parakeet (*Brotogeris chiriri*). In *The Birds of North America*, A. Poole and F. Gill (editors). The Birds of North America, Inc., Philadelphia, PA. Numbers 386–387.

117. Briskie, J. V. 1993. Smith's Longspur (*Calcarius pictus*). In *The Birds of North America*, A. Poole P. Stettenheim, and F. Gill (editors). The Academy of Natural Sciences, Philadelphia, PA, and The American Ornithologists' Union, Washington, DC. Number 34.

118. Britton, P. L. 1967. Weights of the Carmine Bee-eater *Merops nubicoides*. *Ibis* 109:606–614.

119. Britton, P. L. 1969. Weights of the Pennant-winged Nightjar. *Bulletin of the British Ornithologists' Club* 89:21–24.

120. Britton, P. L. 1970. Some non-passerine bird weights from East Africa. *Bulletin of the British Ornithologists' Club* 90:142–144; 90:152–154.

121. Britton, P. L. 1977. Weights of birds in western and coastal Kenya: a comparison. *Scopus* 1:70–73.

122. Britton, P. L. 1978. Seasonality, density and diversity of birds of a papyrus swamp in western Kenya. *Ibis* 120:450–466.

123. Brooke, M. de L. 1978a. Inland observations of Barau's Petrel *Pterodroma baraui* on Reunion. *Bulletin of the British Ornithologists' Club* 98:90–95.

124. Brooke, M. de L. 1978b. Weights and measurements of the Manx shearwater, *Puffinus puffinus*. *Journal of Zoology*, London 186:359–374.

125. Brooke, M. de L. 1983. Ecological segregation of woodcreepers (Dendrocolaptidae) in the state of Rio de Janeiro, Brasil. *Ibis* 125:562–567.

125a. Brooke, M. 2004. *Albatrosses and petrels across the world*. Oxford University Press, Oxford, UK.

126. Brooke, R. K. 1969. *Apus berliozi* Ripley, its races and siblings. *Bulletin of the British Ornithologists' Club* 89:11–16.

127. Brooke, R. K. 1971a. Geographical variation in the Alpine Swift *Apus* (*Tachymarptis*) *melba* (Aves: Apodidae). *Durban Museum Novitates* 9:131–143.

128. Brooke, R. K. 1971b. The eastern and southern populations of the Mottled Spinetail. *Bulletin of the British Ornithologists' Club* 91:134–135.

129. Brooke, R. K. 1971c. Geographical variation in the Little Swift *Apus affinus* (Aves: Apodidae). *Durban Museum Novitates* 9:93–103.

130. Brooke, R. K. 1971d. Taxonomic notes of some lesser known *Apus* swifts. *Bulletin of the British Ornithologists' Club* 91:33–36.

131. Brooker, M. 1996. Morphometrics of the Wedge-tailed Eagle *Aquila audax*. *Corella* 20:129–135.

132. Brooks, T. M. 1993. Project canopy '92. Final unpublished report, Birdlife International.

133. Brosset, A., and C. Erard. 1977. New faunistic records from Gabon. *Bulletin of the British Ornithologists' Club* 97:125–132.

133. Brough, T. 1983. *Average weight of birds*. Avian Bird Unit, Worplesdon Laboratory, Worplesdon, UK.

133a. Brown, C. R. 1997. Purple Martin (*Progne subis*). In *The Birds of North America*, A. Poole and F. Gill (editors). The Academy of Natural Sciences, Philadelphia, PA, and The American Ornithologists' Union. Number 287.

135. Brown, C. R., and M. B. Brown. 1999. Barn Swallow (*Hirundo rustica*). In *The Birds of North America*, A. Poole and F. Gill (editors). The Academy of Natural Sciences, Philadelphia, PA, and The American Ornithologists' Union, Washington, DC. Number 452.

136. Brown, J. H., and M. A. Bowers. 1985. Community organization in hummingbirds: relationships between morphology and ecology. *Auk* 102, 251–269.

137. Brown, J. L. 1994. Mexican Jay (*Aphelocoma ultramarina*). In *The Birds of North America*, A. Poole and F. Gill (editors). The Academy of Natural Sciences, Philadelphia, PA, and The American Ornithologists' Union, Washington, DC. Number 118.

137a. Brown, L. 1976. *Eagles of the world*. Universe Books, New York.

138. Brown, L., and D. Amadon. 1968. *Eagles, hawks and falcons of the world*. McGraw-Hill, New York.

139. Brown, L., E. K. Urban, and K. Newman. 1982. *The birds of Africa*. Academic Press, New York. Volume 1.

140. Brown, P. W., and L. H. Fredrickson. 1997. White-winged Scoter (*Melanitta fusca*). In *The Birds of North America*, A. Poole and F. Gill (editors). The Academy of Natural Sciences, Philadelphia, PA, and The American Ornithologists' Union, Washington, DC. Number 274.

141. Brown, R. G. B., S. P. Barker, D. E. Gaskin, and M. R. Sandeman. 1981. The foods of Great and Sooty Shearwaters *Puffinus gravis* and *P. griseus* in eastern Canadian waters. *Ibis* 123:19–30.
142. Browne, K., and E. Browne. 1956. An analysis of the weights of birds trapped on Skokholm. *British Birds* 49:241–257.
143. Brua, R. B. 2002. Ruddy Duck (*Oxyura jamaicensis*). In *The Birds of North America*, A. Poole and F. Gill (editors). The Birds of North America, Inc., Philadelphia, PA. Number 696.
144. Brumfield, R. T., and J. V. Remsen. 1996. Geographic variation and species limits in *Cinnycerthia* wrens of the Andes. *Wilson Bulletin* 108:205–227.
145. Brush, T. 1998. Olive Sparrow (*Arremonops rufivirigatus*). In *The Birds of North America*, A. Poole and F. Gill (editors). The Birds of North America, Inc., Philadelphia, PA. Number 325.
146. Brush, T. 1999. Couch's Kingbird (*Tyrannus couchii*). In *The Birds of North America*, A. Poole and F. Gill (editors). The Birds of North America, Inc., Philadelphia, PA. Number 437.
147. Brush, T., and J. W. Fitzpatrick.2002. Great Kiskadee (*Pitangus sulphuratus*). In *The Birds of North America*, A. Poole and F. Gill (editors). The Birds of North America, Inc., Philadelphia, PA. Number 622.
148. Buckley, N. J. 1999. Black Vulture (*Coragyps atratus*). In *The Birds of North America*, A. Poole and F. Gill (editors). The Birds of North America, Inc., Philadelphia, PA. Number 411.
149. Bull, E. L., and J. R. Duncan. 1993. Great Gray Owl (*Strix nebulosa*). In *The Birds of North America*, A. Poole and F. Gill (editors). The Academy of Natural Sciences, Philadelphia, PA, and The American Ornithologists' Union, Washington, DC. Number 41.
150. Bull, E. L., and J. A. Jackson. 1995. Pileated Woodpecker (*Dryocopus pileatus*). In *The Birds of North America*, A. Poole and F. Gill (editors). The Academy of Natural Sciences, Philadelphia, PA, and The American Ornithologists' Union, Washington, DC. Number 148.
151. Bump, G., and J. W. Bump. 1969. *A study of the spotted tinamous and the pale spotted tinamous of Argentina*. U.S. Fish and Wildlife Service, Bureau of Sport Fisheries and Wildlife. Special Scientific Report—Wildlife, number 120.
152. Burger, J. 1996. Laughing Gull (*Larus atricilla*). In *The Birds of North America*, A. Poole and F. Gill (editors). The Academy of Natural Sciences, Philadelphia, PA, and The American Ornithologists' Union, Washington, DC. Number 225.
153. Burger, J., and M. Gochfeld. 2002. Bonaparte's Gull (*Larus philadelphia*). In *The Birds of North America*, A. Poole and F. Gill (editors). The Birds of North America, Inc., Philadelphia, PA. Number 634.
154. Burness, G. P., K. Lefevre, and C. T. Collins. 1999. Elegant Tern (*Sterna elegans*). In *The Birds of North America*, A. Poole and F. Gill (editors). The Birds of North America, Inc., Philadelphia, PA. Number 404.
155. Burton, P. J. K. 1973. Non-passerine bird weights from Panama and Colombia. *Bulletin of the British Ornithologists' Club* 93:116–118.
156. Burton, P. J. K. 1975. Passerine bird weights from Panama and Colombia, with some notes on "soft-part" colours. *Bulletin of the British Ornithologists' Club* 95:82–86.
157. Butler, R. G., and D. E. Buckley. 2002. Black Guillemot (*Cepphus grylle*). In *The Birds of North America*, A. Poole and F. Gill (editors). The Birds of North America, Inc., Philadelphia, PA. Number 675.
158. Butler, R. W. 1992. Great Blue Heron (*Ardea herodias*). In *The Birds of North America*, A. Poole, P. Stettenheim, and F. Gill (editors). The Academy of Natural Sciences, Philadelphia, PA, and The American Ornithologists' Union, Washington, DC. Number 25.
159. Butynski, T. M., U. Agenonga, B. Ndera, and J. F. Hart. 1977. Rediscovery of the Congo Bay (Itombwe) Owl *Phodilus prigoginei*. *Bulletin of the African Bird Club* 4:32–35.
159a. Byrd, G. V., and J. C. Williams. 1993. Red-legged Kittiwake (*Rissa brevirostris*). In *The birds of North America*, A. Poole and F. Gill (editors). The Academy of Natural Sciences, Philadelphia, PA, and American Ornithologists' Union, Washington, DC. Number 60.
160. Cabot, J., and T. de Vries. 2003. *Buteo polyosoma* and *Buteo poecilochrous* are two distinct species. *Bulletin of the British Ornithologists' Club* 123:190–207.
161. Cade, T. J., and E. C. Atkinson. 2002. Northern Shrike (*Lanius excubitor*). In *The Birds of North America*, A. Poole and F. Gill (editors) The Birds of North America, Inc., Philadelphia, PA. Number 671.
161a. Calder, W. A. 1984. *Size, function, and life history*. Harvard University Press, Cambridge, MA.
162. Calder, W. A., N. M. Waser, S. M. Hiebert, D. W. Inouye, and S. Miller. 1983. Site-fidelity, longevity, and population dynamics of broad-tailed hummingbirds: a ten year study. *Oecologia* 56:359–364.
163. Calkins, J. D., J. C. Hagelin, and D. F. Lott. 1999. California Quail (*Callipepla californica*). In *The Birds of North America*, A. Poole and F. Gill (editors). The Birds of North America, Inc., Philadelphia, PA. Number 473.
163a. Caldwell, L. D., E. P. Odum, and S. G. Marshall. 1964. Comparison of fat levels in migrating birds killed at a central Michigan and a Florida Gulf Coast television tower. *Wilson Bulletin* 75:428–434.
163b. Canales-Delgadillo, J. C. Unpublished data.
164. Cannings, R. J. 1993. Northern Saw-whet Owl (*Aegolius acadicus*). In *The Birds of North America*, A. Poole and F. Gill (editors). The Academy of Natural Sciences, Philadelphia, PA, and The American Ornithologists' Union, Washington, DC. Number 42.
165. Cannings, R. J., and T. Angell. 2001. Western Screech-Owl (*Otus kennicottii*). In *The Birds of North America*, A. Poole and F. Gill (editors). The Birds of North America, Inc., Philadelphia, PA. Number 597.
166. Cape May Raptor Banding Project, Cape May Bird Observatory, Cape May, NJ, unpublished data (C. Schultz) included in Dunning (1992).
167. Capparella, A. P., and S. M. Lanyon. 1985. Biochemical and morphometric analyses of the sympatric, neotropical, sibling species, *Mionectes macconnelli* and *M. oleagineus*. *Ornithological Monographs* 36:347–355.
168. Cardiff, S. W., and D. L. Dittmann. 2002. Ash-throated Flycatcher (*Myiarchus cinerascens*). In *The Birds of North America*, A. Poole and F. Gill (editors). The Birds of North America, Inc., Philadelphia, PA. Number 664.
169. Cardiff, S. W., and J. V. Remsen. 1981. Three bird species new to Bolivia. *Bulletin of the British Ornithologists' Club* 101:304–305.

170. Carroll, J. P. 1993. Gray Partridge (*Perdix perdix*). In *The Birds of North America*, A. Poole and F. Gill (editors). The Academy of Natural Sciences, Philadelphia, PA, and The American Ornithologists' Union, Washington, DC. Number 58.
171. Casares, M., F. Enders, and J. A. Montoya. 2000. Comparative electrocardiography in four species of macaws (genus *Anodorhynchus* and *Ara*). *Journal of Veterinary Medicine, Series A* 47:277–281.
172. Case, P. R. 1993. European Starling (*Sturnus vulgaris*). In *The Birds of North America*, A. Poole and F. Gill (editors). The Academy of Natural Sciences, Philadelphia, PA, and The American Ornithologists' Union, Washington, DC. Number 48.
173. Castro, I., and A. Phillips. 1996. *A guide to the birds of the Galapagos Islands*. Princeton University Press, Princeton, NJ.
174. Catibog-Sinha, C. 1987. The wild vertebrate fauna of Mount Makiling, Laguna. *Philippine Journal of Science* 116:19–29.
175. Catry, P., R. A. Phillips, and R. W. Furness. 1999. Evolution of reversed sexual size dimorphism in skuas and jaegers. *Auk* 116:158–168.
176. Causey D. 2002. Red-faced Cormorant (*Phalacrocorax urile*). In *The Birds of North America*, A. Poole and F. Gill (editors). The Birds of North America, Inc., Philadelphia, PA. Number 617.
177. Cavalcanti, R. B., and M. A. Marini. 1993. Body masses of birds of the cerrado region, Brazil. *Bulletin of the British Ornithologists' Club* 113:209–212.
178. Caziani, S. M., and J. J. Protomastro. 1991. Nest and eggs of the Stripe-backed Antbird (*Myrmorchilus strigilatus*). *Condor* 93:445–446.
179. Caziani, S. M., and J. J. Protomastro. 1994. Diet of the Chaco Chachalaca. *Wilson Bulletin* 106:640–648.
180. Chardine, J. W., and R. D. Morris. 1996. Brown Noddy (*Anous stolidus*). In *The Birds of North America*, A. Poole and F. Gill (editors). The Academy of Natural Sciences, Philadelphia, PA, and The American Ornithologists' Union, Washington, DC. Number 220.
181. Cheke, A. S. 1987. The ecology of the smaller land-birds of Mauritius. In *Studies of Mascarene Islands birds*, A. W. Diamond (editor). Cambridge University Press, Cambridge, MA, 151–207.
182. Cheke, A. S., and A. W. Diamond. 1986. Birds on Moheli and Grande Comore (Comoro Islands) in February 1975. *Bulletin of the British Ornithologists' Club* 106:138–149.
183. Cheke, A. S., and C. G. Jones. 1987. Measurements and weights of the surviving endemic birds of the Mascarenes and their eggs. In *Studies of Mascarene Islands birds*, A. W. Diamond (editor). Cambridge University Press, Cambridge, MA, 403–422.
184. Cheke, R. A., and C. F. Mann. 2001. *Sunbirds: a guide to the sunbirds, flowerpeckers, spiderhunters and sugarbirds of the world.* Yale University Press, New Haven CT.
184a. Cheng, T.-h. 1993. *Economic birds of China*, second edition. Science Press, Beijing, China.
185. Chesser, R. T. 1995. Comparative diets of obligate ant-following birds at a site in northern Boliva. *Biotropica* 27:382–390.
186. Chilton, G. 1997. Labrador Duck (*Camptorhynchus labradorius*). In *The Birds of North America*, A. Poole and F. Gill (editors). The Academy of Natural Sciences, Philadelphia, PA, and The American Ornithologists' Union, Washington, DC. Number 307.
186a. Christie, M. I., E. J. Ramilo, and M. D. Bettinelli. 2004. *Aves del Noroeaste Patagonico.* Sociedad Naturalista Andino Patagónica. San Carlos de Bariloche, Río Negro, Argentina.
187. Chu, M., and G. Walsberg. 1999. Phainopepla (*Phainopepla nitens*). In *The Birds of North America*, A. Poole and F. Gill (editors). The Birds of North America, Inc., Philadelphia, PA. Number 415.
188. Cicero, C. 2000. Oak Titmouse (*Baeolophus inornatus*) and Juniper Titmouse (*Baeolophus ridgwayi*). In *The Birds of North America*, A. Poole and F. Gill (editors). The Birds of North America, Inc., Philadelphia, PA. Number 485.
189. Cink, C. L. 2002. Whip-poor-will (*Caprimulgus vociferus*). In *The Birds of North America*, A. Poole and F. Gill (editors). The Birds of North America, Inc., Philadelphia, PA. Number 620.
189a. Clark, G. A. 1979. Body weights of birds: a review. *Condor* 81:193–202.
190. Clarkson, K. E., and L. P. Laniawe. 2000. Hawaiian Hawk (*Buteo solitarius*). In *The Birds of North America*, A. Poole and F. Gill (editors). The Birds of North America, Inc., Philadelphia, PA. Number 523.
191. Cleere, N., and D. Nurney. 1998. *Nightjars: a guide to the nightjars, nighthawks and their relatives.* Yale University Press, London.
192. Clement, P. 2000. *Thrushes.* Princeton University Press, Princeton, NJ.
192a. Clements, J. 1998. *Birds of the world: a checklist*, sixth edition. Ibis Publishing, Vista, CA.
193. Clench, M. H., and R .C. Leberman. 1978. Weights of 151 species of Pennsylvania birds analyzed by month, age, and sex. *Bulletin of the Carnegie Museum of Natural History* 5.
194. Cody, M. L. 2000. Antbird guilds in the lowland Caribbean rainforest of southeast Nicaragua. *Condor* 102:784–794.
195. Cohn-Haft, M., unpublished data included in Dunning (1992).
196. Cohn-Haft, M. 1993. Rediscovery of the White-winged Potoo (*Nyctibius leucopterus*). *Auk* 110:391–394.
197. Colbourne, R., and R. Kleinpaste. 1983. A banding study of North Island Brown Kiwis in an exotic forest. *Notornis* 30:109–124.
198. Collins, C. T., unpublished data included in Dunning (1992).
199. Collins, C. T. 1968a. The comparative biology of two species of swifts in Trinidad, West Indies. *Bulletin of the Florida State Museum* 11:257–320.
199a. Collins, C. T. 1968b. *Notes on the biology of Chapman's Swift* Chaetura chapmani *(Aves, Apodidae).* American Museum Novitates 2320.
200. Collins, C. T. 1972a. Weights of some birds of north-central Venezuela. *Bulletin of the British Ornithologists' Club* 92:151–153.
201. Collins, C. T. 1972b. Banding worksheet for western birds—Violet-green Swallow. Western Bird Banding Association.
202. Collins, C. T., and R. A. Bradley. 1971a. Analysis of body weights of spring migrants in southern California. *Western Bird Bander* (July): 39–39.
203. Collins, C. T., and R. A. Bradley 1971b. Analysis of body weights of spring migrants in southern California: part II. *Western Bird Bander* (October): 48–51.
204. Collins, C. T., and R. K. Brooke. 1976. *A review of the swifts of the genus* Hirundapus *(Aves: Apodidae).* Natural History Museum of Los Angeles County Contributions in Science 282.

205. Collister, D. M., and D. Wicklum. 1996. Intraspecific variation in Loggerhead Shrikes: sexual dimorphism and implication for subspecies classification. *Auk* 113:221–223.
206. Colwell, M. A., and J. R. Jehl. 1994. Wilson's Phalarope (*Phalaropus tricolor*). In *The Birds of North America*, A. Poole and F. Gill (editors). The Academy of Natural Sciences, Philadelphia, PA, and The American Ornithologists' Union, Washington, DC. Number 83.
207. Colwell, R. K. 1989. Hummingbirds of the Juan Fernández Islands: natural history, evolution and population status. *Ibis* 131:548–566.
208. Colwell, R. K. 2000. Rensch's rule crosses the line: convergent allometry of sexual size imorphism in hummingbirds and flower mites. *American Naturalist* 156:495–510.
209. Colwell, R. R. 2002. Recommended band size for Spotted Towhees: a suggested revision. *North American Bird Bander* 27:127–129.
209a. Connell, C. E., E. P. Odum, and H. Kale. 1960. Fat-free weights of birds. *Auk* 77:1–9.
210. Contreras, J. R. 1975. Caracteristicas ponderales de las aves del Pargue Nacional Nahuel Huapi y regions adyacentes. *Physis* 34:97–107.
210a. Contreras, J. R. 1979. Bird weights from northeastern Argentina. *Bulletin of the British Ornithological Club* 99:21–24.
211. Contreras, J. R. 1983a. Notas sobre el peso de aves Argentinas. II. *Historia Natural* 3:39–40.
212. Contreras, J. R. 1983b. Notas sobre el peso de aves Argentinas. III. *Historia Natural* 3:95–96.
212a. Contreras, J. R. 1986. Notas sobre el peso de aves Argentinas. V. *Historia Natural* 6:100.
213. Conway, C. J. 1995. Virginia Rail (*Rallus limicola*). In *The Birds of North America*, A. Poole and F. Gill (editors). The Academy of Natural Sciences, Philadelphia, PA, and The American Ornithologists' Union, Washington, DC. Number 173.
214. Cooch, F. G., G. M. Stirrettt, and G. F. Boyer. 1960. Autumn weights of Blue Geese (*Chen caerulescens*). *Auk* 77:460–465.
215. Cooper, J. M. 1994. Least Sandpiper (*Calidris minutilla*). In *The Birds of North America*, A. Poole and F. Gill (editors). The Academy of Natural Sciences, Philadelphia, PA, and The American Ornithologists' Union, Washington, DC. Number 115.
216. Cooper, S. J., and Gessaman, J. A. 2004. Thermoregulation and habitat preference in Mountain Chickadees and Juniper Titmice. *Condor* 106:852–861.
217. Coopmans, P., and N. Krabbe. 2000. A new species of flycatcher (Tyrannidae: *Myiopagis*) from eastern Ecuador and eastern Peru. *Wilson Bulletin* 112:305–312.
217a. Copello, S., F. Quintana, and G. Somoza. 2006. Sex determination and sexual size-dimorphism in Southern Giant-Petrels (*Macronectes giganteus*) from Patagonia, Argentina. *Emu* 106:141–146.
217b. Cornejo, J. Unpublished data from collection of Africam Sahari, Pueblo, Mexico.
218. Cortes, O. 2005. Aspectos ecologicos y de historia natural en *Amazilia castaneiventris* en Soata (Boyaca/Colombia). Trabajo de grado. Departmento de Biologia, Universidad Distrital "Francisco Jose de Caldas."
219. Cotton, P. A. 1998. The hummingbird community of a lowland Amazonian rainforest. *Ibis* 140:512–521.
219a. Cotton, P. A. 2007. Seasonal resource tracking by Amazonian hummingbirds. *Ibis* 149:135–142.
220. Counsell, D. J. R. 1987. The Royal Air Force Ornithological Society expedition to Belize, 1986. Unpublished report to Birdlife International.
221. Counsilman, J. J., K. Nee, A. K. Jalil, and W. L. Keng. 1994. Discriminant analysis of morphometric characters as a means of sexing mynas. *Journal of Field Ornithology* 65:1–7.
222. Craig, R. 1990. Foraging behavior and microhabitat use of two species of white-eyes (Zosteropidae) on Saipan, Micronesia. *Auk* 107:500–505.
223. Cramp, S., and C. M. Perrins (editors). 1977–1994. *Handbook of the birds of Europe, the Middle East, and North Africa*. Oxford University Press, Oxford, UK. Volumes 1–9.
224. Crick, H. Q. P., and C. H. Fry. 1986. Effects of helpers on parental condition in Red-throated Bee-eaters (*Merops bullocki*). *Journal of Animal Ecology* 55:893–905.
225. Crocoll, S. T. 1994. Red-shouldered Hawk (*Buteo lineatus*). In *The Birds of North America*, A. Poole and F. Gill (editors). The Academy of Natural Sciences, Philadelphia, PA, and The American Ornithologists' Union, Washington, DC. Number 107.
226. Crossin, R. S. 1967. The breeding biology of the Tufted Jay. *Proceedings of the Western Foundation of Vertebrate Zoology* 1:265–299.
227. Crossin, R. S. 1974. The storm petrels (Hydrobatidae). In *Pelagic studies of seabirds in the central and eastern Pacific Ocean*, W. B. King (editor). Smithsonian Contributions to Zoology 158, 154–205.
228. Crossin, R. S., and C. A. Ely. 1973. A new race of Sumichrast's Wren from Chiapas, Mexico. *Condor* 75:137–139.
229. Croxall, J. P. 1977. Feeding behaviour and ecology of New Guinea rainforest insectivorous passerines. *Ibis* 119:113–146.
230. Croxall, J. P., and J. E. Furse. 1980. Food of Chinstrap Penguins *Pygoscelis antarctica* and Macaroni Penguins *Eudyptes chrysolophus* at Elephant Island Group, South Shetland Islands. *Ibis* 122:237–245.
231. Cruz, A. 1974. Feeding assemblages of Jamaican birds. *Condor* 76:104–108.
232. Cruz, A. 1976. Distribution, ecology, and breeding biology of the Rufous-throated Solitaire in Jamaica. *Auk* 93:39–45.
233. Cruz, A. 1980. Feeding ecology of the Black-whiskered Vireo and associated gleaning birds in Jamaica. *Wilson Bulletin* 92:40–52.
234. Cruz, A., and D. W. Johnston. 1984. Ecology of the West Indian Red-bellied Woodpecker on Grand Cayman: distribution and foraging. *Wilson Bulletin* 96:366–379.
235. Csada, R. D., and R. M. Brigham. 1992. Common Poorwill (*Phalaenoptilus nuttallii*). In *The Birds of North America*, A. Poole, P. Stettenheim, and F. Gill (editors). The Academy of Natural Sciences, Philadelphia, PA, and The American Ornithologists' Union, Washington, DC. Number 32.
236. Cuervo, A. M., C. D. Cadena, N. Krabbe, and L. M. Renjifo. 2005. *Scytalopus stilesi*, a new species of tapaculo (Rhinocryptidae) from the Cordillera central of Colombia. *Auk* 122:445–463,

237. Cuervo, A. M., P. G. W. Salaman, T. M. Donegan, and J. M. Ochoa. 2001. A new species of piha (Cotingidae: *Lipaugus*) from the Cordillera Central of Colombia. *Ibis* 143:353–368.
238. Cullen, S. A., J. R. Jehl, and G. L. Nuechterlein. 1999. Eared Grebe (*Podiceps nigricollis*). In *The Birds of North America*, A. Poole and F. Gill (editors). The Birds of North America, Inc., Philadelphia, PA. Number 433.
240. Cunningham, J. B. 1984. Differentiating the sexes of the Brown Creeper. *Notornis* 31:19–22.
241. Curio, E., J. Hornbuckle, Y de Soye, P. Aston, and L. L. Lastimoza. 2001. New bird records for the island of Panay, Philippines, including the first record of the Asian Stubtail *Urosphena squameiceps* for the Philippines. *Bulletin of the British Ornithologists' Club* 121:183–197.
242. Currie, D., R. Bristol, J. Millett, M. Hill, U. Bristol, S. J. Parr, and N. J. Shah. 2003. Habitat requirements of the Syechelles Black Paradise Flycatcher *Terpsiphone corvina*: a re-evaluation of translocation priorities. *Ibis* 145:624–636.
243. Curry, R. L., and P. R. Grant. 1990. Galapagos mockingbirds: territorial cooperative breeding in a climatically variable environment. In *Cooperative breeding in birds: long-term studies of ecology and behavior*, P. B. Stacey and W. D. Koenig (editors). Cambridge University Press, Cambridge, MA, 291–332.
244. Curry, R. L., A. T. Peterson, and T. A. Langen. 2002. Western Scrub-Jay (*Aphelocoma californica*). In *The Birds of North America*, A. Poole and F. Gill (editors). The Birds of North America, Inc., Philadelphia, PA. Number 712.
245. Curson, D. R., and C. B. Goguen. 1998. Plumbeous Vireo (*Vireo plumbeus*). In *The Birds of North America*, A. Poole and F. Gill (editors). The Birds of North America, Inc., Philadelphia, PA. Number 366.
246. Czechura, G. V., and S. J. S. Debus. 1985a. The Grey Falcon *Falco hypoleucos*: a summary of information. *Australian Bird Watcher* 11:9–16.
247. Czechura, G. V., and S. J. S. Debus. 1985b. The Black Falcon *Falco subniger*: a summary of information and comparison with the Brown Falcon *Falco berigora*. *Australian Bird Watcher* 11:80–91.
248. Czechura, G. V., and S. J. S. Debus. 1986. The Australian Hobby *Falco longipennis*: a review. *Australian Bird Watcher* 11:185–207.
249. Czechura, G. V., S. J. B. Debus, and N. J. Mooney. 1987. The Collared Sparrowhawk *Accipiter cirrocephalus*: a review and comparison with the Brown Goshawk *Accipiter fasciatus*. *Australian Bird Watcher* 12:35–62.
250. Dahlsten, D. L., L. A. Brennan, D. A. McCallum, and S. L. L. Gaunt. 2002. Chestnut-backed Chickadee (*Poecile rufescens*). In *The Birds of North America*, A. Poole and F. Gill (editors). The Birds of North America, Inc., Philadelphia, PA. Number 689.
251. Dann, P. 1981. Breeding of the Banded and Masked Lapwings in southern Victoria. *Emu* 81:121–127.
252. Darrieu, C. A. 1986. Estudios sobre la avifauna de Corrientes. III. Nuevos registros de Aves Passeriformes (Dendrocolaptidae, Furnariidae, Formicaridae, Cotingidae y Pipridae) y consideraciones sobre su distribucion geografica. *Historia Natural* 6:93–99.
253. Darrieu, C. A. 1987. Estudios sobre la avifauna de Corrientes. IV. Nuevos registros de Aves (Passeriformes, Tyrannidae) y consideraciones sobre su distribucion geografica. *Neotropica* 33:29–36.
254. Darrieu, C. A., and A. R. Camperi. 1993. Estudio de una collección de aves de Corrientes, Argentina. IV. (Phytotomidae a Parulidae). *Neótropica* 39:83–92.
255. Darrieu, C. A., and A. R. Camperi. 1996. Estudio de una collección de aves de Corrientes (Emberizidae y Fringillidae). *Neótropica* 42:69–75.
255a. David, N., and M. Gosselin. 2002. Gender agreement of avian species names. *Bulletin of the British Ornithologists' Club* 122:14–49.
256. Davies, S. J. J. F. 2002. *Ratites and tinamous*. Oxford University Press, Oxford, UK.
257. Davis, J. 1951. Distribution and variation of the brown towhees. *University of California Publications in Zoology* 52:1–120.
258. Davis, J. N. 1995. Hutton's Vireo (*Vireo huttoni*). In *The Birds of North America*, A. Poole and F. Gill (editors). The Academy of Natural Sciences, Philadelphia, PA, and The American Ornithologists' Union. Number 189.
259. Davis, T. J. 1986. Distribution and natural history of some birds from the departments of San Martin and Amazonas, northern Peru. *Condor* 88:50–56.
259a. Daves, W. B. 1945. Notes on Venezuelan birds. *Auk* 62:272–286.
260. Dawson, W. R., and F. C. Evans. 1960. Relation of growth and development to temperature regulation in nestling Vesper Sparrows. *Condor* 62:329–340.
261. Day, D. 1989. *Vanished species*. Gallery Books, W. H. Smith Publ., New York.
262. Day, R. H., and G. V. Byrd. 1989. Food habits of the Whiskered Auklet at Buldir Island, Alaska. *Condor* 91:65–72.
263. Day, R. H., K. J. Kuletz, and D. A. Nigro. 1999. Kittlitz's Murrelet (*Brachyramphus brevirostris*). In *The Birds of North America*, A. Poole and F. Gill (editors). The Birds of North America, Inc., Philadelphia, PA. Number 435.
264. Day, R. H., I. J. Stenhouse, and H. G. Gilchrist. 2001. Sabine's Gull (*Xema sabini*). In *The Birds of North America*, A. Poole and F. Gill (editors). The Birds of North America, Inc., Philadelphia, PA. Number 593.
265. de Vasconcelos, M. F., S. D'Angelo Neto, and M. Rodrigues. 2002. A range extension for the Cipó Castastero *Asthenes luizae* and the consequences for its conservation status. *Bulletin of the British Ornithologists' Club* 122:7–10.
266. Dean, W. R. J. 2000. *The birds of Angola, an annotated checklist*. British Ornithologists' Union, Natural History Museum (Tring), UK.
267. Debus, S. J. S. 1984. A re-appraisal of the dimensions of male Forest Ravens. *Corella* 8:19–20.
268. Debus, S. J. S., and C. Edelstam. 1994. Etymology, type locality and morphology of the Chestnut-shouldered (Burger's) Goshawk *Erythrotriorchis burgersi*. *Australian Bird Watcher* 15:380–382.
269. Degernes, L. A., L. D. Harrison, D. W. Smith, H. M. Newton, C. E. Ross, and D. E. Diaz. 1999. Autologous, homologous, and heterologous red blood cell transfusions in conures of the genus *Aratinga*. *Journal of Avian Medicine and Surgery* 13:10–14.
270. Delacour, J., and D. Amadon. 1973. *Curassows and related birds*. American Museum of Natural History, New York.
271. Delannoy, C. A., and A. Cruz. 1988. Breeding biology of the Puerto Rican Sharp-shinned Hawk (*Accipiter striatus venator*). *Auk* 105:649–662.

272. Delestrade, A. 2001. Sexual size dimorphism and positive assortative mating in Alpine Choughs (*Pyrrhocorax graculus*). *Auk* 118:553–556.
273. Dement'ev, G. P., and N. A. Gladkov (editors). 1951. *Birds of the Soviet Union*. Gosudarstuennae Izdatel'stvo, Moscow, Russia.
274. del Hoyo, J., A. Elliott, and J. Sargatal. 1992–2003. *Handbook of the birds of the world*. Lynx Edicions, Barcelona, Spain. Volumes 1–8.
275. del Hoyo, J., A. Elliott, and D. Christie. 2004–5. *Handbook of the birds of the world*. Lynx Edicions, Barcelona, Spain. Volumes 9–10.
275a. del Hoyo, J. and A. Motis. 2004. Update chapter. In *Curassows and related birds*, 2nd ed., J. Delacour and D. Amadon (editors). Lynx Edicions and The National Museum of Natural History, Barcelona, Spain, 322–476.
276. Deng, H., and X. Zhang. 1990. Standard metabolic rate in several species of passerine birds in alpine meadow. *Acta Zoologica Sinica* 36:377–384.
277. DeSante, D. F., K. M. Burton, P. Velez, and D. Froehlich. 2005. MAPS manual: 2005 protocol. Institute of Bird Populations, Point Reyes Station, CA. Contribution No. 127. Unpublished data collected under the protocols listed within this reference.
278. Devillers, P. 1977. Observations at a breeding colony of *Larus (belcheri) atlanticus*. *Le Gerfaut* 67:22–43.
279. Diamond, A. W. 1975. Biology and behaviour of frigatebirds *Fregata* spp. on Aldabra Atoll. *Ibis* 117:302–323.
280. Diamond, A. W., and E. A. Schreiber. 2002. Magnificent Frigatebird (*Fregata magnificens*). In *The Birds of North America*, A. Poole and F. Gill (editors). The Birds of North America, Inc., Philadelphia, PA. Number 601.
281. Diamond, J. M. 1967. *New subspecies and records of birds from the Karimui Basin, New Guinea*. American Museum Novitates 2284.
282. Diamond, J. M. 1969. *Preliminary results of an ornithological exploration of the North Coastal Range, New Guinea*. American Museum Novitates 2362.
283. Diamond, J. M. 1972. *Avifauna of the Eastern Highlands of New Guinea*. Publications of the Nuttall Ornithological Club, number 12.
284. Diamond, J. M. 1976. Preliminary results of an ornithological exploration of the islands of Vitaiz and Dampier Straits, Papua New Guinea. *Emu* 76:1–7.
285. Diamond, J. M. 1981. Distribution, habits and nest of *Chenorhamphus grayi*, a malurid endemic to New Guinea. *Emu* 81:97–100.
286. Diamond, J. M. 1982. Rediscovery of the Yellow-fronted Gardener Bowerbird. *Science* 216:431–434.
287. Diamond, J. M. 1983. *Melampitta gigantea*: possible relation between feather structure and underground roosting habits. *Condor* 85:89–91.
288. Diamond, J. M. 1985. New distributional records and taxa from the outlying mountain ranges of New Guinea. *Emu* 85:65–91.
289. Diamond, J. M. 1989a. A new subspecies of the Island Thrush *Turdus poliocephalus* from Tolokiwa Island in the Bismarck Archipelago. *Emu* 89:58–60
290. Diamond, J. M. 1989b. Rapid evolution of character displacement in myzomelid honeyeaters. *American Naturalist* 134:675–708.
291. Diamond, J. M., K. D. Bishop, and J. D. Gilardi. 1999. Geophagy in New Guinea birds. *Ibis* 141:181–193.
292. Diamond, J. M., and M. LeCroy. 1979. Birds of Karkar and Bagabag Islands, New Guinea. *Bulletin of the American Museum of Natural History* 164:467–531.
293. Dick, J. A., W. B. McGillivray, and D. J. Brooks. 1984. A list of birds and their weights from Saül, French Guiana. *Wilson Bulletin* 96:347–365.
294. Dickerman, R. W. 1975. Nine new specimen records for Guatemala. *Wilson Bulletin* 87:412–413.
295. Dickerman, R. W., K. C. Phelps, and J. Bell. 1980–81. Notes on the plumages of the Boat-billed Heron. *Living Bird* 19:115–120.
296. Dickerman, R. W., and W. H. Phelps. 1982. *An annotated list of the birds of Cerro Urutani on the border of Estado Bolivar, Venezuela, and Territorio Roraima, Brazil*. American Museum Novitates 2732.
296a. Dickinson, E. C. (editor). 2003. *The Howard and Moore complete checklist of the birds of the world*, third edition. Princeton University Press, Princeton, NJ.
297. Dinesen, L., T. Lehmberg, J. O. Svendsen, L. A. Hansen, and J. Fjeldså. 1994. A new genus and species of perdicine bird (Phasianidae, Perdicini) from Tanzania: a relict form with Indo-Malayan affinities. *Ibis* 136:3–11.
298. Domíguez-Bello, M. G., F. Michelangeli, M. C. Ruiz, A. Garcia, and E. Rodríguez. 1994. Ecology of the folivorous Hoatzin (*Opisthocomus hoazin*) on the Venezuelan plains. *Auk* 111:643–651.
298a. Donegan, T. M., and B. Huertas. 2006. A new brush-finch in the *Atlapetes latinuchus* complex from the Yariguies Mountains and adjacent Eastern Andes of Colombia. *Bulletin of the British Ornithologists' Club* 126:94–116.
298b. Donohue, K. C., and A. M. Dufty. 2006. Sex determination of Red-tailed Hawks (*Buteo jamaicensis calurus*) using DNA analysis and morphometrics. *Journal of Field Ornithology* 77:74–79.
299. Dostine, P. L., and S. R. Morton. 1989. Food of the Black-winged Stilt *Himantopus himantopus* in the Alligator Rivers Region, Northern Territory. *Emu* 89:250–253.
300. Dowling, D. K. 2003. Breeding biology of the red-capped robin. *Australian Journal of Zoology* 51:553–549.
301. Dowsett, R. J. 1970. A collection of birds from the Nyika Plateau, Zambia. *Bulletin of the British Ornithologists' Club* 90:49–53.
302. Dowsett, R. J. 1981. Breeding and other observations on the Slaty Egret *Egretta vinaceigula*. *Bulletin of the British Ornithologists' Club* 101:323–327.
303. Dowsett, R. J. 1983. Sexual size dimorphism in some montane forest passerines from south-central Africa. *Bulletin of the British Ornithologists' Club* 103:59–64.
304. Dowsett, R. J., and C. H. Fry. 1971. Weight losses of trans-Saharan migrants. *Ibis* 113:531–533.
305. Drilling, N., R. Titman, and F. McKinney. 2002. Mallard (*Anas platyrhynchos*). In *The Birds of North America*, A. Poole and F. Gill (editors). The Birds of North America, Inc., Philadelphia, PA. Number 658.
306. Reference deleted.

307. Duffy, D. C. 1980. Patterns of piracy by Peruvian seabirds: a depth hypothesis. *Ibis* 122:521–525.

308. Duffy, D. C. 1987. Ecological implications of intercolony size-variation in Jackass Penguins. *Ostrich* 58:54–57.

309. Duffy, D. C., W. R. Siegfried, and S. Jackson. 1987. Seabirds as consumers in the southern Benguela region. *South African Journal of Marine Science* 5:771–790.

310. Dugger, B. D., K. M. Dugger, and L. H. Fredrickson. 1994. Hooded Merganser (*Lophodytes cucullatus*). In *The Birds of North America*, A. Poole and F. Gill (editors). The Academy of Natural Sciences, Philadelphia, PA, and The American Ornithologists' Union, Washington, DC. Number 98.

311. Duncan, J. R., and P. A. Duncan. 1998. Northern Hawk Owl (*Surnia ulula*). In *The Birds of North America*, A. Poole and F. Gill (editors). The Birds of North America, Inc., Philadelphia, PA. Number 356.

312. Dunk, J. R. 1995. White-tailed Kite (*Elanus leucurus*). In *The Birds of North America*, A. Poole and F. Gill (editors). The Academy of Natural Sciences, Philadelphia, PA, and The American Ornithologists' Union, Washington, DC. Number 178.

313. Dunning, J. B. Unpublished data, included in Dunning (1992).

314. Dunning, J. B. 1984. *Body weights of 686 species of North American birds*. Western Bird Banding Association Monograph 1.

314a. Dunning, J. B. 1985. Owl weights in the literature: a review. *Journal of Raptor Research* 19:113–121.

315. Dunning, J. B. 1988. Yellow-footed Gull kills Eared Grebe. *Colonial Waterbirds* 11:117–118.

315a. Dunning, J. B. 1992. *CRC Handbook of avian body masses*. CRC Press, Boca Raton, FL.

316. Dunning, J. B. 1993. Bachman's Sparrow (*Aimophila aestivalis*). In *The Birds of North America*, A. Poole, P. Stettenheim, and F. Gill (editors). The Academy of Natural Sciences, Philadelphia, PA, and The American Ornithologists' Union, Washington, DC. Number 38.

317. Dymond, J. N. 1996. Biometric data of birds in southern Yemen and Socotra, spring 1993. *Sandgrouse* 17:158–164.

318. Dyrcz, A. Unpublished data.

319. Dyrcz, A. 1990. Understorey bird assemblages in various types of lowland tropical forest in Tambopata Reserve, SE Peru (with faunistic notes). *Acta Zool. Cracov.* 33:215–233.

320. Eadie, J. M., M. L. Mallory, and H. G. Lumsden. 1995. Common Goldeneye (*Bucephala clangula*). In *The Birds of North America*, A. Poole and F. Gill (editors). The Academy of Natural Sciences, Philadelphia, PA, and The American Ornithologists' Union, Washington, DC. Number 170.

321. Eadie, J. M., J.-P. Savard, and M. L. Mallory. 2000. Barrow's Goldeneye (*Bucephala islandica*). In *The Birds of North America*, A. Poole and F. Gill (editors). The Birds of North America, Inc., Philadelphia, PA. Number 548.

322. Earhart, C. M., and N. K. Johnson. 1970. Size dimorphism and food habits of North American owls. *Condor* 72:251–264.

323. Earlé R. A., S. Louw, and J. J. Herholdt. 1989. Notes on the measurements and diet of Ludwig's Bustard *Neotis ludwigii*. *Ostrich* 59:178–179.

324. Eaton, S. W. 1992. Wild Turkey (*Meleagris gallopavo*). In *The Birds of North America*, A. Poole, P. Stettenheim, and F. Gill (editors). The Academy of Natural Sciences, Philadelphia, PA, and The American Ornithologists' Union, Washington, DC. Number 22.

324a. Echeverry-Galvis, M. Á., S. Córdoba-Córdoba, C. A. Peraza, M. P. Baptiste, and J. A. Ahumada. 2006. Body weights of 98 species of Andean cloud-forest birds. *Bulletin of the British Ornithologists' Union* 126:291–298.

325. Eddleman, W. R. 1989. *Biology of the Yuma Clapper Rail in the southwestern U.S. and northwestern Mexico*. Final report, U.S. Bureau of Reclamation and U.S. Fish & Wildlife Service. Inter-Agency Agreement 4-AA-30-02060.

326. Eddleman, W. R., and C. J. Conway. 1998. Clapper Rail (*Rallus longirostris*). In *The Birds of North America*, A. Poole and F. Gill (editors). The Birds of North America, Inc., Philadelphia, PA. Number 340.

327. Eddleman, W. R., R. E. Flores, and M. L. Legare. 1994. Black Rail (*Laterallus jamaicensis*). In *The Birds of North America*, A. Poole and F. Gill (editors). The Academy of Natural Sciences, Philadelphia, PA, and The American Ornithologists' Union, Washington, DC. Number 123.

328. Egli M., G. 1996. Biomorfologia de Algunas Aves de Chile Central. *Boletin Chileno de Ornithologia* 3:2–9.

329. Eisenmann, E., and F. C. Lehmann V. 1962. *A new species of swift of the genus* Cypseloides *from Colombia*. American Museum Novitates 2117.

330. Eisenmann, E., and L. L. Short. 1982. *Systematics of the avian genus* Emberizoides *(Emberizidae)*. American Museum Novitates 2740.

331. Elgood, J. H. 1975. *Malimbus ibadanensis*: a fresh statement of biology and status. *Bulletin of the British Ornithologists' Union* 95:78–80.

332. Elliott, G., K. Walker, and R. Buckingham. 1991. The Auckland Island Rail. *Notornis* 38:199–209.

333. Elphick, C. S., and T. L. Tibbitts. 1998. Greater Yellowlegs (*Tringa melanoleuca*). In *The Birds of North America*, A. Poole and F. Gill (editors). The Birds of North America, Inc., Philadelphia, PA. Number 355.

334. Ely, C. R., and A. X. Dzubin. 1994. Greater White-fronted Goose (*Anser albifrons*). In *The Birds of North America*, A. Poole and F. Gill (editors). The Academy of Natural Sciences, Philadelphia, PA, and the American Ornithologists' Union, Washington, DC. Number 131.

335. Emlen, S. T., and P. H. Wrege. 2004. Size dimorphism, intrasexual competition, and sexual selection in Wattled Jacana (*Jacana jacana*), a sex-role-reversed shorebird in Panama. *Auk* 121:391–403.

336. Engilis, A., T. K. Pratt, C. B. Kepler, A. M. Ecton, and K. M. Fluetsch. 1996. Description of adults, eggshells, nestling, fledgling, and nest of the Poo-uli. *Wilson Bulletin* 108:607–619.

337. England, A. S., M. J. Bechard, and C. S. Houston. 1997. Swainson's Hawk (*Buteo swainsoni*). In *The Birds of North America*, A. Poole and F. Gill (editors). The Academy of Natural Sciences, Philadelphia, PA, and The American Ornithologists' Union, Washington, DC. Number 265.

338. Enkerlin-Hoeflich, E. C., and K. M. Hogan. 1997. Red-crowned Parrot (*Amazona viridigenalis*). In *The Birds of North America*, A. Poole and F. Gill (editors). The Academy of Natural Sciences, Philadelphia, PA, and The American Ornithologists' Union, Washington, DC. Number 292.
339. Enomoto, K. 1941. *A manual of wild birds*. Osaka chapter, Wild Bird Society of Japan, Osaka, Japan.
340. Erritzoe, J., and H. B. Erritzoe. 1998. *Pittas of the world. A monograph on the pitta family*. Lutterworth Press, Cambridge, UK.
341. Ervin, S., unpublished data.
342. Evans, M. R., and P. Barnard. 1995. Variable sexual ornaments in scarlet-tufted malachite sunbirds (*Nectarinia johnstoni*) on Mount Kenya. *Biological Journal of the Linnean Society* 54:371–381.
343. Evans, P. R. 1966. Autumn movements, moult, and measurements of the Lesser Redpoll *Carduelis flammea cabaret*. *Ibis* 108:183–216.
344. Evans, T. D., G. C. L. Dotson, and T. M. Brooks (editors). 1993. *Cambridge Philippines rainforest project 1991*. Final report, International Council of Bird Preservation, 74–77.
345. Ewins, P. J. 1993. Pigeon Guillemot (*Cepphus columba*). In *The Birds of North America*, A. Poole and F. Gill (editors). The Academy of Natural Sciences, Philadelphia, PA, and The American Ornithologists' Union, Washington, DC. Number 49.
346. Eyckerman, R., M. Louette, and M. Becuwe. 1992. On the biometry of birds ringed in Turkey. *Zoology in the Middle East* 6:29–37.
347. Fa, J. E., and M. L. Cavalheiro. 1998. Individual variation in food consumption and food preferences in St Lucia parrots *Amazona versicolor* at Jersey Wildlife Preservation Trust. *International Zoo Yearbook* 36:199–214.
348. Faaborg, J. 1985. Ecological constraints on West Indian bird distributions. *Ornithological Monographs* 36:621–653.
349. Faaborg, J., and J. E. Winters. 1979. Winter resident returns and longevity and weights of Puerto Rican birds. *Bird-Banding* 50:216–223.
350. Fancy, S. G., and C. J. Ralph. 1997. 'Apapane (*Himatione sanguinea*). In *The Birds of North America*, A. Poole and F. Gill (editors). The Academy of Natural Sciences, Philadelphia, PA, and The American Ornithologists' Union, Washington, DC. Number 296.
351. Fancy, S. G., and C. J. Ralph. 1998. 'I'iwi (*Vestiaria coccinea*). In *The Birds of North America*, A. Poole and F. Gill (editors). The Birds of North America, Inc., Philadelphia, PA. Number 327.
352. Farabaugh, S. M. 1982. The ecological and social significance of duetting. In *Acoustic communication in birds*, D. E. Kroodsma, E. H. Miller, and H. Ouellet (editors). Academic Press, New York, 85–125.
353. Reference deleted.
354. Farquhar, C. C., and K. L. Ritchie. 2002. Black-tailed Gnatcatcher (*Polioptila melanura*). In *The Birds of North America*, A. Poole and F. Gill (editors). The Birds of North America, Inc., Philadelphia, PA. Number 690.
355. Favero, M. 2001. Características morfológicas y dimorfismo sexual en la Paloma Antártica (*Chionis alba*). *Ornitologia Neotropical* 12:173–179.
356. Feare, C., and A. Craig. 1999. *Starlings and mynas*. Princeton University Press, Princeton, NJ.
357. Feinsinger, P. 1976. Organization of a tropical guild of nectarivorous birds. *Ecological Monographs* 46:257–291.
358. Ferguson-Lees, J., and D. A. Christie. 2001. *Raptors of the world*. Houghton-Mifflin, Boston, MA.
359. French, R. 1991. *A guide to the birds of Trinidad and Tobago*, second edition. Cornell University Press, Ithaca, NY.
360. Ficken, M., and J. Nocedal. 1992. Mexican Chickadee (*Parus sclateri*). In *The Birds of North America*, A. Poole, P. Stettenheim, and F. Gill (editors). The Academy of Natural Sciences, Philadelphia, PA, and The American Ornithologists' Union, Washington, DC. Number 8.
361. Finlayson, J. C. 1981. Seasonal distribution, weights and fat of passerine migrants at Gibraltar. *Ibis* 123:88–95.
362. Fiora, A. 1933. El peso de las aves. *Hornero* 5:174–188.
363. Fiora, A. 1934. El peso de las aves. *Hornero* 5:358–365.
364. Fisk, E. J. 1979. Fall and winter birds near Homestead, Florida. *Bird-Banding* 50:224–243; 297–419.
365. Fisher, K., and L. Fisher. 1996a. Comparative measurements of the Graceful and Yellow-spotted Honeyeaters. *Corella* 20:102–103.
366. Fisher, K., and L. Fisher. 1996b. Blue-faced Parrotfinch. *Corella* 20:103.
367. Fitzpatrick, J. W., and J. P. O'Neill. 1979. A new tody-tyrant from northern Peru. *Auk* 96:443–447.
368. Fitzpatrick, J. W., and D. F. Stotz. 1997. A new species of tyrannulet (*Phylloscartes*) from the Andean foothills of Peru and Bolivia. *Ornithological Monographs* 48:37–44.
369. Fitzpatrick, J. W., and D. E. Willard. 1990. *Cercomacra manu*, a new species of antbird from southwestern Amazonia. *Auk* 107:239–245.
370. Fitzpatrick, J. W., D. E. Willard, and D. F. Stotz. Weights and wing measurements of birds from southeastern Peru. Unpublished database included in Dunning (1992).
370a. Fjeldså, J. 1981. Biological notes on the Giant Coot, *Fulica gigantea*. *Ibis* 123:423–437.
371. Flammer, K., D. Whitt-Smith, and M. Papich. 2001. Plasma concentrations of doxycycline in selected psittacine birds when administered in water for potential treatment of *Chlamydophila psittaci* infection. *Journal of Avian Medicine and Surgery* 15:276–282.
372. Flood, N. J. 2002. Scott's Oriole (*Icterus parisorum*). In *The Birds of North America*, A. Poole and F. Gill (editors). The Birds of North America, Inc., Philadelphia, PA. Number 608.
373. Ford, H. A., and H. Bell. 1982. Density of birds in eucalypt woodland affected to varying degrees by dieback. *Emu* 82:202–208.
374. Ford, H. A., S. Noske, and L. Bridges. 1986. Foraging of birds in eucalypt woodland in north-eastern New South Wales. *Emu* 86:168–179.
375. Ford, J. 1979. A new subspecies of Grey Butcherbird from the Kimberley, Western Australia. *Emu* 79:191–194.
376. Ford, J. 1983. Evolutionary and ecological relationships between quail-thrushes. *Emu* 83:152–172.
377. Foster, J. T., J. Michael Scott, and P. W. Sykes. 2000. 'Akikiki (*Oreomystis bairdi*). In *The Birds of North America*, A. Poole and F. Gill (editors). The Birds of North America, Inc., Philadelphia, PA. Number 552.
378. Foster, M. S. 1987. Feeding methods and efficiencies of selected frugivorous birds. *Condor* 89:566–680.

379. Foster, M. S., and N. K. Johnson. 1974. Notes on birds of Costa Rica. *Wilson Bulletin* 86:58–62.
380. Fraga, R.M., and A. di Giácomo. 2004. Cooperative breeding of the Yellow-rumped Marshbird (Icteridae) in Argentina and Paraguay. *Condor* 106:671–673.
380a. Francis, C. Unpublished data previously included in Dunning (1992)
380b. Franco, P. Universiteit van Leiden, unpublished data.
381. Franklin, A. B., D. A. Clark, and D. B. Clark. 1979. Ecology and behavior of the Galapagos Rail. *Wilson Bulletin* 91:202–221.
382. Fraser, M. W., and D. J. Briggs. 1992. New information on the *Nesospiza* buntings at Inaccessible Island, Tristan da Cunha, and notes on their conservation. *Bulletin of the British Ornithologists' Club* 112:191–207.
383. Fredrickson, L. H., J. M. Anderson, F. M. Kozlik, and R. A. Ryder. 1977. American coot (*Fulica americana*). In *Management of migratory shore and upland game birds in North America*, G. C. Sanderson (editor). International Association of Fish & Wildlife Agencies, Washington, DC, 123–148.
383a. Freeman, S., and W. M. Jackson. 1990. Univariate metrics are not adequate to measure avian body size. *Auk* 107:69–74.
384. Frings, H., and M. Frings. 1961. Some biometric studies on the albatrosses of Midway Atoll. *Condor* 63:304–312.
385. Frith, C. B., and B. M. Beehler. 1998. *The birds of paradise, Paradisaeidae*. Oxford University Press, Oxford, UK.
386. Frith, C. B., and D. W. Frith. 1990. Nesting biology and relationships of the Lesser Melampitta *Melampitta lugubris*. *Emu* 90:65–73.
387. Frith, C. B., and D. W. Frith. 1993. Results of a preliminary highland bird banding project study at Tari Gap, Southern Highlands, Papua New Guinea. *Corella* 17:5–21.
388. Reference deleted.
389. Reference deleted.
390. Frith, C. B., and D. W. Frith. 2004. *The bowerbirds, Ptilonorhynchidae*. Oxford University Press, Oxford, UK.
390a. Frith, C. B., and D. W. Frith. 2005. A long-term bird banding study in upland tropical rainforest, Paluma Range, north-east Queensland with notes on breeding. *Corella* 29:25–48.
391. Frith, D. W., and C. B. Frith. 2000. The nesting biology of the Grey-headed Robin *Heteromyias albispecularis* (Petroicidae) in Australian upland tropical rainforest. *Emu* 100:81–94.
392. Fry, C. H. 1970. Ecological distribution of birds in north-eastern Mato Grosso State, Brazil. *An. Acad. Brasil. Cienc.* 42:275–318.
393. Fry, C. H., and K. Fry. 1992. *Kingfishers, bee-eaters and rollers*. Princeton University Press, Princeton, NJ.
394. Fry, C. H., S. Keith, and E. K. Urban. 1988. *The birds of Africa*. Academic Press, New York. Volume 3.
395. Fry, C. H., and S. Keith. 2004. *The birds of Africa*. Academic Press, New York. Volume 7.
396. Fry, C. H., S. Keith, and E. K. Urban (editors). 2000. *The birds of Africa*. Academic Press, New York. Volume 6.
396a. Fugler, S. R., S. Hunter, L. P. Newton, and W. K. Steele. 1987. Breeding biology of the Blue Petrels *Halobaena caerulea* at the Prince Edward Islands. *Emu* 87:103–110.
397. Furness, R. W. 1987. *The skuas*. T&AD Poyser, Calton, UK.
398. Gaither, J. C. 1994. Weights of Bornean understory birds. *Bulletin of the British Ornithologists' Club* 114:89–90.
399. Galbraith, I. C. J., and E. H. Galbraith. 1962. Land birds of Guadalcanal and the San Cristoval Group, eastern Solomon Islands. *Bulletin of the British Museum (Natural History): Zoology* 9:1–86.
400. Gamauf, A., M. Preleuthner, and H. Winkler. 1998. Philippines birds of prey: interactions among habitat, morphology, and behavior. *Auk* 115:713–726.
401. Gammonley, J. H. 1996. Cinnamon Teal (*Anas cyanoptera*). In *The Birds of North America*, A. Poole and F. Gill (editors). The Academy of Natural Sciences, Philadelphia, PA, and The American Ornithologists' Union, Washington, DC. Number 209.
401a. Garcia-del-Rey, E., A. G. Gosler, J. Gonzalez, and M. Wink. Sexual size dimorphism and moult in the endemic Plain Swift Apus unicolor—can adults be sexed accurately by discriminant analysis? Unpublished ms.
402. Gardali, T., and G. Ballard. 2000. Warbling Vireo (*Vireo gilvus*). In *The Birds of North America*, A. Poole and F. Gill (editors). The Birds of North America, Inc., Philadelphia, PA. Number 2000.
403. Garrido, O. H. 1983. A new subspecies of *Caprimulgus cubanensis* (Aves: Caprimulgidae) from the Isle of Pines, Cuba. *Auk* 100:988–991.
404. Garrido, O. H., K. C. Parkes, G. B. Reynard, A. Kirkconnell, and R. Sutton. 1997. Taxonomy of the Stripe-headed Tanager, genus *Spindalis* (Aves: Thraupidae) of the West Indies. *Wilson Bulletin* 109:561–594.
404a. Garrido, O. H., J. W. Wiley, and A. Kirkconnell. 2005. The genus *Icterus* in the West Indies. *Ornitologia Neotropical* 16:449–470.
405. Garrison, B. A. 1999. Bank Swallow (*Riparia riparia*). In *The Birds of North America*, A. Poole and F. Gill (editors). The Birds of North America, Inc., Philadelphia, PA. Number 414.
406. Gaston, A. J. 1994. Ancient Murrelet (*Synthliboramphus antiquus*). In *The Birds of North America*, A. Poole and F. Gill (editors). The Academy of Natural Sciences, Philadelphia, PA, and The American Ornithologists' Union, Washington, DC. Number 132.
407. Gaston, A. J., and S. B. C. Dechesne. 1996. Rhinoceros Auklet (*Cerorhinca monocerata*). In *The Birds of North America*, A. Poole and F. Gill (editors). The Academy of Natural Sciences, Philadelphia, PA, and The American Ornithologists' Union, Washington, DC. Number 212.
408. Gauger, V. H. 1999. Black Noddy (*Anous minutus*). In *The Birds of North America*, A. Poole and F. Gill (editors). The Birds of North America, Inc., Philadelphia, PA. Number 412.
409. Gaymer, R., R. A. A. Blackman, P. G. Dawson, M. Penny, and C. M. Penny. 1969. The endemic birds of Seychelles. *Ibis* 111:157–176.
410. Gayou, D. C. 1985. Body weights of south Texas Green Jays (*Cyanocorax yncas*). *Bulletin of the Texas Ornithological Society* 18:30–31.
411. George, T. L. 2000. Varied Thrush (*Ixoreus naevius*). In *The Birds of North America*, A. Poole and F. Gill (editors). The Birds of North America, Inc., Philadelphia, PA. Number 541.

412. Gerhardt, R. P., D. M. Gerhardt, C. J. Flatten, and N. B. Gonzalez. 1994. The food habits of sympatric *Ciccaba* owls in northern Guatemala. *Journal of Field Ornithology* 65:258–264.
413. Geupel, G. R., and G. Ballard. 2002. Wrentit (*Chamaea fasciata*). In *The Birds of North America*, A. Poole and F. Gill (editors). The Birds of North America, Inc., Philadelphia, PA. Number 654.
414. Gibbs, D., E. Barnes, and J. Cox. 2001. *Pigeons and doves: a guide to the pigeons and doves of the world.* Yale University Press, New Haven, CT.
415. Gibson, D. D. 1981. Migrant birds at Shemya Island, Aleutian Islands, Alaska. *Condor* 83:65–77.
416. Gibson, L. 1979. Breeding *Laterallus lecuopyrrhus*. *Avicultural Magazine* 85:63–67.
416a. Giesen, K. M. 1998. Lesser Prairie-Chicken (*Tympanuchus pallidicinctus*). In *The Birds of North America*, A. Poole and F. Gill (editors). The Birds of North America, Inc., Philadelphia, PA. Number 364.
417. Gilchrist, H. G. 2001. Glaucous Gull (*Larus hyperboreus*). In *The Birds of North America*, A. Poole and F. Gill (editors). The Birds of North America, Inc., Philadelphia, PA. Number 573.
418. Gill, B. J., and D. D. Dow. 1983. Morphology and development of nestling Grey-crowned and Hall's Babblers. *Emu* 83:41–43.
419. Gill, B. J., and C. R. Veitch. 1990. Measurements of bush birds on Little Barrier Island, New Zealand. *Notornis* 37:141–145.
420. Gill, F. B., R. A. Canterbury, and J. L. Confer. 2001. Blue-winged Warbler (*Vermivora pinus*). In *The Birds of North America*, A. Poole and F. Gill (editors). The Birds of North America, Inc., Philadelphia, PA. Number 584.
421. Gill, R. E., P. Canevari, and E. H. Iversen. 1998. Eskimo Curlew (*Numenius borealis*). In *The Birds of North America*, A. Poole and F. Gill (editors). The Birds of North America, Inc., Philadelphia, PA. Number 347.
422. Gill, R. E., P. S. Tomkovich, and B. J. McCaffery. 2002. Rock Sandpiper (*Calidris ptilocnemis*). In *The Birds of North America*, A. Poole and F. Gill (editors). The Birds of North America, Inc., Philadelphia, PA. Number 686.
423. Gilliard, E. T. 1961. *Four new birds from the mountains of central New Guinea.* American Museum Novitates 2031.
424. Gilliard, E. T., and M. LeCroy. 1961. Birds of the Victor Emanuel and Hindenburg Mountains, New Guinea. *Bulletin of the American Museum of Natural History* 123:1–86.
425. Gilliard, E. T., and M. LeCroy. 1966. Birds of the Middle Sepik Region, New Guinea. *Bulletin of the American Museum of Natural History* 132:245–276.
426. Gilliard, E. T., and M. LeCroy. 1967a. Results of the 1958–1959 Gilliard New Britain Expedition. *Bulletin of the American Museum of Natural History* 135:173–216.
427. Gilliard, E. T., and M. LeCroy. 1967b. Annotated list of birds of the Adelbert Mountains, New Guinea. *Bulletin of the American Museum of Natural History* 138:51–82.
428. Gilliard, E. T., and M. LeCroy. 1968. *Birds of the Schrader Mountain region, New Guinea. Results of the American Museum of Natural History expedition to New Guinea in 1964.* American Museum Novitates 2343.
429. Reference deleted.
430. Glutz von Blotzheim, U. N., K. M. Bauer, and E. Bezzel. 1977. *Handbuch der Vögel Mitteleuropas.* Akademische Verlagsgesellschaft, Wiesbaden, Germany. Volume 7.
431. Glutz von Blotzheim, U. N., and K. M. Bauer. 1980. *Handbuch der Vögel Mitteleuropas.* Akademische Verlagsgesellschaft, Wiesbaden, Germany. Volume 9.
432. Glutz von Blotzheim, U. N., K. M. Bauer and E. Bezzel. 1984. *Handbuch der Vögel Mitteleuropas*, second edition. Akademische Verlagsgesellschaft, Wiesbaden, Germany. Volume 6.
433. Gochfield, M., J. Burger, and I. C. T. Nisbet. 1998. Roseate Tern (*Sterna dougallii*). In *The Birds of North America*, A. Poole and F. Gill (editors). The Birds of North America, Inc., Philadelphia, PA. Number 370.
434. Gonzaga, L. P., and J. F. Pacheco. 1995. A new species of *Phylloscartes* (Tyrannidae) from the mountains of southern Bahia, Brazil. *Bulletin of the British Ornithologists' Club* 115:88–97.
435. Goodman, S. M., and P. C. Gonzales. 1989. Notes on Philippine birds, 12. Seven species new to Catanduanes Island. *Bulletin of the British Ornithologists' Club* 109:48–50.
436. Goodman, S. M., and P. C. Gonzales. 1990. *The birds of Mt. Isarug National Park, southern Luzon, Philippines, with particular reference to altitudinal distribution.* Fieldiana Zoology (New Series) 60,1–39.
437. Goodman, S. M., D. E. Willard, and P. C. Gonzales. 1995. *The birds of Sibuyan Island, Romblon Province, Philippines, with particular reference to elevational distribution and biogeographic affinities.* Fieldiana Zoology (New Series) 82,1–57.
438. Goodman, S. M., and M. S. Putnam. 1996. The birds of the eastern slopes of the Réserve Naturalle Intégrale d'Andringitra, Madagascar. In *A floral and faunal inventory of the eastern slopes of the Réserve Naturalle Intégrale d'Andringitra, Madagascar: with reference to elevational variation*, S. M. Goodman (editor). Fieldiana Zoology (New Series) 85, 171–190.
439. Gorski, L. J. 1969. Systematics and ecology of sibling species of Traill's Flycatcher. Ph.D. dissertation, University of Connecticut, Storrs, CT.
440. Grant, P. R. 1965. A systematic study of the terrestrial birds of the Tres Marias Islands, Mexico. *Postilla* 90:1–106.
441. Grant, P. R. 1979a. Ecological and morphological variation of Canary Island blue tits, *Parus caeruleus* (Aves: Paridae). *Biological Journal of the Linnean Society* 11:103–129.
442. Grant, P. R. 1979b. Evolution of the chaffinch, *Fringilla coelebs*, on the Atlantic Islands. *Biological Journal of the Linnean Society* 11:301–332.
443. Grant-Mackie, J. A. 1980. Acquisition of a specimen of the New Caledonian Kagu (Cagou). *Notornis* 27:292–293.
444. Gratto-Trevor, C. L. 1992. Semipalmated Sandpiper (*Calidris pusilla*). In *The Birds of North America*, A. Poole and F. Gill (editors). The Academy of Natural Sciences, Philadelphia, PA, and The American Ornithologists' Union, Washington, DC. Number 6.
445. Gratto-Trevor, C. L. 2000. Marbled Godwit (*Limosa fedoa*). In *The Birds of North America*, A. Poole and F. Gill (editors). The Birds of North America, Inc., Philadelphia, PA. Number 492.

446. Graves, G. R. 1992. The endemic land birds of Henderson Island, southeastern Polynesia: notes on natural history and conservation. *Wilson Bulletin* 104:32–43.

447. Graves, G. R., and D. Uribe Restrepo. 1989. A new allopatric taxon in the *Hapalopsittaca amazonina* (Psittacidae) superspecies from Colombia. *Wilson Bulletin* 101:369–524.

448. Graves, G. R., and J. S. Weske. 1987. *Tangara phillipsi*, a new species of tanager from the Cerros del Sira, eastern Peru. *Wilson Bulletin* 99:1–6.

449. Graves, G. R., and R. L. Zusi. 1990. Avian body weights from the lower Rio Xingu, Brazil. *Bulletin of the British Ornithologists' Club* 110:20–25.

450. Green, K. 1997. Biology of the Heard Island Shag *Phalacrocorax nivalis*. 2. Breeding. *Emu* 97:67–75.

451. Green, M. T., P. E. Lowther, S. L. Jones, S. K. Davis, and B. C. Dale. 2002. Baird's Sparrow (*Ammodramus bairdii*). In *The Birds of North America*, A. Poole and F. Gill (editors). The Birds of North America, Inc., Philadelphia, PA.

452. Greene, T. C. 1999. Aspects of the ecology of Antipodes Island Parakeet (*Cyanoramphus unicolor*) and Reischek's Parakeet (*C. novaezelandiae hochstetteri*) on Antipodes Island, October–November 1995. *Notornis* 46:301–310.

453. Greenewalt, C. H. 1962. Dimensional relationships for flying animals. *Smithsonian Miscellaneous Collections* 144(2):1–46.

454. Greenlaw, J. S., and J. D. Rising. Sharp-tailed Sparrow (*Ammodramus caudacutus*). In *The Birds of North America*, A. Poole and F. Gill (editors). The Academy of Natural Sciences, Philadelphia, PA, and The American Ornithologists' Union, Washington, DC. Number 112.

455. Greenway, J. C. 1966. *Birds collected on Batanta, off Western New Guinea, by E. Thomas Gilliard in 1964*. American Museum Novitates 2258.

456. Grinnell, J., J. Dixon, and J. M. Linsdale. 1930. Vertebrate natural history of a section of northern California through the Lassen Peak region. *University of California Publications in Zoology* 35:1–594.

457. Greig-Smith, P. W., and N. C. Davidson. 1977. Weights of West African savannah birds. *Bulletin of the British Ornithologists' Club* 97:96–99.

458. Groschupf, K. 1992. Five-striped Sparrow (*Aimophila quinquestriata*). In *The Birds of North America*, A. Poole, P. Stettenheim, and F. Gill (editors). The Academy of Natural Sciences, Philadelphia, PA, and The American Ornithologists' Union, Washington, DC. Number 21.

459. Groschupf, K. D., and C. W. Thompson. 1998. Varied Bunting (*Passerina versicolor*). In *The Birds of North America*, A. Poole and F. Gill (editors). The Birds of North America, Inc., Philadelphia, PA. Number 351.

460. Gross, A. O. 1937. Birds of the Bowdoin-MacMillan Arctic Expedition, 1934. *Auk* 54:12–42.

461. Gross, D. A., and P. E. Lowther. 2001. Yellow-bellied Flycatcher (*Empidonax flaviventris*). In *The Birds of North America*, A. Poole and F. Gill (editors). The Birds of North America, Inc., Philadelphia, PA. Number 566.

462. Grzybowski, J. A. 1995. Black-capped Vireo (*Vireo atricapillus*). In *The Birds of North America*, A. Poole and F. Gill (editors). The Academy of Natural Sciences, Philadelphia, PA, and The American Ornithologists' Union, Washington, DC. Number 181.

463. Guinan, J. A., P. A. Gowaty, and E. K. Eltzroth. 2000. Western Bluebird (*Sialia mexicana*). In *The Birds of North America*, A. Poole and F. Gill (editors). The Birds of North America, Inc., Philadelphia, PA. Number 510.

464. Gutierrez, R. J., A. B. Franklin, and W. S. Lahaye. 1995. Spotted Owl (*Strix occidentalis*). In *The Birds of North America*, A. Poole and F. Gill (editors). The Academy of Natural Sciences, Philadelphia, PA, and The American Ornithologists' Union, Washington, DC. Number 179.

465. Guzy, M. J. 2002. Antillean Nighthawk (*Chordeiles gundlachii*). In *The Birds of North America*, A. Poole and F. Gill (editors). The Birds of North America, Inc., Philadelphia, PA. Number 619.

466. Guzy, M. J., and B. J. McCaffery. 2002. Bluethroat (*Luscinia svecica*). In *The Birds of North America*, A. Poole and F. Gill (editors). The Birds of North America, Inc., Philadelphia, PA. Number 670.

467. Hackett, S. J., and K. V. Rosenberg. 1990. Comparison of phenotypic and genetic differentiation in South American antwrens (Formicariidae). *Auk* 107:473–489.

467a. Haggerty, T. M. 2006. Sexual size dimorphism and assortative mating in Carolina Wrens. *Journal of Field Ornithology* 77:259–265.

468. Hails, C. J. 1983. The metabolic rate of tropical birds. *Condor* 85:61–65.

469. Hale, A. M. 2004. Predation risk associated with group singing in a Neotropical wood-quail. *Wilson Bulletin* 116:167–171.

470. Hall, B. P. 1974. *Birds of the Harold Hall Australian Expeditions 1962–1970*. British Museum of Natural History, Number 33.

471. Hall, G. A. 1996. Yellow-throated Warbler (*Dendroica dominica*). In *The Birds of North America*, A. Poole and F. Gill (editors). The Academy of Natural Sciences, Philadelphia, PA, and The American Ornithologists' Union. Number 223.

472. Hamel, P. B. 2000. Cerulean Warbler (*Dendroica cerulea*). In *The Birds of North America*, A. Poole and F. Gill (editors). The Birds of North America, Inc., Philadelphia, PA. Number 511.

473. Hancock, J. A., J. A. Kushlan, and M. P. Kahl. 1992. *Storks, ibises, and spoonbills of the world*. Academic Press, New York.

474. Handel, C. M., and R. E. Gill. 2001. Black Turnstone (*Arenaria melanocephala*). In *The Birds of North America*, A. Poole and F. Gill (editors). The Birds of North America, Inc., Philadelphia, PA. Number 585.

475. Haney, J. C., D. S. Lee, and R. D. Morris. 1999. Bridled Tern (*Sterna anaethetus*). In *The Birds of North America*, A. Poole and F. Gill (editors). The Birds of North America, Inc., Philadelphia, PA. Number 468.

476. Haney, J. C., and S. D. McDonald. 1995. Ivory Gull (*Pagophila eburnea*). In *The Birds of North America*, A. Poole and F. Gill (editors). The Academy of Natural Sciences, Philadelphia, PA, and The American Ornithologists' Union, Washington, DC. Number 175.

477. Hanners, L. A., and S. R. Patton. 1998. Worm-eating Warbler (*Helmitheros vermivorus*). In *The Birds of North America*, A. Poole and F. Gill (editors). The Birds of North America, Inc., Philadelphia, PA. Number 367.

478. Hanson, H. C., and C. W. Kossack. 1957. Weight and body-fat relationships of mourning doves in Illinois. *Journal of Wildlife Management* 21:169–181.

479. Harrap, S., and D. Quinn. 1995. *Chickadees, tits, nuthatches and treecreepers*. Princeton University Press. Princeton, NJ.

480. Harrington, B. A. 2001. Red Knot (*Calidris canutus*). In *The Birds of North America*, A. Poole and F. Gill (editors). The Birds of North America, Inc., Philadelphia, PA. Number 563.
481. Harris, M. P. 1969a. Food as a factor controlling the breeding of *Puffinus lherminieri*. *Ibis* 111:140–156.
482. Harris, M. P. 1969b. The biology of storm petrels in the Galápagos Islands. *Proceedings of the California Academy of Sciences* 37:95–166.
483. Harris, M. P. 1970. Breeding ecology of the Swallow-tailed Gull, *Creagrus furcatus*. *Auk* 87:215–243.
484. Harris, M. P. 1979. Measurements and weights of British Puffins. *Bird Study* 26:179–186.
485. Harris, T. 2000. *Shrikes and bush-shrikes*. Princeton University Press, Princeton, NJ.
486. Harrison, C. S., T. S. Hida, and M. P. Seki. 1983. *Hawaiian seabird feeding ecology*. Wildlife Monographs 85.
487. Hartley, I. R., and P. J. K. McGowan. 1991. Moult and biometrics in five birds endemic to Palawan, Philippines. *Forktail* 6:74–77.
488. Hartman, F. A. 1955. Heart weight in birds. *Condor* 57:221–238.
489. Hartman, F. A. 1961. *Locomotor mechanisms of birds*. Smithsonian Miscellaneous Collections 142.
490. Hasegawa, H. 1977. Body temperatures of some autumn migrant birds. *J. Yamashina Institute of Ornithology* 7:97–99.
491. Hashim, I. M. 1994. Group size, weight and some measurements of Clapperton's francolin (*Francolinus clappertoni*). *African Journal of Ecology* 32:334–336.
492. Reference deleted.
493. Hatch, J. J., K. M. Brown, G. G. Hogan, and R. D. Morris. 2000. Great Cormorant (*Phalacrocorax carbo*). In *The Birds of North America*, A. Poole and F. Gill (editors). The Birds of North America, Inc., Philadelphia, PA. Number 553.
494. Hatch, J. J., and D. V. Weseloh. 1999. Double-crested Cormorant (*Phalacrocorax auritus*). In *The Birds of North America*, A. Poole and F. Gill (editors). The Birds of North America, Inc., Philadelphia, PA. Number 297.
495. Hatch, S. A., and D. N. Nettleship. 1998. Northern Fulmar (*Fulmarus glacialis*). In *The Birds of North America*, A. Poole and F. Gill (editors). The Birds of North America, Inc., Philadelphia, PA. Number 361.
496. Haverschmidt, F. 1948. Bird weights from Surinam. *Wilson Bulletin* 60:230–239.
497. Haverschmidt, F. 1952. More bird weights from Surinam. *Wilson Bulletin* 64:234–241.
498. Haverschmidt, F., and G. F. Mees. 1994. *Birds of Suriname*. Paramaribu, Suriname.
499. Hawkins, A., J.-M. Thiollay, and S. M. Goodman. 1998. The birds of the Réserve Spéciale d'Anjanaharibe-Sud, Madagascar. In *A floral and faunal inventory of the eastern slopes of the Réserve Naturalle Intégrale d'Andringitra, Madagascar: with reference to elevational variation*, S. M. Goodman (editor). Fieldiana Zoology (New Series) 85, 93–128.
500. Haug, E. A., B. A. Millsap, and M. S. Martell. 1993. Burrowing Owl (*Speotyto cunicularia*). In *The Birds of North America*. A. Poole and F. Gill (editors). The Academy of Natural Sciences, Philadelphia, PA, and The American Ornithologists' Union, Washington, DC. Number 61.
501. Hayes, F. E. 2001. Geographic variation, hybridization, and the leapfrog pattern of evolution in the Suiriri Flycatcher (*Suiriri suiriri*) complex. *Auk* 118:457–471.
502. Hayward, G. D., and P. H. Hayward. 1993. Boreal Owl (*Aegolius funereus*). In *The Birds of North America* A. Poole and F. Gill (editors). The Academy of Natural Sciences, Philadelphia, PA, and The American Ornithologists' Union, Washington, DC. Number 63.
503. Heather, B. D., and H. A. Robertson. 1997. *The field guide to the birds of New Zealand*. Oxford University Press, Oxford, UK.
504. Hejl, S. J., J. A. Holmes, and D. E. Kroodsma. 2002. Winter Wren (*Troglodytes troglodytes*). In *The Birds of North America*, A. Poole and F. Gill (editors). The Birds of North America, Inc., Philadelphia, PA. Number 623.
505. Henny, C. J., and L. F. VanCamp. 1979. Annual weight cycle in wild screech owls. *Auk* 96:795–796.
506. Henny, C. J., J. L. Carter, and B. J. Carter. 1981. A review of Bufflehead sex and age criteria with notes on weights. *Wildfowl* 32:117–122.
507. Herlugson, C. J. 1983. Growth of nestling Mountain Bluebirds. *Journal of Field Ornithology* 54:259–265.
508. Herholdt, J. J. 1988a. Bird weights from the Orange Free State (Part 1: nonpasserines). *Safring News* 17:3–14.
509. Herholdt, J. J. 1988b. Bird weights from the Orange Free State (Part II: passerines). *Safring News* 17:43–57.
510. Herholdt, J. J., and N. J. Grobler. 1988. Dimensions of Botha's Lark *Spizocorys fringillaris*. *Ostrich* 59:47.
511. Herkert, J. R., D. E. Kroodsma, and J. P. Gibbs. 2001. Sedge Wren (*Cistothorus platensis*). In *The Birds of North America*, A. Poole and F. Gill (editors). The Birds of North America, Inc., Philadelphia, PA. Number 582.
512. Herkert, J. R., P. D. Vickery, and D. E. Kroodsma. 2002. Henslow's Sparrow (*Ammodramus henslowii*). In *The Birds of North America*, A. Poole and F. Gill (editors). The Birds of North America, Inc., Philadelphia, PA. Number 672.
513. Herremans, M., M. Louette, and J. Stevens. 1991. Conservation status and vocal and morphological description of the Grand Comoro Scops Owl *Otus pauliani*. *Bird Conservation International* 1:123–135.
514. Herzog, S., unpublished data.
514a. Hicks, D. L. 1967. Adipose tissue composition and cell size in fall migratory thrushes (Turdidae). *Condor* 69:387–399.
515. Hickling, R. 1983. *Enjoying ornithology*. T & AD Poyser, Calton UK.
515a. Higgins, P. J. and S. J. J. F. Davies. 1996. *Handbook of Australian, New Zealand and Antarctic birds*. Oxford University Press, Melbourne, Australia. Volume 3.
516. Higgins, P. J., et al. 1999. *Handbook of Australian, New Zealand and Antarctic birds*. Oxford University Press, Melbourne, Australia. Volume 4.
517. Higgins, P. J., J. M. Peter, and W. K. Steele. 2001. *Handbook of Australian, New Zealand and Antarctic birds*. Oxford University Press, Melbourne, Australia. Volume 5.
518. Higgins, P. J., and J. M. Peter. 2002. *Handbook of Australian, New Zealand and Antarctic birds*. Oxford University Press, Melbourne, Australia. Volume 6.
518a. Higgins, P. J., J. M. Peter, and S. J. Cowling. 2006. *Handbook of Australian, New Zealand and Antarctic birds*. Oxford University Press, Melbourne, Australia. Volume 7.

519. Hill, D. P., and L. K. Gould. 1997. Chestnut-collared Longspur (*Calcarius ornatus*). In *The Birds of North America*, A. Poole and F. Gill (editors). The Academy of Natural Sciences, Philadelphia, PA, and The American Ornithologists' Union, Washington, DC. Number 288.
520. Hill, G. E. 1995. Black-headed Grosbeak (*Pheucticus melanocephalus*). In *The Birds of North America*, A. Poole and F. Gill (editors). The Academy of Natural Sciences, Philadelphia, PA, and The American Ornithologists' Union, Washington, DC. Number 143.
521. Hilty, S. L. 1977. *Chlorospingus flavovirens* rediscovered, with notes on other Pacific Colombian and Cauca Valley birds. *Auk* 94:44–49.
522. Hilty, S. L. 2003. *Birds of Venezuela*, 2nd ed. Princeton University Press, Princeton, NJ.
523. Hipfner, J. M., and G. Chapdelaine. 2002. Razorbill (*Alca torda*). In *The Birds of North America*, A. Poole and F. Gill (editors). The Birds of North America, Inc., Philadelphia, PA. Number 635.
524. Hitchcock, W. B., and J. L. McKean. 1969. Square-tailed Kite in the Northern Territory. *Emu* 69:115.
525. Hodgdon, K. Y. 1979. Operation Horned Grebe. *North American Bird Bander* 4:11.
526. Hoesch, W., and G. Niethammer. 1940. Die Vogelwelt Deutsch-Südenwestafrikas. *Journal für Ornithologie* 88:1–390.
527. Höglund, J., J. A. Kålås, and L. Løfaldii. 1990. Sexual dimorphism in the lekking Great Snipe. *Ornis Scandinavica* 21:1–6.
528. Hogstad, O. 1976. Sexual dimorphism and divergence in winter foraging behaviour of Three-toed Woodpeckers *Picoides tridactylus*. *Ibis* 118:41–50.
529. Hohman, W. L., and S. A. Lee. 2001. Fulvous Whistling-Duck (*Dendrocygna bicolor*). In *The Birds of North America*, A. Poole and F. Gill (editors). The Birds of North America, Inc., Philadelphia, PA. Number 562.
530. Holder, K., and R. Montgomerie. 1993. Rock Ptarmigan (*Lagopus mutus*). In *The Birds of North America* (Poole, A., and F. Gill, editors). The Academy of Natural Sciences, Philadelphia, PA, and The American Ornithologists' Union, Washington, DC. Number 51.
531. Holt, D. W., and J. L. Petersen. 2000. Northern Pygmy-Owl (*Glaucidium gnoma*). In *The Birds of North America*, A. Poole and F. Gill (editors). The Birds of North America, Inc., Philadelphia, PA. Number 494.
532. Holmes, R. T. Unpublished data.
533. Holmes, R. T., and H. F. Recher. 1986. Determinants of guild structure in forest bird communities: an intercontinental comparison. *Condor* 88:427–439.
534. Holyoak, D. T. 2001. *Nightjars and their allies*. Oxford University Press, Oxford, UK.
535. Hoogerwerf, A. 1971. On a collection of birds from the Vogelkop, near Manokwari, north-western New Guinea. *Emu* 71:1–12, 71:73–83.
536. Horne, R. S. C. 1985. Diet of Royal and Rockhopper Penguins at Macquarie Island. *Emu* 85:150–156.
537. Houston, C. S., and D. E. Bowen. 2001. Upland Sandpiper (*Bartramia longicauda*). In *The Birds of North America*, Poole, A. and F. Gill (editors). The Birds of North America, Inc., Philadelphia, PA. Number 580.
538. Houston, D. C. 1976. Breeding of the White-backed and Ruppell's Griffon Vultures, *Gyps africanus* and *G. rueppellii*. *Ibis* 118:14–50.
539. Howe, H. F. 1977. Bird activity and seed dispersal of a tropical wet forest tree. *Ecology* 58:539–550.
540. Howe, H. F., and G. A. Vande Kerckhove. 1981. Removal of wild nutmeg (*Virola surinamensis*) crops by birds. *Ecology* 62:1093–1106.
541. Howell, C. A., and S. N. G. Howell. 2000. Xantus's Hummingbird (*Hylocharis xantusii*). In *The Birds of North America*, Poole, A. and F. Gill (editors). The Birds of North America, Inc., Philadelphia, PA. Number 554.
542. Howell, T. R. 1972. Birds of the lowland pine savanna of northeastern Nicaragua. *Condor* 74:316–340.
543. Hu, D.-S., L. Joseph, and D. Argo. 2000. Distribution, variation, and taxonomy of *Topaza* hummingbirds (Aves: Trochilidae). *Ornitologia Neotropical* 11:123–142.
544. Hubbard, J. P. 1967. Notes on some Chiapas birds. *Wilson Bulletin* 79:236.
545. Hughes, J. M. 1996. Greater Roadrunner (*Geococcyx californianus*). In *The Birds of North America*, Poole, A. and F. Gill (editors). The Birds of North America, Inc., Philadelphia, PA, and The American Ornithologists' Union, Washington, DC. Number 244.
546. Humphrey, P. S., D. Bridge, P. W. Reynolds, and R. T. Peterson. 1970. *Preliminary Smithsonian manual: Birds of Isla Grande (Tierra del Fuego)*. Smithsonian Institute Press, Washington, D.C.
547. Hunter, F. M., I. L. Jones, J. C. Williams, and G. V. Byrd. 2002. Breeding biology of the Whiskered Auklet (*Aethia pygmaea*) at Buldir Island, Alaska. *Auk* 119:1036–1051.
548. Hussell, D. J. T., and R. Montgomerie. 2002. Lapland Longspur (*Calcarius lapponicus*). In *The Birds of North America*, A. Poole and F. Gill (editors). The Birds of North America, Inc., Philadelphia, PA. Number 656.
549. Illera, J. C. Unpublished data.
550. Illera, J. C., and M. Díaz. Reproduction in an endemic bird of a semiarid island: a food-mediated process. *Journal of Avian Ecology*, in press.
551. Imber, M. J. 1976. Breeding biology of the Grey-faced Petrel *Pterodroma macroptera gouldi*. *Ibis* 118:51–64.
551a. Imber, M. J., and A. J. D. Tennyson. 2001. A new petrel species (Procellariidae) from the south-west Pacific. *Emu* 101:123–127.
552. Indrawan, M., and S. Somadikaria. 2004. A new hawk-owl from the Togian Islands, Gulf of Tomini, central Sulawesi, Indonesia. *Bulletin of the British Ornithologists' Club* 124:160–171.
553. Iñigo E., E., M. Ramos, and F. González. 1987. Two recent records of Neotropical eagles in southern Veracruz, Mexico. *Condor* 89:671–672.
554. Irving, L. 1960. *Birds of Anatuvuk Pass, Kobuk, and Old Crow*. U.S. National Museum Bulletin 217.
555. Ishida, K., Unpublished data.
556. Ishizuka, T. Unpublished data.

557. Islam, K. 1999. Erckel's Francolin (*Francolinus erckelii*), Black Francolin (*Francolinus francolinus*) and Gray Francolin (*Francolinus pondicerianus*). In *The Birds of North America*, A. Poole, A. and F. Gill (editors). The Birds of North America, Inc., Philadelphia, PA. Numbers 394–396.
558. Islam, K., and R. N. Williams. 2000. Red-vented Bulbul (*Pycnonotus cafer*) and Red-whiskered Bulbul (*Haematopus bachmani*). In *The Birds of North America*, A. Poole and F. Gill (editors). The Birds of North America, Inc., Philadelphia, PA. Number 520.
559. Isler, M. L., and P. R. Isler. 1999. *The tanagers: natural history, distribution, and identification*. Smithsonian Institution Press, Washington, D.C.
560. Isler, M. L., P. R. Isler, and B. M. Whitney. 1997. Biogeography and systematics of the *Thamnophilus punctatus* (Thamophilidae) complex. *Ornithological Monographs* 48:355–381.
561. Isler, M. L., J. Alvarez A., P. R. Isler, and B. M. Whitney. 2001. A new species of *Percnostola* antbird (Passeriformes: Thamnophilidae) from Amazonian Peru, and an analysis of species limits within *Percnostola rufifrons*. *Wilson Bulletin* 113:164–176.
562. Jackson, H. D. 1972. Avifaunal survey of the Umtali Municipal Area. 1. The Muneni River collection: a comparison of samples from riparian forest and miombo woodland. *Arnoldia* 6:1–10.
563. Jackson, H. D. 1976. Avifaunal survey of the Umtali Municipal Area. 2. The Cecil Kop collection: a comparison of samples from montane grassland, montane thicket and montane forest. *Arnoldia* 5:1–11.
564. Jackson, H. D. 1986. Avifaunal survey of the Umtali Municipal Area. 3. The Gimboki collection: a comparison of samples from riparian thicket, miombo woodland on sandflats and miombo woodland on rocky slopes. *Arnoldia* 9:325–332.
565. Jackson, H. D. 1987a. Avifaunal survey of the Mutare Municipal Area. V. The Birkley South Collection: a comparison of samples from grassland, woodland and thicket. *Arnoldia Zimbabwe* 9:361–367.
566. Jackson, H. D. 1987b. Avifaunal survey of the Mutare Municipal Area. IV. The Matika Collection: a comparison of samples from grassland, thicket, woodland and riparian forest. *Arnoldia Zimbabwe* 9:353–360.
567. Jackson, H. D. 1989. Weights of birds collected in the Mutare Municipal Area, Zimbabwe. *Bulletin of the British Ornithologists' Club* 109:100–106.
568. Jackson, J. A., and H. R. Ouellet. 2002. Downy Woodpecker (*Picoides pubescens*). In *The Birds of North America*, A. Poole and F. Gill (editors). The Birds of North America, Inc., Philadelphia, PA. Number 613.
569. Jackson, J. A., H. R. Ouellet, and B. J. S. Jackson. 2002. Hairy Woodpecker (*Picoides villosus*). In *The Birds of North America*, A. Poole and F. Gill (editors). The Birds of North America, Inc., Philadelphia, PA. Number 702.
570. Jaksic, F. M., J. A. Iriarte, and J. E. Jiménez. 2002. The raptors of Torres del Paine National Park, Chile: biodiversity and conservation. *Revista Chilena de Historia Natural* 75:449–461.
571. James, J. D., and J. E. Thompson. 2001. Black-bellied Whistling-Duck (*Dendrocygna autumnalis*). In *The Birds of North America*, A. Poole and F. Gill (editors). The Birds of North America, Inc., Philadelphia, PA. Number 578.
572. James, R. D. 1998. Blue-headed Vireo (*Vireo solitarius*). In *The Birds of North America*, A. Poole and F. Gill (editors). The Birds of North America, Inc., Philadelphia, PA. Number 379.
573. Reference deleted.
574. Jeggo, D. J., A. F. Allchurch, and L. Quamie. 1991. A report on surgical sexing of the St. Vincent Parrot *Amazona guildingii* at the in situ captive breeding programme. *Dodo Journal of the Jersey Wildlife Preservation Trust* 27:103–113.
575. Jehl, J. R. 1987. Geographic variation and evolution in the California Gull (*Larus californicus*). *Auk* 104:421–428.
576. Jehl, J. R., and B. G. Murray. 1986. The evolution of normal and reverse sexual size dimorphism in shorebirds and other birds. *Current Ornithology* 3:1–86.
577. Jenkins, A. R. 1995. Morphometrics and flight performance of southern African Peregrine and Lanner Falcons. *Journal of Avian Biology* 26:49–58.
578. Jenkins, J. M. 1979. Natural history of the Guam Rail. *Condor* 81:404–408.
579. Jenkins, J. M. 1983. The native forest birds of Guam. *Ornithological Monographs* 31.
580. Jenkins, P. G. 1982. The Royal Air Force Ornithological Society Belize Expedition, 1981. Unpublished report to Birdlife International.
581. Jensen, F. P. 1983. A new species of sunbird from Tanzania. *Ibis* 125:447–449.
582. Jiménez, J. E., and F. M. Jaksić. 1989. Biology of the Austral Pygmy-Owl. *Wilson Bulletin* 101:377–389.
583. Jiménez, J. E., and F. M. Jaksić. 1990. Historia natural del Aguila *Geranoaetus melanoleucus*: una revision. *El Hornero* 13:97–110.
584. Johnsgard, P. A. 1973. *Grouse and quails of North America*. University of Nebraska Press, Lincoln, NE.
584a. Johnsgard, P. A. 1981. *The plovers, sandpipers, and snipes of the world*. University of Nebraska Press, Lincoln, NE.
585. Johnsgard, P. A. 1983a. *The grouse of the world*. University of Nebraska Press, Lincoln, NE.
586. Johnsgard, P. A. 1983b. *Cranes of the world*. Indiana University Press, Bloomington, IN.
587. Johnsgard, P. A. 1983c. *Hummingbirds of North America*. Smithsonian Institute Press, Washington, DC.
588. Johnsgard, P. A. 1986. *The pheasants of the world*. Oxford University Press, Oxford, UK.
589. Johnsgard, P. A. 1988. *The quails, partridges, and francolins of the world*. Oxford University Press, Oxford, UK.
590. Johnsgard, P. A. 1991. *Bustards, hemipodes, and sandgrouse: birds of dry places*. Oxford University Press, Oxford, UK.
591. Johnsgard, P. A. 1993. *Cormorants, darters, and pelicans of the world*. Smithsonian Institute Press, Washington, D.C.
591a. Johnsgard, P. A. 2000. *Trogons and quetzals of the world*. Smithsonian Institute Press, Washington, D.C.
592. Johnsgard, P. A., and M. Carbonell. 1996. *Ruddy ducks and other stifftails: their behavior and biology*. University of Oklahoma Press, Norman, OK.
593. Johnson, K., and B. D. Peer. 2001. Great-tailed Grackle (*Quiscalus mexicanus*). In *The Birds of North America*, A. Poole and F. Gill (editors). The Birds of North America, Inc., Philadelphia, PA. Number 576.
594. Johnson, N. K. 1965. The breeding avifaunas of the Sheep and Spring Ranges in southern Nevada. *Condor* 67:93–124.

595. Johnson, N. K., and R. E. Jones. 1990. Geographic differentiation and distribution of the Peruvian Screech-Owl. *Wilson Bulletin* 102:199–212.

596. Johnson, N. K., and R. E. Jones. 2001. A new species of tody-tyrant (Tyrannidae: *Poecilotriccus*) from northern Peru. *Auk* 118:334–341.

597. Johnson, O. W., and P. G. Connors. 1996. American Golden-Plover (*Pluvialis dominica*), and Pacific Golden-Plover (*Pluvialis fulva*). In *The Birds of North America*, A. Poole and F. Gill (editors). The Academy of Natural Sciences, Philadelphia, PA, and The American Ornithologists' Union, Washington, DC. Numbers 201–202.

598. Johnson, R. E. 2002. Black Rosy Finch (*Leucosticte atrata*). In *The Birds of North America*, A. Poole and F. Gill (editors). The Birds of North America, Inc., Philadelphia, PA. Number 678.

599. Johnson, R. E., P. Hendricks, D. L. Pattie, and K. B. Hunter. 2000. Brown-capped Rosy-Finch (*Leucosticte australis*). In *The Birds of North America*, A. Poole and F. Gill (editors). The Birds of North America, Inc., Philadelphia, PA, and The American Ornithologists' Union, Washington, DC. Number 536.

600. Johnson, S. R., and R. W. Campbell. 1995. Crested Myna (*Acridotheres cristatellus*). In *The Birds of North America*, A. Poole and F. Gill (editors). The Academy of Natural Sciences, Philadelphia, PA, and The American Ornithologists' Union, Washington, DC. Number 157.

601. Johnson, T. B., and S. Hilty. 1976. Notes on the Sickle-winged Guan in Colombia. *Auk* 93:194–195.

602. Johnston, D. W., and F. S. L. Williamson. 1960. Heart weights of North American crows and ravens. *Wilson Bulletin* 72:248–252.

603. Johnston, R. F., and S. G. Johnson. 1989. Nonrandom mating in feral pigeons. *Condor* 91:23–29.

604. Jones, A., S. Schubel, J. Jolly, M. de L. Brooke, and J. Vickery. 1995. Behaviour, natural history, and annual cycle of the Henderson Island rail *Porzana atra* (Aves: Rallidae). *Biological Journal of the Linnean Society* 56:167–183.

605. Jones, D. N., R. W. Dekker, and C. S. Roselaar. 1995. *The megapodes: Megapodiidae*. Oxford University Press, Oxford.

606. Jones, I. L. 1993. Crested Auklet (*Aethia cristatella*). In *The Birds of North America*, A. Poole and F. Gill (editors). The Academy of Natural Sciences, Philadelphia, PA, and The American Ornithologists' Union, Washington, DC. Number 70.

607. Jones, I. L., N. B. Konyukhov, J. C. Williams, and G. V. Byrd. 2001. Parakeet Auklet (*Aethia psittacula*). In *The Birds of North America*, A. Poole and F. Gill (editors). The Birds of North America, Inc., Philadelphia, PA. Number 594.

608. Jones, P. W., and T. M. Donovan. 1996. Hermit Thrush (*Catharus guttatus*). In *The Birds of North America*, A. Poole and F. Gill (editors). The Academy of Natural Sciences, Philadelphia, PA, and The American Ornithologists' Union, Washington, DC. Number 261.

609. Jones, S. L., and J. S. Dieni. 1995. Canyon Wren (*Catherpes mexicanus*). In *The Birds of North America*, A. Poole and F. Gill (editors). The Academy of Natural Sciences, Philadelphia, PA, and The American Ornithologists' Union, Washington, DC. Number 197.

610. Jouanin, C. 1987. Notes on the nesting of Procellariiformes in Reunion. In *Studies of Mascarene Islands birds*, A. W. Diamond (editor). Cambridge University Press, Cambridge, MA, 359–363.

611. Jouventin, P., J. Martinez, and J. P. Roux. 1989. Breeding biology and current status of the Amsterdam Island Albatross *Diomedea amsterdamensis*. *Ibis* 131:171–182.

611a. Jouventin, P., J.-L. Mougin, J.-C. Stahl, H. Weimerskirch et al. 1985. Comparative biology of the burrowing petrels of the Crozet Islands. *Notornis* 32:157–220.

612. Joyner, K. L., N. de Berger, E. H. Lopez, A. Brice, and P. Nolan. 1992. Health parameters of wild psittacines in Guatemala: a preliminary report. *Proceedings of the Association of Avian Veterinarians*, New Orleans, 287–303.

613. Junge, G. C. A., and G. F. Mees. 1961. The avifauna of Trinidad and Tobago. *Zoologische Verhandelingen* 37:1–172.

613a. Kannan, R., and D. A. James. 1998. Common Peafowl (*Pavo cristatus*). In *The Birds of North America,* A. Poole and F. Gill (editors). The Birds of North America, Inc., Philadelphia, PA. Number 377.

614. Kannan, R., and D. A. James. 2001. Common Myna (*Acridotheres tristis*). In *The Birds of North America*, A. Poole and F. Gill (editors). The Birds of North America, Inc., Philadelphia, PA. Number 583.

615. Karr, J. R. 1971. Structure of avian communities in selected Panama and Illinois habitats. *Ecological Monographs* 41:207–233.

616. Karr, J. R. 1976. Weights of African birds. *Bulletin of the British Ornithologists' Club* 96:92–996.

617. Karr, J. R., S. K. Robinson, J. G. Blake, and R. O. Bierregaard. 1990. *Birds of four neotropical forests*, A. Gentry (editor). Yale University Press, New Haven, CT, 237–269.

618. Karr, J. R., M. F. Willson, and D. J. Moriarty. 1978. Weights of some Central American birds. *Brenesia* 14–15:249–257.

619. Karubian, J. 2001. The social organization and mating system of the Striated Grasswren. *Condor* 103:412–417.

620. Kawakami, K., and H. Higuchi. 2003. Interspecific interactions between the native and introduced white-eyes in the Bonin Islands. *Ibis* 145:583–592.

621. Kazuto, K. 1996. Report on the conservation of endangered Oriental Greenfinch *Carduelis sinica*. Unpublished report [In Japanese].

622. Kazuto, K. Unpublished data.

622a. Kear, J. 2005. *Ducks, geese, and swans*. Oxford University Press, Oxford, UK.

623. Keith, S., E. K. Urban, and C. H. Fry. 1992. *The birds of Africa*. Academic Press, New York. Volume 4.

624. Keitt, B. S., B. R. Tershy, and D. A. Croll. 2000. Black-vented Shearwater (*Puffinus opisthomelas*). In *The Birds of North America*, A. Poole and F. Gill (editors). The Academy of Natural Sciences, Philadelphia, PA, and The American Ornithologists' Union, Washington, DC. Number 521.

625. Kemp, A. C. 1989. Estimation of biological indices for little-known African owls. In *Raptors in the modern world*, B.-U. Meyburg and R. D. Chancellor (editors). World Working Group on Birds of Prey and Owls, Berlin, Germany, 441–449.

626. Kemp, A. 1995. *The hornbills*. Oxford University Press, Oxford, UK.

626a. Kennedy, R. S., T. H. Fisher, S. C. B. Harrap, A. C. Diesmos, and A. S. Manamtam. 2001. A new species of woodcock (Aves: Scolopacidae) from the Philippines and a re-evaluation of other Asian/Papuasian woodcock. *Forktail* 17:1–12.

627. Kenward, B., C. Rutz, A. A. S. Weir, J. Chappell, and A. Kacelnik. 2004. Morphology and sexual dimorphism of the New Caledonian Crow *Corvus moneduloides*, with notes on its behaviour and ecology. *Ibis* 146:652–660.

628. Kennedy, R. S., P. C. Gonzales, and H. C. Miranda. 1997. New *Aethopyga* sunbirds (Aves: Nectariniidae) from the island of Mindanao, Philippines. *Auk* 114:1–10.

629. Kepler, C. B. 1969. *Breeding biology of the Blue-faced Booby on Green Island, Kure Atoll.* Publications of the Nuttall Ornithologists' Club, number 8.

630. Kerlinger, P., and M. R. Lein. 1988. Causes of mortality, fat condition, and weights of wintering Snowy Owls. *Journal of Field Ornithology* 59:7–12.

630a. Kesler, D. C., I. F. Lopes, and S. M. Haig. 2006. Sex determination of Pohnpei Micronesian Kingfishers using morphological and molecular genetic techniques. *Journal of Field Ornithology* 77:229–232.

631. Kessel, B., D. A. Rocque, and J. S. Barclay. 2002. Greater Scaup (*Aythya marila*). In *The Birds of North America*, A. Poole and F. Gill (editors). The Birds of North America, Inc., Philadelphia, PA. Number 650.

632. King, J. R. 1989. Notes on the birds of the Rio Mazan Valley, Azuay Province, Ecuador, with special reference to *Leptosittaca branickii, Hapalopsittaca amazonina pyrrhops* and *Metallura baroni. Bulletin of the British Ornithologists' Club* 109:140–147.

633. King, J. R. 1991. Body weights of some Ecuadorian birds. *Bulletin of the British Ornithologists' Club* 111:46–49.

634. Kirk, D. A., and A. G. Gosler. 1994. Body condition varies with migration and competition in migrant and resident South American vultures. *Auk* 111:933–944.

635. Kirk, D. A., and M. J. Mossman. 1998. Turkey Vulture (*Cathartes aura*). In *The Birds of North America*, (A. Poole and F. Gill (editors). The Birds of North America, Inc., Philadelphia, PA. Number 339.

636. Kirkconnell, A., and O. H. Garrido. 1991. The Thick-billed Vireo, *Vireo crassirostris* (Aves: Vireonidae), a new addition to the Cuban avifauna. *Ornitologia Neotropical* 2:99–100.

637. Kirwan, G. Unpublished data.

638. Kitson, A. R. 1980. *Larus relictus*—a review. *Bulletin of the British Ornithologists' Club* 100:178–185.

639. Kiyosu, Y. 1966. *Encyclopedia of wild birds.* Tokyodo Publ., Tokyo, Japan.

640. Klaas, E. E. 1968. Summer birds from the Yucatan Peninsula, Mexico. *University of Kansas Publications, Museum of Natural History* 17:579–611.

641. Klima, J., and J. R. Jehl. 1998. Stilt Sandpiper (*Calidris himantopus*). In *The Birds of North America*, A. Poole and F. Gill (editors). The Birds of North America, Inc., Philadelphia, PA. Number 341.

642. Knopf, F. L. 1996. Mountain Plover (*Charadrius montanus*). In *The Birds of North America*, A. Poole and F. Gill (editors). The Academy of Natural Sciences, Philadelphia, PA, and The American Ornithologists' Union, Washington, DC. Number 211.

643. Knox, A. G., and P. E. Lowther. 2000. Hoary Redpoll (*Carduelis hornemanni*). In *The Birds of North America*, A. Poole and F. Gill (editors). The Birds of North America, Inc., Philadelphia, PA. Number 544.

644. Kobe Oji Zoo, Kobe, Japan, unpublished data (M. Koichi) included in Dunning (1992).

645. Kochert, M. N., K. Steenhof, C. L. McIntyre, and E. H. Craig. 2002. Golden Eagle (*Aquila chrysaetos*). In *The Birds of North America*, A. Poole and F. Gill (editors). The Birds of North America, Inc., Philadelphia, PA. Number 684.

646. Koenig, W. D., E. L. Walters, J. R. Walters, J. S. Kellam, K. G. Michalek, and M. S. Schrader. 2005. Seasonal body weight variation in five species of woodpeckers. *Condor* 107:810–822.

646a. Komar, O., and W. A. Thurber 2003. Predation on birds by a cuckoo (Cuculidae), mockingbird (Mimidae), and saltator (Cardinalidae). *Wilson Bulletin* 115:205–208.

647. Komen, J. 1991. Energy requirements of nestling Cape Vultures. *Condor* 93:153–158.

648. König, C. 1991. Zur Taxonomie und Ökologie der Sperlingskäuze (*Glaucidium* spp.) des Andenraumes. *Okol. Vögel* 13:15–76.

649. König, C., F. Weick, and J.-H. Becking. 1999. *Owls: a guide to the owls of the world.* Yale University Press, New Haven, CT.

650. Korpimaki, E. 1987. Clutch size, breeding success and brood size experiments in Tengmalm's Owl *Aegolius funereus*: a test of hypotheses. *Ornis Scandinavica* 18:277–284.

650a. Krabbe, N. unpublished data.

651. Krabbe, N. 1992. A new subspecies of the Slender-billed Miner *Geositta tenuirostris* (Furnariidae) from Ecuador. *Bulletin of the British Ornithologists' Club* 112:166–169.

652. Krabbe, N., D. J. Agro, N. H. Rice, M. Jacome, L Navarrete, and F. Sornoza M. 1999a. A new species of antpitta (Formicariidae: *Grallaria*) from the southern Ecuadorian Andes. *Auk* 116:882–890.

653. Krabbe, N., P. Flórez, G. Suárez, J. Castaño, J. D. Arango, P. C. Pulgarin, W. A. Múnera, F. G. Stiles, and P. Salaman. 2005. Rediscovery of the Dusky Starfrontlet *Coeligena orina*, with a description of the adult plumages and a reassessment of its taxonomic status. *Ornitologia Colombiana* 3:28–35.

654. Krabbe, N., M. L. Isler, P. R. Isler, B. M. Whitney, J. Alvarez A., and P. J. Greenfield. 1999b. A new species in the *Myrmotherula haematonota* superspecies (Aves: Thamnophilidae) from the western Amazonian lowlands of Ecuador and Peru. *Wilson Bulletin* 111:157–165.

655. Krabbe, N., P. Salaman, A. Cortés, A. Quevedo, L. A. Ortega, and C. D. Cadena. 2005. A new species of *Scytalopus* tapaculo from the upper Madgalena Valley, Colombia. *Bulletin of the British Ornithologists' Club* 125:93–108.

656. Krabbe, N., and T. S. Schulenberg. 1997. Species limits and natural history of *Scytalopus* tapaculos (Rhinocryptidae), with descriptions of the Ecuadorian taxa, including three new taxa. *Ornithological Monographs* 48:47–88.

656a. Krapu, G. L., J. L. Eldridge, C. L. Gratto-Trevor, and D. A. Buhl. 2006. Fat dynamics of Arctic-nesting sandpipers during spring in mid-continental North America. *Auk* 123:323–334.

656b. Kratter, A. W., J. J. Kirchman, and D. W. Steadman. 2006. Upland bird communities on Santo, Vanuatu, Southwest Pacific. *Wilson Journal of Ornithology* 118:295–308.

657. Kratter, A. W., T. S. Sillett, R. T. Chesser, J. P. O'Neill, T.A. Parker, and A. Castillo. 1993. Avifauna of a chaco locality in Bolivia. *Wilson Bulletin* 105:114–141.

658. Kratter, A. W., D. W. Steadman, C. E. Smith, and C. E. Filardi. 2001. Reproductive condition, moult, and body mass of birds from Isabel, Solomon Islands. *Bulletin of the British Ornithologists' Club* 121:128–144.
659. Krauczuk, E. R. 2000. Presencia de *Chordeiles pusillus* como nidificante en la provincia de Misiones, Argentina. *Ornitologia Neotropical* 11:85–86.
660. Kroodsma, D. E., and J. Verner. 1997. Marsh Wren (*Cistothorus palustris*). In *The Birds of North America*, A. Poole and F. Gill (editors). The Birds of North America, Inc., Philadelphia, PA. Number 308.
661. Kuo, W.-W. 2000. The sex difference and breeding ecology of Grey-cheeked Fulvetta (*Alcippe morrisonia*) at Fushan Experimental Forest. M.S. thesis, National Taiwan University. *Yamashina Institute for Ornithology, Miscellaneous Reports* 5:111–137.
662. Kuroda, N. 1967. Morpho-anatomical analysis of parallel evolution between Diving Petrel and Ancient Auk, with comparative osteological data of other species. Misc. Rpts. Yamashina Institute for Ornithology, Number 27.
663. Kushlan, J. A. 1977. Sexual dimorphism in the White Ibis. *Wilson Bulletin* 89:92–98.
664. LaBastille, A. 1974. *Ecology and management of the Atitlán, Guatemala*. Wildlife Monographs 37.
664. Ladd, C., and L. Gass. 1999. Golden-cheeked Warbler (*Dendroica chrysoparia*). In *The Birds of North America*, A. Poole and F. Gill (editors). The Birds of North America, Inc., Philadelphia, PA. Number 420.
665. Lambert, F. R. 1989. Pigeons as seed predators and dispersers of figs in a Malaysian lowland forest. *Ibis* 131:321–327.
666. Lambert, F. R. 1998. A new subspecies of *Amaurornis* from the Talaud Islands, Indonesia, and a review of taxonomy of bush hens occurring from the Philippines to Australasia. *Bulletin of the British Ornithologists' Club* 118:67–83.
667. Lanctot, R. B., and C. D. Laredo. 1994. Buff-breasted Sandpiper (*Tryngites subruficollis*). In *The Birds of North America*, A. Poole and F. Gill (editors). The Academy of Natural Sciences, Philadelphia, PA, and The American Ornithologists' Union, Washington, DC. Number 91.
667a. Lane, D. F., T. Valqui H., J. Alvarez A., J. Armenta, and K. Eckhardt. 2006. The rediscovery and natural history of the White-masked Antbird (*Pithys castaneus*). *Wilson Journal of Ornithology* 118:13–22.
668. Langham, N. P. E. 1987. Morphometrics and moult in Fijian passerines. *New Zealand Journal of Zoology* 14:463–475.
669. Lanyon, S. M., D. F. Stotz, and D. E. Willard. 1990. *Clytoctantes atrogularis*, a new species of antbird from western Brazil. *Wilson Bulletin* 102:571–580.
670. Lanyon, W. E. 1978. Revision of the *Myiarchus* flycatchers of South America. *Bulletin of the American Museum of Natural History* 161:427–628.
671. Lanyon, W. E. 1995. Eastern Meadowlark (*Sturnella magna*). In *The Birds of North America*, A. Poole and F. Gill (editors). The Academy of Natural Sciences, Philadelphia, PA, and The American Ornithologists' Union, Washington, DC. Number 160.
672. Latta, S. C. Unpublished data.
673. Layne, J. N., and D. R. Smith. 1992. Size comparison of resident and wintering American Kestrels in south-central Florida. *Journal of Field Ornithology* 63:256–263.
674. Leak, J., and S. K. Robinson. 1989. Notes on the social behavior and mating system of the Casqued Oropendula. *Wilson Bulletin* 101:134–137.
674a. Leberman, R. C. 1973. A study of Tufted Titmouse weights. *EBBA News* 36:34–38.
675. Leck, C. F. 1975. Weights of migrants and resident birds in Panama. *Bird-Banding* 46:201–203.
676. LeCroy, M. 1981. *The genus* Paradisaea*—display and evolution*. American Museum Novitates 2714.
677. LeCroy, M., and S. LeCroy. 1974. Growth and fledging in the Common Tern (*Sterna hirundo*). *Bird-Banding* 45:326–340.
678. Lee, D. S., M. Walsh-McGehee. 1998. White-tailed Tropicbird (*Phaethon lepturus*). In *The Birds of North America*, A. Poole and F. Gill (editors). The Birds of North America, Inc., Philadelphia, PA. Number 353.
679. Lee, K.-G., K.-B. Nam, J.-Y. Lee, H.-J. Kim, and J.-C. Yoo. 2002. Morphological characteristics of burrows, adults and eggs of Streaked Shearwaters (*Calonectris leucomelas*). *Korean Journal of Ornithology* 9:23–29.
680. Lee, R. J., and J. Riley. 2001. Morphology, plumage, and habitat of the newly described Cinnabar Hawk-Owl from North Sulawesi, Indonesia. *Wilson Bulletin* 113:17–22.
681. Lee, Y.-F., and L. L. Severinghaus. 2004. Sexual and seasonal differences in the diet of Lanyu scops owls based on fecal analysis. *Journal of Wildlife Management* 68:299–306.
682. Lefranc, N. 1997. *Shrikes: a guide to the shrikes of the world*. Yale University Press, New Haven, CT.
683. Lehmann, V. W. 1941. *Attwater's prairie chicken: its life history and management*. North American Fauna 57.
684. Lentino, M., and R. Restall. 2003. A new species of *Amaurospiza* blue seedeater from Venezuela. *Auk* 120:600–606.
685. Leonard, D. L. 2001. Three-toed Woodpecker (*Picoides tridactylus*). In *The Birds of North America*, A. Poole and F. Gill (editors). The Birds of North America, Inc., Philadelphia, PA. Number 588.
686. Lepson, J. K. 1997. 'Anianiau (*Hemignathus parvus*). In *The Birds of North America*, A. Poole and F. Gill (editors). The Academy of Natural Sciences, Philadelphia, PA, and The American Ornithologists' Union, Washington, DC. Number 312.
687. Lepson, J. K., and L. A. Freed. 1997. 'Akepa (*Loxops coccineus*). In *The Birds of North America*, A. Poole and F. Gill (editors). The Academy of Natural Sciences, Philadelphia, PA, and The American Ornithologists' Union, Washington, DC. Number 294.
688. Lepson, J. K., and S. M. Johnston. 2000. Greater 'Akialoa (*Hemignathus ellisianus*) and Lesser 'Akialoa (*Hemignathus obscurus*). In *The Birds of North America*, A. Poole and F. Gill (editors). The Birds of North America, Inc., Philadelphia, PA. Number 512.
689. Lepson, J. K., and H. D. Pratt. 1997. 'Akeke'e (*Loxops caeruleirostris*). In *The Birds of North America*, A. Poole and F. Gill (editors). The Academy of Natural Sciences, Philadelphia, PA, and The American Ornithologists' Union, Washington, DC. Number 295.
690. Lepson, J. K., and B. L. Woodworth. 2002. Hawai'i Creeper (*Oreomystis mana*). In *The Birds of North America*, A. Poole and F. Gill (editors). The Birds of North America, Inc., Philadelphia, PA. Number 680.
691. Leschack, C. R., S. K. McKnight, and G. R. Hepp. 1997. Gadwall (*Anas strepera*). In *The Birds of North America*, A. Poole and F. Gill (editors). The Academy of Natural Sciences, Philadelphia, PA, and The American Ornithologists' Union, Washington, DC. Number 283.

692. Lessells, C. M., and G. N. Ovenden. 1989. Heritability of wing length and weight in European Bee-eaters (*Merops apiaster*). *Condor* 91:210–214.

693. Ligon, J. D. 1968. Sexual differences in foraging behavior in two species of *Dendrocopos* woodpeckers. *Auk* 85:203–215.

694. Limpert, R. J., and S. L. Earnst. 1994. Tundra Swan (*Cygnus columbianus*). In *The Birds of North America*, A. Poole and F. Gill (editors). The Academy of Natural Sciences, Philadelphia, PA, and The American Ornithologists' Union, Washington, DC. Number 89.

695. Lindsey, G. D., E. A. VanderWerf, H. Baker, and P. E. Baker. 1998. Hawai'I (*Hemignathus virens*), Kaua'I (*Hemignathus kauaiensis*), O'ahu (*Hemignathus chloris*), and Greater 'Amakihi (*Hemignathus sagittirostris*). In *The Birds of North America*, A. Poole and F. Gill (editors). The Birds of North America, Inc., Philadelphia, PA, Inc., Number 360.

696. Linsdale, J. M., and E. L. Sumner. 1934. Winter weights of Golden-crowned and Fox Sparrows. *Condor* 36:107–112.

697. Liversidge, R. 1996. A new species of pipit in southern Africa. *Bulletin of the British Ornithologists' Club* 116:211–215.

698. Liversidge, R., and G. Voelker. 2002. The Kimberley Pipit: a new African species. *Bulletin of the British Ornithologists' Club* 122:93–109.

699. Livezey, B. C. 1989a. Flightlessness in grebes (Aves: Podicipedidae): its independent evolution in three genera. *Evolution* 43:29–54.

700. Livezey, B. C. 1989b. Phylogenetic relationships and incipient flightlessness of the extinct Auckland Islands Merganser. *Wilson Bulletin* 101:410–435.

701. Livezey, B. C. 1990. Evolutionary morphology of flightlessness in the Auckland Islands Teal. *Condor* 92:639–673.

702. Loegering, J. P. 1992. Piping plover breeding biology, foraging ecology and behavior on Assateague Island National Seashore, Maryland. M.S. thesis, Virginia Polytechnic Institute and State University, Blacksburg, VA.

703. Loegering, J. P. 1997. Abundance, habitat association, and foraging ecology of American dippers and other riparian-associated wildlife in the Oregon Coast Range. Ph.D. dissertation, Oregon State University, Corvallis, OR.

704. Long, J. L. 1984. Weights and measurements—Western Rosella. *Data Exchange, Corella* 8:28.

705. López Lamús, B., R. P. Clay, and J. C. Lowen. 1998. A new plumage of the White-winged Nightjar *Caprimulgus candicans* (Aves: Caprimulgidae). *Bulletin of the British Ornithologists' Club* 118:183–187.

706. Louette, M. 1988. Double invasions of birds on the Comoro Islands. Proceedings of the Sixth Pan-African Ornithological Congress. 77–86, Nairobi, Kenya.

707. Louette, M. 2003. Size, plumage, moult and supposed hybrids of African Goshawks (*Accipiter tachiro/toussenelii*) in DR Congo. *Ostrich* 74:18–29.

708. Louette, M., and M. Herremans. 1985. Taxonomy and evolution in the bulbuls (*Hypsipetes*) on the Comoro Islands. In *Proceedings of the International Symposium on African Vertebrates*, K.-L. Schuchmann (editor). Zoologisches Forschungsinstitut, Berlin, 407–423.

709. Louette, M., M. Herremans, L. Bijnens, and L. Janssens. 1988. Taxonomy and evolution in the brush warblers *Nesillas* on the Comoro Islands. *Tauraco* 1:110–129.

710. Louette, M., and J. Prévost. 1987. Passereaux collectés par J. Prévost au Cameroun. *Malimbus* 9:83–96.

710a. Lougheed, S. C., T. W. Arnold, and R. C. Bailey. 1991. Measurement error of external and skeletal variables in birds and its effects on principal components. *Auk* 108:432–436.

711. Low, R. 1985. Breeding the Tahiti blue lory *Vini peruviana*. *Avicultural Magazine* 91:1–14.

712. Low, R. 1991. Pretre's Amazon parrot. *Avicultural Magazine* 97:185–189.

713. Lowe, K. W., A. Clark, and R. A. Clark 1985. Body measurements, plumage and moult of the Sacred Ibis in South Africa. *Ostrich* 56:111–116.

714. Lowery, G. H. Jr., and W. W. Dalquest. 1951. Birds from the state of Veracruz, Mexico. *University of Kansas Publications of the Museum of Natural History* 3:531–649.

715. Lowery, G. H., and J. P. O'Neill. 1969. A new species of antpitta from Peru and a revision of the subfamily Grallariinae. *Auk* 86:1–12.

716. Lowery, G. H., and D. A. Tallman. 1976. A new genus and species of nine-primaried oscine of uncertain affinities from Peru. *Auk* 93:415–429.

717. Lowther, P. E. 1993. Brown-headed Cowbird (*Molothrus ater*). In *The Birds of North America*, A. Poole and F. Gill (editors). The Academy of Natural Sciences, Philadelphia, PA, and The American Ornithologists' Union, Washington, DC. Number 47.

718. Lowther, P. E. 1995. Bronzed Cowbird (*Molothrus aeneus*). In *The Birds of North America*, A. Poole and F. Gill (editors). The Academy of Natural Sciences, Philadelphia, PA, and The American Ornithologists' Union, Washington, DC. Number 144.

719. Lowther, P. E. 2000. Pacific-slope Flycatcher (*Empidonax difficilis*) and Cordilleran Flycatcher (*Empidonax occidentalis*). In *The Birds of North America*, A. Poole and F. Gill (editors). The Birds of North America, Inc., Philadelphia, PA. Number 556.

720. Lowther, P. E. 2002a. Thick-billed Kingbird (*Tyrannus crassirostris*). In *The Birds of North America*, A. Poole and F. Gill (editors). The Birds of North America, Inc., Philadelphia, PA. Number 604.

721. Lowther, P. E. 2002b. Red-billed Pigeon (*Columba flavirostris*). In *The Birds of North America*, A. Poole and F. Gill (editors). The Birds of North America, Inc., Philadelphia, PA. Number 644.

722. Lowther, P. E., and C. T. Collins. 2002. Black Swift (*Cypseloides niger*). In *The Birds of North America*, A. Poole and F. Gill (editors). The Birds of North America, Inc., Philadelphia, PA. Number 676.

723. Lowther, P. E., A. W. Diamond, S. W. Kress, G. J. Robertson, and K. Russell. 2002. Atlantic Puffin (*Fratercula arctica*). In *The Birds of North America*, A. Poole and F. Gill (editors). The Birds of North America, Inc., Philadelphia, PA. Number 709.

724. Lowther, P. E., H. D. Douglas, and C. L. Gratto-Trevor. 2001a. Willet (*Catoptrophorus semipalmatus*). In *The Birds of North America*, A. Poole and F. Gill (editors). The Birds of North America, Inc., Philadelphia, PA. Number 579.

725. Lowther, P. E., and R. T. Paul. 2002. Reddish Egret (*Egretta rufescens*). In *The Birds of North America*, A. Poole and F. Gill (editors). The Birds of North America, Inc., Philadelphia, PA. Number 633.

726. Lowther, P. E., and W. Post. 1999. Shiny Cowbird (*Molothrus bonariensis*). In *The Birds of North America*, A. Poole and F. Gill (editors). The Birds of North America, Inc., Philadelphia, PA. Number 399.

727. Lowther, P. E., C. C. Rimmier, B. Kessel, S. L. Johnson, and W. G. Ellison. 2001b. Gray-cheeked Thrush (*Catharus minimus*). In *The Birds of North America*, A. Poole and F. Gill (editors). The Birds of North America, Inc., Philadelphia, PA. Number 591.

728. Lowther, P. E., and D. F. Stotz. 1999. Sulphur-bellied Flycatcher (*Myiodynastes luteiventris*). In *The Birds of North America*, A. Poole and F. Gill (editors). The Academy of Natural Sciences, Philadelphia, PA, and The American Ornithologists' Union, Washington, DC. Number 475.

729. MacDougall-Shackleton, S. A., R. E. Johnson, and T. P. Hahn. 2000. Gray-crowned Rosy-Finch (*Leucosticte tephrocotis*). In *The Birds of North America*, A. Poole and F. Gill (editors). The Birds of North America, Inc., Philadelphia, PA. Number 559.

730. MacLean, S. F. 1974. Lemming bones as a source of calcium for Arctic sandpipers (*Calidris* spp.). *Ibis* 116:552–557.

731. MacLean, S. F., and R. T. Holmes. 1971. Bill lengths, wintering areas, and taxonomy of North American Dunlins, *Calidris alpina*. *Auk* 88:893–901.

732. MacWirter, B., P. Austin-Smith, and D. Kroodsma. 2002. Sanderling (*Calidris alba*). In *The Birds of North America*, A. Poole and F. Gill (editors). The Birds of North America, Inc., Philadelphia, PA. Number 653.

733. Madge, S., and H. Burn. 1999. *Crows and jays*. Princeton University Press, Princeton, NJ.

734. Madge, S., and P. McGowan. 2002. *Pheasants, partridges, and grouse. A guide to the pheasants, partridges, quails, grouse, guineafowl, buttonquails, and sandgrouse of the world*. Princeton University Press, Princeton, NJ.

735. Magrath, M. J. L., L. Brouwer, A. van Petersen, M. L. Berg, and J. Komdeur. 2003. Breeding behaviour and ecology of the sexually size-dimorphic brown songlark, *Cinclorhampus cruralis*. *Australian Journal of Zoology* 51:429–441.

736. Maguire, G. S., and R. A. Mulder. 2004. Breeding biology and demography of the southern emu-wren (*Stipiturus malachurus*). *Australian Journal of Zoology* 52:583–604.

737. Maher, W. J. 1974. *Ecology of Pomarine, Parasitic, and Long-tailed Jaegers in northern Alaska*. Pacific Coast Avifauna 37.

738. Maher, W. J. 1986. Growth and development of the Brown-backed Honeyeater *Ramsayornis modestus* in North Queensland. *Emu* 86:245–248.

739. Maijer, S., and J. Fjeldså. 1997. Description of a new *Cranioleuca* spinetail from Bolivia and a "leapfrog pattern" of geographic variation in the genus. *Ibis* 139:606–616.

740. Majumdar, N. 1984. *On a collection of birds from Bastar District, Madhya Pradesh*. Records of the Zoological Survey of India, Occasional Paper 59.

741. Male, T. D., S. G. Fancy, and C. J. Ralph. 1998. Red-billed Leiothrix (*Leiothrix lutea*). In *The Birds of North America*, A. Poole and F. Gill (editors). The Birds of North America, Inc., Philadelphia, PA. Number 359.

742. Mann, C. F. 1985. An avifaunal study in Kakamega Forest, Kenya, with particular reference to species diversity, weight and moult. *Ostrich* 56:236–262.

743. Mann, R. et al. 1982. Breeding the red-topped amazon parrot *Amazona dufresnia rhodocorytha*. *Avicultural Magazine* 88:12–14.

744. Manuwal, D. A., and A. C. Thoresen. 1993. Cassin's Auklet (*Ptychoramphus aleuticus*). In *The Birds of North America*, A. Poole and F. Gill (editors). The Academy of Natural Sciences, Philadelphia, PA, and The American Ornithologists' Union, Washington, DC. Number 50.

745. Marchant, S., and P. J. Higgins (editors). 1990. *Handbook of Australian, New Zealand and Antarctic birds*. Oxford University Press, Oxford. Volume 1.

746. Marchant, S., and P. J. Higgins (editors). 1993. *Handbook of Australian, New Zealand and Antarctic birds*. Oxford University Press, Oxford. Volume 2.

747. Marín A., J. M. Carrión B., and F. C. Sibley. 1992. New distributional records for Ecuadorian birds. *Ornithología Neotropical* 3:27–34.

748. Marín A., M., L. F. Kiff, and L. Peña G. 1989. Notes on Chilean birds, with descriptions of two new subspecies. *Bulletin of the British Ornithologists' Club* 109:66–82.

749. Marin, M. 1997. Species limits and distribution of some New World spine-tailed swifts (*Chaetura* spp.). *Ornithological Monographs* 48:431–443.

750. Marin, M. 2000. Species limits, distribution, and biogeography of some New World gray-rumped spine-tailed swifts (*Chaetura*, Apodidae). *Ornitología Neotropical* 11:93–107.

751. Marini, M. A., and R. B. Cavalcanti. Unpublished data included in Dunning (1992).

752. Marini, M. A., J. C. Motta-Junior, L. A. S. Vasconcellos, and R. B. Cavalcanti. 1997. Avian body masses from the *cerrado* region of central Brazil. *Ornithologia Neotropical* 8:93–99.

753. Marini, M. A., and S. J. Hackett. 2002. A multifaceted approach to the characterization of an intergeneric hybrid manakin (Pipridae) from Brazil. *Auk* 119:1114–1120.

754. Marion, W. R. 1977. Growth and development of the Plain Chachalaca in south Texas. *Wilson Bulletin* 89:47–56.

755. Marks, J. S., D. L. Evans, and D. W. Holt. 1994 Long-eared Owl (*Asio otus*). In *The Birds of North America*, A. Poole and F. Gill (editors). The Academy of Natural Sciences, Philadelphia, PA, and The American Ornithologists' Union, Washington, DC. Number 133.

756. Marks, J. S., and S. M. Leasure. 1992. Breeding biology of Tristram's Storm-Petrel on Laysan Island. *Wilson Bulletin* 104:719–731.

757. Marks, J. S., T. L. Tibbitts, R. E. Gill, and B. J. McCaffery. 2002. Bristle-thighed Curlew (*Numenius tahitiensis*). In *The Birds of North America*, A. Poole and F. Gill (editors). The Birds of North America, Inc., Philadelphia, PA. Number 705.

758. Marshall, J. T. 1948. Ecologic races of Song Sparrows in the San Francisco Bay region. *Condor* 50:233–256.

759. Marshall, J. T. 1978. Systematics of smaller Asian night birds based on voice. *Ornithological Monographs* 25.

759a. Marshall, S. G., and V. B. Baker. 1964. The role of fat in bird migration. *Oriole* 29:35–39.

760. Martens, J., and S. Eck. 1991. *Pnoepyga immaculate* n. sp., eine neue bodenbewohnende Timalie aus dem Nepal-Himalaya. *Journal für Ornithologie* 132:179–198.

761. Marti, C. D. 1992. Barn Owl (*Tyto alba*). In *The Birds of North America*, A. Poole, P. Stettenheim, and F. Gill (editors). The Academy of Natural Sciences, Philadelphia, PA, and The American Ornithologists' Union, Washington, DC.
762. Martin, J. W., and B. A. Carlson. 1998. Sage Sparrow (*Amphispiza belli*). In *The Birds of North America*, A. Poole and F. Gill (editors). The Birds of North America, Inc., Philadelphia, PA. Number 326.
763. Martin, P. S., C. R. Robins, and W. B. Heed. 1954. Birds and biogeography of the Sierra de Tamaulipas, an isolated pine-oak habitat. *Wilson Bulletin* 66:38–57.
764. Martin, S. G., and T. A. Gavin. 1995. Bobolink (*Dolichonyx oryzivorus*). In *The Birds of North America*, A. Poole and F. Gill (editors). The Academy of Natural Sciences, Philadelphia, PA, and The American Ornithologists' Union, Washington, DC. Number 176.
765. Martinez-Gómez, J. E., and R. L. Curry. 1998. Distinguishing sex of Socorro Mockingbirds by body measurements. *Ornitologia Neotropical* 9:103–110.
766. Mascitti, V. I., Unpublished data.
766a. Masello, J. F., and P. Quillfeldt. 2003. Body size, body condition and ornamental feathers of Burrowing Parrots: variation between years and sexes, assortative mating and influences on breeding success. *Emu* 103:149–161.
766b. Massoni, V., V. Loetti, and J. C. Reboreda. 2005. Sexual dimorphism and parental care in the Yellow-winged Blackbird (*Agelaius thilius*). *Ornitologia Neotropical* 16:551–555.
767. Mayfield, H. F. 1992. Kirtland's Warbler (*Dendroica kirtlandii*). In *The Birds of North America*, A. Poole, P. Stettenheim, and F. Gill (editors). The Academy of Natural Sciences, Philadelphia, PA, and The American Ornithologists' Union, Washington, DC. Number 19.
768. Mayr, E. 1938. *Birds collected during the Whitney South Sea Expedition. XL.* American Museum Novitates 1007.
769. Mayr, E. 1944. *Birds collected during the Whitney South Sea Expedition. 54.* American Museum Novitates 1269.
770. Mayr, E., and E. T. Gilliard. 1954. Birds of central New Guinea. *Bulletin of the American Museum of Natural History* 103:311–374.
771. Mayr, E., and J. M. Diamond. 2001. *The birds of northern Melanesia: speciation, ecology, and biogeography.* Oxford University Press, Oxford, UK.
771a. Mays, H. L., S. M. Doucet, C.-T. Yao, and H.-W. Huan. 2006. Sexual dimorphism and dichromatism in Steere's Liocichla (*Liocichla steerii*). *Journal of Field Ornithology* 77:437–443.
772. McCaffery, B., and R. Gill. 2001. Bar-tailed Godwit (*Limosa lapponica*). In *The Birds of North America*, A. Poole and F. Gill (editors). The Birds of North America, Inc., Philadelphia, PA. Number 581.
773. McCallum, D. A. 1994. Flammulated Owl (*Otus flammeolus*). In *The Birds of North America*, A. Poole and F. Gill (editors). The Academy of Natural Sciences, Philadelphia, PA, and The American Ornithologists' Union, Washington, DC. Number 93.
774. McCallum, D. A., F. B. Gill, and S. L. L. Gaunt. 2001. Community assembly patterns of parids along an elevational gradient in western China. *Wilson Bulletin* 113:53–64.
775. McClure, H. E. 1964. Avian bionomics in Malaya: 1. The avifauna above 5000 feet altitude at Mount Brinchang, Pahang. *Bird-Banding* 35:141–183.
776. McCracken, K. G. 2000. The 20-cm spiny penis of the Argentine Lake Duck (*Oxyura vittata*). *Auk* 117:820–825.
776a. McFarlane, R. W. Unpublished data.
777. McGregor, R. Unpublished data.
778. McIntosh, R. R., R. Kats, M. Berg, J. Komdeur, and M. A. Elgar. 2003. Breeding ecology and bias in offspring sex ratio in little grassbirds (*Megalurus gramineus*). *Australian Journal of Zoology* 51:505–514.
779. McIntyre, J. W., and J. F. Barr. 1997. Common Loon (*Gavia immer*). In *The Birds of North America*, A. Poole and F. Gill (editors). The Academy of Natural Sciences, Philadelphia, PA, and The American Ornithologists' Union, Washington, DC. Number 313.
780. McLandress, M. R., and D. G. Raveling. 1981. Changes in diet and body composition of Canada Geese before spring migration. *Auk* 98:65–79.
781. McNab, B. K. 2001. Energetics of toucans, a barbet, and a hornbill: implications for avian frugivory. *Auk* 118:916–933.
782. McNicholl, M. K., unpublished data included in Dunning (1992).
783. Meanley, B. 1969. *Natural history of the King Rail.* North American Fauna 67.
784. Meanley, B. 1971. *Natural history of the Swainson's Warbler.* North American Fauna 69.
785. Medway, Lord. 1972. The Gunong Benom Expedition 1967. Part 6. The distribution and altitudinal zonation of birds and mammals on Gunong Benom. *Bulletin of the British Museum (Natural History) Zoology* 23:103–154.
785a. Mee, A. Unpublished data.
786. Mees, G. F. 1982. Birds from the lowlands of southern New Guinea (Merauke and Koembe). *Zoologische Verhandelingen* 191.
787. Melville, D. S., and P. D. Round. 1984. Weights and gonad condition of some Thai birds. *Bulletin of the British Ornithologists' Club* 104:127–138.
788. Reference deleted.
789. Mendelssohn, H., Y. Yom-Tov, and U. Safriel. 1975. Hume's Tawny Owl *Strix butleri* in the Judean, Negev and Sinai deserts. *Ibis* 117:110–111.
790. Merritt, R. E., P. A. Bell, and V. Laboudallon. 1986. Breeding biology of the Seychelles Black Parrot (*Coracopsis nigra barklyi*). *Wilson Bulletin* 98:160–163.
791. Mermoz, M. E., and J. C. Reboreda. 2003. Reproductive success of Shiny Cowbird (*Molothrus bonariensis*) parasitizing the larger Brown-and-yellow Marshbird (*Pseudoleistes virescens*) in Argentina. *Auk* 120:1128–1139.
792. Metz, V. G., and E. A. Schreiber. 2002. Great Frigatebird (Fregata minor). In *The Birds of North America*, A. Poole and F. Gill (editors). The Birds of North America, Inc., Philadelphia, PA. Number 681.
793. Meyburg, B.-U., P. Paillat, and C. Meyburg. 2003. Migration routes of Steppe Eagles between Asia and Africa: a study by means of satellite telemetry. *Condor* 105:219–227.

794. Mezquida, E. T., and L. Marone. 2001. Factors affecting nesting success of a bird assembly in the central Monte Desert, Argentina. *Journal of Avian Biology* 32:287–296.
795. Miller, A. H. 1947. The tropical avifauna of the Upper Magdalena Valley, Colombia. *Auk* 64:351–381.
796. Miller, A. H. 1952. Supplemental data on the tropical avifauna of the arid Upper Magdalena Valley of Colombia. *Auk* 69:450–457.
797. Miller, A. H. 1955. The avifauna of the Sierra del Carmen of Coahuila, Mexico. *Condor* 57:154–178.
798. Miller, A. H. 1963. Seasonal activity and ecology of the avifauna of an American equatorial cloud forest. *University of California Publications in Zoology* 66:1–78.
799. Miller, K. E., and K. D. Meyer. 2002. Short-tailed Hawk (*Buteo brachyurus*). In *The Birds of North America*, A. Poole and F. Gill (editors). The Birds of North America, Inc., Philadelphia, PA. Number 674.
800. Milwaukee County Zoo, WI, unpublished data (E. Diebold) included in Dunning (1992).
801. Mitchell, A., unpublished data.
802. Mitchell, C. D. 1994. Trumpeter Swan (*Cygnus buccinator*). In *The Birds of North America*, A. Poole and F. Gill (editors). The Academy of Natural Sciences, Philadelphia, PA, and The American Ornithologists' Union. Number 105.
803. Moeliker, C. W., and C. J. Heij. 1995. The rediscovery of *Monarcha boanensis* (Aves: Monarchidae) from Boano Island, Indonesia. *Deinsea* 2:123–143.
804. Montalti, D., and R. Maragliano. 1996. Nueva localidad para *Phimosus infuscatus* (Aves: Threskiornithidae). *Neotrópica* 42:124.
805. Montevecchi, W. A., and I. J. Stenhouse. 2002. Dovekie (*Alle alle*). In *The Birds of North America*, A. Poole and F. Gill (editors). The Birds of North America, Inc., Philadelphia, PA. Number 701.
806. Montoya, A. B., P. J. Zwank, and M. Cardenas. 1997. Breeding biology of Aplomado Falcons in desert grasslands of Chihuahua, Mexico. *Journal of Field Ornithology* 68:135–144.
807. Moorman, T. E., and P. N. Gray. 1994. Mottled Duck (*Anas fulvigula*). In *The Birds of North America*, A. Poole and F. Gill (editors). The Academy of Natural Sciences, Philadelphia, PA, and The American Ornithologists' Union, Washington, DC. Number 81.
808. Moraes, U. S., and R. Krul. 1997. Notes on the Black-backed Tanager *Tangara peruviana* (Desmarest 1806). *Bulletin of the British Ornithologists' Club* 117:316–318.
809. Moreau, R. E., and R. M. Dolp. 1970. Fat, water, weights and wing-lengths of autumn migrants in transit on the northwest coast of Egypt. *Ibis* 112:209–228.
810. Morin, M. P., and S. Conant. 2002. Laysan Finch (*Telespiza cantans*) and Nihoa Finch (*Telespiza ultima*). In *The Birds of North America*, A. Poole and F. Gill (editors). The Birds of North America, Inc., Philadelphia, PA. Number 639.
811. Morin, M. P., S. Conant, and P. Conant. 1997. Laysan and Nihoa Millerbird (*Acrocephalus familiaris*). In *The Birds of North America*, A. Poole and F. Gill (editors). The Academy of Natural Sciences, Philadelphia, PA, and The American Ornithologists' Union, Washington, DC. Number 302.
812. Morris, A. K. 1971. Distribution of the Black Bittern in New South Wales. *Emu* 71:175–176.
813. Morrison, J. L. 1996. Crested Caracara (*Caracara plancus*). In *The Birds of North America*, A. Poole and F. Gill (editors). The Academy of Natural Sciences, Philadelphia, PA, and The American Ornithologists' Union, Washington, DC. Number 249.
814. Reference deleted.
815. Morton, E. S. 1977. Intratropical migration in the Yellow-green Vireo and Piratic Flycatcher. *Auk* 94:97–106.
815a. Moser, T. D., and D. H. Rusch. 1988. Indices of structural size and condition of Canada Geese. *Journal of Wildlife Management* 52:202–208.
816. Moskoff, W. 2002. Green Kingfisher (*Chloroceryle americana*). In *The Birds of North America*, A. Poole and F. Gill (editors). The Birds of North America, Inc., Philadelphia, PA. Number 621.
817. Moskoff, W., and L. R. Bevier. 2002. Mew Gull (*Larus canus*). In *The Birds of North America*, A. Poole and F. Gill (editors). The Birds of North America, Inc., Philadelphia, PA. Number 687.
818. Mostrum, A. M., R. L. Curry, and B. Lohr. 2002. Carolina Chickadee (*Poecile carolinensis*). In *The Birds of North America*, A. Poole and F. Gill (editors). The Birds of North America, Inc., Philadelphia, PA. Number 636.
819. Moulton, D. W., and A. P. Marshall. 1996. Laysan Duck (*Anas laysanensis*). In *The Birds of North America*, A. Poole and F. Gill (editors). The Academy of Natural Sciences, Philadelphia, PA, and The American Ornithologists' Union, Washington, DC. Number 242.
820. Mountainspring, S., and Scott, J. M. 1985. Interspecific competition among Hawaiian forest birds. *Ecological Monographs* 55:219–239.
821. Mowbray, T. B. 2002. Canvasback (*Aythya valisineria*). In *The Birds of North America*, A. Poole and F. Gill (editors). The Birds of North America, Inc., Philadelphia, PA. Number 659.
822. Mowbray, T. B., F. Cooke, and B. Ganter. 2000. Snow Goose (*Chen caerulescens*). In *The Birds of North America*, A. Poole and F. Gill (editors). The Birds of North America, Inc., Philadelphia, PA. Number 514.
823. Mowbray, T. B., C. R. Ely, J. S. Sedinger, and R. E. Trost. 2002. Canada Goose (*Branta canadensis*). In *The Birds of North America*, A. Poole and F. Gill (editors). The Birds of North America, Inc., Philadelphia, PA. Number 682.
824. Mueller, H. C., D. D. Berger, and G. Allez. 1976. Age and sex variation in the size of Goshawks. *Bird-Banding* 47:310–318.
825. Mueller, H. C., D. D. Berger, and G. Allez. 1979. Age and sex variation in the size of Sharp-shinned Hawks. *Bird-Banding* 50:34–44.
826. Mueller, H. C., D. D. Berger, and G. Allez. 1981. Age, sex, and seasonal differences in size of Cooper's Hawks. *Journal of Field Ornithology* 52:112–126.
827. Muller, M. J., and R. W. Storer. 1999. Pied-billed Grebe (*Podilymbus podiceps*). In *The Birds of North America*, A. Poole and F. Gill (editors). The Birds of North America, Inc., Philadelphia, PA. Number 410.
828. Mulvihill, R. S., R. C. Leberman, and A. J. Leppold. 2004. *Relationships among body mass, fat, wing length, age, and sex for 170 species of birds banded at Powdermill Nature Reserve*. Eastern Bird Banding Association Monograph 1.
828a. Murata, K. Unpublished data.

829. Murie, J. O., L. S. Davis, and I. G. McLean. 1991. Identifying the sex of Fiordland Crested Penguins by morphometric characters. *Notornis* 38:233–238.
830. Murphy, M. T. 1996. Eastern Kingbird (*Tyrannus tyrannus*). In *The Birds of North America*, A. Poole and F. Gill (editors). The Academy of Natural Sciences, Philadelphia, PA, and The American Ornithologists' Union, Washington, DC. Number 253.
831. Murray, B. G., and J. W. Hardy. 1981. Behavior and ecology of four syntopic species of finches in Mexico. *Zeitschrifte Tierpsychologie* 57:51–72.
832. Murray, B. G., and J. R. Jehl. 1964. Weights of autumn migrants from coastal New Jersey. *Bird-Banding* 35:251–263.
833. Murray, K. G., K. Winnett-Murray, Z. A. Eppley, G. L. Hunt, and D. B. Schwartz. 1983. Breeding biology of the Xanthus' Murrelet. *Condor* 85:12–21.
834. Museum of Taiwan, Endemic Species Research Institute, unpublished database.
835. Nacinovic, J. B., and D. M. Teixeira. 1989. As aves de Fernando de Noronha: uma lista sistemática anotada. *Revista Brasiliera de Biologie* 49:709–729.
835a. Nagata, H. Unpublished data.
836. Naik, S., and R. M. Naik. 1966. Studies on the House Swift, *Apus affinis* (G. E. Gray). 6. Body weight. *Pavo* 4:84–91.
837. Nakamura, M. 1990. Cloacal protuberance and copulatory behavior of the Alpine Accentor (*Prunella collaris*). *Auk* 107:284–295.
837a. Naoki, K. 2007. Arthropod resource partitioning among omnivorous tanagers (*Tangara* spp.) in western Ecuador. *Auk* 124:197–209.
838. Naugler, C. T. 1993. American Tree Sparrow (*Spizella arborea*). In *The Birds of North America*, A. Poole, P. Stettenheim, and F. Gill (editors). The Academy of Natural Sciences, Philadelphia, PA, and The American Ornithologists' Union, Washington, DC. Number 37.
839. Navarro S., A. T. Peterson, B. P. Escalante P., and H. Benitez D. 1992. *Cypseloides storeri*, a new species of swift from Mexico. *Wilson Bulletin* 104:55–64.
840. Navas, J. R. 1971. Estudios sobre la avifauna Andinopatagonica. 1. Generos *Upucerthia*, *Ochetorhynchus* y *Eremobius* (Furnariidae). *Revista del Museo Argentino de Ciencias Naturales "Bernardino Rivadavia"* 7:267–304.
841. Navas, J. R., and N. A. Bó. 1986. Notas sobre una coleccion de aves del Parque Nacional Lihue Calel, La Pampa, Argentina. *Hornero* 12:250–261.
842. Navas, J. R., and N. A. Bó. 1991. Anotaciones taxionomicas sobre Emberizidae y Fringillidae de la Argentina (Aves: Passeriformes). *Revista del Museo de la Plata (Nueva Sorie)* 158:119–134.
843. Navas, J. R., and N. A. Bó. 1994. Apuntes sobre distribucion y cria de algunas especies de los generos *Agriornis, Xolmis, Neoxolmis* y *Muscisaxicola* (Aves, Tyrannidae). *Revista del Museo Argentino de Ciencias Naturales "Bernardino Rivadavia"* 16:81–96.
844. Nelson, A. L., and A. C. Martin. 1953. Gamebird weights. *Journal of Wildlife Management* 17:36–42.
845. Nelson, J. B. 1978. *The Sulidae: gannets and boobies.* Oxford University Press, Oxford, UK.
846. Nelson, S. K. 1997. Marbled Murrelet (*Brachyramphus marmoratus*). In *The Birds of North America*, A. Poole and F. Gill (editors). The Academy of Natural Sciences, Philadelphia, PA, and The American Ornithologists' Union, Washington, DC. Number 276.
847. Nesbitt, S. A., D. T. Gilbert, and D. B. Barbour. 1976. Capturing and banding Limpkins in Florida. *Bird-Banding* 47:164–165.
848. Nesbitt, S. A., C. T. Moore, and K. S. Williams. 1992. Gender prediction from body measurements of two subspecies of sandhill cranes. *Proceedings of the North American Crane Workshop* 6:38–42.
849. Neufeldt, I. 1966. Life history of the Indian Cuckoo, *Cuculus micropterus micropterus* Gould, in the Soviet Union. *Journal of the Bombay Natural History Society* 63:399–419.
850. Newton, I., M. Marquiss, and A. Village. 1983. Weights, breeding, and survival in European Sparrowhawks. *Auk* 100:344–354.
851. New York Zoological Society, unpublished data (A. Lyles) included in Dunning (1992).
852. Niethammer, G. 1955. Zur Vogelwelt des Ennedi-Gebirges (Franzósisch Áquatorial-Afrika. *Bonner zoologische Beitrage* 6:29–80.
853. Nisbet, I. C. T. 1968. Weights of birds caught at night at a Malayan radio tower. *Ibis* 110:352–354.
854. Nol, E., and R. C. Humphrey. 1994. American Oystercatcher (*Haematopus palliatus*). In *The Birds of North America*, A. Poole and F. Gill (editors). The Academy of Natural Sciences, Philadelphia, PA, and The American Ornithologists' Union, Washington, DC. Number 82.
855. Nolan, V. 1978. The ecology and behavior of the Prairie Warbler *Dendroica discolor. Ornithological Monographs* 26.
856. Nolan, V., E. D. Ketterson, D. A. Cristol, C. M. Rogers, E. D. Clotfelter, R. C. Titus, S.J. Schoech, and E. Snajdr. 2002. Dark-eyed Junco (*Junco hyemalis*). In *The Birds of North America*, A. Poole and F. Gill (editors). The Birds of North America, Inc., Philadelphia, PA. Number 716.
857. Norment, C. J., P. Hendricks, and R. Santonocito. 1998. Golden-crowned Sparrow (*Zonotrichia atricapilla*). In *The Birds of North America*, A. Poole and F. Gill (editors). The Birds of North America, Inc., Philadelphia, PA. Number 352.
858. Norment, C. J., and S. A. Shackleton. 1993. Harris' Sparrow (*Zonotrichia querula*). In *The Birds of North America*, A. Poole and F. Gill (editors). The Academy of Natural Sciences, Philadelphia, PA, and The American Ornithologists' Union, Washington, DC. Number 64.
859. Norris, R. A. 1958. Comparative biosystematics and life history of the nuthatches *Sitta pygmaea* and *Sitta pusilla*. *University of California Publications in Zoology* 56:1–300.
860. North, M. R. 1994. Yellow-billed Loon (*Gavia adamsii*). In *The Birds of North America*, A. Poole and F. Gill (editors). The Academy of Natural Sciences, Philadelphia, PA, and The American Ornithologists' Union, Washington, DC. Number 121.
861. Novaes, F. C. 1976. As aves do rio Aripuanã, Estados de Mato Grosso e Amazonas. *Acta Amazonica, Suplemento* 6:61–85.
861a. Nyuyen Quang, P., Y. Vo Quang, and J.-F. Voisin. 2002. *The White-nest Swiftlet and the Black-nest Swiftlet: A Monograph.* Societe Nouvelle des Editions Boubee, Paris.
861b. Ochoa-acuña, H., M. S. Sepúlveda, and T. S. Gross. 2002. Mercury in feathers from Chilean birds: influence of location, feeding strategy, and taxonomic affiliation. *Marine Pollution Bulletin* 44:340–349.

862c. Odum, E. P. 1960. Lipid deposition in nocturnal migrant birds. *Proceedings of the International Ornithological Congress* 12:563–576.
862d. Odum, E. P., C. E. Connell, and H. L. Stoddard. 1961. Flight energy and estimated flight ranges of some migratory birds. *Auk* 78:515–527.
862e. Odum, E. P., S. G. Marshall, and T. G. Marples. 1965. The caloric content of migrating birds. *Ecology* 46:901–904.
862. Olsen, J., M. Wink, H. Sauer-Gürth, and S. Trost. 2002. A new *Ninox* owl from Sumba, Indonesia. *Emu* 102:223–231.
863. Olsen, P. D., S. J. S. Debus, G. V. Czechura, and N. J. Mooney. 1990. Comparative feeding ecology of the Grey Goshawk *Accipiter novaehollandiae* and Brown Goshawk *Accipiter fasciatus*. *Australian Bird Watcher* 13:178–192.
864. Olsen, P. D., and J. Olsen. 1987. Movements and measurements of the Australian Kestrel *Falco cenchroides*. *Emu* 87:35–41.
865. Olson, C. R., and T. E. Martin. 1999. Virginia's Warbler (*Vermivora virginiae*). In *The Birds of North America*, A. Poole and F. Gill (editors). The Birds of North America, Inc., Philadelphia, PA. Number 477.
866. Olson, S. L. 1985. Weights of some Cuban birds. *Bulletin of the British Ornithologists' Club* 105:68–69.
867. Olson, S. L., and J. P. Angle. 1977. Weights of some Puerto Rican birds. *Bulletin of the British Ornithologists' Club* 97:105–107.
868. Olson, S. L., H. F. James, and C. A. Meister. 1981. Winter field notes and specimen weights of Cayman Island birds. *Bulletin of the British Ornithologists' Club* 101:339–346.
869. O'Neill, J. P., and G. R. Graves. 1977. A new genus and species of owl (Aves: Strigidae) from Peru. *Auk* 94:409–416.
870. O'Neill, J. P., D. F. Lane, A. W. Kratter, A. P. Capparella, and C. F. Joo. 2000. A striking new species of barbet (Capitoninae: *Capito*) from the eastern Andes of Peru. *Auk* 117:569–577.
871. O'Neill, J. P., C. A. Munn, and I. Franke. 1991. *Nannopsittaca dachilleae*, a new species of parrotlet from eastern Peru. *Auk* 108:225–229.
872. O'Neill, J. P., and T. A. Parker. 1997. New subspecies of *Myrmoborus leucophrys* (Formicariidae) and *Phrygilus alaudinus* (Emberizidae) from the Upper Huallaga Valley, Peru. *Ornithological Monographs* 48: 485–491.
873. O'Neill, J. P., and T. S. Schulenberg. 1979. Notes on the Masked Saltator, *Saltator cinctus*, in Peru. *Auk* 96:610–613.
874. Oniki, Y. 1972. Some temperatures of Panamanian birds. *Condor* 74:209–215.
875. Oniki, Y. 1974. Some temperatures of birds of Belém, Brazil. *Acta Amazonica* 4:63–68.
876. Oniki, Y. 1978. Weights, digestive tracts and gonadal conditions of some Amazonian birds. *Revista Brasiliera de Biologia*. 38:679–681.
877. Oniki, Y. 1980. Weights and cloacal temperatures of some birds of Minas Gerais, Brazil. *Revista Brasileira de Biologia* 40:1–4.
878. Oniki, Y. 1981. Weights, cloacal temperatures, plumage and molt condition of birds in the state of Sáo Paulo. *Revista Brasileira de Biologia* 41:451–460.
879. Oniki, Y. 1986. Pesos, medidas, temperatures cloaçãis, condição da plumagem e de reprodução de aves da região de Balbina, Amazonas, Brasil. *Anals Soc. Sul-Riograndense de Ornitologia* 7:7–9
880. Oniki, Y. 1996. Band sizes of southeastern Brazilian hummingbirds. *Journal of Field Ornithology* 67:387–391.
881. Oniki, Y., and E. O. Willis. 1991. Morphometrics, molt, cloacal temperatures and ectoparasites in Colombian birds. *Caldasia* 16:519–524.
882. Oniki, Y., and E. O. Willis. 1993. Pesos, medidas, mudas, temperatures cloacais e ectoparasitos de aves da Reserva Ecológica do Panga, Minas Gerias, Brasil. *Bol. CEO* 9:2–10.
883. Oniki, Y., and E. O. Willis. 1999. Body mass, cloacal temperature, morphometrics, breeding and molt of birds of the Serra das Araras region, Mato Grosso, Brazil. *Ararajuba* 7:17–21.
884. O'Reilly, K. M., and J. C. Wingfield. 2003. Seasonal, age, and sex differences in weight, fat reserves, and plasma corticosterone in Western Sandpipers. *Condor* 105:13–26.
885. Oren, D. C., and T. A. Parker 1997. Avifauna of the Tapajó National Park and vicinity, Amazonian Brazil. *Ornithological Monographs* 48:493–525.
886. Ornelas, J. F., C. González, and J. Uribe. 2002. Complex vocalizations and aerial displays of the Amethyst-throated Hummingbird (*Lampornis amethystinus*). *Auk* 119:1141–1149.
887. Reference deleted.
888. Ottenwalder, J. A., and T. Vargas. 1979. *Nueva localidad para el Diablotin en la Republica Dominicana*. Naturu334ralista Postal, Universidad Autonoma De Santo Domingo, number 36.
889. Owen, M., and W. A. Cook. 1977. Variations in body weight, wing length and condition of Mallard *Anas platyrhynchos platyrhynchos* and their relationship to environmental changes. *Journal of Zoology, London* 183:377–395.
889a. Owen, M., and M. A. Ogilvie. 1979. Wing molt and weights of Barnacle Geese in Spitsbergen. *Condor* 81:42–52.
890. Pacheco, J. F., B. M. Whitney, and L. P. Gonzaga. 1996. A new genus and species of furnariid (Aves: Furnariidae) from the cocoa-growing region of southeastern Bahia, Brazil. *Wilson Bulletin* 108:397–406.
891. Pacheco, J. F., and L. P. Gonzaga. 1995. A new species of *Synallaxis* of the *ruficapilla/infuscata* complex from eastern Brazil (Passeriformes: Furnariidae). *Ararajuba* 3:3–11.
892. Page, G. W., J. S. Warriner, J. C. Warriner, and P. W. C. Paton. 1995. Snowy Plover (*Charadrius alexandrinus*). In *The Birds of North America*, A. Poole and F. Gill (editors). The Academy of Natural Sciences, Philadelphia, PA, and The American Ornithologists' Union, Washington, DC. Number 154.
893. Reference deleted.
894. Palmer, R. S. 1962–1988. *Handbook of North American Birds*. Yale University Press, New Haven CT. Volumes 1–5.
895. Pardieck, K., and R. B. Waide. 1992. Mesh size as a factor in avian community studies using mist nets. *Journal of Field Ornithology* 63:250–255.
896. Parker, T. A. 1981. Distribution and biology of the White-cheeked Cotinga *Zaratornis stresemanni*, a high Andean frugivore. *Bulletin of the British Ornithologists' Club* 101:256–265.

897. Parker, T. A. 1984. Notes on the behavior of *Ramphotrigon* flycatchers. *Auk* 101:186–188.
898. Parker, T. A., and J. P. O'Neill. 1985. A new species and a new subspecies of *Thryothorus* wren from Peru. *Ornithological Monographs* 36:9–15.
899. Parker, T. A., and S. A. Parker. 1982. Behavioural and distributional notes on some unusual birds of a lower montane cloud forest in Peru. *Bulletin of the British Ornithologists' Club* 102:63–70.
900. Parker, T. A., and T. S. Schulenberg, G. R. Graves, and M. J. Braun. 1985. The avifauna of the Huancabamba region, northern Peru. *Ornithological Monographs* 46:169–198.
901. Parker, T. A., and E. O. Willis. 1997. Notes on three tiny grassland flycatchers, with comments on disappearance of South American fire-diversified savannas. *Ornithological Monographs* 48:549–555.
902. Paton, D. C., and B. G. Collins. 1989. Bills and tongues of nectar-feeding birds: a review of morphology, function and performance, with intercontinental comparison. *Australian Journal of Ecology* 14:473–506.
903. Payne, L. X., and E. P. Pierce. 2002. Purple Sandpiper (*Calidris maritima*). In *The Birds of North America*, A. Poole and F. Gill (editors). The Birds of North America, Inc., Philadelphia, PA. Number 706.
904. Payne, R. B. 1977. Clutch size, egg size, and the consequences of single vs. multiple parasitism in parasitic finches. *Ecology* 58:500–513.
905. Reference deleted.
906. Payne, R. B. 1984. Sexual selection, lek and arena behavior, and sexual size dimorphism in birds. *Ornithological Monographs* 33.
907. Payne, R. B. 1992. Indigo Bunting (*Passerina cyanea*). In *The Birds of North America*, A. Poole, P. Stettenheim, and F. Gill (editors). The Academy of Natural Sciences, Philadelphia, PA, and The American Ornithologists' Union, Washington, DC. Number 4.
908. Payne, R. B. 2005. *The cuckoos*. Oxford University Press, Oxford, UK.
909. Payne, R. B., L. L. Payne, and M. E. D. Nhlane. 1992. Song mimicry and species status of the Green Widowfinch *Vidua codringtoni*. *Ostrich* 63:86–97.
910. Paynter, R. A. 1952. Birds from Popocatépetl and Ixtaccíhuatl, Mexico. *Auk* 69:293–301.
910a. Paynter, R. A. 1955. The ornithogeography of the Yucatan Peninsula. *Bulletin of the Peabody Museum of Natural History* 9:1–347.
911. Paynter, R. A. 1956. Avifauna of the Jorullo region, Michoacán, Mexico. *Postilla* 25:1–11.
912. Pearson, D. L. 1977. Ecological relationships of small antbirds in Amazonian bird communities. *Auk* 94:283–292.
913. Pearson, S. F. 1997. Hermit Warbler (*Dendroica occidentalis*). In *The Birds of North America*, A. Poole and F. Gill (editors). The Academy of Natural Sciences, Philadelphia, PA, and The American Ornithologists' Union, Washington, DC. Number 303.
914. Peer, B. D., and E. K. Bollinger. 1997. Common Grackle (*Quiscalus quiscula*). In *The Birds of North America*, A. Poole and F. Gill (editors). The Academy of Natural Sciences, Philadelphia, PA, and The American Ornithologists' Union, Washington, DC. Number 271.
915. Peirce, M. A. 1984. Weights of birds from Balmoral, Zambia. *Bulletin of the British Ornithologists' Club* 104:84–85.
916. Peña, L. E. 1961. Explorations in the Antofagasta Range, with observations on the fauna and flora. *Postilla* 49:3–42.
917. Pérez-Emán, J., C. J. Sharpe, M. Lentino, R. O. Prum, and I. J. Carreño F. 2003. New records of birds from the summit of Cerro Guaiquinima, Estado Bolivar, Venezuela. *Bulletin of the British Ornithologists' Club* 123:78–90.
918. Peris, S. J. 1990. Peso y biometria de algunas aves de chaco humedo (Presidente Hayes, Paraguay). *Ornitologia Neotropical* 1:31–32.
919. Perrins, C. M. et al. 1994. *Handbook of the birds of Europe, the Middle East, and North Africa*. Oxford University Press, Oxford, UK. Volume 8.
920. Petersen, K. L., L. B. Best, and B. M. Winter. 1986. Growth of nestling Sage Sparrows and Brewer's Sparrows. *Wilson Bulletin* 98:535–546.
921. Peterson, M. R., J. B. Grand, and C. P. Dau. 2000. Spectacled Eider (*Somateria fischeri*). In *The Birds of North America*, A. Poole and F. Gill (editors). The Birds of North America, Inc., Philadelphia, PA. Number 547.
921a. Peterson, M. R., J. A. Schmutz, and R. F. Rockwell. 1994. Emperor Goose (*Chen canagica*). In *The Birds of North America*, A. Poole and F. Gill (editors). The Academy of Natural Sciences, Philadelphia, PA, and The American Ornithologists' Union, Washington, DC. Number 174.
922. Petit, L. Unpublished banding data.
923. Philadelphia Zoological Gardens, unpublished data included in Dunning (1992).
924. Piatt, J. F. and A. S. Kitaysky. 2002. Horned Puffin (*Fratercula corniculata*). In *The Birds of North America*, A. Poole and F. Gill (editors). The Birds of North America, Inc., Philadelphia, PA. Number 603.
924a. Piatt, J. F., and A. S. Kitaysky. 2002. Tufted Puffin (*Fratercula cirrhata*). In *The Birds of North America*, A. Poole and F. Gill (editors). The Birds of North America, Inc., Philadelphia, PA. Number 708.
924b. Pichorim, M. 2006. Reproduction of the Mottled Piculet in southern Brazil. *Journal of Field Ornithology* 77:244–249.
925. Piechocki, R. 1958. Beiträge zur Avifauna Nord- und Nordost-China (Mandschurei). *Abhandlungen und Berichte aus dem staatlichen Museum für Tierkunde in Dresden* 24:105–203.
926. Piechocki, R. 1968. Beitrage zur Avifauna der Mongolei. Teil I. Non-passeriformes. *Mittleilungen aus dem Zoologischen Museum in Berlin. Akademie-Verlag, Berlin* 44: 149–292.
927. Piechocki, R., and A. Bolod. 1972. Beiträge zur Avifauna der Mongolei Teil II. Passeriformes. *Mitteilungen aus dem Zoologischen Museum in Berlin. Akademie-Verlag, Berlin* 48:41–176.
928. Pierce, R. J. 1984. Plumage, morphology and hybridisation of New Zealand stilts *Himantopus* spp. *Notornis* 31:106–130.
929. Perió, I. G. 1994. Biometría del Bigotudo *Panurus biarmicus* en una localidad del Sureste de España. *Butll. GCA* 11:51–55.
930. Pierotti, R. J., and C. A. Annett. 1995. Western Gull (*Larus occidentalis*). In *The Birds of North America*, A. Poole and F. Gill (editors). The Academy of Natural Sciences, Philadelphia, PA, and The American Ornithologists' Union, Washington, DC. Number 174.

931. Pierotti, R. J., and T. P. Good. 1994. Herring Gull (*Larus argentatus*). In *The Birds of North America,* A. Poole and F. Gill (editors). The Academy of Natural Sciences, Philadelphia, PA, and The American Ornithologists' Union, Washington, DC. Number 124.

931a. Piersma, T., and N. C. Davidson. 1991. Confusions of mass and size. *Auk* 108:441–444.

932. Post, W., and J. S. Greenlaw. 1994. Seaside Sparrow (*Ammodramus maritimus*). In *The Birds of North America*, A. Poole and F. Gill (editors). The Academy of Natural Sciences, Philadelphia, PA, and The American Ornithologists' Union, Washington, DC. Number 127.

933. Post, W., J. P. Poston, and G. T. Bancroft. 1996. Boat-tailed Grackle (*Quiscalus major*). In *The Birds of North America*, A. Poole and F. Gill (editors). The Academy of Natural Sciences, Philadelphia, PA, and The American Ornithologists' Union, Washington, DC. Number 207.

934. Poulin, R. G., S. D. Grindal, and R. M. Brigham. 1994. Common Nighthawk (*Chordeiles minor*). In *The Birds of North America*, A. Poole and F. Gill (editors). The Academy of Natural Sciences, Philadelphia, PA, and The American Ornithologists' Union, Washington, DC. Number 213.

935. Power, D. M. 1980. Evolution of land birds on the California islands. In *The California Islands: Proceedings of a Multidisciplinary symposium*, D. M. Power (editor). Santa Barbara Museum of Natural History, Santa Barbara, CA, 613–649.

936. Powers, D. R. 1996. Magnificent Hummingbird (*Eugenes fulgens*). In *The Birds of North America*, A. Poole and F. Gill (editors). The Academy of Natural Sciences, Philadelphia, PA, and The American Ornithologists' Union, Washington, DC. Number 221.

937. Powlesland, R. G., B. D. Lloyd, H. A. Best, and D. V. Merton. 1992. Breeding biology of the Kakapo *Strigops habroptilus* on Stewart Island, New Zealand. *Ibis* 134:361–373.

938. Pratt, T. K. 2000. Evidence for a previously unrecognized species of owlet-nightjar. *Auk* 117:1–11.

939. Pratt, T. K., S. G. Fancy, and C. J. Ralph. 2001. 'Akiapola'au (*Hemignathus munroi*) and Nukupu'u (*Hemignathus lucidus*). In *The Birds of North America*, A. Poole and F. Gill (editors). The Birds of North America, Inc., Philadelphia, PA. Number 600.

940. Price, T. D. 1979. The seasonality and occurrence of birds in the Eastern Ghats of Andhra Pradesh. *Journal of the Bombay Natural History Society* 76:378–422.

941. Price, T. D., A. J. Helbig, and A. D. Richman. 1997. Evolution of breeding distributions in the Old World leaf warblers (genus *Phylloscopus*). *Evolution* 51:552–561.

942. Price, T. D., and N. Jamdar. 1990. The breeding birds of Overa Wildlfie Sanctuary, Kashmir. *Journal of the Bombay Natural History Society* 87:1–15.

943. Prince, P. A., C. Ricketts, and G. Thomas. 1981. Weight loss in incubating albatrosses and its implications for their energy and food requirements. *Condor* 83:238–242.

944. Prinzinger, R., K. Krüger, and K. L. Schuchmann. 1981. Metabolism-weight relationship in 17 humming-bird species at different temperatures during day and night. *Experientia* 37:1307–1309.

945. Prinzinger, R., and E. Schleucher. 1998. Fruits of tall saltbush *Rhagodia eremaea* as an important source of energy and water for arid zone honeyeaters. *Emu* 98:236–240.

946. Proudfoot, G. A., and R. R. Johnson. 2000. Ferruginous Pygmy-Owl (*Glaucidium brazilianum*). In *The Birds of North America*, A. Poole and F. Gill (editors). The Birds of North America, Inc., Philadelphia, PA. Number 498.

947. Prum, R. O., J. D. Kaplan, and J. E. Pierson. 1996. Display behavior and natural history of the Yellow-crowned Manakin (*Heterocercus flavivertex*: Pipridae). *Condor* 98:722–735.

948. Prŷs-Jones, R. P. 1979. The ecology and conservation of the Aldabran brush warbler *Nesillas aldabranus. Philosophical Transactions fo the Royal Society of London B* 286:211–224.

949. Prŷs-Jones, R. P. 1982. Molt and weight of some land-birds on Dominica, West Indies. *Journal of Field Ornithology* 53:32–362.

950. Puerta, M. L., J. C. Alonso, V. Huecas, J. A. Alonso, M. Abelenda, and R. Munoz-Pulido. 1990. Hematology and blood chemistry of wintering Common Cranes. *Condor* 92:210–214.

950a. Pulgarin, P. Unpublished data.

951. Quinn, J. S. 1990. Sexual size dimorphism and parental care patterns in a monomorphic and a dimorphic larid. *Auk* 107:260–274.

952. Quinn, J. S., and Startek-Foote, J. M. 2000. Smooth-billed Ani (*Crotophaga ani*). In *The Birds of North America*, A. Poole and F. Gill (editors). The Birds of North America, Inc., Philadelphia, PA. Number 539.

953. Rahbek, C., H. Bloch, M. K. Poulsen, and J. F. Rasmussen. 1993. Avian body weights from southern Ecuador. *Bulletin of the British Ornithologists' Club* 113:103–108.

954. Rand, A. L., and D. S. Rabor. 1960. *Birds of the Philippines Islands: Siquijor, Mount Malindang, Bohol, and Samar.* Fieldiana Zoology 35, 221–441.

954a. Rappole, J. H., S. C. Rimmer, N. M. Shwe, and P. R. Sweet. 2005. A new species of scimitar-babbler (Timaliidae: *Jabouilleia*) from the sub-Himalayan region of Myanmar. *Auk* 122:1055–1063.

955. Raveling, D. G. 1968. Weights of *Branta canadensis interior* during winter. *Journal of Wildlife Management* 32:412–414.

956. Raveling, D. G. 1978. Morphology of the cackling Canada goose. *Journal of Wildlife Management* 42:897–900.

957. Ravokatra, M., L. Wilmé, and S. M. Goodman. 2003. Bird weights. In *The Natural History of Madagascar*, S. M. Goodman and J. P. Benstead (editors). University of Chicago Press, Chicago IL, 1059–1063.

958. Redman, N. J. 1987. The Arabian Accentor in North Yemen. *Sandgrouse* 9:78–81.

959. Reed, A., D. H. Ward, D. V. Derksen, and J. S. Sedinger. 1998. Brant (*Anas acuta*). In *The Birds of North America*, A. Poole and F. Gill (editors). The Birds of North America, Inc., Philadelphia, PA. Number 337.

960. Regosin, J. V. 1998. Scissor-tailed Flycatcher (*Tyrannus forficatus*). In *The Birds of North America*, A. Poole and F. Gill (editors). The Birds of North America, Inc., Philadelphia, PA. Number 342.

960a. Reichel, J. Unpublished data from Yamashima Institute of Ornithology database.

961. Reinert, B. L., J. C. Pinto, M. R. Bornschein, M. Pichorim, and M. A. Marini. 1996. Body masses and measurements of birds from southern Atlantic Forest, Brazil. *Revista Brasileira de Zoologia* 13:815–820.

962. Remsen, J. V. 1977. Five bird species new to Colombia. *Auk* 94:363.
963. Remsen, J. V. 1985. Community organization and ecology of birds of high elevation humid forest of the Bolivian Andes. *Ornithological Monographs* 36:733–756.
964. Remsen, J. V. (editor). 1997. Studies in Neotropical ornithology honoring Ted Parker. *Ornithological Monographs* 48.
965. Remsen, J. V., and M. A. Traylor. 1983. Additions to the avifauna of Bolivia, part 2. *Condor* 85:95–98.
966. Renner, Swen C., unpublished data.
967. Reynolds, C. M. 1972. Mute Swan weights in relation to breeding. *Wildfowl* 23:111–118.
968. Reynolds, T. D., T. D. Rich, and D. A. Stephens. 1999. Sage Thrasher (*Oreoscoptes montanus*). In *The Birds of North America*, A. Poole and F. Gill (editors). The Academy of Natural Sciences, Philadelphia, PA, and The American Ornithologists' Union, Washington, DC. Number 463.
969. Richdale, L. E. 1964. Notes on the Mottled Petrel *Pterodroma inexpectata* and other petrels. *Ibis* 106:110–114.
970. Ricklefs, R. E. Unpublished compilation of papers with body mass data, included in Dunning (1992).
971. Ricklefs, R. E. 1968. Weight recession in nestling birds. *Auk* 85:30–35.
972. Ricklefs, R. E., D. F. Bruning, and G. W. Archibald. 1986. Growth rates of cranes reared in captivity. *Auk* 103:125–134.
973. Ridgely, R. S., and M. B. Robbins. 1988. *Pyrrhura orcesi*, a new parakeet from southwestern Ecuador, with systematic notes on the *P. melanura* complex. *Wilson Bulletin* 100:173–182.
974. Rimmer, C. C., J. Almonte, E. Garrido G., D. A. Mejia, M. Milagros P., and P. R. Wieczoreck. 2003 Bird records in a montane forest fragment of western Sierra de Neiba, Dominican Republic. *Journal of Caribbean Ornithology* 16:55–60.
975. Rimmer, C. C., K. P. McFarland, W. G. Ellison, and J. E. Goetz. 2001. Bicknell's Thrush (*Catharus bicknelli*). In *The Birds of North America*, A. Poole and F. Gill (editors). The Birds of North America, Inc., Philadelphia, PA. Number 592.
976. Rinke, D. 1986. Notes on the avifauna of Niuafo'ou Island, Kingdom of Tonga. *Emu* 86:82–86.
977. Rinke, D. 1987. The avifauna of 'Eua and its off-shore islet Kalau, Kingdom of Tonga. *Emu* 87:26–34.
978. Rinke, D. 1989. The reproductive biology of the Red Shining Parrot *Prosopeia tabuensis* on the island of 'Eua, Kingdom of Tonga. *Ibis* 131:238–149.
978a. Ripley, S. D. 1950. A small collection of birds from Argentine Tierra del Fuego. *Postilla* 3:1–11.
979. Ripley, S. D. 1959. Birds from Djailolo, Halmahera. *Postilla* 41:1–8.
980. Ripley, S. D. 1985. Relationships of the Pacific warbler *Cichlornis* and its allies. *Bulletin of the British Ornithologists' Club* 105:109–112.
981. Ripley, S. D., and G. H. Heinrich. 1960. Additions to the avifauna of northern Angola I. *Postilla* 47:1–7.
982. Ripley, S. D., and G. H. Heinrich. 1966. Additions to the avifauna of northern Angola II. *Postilla* 95:1–29.
983. Ripley, S. D., and D. S. Rabor. 1958. Notes on a collection of birds from Mindoro Island, Philippines. *Bulletin of the Peabody Museum of Natural History*, number 15.
984. Ripley, S. D., and G. E. Watson. 1956. Cuban bird notes. *Postilla* 26:1–6.
984a. Rising, J. D., and K. M. Somers. 1989. The measurement of overall body size in birds. *Auk* 106:666–674.
985. Rising, J. D., and P. L. Williams. 1999. Bullock's Oriole (*Icterus bullockii*). In *The Birds of North America*, A. Poole and F. Gill (editors). The Birds of North America, Inc., Philadelphia, PA. Number 416.
986. Rivera-Milan, F. Unpublished data.
987. Robbins, M. B., M. J. Braun, C. J. Huddleston, D. W. Finch, and C. M. Milensky. 2005. First Guyana records, natural history and systematics of the White-naped Seedeater *Dolospingus fringilloides*. *Ibis* 147:334–341.
988. Robbins, M. B., A. P. Capparella, R. S. Ridgely, and S. W. Cardiff. 1991. Avifauna of the Rio Manití and Quebrada Vainilla, Peru. *Proceedings of the Academy of Natural Sciences, Philadelphia* 143:145–159.
989. Robbins, M. B., N. Krabbe, G. H. Rosenberg, R. S. Ridgely, and F. Sornoza Molina. 1994a. Notes on the natural history of the Crescent-faced Antpitta. *Wilson Bulletin* 106:169–173.
990. Robbins, M. B., T. A. Parker, and S. E. Allen. 1985. The avifauna of Cerro Pirre, Darién, eastern Panama. *Ornithological Monographs* 36:198–232.
991. Robbins, M. B., and R. S. Ridgely. 1990. The avifauna of an upper tropical cloud forest in southwestern Ecuador. *Proceedings of the Academy of Natural Sciences, Philadelphia* 142:59–71.
992. Robbins, M. B., and R. S. Ridgely. 1992. Taxonomy and natural history of *Nyctiphrynus rosenbergi* (Caprimulgidae). *Condor* 94:984–987.
993. Robbins, M. B., R. S. Ridgely, and S. W. Cardiff. 1994. Voice, plumage and natural history of Anthony's Nightjar (*Caprimulgus anthonyi*). *Condor* 96:224–228.
994. Robbins, M. B., G. H. Rosenberg, and F. Sornoza Molino. 1994b. A new species of cotinga (Cotingidae: *Doliornis*) from the Ecuadorian Andes, with comments on plumage sequences in *Doliornis* and *Ampelion*. *Auk* 111:1–7.
995. Robbins, M. B., G. H. Rosenberg, F. Sornoza Molino, and M. A. Jácome. 1997. Taxonomy and nest description of the Tumbes Swallow (*Tachycineta* [*albilinea*] *stolzmanni*). *Ornithological Monographs* 48:609–612.
996. Robbins, M. B., and F. G. Stiles. 1999. A new species of pygmy-owl (Strigidae: *Glaucidium*) from the Pacific slope of the northern Andes. *Auk* 116:305–315.
997. Roberts, G. E., T. D. Male, and S. Conant. 1998. White-rumped Shama (*Copsychus malabaricus*). In *The Birds of North America*, A. Poole and F. Gill (editors). The Birds of North America, Inc., Philadelphia, PA. Number 378.
998. Roberts, M., L. Cloutier, and P. Laporte. 1997. The summer diet of the Yellow Rail in southern Quebec. *Wilson Bulletin* 109:702–710.
999. Roberts, T. J. 1991. *The birds of Pakistan*. Oxford University Press, Oxford UK. Volume 1.
1000. Robertson, G. J., and R. I. Goudie. 1999. Harlequin Duck (*Histrionicus histrionicus*). In *The Birds of North America*, A. Poole and F. Gill (editors). The Birds of North America, Inc., Philadelphia, PA. Number 274.

1001. Robertson, H. A., and M. D. Dennison. 1984. Sexual dimorphism of the Chatham Island Warbler *Gerygone albofrontata*. *Emu* 84:103–107.

1002. Robertson, H. A., A. H. Whitaker, and B. M. Fitzgerald. 1983. Morphometrics of forest birds in the Orongorongo Valley, Wellington, New Zealand. *New Zealand Journal of Zoology* 10:87–98.

1003. Robertson, R. J., B. J. Stutchbury, and R. R. Cohen. 1992. Tree Swallow (*Tachycineta bicolor*). In *The Birds of North America*, A. Poole, P. Stettenheim, and F. Gill (editors). The Academy of Natural Sciences, Philadelphia, PA, and The American Ornithologists' Union, Washington, DC. Number 11.

1004. Robinet, O., and M. Salas. 1999. Reproductive biology of the endangered Ouvea Parakeet *Eunymphicus cornutus uvaeensis*. *Ibis* 141:660–669.

1005. Robinson, J. A., L. W. Oring, J. P. Skorupa, and R. Boettcher. 1997. American Avocet (*Recurvirostra americana*). In *The Birds of North America*, A. Poole and F. Gill (editors). The Academy of Natural Sciences, Philadelphia, PA, and The American Ornithologists' Union, Washington, DC. Number 275.

1006. Robinson, J. A., J. M. Reed, J. P. Skorupa, and L. W. Oring. 1999. Black-necked Stilt (*Himantopus mexicanus*). In *The Birds of North America*, A. Poole and F. Gill (editors). The Birds of North America, Inc., Philadelphia, PA. Number 449.

1007. Robinson, S. K. 1985. The Yellow-rumped Cacique and its associated nest pirates. *Ornithological Monographs* 36:898–907.

1008. Rodriguez de los Santos, M., and J. C. Rubio Garcia. 1986. Biology and biometry of the Pallid Swift (*Apus pallidus*) in southern Spain. *Le Gerfaut* 76:19–30.

1009. Rodriguez-Flores, C. I., and F. G. Stiles. 2005. Análisis ecomorfológico de una comunidad de colibríes ermitaños (Trochildidae, Phaethorninae) y sus flores en la Amazona Colombiana. *Ornitologia Columbiana* 3:7–27.

1010. Reference deleted.

1011. Rogers, D. T., and E. P. Odum. 1966. A study of autumnal postmigrant weights and vernal fattening of North American migrants in the tropics. *Wilson Bulletin* 78:415–433.

1012. Rohwer, F. C. 1988. Inter- and intraspecific relationships between egg size and clutch size in waterfowl. *Auk* 105:161–176.

1013. Rohwer, F. C., W. P. Johnson, and E. R. Loos. 2002. Blue-winged Teal (*Anas discors*). In *The Birds of North America*, A. Poole and F. Gill (editors). The Birds of North America, Inc., Philadelphia, PA. Number 625

1014. Romdal, Thomas Skovlund. Unpublished data.

1015. Roper, J. J. 2005. Sexually distinct songs in the duet of the sexually monomorphic Rufous Hornero. *Journal of Field Ornithology* 76:234–236.

1016. Rosenberg, K. V. 1997. Ecology of dead-leaf foraging specialists and their contribution to Amazonian bird diversity. *Ornithological Monographs* 48:673–700.

1017. Reference deleted.

1018. Ross, C. A. 1988. Weights of some New Caledonian birds. *Bulletin of the British Ornithologists' Club* 108:91–93.

1018a. Rosselli, L., P. Vasquez, and I. Ayub. 2002. The courtship displays and social system of the White-ruffed Manakin in Costa Rica. *Wilson Bulletin* 114:165–178.

1019. Roth, R. R., M. S. Johnson, and T. J. Underwood. 1996. Wood Thrush (*Hylocichla mustelina*). In *The Birds of North America*, A. Poole and F. Gill (editors). The Academy of Natural Sciences, Philadelphia, PA, and The American Ornithologists' Union, Washington, DC. Number 246.

1020. Round, P. D., and R. L. Swan. 1977. Aspects of the breeding of Cory's Shearwater *Calonectris diomedea* in Crete. *Ibis* 119:350–353.

1021. Rowley, I. 1978. Communal activities among White-winged Choughs *Corcorax melanorhamphus*. *Ibis* 120:178–197.

1022. Rowley, I., and E. Russell. 1997. *Fairy-wrens and grasswrens*. Oxford University Press, Oxford, UK.

1023. Ruiz, X., J. Gonzales-Solis, D. Oro, and L. Jover. 1998. Body size variation in Audouin's Gull *Larus audouinii*: a density-dependent effect? *Ibis* 140:411–438.

1024. Rusch, D. H., S. Destefano, M. C. Reynolds, and D. Lauten. 2000. Ruffed Grouse (*Bonasa umbellus*). In *The Birds of North America*, A. Poole and F. Gill (editors). The Birds of North America, Inc., Philadelphia, PA. Number 515.

1025. Russell, R. W. 2002. Pacific Loon (*Gavia pacifica*) and Arctic Loon (*Gavia arctica*). In *The Birds of North America*, A. Poole and F. Gill (editors). The Birds of North America, Inc., Philadelphia, PA. Number 657.

1026. Russell, S. M. Unpublished data included in Dunning (1992).

1027. Russell, S. M. 1964. A distributional study of the birds of British Honduras. *Ornithological Monographs* 1.

1028. Russell, S. M. 1996. Anna's Hummingbird (*Calypte anna*). In *The Birds of North America*, A. Poole and F. Gill (editors). The Academy of Natural Sciences, Philadelphia, PA, and The American Ornithologists' Union, Washington, DC. Number 226.

1029. Russell, S. M., J. C. Barlow, and D. W. Lamm. 1979. Status of some birds on Isla San Andres and Isla Providencia, Colombia. *Condor* 81:98–100.

1030. Ryan, P. G. Unpublished data.

1030a. Ryan, P. G., C. Bento, C. Cohen, J. Graham, V. Parker, and C. Spttiswoode. 1999. The avifauna and conservation status of the Namuli Massif, northern Mozambique. *Bird Conservation International* 9:315–331.

1031. Ryan, P. G., and C. L. Moloney. 1991. Tristan Thrushes kill adult White-bellied Storm-Petrels. *Wilson Bulletin* 103:130–132.

1032. Ryan, P. G., and B. P. Watkins. 1989. Snow Petrel breeding biology at an inland site in continental Antarctica. *Colonial Waterbirds* 12:176–184.

1033. Ryan, P. G., B. P. Watkins, and W. R. Siegfried. 1989. Morphometrics, metabolic rate and body temperature of the smallest flightless bird: the Inaccessible Island Rail. *Condor* 91:465–467.

1034. Ryder, J. P.. 1993. Ring-billed Gull (*Larus delawarensis*). In *The Birds of North America*, A. Poole, P. Stettenheim, and F. Gill (editors). The Academy of Natural Sciences, Philadelphia, PA, and The American Ornithologists' Union, Washington, DC. Number 33.

1035. Ryder, J. P., and R. T. Alisauskas. 1995. Ross' Goose (*Chen rossii*). In *The Birds of North America*, A. Poole and F. Gill (editors). The Academy of Natural Sciences, Philadelphia, PA, and The American Ornithologists' Union, Washington, DC. Number 162.

1036. SAFRING ringing project, Avian Demographic Unit, University of Cape Town, South Africa. Unpublished online database.
1037. Salaman, P. 2001. The study of an understorey avifauna community in an Andean premontane pluvial forest. Ph.D. dissertation, Oxford University, Oxford, UK.
1038. Salaman, P., P. Coopmans, T. M. Donegan, M. Mulligan, A. Cortés, S. L. Hilty, L. A. Ortega. 2003. A new species of wood-wren (Troglodytidae: *Henicorhina*) from the western Andes of Colombia. *Ornitología Colombiana* 1:4–21.
1038a. Salaman, P., and T. M. Donegan. Biodiversity surveys and conservation in Serrania de los Churumbelos, Colombia. Unpublished report, Conservation International, Washington, DC.
1039. Salaman, P. G. W., and F. G. Stiles. 1996. A distinctive new species of vireo (*Vireo masteri*) from the Western Andes of Colombia. *Ibis* 138:610–619.
1040. Salvador, S. A. 1988. Datos de peso de aves argentines. *Hornero* 13:78–83.
1041. Salvador, S. A. 1990. Datos de peso de aves argentines. 2. *Hornero* 13:169–171.
1041a. SalvaNATURA Conservation Science Program database, El Salvador.
1042. Sanft, V. K. 1970. Gewichte südamerikanischer Vögel—Nonpasseres. *Beiträge zur Vogelkunde* 16:344–354.
1043. Sanft, V. K. 1973. Gewichte südamerikanischer Vögel—Passeres. *Beiträge zurVogelkunde* 19:406–423.
1044. Santiago, D. Unpublished data.
1045. Saul, E. K., H. A. Robertson, and A. Tiraa. 1998. Breeding biology of the Kakerori (*Pomarea dimidiata*) on Rarotonga, Cook Islands. *Notornis* 45:255–269.
1046. Saunders, D. A. 1978. Measurements of the Little Corella from Kununurra, WA. *Emu* 78:37–39.
1047. Saunders, D. A. 1979. Distribution and taxonomy of the white-tailed and yellow-tailed black-cockatoos *Calyptorhynchus* spp. *Emu* 79:215–227.
1048. Saunders, D. A., G. T. Smith, and N. A. Campbell. 1984. The relationship between body weight, egg weight, incubation period, nestling period and nest site in the Psittaciformes, Falconiformes, Strigiformes and Columbiformes. *Australian Journal of Zoology* 32:57–65.
1049. Savard, J.-P., D. Bordage, and A. Reed. 1998. Surf Scoter (*Melanitta perspicillata*). In *The Birds of North America*, A. Poole and F. Gill (editors). The Birds of North America, Inc., Philadelphia, PA. Number 363.
1050. Scharf, W. C., and J. Kren. 1996. Orchard Oriole (*Icterus spurius*). In *The Birds of North America*, A. Poole and F. Gill (editors). The Academy of Natural Sciences, Philadelphia, PA, and The American Ornithologists' Union, Washington, DC. Number 255.
1051. Schillaci, J., unpublished data from Ft. Bragg, NC, MAPS station, as collected by protocols in DeSante et al. (2000).
1052. Schluter, D., unpublished data included in Dunning (1992).
1053. Schluter, D., and R. R. Repasky. 1991. Worldwide limitation of finch densities by food and other factors. *Ecology* 72:1763–1774.
1054. Schmitt, C. G., and D. C. Cole. 1981. First records of Black-legged Seriema (*Chunga burmeisteri*) in Bolivia. *Condor* 83:182–183.
1055. Schmitt, C. G., D. C. Schmitt, and J. V. Remsen. 1997. Birds of the Tambo area, an arid valley in the Bolivian Andes. *Ornithological Monographs* 48:701–716.
1056. Schodde, R. 1978. The identity of five type-specimens of New Guinean birds. *Emu* 78:1–6.
1057. Schodde, R., and I. J. Mason. 1979. Revision of the Zitting Cisticola *Cisticola juncidis* (Rafinesque) in Australia, with description of a new subspecies. *Emu* 79:49–53.
1058. Schodde, R., and J. L. McKean. 1973. Distribution, taxonomy and evolution of the Gardener Bowerbirds, *Ambyornis* spp. in eastern New Guinea with descriptions of two new subspecies. *Emu* 73:51–60.
1059. Schramm, M. 1983. The breeding biologies of the petrels *Pterodroma macroptera, P. brevirostris* and *P. mollis* at Marion Island. *Emu* 83:75–81.
1060. Reference deleted.
1061. Schreiber, E. A., C. J. Feare, B. A. Harrington, B. G. Murray, W. B. Robertson, M. J. Robertson, and G. E. Woolfenden. 2002. Sooty Tern (*Sterna fuscata*). In *The Birds of North America*, A. Poole and F. Gill (editors). The Birds of North America, Inc., Philadelphia, PA. Number 665.
1062. Schreiber, E. A., and R. L. Norton. 2002. Brown Booby (*Sula leucogaster*). In *The Birds of North America*, A. Poole and F. Gill (editors). The Birds of North America, Inc., Philadelphia, PA. Number 649.
1062a. Schreiber, E. A., and R. W. Schreiber. 1988. Great Frigatebird size dimorphism on two Central Pacific atolls. *Condor* 90:90–99.
1063. Schreiber, E. A., and R. W. Schreiber. 1993. Red-tailed Tropicbird (*Phaethon rubricauda*). In *The Birds of North America*, A. Poole and F. Gill (editors). The Academy of Natural Sciences, Philadelphia, PA, and The American Ornithologists' Union, Washington, DC. Number 43.
1064. Schreiber, E. A., R. W. Schreiber, and G. A. Schenk. 1996. Red-footed Booby (*Sula sula*). In *The Birds of North America*, A. Poole and F. Gill (editors). The Academy of Natural Sciences, Philadelphia, PA, and The American Ornithologists' Union, Washington, DC. Number 241.
1065. Schreiber, R. W., and E. A. Schreiber. 1984. Mensural and moult data for some birds of Martinique, French West Indies. *Bulletin of the British Ornithologists' Club* 104:62–68.
1066. Schroeder, M. A., and L. A. Robb. 1993. Greater Prairie-Chicken (*Tympanuchus cupido*). In *The Birds of North America*, A. Poole, P. Stettenheim, and F. Gill (editors). The Academy of Natural Sciences, Philadelphia, PA, and The American Ornithologists' Union, Washington, DC. Number 36.
1067. Schroeder, M. A., J. R. Young, and C. E. Braun. 1999. Sage Grouse (*Centrocercus urophasianus*). In *The Birds of North America*, A. Poole and F. Gill (editors). The Birds of North America, Inc., Philadelphia, PA. Number 425.
1068. Schuchmann, K.-L. 1979. Notes on sexual dimorphism and the nest of the Greenish Puffleg *Haplophaedia aureliae caucensis*. *Bulletin of the British Ornithologists' Club* 99:59–61.
1069. Schuchmann, K.-L., A.-A. Weller, and I. Heynen. 2001. Systematics and biogeography of the Andean genus *Eriocnemis* (Aves: Trochilidae). *Journal für Ornithologie* 142:433–481.

1070. Schulenberg, T. S. 1985. An intergeneric hybrid conebill (*Conirostrum* x *Oreomanes*) from Peru. *Ornithological Monographs* 36:390–395.

1071. Schulenberg, T. S. 1987. Observations on two rare birds, *Upucerthia albigula* and *Conirostrum tamarugense*, from the Andes of southwestern Peru. *Condor* 89:654–658.

1072. Schulenberg, T. S., and L. C. Binford. 1985. A new species of tanager (Emberizidae: Thraupinae, *Tangara*) from southern Peru. *Wilson Bulletin* 97:413–420.

1073. Schulenberg, T. S., and T. A. Parker. 1997a. A new species of tyrant-flycatcher (Tyrannidae: *Tolmomyias*) from the western Amazon basin. *Ornithological Monographs* 48:723–731.

1074. Schulenberg, T. S., and T. A. Parker. 1997b. Notes on the Yellow-browed Toucanet *Aulacorhynchus huallagae*. *Ornithological Monographs* 48:717–720.

1075. Scott, P. and The Wildfowl Trust. 1972. *The Swans*. Houghton-Mifflin, Boston, MA.

1076. Seddon, N., unpublished data.

1077. Sedgwick, J. A. 1994. Hammond's Flycatcher (*Empidonax hammondii*). In *The Birds of North America*, A. Poole and F. Gill (editors). The Academy of Natural Sciences, Philadelphia, PA, and The American Ornithologists' Union, Washington, DC. Number 109.

1078. Sekercioglu, C. H. Unpublished data.

1078a. Seki, S.-I. Unpublished data.

1079. Selander, R. K. 1966. Sexual dimorphism and differential niche utilization in birds. *Condor* 68:113–151.

1080. Selander, R. K., and D. R. Giller. 1959. The avifauna of the Barranca de Oblatos, Jalisco, Mexico. *Condor* 61:210–222.

1081. Senner, S. E., and B. J. McCaffery. 1997. Surfbird (*Aprhiza virgata*). In *The Birds of North America*, A. Poole and F. Gill (editors). The Academy of Natural Sciences, Philadelphia, PA, and The American Ornithologists' Union, Washington, DC. Number 266.

1082. Sepúlveda, M., and H. Ochoa-acuña. Unpublished data.

1083. Servat, G., and D. L. Pearson. 1991. Natural history notes and records for seven poorly known bird species from Amazonian Peru. *Bulletin of the British Ornithologists' Club* 111:92–95.

1084. Serventy, D. L., and H. M. Whittell. 1967. *Birds of Western Australia*. Lamb Publications, Perth, Western Australia.

1085. Severinghaus, L. L. 1987. Social behavior of the Vinous-throated Parrotbill during the non-breeding season. *Bulletin of the Institute of Zoology, Academia Sinica* 26:231–244.

1085a. Shannon, P. Unpublished data included in Dunning (1992).

1086. Shaw, P., E. Mungaya, N. Mbilinyi, and M. Mbilinyi. 2005. The voice and bill length of Karamoja Apalis *Apalis karamojae* are atypical of the genus. *Bulletin of the British Ornithologists' Club* 125:122–129.

1087. Shaw, T.-H. 1936. *The birds of Hopei Province*. Zoologia Sinica, Series B. *The Vertebrates of China*, volume 15, fascicle 1. Fan Museum Institute of Biology Peiping (Peking), China.

1088. Shaw, T.-H. 1938. The avifauna of Tsingtao and neighboring districts. *Bulletin of the Fan Memorial Institute of Biology, Zoological Series* 8:133–222.

1089. Shelley, E. L., M. Y. U. Tanaka, A. R. Ratnathicam, and D. T. Blumstein. 2004. Can Lanchester's laws help explain interspecific dominance in birds? *Condor* 106:395–400.

1090. Sheppard, J. M. 1972a. Banding worksheet for western birds—Lesser Goldfinch. Western Bird Banding Association.

1091. Sheppard, J. M. 1972b. Banding worksheet for western birds—Lazuli Bunting. Western Bird Banding Association.

1092. Sheppard, J. M. 1996. LeConte's Thrasher (*Toxostoma lecontei*). In *The Birds of North America*, A. Poole and F. Gill (editors). The Academy of Natural Sciences, Philadelphia, PA, and The American Ornithologists' Union, Washington, DC. Number 230.

1093. Sheppard, J. M., and C. T. Collins. 1971. Banding worksheet for western birds—Western Tanager. Western Bird Banding Association.

1094. Sherman, P. T. 1995. Breeding biology of White-winged Trumpeters (*Psophia leucoptera*) in Peru. *Auk* 112:285–295.

1095. Shin-Ichi, S. Unpublished data.

1096. Shirihai, H., G. Gargallo, and A. J. Helbig. 2001. *Sylvia warblers*. Princeton University Press, Princeton, NJ.

1097. Short, L. L. 1968. *Sympatry of Red-breasted Meadowlarks in Argentina, and the taxonomy of meadowlarks (Aves:* Leistes, Pezites, *and* Sturnella). American Museum Novitates 2349.

1098. Short, L. L. 1969. Observations on three sympatric species of tapaculos (Rhinocryptidae) in Argentina. *Ibis* 111:239–240.

1099. Short, L. L. 1971. Aves nuevas o poco comunes de Corrientes, Republica Argentina. *Revista del Museo Argentino de Ciencias Naturales "Bernardino Rivadavia," Zoologia* 9:283–309.

1100. Short, L. L. 1982. *Woodpeckers of the world*. Delaware Museum of Natural History, Greenville, DE.

1101. Short, L. L., and R. C. Banks. 1965. Notes on birds of northwestern Baja California. *Transactions of the San Diego Society of Natural History* 14:41–52.

1102. Short, L. L. and J. F. M. Horne. 2001. *Toucans, barbets and honeyguides*. Oxford University Press, Oxford, UK.

1103. Short, L. L., and J. J. Morony. 1969. Notes on some birds of central Peru. *Bulletin of the British Ornithologists' Club* 89:112–115.

1104. Sick, H. 1991. Distribution and subspeciation of the Biscutate Swift *Streptoprocne biscutata*. *Bulletin of the British Ornithologists' Club* 111:38–40.

1105. Sick, H. 1993. *Birds in Brazil*. Princeton University Press, Princeton, NJ.

1106. Silva, J. M. C., G. Coelho, and L. P. Gonzaga. 2002. Discovered on the brink of extinction: a new species of pygmy-owl (Strigidae: *Glaucidium*) from Atlantic Forest of northeasten Brazil. *Ararajuba* 10:123–130.

1107. Silva, J. M. C., M. de F. Cunha L., and M. L. Videira Marceliano. 1990. Pesos de aves de duas localidades na Amazonia Oriental. *Ararajuba* 1:99–104.

1108. Silva, J. M. C., F. C. Novaes, and D. C. Oren. 2002. Differentiation of *Xiphocolaptes* (Dendrocolaptidae) across the river Xingu, Brazilian Amazonia: recognition of a new phylogenetic species and biogeographic implications. *Bulletin of the British Ornithologists' Club* 122:185–194.

1109. Silva, J. M. C., D. C. Oren, J. César Roma, and L. M. P. Henriques. 1997. Composition and distribution patterns of the avifauna of an Amazonian upland savanna, Amapa, Brazil. *Ornithological Monographs* 48:743–762.
1110. Silveiro, L. F., F. C. Thadeo de Lima, E. Höfling. 2005. A new species of *Aratinga* parakeet (Psittaciformes: Psittacidae) from Brazil, with taxonomic remarks on the *Aratinga solstitialis* complex. *Auk* 122:292–305.
1111. Simmons, R. E. 2000. *Harriers of the world: their behaviour and ecology*. Oxford University Press, Oxford, UK.
1112. Simon, J. C., P. E. Baker, and H. Baker. 1997. Maui Parrotbill (*Pseudonestor xanthophrys*). In *The Birds of North America*, A. Poole and F. Gill (editors). The Academy of Natural Sciences, Philadelphia, PA, and The American Ornithologists' Union, Washington, DC. Number 311.
1113. Simons, T. R. 1985. Biology and behavior of the endangered Hawaiian Dark-rumped Petrel. *Condor* 87:229–243.
1114. Sirén, A., P. Hambäck, and J. Machoa. 2004. Including spatial heterogeneity and animal dispersal when evaluating hunting: a model analysis and an empirical assessment in an Amazonian community. *Conservation Biology* 18:1315–1329.
1115. Skeel, M. A. 1982. Sex determination of adult Whimbrels. *Journal of Field Ornithology* 53:414–416.
1116. Slotterback, J. W. 2002. Band-rumped Storm-Petrel (*Oceanodroma castro*) and Tristram's Storm-Petrel (*Oceanodroma tristrami*). In *The Birds of North America*, A. Poole and F. Gill (editors). The Birds of North America, Inc., Philadelphia, PA. Number 673.
1117. Smith, G. A. 1975. Systematics of parrots. *Ibis* 117:18–68.
1118. Smith, G. A. 1976. Notes on some species of parrot in captivity. *Avicultural Magazine* 82:143–150.
1119. Smith, G. A. 1977a. Notes on some species of parrot in captivity. *Avicultural Magazine* 83:21–27.
1120. Smith, G. A. 1977b. Notes on some species of parrot in captivity. *Avicultural Magazine* 83:119–123.
1121. Smith, G. A. 1977c. Notes on some species of parrot in captivity. *Avicultural Magazine* 83:160–166.
1122. Smith, G. A. 1978. Notes on some species of parrot in captivity. *Avicultural Magazine* 84:200–201.
1123. Smith, G. T. 1991. Ecology of the Western Whipbird *Psophodes nigrogularis* in Western Australia. *Emu* 91:145–157.
1124. Smith, J. N. M., and H. P. A. Sweatman. 1976. Feeding habits and morphological variation in Cocos Finches. *Condor* 78:244–248.
1125. Smith, T. B. 1993. Patterns of morphological and geographic variation in trophic bill morphs of the African finch *Pyrenestes*. *Biological Journal of the Linnean Society* 41:381–414.
1126. Snell, R. R. 2002. Iceland Gull (*Larus glaucoides*) and Thayer's Gull (*Larus thayeri*). In *The Birds of North America*, A. Poole and F. Gill (editors). The Birds of North America, Inc., Philadelphia, PA. Number 699.
1127. Snetsinger, T. J., K. M. Wakelee, and S. G. Fancy. 1999. Puaiohi (*Myadestes palmeri*). In *The Birds of North America*, A. Poole and F. Gill (editors). The Birds of North America, Inc., Philadelphia, PA. Number 461.
1128. Snow, B. K. 1966. Observations on the behaviour and ecology of the Flightless Cormorant *Nannopterum harrisi*. *Ibis* 108:265–280.
1129. Snow, D. 1982. *The cotingas*. Cornell University Press, Ithaca, NY.
1130. Snow, D. W., and C. M. Perrins. 1998. *The birds of the Western Palearctic, concise edition*. Oxford University Press, Oxford, UK.
1131. Snow, D. W., and B. K. Snow. 1963. Weights and wing-lengths of some Trinidad birds. *Zoologica* 48:1–12.
1132. Snyder, N. F. R., E. C. Enkerlin-Hoeflich, and M. A. Cruz-Nieto. 1999. Thick-billed Parrot (*Rhynchopsitta pachyrhyncha*). In *The Birds of North America*, A. Poole and F. Gill (editors). The Birds of North America, Inc., Philadelphia, PA. Number 406.
1133. Snyder, N. F. R., and K. Russell. 2002. Carolina Parakeet (*Conuropsis carolinensis*). In *The Birds of North America*, A. Poole and F. Gill (editors). The Birds of North America, Inc., Philadelphia, PA. Number 667.
1134. Snyder, N. F. R., and N. J. Schmitt. 2002. California Condor (*Gymnogyps californianus*). In *The Birds of North America*, A. Poole and F. Gill (editors). The Birds of North America, Inc., Philadelphia, PA. Number 610.
1135. Snyder, N. F. R., and J. W. Wiley. 1976. Sexual size dimorphism in hawks and owls of North America. *Ornithological Monographs* 20.
1136. Snyder, N. F. R., J. W. Wiley, and C. B. Kepler. 1987. *The parrots of Luquillo: natural history and conservation of the Puerto Rican Parrot*. Western Foundation of Vertebrate Zoology, Los Angeles, CA.
1137. Sogge, M. K., W. M. Gilbert, and C. Van Riper. 1994. Orange-crowned Warbler (*Vermivora celata*). In *The Birds of North America*, A. Poole and F. Gill (editors). The Academy of Natural Sciences, Philadelphia, PA, and The American Ornithologists' Union, Washington, DC. Number 101.
1138. Sogge, M. K., M. D. Kern, R. Kern, and C. van Riper. 1991. Growth and development of thermoregulation in nestling San Miguel Island Song Sparrows. *Condor* 93:773–776.
1139. Sözer, R., C. R. Shepherd, and Darjono. 2006. First description of male Hoogerwerf's Pheasant *Lophura* (*inornata*) *hoogerwerfi* (Chasen, 1939), with notes on distribution. *Bulletin of the British Ornithologists' Club* 126:207–211.
1140. Spear, L. B., and D. G. Ainley. 1997. Flight behaviour of seabirds in relation to wind direction and wing morphology. *Ibis* 139:221–233.
1141. Spear, L. B., and D. G. Ainley. 1998. Morphological differences relative to ecological separation in petrels (Family Procellariidae) of the southern ocean and tropic Pacific. *Auk* 15:1017–1033.
1141a. Spear, L. B., and D. G. Ainley. 2007. Storm-petrels of the Eastern Pacific Ocean: species assembly and diversity along marine habitat gradients. *Ornithological Monographs* 62.
1142. Specimens from the British Museum of Natural History.
1142a. Specimens from the Delaware Museum of Natural History, Dover, DE.
1142b. Specimens from the Departamento de Zoologia da Universidade Federal de Minas Gerais, Belo Horizonte, Minas Gerais, Brazil.
1143. Specimens from the Field Museum of Natural History, Chicago, IL.
1144. Specimens from Instituto de Ciencias Naturales, Universidad Nacional de Colombia.
1145. Specimens from Louisiana State University, Baton Rouge, LA.
1146. Specimens from the Museo Nacional, Rio de Janiero, Brazil.
1147. Specimens from the Philadelphia Academy of Natural Sciences, Philadelphia, PA.
1147a. Specimens from the Royal Ontario Museum, Toronto, Ontario, Canada.

1147b. Specimens from the San Diego Museum of Natural History, San Diego, CA.

1148. Specimens from the Smithsonian Institution, Washington, DC.

1148a. Specimens from the Texas Cooperative Wildlife Collection, Texas A&M University, College Station, TX.

1148b. Specimens from the United States National Museum, Washington, DC.

1149. Specimens from the University of Alaska, Fairbanks, AK.

1150. Specimens from the University of Arizona, Tucson, AZ.

1151. Specimens from the Cowan Vertebrate Museum, University of British Columbia, Vancouver, British Columbia, Canada.

1151a. Specimens from the Museum of Vertebrate Zoology, University of California, Berkeley, CA.

1152. Specimens from the University of Florida, Gainesville, FL.

1153. Specimens from the University of Kansas, Lawrence, KS.

1153a. Specimens from the University of Oklahoma, Sam Noble Oklahoma Museum of Natural History, Norman, OK.

1153b. Specimens from the University of Washington, Seattle, WA.

1154. Specimens from the Western Foundation of Vertebrate Zoology, Camarillo, CA.

1155. Specimens from the Yale Peabody Museum, New Haven, CT.

1156. Specimens from the Zimbabwe Natural History Museum, Bulawayo, Zimbabwe.

1157. Spring, L. 1971. A comparison of functional and morphological adaptations in the Common Murre (*Uria aalge*) and Thick-billed Murre (*Uria lomvia*). *Condor* 73:1–27.

1158. Squires, J. R., and R. T. Reynolds. 1997. Northern Goshawk (*Accipiter gentilis*). In *The Birds of North America*, A. Poole and F. Gill (editors). The Academy of Natural Sciences, Philadelphia, PA, and The American Ornithologists' Union, Washington, DC. Number 298.

1159. Steadman, D. W., S. L. Olson, J. C. Barber, C. A. Meister, and M. E. Melville. 1980. Weights of some West Indian birds. *Bulletin of the British Ornithologists' Club* 100:155–158.

1159a. Steenhof, K., and J. O. McKinley. 2006. Size dimorphism, molt status, and body mass variation of Prairie Falcons nesting in the Snake River Birds of Prey National Conservation Area. *Journal of Raptor Research* 40:71–75.

1160. Stiles, F. G. 1978. Possible specialization for hummingbird-hunting in the Tiny Hawk. *Auk* 95:550–553.

1161. Stiles, F. G. 1983a. Systematics of the southern forms of *Selasphorus* (Trochilidae). *Auk* 100:311–325.

1162. Stiles, F. G. 1983b. The taxonomy of *Microcerculus* wrens (Troglodytidae in Central America. *Wilson Bulletin* 95:169–183.

1163. Stiles, F. G. 1985. Seasonal patterns and coevolution in the hummingbird-flower community of a Costa Rican subtropical forest. *Ornithological Monographs* 36:757–787.

1164. Stiles, F. G. 1992. A new species of antpitta (Formicariidae: *Grallaria*) from the eastern Andes of Colombia. *Wilson Bulletin* 104:389–399.

1165. Stiles, F. G. 1996. A new species of emerald hummingbird (Trochilidae, *Chlorostilbon*) from the Sierra de Chiribiquete, southeastern Colombia, with a review of the *C. mellisugus* complex. *Wilson Bulletin* 108:1–27.

1166. Stiles, F. G. 1998. Notes on the biology of two threatened species of *Bangsia* tanagers in northwest Colombia. *Bulletin of the British Ornithologists' Club* 118:25–31.

1167. Stiles, F. G. 2003. Notas taxonómicas sobre aves Columbianas I: El rango taxonómico de *Hylocharis humboldtii* (Trochilidae). *Ornitologia Columbiana* 1:68–70.

1168. Stiles, F. G. 2005. El Trepatroncos de Zimmer *Xiphorhynchus kieneri* (Dendrocolaptidae) en la Amazonia Colombiana. *Ornitologia Columbiana* 3:104–106.

1169. Stiles, F. G., and P. Caycedo. 2002. A new subspecies of Apolinar's Wren (*Cistothorus apolinari*, Aves: Troglodytidae), an endangered Colombian endemic. *Caldasia* 24:191–199.

1170. Stiles, F. G., and A. F. Skutch. 1989. *A guide to the birds of Costa Rica*. Cornell University Press, Ithaca, NY.

1171. Stober, J. M. 1996. Territory dynamics and basic biology of the Bachman's Sparrow (*Aimophila aestivalis*) at the Savannah River Site, South Carolina. M.S. thesis, University of Georgia, Athens, GA.

1172. Stokes, T. 1980. Notes on the landbirds of New Caledonia. *Emu* 80:81–86.

1172a. Stonehouse, B. 1970. Geographic variation in Gentoo Penguins *Pygoscelis papua*. *Ibis* 112:52–57.

1173. Storer, R. W. 1980–81. The Hooded Grebe on Laguna de los Escarchardos: ecology and behavior. *Living Bird* 19:51–68.

1174. Storer, R. W. 1987. Morphology and relationships of the Hoary-headed Grebe and the New Zealand Dabchick. *Emu* 87:150–158.

1175. Storer, R. W. 1989. *Notes on Paraguayan birds*. Occasional papers of the Museum of Zoology, University of Michigan 719.

1176. Storer, R. W. 1992. Least Grebe (*Tachybaptus dominicus*). In *The Birds of North America*, A. Poole, P. Stettenheim, and F. Gill (editors). The Academy of Natural Sciences, Philadelphia, PA, and The American Ornithologists' Union, Washington, DC. Number 24.

1177. Storer, R. W., and G. L. Nuechterlein. 1992. Western Grebe and Clark's Grebe (*Aechmophorus occidentalis* and *A. clarkii*). In *The Birds of North America*, A. Poole, P. Stettenheim, and F. Gill (editors). The Academy of Natural Sciences, Philadelphia, PA, and The American Ornithologists' Union, Washington, DC. Number 26.

1178. Stotz, D. Unpublished data included in Dunning (1992).

1179. Straight, C. A., and R. J. Cooper. 2000. Chuck-will's-widow (*Caprimulgus carolinensis*). In *The Birds of North America*, A. Poole and F. Gill (editors). The Birds of North America, Inc., Philadelphia, PA. Number 499.

1180. Strauch, J. G. 1977. Further bird weights from Panama. *Bulletin of the British Ornithologists' Club* 97:61–66.

1181. Stresemann, E. 1914a. Die Vogel von Seran (Ceram). *Novitates Zoologicae* 21:25–153.

1182. Stresemann, E. 1914b. Beiträge zur Kentnis der avifauna von Buru. *Novitates Zoologicae* 21:358–400.

1183. Stresemann, E., W. Meise, and M. Schönwetter. 1937. Aves Beickianae. *Journal für Ornithologie* 85:376–576.

1184. Strickland, D., and H. Ouellet. 1993. Gray Jay (*Perisoreus canadensis*). In *The Birds of North America*, A. Poole, P. Stettenheim, and F. Gill (editors). The Academy of Natural Sciences, Philadelphia, PA, and The American Ornithologists' Union, Washington, DC. Number 40.

1185. Stromberg, M. R. 2000. Montezuma Quail (*Cyrtonyx montezumae*). In *The Birds of North America*, A. Poole and F. Gill (editors). The Birds of North America, Inc., Philadelphia, PA. Number 524.

1186. Reference deleted.

1187. Stýblo, P., and J. Mlíkovský. 1992. Weights and measurements of birds from the Svjatoj Nos wetlands, Lake Baikal. In *Ecology of the Svjatoj Nos wetlands, Lake Baikal*, J., Mlíkovský and P. Stýblo (editors). Ninox Press, Praha, 89–100.

1188. Suedmeyer, W. K., A. Bermudez, and L. Shaiken. 1998. Osteolysis and hepatomegaly caused by *Salmonella typhimurium* in a Temminck's fruit-dove (*Ptilinopus porphyrea*). *Journal of Avian Medicine and Surgery* 12:184–189.

1189. Summers-Smith, J. D. 1988. *The sparrows. A study of the genus* Passer. T&AD Poyser, Calton, UK.

1190. Reference deleted.

1191. Sutton, R. Unpublished data included in Dunning (1992).

1192. Swales, M. K. 1965. The sea-birds of Gough Island. *Ibis* 107:17–42.

1193. Sykes, P. W., A. K. Kepler, C. B. Kepler, and J. M. Scott. 2000. Kaua'i O'o (*Moho braccatus*), O'ahu O'o (*Moho apicalis*), Bishop's O'o (*Moho bishopi*), Hawai'i O'o (*Moho nobilis*) and Kioea (*Chaetoptila angustipluma*). In *The Birds of North America*, A. Poole and F. Gill (editors). The Birds of North America, Inc., Philadelphia, PA. Number 535.

1194. Sykes, P. W., J. A. Rodgers, and R. E. Bennetts. 1995. Snail Kite (*Rostrhamus sociabilis*). In *The Birds of North America*, A. Poole and F. Gill (editors). The Academy of Natural Sciences, Philadelphia, PA, and The American Ornithologists' Union, Washington, DC. Number 171.

1195. Tachia, T. C., S. A. Nesbitt, and P. A Vohs. 1992. Sandhill Crane (*Grus canadensis*). In *The Birds of North America*, A. Poole and F. Gill (editors). The Academy of Natural Sciences, Philadelphia, PA, and The American Ornithologists' Union, Washington, DC. Number 31.

1196. Tanner, J. T. 1942. *The Ivory-billed Woodpecker.* National Audubon Society Research Report 1.

1196a. Tarburton, M. K. Unpublished data.

1196b. Tarburton, M. K. 1990. Breeding biology of the Atiu Swiftlet. *Emu* 90:175–179.

1196c. Tarburton, M. K. 1992. Weights of some birds from Fiji. *Bulletin of the British Ornithologists' Club* 112:34–36.

1196d. Tarburton, M. K. 1988. What swiftlets do in the caves at Chillagoe and some ideas on how cavers can share those caves without interference. *17th Biennial Conference of the Australian Speleological Federation*, 98–107.

1196e. Tarburton, M. K. 2003. The breeding biology of the Mountain Swiftlet, *Aerodramus hirundinaceus*, in Irian Jaya. *Emu* 103:177–182.

1197. Tashian, R. E. 1952. Some birds from the Palenque region of northeastern Chiapas, Mexico. *Auk* 69:60–66.

1198. Taylor, B. 1998. *Rails: a guide to the rails, crakes, gallinules and coots of the world.* Yale University Press, New Haven CT.

1199. Taylor, I. 1994. *Barn owls.* Cambridge University Press, Cambridge, UK.

1200. Teixeira, D. M. 1987. Notas sobre o "Gravatazeiro," *Rhopornis ardesiaca* (Wied, 1831) (Aves, Fromicariidae). *Revista Brasiliera de Biologie* 47:409–414.

1201. Teixeira, D. M., and N. Carnevalli. 1989. Nova espêcie de Scytalopus Gould, 1837, do nordeste do Brasil (Passeriformes, Rhinocryptidae). *Boletim do Museu Nacional, Nova Sêrie, Rio de Janeiro, Brazil* 331:1–11.

1202. Teixeira, D. M., J. B. Nacinovic, and M. S. Tavares. 1986. Notes on some birds of northeastern Brazil. *Bulletin of the British Ornithologists' Club* 106:70–74.

1203. Teixeira, D. M., J. B. Nacinovic, and F. B. Pontual. 1987b. Notes on some birds of northeastern Brazil (2). *Bulletin of the British Ornithologists' Club* 107:151–157.

1204. Teixeira, D. M., J. B. Nacinovic, and G. Luigi. 1988. Notes on some birds of northeastern Brazil (3). *Bulletin of the British Ornithologists' Club* 108:75–79.

1205. Teixeira, D. M., J. B. Nacinovic, and G. Luigi. 1989. Notes on some birds of northeastern Brazil (4). *Bulletin of the British Ornithologists' Club* 109:152–157.

1206. Teixeira, D. M., R. Otoch, G. Luigi, M. A. Raposo, and A. C. C. de Almeida. 1993. Notes on some birds of northeastern Brazil (5). *Bulletin of the British Ornithologists' Club* 113:48–52.

1207. Telfair, R. C. 1994. Cattle Egret (*Bubulcus ibis*). In *The Birds of North America*, A. Poole and F. Gill (editors). The Birds of North America, Inc., Philadelphia, PA, and The American Ornithologists' Union, Washington, DC. Number 113.

1208. Temple, S. A. 2002. Dickcissel (*Spiza americana*). In *The Birds of North America*, A. Poole and F. Gill (editors). The Birds of North America, Inc., Philadelphia, PA. Number 703.

1209. Tenney, C. R. 1997. Black-chinned Sparrow (*Spizella atrogularis*). In *The Birds of North America*, A. Poole and F. Gill (editors). The Academy of Natural Sciences, Philadelphia, PA, and The American Ornithologists' Union, Washington, DC. Number 270.

1210. Terborgh, J., S. K. Robinson, T. A. Parker, C. A. Munn,, and N. Pierpont. 1990. Structure and organization of an Amazonian forest bird community. *Ecological Monographs* 60:213–238.

1210a. Thibault, J.-C. Unpublished data.

1210b. Thibault, J.-C., A. Cibois, and E. Pasquet. Weights of Polynesian monarchs, unpublished data.

1211. Thiollay, J.-M. 1970. Recherches ecologiques dans la savane de lamto (Cote D'Ivoire): le peuplement avien. *Essai d'stude quantitative. Le Terre et al Vie Nó.* 1:108–144.

1212. Thiollay, J.-M. 1974. Le peuplement avien de la savane de Lamto. *Bulletin de Liaison des chercheurs de lamto. No. Spec.* 4:39–68.

1213. Thiollay, J.-M. 1977. Les rapaces d'une zone de contact savane-foret en Cote-D'Ivoire: modes d'exploitation du milieu. *Alauda* 45:197–218.

1214. Thiollay, J.-M. 1988. Comparative foraging success of insectivorus birds in tropical and temperate forests: ecological implications. *Oikos* 53:17–30.

1215. Thomas, B. T. 1982. Weights of some Venezuelan birds. *Bulletin of the British Ornithologists' Club* 102:48–52.

1216. Thomas, B. T. 1990. Additional weights of Venezuelan birds. *Bulletin of the British Ornithologists' Club* 110:48–51.

1217. Thompson, B. C., J. A. Jackson, J. Burger, L. A. Hill, E. M. Kirsch, and J. L. Atwood. 1997. Least Tern (*Sterna antillarum*). In *The Birds of North America*, A. Poole and F. Gill (editors). The Academy of Natural Sciences, Philadelphia, PA, and The American Ornithologists' Union, Washington, DC. Number 290.

1218. Thompson, M. C. 1966. Birds from North Borneo. *University of Kansas Publications, Museum of Natural History* 17:377–433.

1219. Thorstrom, R. 1997a. A description of nests and behavior of the Gray-headed Kite. *Wilson Bulletin* 109:173–177.

1220. Thorstrom, R. 1997b. New record, ranging behaviour, vocalization and food of the Madagascar Red Owl *Tyto soumagnei*. *Ibis* 139:477–481.

1221. Reference deleted.

1222. Threlfall, W., and S. P. Mahoney. 1980. The use of measurements in sexing Common Murres from Newfoundland. *Wilson Bulletin* 92:266–268.

1223. Tickell, W. L. N. 2000. *Albatrosses*. Yale University Press, New Haven, CT.

1224. Tielman, B. I., J. B. Williams, M. E. Buschur, and C. R. Brown. 2003. Phenotypic variation of larks along an aridity gradient: are desert birds more flexible? *Ecology* 84:1800–1815.

1225. Tobalske, B. W. 1997. Lewis' Woodpecker (*Melanerpes lewis*). In *The Birds of North America*, A. Poole and F. Gill (editors). The Academy of Natural Sciences, Philadelphia, PA, and The American Ornithologists' Union, Washington, DC. Number 284.

1226. Tomkins, R. J., and B. J. Milne. 1991. Differences among Dark-rumped Petrel (*Pterodroma phaeopygia*) populations within the Galapagos Archipelago. *Notornis* 38:1–35.

1227. Tomlinson, R. E. 1975. 1975. Weights and wing lengths of wild Sonoran Masked Bobwhites during fall and winter. *Wilson Bulletin* 87:180–186.

1227a. Tompkins, D. M. Unpublished data.

1228. Tracy, D. M., D. Schamel, and J. Dale. 2002. Red Phalarope (*Phalaropus fulicarius*). In *The Birds of North America*, A. Poole and F. Gill (editors). The Birds of North America, Inc., Philadelphia, PA. Number 698.

1229. Trail, P. W., and L. Baptista. Unpublished data included in Dunning (1992).

1230. Trail, P. W., and P. Donahue. 1991. Notes on the behavior and ecology of the red-cotingas (Cotingidae: *Phoenicircus*). *Wilson Bulletin* 103:539–551.

1231. Reference deleted.

1232. Tubaro, P. L., and B. Mahler. 1998. Acoustic frequencies and body mass in New World doves. *Condor* 100:54–61.

1233. Turner, A., and C. Rose. 1989. *Swallows and martins: an identification guide and handbook*. Houghton Mifflin, Boston, MA.

1234. Tweit, R. C. 1996. Curve-billed Thrasher (*Toxostoma curivrostre*). In *The Birds of North America*, A. Poole and F. Gill (editors). The Academy of Natural Sciences, Philadelphia, PA, and The American Ornithologists' Union, Washington, DC. Number 235.

1235. Tweit, R. C. 1997. Long-billed Thrasher (*Toxostoma longirostre*). In *The Birds of North America*, A. Poole and F. Gill (editors). The Academy of Natural Sciences, Philadelphia, PA, and The American Ornithologists' Union, Washington, DC. Number 317.

1236. Tweit, R. C., and C. W. Thompson. 1999. Pyrrhuloxia (*Cardinalis sinuatus*). In *The Birds of North America*, A. Poole and F. Gill (editors). The Birds of North America, Inc., Philadelphia, PA. Number 391.

1237. Tweit, R. C., and J. C. Tweit. 2002. Dusky-capped Flycatcher (*Myiarchus tuberculifer*). In *The Birds of North America*, A. Poole and F. Gill (editors). The Birds of North America, Inc., Philadelphia, PA. Number 631.

1238. Tyler, S. 1979. Bird ringing in an Addis Ababa garden. *Scopus* 3:1–8.

1239. Uchida, H. 1989. Sexing and ageing of Japanese Wagtail (*Motacilla grandis*) in winter. *Bulletin of Japanese Bird Banding Association* 4:48–52.

1240. Uezu, A., Metzger, J. P., and J. M. E. Vielliard. 2005. Effects of structural and functional connectivity and patch size on the abundance of seven Atlantic Forest bird species. *Biological Conservation* 123:507–519.

1241. Underhill, L. G., B. Erni, and F. X. Mashinini. 1998. Ecological differentiation of canaries (Fringillidae) in the Western Cape. *Durban Museum Novitates* 23:56–60.

1242. Uramoto, M. 1961. Ecological study of the bird community of the broad-leaved deciduous forest of central Japan. *Miscellaneous Reports of the Yamashina Institute for Ornithology and Zoology* 3:1–32.

1243. Urban, E. K. 1959. Birds from Coahuila, Mexico. *Publications of the University of Kansas Museum of Natural History* 11:443–516.

1243a. Urban, E. K. 1975. Weights and longevity of some birds from Addis Ababa, Ethiopia. *Bulletin of the British Ornithologists' Club* 95:96–98.

1244. Urban, E. K., C. H. Fry, and S. Keith. 1986. *The birds of Africa*. Academic Press, New York. Volume 3.

1245. Urban, E. K., C. H. Fry, and S. Keith. 1997. *The birds of Africa*. Academic Press, New York. Volume 5.

1246. Vanderwerf, E. A. 1998. 'Elepaio (*Chasiempis sandwichensis*). In *The Birds of North America*, Poole, A., and F. Gill, (editors). The Birds of North America, Inc., Philadelphia, PA. Number 344.

1247. van Riper, C. 1987. Breeding ecology of the Hawaii Common Amakihi. *Condor* 89:85–102.

1248. van Riper, S. G. 2000. Japanese White-eye (*Zosterops japonicus*). In *The Birds of North America*, A. Poole and F. Gill (editors). The Birds of North America, Inc., Philadelphia, PA. Number 487.

1249. Velando, A., Lessells, C. M., and J. C. Márquez. 2001. The function of female and male ornaments in the Inca Tern: evidence for links between ornament expression and both adult condition and reproductive performance. *Journal of Avian Biology* 32:311–318.

1250. Verbeek, N. A. M. 1976. Banding worksheet for western birds—Yellow-billed Magpie. Western Bird Banding Association.

1251. Verbeek, N. A. M. 1993. Glaucous-winged Gull (*Larus glaucescens*). In *The Birds of North America*, A. Poole and F. Gill (editors). The Academy of Natural Sciences, Philadelphia, PA, and The American Ornithologists' Union, Washington, DC. Number 59.

1252. Verbeek, N. A. M., and C. Caffrey. 2002. American Crow (*Corvus brachyrhynchos*). In *The Birds of North America*, A. Poole and F. Gill (editors). The Birds of North America, Inc., Philadelphia, PA. Number 647.

1253. Verbeek, N. A. M., and P. Hendricks. 1994. American Pipit (*Anthus rubescens*). In *The Birds of North America*, A. Poole and F. Gill (editors). The Academy of Natural Sciences, Philadelphia, PA, and The American Ornithologists' Union, Washington, DC. Number 95.

1253a. Verea, C., 2007. Algunas notia sobre las caracteristicas externas del Ponchito Pechiescamado (Grallaricula loricata), una especie endemica del norte de Venezuela. *Ornitologia Neotropical* 18:1–9.

1254. Verea, C., A. Solorzano, and A. F. Badillo. 1999. Pesos y distribución de aves del sotobosque del Parque Nacional Henri Pittier al norte de Venezuela. *Ornitologie Neotropical* 10:217–231.

1255. Vickery, P. D. 1996. Grasshopper Sparrow (*Ammodramus savannarum*). In *The Birds of North America*, A. Poole and F. Gill (editors). The Academy of Natural Sciences, Philadelphia, PA, and The American Ornithologists' Union, Washington, DC. Number 239.

1256. Vielliard, J. 1990. Uma nova espécie de *Asthenes* da serra do Cipó, Minas Gerias, Brasil. *Ararajuba* 1:121–122.

1257. Vink-Nooteboom, M., N. J. Schoemaker, M. J. L. Kik, J. T. Lumeij, and W. Th. C. Wolvekamp. 1998. Clinical diagnosis of aneurysm of the right coronary artery in a white cockatoo (*Cacatua alba*). *Journal of Small Animal Practice* 39:533–537.

1258. Voelker, G. 1999. Body mass and moult data for South African motacillids, with body mass data for 47 additional South African species. *Ostrich* 70:233–235.

1259. Voelker, G. 2001. Morphological correlates of migratory distance and flight display in the avian genus *Anthus*. *Biological Journal of the Linnean Society* 73:425–435.

1260. Voous, K. H. 1969. Predation potential in birds of prey from Surinam. *Ardea* 57:117–148.

1261. Vuilleumier, F. 1969. Field notes on some birds from the Bolivian Andes. *Ibis* 111:599–608.

1262. Vuilleumier, F. 1997. Status and distribution of *Asthenes anthoides* (Furnariidae), a species endemic to Fuego-Patagonia, with notes on its systematic relationships and conservation. *Ornithological Monographs* 48:791–808.

1263. Vuilleumier, F., and D. N. Ewert. 1978. The distribution of birds in Venezuelan páramos. *Bulletin of the American Museum of Natural History* 162:47–90.

1264. Vuilleumier, F., M. LeCroy, and E. Mayr. 1992. New species of birds described from 1981 to 1990. *Bulletin of the British Ornithologists' Club* 112A:267–309.

1265. Wakelee, K. M., and S. G. Fancy. 1999. 'Oma'o (*Myadestes obscurus*), Kama'o (*Myadestes myadestinus*), Oloma'o (*Myadestes lanaiensis*), and 'Amaui (*Myadestes woahensis*). In *The Birds of North America*, A. Poole and F. Gill (editors). The Birds of North America, Inc., Philadelphia, PA. Number 460.

1266. Reference deleted.

1267. Wallace, G. E., unpublished data included in Dunning (1992).

1268. Wallace, G. J. 1965. Studies on Neotropical thrushes in Colombia. *Publications of the Museum, Michigan State University. Biological Series* 3(1):1–48.

1269. Wallace, M. P., and S. A. Temple. 1987. Competitive interactions within and between species in a guild of avian scavengers. *Auk* 104:290–295.

1270. Walters, E. L., E. H. Miller, and P. E. Lowther. 2002. Red-breasted Sapsucker (*Sphyrapicus ruber*) and Red-naped Sapsucker (*Sphyrapicus nuchalis*). In *The Birds of North America*, A. Poole and F. Gill (editors). The Birds of North America, Inc., Philadelphia, PA. Number 663.

1271. Walters, P. M. 1981. Notes on the body weight and molt of the Elf Owl (*Micrathene whitneyi*) in southeastern Arizona. *North American Bird Bander* 6:104–105.

1272. Waltert, M. unpublished data.

1273. Reference deleted.

1274. Warner, D. W. 1959. The song, nest, eggs, and young of the Long-tailed Partridge. *Wilson Bulletin* 71:307–312.

1275. Watanuki, Y. 1986. Moonlight avoidance behavior in Leach's Storm-Petrels as a defense against Slaty-backed Gulls. *Auk* 103:14–22.

1276. Watanuki, Y., Kato, A., and Naito, Y. 1996. Diving performance of male and female Japanese Cormorants. *Canadian Journal of Zoology* 74:1098–1109.

1277. Watling, D. 1986. Notes on the Collared Petrel *Pterodroma (leucoptera) brevipes*. *Bulletin of the British Ornithologists' Club* 106:63–70.

1278. Watling, D., and R. F. Lewanavanua. 1985. A note to record the continuing survival of the Fiji (MacGillivray's) Petrel *Pseudobulweria macgillivrayi*. *Ibis* 127:230–233.

1279. Watson, R. T., S. Thomsett, D. O'Daniel, and R. Lewis. 1996. Breeding, growth, development, and management of the Madagascar Fish-Eagle (*Haliaeetus vociferoides*). *Journal of Raptor Research* 30:21–27.

1280. Weatherhead, P. J. 1980. Sexual dimorphism in two savannah sparrow populations. *Canadian Journal of Zoology* 8:412–415.

1281. Weathers, W. W. 1977. Temperature regulation in the Dusky Munia, *Lonchura fuscans* (Cassin) (Estrildidae). *Australian Journal of Zoology* 25:193–199.

1282. Weckstein, J. D., D. E. Kroodsma, and R. C. Faucett. 2002. Fox Sparrow (*Passerella iliaca*). In *The Birds of North America*, A. Poole and F. Gill (editors). The Birds of North America, Inc., Philadelphia, PA. Number 715.

1283. Weeden, R. B. 1979. Relative heart size in Alaskan Tetraonidae. *Auk* 96:306–318.

1284. Weimerskirch, H., R. Zotier, and P. Jouventin. 1989. The avifauna of the Kerguelen Islands. *Emu* 89:15–29.

1285. Weller, M. W. 1975. Ecology and behaviour of the South Georgia Pintail *Anas g. georgica*. *Ibis* 117:217–231.

1286. Weller, M. W. 1980. *The island waterfowl*. Iowa State University Press, Ames, IA.

1287. Wells, D. R. 1999. *The birds of the Thai-Malay Peninsula*. Academic Press, New York.

1288. Weske, J. S. 1972. The distribution of the avifauna in the Apurimac Valley of Peru with respect to environmental gradients, habitat, and related species. Ph.D. dissertation, University of Oklahoma, Norman, OK.

1289. Weske, J. S., and J. W. Terborgh. 1981. *Otus marshalli*, a new species of screech-owl from Peru. *Auk* 98:1–7.

1290. West, G. C., R. B. Weeden, L. Irving, and L. J. Peyton. 1970. Geographic variation in body size and weight of willow ptarmigan. *Arctic* 23:240–253.

1291. West, S. 1995. Cave Swallow (*Hirundo fulva*). In *The Birds of North America*, A. Poole and F. Gill (editors). The Academy of Natural Sciences, Philadelphia, PA, and The American Ornithologists' Union, Washington, DC. Number 141.

1292. Wethington, S. M. 2002. Violet-crowned Hummingbird (*Amazilia violiceps*). In *The Birds of North America*, A. Poole and F. Gill (editors). The Birds of North America, Inc., Philadelphia, PA. Number 688.

1292a. Wheelwright, N. T., and J. D. Rising. 1993. Savannah Sparrow (*Passerculus sandwichensis*). In *The Birds of North America*, A. Poole and F. Gill (editors). The Academy of Natural Sciences, Philadelphia, PA, and The American Ornithologists' Union, Washington, DC. Number 45.

1292a. Whitacre, D. F. 1992. Studies of the ecology of the White-collared Swift (*Streptoprocne zonaris*) and White-naped Swift (*S. semicollaris*) and patterns of adaptation among the swifts (Aves: Apodidae). Ph.D. dissertation, University of California, Davis, CA.

1293. White, C. M., N. J. Clum, T. J. Cade, and W. G. Hunt. 2002. Peregrine Falcon (*Falco peregrinus*). In *The Birds of North America*, A. Poole and F. Gill (editors). The Birds of North America, Inc., Philadelphia, PA. Number 660.

1294. Whitney, B. M. 1994. A new *Scytalopus* tapaculo (Rhinocryptidae) from Bolivia, with notes on other Bolivian members of the genus and the *magellanicus* complex. *Wilson Bulletin* 106:585–614.

1295. Whitney, B. M., and J. Alvarez Alonso. 1998. A new *Herpsilochmus* antwren (Aves: Thamnophilidae) from northern Amazonian Peru and adjacent Ecuador: the role of edaphic heterogeneity of *terre firme* forest. *Auk* 115:559–576.

1296. Whitney, B. M., and J. Alvarez Alonso. 2005. A new species of gnatcatcher from white-sand forests of northern Amazonian Peru with revision of the *Polioptila guianensis* complex. *Wilson Bulletin* 117:113–127.

1297. Whitney, B. M., D. C. Oren, and R. T. Brumfield. 2004. A new species of *Thamnophilus* antshrike (Aves: Thamnophilidae) form the Serra do Divisor, Acre, Brazil. *Auk* 121:1031–1039.

1298. Whitney, B. M., J. F. Pacheco, D. R. C. Buzzetti, and R. Parrini. 2000. Systematic revision and biogeography of the *Herpsilochmus pileatus* complex, with description of a new species from northeastern Brazil. *Wilson Bulletin* 117:869–891.

1299. Whittaker, A. 2002. A new species of forest-falcon (Falconidae: *Micrastur*) from southeastern Amazonia and the Atlantic rainforests of Brazil. *Wilson Bulletin* 114:421–445.

1300. Whittingham, L. A., A. Kirkconnell, and L. M. Ratcliffe. 1996. Breeding behavior, social organization and morphology of Red-shouldered (*Agelaius assimilis*) and Tawny-shouldered (*A. humeralis*) Blackbirds. *Auk* 98:832–836.

1301. Whittow, G. C. 1997. Factors affecting egg mass loss in the Phoenix Petrel on Christmas Island. *Journal of Field Ornithology* 68:376–381.

1302. Wiedenfeld, D. A., T. S. Schulenberg, and M. B. Robbins. 1985. Birds of a tropical deciduous forest in extreme northwestern Peru. *Ornithological Monographs* 36:305–315.

1303. Wiens, J. A., and J. T. Rotenberry. 1980. Patterns of morphology and ecology in grassland and shrubsteppe bird populations. *Ecological Monographs* 50:287–308.

1304. Wilcox, L. 1959. A twenty-year banding study of the Piping Plover. *Auk* 76:129–152.

1305. Wiles, G. J., J. Bart, R. E. Beck, and C. F. Aguon. 2003. Impacts of the brown tree snake: patterns of decline and species persistence in Guam's avifauna. *Conservation Biology* 17:1350–1360.

1306. Wiley, J. W. Unpublished data.

1307. Wiley, R. H., and D. S. Lee. 1998. Long-tailed Jaeger (*Stercorarius longicaudus*). In *The Birds of North America*, A. Poole and F. Gill (editors). The Academy of Natural Sciences, Philadelphia, PA, and The American Ornithologists' Union, Washington, DC. Number 365.

1308. Willard, D. E., M. S. Foster, G. F. Barrowclough, R. W. Dickerman, P. F. Connell, S. L. Coats, J. L. Cracraft, and J. P. O'Neill. 1991. *The birds of Cerro de la Neblina, Territorio Federal Amazonas, Venezuela*. Fieldiana Zoology (New Series) 65,1–80.

1309. Williams, G. R. 1960. The birds of the Pitcairn Islands, central South Pacific Ocean. *Ibis* 102:58–70.

1310. Williams, J. B., and K. A. Nagy. 1985. Daily energy expenditure by female Savannah Sparrows feeding nestlings. *Auk* 102:187–190.

1311. Williams, R. S. R., and J. A. Tobias. 1993. The conservation of southern Ecuador's threatened avifauna. Unpublished final report, Birdlife International.

1312. Williams, T. D. 1995. *The penguins, Spheniscidae*. Oxford University Press, Oxford, UK.

1313. Willis, E. O. 1982. The behavior of Black-banded Woodcreepers (*Dendrocolaptes picumnus*). *Condor* 84:272–285.

1314. Willis, E. O. 1992. Three *Chamaeza* antthrushes in eastern Brazil (Formicariidae). *Condor* 94:110–116.

1315. Willis, E. O., and Y. Oniki. 1992. A new *Phylloscartes* (Tyrannidae) from southeastern Brazil. *Bulletin of the British Ornithologists' Club* 112:158–165.

1316. Willson, S. K. 2004. Obligate army-ant-following birds: a study of ecology, spatial movement patterns, and behavior in Amazonian Peru. *Ornithological Monographs* 55.

1317. Reference deleted.

1318. Wilson, R. T. 1981. The Corvidae in the Sudan Republic, with special reference to Darfur. *African Journal of Ecology* 19:285–294.

1319. Wilson, R. T., and D. M. Ball. 1979. Morphology, wing loading and food of western Darfur birds. *Bulletin of the British Ornithologists' Club* 99:15–20.

1320. Wilson, R. T., and J. G. Lewis. 1977. Observations on the Speckled Pigeon *Columba guinea* in Tigrai, Ethiopia. *Ibis* 119:195–198.

1321. Wingate, D. B. 1972. First successful hand-rearing of an abandoned Bermuda Petrel chick. *Ibis* 114:97–101.

1322. Winker, K., M. A. Ramos, J. H. Rappole, and D. W. Warner. 1992. A note on *Campylopterus excellens* in southern Veracruz, with a guide to sexing captured individuals. *Journal of Field Ornithology* 63:339–343.

1323. Winker, K., J. T. Klicka, and G. Voelker. 1996. Sexual size dimorphism in birds from southern Veracruz, Mexico. II. *Thryothorus maculipectus* and *Henicorhina* [*leucosticta*] *prostheleuca*. *Journal of Field Ornithology* 67:236–251.

1324. Winter, M. Unpublished banding data.

1325. Winterstein, S. R., and R. J. Raitt. 1983. Nestling growth and development and the breeding ecology of the Beechey Jay. *Wilson Bulletin* 95:256–268.
1326. Wishart, R. A. 1979. Indices of structural size and condition of American wigeon (*Anas americana*). *Canadian Journal of Zoology* 57:2369–2374.
1327. With, K. A. 1994. McCown's Longspur (*Calcarius mccownii*). In *The Birds of North America*, A. Poole and F. Gill (editors). The Academy of Natural Sciences, Philadelphia, PA, and The American Ornithologists' Union, Washington, DC. Number 96.
1328. Withers, P. C., R. B. Forbes, and M. S. Hedrick. 1987. Metabolic, water and thermal relations of the Chilean Tinamou. *Condor* 89:424–426.
1329. Woinarski, J. C. Z. Unpublished data.
1330. Woinarski, J. C. Z., and C. Bulman. 1985. Ecology and breeding biology of the Forty-spotted Pardalote and other pardalotes on North Bruny Island. *Emu* 85:106–120.
1331. Woinarski, J. C. Z., D. F. Dorward, and J. M. Cullen. 1983. Variation in the *Pardalotus striatus* complex in south-eastern Australia. *Emu* 83:82–93.
1332. Wolf, L. L. 1977. Species relationships in the avian genus *Aimophila*. *Ornithological Monographs* 23.
1333. Wong, M. 1986. Trophic organization of understory birds in a Malaysian dipterocarp forest. *Auk* 103:100–116.
1334. Wood, B., S. C. Madge, and C. S. Waller. 1978. Description, moult and measurements of *Montifringilla theresae*. *Bulletin of the British Ornithologists' Club* 98:55–59.
1335. Wood, B. 1989. Biometrics, iris and bill colouration, and moult of Somali forest birds. *Bulletin of the British Ornithologists' Club* 109:11–22.
1336. Woodin, M. C., and T. C. Michot. 2002. Redhead (*Aythya americana*). In *The Birds of North America*. A. Poole and F. Gill (editors). The Birds of North America, Inc., Philadelphia, PA. Number 695.
1337. Woods, J. G. Unpublished data included in Dunning (1992).
1338. Woods, R. W. 1993. Cobb's Wren *Troglodytes (aedon) cobbi* of the Falkland Islands. *Bulletin of the British Ornithologists' Club* 113:195–207.
1339. Woolfenden, G. E. 1956. Comparative breeding behavior of *Ammospiza caudacuta* and *A. maritima*. *University of Kansas Publications of the Museum of Natural History* 10:45–75.
1340. Woolfenden, G. E., and J. W. Fitzpatrick. 1996. Florida Scrub-Jay (*Aphelocoma coerulescens*). In *The Birds of North America*, A. Poole and F. Gill (editors). The Academy of Natural Sciences, Philadelphia, PA, and The American Ornithologists' Union, Washington, DC. Number 228.
1341. Wooller, R. D., and M. C. Calver. 1981. Diet of three insectivorous birds on Barrow Island, WA. *Emu* 81:48–50.
1342. Woxvold, I. A. 2004. Breeding ecology and group dynamics of the apostlebird. *Australian Journal of Zoology* 52:561–581.
1342a. Wu, H.-J., Y.-H. Sun, Y. Wang, and Y.-S. Tseng. 2006. Food habits of Tawny Fish-Owls in Sakatang Stream, Taiwan. *Journal of Raptor Research* 40:111–119.
1343. Wyndham, E. 1980. Diurnal cycle, behaviour and social organization of the Budgerigar *Melopsittacus undulatus*. *Emu* 80:25–33.
1344. Reference deleted.
1345. Yamashina, Y., and T. Mano. 1981. A new species of rail from Okinawa Island. *Journal of the Yamashina Institute of Ornithology* 12:1–6.
1346. Yañez, J. L., H. Nuñuz, and F. M. Jaksić. 1982. Food habits and weight of Chimango Caracaras in central Chile. *Auk* 99:170–171.
1347. Reference deleted.
1348. Yogev, A., A. Ar, and Y. Yom-Tov. 1996. Determination of clutch size and the breeding biology of the Spur-winged Plover (*Vanellus spinosus*) in Israel. *Auk* 113:68–73.
1349. Yorio, P., P. D. Boersma, and S. Swann. 1996. Breeding biology of the Dolphin Gull at Punta Tomba, Argentina. *Condor* 98:208–215.
1349. Yosef, R. Unpublished data.
1350. Yosef, R. 1996. Loggerhead Shrike (*Lanius ludovicianus*). In *The Birds of North America*, A. Poole and F. Gill (editors). The Academy of Natural Sciences, Philadelphia, PA, and The American Ornithologists' Union, Washington, DC. Number 231.
1351. Yosef, R., and L. Fornasari. 2000. Biometric differences between age and sex classes of the Levant Sparrowhawk *Accipiter brevipes* on migration at Eilat, Israel. *Israel Journal of Zoology* 46:207–214.
1352. Yosef, R., and P. Tryjanowski. 2002a. Migratory Masked Shrikes *Lanius nubicus* staging at the desert edge: phenology, and sex- and age-related differences in body mass. *Ostrich* 73:1–4.
1353. Yosef, R., and P. Tryjanowski. 2002b. Spring migration ecology of Cretzschmar's Bunting (*Emberiza caesia*) through Elat, Israel. *Israel Journal of Zoology* 48:149–153.
1354. Yosef, R., and P. Tryjanowski. 2002c. Differential spring migration of Ortolan Bunting *Emberiza hortulana* by sex and age at Eilat, Israel. *Ornis Fennica* 79:173–180.
1355. Yosef, R., P. Tryjanowski, and K. L. Bildstein. 2002a. Spring migration of adult and immature buzzards (*Buteo buteo*) through Elat, Israel: timing and body size. *Journal of Raptor Research* 36:115–120.
1356. Yosef, R., P. Tryjanowski, and M. Remisiewicz. 2002b. Migration characteristics of the Wood Sandpiper (*Tringa glareola*) at Eilat (Israel). *Ring* 24:61–69.
1357. Young, G. 1991. Sexual dimorphism in Meller's Duck *Anas melleri*. *Bulletin of the British Ornithologists' Club* 111:225–228.
1358. Young, H. G., M. D. Sorenson, and K. P. Johnson. 1997. A description of the Madagascar Teal *Anas bernieri* and an examination of its relationships with the Grey Teal *A. gracilis*. *Wildfowl* 48:174–180.
1359. Yu, S. 1999. The breeding and foraging ecology of the Taiwan laughing thrushes, *Garrulax morrisonianus*, in the Tatachia area, Taiwan. M.S. thesis, National Taiwan University.
1360. Yuan, H.-W., M. Liu, and S.-F. Shen. 2004. Joint nesting in Taiwan Yuhinas: a rare passerine case. *Condor* 106:862–872.

1360a. Yue-Hua, S. Unpublished data.

1361. Yue-Hua, S., J. Chen-Xi, and F. Yun. 2001. The distribution and status of Sichuan Grey Jay (*Perisoreus internigrans*). *Journal für Ornithologie* 142:93–98.

1362. Zammuto, R. M., and E. C. Franks. 1979. Trapping flocks of Chimney Swifts in Illinois. *Bird-banding* 50:201–209.

1363. Zavalaga, C. B., and J. Jahncke. 1997. Maximum dive depths of the Peruvian Diving-Petrel. *Condor* 99:1002–1004.

1364. Zavalaga, C. B., and R. Paredes. 1997. Sex determination of adult Humboldt Penguins using morphometric characters. *Journal of Field Ornithology* 68:102–112.

1365. Zimmer, K. J., A. Whittaker, and D. C. Oren. 2001. A cryptic new species of flycatcher (Tyrannidae: *Suiriri*) from the cerrado region of central South America. *Auk* 118:56–78.

1366. Zwickel, F. C. 1992. Blue Grouse (*Dendragapus obscurus*). In *The Birds of North America*, A. Poole, P. Stettenheim, and F. Gill (editors). The Academy of Natural Sciences, Philadelphia, PA, and The American Ornithologists' Union, Washington, DC. Number 15.

1367. Zwickel, F. C., and J. H. Brigham. 1974. Autumn weights of spruce grouse in north-central Washington. *Journal of Wildlife Management* 38:315–319.

1368. Zwickel, F. C., and J. H. Brigham, and I. O. Buss. 1966. Autumn weights of Blue Grouse in north-central Washington, 1954 to 1963. *Condor* 68:488–496.

1369. Karubian, J., L. Carrasco, D. Cabrera, A. Coole, and J. Oliva, 2007. Nesting biology of the Banded Ground-cuckoo (*Neomorphus radiolosus*), *Wilson Journal of Ornithology* 119:221–227.

Index

D

E

M

Q

R

S

T

U

V

W

X

Y

Z